THE DEVELOPMENT AND FUNCTION OF ROOTS

THE DEVELOPMENT AND FUNCTION OF ROOTS

Third Cabot Symposium

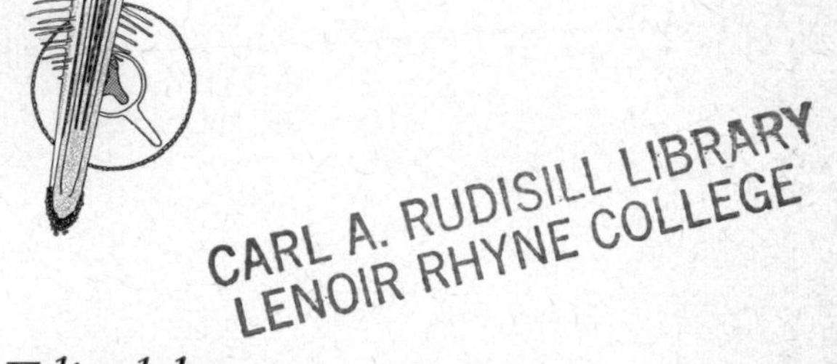

Edited by

J. G. TORREY

Cabot Foundation Harvard University Harvard Forest
Petersham Massachusetts U.S.A.

and

D. T. CLARKSON

Agricultural Research Council Letcombe Laboratory
Wantage Berkshire England

1975

ACADEMIC PRESS

LONDON · NEW YORK · SAN FRANCISCO

A Subsidiary of Harcourt Brace Jovanovich, Publishers

ACADEMIC PRESS INC. (LONDON) LTD.
24/28 Oval Road,
London NW1

United States Edition published by
ACADEMIC PRESS INC.
111 Fifth Avenue,
New York, New York 10003

Library of Congress Catalog Card Number: 75 153 51
ISBN: 0 12 695750 9

Text set in 11/12 pt. Monotype Bembo, printed by letterpress, and bound in Great Britain at The Pitman Press, Bath

Preface

During the past decade roots and root systems of vascular plants have come to be much more appreciated by botanists, agriculturalists, foresters and other plant scientists than formerly. Roots have long been known as absorptive structures for water and mineral nutrients and as mechanical devices for supporting shoot systems in a soil substrate. Only in recent years has it been recognized that roots are sites of synthesis for hormones required by the shoot system, that roots accomplish other biosynthetic activities for the plant as a whole, and that roots serve as model systems within the plant for studying problems of organogenesis, cytodifferentiation and ultrastructural organelle function. For large groups of plants, roots, together with invading infective organisms or symbiotic associates, serve as sites for fixation of atmospheric nitrogen, for accelerated and facilitated ion uptake and for extended absorptive surfaces for moisture accumulation and transport.

This volume attempts to document through comprehensive accounts by recognized specialists in their respective fields the anatomy and cytology of root structure, the physiology and biochemistry of root function and the ecological importance of root distribution and function.

In the course of reading this book, the professional botanist will learn new information and gain new insight into older ideas, now modified, as to how a root works. He will learn, for example, that the average mature red oak tree possesses a root system with perhaps five hundred million living root tips. He will be urged to discard the classical text-book idea of an "absorption zone" near the root apex and to believe rather that water and some cations may be absorbed along the length of the root, despite the presence of a suberized endodermis. New evidence is discussed concerning the role of auxin, its positive identification in roots, its predominantly acropetal movement toward the root apex and the probability of its role in root cell elongation. A comprehensive review of recent literature is made which points to the importance of the root apex as the site of cytokinin formation, to the important role of these root-originating hormones in normal shoot development, and to the striking deleterious effect of stresses imposed upon the root system, such as drought or water stress, whose major impact can be interpreted as acting via interference with hormone biosynthesis in the root.

Root structure and morphogenesis receive considerable attention. Accumulating evidence is reviewed showing that lateral roots originate along the root in a more precise order than is usually appreciated—precise in relation to the radial dimension of the root, opposite or in relation to the primary xylem poles, but also spaced in a non-random order along the longitudinal axis. Formation of root buds is shown to be not at all uncommon and *bona fide* cases of the conversion of a root apex to a shoot apex and, perhaps, vice versa are discussed. Aerial roots, with their great variety of structure and responsiveness, roots of aquatic plants with highly modified internal tissues, reduced vascular systems and lack of root hairs are described as essentially unexplored fields for research.

Some of the world's experts on the organization and function of the root apical meristem discuss their ideas and the new evidence about the quiescent center, eliciting positive evidence, rather than only negative, about the role this population of relatively inert and slowly cycling cells in the center of the root tip may play in root development. The root cap receives a fair share of attention and a new group of chemical growth regulators, related to abscisic acid, is brought into play in interpreting the activity of the root cap in influencing quiescent center behavior and, of more general interest perhaps, in exerting control over the geotropic sensitivity of the growing root.

The book conveys convincingly the idea that roots in nature develop and function in a complex environment, made manifestly more complicated and subtle by the presence of a vast array of soil organisms, which range all the way from mice and voles which chew on roots and change their branching patterns, through ants and worms which actively rearrange the substrate in which roots grow, to the soil microorganisms which surround and enter the root systems in a myriad varied ways. The almost universal occurrence of mycorrhizal fungi in and around roots is stressed and their important beneficial effects are reemphasized. The increasingly important subject of symbiotic nitrogen fixation by nodulated roots of legumes and non-legumes is summarized and new insight brought to the structure and function of these associations.

These lessons and many more are here for the practical agriculturalist, for the horticulturalist, for the forester, perhaps even for the home gardener, or for any one whose focus is on the growth of vascular plants. The point is clear that roots are an integral part of plant growth and development and it follows that the more we know or can learn about roots and how they work, the better able we will be to understand and control plant development and plant productivity in the field.

This book had its inception in a symposium organized by the Director of the Maria Moors Cabot Foundation of Harvard University and held in April, 1974 at the Harvard Forest in Petersham, Massachusetts, U.S.A. All participants extended their thanks to Harvard University and to the Harvard Forest for acting as hosts for the week-long meeting. The editors are indebted to the authors of these collected papers for their interest and willingness to share their enthusiasms in written form, making them available to the reader. Acknowledgement and thanks are here expressed to the Cabot Foundation for supporting the symposium and making the publication of this volume possible.

JOHN G. TORREY
DAVID T. CLARKSON

April, 1975.

Contributors

ANDERSON, W. P., *Department of Botany, University of Liverpool, Liverpool, England.*

AUDUS, L. J., *Department of Botany, University of London, Bedford College, London, NW1, England.*

BARLOW, PETER W., *Agricultural Research Council Unit of Developmental Biology, University of Cambridge, Cambridge, England.*

BATRA, M. W., *Department of Botany, University of North Carolina, Chapel Hill, North Carolina, USA.*

BECKING, J. H., *Institute for Atomic Sciences in Agriculture, Wageningen, The Netherlands.*

BRISTOW, J. M., *Biology Department, Queens University, Kingston, Ontario, Canada.*

CLARKSON, D. T., *Agricultural Research Council, Letcombe Laboratory, Wantage, Berkshire England.*

CLOWES, F. A. L., *Botany School, Oxford University, Oxford, England.*

DART, P. J., *Soil Microbiology Department, Rothamsted Experimental Station, Harpenden, Hertfordshire, England.*

EDWARDS, K. L., *Department of Biology, Kline Tower, Yale University, New Haven, Connecticut, USA.*

FELDMAN, LEWIS J., *Biological Laboratories, Harvard University, Cambridge, Massachusetts, USA.*

GERDEMANN, J. W., *Department of Plant Pathology, University of Illinois, Urbana Illinois, USA.*

GILL, A. M., *CSIRO Division of Plant Industry, P.O. Box 1600, Canberra A.C.T., Australia.*

LUXOVA, M., *Institute of Botany, Slovak Academy of Sciences, Bratislava, Czechoslovakia.*

LYFORD, W. H., *Harvard Forest, Harvard University, Petersham, Massachusetts, USA.*

MCCULLY, MARGARET E., *Biology Department, Carleton University, Ottawa, Canada.*

MASON, P., *Institute of Terrestrial Ecology, Unit of Tree Biology, Bush Estate, Penicuik, Midlothian, Scotland.*

OBROUCHEVA, NATALIE V., *Institute of Plant Physiology, USSR Academy of Sciences, Moscow, USSR.*

PETERSON, R. L., *Department of Botany and Genetics, University of Guelph, Guelph, Ontario, Canada.*

REYNOLDS, E. R. C., *Department of Forestry, University of Oxford, Oxford, England.*

ROBARDS, A. W., *Department of Biology, University of York, York, England.*

SCOTT, T. K., *Department of Botany, University of North Carolina, Chapel Hill, North Carolina, USA.*

SKENE, K. G. M., *CSIRO Division of Horticultural Research, Adelaide, South Australia.*

TOMLINSON, P. B., *Harvard University, Harvard Forest, Petersham, Massachusetts, USA.*

TORREY, JOHN G., *Cabot Foundation, Harvard University, Petersham, Massachusetts, USA.*

WALLACE, WILLIAM D., *Cabot Foundation, Harvard University, Petersham, Massachusetts, USA.*

WEATHERLEY, P. E., *Botany Department, University of Aberdeen, Aberdeen, Scotland.*

WILSON, B. F., *Department of Forestry and Wildlife Management, University of Massachusetts, Amherst, Massachusetts, USA.*

ZOBEL, R. W., *Cabot Foundation, Harvard University, Petersham, Massachusetts, USA.*

Contents

Part II PHYSIOLOGICAL ASPECTS OF ROOT FUNCTION

Part III ROOTS IN RELATION TO SOIL MICROFLORA

Part I

THE ORGANIZATION AND STRUCTURE OF ROOTS

Chapter 1

The Quiescent Centre

F. A. L. CLOWES

Botany School, Oxford University, England

I. Introduction

The quiescent centre was discovered as a result of a purely geometrical analysis of the cell lineages in the root meristems of *Zea* (Clowes, 1954). In roots other than those of the grass type it is not possible to derive an unambiguous result from such an analysis, but the development of methods of studying cell cycles in meristems has shown that all root apices have a quiescent centre in normal growth. Its cells proliferate much more slowly than any of the surrounding cells of the meristem and it is always situated at the pole of the stelar and cortical complexes of cells adjacent to the initials of the central part of the root cap (Clowes, 1956a, b, 1961a, 1965; Phillips and Torrey, 1972). Pulse-labelling with ^{3}H-thymidine and microdensitometry show that the small contribution that the region makes to cell production is due partly to most of the cells being out of cycle (in the G_0 condition) and partly to the extension of G_1 in the cycling cells (Clowes, 1968, 1971, 1972a).

There are interesting consequences in the presence of such a region surrounded by actively cycling cells, and root tips have been exploited in recent work on the control of cycling. The quiescent centre acts as a reservoir of cells relatively immune from perturbations that damage cycling cells. After severe X-irradiation, for example, the normally rapidly cycling cells stop cycling and the root recovers its growth by the repopulation of the meristem with cells derived from renewed meristematic activity within the quiescent centre (Clowes, 1959, 1963a, 1970a). The same kind of response occurs after many kinds of treatment and we should expect small environmental changes to elicit differential responses in the same directions even where cycling is not stopped. Thus the quiescent centre cells are by no means inherently quiescent. They can be induced to divide by experimental treatments of the meristem and they can approach normal rates of division for the meristem.

The maintenance of cell cycles with different parameters within an organism is an important biological and practical problem and in the root tip we probably have the best system for solving it. Variations in cycling are known also to exist in animal tissues, in the intestinal crypts and in tumours, for example, but it is more difficult to acquire the basic information because of the practical difficulties involved in measurements and experimentation on animal cells *in vivo*. In this paper I shall contrast the quiescent centre with its neighbouring tissues in an attempt to produce ideas about the control of cycling in the maintenance of an organized meristem.

II. Experimental Methods

The roots used in most of these investigations are primary ones of seedlings of *Zea mays* (cv. Golden Bantam) and in each batch of plants those with roots 20–40 mm long at the start of treatment have been selected. The plants are grown in constant darkness in damp *Sphagnum* or water so that they rely entirely on the seed reserves. Such roots have quiescent centres containing 500–600 cells. The temperature is maintained at a constant level, 23°C, for most experiments described here.

The techniques used are those of cell kinetics: pulse-labelling with ^{3}H-thymidine and the use of colchicine as a cycle-blocker. *Zea* roots readily incorporate tritium into DNA from thymidine labelled either at the 6-position or in the methyl group and 0·05 or 0·1% colchicine effectively stops the cell cycle at metaphase without altering the rate of entry into prophase over at least three hours. Three rates of mitosis have been measured: (1) the rate for the fast cycling cells displayed by the pulse-labelled mitosis (PLM) curve; (2) the rate for all cycling cells derived from the duration of the DNA-synthetic phase (S) and the fraction of mitoses labelled integrated over the first mitotic cycle; and (3) the cell-doubling time derived either from the accumulation of metaphases during colchicine treatment or from the duration of S and the fraction of all cells labelled by a pulse or by measurement of cell-flow after multiple pulses of ^{3}H-thymidine (Clowes, 1971; De la Torre and Clowes, 1974). When colchicine is used the cell-doubling time may be calculated by a simple cell flow procedure or by the more complex method of Evans *et al.* (1957). Cell flow measurement involves assessing, by sampling, the increase in some recognizable phase of the cell cycle either when the cycle is blocked in continuous treatments or with interrupted pulses of a marker in unblocked cycles.

The fraction of cells cycling (the proliferative fraction) is an important parameter (see also p. 6) and has been measured from pulse-labelled roots where it was reasonably certain that labelled cells were in a mitotic rather than an endomitotic cycle. This parameter, usually called the growth fraction in medical literature, has been calculated as the ratio of the fraction of all cells labelled by the pulse to the fraction of mitoses labelled (Clowes, 1971, 1972a). The denominator here varies cyclically and has to be integrated over one mitotic cycle. The numerator is constant over at least two cycles. Where there is a possibility that labelled cells include some which were preparing for endomitosis during the pulse, the proliferative fraction is obtained by comparing the average cycle duration of all cycling cells with the cell-doubling time determined independently.

Median longitudinal sections of appropriate thickness have been used throughout so that various regions of the meristem could be separately scored. The regions include the quiescent centre whose boundaries are determined by the thick wall between the cap and the rest of the root, the appearance of the cells and experience of its size from continuously labelled roots. The cap initials are taken to be the proximal three tiers of meristematic cap cells. Parts of the stelar meristem have also been scored as bands of cells across the sections at various distances from the cap boundary.

The sections have been stained in Feulgen except that in investigations of the nucleolar cycle the ability of nucleoli to reduce silver salts as in photography has been exploited to display details of nucleolar structure for light microscopy (De la Torre and Clowes, 1972).

III. Results

A. Termination of Cycling

In the quiescent centre the transition to quiescence occurs at early developmental stages that vary from species to species and, no doubt, also with rates of development (Clowes, 1958a; Byrne and Heimsch, 1970; Byrne, 1973). The quiescent centre shares with the cells at the margin of the meristem a transition from short to long cell-doubling times i.e. from high to low average rates of mitosis. In one batch of roots, for example, the mean rate of mitosis was 0·16 per day in the quiescent centre and 0·17 per day in the stele 1200 μm from the cap boundary at the proximal margin of the meristem. These figures compare with over one division per day within the meristem proper. But the similarity between the two regions is superficial. The mitotic cycle of the cells that are cycling, or are displayed by the PLM curve, is longer in the quiescent centre than at the margin of the stelar meristem and the difference is accounted for by the lower proliferative fraction in the latter region (Fig. 1). An interesting point here is that the cycle of the

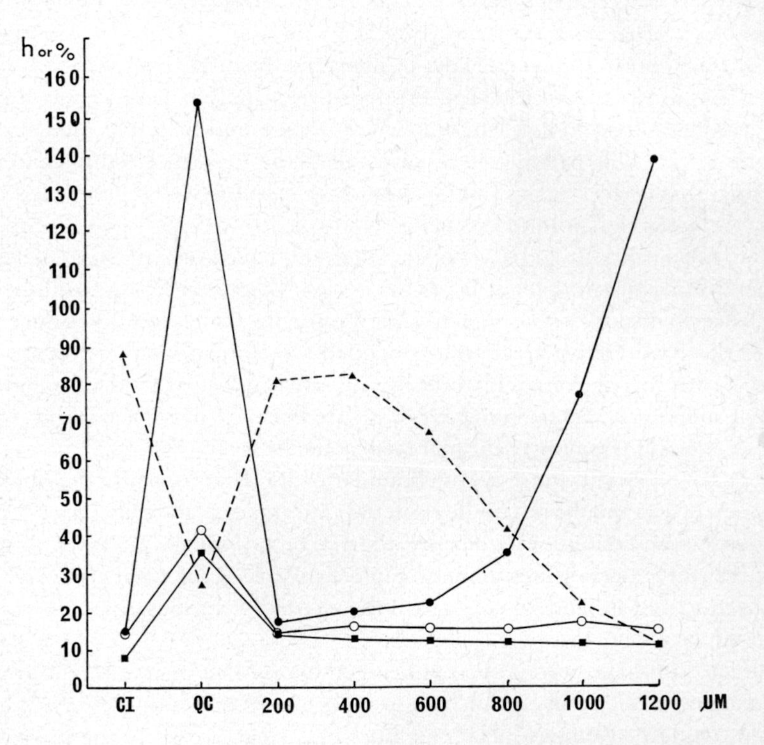

FIG. 1. Durations of the mitotic cycle (hours) and the proliferative fraction (%) at various levels in the root apex of *Zea mays* growing at 23°C. ●—● cell-doubling time determined by metaphase accumulation; ○—○ average duration of the mitotic cycle in all cycling cells; ■—■ average duration of the mitotic cycle displayed by the pulse-labelled mitosis curve; ▲—▲ proliferative fraction. CI: cap initials; QC: quiescent centre; other points are for the stele at the stated distance above the cap-quiescent centre boundary.

cells displayed by the PLM curve actually decreases slightly in length in the stele as the rate of cell proliferation goes down. This work therefore confirms the finding of Barlow and Macdonald (1973) on the fast cycling cells of *Zea* and provides the mode of transition from the meristem to the zone of differentiation. Cycling continues at normal rates and with normal phase durations, including a short G_1 phase, and the reduction of proliferation is effected by cells going out of cycle. In the quiescent centre, on the other hand, although the proliferative fraction is low, the mitotic cycle is also lengthened especially by the extension of G_1.

A further difference between the quiescent centre and the cells going out of cycle at the proximal margin of the meristem lies in what happens after cycling stops. In the quiescent centre the ending of cycling signals the end of DNA synthesis, but this need not be so at the margin of the meristem. In *Zea* DNA synthesis continues after the mitotic cycle stops and cells enter the zone of differentiation resulting in endoreduplication. This is seen in the increase in DNA levels beyond 2C or 4C amounts and also in the discrepancy between the ratio of S to the labelling index and the cell-doubling time. This discrepancy is small or non-existent in the quiescent centre and other distal regions of the meristem in *Zea*. Judging from the earlier literature on endoreduplication, roots of some plants probably would not show this increase in the DNA content of nuclei beyond the 2C level at the end of normal cycling, and the labelling index should then indicate only cells in normal S. In other species endoreduplication seems to be confined to certain tissues or is most prominent in certain tissues.

The cessation of DNA synthesis at the end of cycling appears from autoradiography to be the general rule in quiescent centres of angiosperms, but not in pteridophytes with apical cells. In these roots the apical cell and its immediate derivatives are in the position occupied by the quiescent centre in other roots. It was shown that these cells do synthesize DNA (Clowes, 1956b), but D'Amato and Avanzi also showed that this synthesis prepared for mitosis only in the early stages of the root's existence, as in the development of an angiosperm quiescent centre, and that, in later stages, it preceded endoreduplication (D'Amato and Avanzi, 1965; Avanzi and D'Amato, 1970).

In measuring the time parameters of the cell cycle and the proliferative fractions we are almost wholly dependent upon radioactive precursors of DNA and work on pulses of ^{3}H-thymidine shows how damaging these can be to the meristem (De la Torre and Clowes, 1974). The damage to the meristem is obscured by treating whole meristems as homogeneous units, but revealed when the reactions of the various regions are looked at separately (Clowes, 1961b; Phillips and Torrey, 1971).

The main experiment in which the damage was evident was a series of three 20-minute pulses of ^{3}H-thymidine given to roots of *Zea* at 0·5 μCi/ml and 23·3 Ci/mM at intervals of three hours. The labelling index was measured immediately after each of the three pulses and rates of mitosis were calculated from cell flow between pulses. The difference between two adjacent labelling indices is then equal to the interval between the start of the first and second pulses divided by the mean duration of the cycle. This fraction is multiplied by $\log_e 2$ if we assume exponential growth. The experiment is controlled by measuring the cell flow with colchicine as a blocker of the cycle in the presence and absence of non-radioactive thymidine at the labelling concentration, here $2{\cdot}15 \times 10^{-8}$ M.

The mean duration of the cell cycle measured by both cell flow methods is the cell-doubling time and, of course, differs from that obtained from the PLM curve.

The main results are displayed in Fig. 2. Tritium-labelled thymidine, even when supplied as pulses of 20 minutes, has a profound effect upon the organization

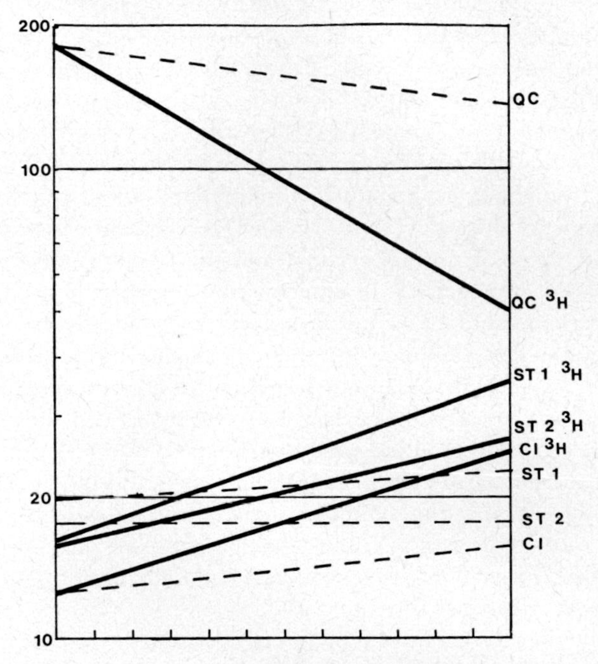

FIG. 2. Changes in the cell-doubling time (hours) measured between the first and second and between the second and third pulses of thymidine ($2{\cdot}15 \times 10^{-8}$ M) in root apices of *Zea mays* at 23°C. Each pulse lasted 20 min and the intervals between pulses over which the cell-doubling time was calculated by cell flow were 3 h. Continuous lines, ^{3}H-thymidine ($0{\cdot}5\mu$Ci/ml); dashed lines, non-radioactive thymidine at the same concentration. QC: quiescent centre; ST1: stele just above QC; ST2: stele 200μm above QC; CI: cap initials. Note log scale of the ordinate. Data from De la Torre and Clowes (1974).

of the meristem. It lengthens the cell-doubling time in the rapidly dividing regions and shortens it in the quiescent centre.

Non-radioactive thymidine has only a small or negligible effect at the labelling concentration. We can measure the changes in organization only because we can compare results after different intervals, but there is no reason to believe that similar changes do not also take place immediately after a single pulse of tritium as they do after acute X-irradiation. The results reported above and elsewhere from experiments that utilize labelled precursors of DNA should, therefore, always be treated with some reservation especially because the radiation effect is enhanced by the differences in incorporation between cycling and non-cycling cells of a meristem.

B. Rates of Cycling in Quiescent Centres

The treatments which we know to stimulate cycling in the quiescent centre include acute X-irradiation, chronic γ-irradiation, β-decay from isotopes incorporated in cycling nuclei, recovery from cold-induced dormancy, treatment with 5-amino uracil, triiodobenzoic acid, colchicine, non-labelled thymidine at $3{\cdot}3 \times 10^{-6}$ M (but not at $2{\cdot}2 \times 10^{-8}$ M), cyclic AMP and indole acetic acid. The impression given is that anything that reduces the rate of cycling in the normally meristematic cells will increase the rate in the quiescent centre.

In the course of determining rates of mitosis in a large number of experiments I have found a wide variation in cell-doubling times in untreated roots of *Zea*. Variations may be due to temperature, age of seeds and age of roots. In order to see if there are correlations between rates of proliferation in various parts of the meristem, cell-doubling times determined by metaphase accumulation in the cap initials have been compared with cell-doubling times in three other regions of the same meristems: the adjacent quiescent centre, the stele just above it and at 200 μm above it. All these regions are well within the meristem. The best fits chosen by computer from twelve models are given in Fig. 3 with their confidence limits. They show that, whereas there is a positive correlation between the mean cell cycle durations of the cap initials and the stelar regions, there is a negative correlation with the quiescent centre.

The range of cell-doubling times shown in Fig. 3 is that for roots growing in a small temperature range of 23–30°C and with no experimental treatments. If we extend the temperature range or introduce chemical or irradiation treatments, a different picture emerges. Some of the irradiated samples shorten the cell-doubling time in the quiescent centre beyond that shown in Fig. 3 and lengthen the cell-doubling time in the other regions far beyond the range in Fig. 3. Extension is especially prominent in the cap initials which, in *Zea*, are the most sensitive cells of the meristem and often stop dividing when other regions of the meristem still have measurable rates of mitosis.

When the temperature range is extended beyond the optimal range, cell-doubling times may be prolonged in all regions. A study of this effect by A. T. Taylor in *Allium sativum* is given in Fig. 4. The high sensitivity of the cap initials of this species also is shown by the shape of its curve for rates of mitosis.

C. Nucleolar Cycles and RNA Synthesis

Cells of the quiescent centre differ from their neighbours in the meristem in several cytological respects (Clowes, 1956a, 1958b; Clowes and Juniper, 1964; Pilet and Lance-Nougarède, 1965; Hyde, 1967; Barlow, 1970a, b). Among these there is a conspicuous difference in the size of the nucleoli. Cap initials have nucleoli of twice the volume, and stelar cells just above the quiescent centre have nucleoli of eight times the volume, found in the quiescent centre cells in *Zea*. There are also differences in the numbers of nucleolar vacuoles and in the organization of the fibrillar and granular components (De la Torre and Clowes, 1972).

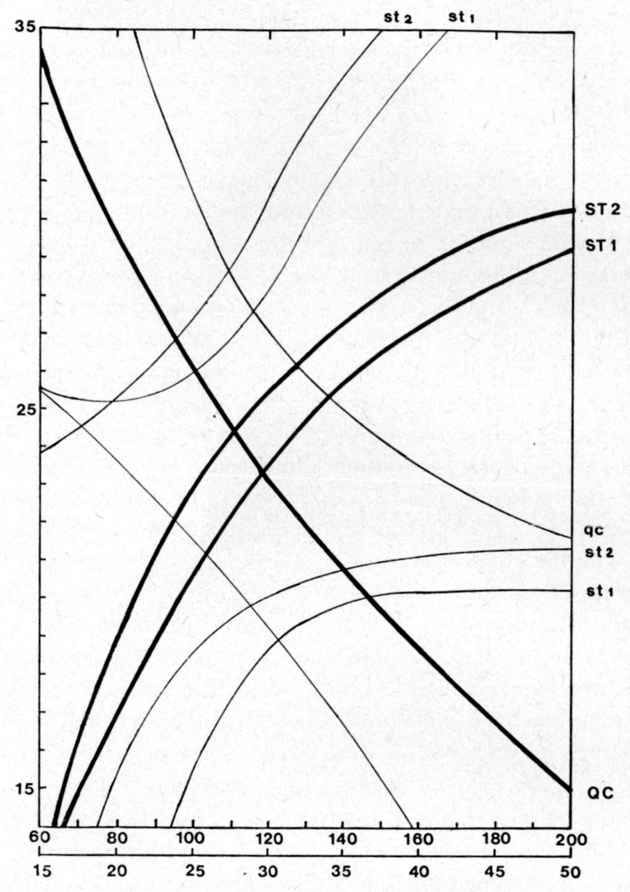

FIG. 3. Cell-doubling times in the cap initials plotted against cell-doubling times in either the stele or the quiescent centre of batches of roots of *Zea mays* growing in various conditions (see text). The thick lines are plots of the best fit to the data chosen by computer and the thin lines are the confidence limits. The upper scale of the abscissa refers to the quiescent centre, QC the lower to the two stelar regions, ST1 and ST2, defined in Fig. 2.

Nucleoli of meristematic cells exhibit cycles of disorganization, dispersion and reorganization which normally fit the mitotic cycle in the following manner. Disorganization starts during prophase and ends with the dispersion of the nucleolar material at the prophase–metaphase transition. Reorganization starts during telophase and ends during G_1. This is what happens in the stelar regions in *Zea*. In the cap initials, the rapid cycle is accommodated differently by the nucleolar cycle. Disorganization starts before the onset of prophase and re-organization starts at the beginning of telophase and is completed half way

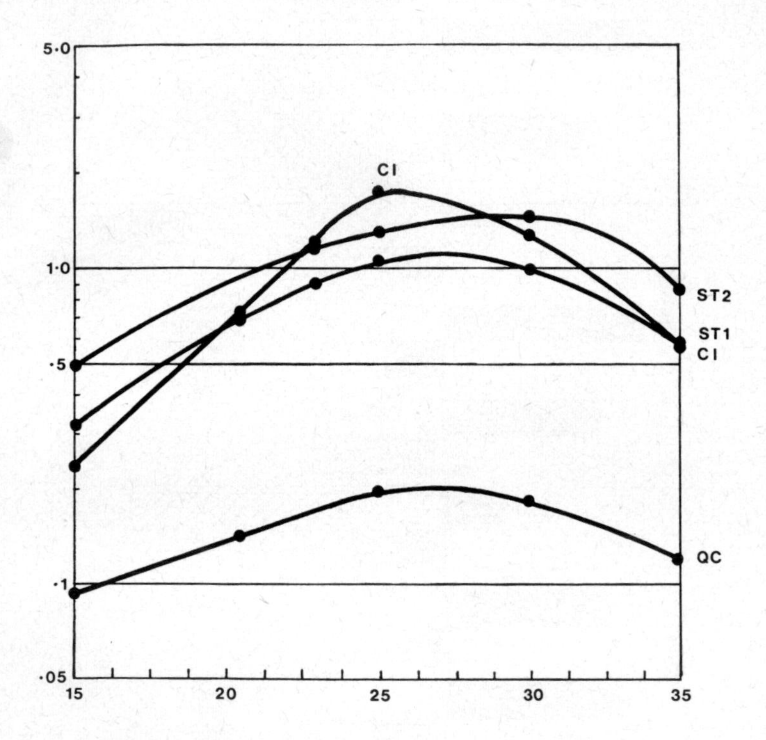

FIG. 4. Rates of mitosis (divisions per cell per day) of four regions of the root meristem of *Allium sativum* between 15 and 35°C. QC: quiescent centre; CI: cap initials; ST1: stele just above QC; ST2: stele 150μm above QC. Note log scale of ordinate. Data provided by A. T. Taylor.

through telophase thus allowing for the fact that G_1 is here eliminated by advancing S into telophase (Clowes, 1967). In the quiescent centre the nucleoli remain dispersed into G_1 and take longer to reorganize than in other regions (Fig. 5).

When roots are X-rayed the relationships between the nucleolar and mitotic cycles are altered in the quiescent centre and cap initials, though not in the stele. In the quiescent centre, after an acute dose of 1700 rads, reorganization of the nucleolus is immediately accelerated so that, for a short time, fully organized nucleoli are available in the last half of telophase. This condition is lost forever in the cap initials (Clowes and De la Torre, 1972). This is true also of roots treated with ethidium bromide, an inhibitor of DNA- and RNA-polymerases though the

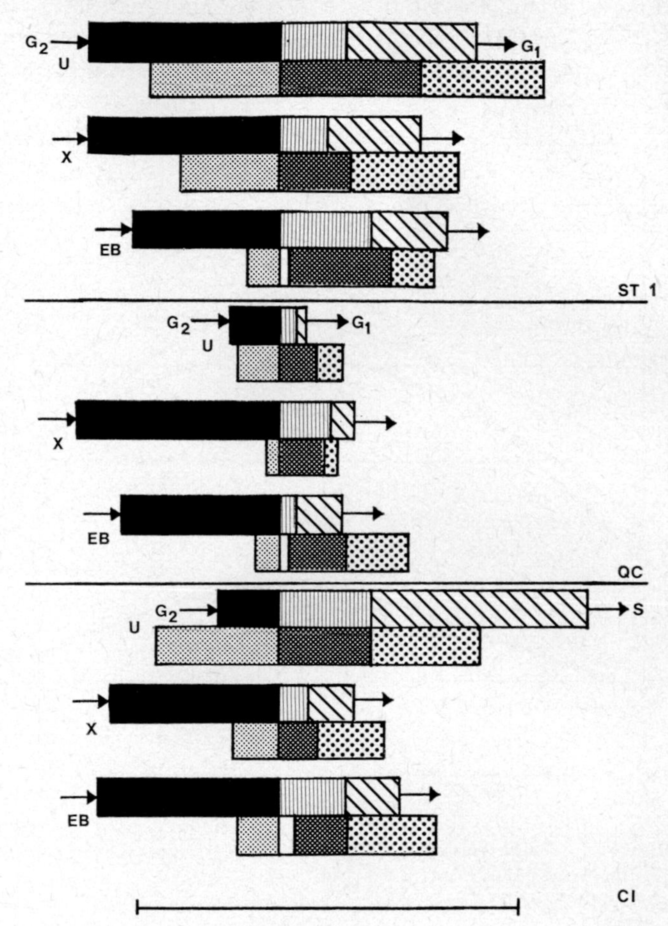

FIG. 5. The relationship between nucleolar and mitotic cycles in three regions of the meristem of *Zea mays* defined in Fig. 2. The top band of each pair represents the fraction of the cell cycle occupied by mitosis and the bottom band the fraction occupied by phases of the nucleolar cycle. Black—prophase; vertical shading—meta-anaphase; diagonal shading—telophase; small black dots—disorganizing; white dots—dispersed; large black dots—reorganizing; white—nucleoli persisting into metaphase. U: untreated roots; X: 1 h after acute dose of 1700 rads of X-irradiation; EB: after 1 h of treatment with ethidium bromide. The scale represents 10% of a cell cycle. Data from Clowes and De la Torre (1972, 1974).

drug, of course, does not have the same effect upon the quiescent centre as X-rays (Clowes and De la Torre, 1974).

IV. Discussion

There are three points about the quiescent centre especially worth discussing in relation to the development and function of roots namely its significance, its maintenance and its control at the cell level.

A. Significance

I have shown that a quiescent centre is a geometrical necessity for a meristem whose planes of cell division are organized in the way that exists in some kinds of roots. This appears at first sight not to be so in other kinds of roots and yet all roots have a quiescent centre. The key to resolving this paradox is a proper understanding of root apex structure in relation to the behaviour of the component cells. Two extreme examples of types of roots, which have caused much trouble in interpretation, may be considered.

One of these is the kind of root with a tetrahedral apical cell. The pattern of cells around the apical cell clearly shows their derivation by divisions within the tetrahedron by walls parallel to each facet in turn (or from the three basiscopic facets in some species). This fact has been used to prove that the apical cell does divide, but it provides no such proof. All that the pattern proves is that the apical cell did divide in this regular fashion in some past epoch. Exactly the same kind of argument can be presented for ordinary roots without apical cells and the dilemma of the observer who wants to accept the existence of a quiescent centre has been presented by Kadej (1963) and resolved (Clowes, 1964; D'Amato and Avanzi, 1965). There is a time in the development of the root primordium when there is no quiescent centre. This is the time when the cell pattern at the apex of the future root is decided. Guttenberg (1960) never conceded this point. He believed that the pattern of cells implies current activity and this view and its consequences continue to mislead students of root structure.

The other troublesome kind of root apex is the "open" type, such as *Vicia faba* and *Pisum sativum* possess. This was formerly interpreted as having a group of initials common to the cap, stele and cortex because the files of cells in the centre sometimes appear to be continuous from the stele into the cap. Autoradiography shows such roots to have quiescent centres just like other roots. Moreover, careful observation and the use of adequate cell wall stains show that there is a discontinuity in the files of cells at the boundary between the quiescent centre and the cap. It is true that this boundary is not inviolate as it is in roots of the "closed" type and the cap initials are renewed periodically by the mitotic activity of the quiescent centre. This may be seen in the pattern of cap cells, but does not invalidate the concept of the quiescent centre which still has an average rate of mitosis some ten times less than in the surrounding cells. The difference between this kind of root and the kind found in *Zea* is that mitosis in the quiescent centre contributes cells both to the cap and to the rest of the root whereas in *Zea* it contributes no cells to the cap because of the way the cells are polarized and consequently the way the files of cells are orientated (Fig. 6).

A further point about the significance of the quiescent centre concerns the concept of initials. This is largely a matter of semantics and not of science. Before

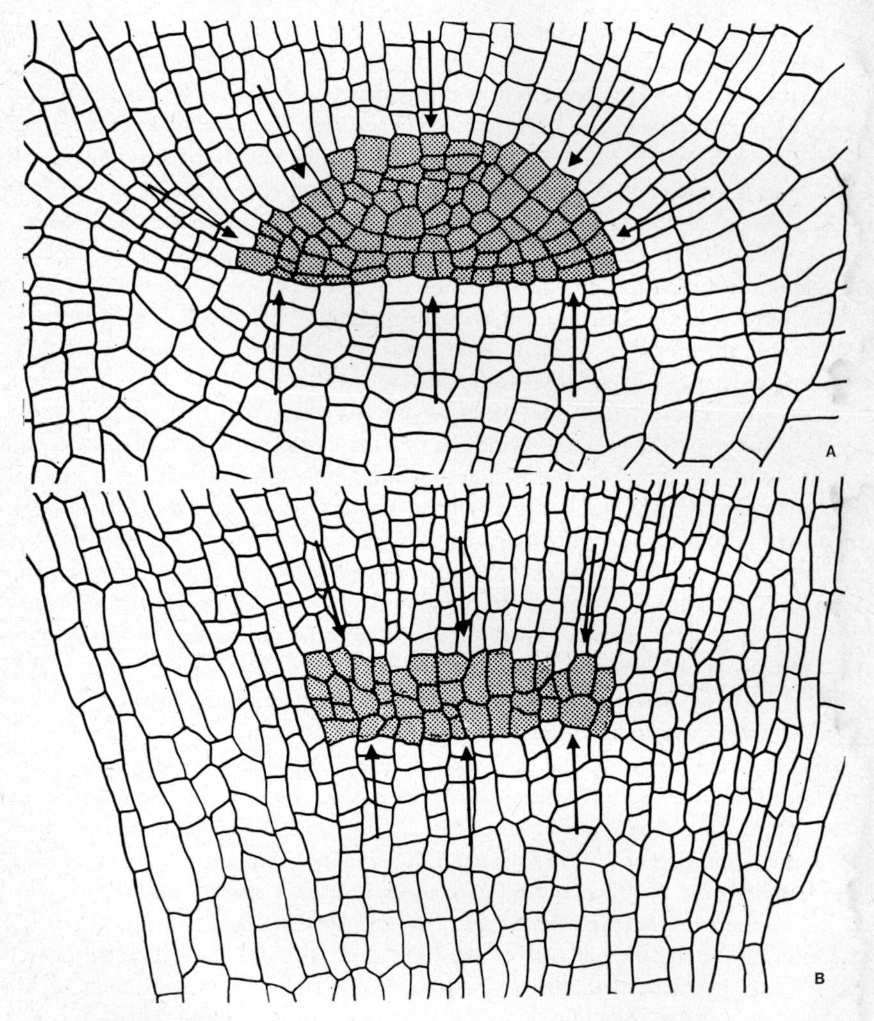

Fig. 6. Median sections of root apices of *Zea mays* (A) and *Pisum sativum* (B) to show the directions of restraint imposed by the dividing cells on the quiescent centres in different kinds of meristems.

1954 it was thought that angiosperm roots had a small number of initial cells situated near the axis of the root. The initials for the main body of the root would occupy the centre or the pole of the quiescent centre. With the discovery of the quiescent centre it may be thought more useful to designate the cells over its surface as the initials, but some people have preferred to retain the previous site on the grounds that the quiescent centre is not completely inert mitotically. Steeves and Sussex (1972) have then brought up the problem of where within the quiescent centre the mitosis occurs. If only the peripheral cells divide, then we merely think of the quiescent centre fluctuating in size. But, if mitosis is not so confined, they consider the centre of the quiescent centre to constitute the initials. A minor objection to this view is that this region includes no progenitors of the cap in roots of the "closed" type. A serious objection is that it is almost certainly pointless to look for permanent initials in any meristem. The flux of cells is such that we ought not to seek such cells either by their position or by their characters.

Everybody who has examined the quiescent centre finds that cells in mitosis or S occur more often near its periphery than at its pole. These facts provide no grounds for saying that the rate of mitosis is higher at the periphery. It would be laborious, though not impossible, to determine rates of mitosis at the pole of the quiescent centre, but our purpose is served merely by noting that cycling cells do occasionally occur at the pole. They occur there with so low a frequency in untreated roots that it is not possible to deny that their presence is due to some accident in the meristem that has not revealed its full consequence at the time of fixation.

B. Maintenance

Several theories have been put forward to explain the existence of the quiescent centre. One of these is that quiescence is due to lack of oxygen or perhaps some other substance obtained by the apex directly from the environment. The oxygen theory is attractive because it could explain directly the low sensitivity to ionizing radiation as well as failure to reach S or mitosis, but radiation experiments show that it is unlikely to be the reason for resistance to radiation (Hall *et al.*, 1962) and probably therefore can be neglected.

Another view is that the quiescent centre is starved of materials supplied from the shoot or seed by the prior demand of the proximal cells. The existence of fast cycling cap initials on the distal side could be explained here by the possibility of the effete cells of the cap returning substances to its meristem before being sloughed off. Feeding experiments, however, show that demand for nutrients is more likely to determine supply than the opposite, although supply may be attentuated by passage through demanding cells (Clowes, 1970b).

There have been many suggestions that the quiescent centre is the site of synthesis or of accumulation of hormones, (Clowes, 1961c; Torrey, 1963, 1972; González-Fernández *et al.*, 1968). This view has so far proved to be untestable and

I propose not to discuss it here partly because it is to be presented elsewhere in this volume in relation to cytokinins (see p. 55) and partly because my own approach to the subject is different from that of a hormone physiologist. It is important to note, however, that when root meristems are released from a general inhibition of mitosis as during germination of the seed (Clowes, 1958a; Byrne and Heimsch, 1970) or after dormancy imposed by cold (Clowes and Stewart, 1967) or after dormancy imposed by starvation (Webster and Langenauer, 1973) all the cells, including those of the quiescent centre, start cycling and quiescence is acquired only at a later stage of recovery. This suggests that quiescence is imposed by the proliferating cells primarily, rather than that the quiescent centre imposes quiescence on its own cells and cycling on the surrounding cells.

The view that I find most attractive is that the quiescent centre is maintained by the pressure exerted by the growth of the surrounding cells. The critical fact here is that the quiescent centres of different kinds of roots are of different shape and their shape fits into the pattern of the files of dividing cells. The files of dividing cells abut on to the quiescent centre from all round so that, in roots of *Zea* for example with very curved files in the cortex, the quiescent centre is more or less hemispherical with the flat side facing the tip. Here the restraint exerted by the cap initials on the flat surface of the quiescent centre is not balanced directly by the restraint exerted by the stele and cortex initials on the curved surface and the quiescent centre never contributes cells to the cap in normal growth whereas it does give cells to the stele and cortex. In species, such as *Pisum*, where the cortical files curve inwards very little, the quiescent centre forms a flat disc sandwiched between the cap initials on one side and the stelar and cortical meristem on the other. Here the restraints exerted on the two flat surfaces are directly opposed and, as we have seen, the quiescent centre's low rate of cell proliferation contributes cells both to the cap and to the rest of the root (Fig. 5). Similarly, in other species, quiescent centres of different shape can be explained on the cell pattern and so probably could their contributions to the different regions of the meristem.

The pattern of cells is decided in the early ontogeny of the root primordium by the planes and rates of cell division. The quiescent centre develops with the development of the cell pattern and with increasing mass and activity of the meristem.

This view and a hormonal theory are not mutually exclusive. Cell growth, cell division and hormonal levels are all interrelated and which one we consider as causal depends not on the known facts, but upon whether we are seeking a morphological or a physiological explanation. Radiation-induced stimulation of cycling in the quiescent centre, for example, could be considered to be due to relief from pressure exerted by the surrounding sensitive cells or by the availability of hormones derived from moribund cells or indeed to relief from competition for hormones or nutrients or the inactivation of certain enzymes. The very rapid response of the quiescent centre to irradiation (Clowes, 1970a, 1972b) suggests that the overall control of cycling throughout the meristem is a fine balance of forces and it may be a very complex balance. The complexity of the reaction is also shown when the meristem is interfered with in any other way. Surgical operations in recent years have concentrated on the control exerted by the cap (Pilet, 1971, 1973; Clowes, 1972c; Barlow, this volume, p. 21) because of the interesting effect upon geotropism, but there is no need to ascribe a special hormonal role for this tissue yet.

C. Cellular Control

At the cell level, views on the control of cycling are now centred round the principle control point hypothesis. Some kinds of cells seem to proceed through the cycle as if they have to pass stop–go signals at fixed points. These points are often in G_1 and G_2 and, once past them, the cell can proceed through S or mitosis to the next control point. On the basis of some experiments on whole root meristems of *Pisum* with non-ionizing radiations, Brown and Klein (1973) considered that the control point in G_1 may be associated with protein synthesis while that in G_2 may be associated with RNA synthesis. Webster and Van't Hof (1970) have shown that, in carbohydrate-starved roots of *Pisum*, while inhibitors of protein synthesis prevent recycling when the starvation is stopped, actinomycin-D has no immediate effect on release of G_1 cells to S or G_2 cells to mitosis. This resistance to actinomycin-D is lost slowly during the stationary phase suggesting the necessity for an RNA which is at first supplied from stores. If the quiescent centre is maintained by starvation we should expect this RNA to be depleted whereas, as we have seen, return to cycling in the quiescent centre can occur very quickly. Van't Hof and Kovacs (1971) have shown that in *Pisum* roots the cells whose cycles can be stopped in G_1 can also be stopped in G_2. There also appears to be a G_2 control factor produced in *Pisum* cotyledons which affects the roots (Evans and Van't Hof, 1973). There is therefore no distinction between different populations of cells as Gelfants (1963) thought there was in some mammalian cells.

Experiments are only now being done on the principal control point hypothesis in relation to the quiescent centre and only one important fact exists. This is that the cycle stops in G_1 in the quiescent centres of angiosperms, but allows cells into S and endoreduplication in ferns with apical cells. Clearly the ideas generated by the principal control point hypothesis ought to be examined using the quiescent centre and we ought to exploit the differences between angiosperms and ferns.

References

Avanzi, S. and D'Amato, F. (1970). Cytochemical and autoradiographic analyses on root primordia and root apices of *Marsilea strigosa*. *Caryologia* **23**; 335–345.

BARLOW, P. W. (1970a). Mitotic spindle and mitotic cell volumes in the root meristem of *Zea mays*. *Planta* (*Berl.*) **91**; 169–172.

BARLOW, P. W. (1970b). Vacuoles in the nucleoli of *Zea mays* root apices and their possible significance in nucleolar physiology. *Caryologia* **23**; 61–70.

BARLOW, P. W. and MACDONALD, P. D. M. (1973). An analysis of the mitotic cell cycle in the root meristem of *Zea mays*. *Proc. R. Soc. Lond.* **183**; 385–398.

BROWN, S. J. and KLEIN, R. M. (1973). Effects of near ultraviolet and visible radiations on cell cycle kinetics in excised root meristems of *Pisum sativum*. *Am. J. Bot.* **60**; 554–560.

BYRNE, J. M. (1973). The root apex of *Malva sylvestris*. III Lateral root development and the quiescent centre. *Am. J. Bot.* **60**; 657–662.

BYRNE, J. M. and HEIMSCH, C. (1970). The root apex of *Malva sylvestris*. II The quiescent centre. *Am. J. Bot.* **57**; 1179–1184.

CLOWES, F. A. L. (1954). The promeristem and the minimal constructional centre in grass root apices. *New Phytol.* **53**; 108–116.

CLOWES, F. A. L. (1956a). Nucleic acids in root apical meristems of *Zea mays*. *New Phytol.* **55**; 29–34.

CLOWES, F. A. L. (1956b). Localization of nucleic acid synthesis in root meristems. *J. exp. Bot.* **7**; 307–312.

CLOWES, F. A. L. (1958a). Development of quiescent centres in root meristems. *New Phytol.* **57**; 85–88.

CLOWES, F. A. L. (1958b). Protein synthesis in root meristems. *J. exp. Bot.* **9**; 229–238.

CLOWES, F. A. L. (1959). Reorganization of root apices after irradiation. *Ann. Bot.* **23**; 205–210.

CLOWES, F. A. L. (1961a). Duration of the mitotic cycle in a meristem. *J. exp. Bot.* **12**; 283–293.

CLOWES, F. A. L. (1961b). Effects of β-irradiation on meristems. *Expl. Cell Res.* **25**; 529–534.

CLOWES, F. A. L. (1961c). *Apical Meristems*, Blackwell Scientific Publications, Oxford.

CLOWES, F. A. L. (1963). X-irradiation of root meristems. *Ann. Bot.* **26**; 343–352.

CLOWES, F. A. L. (1964). Segmentation patterns in root apices. *Acta Soc. Bot. Pol.* **33**; 393–395.

CLOWES, F. A. L. (1965). The duration of the G_1 phase of the mitotic cycle and its relation to radiosensitivity. *New Phytol.* **64**; 355–359.

CLOWES, F. A. L. (1967). Synthesis of DNA during mitosis. *J. exp. Bot.* **18**; 740–745.

CLOWES, F. A. L. (1968). The DNA content of the cells of the quiescent centre and root cap of *Zea mays*. *New Phytol.* **67**; 631–639.

CLOWES, F. A. L. (1970a). The immediate response of the quiescent centre to X-rays. *New Phytol.* **69**; 1–18.

CLOWES, F. A .L. (1970b). Nutrition and the quiescent centre of root meristems. *Planta* (*Berl.*) **90**; 340–348.

CLOWES, F. A. L. (1971). The proportion of cells that divide in root meristems of *Zea mays*. L. *Ann. Bot.* **35**; 249–261.

CLOWES, F. A. L. (1972a). In *Chromosomes Today* (C. D. Darlington and K. R. Lewis, eds) Vol. **3**; 110–117. Oliver and Boyd, Edinburgh and London.

CLOWES, F. A. L. (1972b). Cell cycles in a complex meristem after X-irradiation. *New Phytol.* **71**; 891–897.

CLOWES, F. A. L. (1972c). Regulation of mitosis in roots by their caps. *Nature, Lond.* **235**; 143–144.

CLOWES, F. A. L. and DE LA TORRE, C. (1972). Nucleoli in X-rayed meristems. *Cytobiologie* **6**; 318–326.

CLOWES, F. A. L. and DE LA TORRE, C. (1974). Inhibition of RNA synthesis and the relationship between nucleolar and mitotic cycles in root meristems. *Ann. Bot.* **38**; 961–966.

CLOWES, F. A. L. and JUNIPER, B. E. (1964). The fine structure of the quiescent centre and neighbouring tissues in root meristems. *J. exp. Bot.* **15**; 622–630.

CLOWES, F. A. L. and STEWART, H. E. (1967). Recovery from dormancy in roots. *New Phytol.* **66**; 115–123.

D'AMATO, F. and AVANZI, S. (1965). DNA content, DNA synthesis and mitosis in the root apical cell of *Marsilea strigosa. Caryologia* **18**; 383–394.

DE LA TORRE, C. and CLOWES, F. A. L. (1972). Timing of nucleolar activity in meristems. *J. Cell. Sci.* **11**; 713–721.

DE LA TORRE, C. and CLOWES, F. .A L. (1974). Thymidine and the measurement of rates of mitosis in meristems. *New Phytol.* **73**; 919–925.

EVANS, H, J., NEARY, G. J. and TONKINSON, S. M. (1957). The use of colchicine as an indicator of mitotic rate in broad bean root meristems. *J. Genet.* **55**; 487–502.

EVANS, L. S. and VAN'T HOF, J. (1973). Cell arrest in G_2 in root meristems; a control factor from the cotyledons. *Expl. Cell Res.* **82**; 471–473

GELFANTS, S. (1963). Patterns of epidermal cell division. *Expl. Cell Res.* **32**; 521–528.

GONZÁLEZ-FERNÁNDEZ, A., LÓPEZ-SÁEZ, J. F., MORENO, P. and GIMÉNEZ-MARTÍN, G. (1968). A model for dynamics of cell division cycle in onion roots. *Protoplasma* **65**; 263–276.

GUTTENBERG, H. VON (1960). "Grundzüge der Histogenese Höherer Pflanzen I. Die Angiospermen." Gebrüder Borntraeger, Berlin.

HALL, E. J., LAJTHA, L. G. and CLOWES, F. A. L. (1962). The role of the quiescent centre in the recovery of *Vicia faba* roots from irradiation. *Radiat. Bot.* **2**; 189–194.

HYDE, B. B. (1967). Changes in nucleolar ultrastructure associated with differentiation in the root tip. *J. Ultrastruct. Res.* **18**; 25–54.

KADEJ, F. (1963). Interpretation of the pattern of the cell arrangement in the root apical meristem of *Cyperus gracilis* var. alternifolius. *Acta Soc. Bot. Pol.* **32**; 295–301.

PHILLIPS, H. L. JR. and TORREY, J. G. (1971). The quiescent centre in cultured roots of *Convolvulus arvensis. Am. J. Bot.* **58**; 665–671.

PHILLIPS, H. L. JR. and TORREY, J. G. (1972). Duration of cell cycles in cultured roots of *Convolvulus. Am. J. Bot.* **59**; 183–188.

PILET, P-E. (1971). Rôle de l'apex radiculaire dans la croissance, le géotropisme et le transport des auxines. *Bul. Soc. bot. Suisse* **81**; 52–65.

PILET, P-E. (1973). Growth inhibitor from the root cap of *Zea mays. Planta (Berl.)* **111**; 275–278.

PILET, P-E. and LANCE-NOUGARÈDE, A. (1965). Quelques characteristiques structurales et physiologiques du méristème radiculaire du *Lens culinaris. Bull. Soc. fr. Physiol. vég.* **11**; 187–201.

STEEVES, T. A. and SUSSEX, I. M. (1972). *Patterns in plant development.* Prentice-Hall, Englewood Cliffs.

TORREY, J. G. (1963). Cellular patterns in developing roots. *S.E.B. symp.* **17**; 285–314.

TORREY, J. G. (1972). In *The Dynamics of Meristem Cell Populations.* (M. W. Miller and C. C. Kuehnert eds.) Plenum Press, New York and London.

VAN'T HOF, J. and KOVACS, C. J. (1971). In *The Dynamics of Meristem Cell Populations* (M. W. Miller and C. C. Kuehnert eds.) Plenum Press, New York and London.

WEBSTER, P. L. and LANGENAUER, H. D. (1973). Experimental control of the activity of the quiescent centre in excised root tips of *Zea mays. Planta (Berl.)* **112**; 91–100.

WEBSTER, P. L. and VAN'T HOF, J. (1970). DNA synthesis and mitosis in meristems: requirements for RNA and protein synthesis. *Am. J. Bot.* **57**; 130–139.

Chapter 2

The Root Cap

PETER W. BARLOW

Agricultural Research Council Unit of Developmental Botany, University of Cambridge, England.

I. Introduction

The roots of all plants have at their tip a group of cells known as the cap. In many species with roots of the so-called "open" type of construction it is not always easy to see exactly where the cap is delimited from the root proper, but in roots with the "closed" type of construction, such as grasses, the cap can be distinguished from the root proper by a distinct cell wall that forms a common boundary to the two cell populations. The apex of roots with open and closed constructions are illustrated in Fig. 1, taking *Pisum sativum* and *Zea mays* as the respective examples. The reason for the difference between the open and closed apices lies in the caryopsis, showing the relation of the root tip and cap to the coleorhiza, is illustrated in Fig. 2.

Although Randolph did not pay particular attention to the development of the root cap of *Zea*, the embryogeny of this species is very similar to that of *Poa annua*. In this latter species Souèges (1924) and Guttenberg *et al.* (1954) have shown

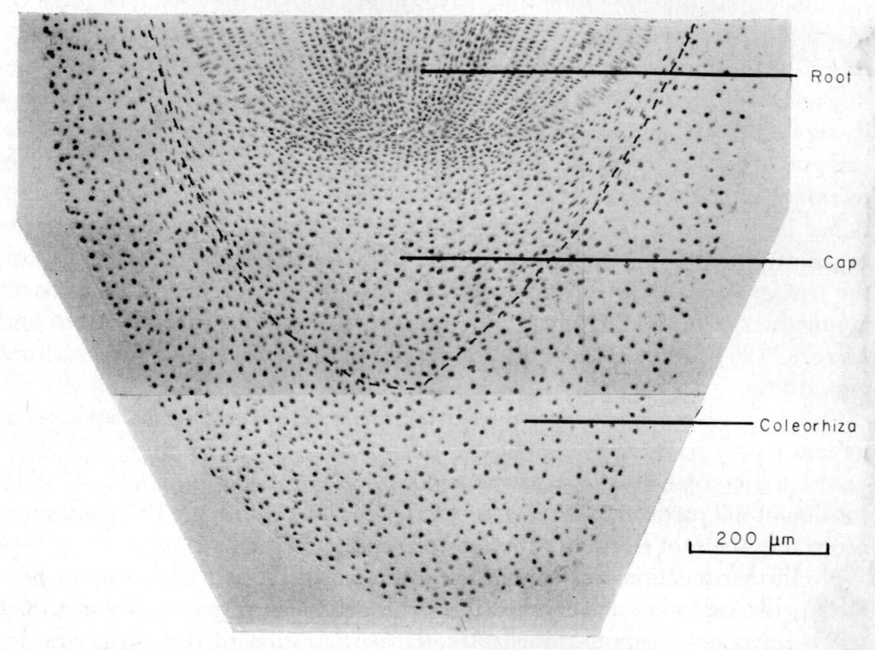

FIG. 2. Median longitudinal section of the primary root apex of *Zea mays* within the dry caryopsis. The boundary of cap and coleorhiza (broken line) can be recognized by the smaller size of the cells there. The suspensor is absent, it has probably withered away, but would have been joined to the region of the coleorhiza that is at the bottom of this photograph. Stained by Feulgen's reaction.

how the cap initials originate as a few cells that establish their independence from the primordium of the root proper at an early stage of embryogenesis. The position of the cap initials in the mass of embryonic cells becomes clearly delimited by an increase in the thickness of the cell wall that separates them from the more internally located root meristem cells (Fig. 3). The presence of this thicker cell wall indicates that once the two meristematic groups are established within the root primordium there is no interchange of cells between them. This course of development accounts for the origin of the "closed" construction of the apex and is maintained in the root apex of the germinated plant.

The development of the root cap during embryogeny of many other species of angiosperms is discussed by Guttenberg (1968). This author also describes the

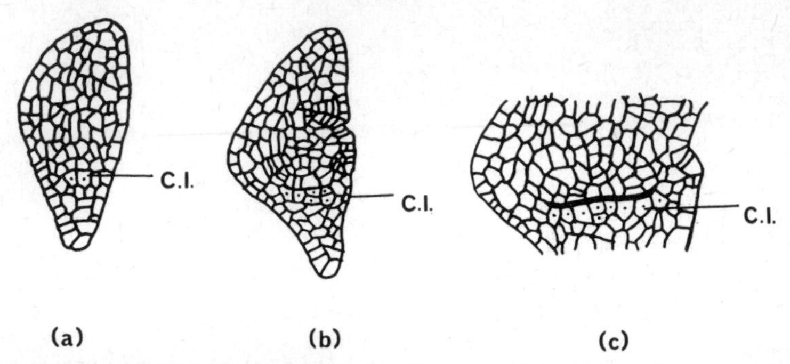

FIG. 3. (a), (b) and (c) are stages in the development of the cap meristem in embryos of *Poa annua*. The initials (C.I.) are marked with a dot. Note the formation of a cap–root boundary: it is the heavy line in (c). Redrawn from Guttenberg *et al.* (1954).

origin of cap tissue in the primordia of lateral roots, including *Zea mays* (Guttenberg, 1968; Schade and Guttenberg, 1951; see also Bell and McCully, 1970; and McCully, this volume, Chapter 6.)

IV. Cytology of the Root Cap

A. The Nucleus

1. *DNA Contents*

The DNA content of nuclei in cells of the dry, ungerminated, seed reflects the position in the cell cycle at which the nuclei were arrested when the embryo entered the dormant state at the end of its development. Nuclear DNA contents ("C" values: 1C is the DNA content of a haploid nucleus in G_1) in cells of root cap and coleorhiza in dry caryopses of *Zea mays* are shown in Fig. 4. In the region of the cap immediately distal to the initials most nuclei are held with the 2C and 4C values with a few nuclei reaching 8C. Peripheral cells of the cap and cells of the coleorhiza have most of their nuclei with the 4C and 8C values and a few with the 2C and 16C value. These data show that during cap and coleorhiza development some of the nuclei are participating in endomitotic cycles which result in DNA values higher than those found in mitotic cycling cells (i.e. 2C–4C). As this situation is also found in the caps of actively growing root tips after germination, we may conclude that cap cell differentiation, at the level of the nucleus, has occurred by the time the embryonic root becomes dormant.

When caryopses are set to germinate at 25°C the nuclei swell and 48 hours after the start of imbibition the first mitoses seen in the cap are in the most proximal four or five tiers of cells; no divisions are found in the more distal cells, thus the distribution of mitoses is similar to that seen in the cap of an emerged root (see later). Mitoses in the most proximal cell-tiers were also found at this time by Berjak and Villiers (1970) who, in addition, found these cells to be the only site of nuclear DNA synthesis in the cap. These results suggest that during the dormant period the tissue retains a "memory" of which cells were previously in a mitotic cycle and must resume cycling upon germination. Although the sequence of reactivation of nuclear cycles in the cap remains to be worked out in more detail, if the initial cells of the cap behave during germination as do cells in the meristem of the root proper (Deltour, 1971), it is likely that the onset of DNA

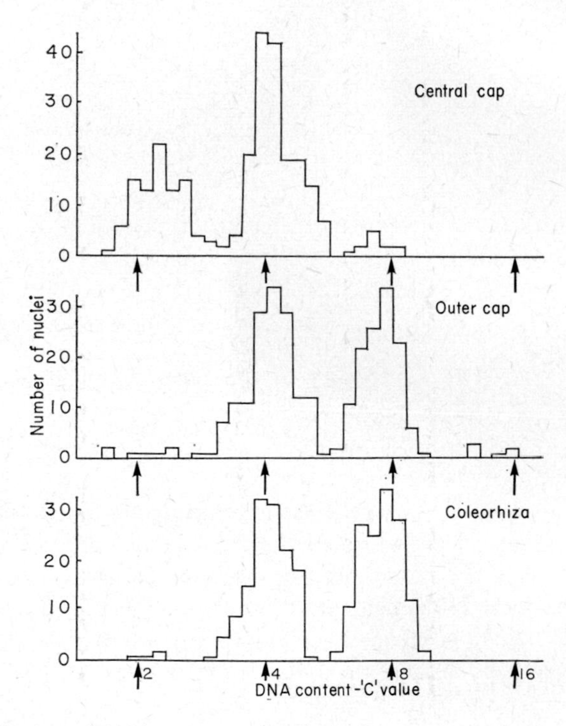

Fig. 4. Distribution of DNA contents of nuclei in three regions of a dry caryopsis such as the one shown in Fig. 2.

synthesis precedes mitosis by some hours. Whether the endopolyploid nuclei present in the dormant cap also resume DNA synthesis is not yet known. Berjak and Villiers (1970) do not record any of these nuclei as synthesizing DNA 44–48 hours after the onset of imbibition (the nuclei may have already done so), but DNA synthesis was found in these nuclei at this time after the seeds had been artificially aged by heat treatment (Berjak and Villiers, 1972).

In the caps of young, actively growing primary roots of *Zea* the pattern of cells allows two regions to be distinguished: columella cells, which form a core of the cap, and peripheral cells. The initials of the former group of cells divide predominantly transversely with respect to the axis of the root, while meristematic peripheral cells divide both transversely and longitudinally. Because of the rarity of longitudinal divisions in the columella, the arrangement of its cells is in lineages. This facilitates a detailed analysis of the columella cells as each tier of cells can be given a number to identify its position in the lineage. Tier number 1 refers to the cells that abut the wall that separates cap from quiescent centre, and then the more distal the cell the higher its tier number. Using this method of identification, an attempt has been made to locate cells engaged in

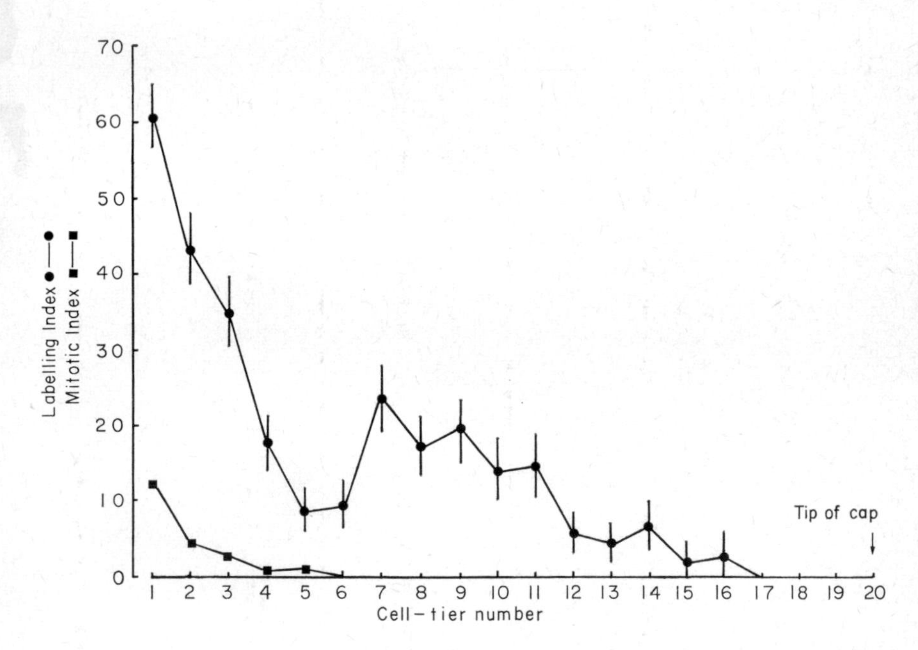

FIG. 5. Labelling and mitotic indices (%) of cells of the cap columella. Cell-tier number 1 is the most proximal layer of cells, the tip of the cap is at about cell-tier number 20. The error bars on the labelling index is the standard deviation of a binomial distribution.

DNA synthesis and mitosis in the columella with a view to gaining a better understanding of the relation of these two processes to each other during the life of a cap cell. In one experiment, root tips 2–3 cm long were exposed to a solution of tritiated thymidine (^{3}H-TdR) for 20 minutes and autoradiographs made of longitudinal sections. The nuclei of the cells in the different tiers were scored as mitotic or interphase and, if the latter, whether labelled or unlabelled. The result of this analysis is shown in Fig. 5. Mitoses are most frequent in cell-tier number 1, but are absent from all cells distal to tier number 5. Labelled nuclei are most frequent in tier 1, but their frequency falls to a lower value in tiers 5 and 6. Labelled nuclei increase in frequency in tiers 7–11 before declining and becoming absent from the cells at the tip of the cap (about tier 20–22). The interpretation of these findings is as follows. In these young roots the zone of cell division in the cap columella is confined to the five most proximal tiers of cells; the labelled and mitotic nuclei are indicators of cells in a mitotic cycle. The cells immediately distal to the initials are participating in an endomitotic cycle; the nuclei of tiers 7–16 labelled by the ^{3}H-TdR are in the endoS phase of this cycle.

To confirm this interpretation, the DNA contents of nuclei in cells in different regions of the caps of roots 2–3 cm long were determined by microdensitometry (Fig. 6). As expected in a mitotically-cycling population, the nuclear DNA

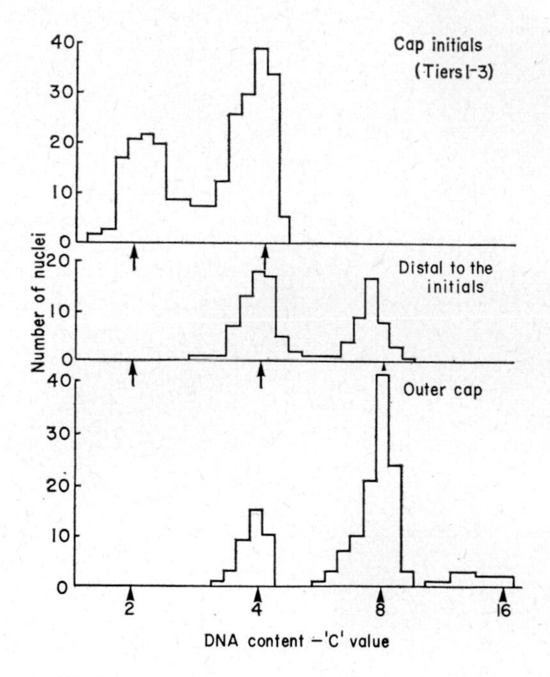

FIG. 6. Distribution of DNA contents of nuclei in three regions of the cap of a root 2–3 cm long, such as the one shown in Fig. 1b. The region called "distal to the initials" corresponds to cells of the columella in the 5th–9th tiers distal to the cap-quiescent centre junction.

contents in cells of tiers 1–3 are distributed between the 2C and 4C values. In the cells immediately distal to the initials, nuclei are found with 4–8C DNA contents; in the outer cells of the cap, the majority of nuclei have an 8C content, but a few nuclei reach a 16C content. Thus, microdensitometry and autoradiography complement each other perfectly and it seems clear that the non-dividing cells of the cap marked in an endoS phase were synthesizing DNA to go from a 4C to an 8C DNA content.

In these young roots, no nuclei with a 2C DNA content were found in cells distal to the initial zone and this raises the possibility that cells leave the mitotic cycle to enter the endomitotic cycle having replicated their DNA to the 4C level. This point was checked by carrying out further nuclear DNA measurements over the most proximal tiers of cells. The roots used in this study had been grown for an

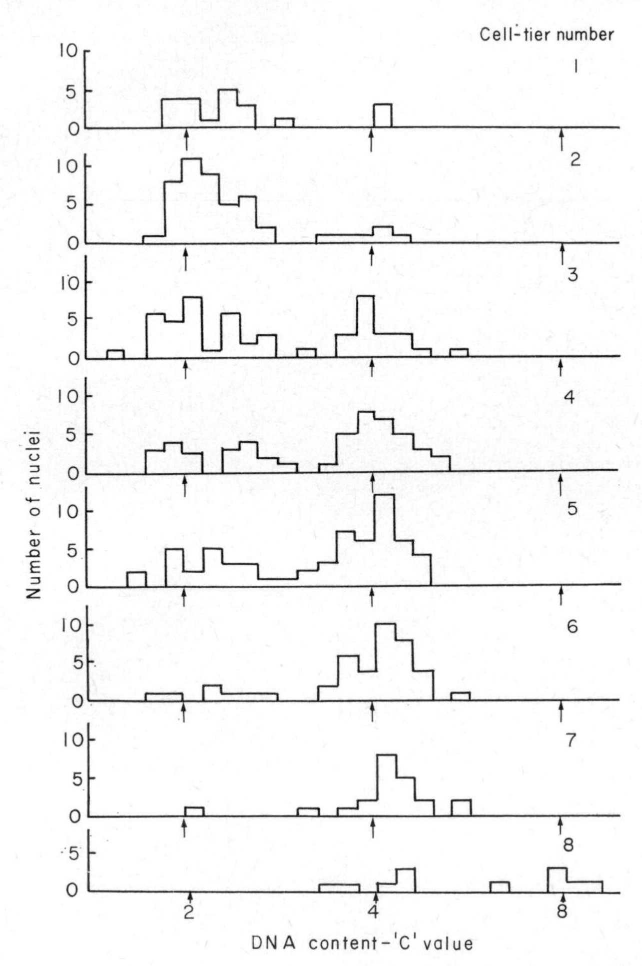

FIG. 7. Distribution of DNA contents of nuclei in different cell-tiers of the columella. Cell-tier 1 is the most proximal, then the more distal the cell the higher its tier number. Mitoses were found in tiers 1–3 only.

extra day in water. In such roots, mitoses are present only in the first three tiers of cells and here the nuclei have a DNA content close to the 2C value; nuclei with a 4C DNA content do not predominate until about the 5–7th tiers (Fig. 7). Thus, unlike the situation in the root caps described in the first experiment, which were a day younger, the nuclei in the caps of water-grown roots were passing from a mitotic into an endomitotic cycle with a 2C DNA content. It is not possible, therefore, to conclude that the "decision" of a cell to enter the endomitotic cycle is made at a particular phase of the nuclear cycle: probably cells leave the mitotic cycle with nuclei possessing either a 2C or a 4C DNA content.

In the caps of the water-grown roots and of roots sampled earlier in their growth, the nuclei never have a DNA content in excess of 16C. Such nuclei must have completed 2 or 3 endoS phases (the exact number of phases completed

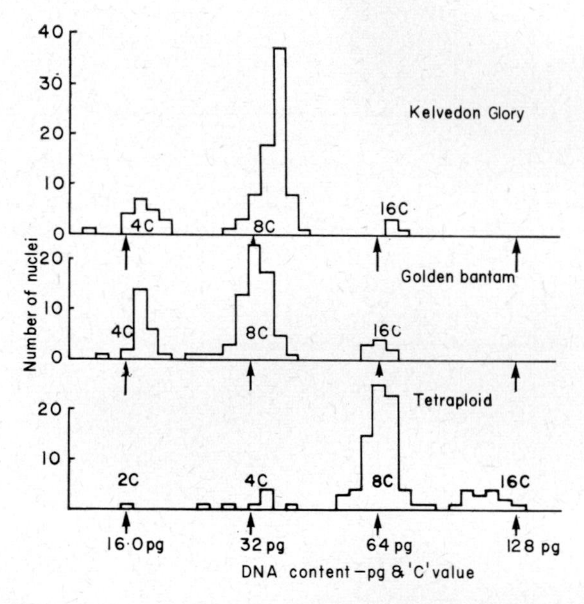

FIG. 8. Distribution of DNA contents of nuclei in the outer cap cells of two diploid cultivars of *Zea*, Kelvedon Glory and Golden Bantam, and a tetraploid strain (cf. Fig. 6). The roots were 5 cm long when fixed. The highest "C" value in each case is 16C. The DNA content of a 2C nucleus of c.v. Golden Bantam was calculated as 7·96 pg using *Hordeum vulgare* c.v. Sultan as a standard (2C = 13·5 pg).

depends on whether the cells left the initial zone with nuclei of a 4C or 2C DNA content respectively). The question now arises as to whether there is a limit beyond which the DNA content of a cap nucleus cannot increase. A partial answer is provided by comparing the DNA contents of nuclei in caps of diploid and tetraploid maize (Fig. 8). In two diploid cultivars, Golden Bantam and Kelvedon Glory, and in a tetraploid strain,* nuclei with a 16C DNA content are found in outer cap cells. As a tetraploid nucleus contains twice as much DNA as a

* Seeds of this tetraploid strain were supplied by Dr. J. W. Dudley, Illinois University, to The Botany School, Oxford University, England where stocks have since been maintained.

diploid nucleus of similar C value, and because the cap nuclei of all three varieties reached the 16C value, the quantity of DNA per nucleus would not seem to set the limit to the C value attained by cap nuclei. More probably, the nuclear DNA content of outer cap cells is determined by the duration of the endomitotic cycle and the time spent before the outer cells detach from the cap.

Cells with endopolyploid nuclei are found in the caps of some other grasses besides *Zea*, e.g. *Secale cereale* (8C) and *Hordeum vulgare* (8C), but not in *Spartina anglica*, where some nuclei in the outer cap reach only a 4C DNA content, or *Zizania aquatica* in whose cap all the nuclei of non-meristematic cells have a 2C content (personal observations).

2. *The Duration of Cell Cycles*

(*a*) *The Mitotic Cycle*. An estimate of the duration of the mitotic cell cycle for the cap initials of *Zea mays* was first made by Clowes (1961). A value of 12 hours was calculated using the results obtained from two independent methods, one relying on the accumulation of colchicine-metaphases, the other on continuous labelling of nuclei with ^{3}H-TdR. These two methods yield limited information about the mitotic cycle and, in addition, both require an assumption to be made concerning the proportion of cells that are participating in the cycle. Nevertheless, this value of 12 hours for the cycle duration is shorter than for other cells of the *Zea* root meristem, particularly the quiescent centre. The pulse labelling method, devised by Quastler and Sherman (1959), provides information about the duration of the component phases of the mitotic cycle and, further, only cells that are actually proliferating are considered. Using this method, Clowes (1965) confirmed a rapid rate of cycling in the cap initials and he also observed that some nuclei entered the DNA synthesis phase (S phase) at the telophase of mitosis (Clowes, 1967), that is, there was no G_1 phase in the daughter cells. The absence of a G_1 phase is only found in the columella initials; dividing cells at the periphery of the cap, and all other cells in the root meristem, have a G_1 phase in their mitotic cycle. The cells of the quiescent centre, a population that adjoins the cap columella initials, have a particularly long G_1 which occupies about 80% of the total cycle duration (Clowes, 1965; Barlow and Macdonald, 1973). Absence of a G_1 phase is unusual in cycling cells of both plants and animals and its omission from the cycle of the columella initials must indicate some difference in the biochemistry of these cells compared with other meristematic cells.

It is not known whether the cap initials of other species can dispense with a G_1 phase in their cycles; *Allium sativum* and *Avena sativa* are the only other plants, in addition to *Zea*, in which the duration of the component phases of the mitotic cycle of the cap have been timed. The cap initials of both species have a G_1 phase; in *Avena* it is the G_2 phase that has a very short duration. Table I summarizes what is known about the duration of the mitotic cycle in root cap cells.

One additional parameter that needs to be known about any meristematic

TABLE I. Duration of the mitotic cycle and its component phases in root cap initial cells.

Species	Mean Duration ± S.E. (Hours)					Temp °C	Author
	Cycle	G_1	S	G_2	M		
Zea mays	14	− 1[a]	8	5	2	18	Clowes (1965)
Columella initials	14·0 ± 0·7	− 0·3[a] ± 0·6	7·6 ± 0·9	5·3 ± 0·5	1·4	21	Barlow and Macdonald (1973)
Peripheral initials	22·5 ± 2·4	6·3 ± 1·8	6·7 ± 0·8	7·3 ± 0·4	2·2	21	Barlow and Macdonald (1973)
Allium sativum	27	3	13	6	5	20	Thompson and Clowes (1968)
Avena sativa	11·1 ± 0·4	3·1 ± 0·5	4·7 ± 0·3	0·5 ± 0·2	2·8	25	Harkes[b]
Convolvulus arvensis	13					23	Phillips and Torrey (1972)
Sinapis alba	25					22	Clowes (1962)
Vicia faba	44					19	Clowes and Hall (1962)

The results for *Zea*, *Allium* and *Avena* were calculated from the pulse labelled mitosis method, those for the other species were calculated from the rate of accumulation of metaphases by colchicine. Each method requires certain assumptions to be made about the behaviour of the cell population under study—see text and Barlow and Macdonald (1973) for further details. [a] A negative value for the duration of the G_1 phase was obtained from pulse labelling method and analysis of the resulting fraction labelled mitoses curves. It means that DNA synthesis starts before the phase scored as "mitosis" has finished. [b] These values were calculated by applying the methods referred to by Barlow and Macdonald (1973) and Macdonald (1974) to data supplied by Dr. P. A. A. Harkes.

cell population is the value of the proliferative fraction (P.F.) held by that population. Proliferative fraction, or as some people call it, growth fraction, is defined as the proportion of cells in a population which are progressing towards mitosis. When P.F. = 1·0, all the cells are participating in the cycle and will divide; when P.F. < 1·0, some cells are not capable of dividing, being arrested, either temporarily or permanently, in what some authors call a G_0 phase. It is difficult to determine the value of P.F. in most meristematic populations by direct experimentation. Clowes (1971) and Macdonald (1974) have attempted to estimate P.F. in the meristem of *Zea* using data obtained from pulse labelling experiments, as theoretically it is possible to calculate the fraction of non-cycling cells in the population by comparing the observed proportion of nuclei in S phase with the proportion expected in this phase after analysis of the fraction of labelled mitoses curve. However, this method of calculating P.F. uses the assumption that the nuclei which synthesize DNA will eventually divide but, in the case of the cap initials (described on p. 29) and perhaps other regions of the meristem too, cells may cease to progress towards mitosis, even after having synthesized nuclear DNA. With this reservation in mind, the estimates of P.F. made by Clowes and by Macdonald (opp. cit.) for cap initials of *Zea* are given in Table II.

A further uncertainty is introduced into the calculation of P.F. for cap initials by the fact that each tier of cells has its own characteristic pattern of proliferation (c.f. Fig. 5). Therefore, it is most important that the exact location of the cells sampled should be specified. The value of P.F. calculated by Macdonald (1974) used data that I had gathered from the first 3 or 4 tiers of cells of the columella initials and peripheral cap initials (presented in Barlow and Macdonald, 1973). This latter region very probably contained some cells that had entered an endomitotic cycle. As there is no way of distinguishing a nucleus in an endomitotic S phase from one that is in a mitotic S phase, the value of P.F. calculated for this region (given in Table II) is almost certainly an overestimate. Changes in P.F. from tier to tier in the cap initials are very likely to be the reason for Phillips and Torrey (1972) estimating the average mitotic cycle duration of cell-tier 1 of *Convolvulus arvensis* root caps to be twelve times shorter than the average cycle duration of cell-tiers 2–4. In this case cycle durations were calculated from the rate at which nuclei accumulated in metaphase in the presence of colchicine; obviously the rate of accumulation, and therefore the average rate of cycling, will appear to decrease if a proportion of cells is incapable of proceeding to mitosis (i.e. P.F. < 1·0). If it is assumed that all cells in tier 1 of the *Convolvulus* root caps are able to divide (P.F. = 1·0), and that cells in the more distal tiers do not slow down their rate of cycling before ceasing to proliferate (for which there is some evidence, see Barlow and Macdonald, 1973), then the rate of metaphase assumulation in cell-tiers 2–4 would be identical to that of tier 1 if only about 20% of all the cells in tiers 2–4 are in cycle, that is, their P.F. = 0·2 (Table II). This is not an unreasonable

Table II. Proliferative fraction (P.F.) of root cap initials.

Species	Location	P.F.	Temp °C	Author
Zea mays c.v. Golden Bantam	Peripheral cap initials	0·79	21	Macdonald (1974)
Zea mays c.v. Golden Bantam	Columella initials (tiers 1–4)	0·40	21	Macdonald (1974)
Zea mays c.v. Golden Bantam	"Cap initials"	0·83–0·89	23	Clowes (1971)
Avena sativa		1·0	25	Harkes[a]
Convolvulus arvensis	Columella initials cell-tiers 1	assumed 1·0	23	from Phillips and Torrey (1972)
Convolvulus arvensis	Columella initials cell-tiers 2–4	0·2	23	from Phillips and Torrey (1972)

[a] See footnote [b] of Table I.

value, as at some point towards the distal limit of the cap initial zone the P.F. must fall to zero.

Some readers may feel that I have discussed the proliferative fraction at undue length when so few facts about it are available, but I believe that any discussion of the behaviour of meristematic cells is incomplete unless it is recognized that some cells in the population may be destined not to divide. And further, if one wishes to understand the mechanism that brings about the transition of cells from a proliferative to a non-proliferative state, it is important to know exactly where in the tissue the cells make this transition so that they can be studied in more detail.

(*b*) *The Endomitotic Cycle.* As mentioned earlier (see p. 28), cells that have ceased dividing are still able to synthesize nuclear DNA in the course of an endomitotic cycle. It is possible to determine the duration of the endoS phase in such a cycle by a double labelling technique using two different radioactively-marked precursors of DNA, ^{14}C- and ^{3}H-TdR (Wimber and Quastler, 1963). In this way I have found (Barlow, 1974a) that in the cells immediately distal to the cap initials, whose nuclei are synthesizing DNA to take them from the 4C to the 8C DNA content, the endoS phase lasts 8·1 h; this value is similar to the S phase duration of the mitotically-cycling nuclei of the initials. Once again, it is necessary to know what fraction of the cell population is capable of participating in the endomitotic cycle (the endo-cycling fraction) before the duration of the remaining phases of the endo-cycle can be calculated. In these roots, which were 2–3 cm long, the endo-cycling fraction of the cells studied is probably close to 1·0, as microdensitometry shows that the majority of nuclei in the outer cap cells of such roots have the 8C DNA content or more (Fig. 6). If this is the case, then the duration of the endocycle is 29·3 h and the endoG phases, in which no DNA synthesis takes place, last 21·2 h. However, the endo-cycling fraction in older roots may be less than 1·0 because some nuclei reach the outside of the cap with a 2C DNA content, that is, they have not synthesized DNA since they completed their last mitosis.

(*c*) *Cell Turn-over.* Cells of the cap, in particular those of the columella, may be considered in the following way with regard to their nuclear and cell division cycles. Cells in tier 1 of the columella are a stem-cell population and generate all the cells of the columella. When a cell in tier 1 divides transversely, one daughter cell remains in tier 1 and the other lies in tier 2. Cells distal to tier 1 may divide, but do so with a probability that decreases the further they become displaced from tier 1; by about tier 5 the probability of division is zero (see Fig. 13). In contrast, the probability that a nucleus will synthesize DNA does not decrease so sharply on the displacement of a cell from the stem-cell tier, indeed a nucleus may still synthesize DNA even within a cell that has reached the distal margin of the cap. The transition of a cell from a dividing state to a non-dividing state may be due to the inability of the chromatin to be organized into chromosomes.

Cells are constantly being sloughed from the cap, thus the cap depends on the

TABLE III. Cap renewal times: the time taken for all cells in the cap of a primary root to be completely replaced by meristematic activity.

Species	Manner of calculation	Renewal time	Temp °C	Author
Zea mays c.v. Golden Bantam 3–5 cm roots	There are 5,600 dividing cells in a cap with a total of 10,000 cells. Mitosis replaces all these cells in . . . ∴ 10,000 cells sloughed off from cap each day.	1 day	23	Clowes (1971)
Zea mays c.v. Golden Bantam	When the cap is removed cells of the quiescent centre are stimulated to divide and regenerate a new cap in . . .	3–4 days	23	Barlow (1974b)
Avena sativa c.v. Seger I	Cap initials were labelled with ^{3}H-TdR[a] for 30 minutes. Cells with labelled nuclei were displaced from the meristem and sloughed from the cap after . . .	5–6 days	25	Harkes (1973)
Convolvulus arvensis cloned roots in liquid culture	Cap initials were labelled with ^{3}H-TdR[b] for 14 h. Columella cells with labelled nuclei were displaced from the meristem and sloughed from the cap after about . . .	6–9 days	23	Phillips and Torrey (1971)

[a, b] Methyl-^{3}H-thymidine used at 5μCi/ml and 0·2μCi/ml respectively. β-irradiation from the incorporated ^{3}H may impair the proliferation of cells that contain it and so may exaggerate renewal times.

activity of the meristem to replace the lost cells. The renewal time for caps of primary roots, that is, the time taken to renew all the cells of the cap, varies depending on the number of meristematic cells, the duration of their mitotic cycle and the total number of cap cells. Renewal times for the caps of three species, together with the manner of their calculation, are given in Table III.

B. The Cytoplasm

When cells of the cap leave the initial zone they may be said to commence to differentiate. Differentiation can be recognized by alterations in the structure of

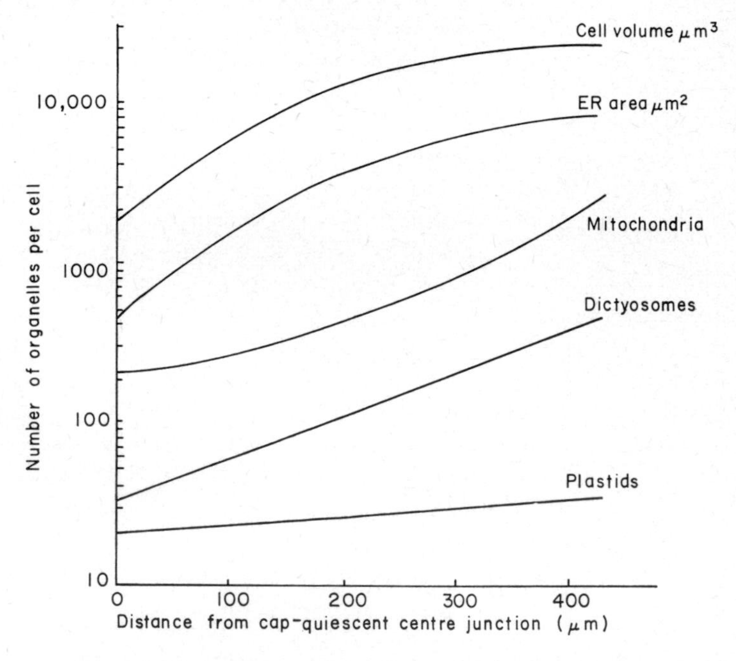

FIG. 9. Number of organelles per cell, total area of endoplasmic reticulum (E.R.) per cell and cell volume in different regions of the cap. Redrawn from Juniper and Clowes (1965).

the cytoplasmic organelles which, in turn, may lead to the changing metabolic activities of the cap cells (see Section V). A detailed analysis can be made of cytoplasmic differentiation with the electron microscope. To this end Juniper and Clowes (1965) counted the number of the various organelles in cells throughout the columella of the root cap of *Zea mays* (Fig. 9); from their quantitative analysis it is possible to draw certain conclusions about the development of the organelles in the cells as they differentiate during the period in which they are displaced from the initial zone to the outside of the cap. However, the conclusions only apply if the population of cells that constitutes the cap is in a "steady-state", that is, the rate of cell loss equals the rate of cell production. As Juniper and Clowes examined roots 3–5 cm long this was probably the case. Using the quantitative data of these authors, together with observations made with the electron microscope in collaboration with Dr. J. A. Sargent, the cytoplasmic organelles in cap cells of *Zea* show the following developmental changes.

1. *Plastids*

In *Zea* there are about 20 plastids in each cell of the cap (Fig. 9). The constant number, but changing appearance, of plastids leads to the following inference about their behaviour. In dividing cells the plastid population is in an immature, or proplastid, state (Fig. 10a) and in order to maintain a constant number the proplastids must double their number by division in the period between successive mitoses. As the cells make the transition from a proliferating to a non-proliferating state, the proplastids develop into amyloplasts (Fig. 10b) and must no longer be able to divide as their number remains constant in the non-dividing cells distal to the initials. Therefore, it seems that there might be some connexion between division of a cell and division of its plastids. However, plastids with small starch grains can sometimes be seen in mitotic cells of the cap showing that amyloplast development is not necessarily incompatible with cell division. Amyloplasts continue to grow until, in the central cells of the cap, they reach about 3 μm in diameter and contain up to about 10 starch grains. In the cells at the outside of the cap the starch in the amyloplasts is broken down (Fig. 10c). The build-up and the subsequent break-down of starch grains may reflect a balance between the enzymic activities concerned with each of these processes (Huber *et al.*, 1973). Maitra and De (1972) have suggested that as the starch contained in the amyloplasts of central cap cells of *Medicago sativa* is broken down, so the products are converted into a lipid that is highly osmiophilic and therefore very electron-dense when seen in the electron microscope. In the outermost cells these lipids are themselves broken down to leave a vacuole. The authors suggest that the lipids are used as a source of energy for the enlargement of the cell and the synthesis of the polysaccharide slime. In the cap of *Zea* lipid bodies are also seen in the cytoplasm but these are not particularly osmiophilic (Fig. 11a, b). Unlike the situation in *Medicago* there does not seem to be a correlation between the presence of lipid bodies and the decreasing amount of starch.

2. *Mitochondria*

In the cap initials there are about 200 mitochondria per cell and their number rises to about 2000 per cell in the most distal cap cells (Fig. 9). Thus, mitochondria

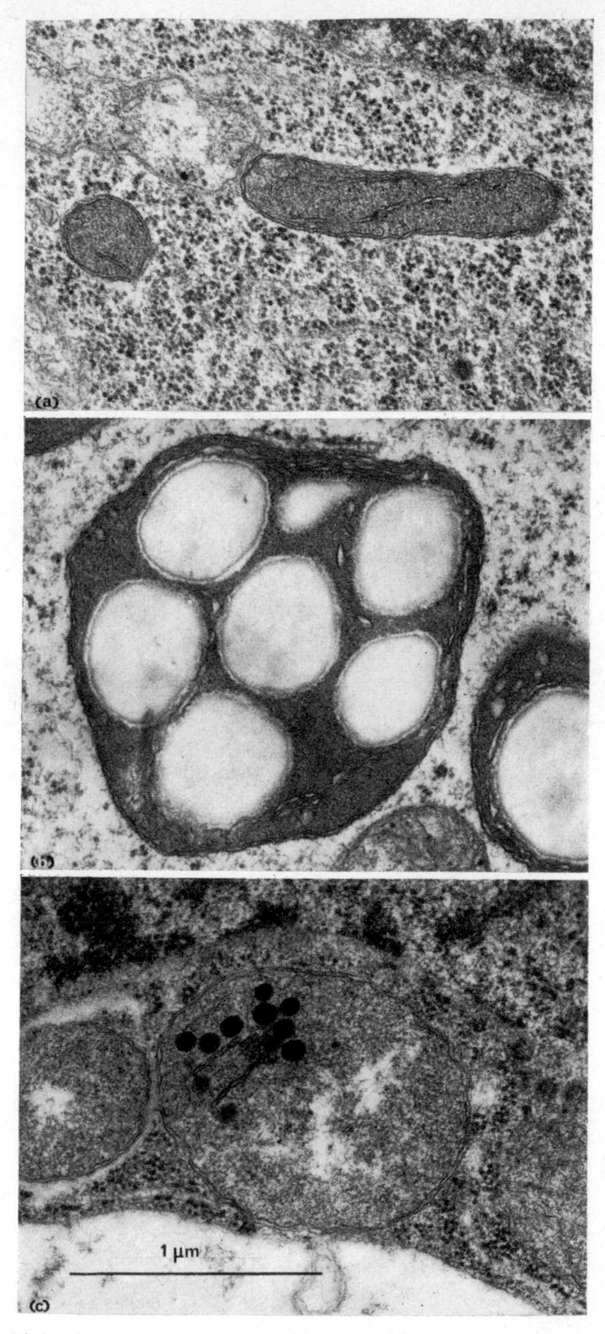

FIG. 10. Plastid development in cap columella cells of *Zea mays*. (a) Proplastids in an initial cell. (b) Plastid containing starch in a cell distal to the initials, 90 μm from the cap–Q.C. junction. (c) Plastid in a cell 300 μm from the cap–Q.C. junction. Starch has disappeared and osmiophilic globules are now present. The magnification is the same in (a)–(c). Glut/OsO_4 fixation: U/Pb staining in Figs 10, 11 and 12.

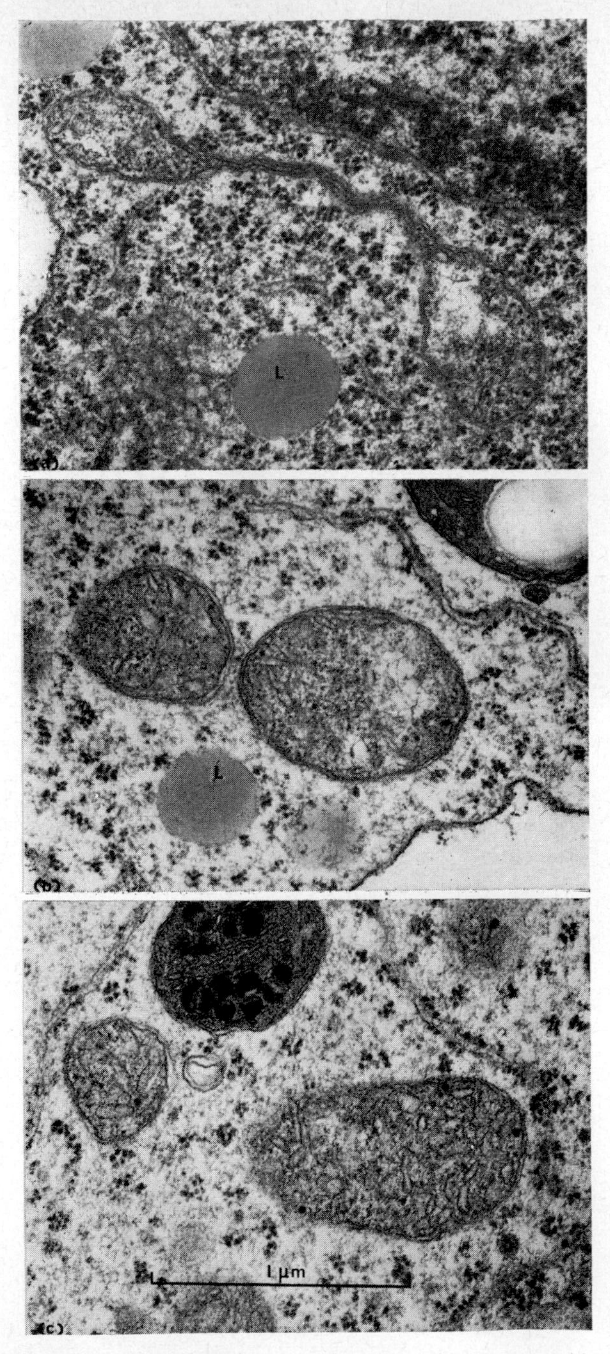

FIG. 11. Mitochondrion development in cap columella cells of *Zea mays*. (a) Mitochondrion in an initial cell. The dumb-bell shape of the mitochondrion may indicate that it is about to divide. Note that cristae are not evident. (b) Mitochondria in a cell distal to the initials, 90 μm from the cap–Q.C. junction. Cristae are more evident. (c) Mitochondria in a cell 240 μm from the cap–Q.C. junction. Cristae are well developed and small osmiophilic globules are also present. In (b) and (c) note the profiles of E.R. with clusters of ribosomes at intervals along their surfaces. L is a lipid droplet. The magnification is the same in (a)–(c).

continue to be formed throughout the life of a cap cell. The most probable means of their formation is by division (Fig. 11a). If it takes a cell in the columella 3 days to pass from the meristematic zone to the outside of the cap, the increase in the number of mitochondria can be accounted for if they double their number each day. But if the renewal time of 1 day, calculated for the cap by Clowes, is accepted (Table III) then their rate of increase must be three times faster (i.e. 8 hours). As the cells pass down the cap from the initial zone, the cristae of the mitochondria become more noticeably developed and seem to have a somewhat swollen appearance; in addition, the interior of the mitochondria contains osmiophilic droplets that are absent in the mitochondria of the initials (Fig. 11).

3. *Membranes*

The total amount of endoplasmic reticulum (E.R.), as judged from the length of E.R. profiles, increases as the cap cells mature (Fig. 9). In the cap cells described by Juniper and Clowes (1965) there was little if any vacuolation, so the total amount of E.R. per unit volume of cytoplasm was constant. The E.R. is covered by groups of polysomes in all cells of the cap; the grouping is dense in the initials, but in mature cells the groups are separated by stretches of E.R. free of polysomes.

The distribution of E.R. changes within cells throughout the cap (Juniper and French, 1973). In the initial cells profiles of E.R. are apparently scattered at random in the cytoplasm, perhaps because it is churned about at each mitosis. Once the cells leave the initial zone many of the profiles of E.R. come to lie parallel to the plasmalemma. In the outermost cells, the profiles of E.R. may be found thrown into loops and whorls (Clowes and Juniper, 1968).

Meristematic cells of the cap each contain about 30 dictyosomes; this number rises to about 200 per cell in the outer cap (Fig. 9). The magnitude of this increase (about 3 doublings over the length of the cap) is similar to the increases in the number of mitochondria and the length of E.R. profiles and all may reflect coordinated rates of production.

The structure of the dictyosomes in cells at various sites in the cap shows marked differences (Whaley *et al.*, 1959; Mollenhauer *et al.*, 1961; Clowes and Juniper, 1968). In the meristematic cells the dictyosomes have small vesicles in association with the cisternae, while in cells towards the outside of the cap some of the dictyosomal cisternae have a hypertrophied appearance and pinch off much larger vesicles whose contents are discharged through the plasmalemma into the cell wall. Cells that have detached from the cap show dictyosomes that are no longer hypertrophied (Clowes and Juniper, 1968) but dictyosomes with both hypertrophied and non-hypertrophied vesicles can be found in one and the same cell elsewhere towards the outside of the cap. These differences in structure, illustrated in Fig. 12, relate with differences in function and confer on the cells different biochemical properties, as judged by the pattern of polysaccharide synthesis in the cap.

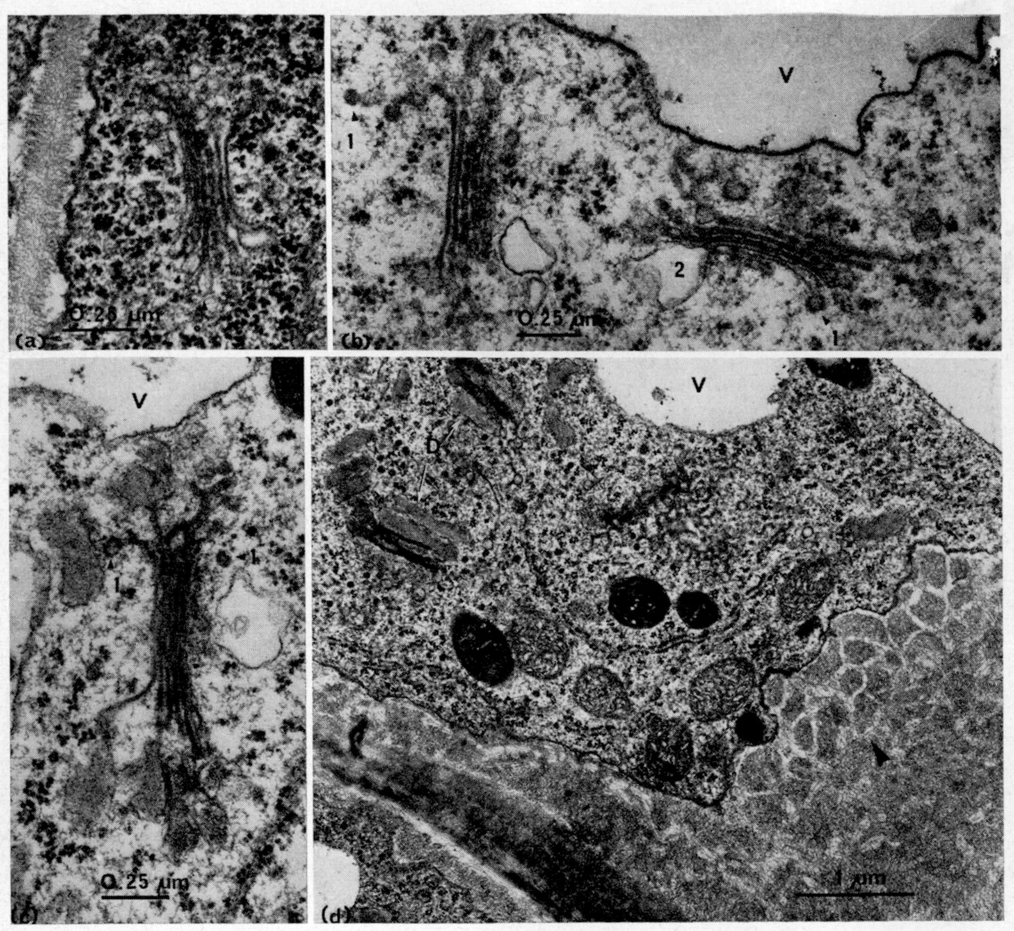

FIG. 12. Dictyosome development in cap columella cells of *Zea mays*. (a) Dictyosome in an initial cell. The dictyosome is producing electron-lucent vesicles (arrow). (b) Dictyosomes in a cell distal to the initials about 90 μm from the cap–Q.C. junction. Two types of dictyosome vesicles are apparent: type 1 which are small and electron dense, and type 2 which are larger and electron-lucent. The latter may contribute to the vacuoles, V, which are present in all the cap cells of this root. (c) Dictyosome in a cell 175 μm from the cap–Q.C. junction. The dictyosome is producing another type of vesicle that contains a fibrous material (probably polysaccharide slime). Vesicles of type 1 in Fig. 12b may also be being formed. (d) From a cell 125 μm distal to that in (c). Here the dictyosomes (D) have the entire cisterna filled with fibrous material. These are believed to detach from the dictyosomes as vesicles, the membrane of which fuses with the plasmalemma and the vesicle contents pass into the space between wall and plasmalemma. In this photograph the discharged products of the dictyosomes seem to be aggregated as separate packets before coalescing (arrow).

V. Biochemical Properties of Root Cap Cells

In Section II mention was made of the slime produced by cap cells and its possible role in the growth of the root. The slime contains an acidic polysaccharide that has properties of pectin (Harris and Northcote, 1970; Rougier, 1971; Wright and Northcote, 1974). As slime is produced in such abundance by the cap, it has been relatively easy, using autoradiography, in conjunction with electron microscopy, to establish which cells of the cap are responsible for slime production and even whereabouts in the cell slime is made. All the evidence points to the hypertrophied dictyosomes in mature cap cells as being concerned with both the synthesis of slime and its transport to the exterior of the cell.

It was Juniper and Roberts (1966) who first noticed a correlation between the hypertrophied dictyosomes in the outer cap cells of *Zea mays* and a high rate of incorporation of ^{14}C-glucose and ^{14}C-arabinose by these cells. A more direct relation between these two observations was established in the same year by Northcote and Pickett-Heaps (1966) who, using *Triticum vulgare*, showed that the hypertrophied dictyosomes of the outer cap cells were rapidly labelled when D-6-^{3}H-glucose was fed to the roots. On returning the roots to non-radioactive medium, radioactivity was then found over free dictyosome vesicles and later over the cell wall and slime. The transfer of label from the dictyosomes to the cell wall took only 15–20 minutes to accomplish. This sequence demonstrated the dictyosome to be both a dynamic organelle continually budding off vesicles that move across the cytoplasm to fuse with the plasmalemma, and directly involved in polysaccharide synthesis. These observations, together with a chemical analysis of the slime (Harris and Northcote, 1970) and the results of histochemical studies on the contents of the dictyosome vesicles (Coulomb and Coulon, 1970; Rougier, 1971), make it likely that the hypertrophied dictyosomes are the site of synthesis of the slime. Additional evidence for this conclusion comes from results of Bowles and Northcote (1972) who showed that when roots of *Zea mays* were fed with D-^{14}C-U-glucose, and the subcellular fractions examined for radioactivity, the dictyosome-membrane fraction of the cap was found to contain radioactive fucose. Fucose is a sugar that is much more prevalent in the cap of *Zea mays* than it is elsewhere in the root, and further, it is present in relatively large amounts in the weakly-acidic component of slime (Wright and Northcote, 1974). Thus fucose can be considered to be a specific marker for slime and finding it in the dictyosomes is good evidence that these organelles are instrumental in its production. The constituent sugars of the slime of *Zea* are given in Table IV. Up to the present, chemical analyses of root tips shows fucose to be found in substantial amounts only in the slime of *Zea*; it is found in small amounts in the polysaccharides extracted from root tips of *Acer pseudoplatanus*, *Pisum sativum* and *Triticum vulgare* (Wright and Northcote, 1974).

When roots are fed radioactive sugars, such as glucose, galactose or arabinose,

TABLE IV. Composition of the slime secreted by cells of *Zea mays*. The amounts given are the percentages, by weight, of the constituent sugars.

Constituent	1	2	3	4 a	4 b
Galacturonic Acid	Present	12			
Glucuronic Acid	Present				
Uronic Acids				—	6·9
Galactose	21	39	37·7	18·4	22·1
Glucose	22	22	16·7	27·3	—
Mannose	6	6	7·5		
Fucose	32	8	16·7	16·6	33·1
Rhamnose		1			
Arabinose	15	7	6·7	15·1	23·5
Xylose	4	5	12·8	6·2	14·3
Fructose				16·4	—

(1) *Z. mays* c.v. Orla 266. The slime was wiped from the *root cap* of surface-sterilized seedlings, hydrolysed and the neutral sugars analysed by gas-liquid chromatography. From Harris and Northcote, 1970. (2) *Z. mays* hybrid WF-9 X M-14. The slime was collected from the *root cap* of surface-sterilized seedlings, precipitated in 80% ethanol and the neutral sugars in this fraction analysed by gas-liquid chromatography. Uronic acids were determined by the carbazole procedure. From Jones and Morré, 1973. (3) *Z. mays* c.v. Caldera. The slime was wiped from sterile *callus* derived from *coleoptile* tissue, hydrolysed and the neutral sugars analysed by gas-liquid chromatography. From Wright and Northcote, 1974. (4) *Z. mays* hybrid WF9 X 38-11. Slime was collected by suction from *nodal roots* 0·5–2·5 cm long, precipitated, hydrolysed and the sugars analysed from paper chromatograms. Uronic acids were determined by the carbazole method. Column (a) is the analysis of the slime from the 1st and 2nd nodal roots; column (b) is for the 4th–6th nodal roots. The roots were not sterile and therefore bacteria could contribute an unknown amount of carbohydrate to the results. Acid phosphatase and ATPase were present in both slime samples. From Floyd and Ohlrogge, 1970.

much less label is found in the cells of the cap interior than in cells towards the exterior (Juniper and Roberts, 1966; Northcote and Pickett-Heaps, 1966; Barlow, 1974b; Dauwalder and Whaley, 1974). For example, after feeding root tips with D-6-^{3}H-glucose for 4 hours, autoradiographs of sectioned caps show over twice as many silver grains per unit area over the cytoplasm and walls of outer cap cells compared with the inner cells (Barlow, 1974b). Although a differential penetration of the radioactive sugars cannot be ruled out, it would seem probable that they are utilized differently by the inner cells than by the outer cells, where, in the latter, the label is incorporated into the slime. Northcote and Pickett-Heaps (1966) found that in the inner cells ^{3}H-glucose found its way to the starch grains within the amyloplasts, but the starch in the outer cells was unlabelled; this is possibly because the starch in the outer cells is being degraded faster than it is being synthesized. Further, the disappearance of starch from the amyloplasts occurs in the same cells that contain hypertrophied dictyosomes. Therefore, the sugars released from the starch could enter a pool available for utilization by the dictyosomes. Thus, the patterns of polysaccharide (starch, slime, etc.) metabolism within the cap tissue, besides showing regional differences, may also be interdependent processes.

The extent to which the differences are due to changes in enzyme complements within the various compartments and organelles in the cytoplasm is not known for certain, nevertheless it should be seriously considered that the pattern of slime synthesis in cap tissue may be controlled by differentiation of the dictyosomes. In support of a thesis of organelle enzyme differentiation, Dauwalder *et al.* (1969) have found a correlation between the amount of inosine diphosphatase detectable in dictyosomes in different regions of the cap and the pattern of ^{3}H-glucose uptake. Thiamine pyrophosphatase could also be detected in dictyosomes, and for some reason the reaction for this enzyme was stronger in peripheral cap cells surrounding the epidermis than elsewhere in the cap.

A cytoplasmic organelle that may originate from a particular class of dictyosome vesicles is the lysosome. This was suggested by Coulomb and Coulon (1970), who found that dictyosomes, besides producing large vesicles containing polysaccharides, also produced a class of smaller vesicles that contained acid phosphatases and these were presumed to be lysosomes. The two differently-sized dictyosome vesicles were found by Coulomb and Coulon in root tips of *Cucumis pepo* and have also been seen in *Zea* root caps (see Fig. 12). Berjak (1968) equated lysosomes with electron-dense bodies seen after permanganate and lead citrate staining, and which also showed acid phosphatase and esterase activities. She found that in *Zea mays* the number of these bodies per cell section increased nearly fourfold over the distance from the region of the cap initials to the outer cells. In the latter cells the dense bodies disintegrate and Berjak believes the release of the hydrolytic enzymes contained within the bodies may result in autodigestion of the outer cap cells. This event could contribute to the detachment of the cells from the cap and, if it also occurs in the cap cells of lateral root primordia, may assist in the penetration of the laterals through the cortex of the primary root by digesting, or softening, some component of these cells (see also Bell and McCully, 1970). Another consequence of the release of hydrolytic enzymes from outer cap cells could be to assist the emergence of the root during germination. In dry caryopses of *Zea*, cells of the cap and coleorhiza are often continuous and, during imbibition, activation of the dictyosomes and lysosomal vesicles occurs (Berjak and Villiers, 1970) which may result in the split that appears between the two groups of cells and the penetration of the root tip through the coleorhiza.

Other biochemical differences between cells in different locations in the cap are incompletely understood and their significance can only be guessed at. Ashford and McCully (1970) found regional differences in the activity of β-glucosidase in the cap, but it is not known how this relates to the biochemistry of the cells *in vivo*. Mitochondrial activities may vary within the cap; Huber *et al.* (1973) have suggested that such changes in these organelles influence the metabolism of starch and therefore may effect amyloplast development.

VI. Towards an Understanding of the Control of Cap Cell Differentiation

As we have seen (Section IV.A.1), cell proliferation in the cap is limited to a zone of cells around the root–cap junction. The number of proliferating cells remains approximately constant, or may decline slightly, with continued root growth. And as proposed earlier (Section IV.A.3), this may be explained by the decreasing probability of cell division occurring the further a cell is displaced from the tier of stem cells (tier 1). Continuous meristematic activity of the cap initials constantly generates cells which, when they cease dividing, are displaced from the initial zone towards the outside of the cap where they slough-off. As the total number of cells in the cap also remains approximately constant the cap is a population of cells in "steady-state", that is, cells are being lost at approximately the same rate as they are being generated. The structure and properties of cells in their transit through the cap were described in Sections IV and V and represent a sequential differentiation, or development, of the cells. The problem that remains to be tackled is one that concerns the control of differentiation.

Two propositions that help delimit a view of developmental control are that the properties of a cell in a developing system are determined by (1) its position within the system, and/or (2) the time that has elapsed since its generation at mitosis (Wolpert, 1969; Summerbell *et al.*, 1973). With reference to the root cap, the cellular properties that may be most easily monitored and perhaps experimentally manipulated, are indicated in Fig. 13. In this scheme, which also summarizes much of what has been described in the foregoing sections, a cell is considered to have a probability that it will do certain things in a particular location. For instance, a cell in tier 1 almost invariably synthesizes nuclear DNA and then divides; in a more distal location the probability of division falls off sharply and the cell then becomes a non-dividing cell and acquires other properties. Because a cell distal to the meristematic zone can still synthesize nuclear DNA it may become endopolyploid. As a cell is displaced towards the tip of the cap, so the probability increases that starch will be deposited in the plastids, slime will be synthesized and lysosomes will become active. Although probability is plotted against an axis marked "cell-number", the question (in terms of the two propositions stated above) is whether this axis is the positional co-ordinate of the cell or the time that has elapsed since the generation of the cell. Similarly, if temporal cues determine differentiation, a time axis could replace the positional axis of Fig. 9. I believe that it is feasible to perform experiments that could decide whether a cell differentiates in response to positional and/or temporal cues, or to neither. However, such theories and hypotheses are useful as concepts only and should

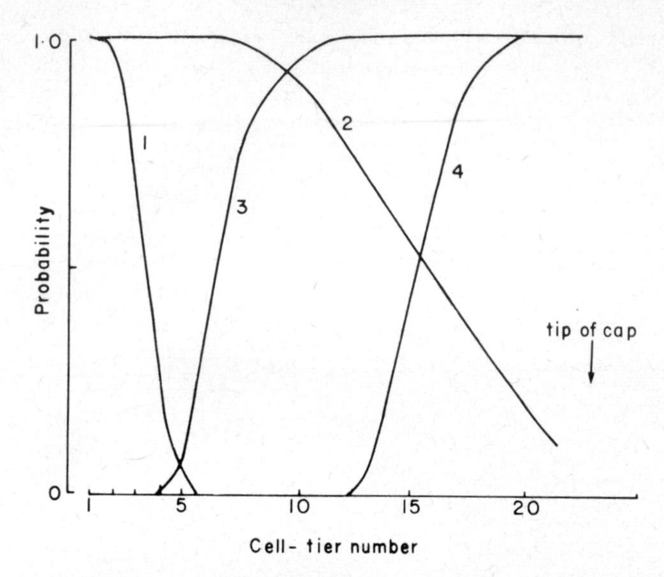

FIG. 13. A model for cap cell differentiation. Probability of an event is plotted against cell position in the cap. Probability line 1: that a cell will enter mitosis; 2: that the nucleus will continue to synthesize DNA; 3: that starch will be deposited in the plastids; 4: that slime will be synthesized and starch will be degraded.

not be considered substitutes for the realities of differentiation; these realities reside in a matrix of stochastic changes of molecular activity.

VII. The Influence of the Cap on the Behaviour of Cells Elsewhere in the Root

A. The Quiescent Centre

Elsewhere in this volume (pp. 16 and 67) reference is made to the possible role of the cap in maintaining the group of cells known as the quiescent centre in their inactive state. This possibility was first suggested by Clowes (1954) and later by myself (Barlow, 1971, 1973, 1974b). The hypothesis is mechanistic rather than hormonal (cf. Feldman, Chapter 3) and may be stated as follows, as it applies to the root apex of *Zea*. The meristematic cells of the cap columella grow in a direction parallel to the axis of the root. These initial cells are attached to cells of the epidermal lineage at the apex of the root proper whose preferred direction of growth is at right angles to the direction of growth of the columella cells. Because cell walls do not slide over one another, and because the preferred directions of growth of the two groups of cells are not the same, the epidermal cells must be prevented from growing by the constraint due to the attached columella cells. It follows that if the epidermal cells do not grow, then the cells in the adjoined lineage of cortical cells cannot grow either; these latter cells will in turn prevent growth of the cells in the neighbouring lineage, and so on. The orientation of cell lineages at the root apex is such that the zone where these non-growing cells is located is hemispherical and coincides with the quiescent centre. It follows from this hypothesis that constraint to cell growth must be translated into a constraint to cell division and other metabolic activities, but how this comes about is not known.

The hypothesis that the cap imposes quiescence on cells at the pole of the root can be put to the test by removing the root cap and then looking to see how the quiescent centre behaves. As expected, decapping stimulates the cells of the

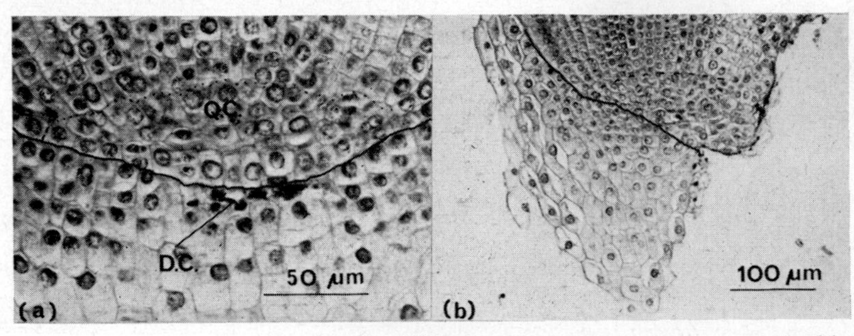

FIG. 14. To show the reaction of cells of the quiescent centre to damage sustained by the cap. (a) Cells of the cap initials have died as a consequence of unsuccessful decapping 18 hours earlier. Note that cells of the quiescent centre (Q.C.) adjacent to the dead cells (D.C.) have enlarged and divided. (b) Section of a root tip fixed 28 h after cutting the cap longitudinally and removing one half. Note that cells of the Q.C. exposed by the half-decapping have grown and divided, those in contact with the remaining cap have not. The solid line traces the cell wall between cap and root proper, the broken line traces the basiscopic face of the original quiescent centre.

quiescent centre to grow and divide with the consequence that a new root cap is eventually regenerated with planes of cell division and growth similar to unviolated roots. Once the new cap is complete a quiescent centre is re-established (Barlow, 1974b). These observations can be interpreted as providing some support for the hypothesis. Additional favourable evidence comes from two other observations. First, occasionally in normal roots and in roots where decapping was unsuccessful, a few of the cap initials die and collapse. The cells of the quiescent centre adjacent to these collapsed cells are no longer under constraint and are free to enlarge and divide (Fig. 14a). Second, it is possible to bisect the cap longitudinally and remove one half. When this is done, only cells exposed by the half-decapping, i.e. quiescent centre cells, are stimulated to grow and divide; cells of the quiescent centre in contact with the remaining cap are not stimulated (Fig. 14b). Although all these observations are open to alternative interpretations in terms of hormone action, using arguments that would have to presume that the inhibition of growth and division of quiescent centre cells is mediated by a hormone whose concentration here is regulated by the cap (see Barlow, 1974b), the interpretation that requires the fewest assumptions is that the cap constrains growth and division of the cells that constitute the quiescent centre by its physical presence rather than by its metabolic activities.

An observation that also indicates an interaction between the cap and quiescent centre of *Zea* was made by Clowes (1972) when he cut off the distal half of the cap. Five hours after the removal of these mature cap cells the average rate of cell division declined in the cap initials but increased in the quiescent centre. As the rates of division were calculated from the number of metaphases accumulated by colchicine, any change assumed to occur in the average duration of the mitotic cycle of the two cell populations may be more apparent than real because the result would by influenced by any change in the proliferative fraction (P.F.). Changes in the P.F. are the most likely explanation for Clowes' results: the P.F. being increased in the quiescent centre and decreased in the cap initials by cutting off part of the cap. It would be most interesting to know how removal of the mature cap cells brings about changes in the activity of the two meristematic groups of cells.

Because the removal of cap cells leads to their replacement from the quiescent centre, we are given an indication of how the cap can continue to fulfil its role as protector of the root meristem even after sustaining damage. As cells of the cap detach naturally or are abraded away by soil, so they are replaced from the initials. But if many cells are lost, then new cap cells can be formed from the quiescent centre. Thus, in a sense the quiescent centre protects the cap, for the cap will never be irretrievably lost if the quiescent centre is present. However, the lubricant slime sheath of the cap probably prevents excessive damage to the cap during the passage of the root through soil.

B. Geotropism

Since the turn of the century the root cap has been considered to be the most likely site of gravity perception (Haberlandt, 1914; Němec, 1964). The observation that removal of the root cap causes a temporary loss of the root's ability to express a response to a gravitational stimulus (Juniper *et al.*, 1966; Shachar, 1967; Barlow, 1974c) gives added weight to this proposition and leads to the hypothesis that the cap is either the site of gravity perception, or in some other way mediates the geotropic response, perhaps by providing a necessary hormone. Of course, both possibilities may hold (Barlow, 1974c). The relevance of hormones to geotropism in roots is more fully discussed elsewhere in this volume by L. J. Audus (Chapter 16).

While removal of the cap clearly implicates this organ in geotropism, it is not known for certain which cells in the cap are important in mediating this growth response. The most favoured cells for this role are those that contain both amyloplasts and a well developed endoplasmic reticulum (Juniper and French,

1970; Volkmann, 1974). But, in certain circumstances, the cap seems to be dispensable in mediating the geotropic response. This conclusion is based on the following observations. When the cap is removed from the primary root of *Zea mays* and *Triticum aestivum*, a positive orthogeotropic response is immediately abolished, but returns after about 14 hours in both species (Barlow, 1974c). At the time that geotropism returns no new cap has regenerated (Barlow, 1974b, c); however, during the ageotropic period amyloplasts develop throughout the meristem (Barlow, 1974c; Barlow and Grundwag, 1974). Thus, a cap is not required for geotropism, but amyloplasts probably are. These same observations have been made on *Pisum sativum*, *Hordeum vulgare* and *Secale cereale* (unpublished personal observations). As amyloplasts are particularly well developed in the cap of normal roots, it is difficult to escape the conclusion that it is by virtue of their presence that the cap plays a vital part in the response to gravity.

A completely different role for the cap in geotropism has been put forward by Kutschera-Mitter (1972). She believes that the slime secreted by the cap regulates the extensibility of cells in the region in which the geotropic bending occurs, and that geotropism is a response to different water relations, mediated by the presence of the slime, between cells at the upper and lower surfaces of the root.

VIII. Conclusion

Caps are present on the roots of all plants. Although much of the foregoing discussion has centred around the root cap of *Zea mays*, this is only because more observation and experimentation has been made using this species; there is no reason to think that the properties of the cap of *Zea* are not shared by the caps of other species.

The cap has a meristematic zone that continually produces cells, some of which remain meristematic, and others which cease to divide. These latter are displaced by the continued activity of the meristem until they reach the edge of the cap from which they are sloughed off. During their passage to the outside of the cap, cells change their structure and metabolism in a predictable sequence. This pattern of differentiation is recognizable at the levels of both nucleus and cytoplasm. The nuclei become endopolyploid, while the most noticeable cytoplasmic changes relate to the metabolism of starch and polysaccharide slime. These two constituents are important for the role they play in root growth. The amyloplasts, which contain the starch, may be the gravity sensors that trigger geotropism, the slime may assist the passage of the root through soil and influence the rhizosphere.

Acknowledgement

I thank Dr. J. A. Sargent for his assistance with electron microscopy and for preparing the electron micrographs. I am also grateful to Drs. P. A. A. Harkes

(Leiden University) and P. D. M. Macdonald (McMaster University) for supplying me with unpublished results.

References

ASHFORD, A. E. and MCCULLY, M. E. (1970). Histochemical localization of β-glycosidases in roots of *Zea mays*. II. Changes in localization and activity of β-glucosidase in the main root apex. *Protoplasma* **71**; 389–402.

BARLOW, P. W. (1971). Properties of cells in the root apex. *Revta Fac. Agron. Univ. nac. La Plata (3a época)* **47**; 275–301.

BARLOW, P. W. (1973). *In* "The Cell Cycle in Development and Differentiation" (M. Balls and F. S. Billett, eds.), pp. 135–165. University Press, Cambridge.

BARLOW, P. W. (1974a). The duration of endomitotic cycles in cells of the root apex of *Zea mays*. (MS in preparation).

BARLOW, P. W. (1974b). Regeneration of the cap of primary roots of *Zea mays*. *New Phytol.* **73;** 937–954.

BARLOW, P. W. (1974c). Recovery of geotropism after removal of the root cap. *J. exp. Bot.* **25;** 1137–1146.

BARLOW, P. W. and GRUNDWAG, M. (1974). The development of amyloplasts in cells of the quiescent centre of *Zea* roots in response to removal of the root cap. *Z. Pfl Physiol.* **73;** 56–64.

BARLOW, P. W. and MACDONALD, P. D. M. (1973). An analysis of the mitotic cell cycle in the root meristem of *Zea mays*. *Proc. R. Soc. Ser. B.* **183**; 385–398

BELL, J. K. and MCCULLY, M. E. (1970). A histological study of lateral root initiation and development in *Zea mays*. *Protoplasma* **70**; 179–205.

BERJAK, P. (1968). A lysosome-like organelle in the root cap of *Zea mays*. *J. Ultrastruct. Res.* **23**; 233–242.

BERJAK, P. and VILLIERS, T. A. (1970). Ageing in plant embryos. I. The establishment of the sequence of development and senescence in the root cap during germination. *New Phytol.* **69**; 929–938.

BERJAK, P. and VILLIERS, T. A. (1972). Ageing in plant embryos. II. Age-induced damage and its repair during early germination. *New Phytol.* **71**; 135–144.

BOWLES, D. J. and NORTHCOTE, D. H. (1972). The sites of synthesis and transport of extra-cellular polysaccharides in the root tissues of maize. *Biochem. J.* **130**; 1135–1145.

CLARKSON, D. T. and SANDERSON, J. (1969). The uptake of a polyvalent cation and its distribution in the root apices of *Allium cepa:* Tracer and autoradiographic studies. *Planta (Berl.)* **89**; 136–154.

CLOWES, F. A. L. (1954). The promeristem and the minimal constructional centre in grass root apices. *New Phytol.* **53**; 108–116.

CLOWES, F. A. L. (1961). Duration of the mitotic cycle in a meristem. *J. exp. Bot.* **12;** 283–293.

CLOWES, F. A. L. (1962). Rates of mitosis in a partially synchronous meristem. *New Phytol.* **61**; 111–118.

CLOWES, F. A. L. (1965). The duration of the G_1 phase of the mitotic cycle and its relation to radiosensitivity. *New Phytol.* **64**; 355–359.

CLOWES, F. A. L. (1967). Synthesis of DNA during mitosis. *J. exp. Bot.* **18**; 740–745.

CLOWES, F. A. L. (1971). The proportion of cells that divide in root meristems of *Zea mays* L. *Ann. Bot.* **35**; 249–261.

CLOWES, F. A. L. (1972). Regulation of mitosis in roots by their caps. *Nature New Biol.* **235**; 143–144.

Clowes, F. A. L. and Hall, E. J. (1962). The quiescent centre in root meristems of *Vicia faba* and its behaviour after acute X-irradiation and chronic γ-irradiation. *Radiat. Bot.* **3**; 45–53.

Clowes, F. A. L. and Juniper, B. E. (1968). "Plant Cells." Blackwell Scientific Publications, Oxford and Edinburgh.

Coulomb, P. and Coulon, J. (1970). Fonctions de l'appareil de Golgi dans les méristèmes radiculaires de la courge (*Cucurbita pepo* L. Cucurbitacée) *J. Microscop.* **10**; 203–214.

Darbyshire, J. F. and Greaves, M. P. (1971). The invasion of pea roots, *Pisum sativum* L., by soil microorganisms, *Acanthamoeba palestinensis* (Reich) and *Pseudomonas* sp. *Soil Biol. Biochem.* **3**; 151–155.

Dart, P. J. and Mercer, F. V. (1964). The legume rhizosphere. *Arch. Mikrobiol.* **47**; 344–378.

Dauwalder, M. and Whaley, W. G. (1974). Patterns of incorporation of [^{3}H] galactose by cells of *Zea mays* root tips. *J. Cell Sci.* **14**; 11–27.

Dauwalder, M., Whaley, W. G. and Kephart, J. E. (1969). Phosphatases and differentiation of the Golgi apparatus. *J. Cell Sci.* **4**; 455–497.

Deltour, R. (1971). *Etude cytologique, autoradiographique et ultrastructurale du passage de l'état de vie quiescente a l'état de vie active des cellules radiculaires de l'embryon de* Zea mays L. Thesis: Mémoire de Doctorat en Sciences Botanique, Universite de Liège.

Floyd, K. A. and Ohlrogge, A. J. (1970). Gel formation on nodal root surfaces of *Zea mays.* I. Investigation of the gel's composition. *Pl. Soil* **33**; 331–343.

Floyd, K. A. and Ohlrogge, A. J. (1971). Gel formation on nodal root surfaces of *Zea mays.* Some observations relevant to understanding its action at the root–soil interface. *Pl. Soil* **34**; 595–606.

Greaves, M. P. and Darbyshire, J. F. (1972). The ultrastructure of the mucilaginous layer on plant roots. *Soil Biol. Biochem.* **4**; 443–449.

Guttenberg, H. von (1968). *In* "Handbuch der Pflanzenanatomie" Vol. 8, part 5. Gebruder Borntraeger, Berlin and Stuttgart.

Guttenberg, H. von, Heydal, H.-R. and Pankow, H. (1954). Embryologische Studien an Monokotyledonen I. Die Enstehung der Primarwurzel bei *Poa annua* L. *Flora, Jena* **141**; 298–311.

Haberlandt, G. (1914). "Physiological Plant Anatomy" (translation of 4th German edition by M. Drummond). MacMillan, London.

Harkes, P. A. A. (1973). Structure and dynamics of the root cap of *Avena sativa* L. *Acta bot. neerl.* **22**; 321–328.

Harris, P. J. and Northcote, D. H. (1970). Patterns of polysaccharide biosynthesis in differentiating cells of maize root tips. *Biochem. J.* **120**; 479–491.

Huber, W., Fekete, M. A. R. de and Ziegler, H. (1973). Enzyme des Stärkeumsatzes in den Wurzelhaubenzellen von *Zea mays* L. *Planta* **112**; 343–356.

Jones, D. D. and Morré, D. J. (1973) Golgi apparatus mediated polysaccharide secretion by outer root cap cells of *Zea mays.* III. Control by exogenous sugars. *Physiologia Pl.* **29**; 68–75.

Juniper, B. E. and Clowes, F. A. L. (1965). Cytoplasmic organelles and cell growth in root caps. *Nature, Lond.* **208**; 864–865.

Juniper, B. E. and French, A. (1970). The fine structure of cells that perceive gravity in the root tip of maize. *Planta* **95**; 314–329.

Juniper, B. E. and French, A. (1973). The distribution and redistribution of endoplasmic reticulum (ER) in geoperceptive cells. *Planta* **109**; 211–224.

Juniper, B. E. and Roberts, R. M. (1966). Polysaccharide synthesis and the fine structure of root cells. *Jl R. Microsc. Soc.* **85**; 63–72.

Juniper, B. E., Groves, S., Landau-Schacher, B. and Audus, L. J. (1966). Root cap and the perception of gravity. *Nature, Lond.* **209**; 93–94.

Kutschera-Mitter, L. (1972). Erklarung des geotropen wachstums aus Standort und Bau der Pflanzen. *Land- und Forstwirtschaftliche Forschung in Österreich* **5**; 35–89.
Leiser, A. T. (1968). A mucilaginous root sheath in Ericaceae. *Am. J. Bot.* **55**; 391–398.
Macdonald, P. D. M. (1974). *In* "Mathematical Problems in Biology, Victoria Conference"; Lecture notes in biomathematics. (P. van den Driessche, ed.), pp. 153–163. Springer, Berlin–Heidelberg–New York.
Maitra, S. C. and De, D. N. (1972). Ultrastructure of root cap cells: formation and utilization of lipid. *Cytobios* **5**; 111–118.
Mollenhauer, H. H., Whaley, W. G. and Leech, J. H. (1961). A function of the Golgi apparatus in outer root cap cells. *J. Ultrastruct. Res.* **5**; 193–200.
Němec, B. (1964). Uber Georezeptoren in Wurzeln. *Biologia Pl.* **6**; 243–249.
Northcote, D. H. and Pickett-Heaps, J. D. (1966). A function of the Golgi apparatus in polysaccharide synthesis and transport in the root-cap cells of wheat. *Biochem. J.* **98**; 159–167.
Phillips, H. L. and Torrey, J. G. (1971). Deoxyribonucleic acid synthesis in root cap cells of cultured roots of *Convolvulus*. *Pl. Physiol.* **48**; 213–218.
Phillips, H. L. and Torrey, J. G. (1972). Duration of cell cycles in cultured roots of *Convolvulus*. *Am. J. Bot.* **59**; 183–188.
Quastler, H. and Sherman, F. G. (1959). Cell population kinetics in the intestinal epithelium of the mouse. *Expl. Cell Res.* **17**; 420–438.
Randolph, L. F. (1936). Developmental morphology of the caryopsis in maize. *J. Agric. Res.* **53**; 881–916.
Rougier, M. (1971). Etude cytochemique de la secretion des polysaccharides végétaux a l'aide d'un material de choix: les cellules de la coiffe de *Zea mays*. *J. Microscop.* **10**; 67–81.
Schade, C. and Guttenberg, H. von (1951). Uber die Entwicklung des Wurzelvegetationspunktes der Monokotyledonen. *Planta* **40**; 170–198.
Shachar, B. (1967). *Studies on the root cap and its significance in graviperception.* Ph.D. Thesis, University of London.
Soueges, R. (1924). Embryogenie des Gramines. Developpement de l'embryon chez le *Poa annua* L. *C. R. hebd. séanc. Acad. Sci., Paris* **178**; 860–862.
Summerbell, D., Lewis, J. H. and Wolpert, L. (1973). Positional information in chick limb morphogenesis. *Nature, Lond.* **244**; 492–496.
Thompson, J. and Clowes, F. A. L. (1968). The quiescent centre and rates of mitosis in the root meristem of *Allium sativum*. *Ann. Bot.* **32**; 1–13.
Volkmann, D. (1974). Amyloplasten und Endomembranen: Das Geoperzeptionssystem der Primarwurzel. *Protoplasma* **79**; 159–183.
Whaley, W. G., Kephart, J. E. and Mollenhauer, H. H. (1959). Developmental changes in the Golgi-apparatus of maize root cells. *Am. J. Bot.* **46**; 743–751.
Wimber, D. E. and Quastler, H. (1963). A ^{14}C- and ^{3}H-thymidine double labelling technique in the study of cell proliferation in *Tradescantia* root tips. *Expl. Cell Res.* **30**; 8–22.
Wolpert, L. (1969). Positional information and the spatial pattern of cellular differentiation. *J. Theoret. Biol.* **25**; 1–47.
Wright, K. and Northcote, D. H. (1974). The relationship of root-cap slimes to pectins. *Biochem. J.* **139**, 525–534.

Added in proof

Following the submission of this article for printing, additional information on some of the points covered has been published.

designated the promeristem (Clowes, 1961). From a careful study of cell lineages Clowes (1950) was able initially to deduce that the promeristem ought to be cup-shaped and more importantly, that the cells at the centre of this cup should divide rarely or not at all. He further suggested that cells on the periphery of this hemispherical region are meristematic and thus ought to be considered part of the promeristem. By various techniques the existence of this central, inactive, region (the quiescent center) has been shown clearly in root apices of numerous species (Clowes, 1956a, b; Byrne and Heimsch, 1970; Miksche and Greenwood, 1966; Phillips and Torrey, 1971; Wilcox, 1962).

The quiescent center (QC) thus originates sometime during the ontogeny of the root. Why it develops is not known, but because it is situated within this hemisphere of activity, within a region which was formerly meristematic, one may wonder how important it is for the establishment and maintenance of normal root growth. Must a QC be maintained for normal root development? How does a QC come to originate, and further, how is it perpetuated?

Torrey suggested that the organization of the root apex may depend largely upon or be maintained by a "root organizer substance" (Torrey, 1962, 1972). The nature of this hypothetical substance remains obscure, but Torrey proposed that it may be a hormone and advanced a cytokinin or cytokinins as candidates (Torrey, 1972). Strong, but nevertheless indirect evidence favours roots as one site of cytokinin biosynthesis (Burrows and Carr, 1969; Kende, 1965, 1971; Short and Torrey, 1972; Weiss and Vaadia, 1965). From work on pea, it was suggested by Short and Torrey (1972) that the QC and the surrounding meristematic tissue of the root apex are probable sites of cytokinin biosynthesis.

In this paper the distribution of cytokinins within the terminal millimeter region of *Zea* roots is presented. Further, changes of cytokinin levels associated with increasing and decreasing activity of the QC will be considered.

II. Materials and Methods

A. Materials

Corn seed (*Zea mays*, c.v. Kelvedon 33) was obtained from Hurst Ltd., Witham, Essex, England. The seed was surface sterilized in half strength commercial chlorox, rinsed 3 times in sterile distilled water, and aseptically germinated in the dark at 23°C. After 3 days, seedlings with primary roots 25–35 mm in length were selected and used in all experiments.

B. Surgical Techniques

With the aid of a dissecting microscope and sterile instruments, 3-day-old roots were decapped. Removal of the root cap was accomplished by holding a blade at a 45° angle and placing the cutting edge at the junction of the root cap and the main body of the root. By illuminating the root from beneath, this junction was usually readily evident. Gentle pressure on the blade yielded a clean separation of the root cap from the remaining portion of the root. Experimentation was carried out either directly on these decapped 72 h roots or the decapped roots were returned to the dark, at 23°C, for varying periods of time and then experimentation begun.

C. Histoautoradiography

Sterile roots were immersed in half-strength Knop's solution (Cutter and Feldman, 1970) containing methyl-^{3}H-thymidine (specific activity 6·0 ci/mM) at a concentration of 0·3μci/ml for 12 h, washed, excised, fixed in fresh FAA and embedded in paraplast. Serial sections, 10μm in thickness, were mounted on subbed slides (Boyd, 1955), taken down to water and coated with Kodak Nuclear Track Emulsion, NTB3, diluted 1:1 with distilled water. The dried slides were stored in boxes containing Drierite, at 4°C for 3–6 weeks. The slides were developed in Kodak Dektol Developer diluted with water 1:2, rinsed and washed for 20 minutes. The sections were stained with freshly prepared aqueous 0·1% Toluidine Blue 0, dehydrated and mounted in damar.

D. Extraction of Free Cytokinins

Tissues from various portions and treatments were harvested and immediately immersed separately in precooled 95% ethanol at —72°C. If the tissue was not to be extracted immediately, it was stored in 95% ethanol, under liquid nitrogen. The frozen segments were collected, extracted and the extracts purified according to the methods of Short and Torrey (1972), with the following modifications. The Dowex ion exchange resin 50W-X8, H+ (200–400 mesh) was purified and regenerated prior to packing. In place of paper chromatography thin layer chromatography (TLC) was employed. Residues were streaked either on plates coated with Silica Gel HF-254, at 400 μm, or on Cellulosepulver MN 300 UV_{254} at 400 μm. The developing solvents, respectively, were chloroform: methanol, 8:2 and 0·03 M borate buffered at pH 8·4. Following development, the plates were divided into ten equal bands, parallel to the origin. Each band was scraped off the plate and eluted 3 times with water-saturated ethyl acetate followed by two elutions with 95% ethanol. The eluate was reduced in volume and assayed for cytokinins by the soybean callus bioassay (Miller, 1963).

E. Cytokinin Bioassay

To the aqueous extract of each band was added 40 ml of SCF agar medium (Phillips and Torrey, 1970) which prior to autoclaving was divided equally into each of four 125 ml flasks. Four pieces of soybean callus, each approximately $1\frac{1}{2}$–2 mm square, were inoculated into each flask. The cultures were maintained

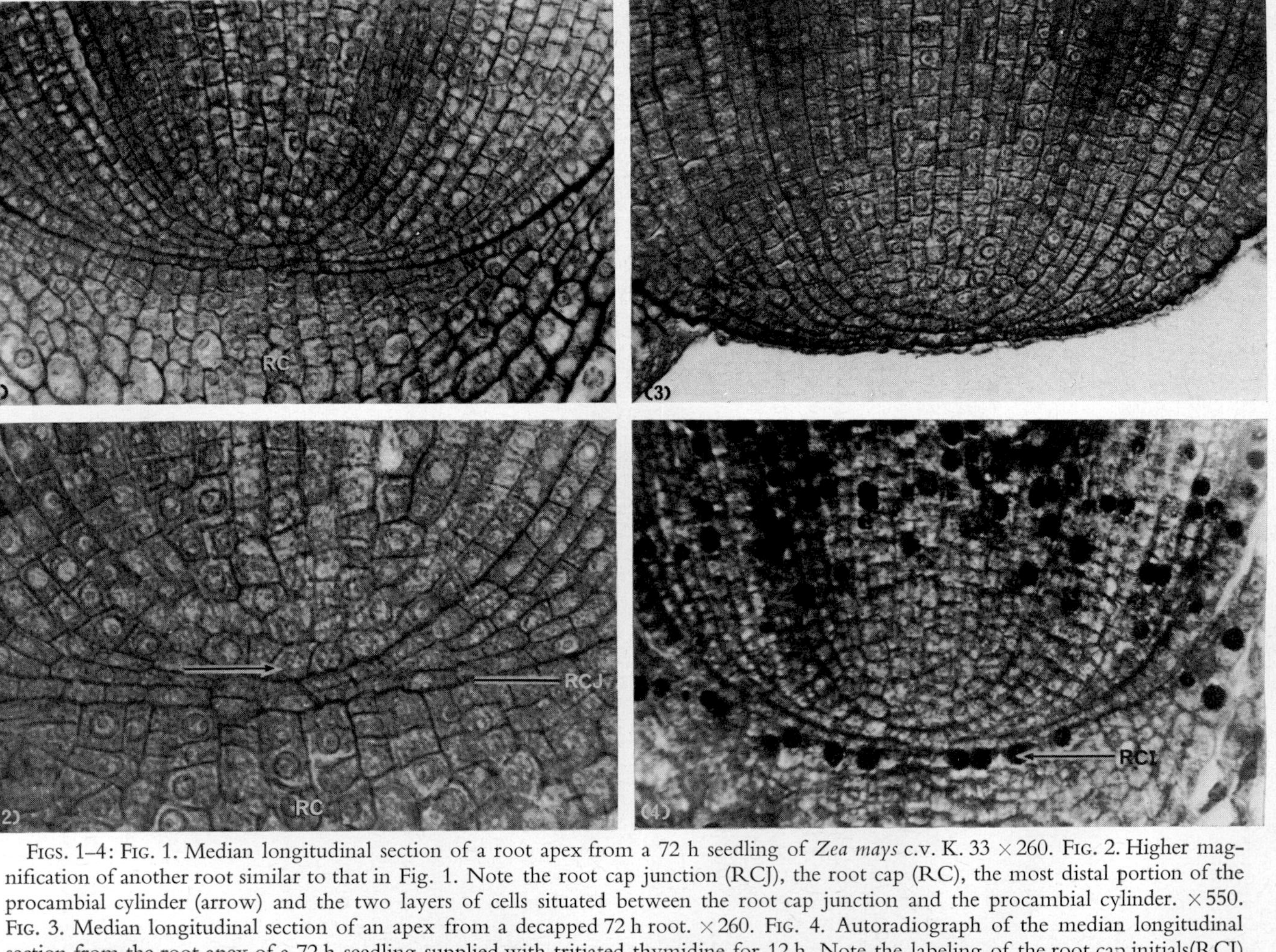

FIGS. 1–4: FIG. 1. Median longitudinal section of a root apex from a 72 h seedling of *Zea mays* c.v. K. 33 ×260. FIG. 2. Higher magnification of another root similar to that in Fig. 1. Note the root cap junction (RCJ), the root cap (RC), the most distal portion of the procambial cylinder (arrow) and the two layers of cells situated between the root cap junction and the procambial cylinder. ×550. FIG. 3. Median longitudinal section of an apex from a decapped 72 h root. ×260. FIG. 4. Autoradiograph of the median longitudinal section from the root apex of a 72 h seedling supplied with tritiated thymidine for 12 h. Note the labeling of the root cap initials(RCI) and the lack of label in the cells of the QC. ×260.

at 23°C and received 12 h of diffuse, warm-white fluorescent light per day. After 4 weeks the average fresh weight per flask was determined.

III. Results

A. Microsurgery

Surgical removal of the root cap, as noted by Juniper *et al.* (1966), occurred by a break through the walls of the cap initial cells. With careful technique, only the root cap initials were destroyed, leaving cells proximal to this break intact and uninjured. Distal to the stele, one or two layers of intact cells remained (Figs 1, 2 and 3).

B. Autoradiographs

Autoradiographs of intact 72 h roots revealed a distinct QC at the distal end of the procambial cylinder (Fig. 4). The layer of root cap initials was usually clearly defined since these cells were actively synthesizing DNA and thus incorporated the labeled thymidine. The size of the QC varied somewhat from one intact root to another, with the shifting of the proximal boundary accounting for this size fluctuation. In order to describe the QC, measurements of height and width were obtained from median longitudinal sections. The distal boundary was readily observed from median sections, whereas the proximal limit of individual QC's was best obtained by noting labeling patterns of the median as well as adjacent sections. Measurements of the width denote dimensions of the QC at its widest point (Table I).

As can be seen from Table I and Fig. 5, if 72 h roots were decapped, provided

TABLE I. Measurements of the dimensions of the quiescent center, taken from the most median longitudinal section.

Treatment	Height in microns	Average height	Width
Intact 72 h roots	100		106
	105	103	100
	103		110
Decap, then regeneration	95		110
for 12 h	90	93	100
Decap, then regeneration	58		100
for 24 h	73	65	115
	63		110
Decap, then regeneration	100		110
for 36 h	95	98	110
	100		105
Decap, then regeneration	111		121
for 48 h	100	105	116
	105		116

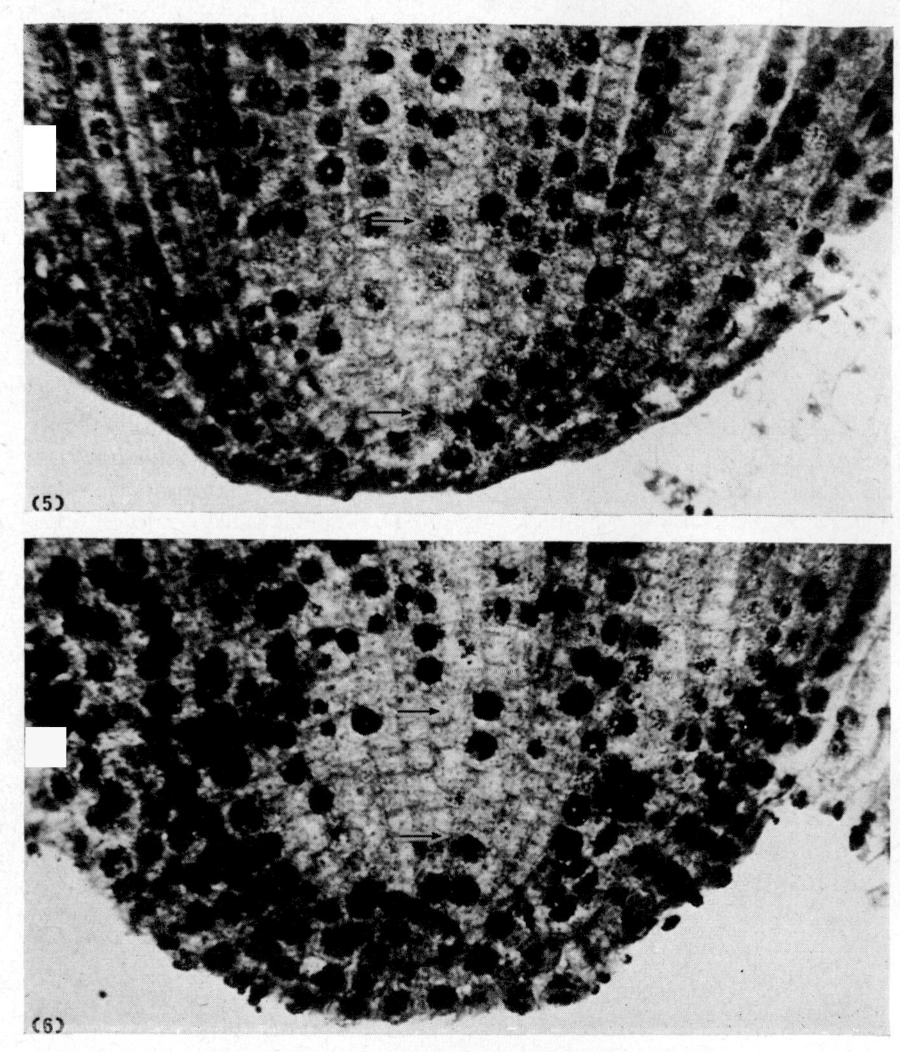

FIGS. 5 and 6. Autoradiographs of median longitudinal sections of root apices of *Zea mays*. All roots were decapped at an age of 72 h and labeled for 12 h periods with tritiated thymidine, as indicated. The QC is taken as that area between the arrows. ×350. FIG. 5. Autoradiograph of a root supplied with label for 12 h immediately subsequent to decapping. The root was 84 h old when fixed. FIG. 6. Autoradiograph of a root labeled for the time period 12–24 h after decapping. The root was 96 h old when fixed.

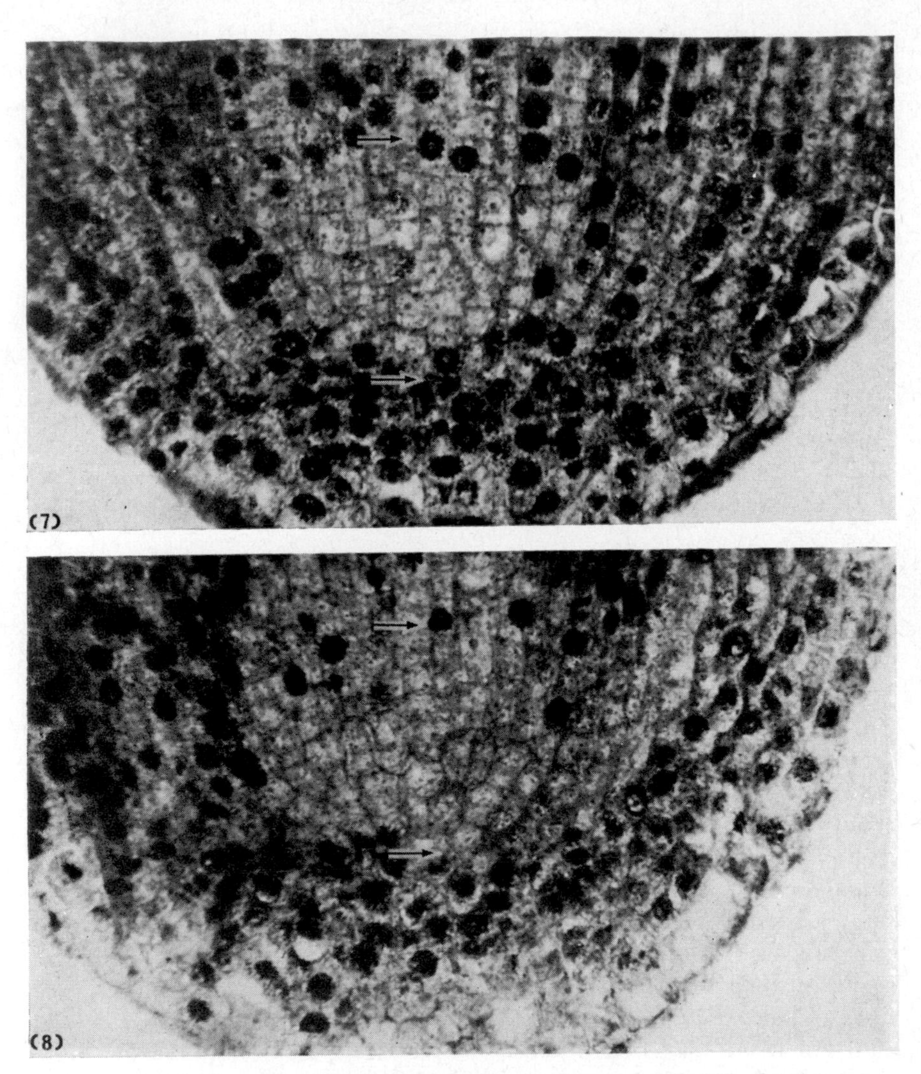

FIG. 7. Autoradiograph of a root labeled for the time period 24–36 h after decapping. The root was 108 h old when fixed. FIG. 8. Autoradiograph of a root labeled for the time period 36–48 h after decapping. The root was 120 h old when fixed.

with ^{3}H-thymidine for an additional 12 h and then examined, activation of the distal cells of the QC was evident. Incorporation of label was slight (relative to surrounding nuclei) but nevertheless the onset of DNA synthesis was indicated. Relative to intact controls the QC has already begun to diminish in height (93 μm versus 103 μm). Decapping a 72 h root, returning it to the dark for 12 h, with subsequent labelling 12–24 h after decapping typically showed the pattern of label incorporation seen in Fig. 6. On the average the QC was 65 μm in height, compared to 103 μm for intact controls. Between 12 and 24 h the QC reached its smallest recorded size, with this decrease attributed to an activation and subsequent label incorporation into the now active distal QC cells. Twenty-four hours after decapping, a new root cap initial layer was evident as well as usually 1 layer of derivative "root cap" cells.

Decapping 72 h roots, returning them to the dark for 24 h, with subsequent labeling from 24 to 36 h yielded an average QC height of 98 μm (Fig. 7). Several layers of new root cap cells were found, though the root cap was not as large as that of the intact control. Finally, labeling 36 to 48 hours after decapping resulted in a QC with an average height of 105 μm as compared to the average of 103 μm for intact controls (Fig. 8).

The distance to the proximal boundary relative to the apex of the root remained relatively constant. Decreases in height of the QC resulted from an activation of distal and not proximal QC cells.

C. Cytokinins

The extraction, purification and bioassay for cytokinins showed the presence of three distinct regions of cytokinin activity. Extracts from intact terminal millimeter tips from 72 h roots, chromatographed on silica gel, yielded peaks of activity at R_f's 0·1–0·2, 0·35–0·55 and 0·7–0·9 (Fig. 9a). As in the work of Radin and Loomis (1971), these peaks will be designated cytokinin I, cytokinin II and cytokinin III respectively. Attempts were made to characterize the cytokinins in these peaks.

Cytokinin II chromatographed with the same R_f as a mixture of zeatin and zeatin riboside. On silica gel, commercial zeatin and zeatin riboside (obtained from Calbiochem) chromatographed at R_f's 0·39–0·51 and 0·3–0·42 respectively, and thus it was suspected that peak II might consist of a mixture of these cytokinins. In an attempt to further sub-divide cytokinin II, the purified extract was applied to cellulose TLC plates, run in a borate buffer and then eluted and bio-assayed with the soybean callus. Three peaks of activity were observed; at R_f's 0·5–0·6, 0·7–0·8 and 0·9–1·0 (Fig. 10). Commercial zeatin and its riboside, in this system, chromatographed at R_f's of 0·5–0·6 and 0·7–0·8 respectively.

From other work (Radin and Loomis, 1971; Short and Torrey, 1972) peak number I on silica gel was suspected of being a nucleotide. In *Zea* roots,

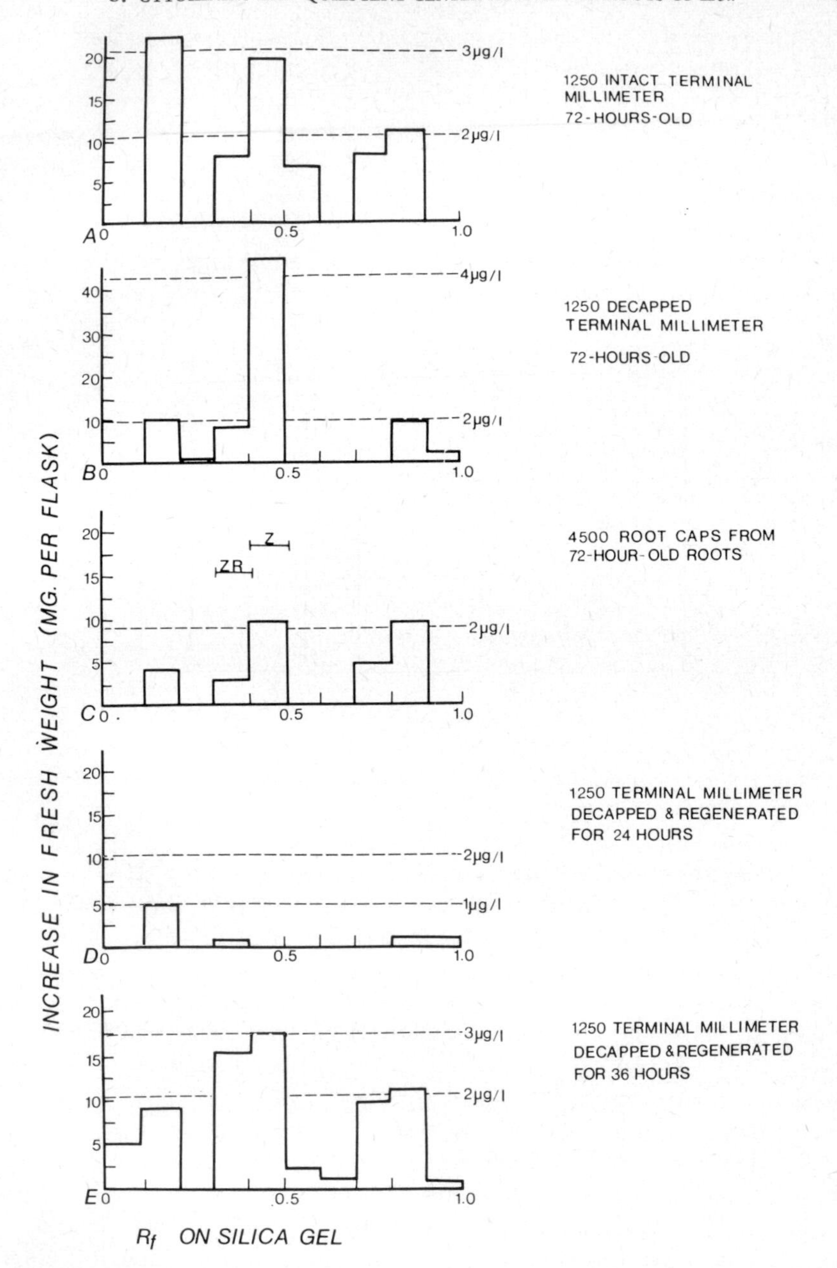

FIG. 9A, B, C, D, and E. Histograms of soybean callus bioassay of extracts chromatographed on silica gel. Sources of cytokinins for extractions are indicated to the right of each histogram. The growth response of kinetin standards is illustrated by the dashed lines.

preliminary evidence strongly suggested that cytokinin I is at least partly composed of zeatin ribotide.

Identification of cytokinin III, R_f 0·7–0·9 on silica gel, was not attempted. Isopentenyl adenine, 2iP, chromatographed with an R_f of 0·62–0·72 on silica gel, always slightly behind cytokinin III. The chemical properties of peak III were not investigated any further.

Thus the evidence supports the view that in *Zea* root tips there are at least 4 cytokinins: zeatin and zeatin riboside (Peak II on silica gel), zeatin ribotide (Peak I) and one other unidentified cytokinin (Peak III).

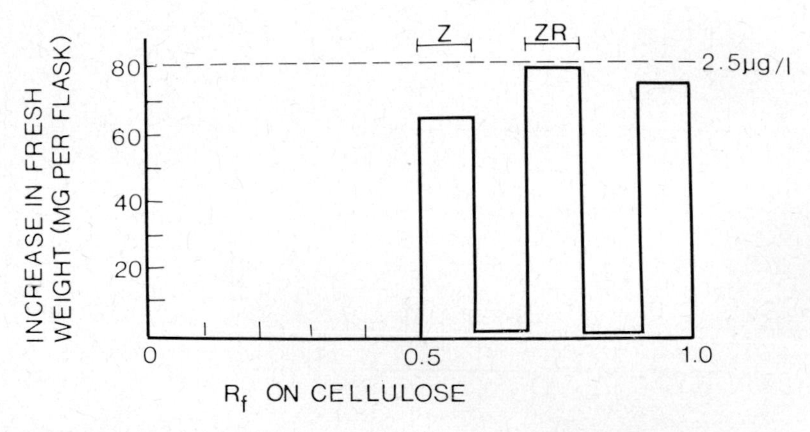

FIG. 10. Histogram of the soybean callus bioassay of the purified extract from 1,250 intact, 72 h terminal millimeter root segments, on cellulose TLC.

Following the partial characterization of cytokinins in intact terminal millimeter tips from 72 h seedling roots, an attempt was made to localize cytokinins in the apical region more precisely. Roots were decapped and the caps plus the remaining portion of the terminal millimeter were collected separately, weighed (Table II) and assayed for cytokinins. As seen in Figs. 9b and c, the three cytokinin peaks noted in the intact terminal millimeter segments were present in extracts of root caps alone and also in extracts of the terminal millimeter minus the caps.

In order to compare cytokinin amounts obtained from the various tissues (indicated by non-bracketed numbers in Table III), it was necessary, for purposes of calculation, to assume each bioassay was run on identical weights of tissue. Ten grams was chosen as this arbitrary weight of tissue compared in each bioassay. The bracketed kinetin equivalents in Table III were obtained by dividing individual fresh weights in Table II into 10, yielding a resultant number designated the "multiplication factor" (Table II). This factor, when multiplied by the non-bracketed kinetin equivalents yielded the corresponding bracketed equivalents

TABLE II. Measurements of fresh weight of tissues, from the terminal millimeter of *Zea* roots.

Weight of an individual		Number extracted	Total fresh weight extracted	Multiplication factor (used in Table III)
Intact 72 h root	4×10^{-4} gm	1,250	0·5 grams	10/0·5 = 20
Decapped 72 h root	3×10^{-4} gm	1,250	0·375	10/0·375 = 26·6
Root cap	1×10^{-4} gm	4,500	0·45	10/0·45 = 22·2
Decapped then regenerated 24 h	$3{\cdot}2 \times 10^{-4}$ gm	1,250	0·4	10/0·4 = 25
Decapped then regenerated 36 h	$3{\cdot}5 \times 10^{-4}$ gm	1,250	0·44	10/0·44 = 22·7

Except for the intact 72 h root tips, which were 1 mm in length, other tissues were some portion of a millimeter in length.

seen in Table III. Then kinetin equivalents can be compared as if identical weights of tissues were extracted for each bioassay.

On a basis of tissue weight total cytokinin was lower in the root caps than in intact 72 h or decapped 72 h tissue. Cytokinin III was fairly constant in the three treatments, whereas cytokinins I and II were about half as concentrated in the root cap alone when compared to the intact 72 h root or to the just decapped 72 h root (Table III). On a basis of tissue weight the decapped 72 h root had the highest

TABLE III. Kinetin equivalents from the terminal millimeter of *Zea* roots.

	Cytokinin concentration (μg.l^{-1}) Cytokinin I	Cytokinin II	Cytokinin III	Total of I, II and III
Intact 72 h roots	3·1 (62)	5·6 (112)	3·8 (76)	12·5 (248)
Decapped 72 h roots	2·0 (53)	5·7 (152)	2·7 (72)	10·4 (290)
Root caps	1·2 (26)	3·1 (69)	3·8 (84)	8·1 (206)
Decapped then regenerated 24 h	1·2 (30)	0·3 (7·5)	0·3 (7·5)	1·8 (45)
Decapped then regenerated 36 h	2·6 (57)	5·8 (132)	4·3 (97)	12·7 (286)

The non-bracketed values for cytokinins I, II and III were obtained from the corresponding histograms in Fig. 9. Bracketed numbers represent normalized kinetin equivalents per 10 grams fresh weight tissue. Except for the intact 72 h root tips, which were 1 mm in length, other tissues were some portion of a millimeter in length.

total amount of kinetin equivalents, 1·4 (290/206) times as much as root caps and 1·17 (290/248) times as much as intact 72 h root tissue.

In order to secure measurable quantities of cytokinins from root caps, 4500 caps were collected and extracted *en masse* each time this tissue was assayed for cytokinins. The histogram in Fig. 9c was obtained by extracting 4500 caps, whereas the histograms for either intact or decapped roots were from extracting 1250 roots. Dividing 1250 by 4500 yields 0·28, which, when multiplied by the figures for root caps in Table III yielded cytokinin equivalents from 1250 caps (Table IV). Adding this figure to that from 1250 decapped 72 h roots, as seen in

TABLE IV. Distribution of cytokinins I, II and III in the terminal millimeter of *Zea* roots; expressed in μgrams/liter kinetin equivalents.

	Cytokinin I	Cytokinin II	Cytokinin III	Total I, II and III
(A) 1,250 decapped 72 h roots	2·00	5·70	2·80	10·50
(B) Equivalent number of root caps[a]	0·33	0·87	1·05	2·25
Sum of (A) + (B)	2·33	6·57	3·85	12·75
1,250 intact 72 h roots	3·10	5·60	3·80	12·50

Compare the μgram/l equivalents from intact roots to the corresponding total values obtained by adding equivalents from root caps plus equivalents from decapped roots.

[a]Derived by multiplying observed value by 0·28—see text.

Table IV yielded a total amount of kinetin equivalents nearly identical to that from intact 72 h roots. Furthermore, this table indicates that in an intact terminal millimeter about 1/5 (2·25) of the total extractable cytokinins were in the root cap, while the remainder of the terminal millimeter contained the other 4/5 (10·5).

When decapped 72 h roots were returned to the dark for 24 h, and then assayed for cytokinins, the histogram in Fig. 9d was obtained. A sharp decrease was noted in all three peaks. On a basis of tissue weight approximately 1/5 to 1/6 of the *total* cytokinin present in the intact or decapped 72 h root tips was present in the decapped plus regenerated 24 h roots. In cytokinin II, which is thought to be at least partly composed of zeatin and its riboside, a 15–20 fold drop was noted in the decapped plus regenerated 24 h roots, compared to the intact or decapped 72 h roots. Other ratio values are seen in Table V. These figures were obtained by taking the bracketed results in Table III and dividing them by the bracketed μgram/liter equivalents of the decapped plus regenerated 24 h. So, for example, comparing cytokinin II in the decapped 72 h roots to the decapped plus regenerated 24 h, one obtains 152/7·5 = 20.

TABLE V. Ratios of the relative amounts of the three cytokinins in the terminal millimeter of roots of *Zea*.

	Cytokinin I	Cytokinin II	Cytokinin III
Intact 72 h roots	62/30 = 2·0	112/7·5 = 15·0	76/7·5 = 10·0
Decapped 72 h roots	53/30 = 1·8	152/7·5 = 20·0	69/7·5 = 9·2
Roots decapped and then regenerated 24 h	30/30 = 1·0	7·5/7·5 = 1·0	7·5/7·5 = 1·0
Roots decapped and then regenerated 36 h	57/50 = 1·9	132/7·5 = 17·6	97/7·5 = 12·9

The value for decapped roots which have regenerated for 24 h was set equal to 1.0.

Finally, if one decapped, returned the roots to the dark for 36 h and then extracted for cytokinins, the histogram in Fig. 9e was obtained. The three peaks were again evident, and from Table V the values and the ratios obtained for roots decapped plus regenerated 36 h were similar though somewhat higher than for the intact 72 h roots.

IV. Discussion and Conclusions

It is not understood how a quiescent center arises nor how it is perpetuated. From limited anatomical work and histoautoradiography, it is known that the first cells to become quiescent in developing roots are those situated immediately proximal to the root-cap-initiating region (Clowes, 1958). That the QC can initiate early in root development is supported by the recent work of Byrne and Heimsch (1970) who concluded that, in roots of *Malva*, the QC was established during embryogeny and continued to exist during germination. The lateral roots of *Eichornia* and *Pistia* first develop a QC at the apex of the procambial cylinder (Clowes, 1958), and then the QC enlarges in a proximal direction. According to Clowes (1958), the proximal boundary of the QC is indefinite. In *Malva*, the developing QC at its proximal surface includes portions of undifferentiated cortex, the apical parts of which are quiescent (Byrne and Hemisch, 1970). Thus, the existing evidence indicates that the QC initiates in roots immediately basal to the root cap junction and then spreads in a proximal direction.

In *Zea*, decapping the root causes the QC to begin cell divisions at the distal surface with little or no activation at the proximal end. This distal cell activation in the QC occurs between 12 and 24 h. Concomitantly, cytokinin concentrations measured in roots which have been decapped and returned to the dark for 24 h, reach their lowest measured levels. In sectioned material, 24 h after decapping new root cap initials are already evident, associated with a layer of new root cap cells. At this time the QC reaches its smallest recorded size, but never totally disappears. From 24 to 36 h new root cap cells are formed, the QC enlarges to nearly normal size and the total cytokinin levels in the tip reach pre-decapping levels.

How might one account for these observed activities on the part of the QC? One explanation suggested by Torrey (1972), proposed that the QC is the site of synthesis of a substance which stimulates cell-division; its formation in supra-optimal concentrations at the center of the QC would result ultimately in the quiescence of these cells.

An alternative theory, proposed by Webster and Langenauer (1973), holds that the meristematic cells surrounding the QC produce a substance which, when present in sufficiently high concentrations, inhibits cell division or greatly increases the cycle times in the QC. From observations on *Zea* and from the work of others (Clowes and Stewart, 1967; Webster and Langenauer, 1973; Wilcox, 1962), it is noted that activation of the QC often ensues after an impairment in function of the meristematic cells encircling the QC. Taken together, these observations suggest that these meristematic initials in some manner exert a controlling influence on the maintenance of a reduced metabolic state in cells comprising the QC. Such a regulatory effect, as Webster and Langenauer (1973) note, could be ascribed to substances produced by both the distal and proximal initials.

Except for roots with a single apical cell as in ferns, meristematic regions can be described in terms of layers of initial cells, with *at least* 2 discrete meristematic layers evident early during initiation. As adjacent layers continue to divide and synthesize substances, it is suggested that localized areas of more highly concentrated growth substances would result. The accumulation of relatively high concentrations could lead to these cells eventually inhibiting themselves. In order to avoid this self or mutual inhibition, these meristematic layers must either physically move apart from each other, or meristematic function must be transferred to *new* layers of cells. Such a transfer of function could occur only to the proximal initials. Separating the meristematic layers from each other by interposing a zone of non-dividing cells creates a buffer between the two suggested sites of origin of various substances. The interposed cells, which form the QC, serve to dilute potentially inhibitory concentrations of growth substances from the proximal meristem, before these substances can reach the distal meristem, and vice versa. This dilution could occur in a number of ways. For example, compounds produced by either the distal or the proximal meristem, in order to affect the opposing meristem, must first traverse the cells of the QC. Such cells, though greatly reduced in metabolic activity, nevertheless would still require various substances, including hormones, to continue to function. Thus it is suggested that, as compounds from either meristem move through the QC, some are retained or metabolized by cells composing the QC, so that in effect, a compound originating at one meristem is diluted or reduced in concentration upon arriving at the opposite meristem. Further, when separated, each meristem maintains both a proximal and a distal exit to facilitate the removal of potentially inhibitory concentrations of growth substances. The number of cells interposed would be dictated by the capacity of the meristematic cells to produce and ultimately to maintain now inhibitory concentrations of substances within cells composing the QC. Cells situated within this meristematic hemisphere would be continually provided with an inhibitory amount of some growth substance(s) with the only available exit for these compounds through the meristematic cells themselves. Fluctuations in the metabolic activity of the meristematic layers would lead to concomitant changes in inhibitor concentration resulting in corresponding shifts in QC size.

If the meristematic initials surrounding the QC are indeed sites of production of some factor which maintains the QC in a state of reduced activity, certain features could be ascribed to this regulatory substance. First, and obviously, it must be synthesized in the meristematic cells. Second, this substance, at one concentration, must have the capacity to maintain low rates of mitoses within the QC, and at another concentration have no effect, or perhaps a stimulatory effect, on divisions in these same, formerly inhibited QC cells. Third, changes in concentration of this substance within the QC could be reflected by concomitant changes in QC activity.

Whether cytokinins, in varying concentrations, could act as this proposed regulator of QC activity is not yet established. In young, developing fruitlets Letham (1963) and others have shown that meristematic cells can serve as sites of cytokinin biosynthesis. Furthermore, in soybean callus, zeatin at one concentration may stimulate cell division and at a slightly higher concentration becomes supra-optimal and the number of divisions are reduced (unpublished data). That changes in cytokinin concentration within the terminal millimeter of roots (not within the QC only) may also accompany changes in QC activity is clearly shown from this work. Whether this parallel is real or only co-incidental remains to be established. Nevertheless, since cytokinins seem to possess the proposed characteristics of this regulatory factor, let us assume, as Short and Torrey (1972) have suggested, that cytokinins, at least in part, are produced by the meristematic cells encircling the QC.

In roots of *Zea*, the argument could be made that at the time of lowest cytokinin levels, except for the absence of a root cap and its initials, the decapped seedling roots ought to be identical morphologically to intact seedling roots. Nevertheless, cytokinin levels can differ greatly between intact and decapped roots. These facts offer strong evidence that the root apical meristem is a site of cytokinin biosynthesis. If cytokinins extracted from the root apex came only from the shoot or more mature regions of the root, cytokinin levels in the terminal millimeter following decapping would be expected to remain constant and not fluctuate.

Surgical removal of the root cap destroys the distal initials, thus removing a suggested site of cytokinin biosynthesis. Distal QC cells would no longer be inhibited and thus ought to become activated, as indeed they do. Proximal initials remain present and functional at this time implying that cytokinin levels in the proximal QC cells remain high. This seems to be the case, for proximal QC cells remain quiescent. Between 12 and 24 h cytokinin levels reach 1/6 of the total amount of that in intact roots. However, since the proximal initials are still present one might expect cytokinin levels to be closer to 1/2. The observed difference suggests that the distal initials and perhaps the root cap itself may be more important for cytokinin production than the proximal initials. Work of Clowes (1972) suggests that this conclusion might fit, since the distal cap initials cycle much faster than any other cells in the meristem, largely eliminating the G_1 phase.

Irradiation destroys the root initials, both proximal and distal, ultimately resulting in the regeneration of new initials from the now activated QC. Similarly, artificially induced dormancy, produced with cold treatments, most often causes the activation of the QC with only the distal initials for certain being replaced (Clowes and Stewart, 1967). In both of these treatments Clowes and Stewart (1967) note that cells of the meristem which normally divide rapidly suffer a reduction in their reproductive capacity and presumably in their overall metabolism. It is here suggested that the reduction in number of cells remaining quiescent reflects a concomitant drop in the amounts of growth substances produced.

In *Libocedrus* roots, in which dormancy is induced naturally, Wilcox (1962) implies that as roots become dormant, the QC shrinks in size. This presumably would occur because the meristematic initials are reduced in cellular activity with an attendant reduction in the area in which supra-optimal concentrations of growth substances could be maintained. New growth is resumed by cells within the QC, recapitulating the process of QC development. Recently Webster and Langenauer (1973) noted that starving roots results in a cessation of DNA synthesis in both distal and proximal initials. When resupplied with nutrients no QC was evident. Several comparisons can be drawn from the starvation, decapping and natural dormancy work; first, the source of the proposed inhibitor is either removed or its metabolic effect is impaired, and second, the QC is lost or greatly reduced in size.

Because changes in extractable cytokinins correspond with shifts in QC activity, it is suggested that this class of hormones may, at least in part, influence the establishment and maintenance of the QC in roots.

Acknowledgements

With great pleasure, I wish to thank Dr. John G. Torrey for the numerous stimulating discussions, from which much of this work is derived.

References

BOYD, G. A. (1955). "Autoradiography in Biology and Medicine". Academic Press, New York, London and San Francisco.

BURROWS, W. J. and CARR, D. J. (1969). Effects of flooding the root system of sunflower plants on the cytokinin content in the xylem sap. *Physiol. Pl.* **22**; 1105–1112.

BYRNE, J. M. and HEIMSCH, C. (1970). The root apex of *Malva sylvestris*. II. The quiescent center. *Am. J. Bot.* **57**; 1179–1184.

CLOWES, F. A. L. (1950). Root apical meristems of *Fagus sylvatica*. *New Phytol.* **49**; 248–268.

CLOWES, F. A. L. (1956a). Nucleic acids in root apical meristems of *Zea*. *New Phytol.* **55**; 29–34.

CLOWES, F. A. L. (1956b). Localization of nucleic acid synthesis in root meristems. *J. exp. Bot.* **7**; 307–312.

CLOWES, F. A. L. (1958). Development of quiescent centres in root meristems. *New Phytol.* **57**; 85–88.

CLOWES, F. A. L. (1961). *Apical Meristems*. Blackwells, Oxford.

CLOWES, F. A. L. (1972). The control of cell proliferation within root meristems. *In* "The Dynamics of Meristem Populations" (M. W. Miller and C. C. Kuehnert, eds.) Vol. 18, pp. 133–147. Plenum Press, New York.

CLOWES, F. A. L. and STEWART, H. E. (1967). Recovery from dormancy in roots. *New Phytol.* **66**; 115–123.

CUTTER, E. G. and FELDMAN, L. J. (1970). Trichoblasts in *Hydrocharis*. I. Origin, differentiation, dimensions and growth. *Am. J. Bot.* **57**; 190–201.

JUNIPER, B. E., GROVES, S., LANDAU-SCHACHAR, B. and AUDUS, L. J. (1966). Root cap and the perception of gravity. *Nature, Lond.* **209**; 93–94.

KENDE, H. (1965). Kinetin-like factors in the root exudate of sunflowers. *Proc. Natl. Acad. Sci. U.S.A.* **53**; 1302–1307.

KENDE, H. (1971). The cytokinins. *In:* "International Review of Cytology" (G. A. Bourne and J. F. Danielli, eds). Vol. 31, pp. 301–338. Academic Press, New York, London and San Francisco.

LETHAM, D. S. (1963). Purification of factors inducing cell division extracted from plum fruitlets. *Life Sci.* **3**; 152–157.

MIKSCHE, J. P. and GREENWOOD, M. (1966). Quiescent center of the primary root of *Glycine max*. *New Phytol.* **15**; 1–4.

MILLER, C. O. (1963). Kinetin and kinetin-like compounds. *In:* "Modern Methods of Plant Analysis" (H. F. Linskens and M. V. Tracey, eds.) Vol. 6, pp. 194–202. Springer, Berlin.

PHILLIPS, D. A. and TORREY, J. G. (1970). Cytokinin production by *Rhizobium japonicum*. *Physiol. Pl.* **23**; 1057–1063.

PHILLIPS, H. L. and TORREY, J. G. (1971). The quiescent center in cultured roots of *Convolvulus arvensis* L. *Am. J. Bot.* **58**; 665–671.

RADIN, J. W. and LOOMIS, R. S. (1971). Changes in the cytokinins of radish roots during maturation. *Physiol. Pl.* **25**; 240–244.

SHORT, K. C. and TORREY, J. G. (1972). Cytokinins in seedling roots of pea. *Plant Physiol.* **49**; 155–160.

Torrey, J. G. (1962). Auxin and purine interaction in lateral root initiation in isolated pea segments. *Physiol. Pl.* **15**; 177–185.

Torrey, J. G. (1972). On the initiation of organization in the root apex. *In:* "The Dynamics of Meristem Populations" (M. W. Miller and C. C. Kuehnert, eds.) Vol. 18, pp. 1–13. Plenum Press, New York.

Webster, P. L. and Langenauer, H. D. (1973). Experimental control of the quiescent centre in excised root tips of *Zea mays*. *Planta (Berl.)* **112**; 91–100.

Weiss, C. and Vaadia, Y. (1965). Kinetin-like activity in root apices of sunflower plants. *Life Sci.* **4**; 1323–1326.

Wilcox, H. (1962). Growth studies of the root of incense cedar. *Libocedrus decurrens*. I. The origin and development of primary tissues. *Am. J. Bot.* **49**; 221–236.

Chapter 4

Some Aspects of the Differentiation of Primary Root Tissues

M. LUXOVÁ

Institute of Botany, Slovak Academy of Sciences, Bratislava, Czechoslovakia

I. Introduction

Cell differentiation proceeds through a series of partial processes which culminate in an adult, mature cell. The study of differentiation may be approached from various standpoints, e.g. from cytogenetics, histology, cyto- and histo-chemistry, physiology and ultrastructure. In this paper I shall deal with certain histological problems.

The root promeristem is formed by a group of initial cells, except in the leptosporangiate ferns (Guttenberg, 1947). There are several ways in which these initials can be organized. For general purposes of description the apical meristem may be characterized as a region of rapidly cycling initial cells which surround the group of non-cycling, or long cycling cells of the quiescent centre (Clowes, 1958).

The derivative region is immediately proximal to the apical meristem and I would like to focus my attention on this region in which the differentiation pattern is formed. My paper will deal with three topics, with the formation of the pattern and the extent of the mitotically active region, with variations of the pattern, and with certain cases of asynchronous differentiation.

Differences in ontogenesis can be observed in the various cell types which form the tissues of the mature root. Each cell type has a characteristic course of differentiation which results from the interaction of internal and external factors.

II. Material and Methods

A. Roots from Seedlings

Root tips of *Hordeum vulgare*, cv. Slovenský dunajský trh, the two-row spring barley, were used for the analyses. The seeds were soaked in running tap water for 4 h and then allowed to germinate on sheets of moist filter paper in an incubator at 25°C. After 30 h tips from primary roots, 10 mm in length, were fixed in Navashin's fluid and were passed into paraffin through methylsalicylate. Transverse sections 10 μm thick and longitudinal sections of 5 μm were stained with Heidenhain's hematoxylin. The results obtained from the analyses of a series of transverse sections from ten root tips were complemented by the analysis of longitudinal median sections. An analysis was made of the course of cell division paying special attention to the origin of the individual cell types, as well as to the extent of cell division in the periclinal, anticlinal and transverse planes. In addition, the pattern of the inner root organization was obtained from analyses of hand cut transverse sections with longer roots (50 mm); this served as a basis for the delimitation of the extent of longitudinal divisions. Mean values of the cell numbers in the individual layers, together with the variation in the data were obtained from the analyses of twenty roots. The extent of transverse division was estimated from appearance of the last mitoses in the proximal part of the root.

For the analysis of *Zea* root tips, the inbred line A15 was used. The tips of primary roots 50 mm long were processed and analysed using the same technique as described for barley.

B. Roots from Excised Embryos

For cultivation *in vitro*, *Zea mays* embryos, of the inbred line A15 were used. The plants were artificially pollinated. The cultivation of embryos excised 14, 20 and 30 days after pollination was performed in test tubes in White's agar medium containing 2, 5, 10 and 15% sucrose, at 26°C under daylight illumination. The embryos of kernels developed under natural conditions served as controls. After growing for 1 month, the bases of 20 primary seminal roots from each treatment were analysed. The material for anatomical analyses was prepared as described above. (Preliminary results of these experiments were presented at The International Symposium on Morphogenesis in Plant Cell, Tissue and Organ Cultures, Delhi in 1971.)

In a second experiment larger numbers of embryos were excised 14–16 days after pollination to provide sufficient material for detailed analysis of the histological pattern in the stelar regions. Out of a total of 250 plantlets, detailed observations were made on individuals with 5 large metaxylem vessels in the stele; in this respect they were comparable with control plants where the embryo had matured in the seed.

C. Roots Showing Geotropic Curvature

The asynchronous formation of the differentiation pattern was studied using geotropically curved roots of *Hordeum*, *Sorghum*, *Zea* and other species. The preparation of the material was as described above.

III. Results and Discussion

A. Pattern Formation : the Extent of the Meristematic Region

The histological pattern in the root apex is determined by two factors, the frequency of cell division, and the orientation of the divisions. Periclinal divisions increase the number of layers in the root and thus enlarge its diameter. Anticlinal divisions increase the number of cells in an individual cell layer. Transverse division, together with cell elongation, determines root elongation. The distribution of these three types of cell division has been observed in the primary seminal root of barley (Luxová and Lux, 1974). The anatomy of this species has been studied frequently (e.g. Jackson, 1922; Heimsch, 1951; Hagemann, 1957). The pattern of differentiation is as follows: under the epidermis there are 5 to 6 layers of cortex beneath which the polyarch protostele is found. The stele has 7 to 8 xylem poles which alternate with the same number of phloem poles. The large metaxylem vessel is situated at the centre of the stele. The uniseriate layers of the epidermis and pericycle have approximately twice as many cells in them in comparison with the layers adjacent to them, i.e. the hypodermis, and the endodermis. The number of cells in the individual layers is shown in Fig. 1.

Periclinal divisions are the least numerous. In the epidermis and the one-layer pericycle they do not occur. The outer cortical layers are almost completely derived from the repeated periclinal divisions of the innermost layer and when this has been done this layer forms the endodermis. In the central cylinder periclinal divisions cease centripetally.

Anticlinal divisions in the peripheral part of the root cease at a very early stage; the decline in frequency being from the outside towards the centre, i.e. centripetally, whereas in the stele they cease centrifugally. The doubling-up of cell numbers in the uniseriate epidermis and pericycle indicates that anticlinal division takes place in these tissues. The longitudinal divisions, both periclinal and anticlinal, are thus limited to the distal, relatively short, part of the meristematic region (Figs. 2a and b). It is in this part that the pattern formation occurs.

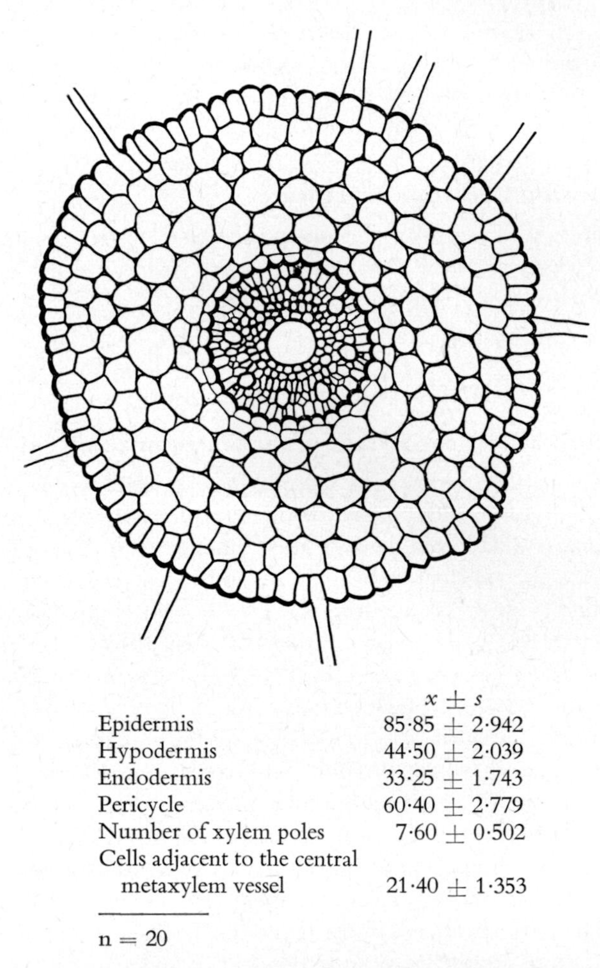

	$x \pm s$
Epidermis	85·85 ± 2·942
Hypodermis	44·50 ± 2·039
Endodermis	33·25 ± 1·743
Pericycle	60·40 ± 2·779
Number of xylem poles	7·60 ± 0·502
Cells adjacent to the central metaxylem vessel	21·40 ± 1·353

n = 20

Fig. 1. Transection of a primary root of *Hordeum vulgare* with the average numbers of the cells in individual layers.

Meristematic cells of the derivative region are morphologically non-uniform. The first changes can be observed immediately above the initials of the stele and are associated with the origin of the large single central metaxylem vessel (Fig. 3A). The pericycle surrounding the central cylinder can be identified at quite

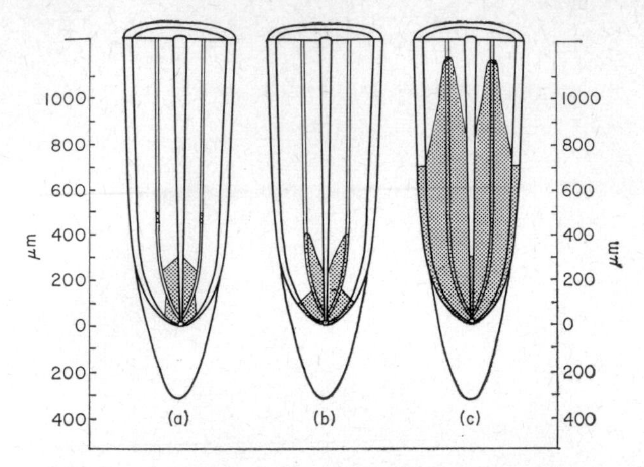

FIG. 2. Longitudinal sections from a root tip of *Hordeum vulgare* showing the extent of periclinal (a), anticlinal (b) and transverse cell divisions (c).

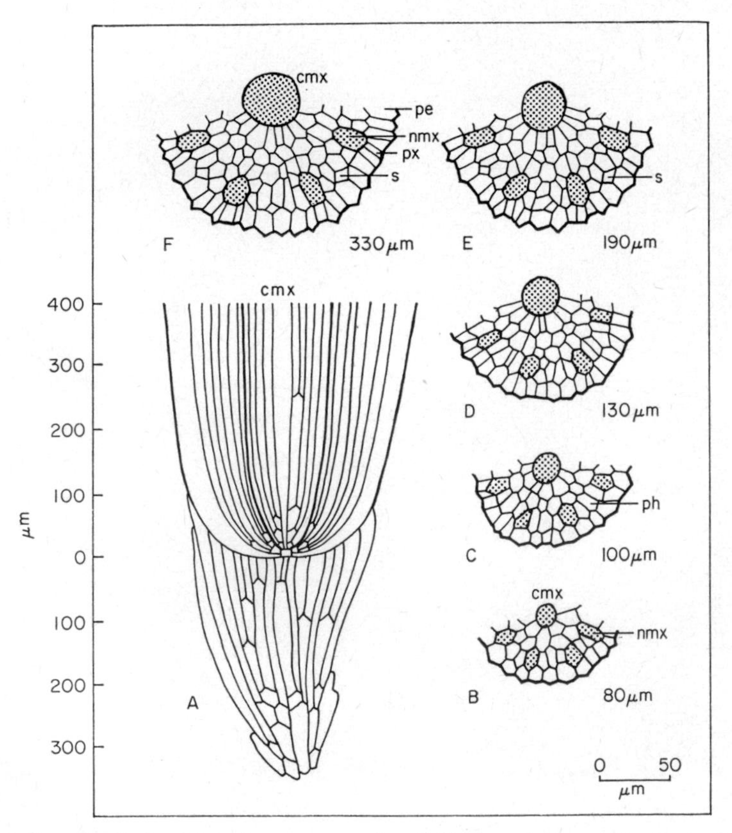

FIG. 3. Longitudinal section from a root tip of *Hordeum vulgare* (A) and partial transections of the vascular cylinder in successive stages of differentiation (B–F). cmx, central metaxylem vessel; nmx, narrow metaxylem; ph, protophloem mother cells; s, sieve tube.

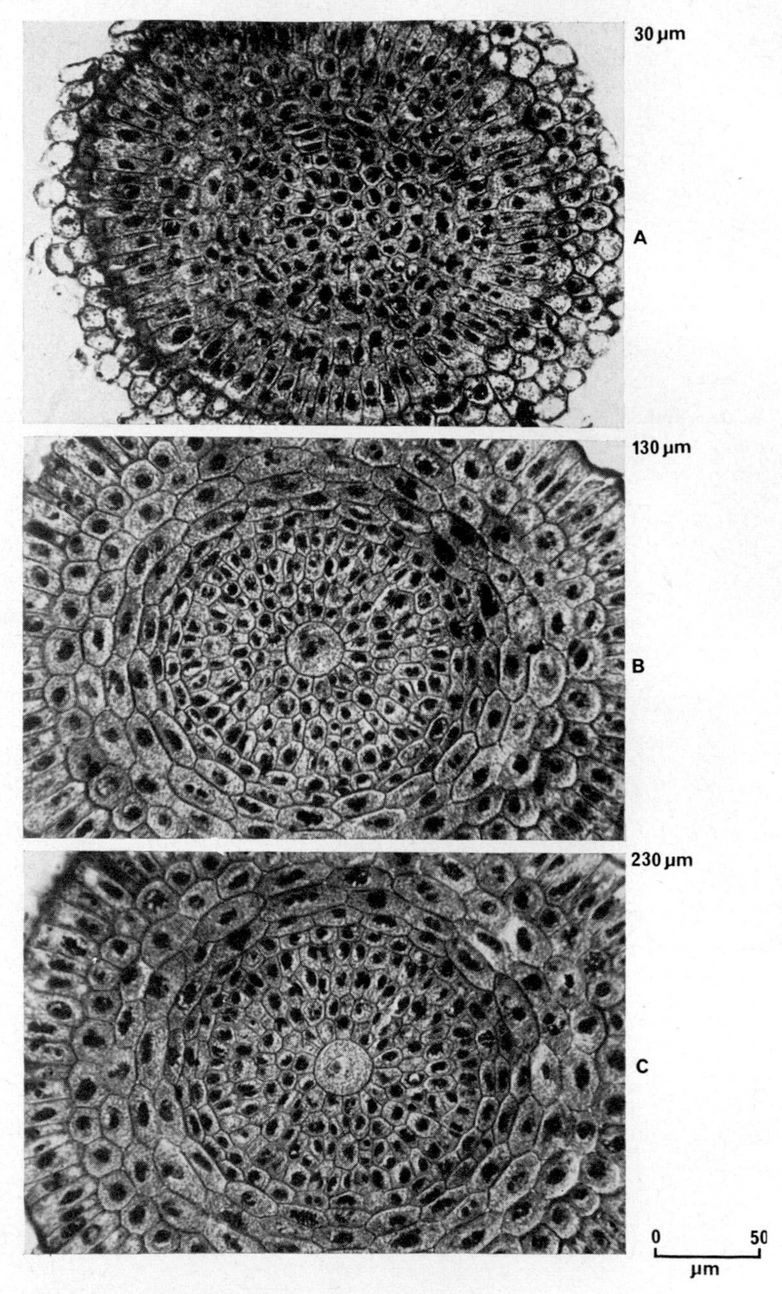

FIG. 4. Transection of a root tip of *Hordeum vulgare* showing the stages of determina-
narrow metaxylem; cmx, central metaxylem.

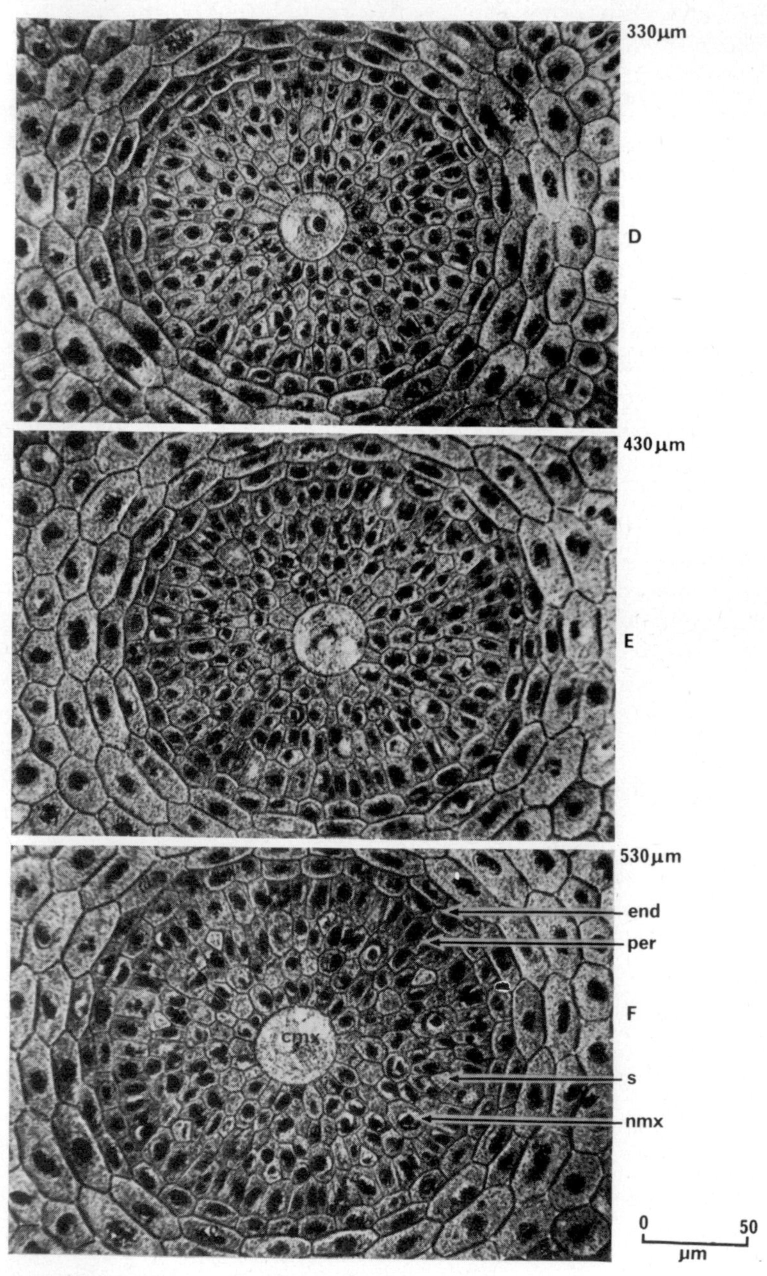

tion of the central cylinder. end, endodermis; per, pericycle; s, sieve tube; nmx,

a short distance from the apical initials. At a distance of 70–80 μm, the differentiation takes place, in the layer immediately interior to the pericycle, of 7–8 cells which become the mother cells of the narrow metaxylem (Fig. 3B); thus, the position of the xylem poles is determined. At a distance of approximately 100 μm the protophloem mother cells appear alternately between them, thus determining the position of the phloem poles (Fig. 3C). The formation of the root pattern continues with successive longitudinal cell divisions (Fig. 3D). The origin of sieve tubes may be observed at a distance of 170–230 μm from the initial cells (Fig. 3E). The protoxylem, the last constituent of the xylem to be differentiated, appears 300–500 μm from the initials (Fig. 3F). This delayed appearance of the protoxylem is connected with its derivation from the pericycle by periclinal division. The formation of xylem proceeds centrifugally. Figures 4A–F demonstrate the development of the radial pattern in the central cylinder.

Transverse divisions extend further back along the axis; thus, they determine the extent of the meristematic region (Fig. 2c). Transverse divisions occur over a greater length of axis in some tissues than in others; in the peripheral part of the root they cease centripetally, but in the central cylinder they cease centrifugally. The large central metaxylem, which divides only transversely, stops dividing at a distance of 300–350 μm from the initials. In the pericycle, i.e. in the layer with the longest duration of division, the cells divide transversely up to the distance of 1000–1150 μm, the capacity for division being maintained longest by those cells of the pericycle which are adjacent to the xylem poles. A comparison of the extent of cell division in the various planes and in the various tissues of the root is shown in Table I.

TABLE I. The extent of the cell division in barley root tip.

Division	Periclinal	Anticlinal	Transverse
Epidermis	—	50–100 μm	600–750 μm
Hypodermis	it ceases very early	50–100 μm	650–800 μm
Central cortical region	it ceases very early	50–100 μm	750–850 μm
Endodermis	100–150 μm	100–150 μm	850–900 μm
Pericycle	only a few cells, more distant from the initials (450–500 μm), during the initiation of protoxylem.	350–450 μm	1,000–1,150 μm
Layer beneath the pericycle	200–250 μm	—	900–1,000 μm
Cells adjacent to the central metaxylem	300–350 μm	150–200 μm	750 800 μm
Central metaxylem	—	—	300–350 μm

When comparing the pattern of tissue determination in other species with these observations, substantially similar patterns are revealed but interesting differences in detail are found. With the *Zea* root tip (Fig. 5A), the prolonged mitotic activity of the epidermis is conspicuous. The last mitoses were observed in this tissue at a distance of 1400 μm from the apex, whereas in the hypodermis, the mitotic activity ceased at a distance of 950 μm from the apex. In the inner part of the cortex and the outer part of the stele the cell division stopped approximately at

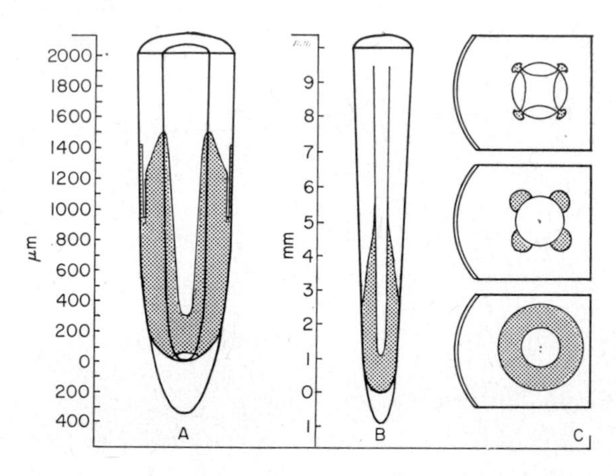

FIG. 5. Central longitudinal sections from root tips of *Zea mays* (A) and *Vicia faba* (B) showing the extent of the meristematic region. The progressive reduction of *Vicia* apical meristem to groups of dividing cells adjacent to protoxylem is illustrated in transverse sections (C). (B, C by Luxová and Murín, 1973, modified).

the same point, at about 1500 μm. With the root of *Vicia faba* the cells of the protophloem were the first to cease dividing, the sieve tubes of the protophloem being mature at a distance of 600–700 μm from the apex (Luxová and Murín 1973). At about 1000 μm, all division in the central stele stopped centrifugally (Fig. 5B). In the peripheral parts of the root apex, the apidermis ceased dividing first, at a distance of about 2000 μm from the apex, followed by the centripetal cessation of division in the cortex. In this way the region of continuing cell division is formed from the inner cortical layers, including the endodermis, and by the pericycle and the few layers adjacent to it. This region, originally cylindrical, is gradually reduced to isolated strands in which cell division occurs and these lie adjacent to the xylem poles (Fig. 5C). At a distance of 900–1000 μm from the initials, these isolated groups of cells in the pericycle cease to divide.

Thus the main region where cells are to be found in mitosis is a restricted part of the root apex. This is also the case in other species where the latest cell divisions were observed at the following distances from the root tip; *Phleum pratense*—425 μm (Goodwin and Stepka, 1945); *Melilotus alba*—700 μm (Alfieri and

Evert, 1968); *Triticum*—900–1300 μm; *Allium cepa*—1500–2100 μm; *Helianthus annuus*—1100–2000 μm (Balodis, 1968). As stated by Němec (1930), the number of division cycles characteristic of individual meristematic cells prior to their transition to the non-active state is generally low.

In accordance with the general finding we have seen that the cells at the centre of the stele undergo the lowest number of transverse divisions, e.g. with *Vicia*, the cells in the central stele divide transversely approximately five times (2^5), whereas the cells of the cortex, as well as those of the epidermis divide transversely approximately seven times (2^7) (Luxová and Murín, 1973). The centrifugal course of determination of the tissues of the central cylinder applies to the roots which have the metaxylem at the centre of the stele, as well as to those which have pith in this position. On the periphery of the central cylinder, the regions of the prospective xylem alternate with those of the prospective phloem. The differentiation of phloem is faster than that of xylem; the sieve tubes of the protophloem being mature at a distance where the other cells are still dividing. However, the asynchronous development is not limited to regions of the phloem and xylem, or to the adjoining parts of the pericycle, but is characteristic also of the adjacent endodermal cells and more distant cells in the inner cortex.

B. Changes in the Pattern

The pattern of differentiation and the extent of the meristematic region is characteristic of a given species, or a given variety or cultivar of a species. In the vascular system of *Zea mays*, v. *dentiformis* had an average of 12 to 15 xylem poles in the radicle, whereas variety *tunicata* had only 8 to 11 poles (Luxová, 1967).

According to the morphological root type involved, quantitative differences may occur during its formation. A suitable model subject for this purpose is again *Zea*. The inbred line A15 has 3 roots preformed in the embryo, i.e. the main, or primary, seminal root and two further seminal roots. In addition, whorls of adventitious roots are formed around the 6–7 basal nodes of the plant. Quantitative analyses of the number of xylem vessels and protoxylem poles were made by examining transverse sections near the base of these various root types (Table II). The results obtained demonstrate that the site of origin affects the "architecture" of the stele.

Quantitative changes in pattern formation can also occur along the length of a root (see also p. 95). When analysing the individual root types of *Zea* (Lux and Luxová, unpublished data), it was observed that both the number of cell layers and the number of cells in an individual layer were reduced as the root grew longer. There were also reductions in the number of large metaxylem vessels, and xylem and phloem poles. As the root became longer the frequency of the periclinal and anticlinal divisions of the meristematic zone decreased.

Turning to the question of the histological determination of the individual root regions, I shall refer to some experiments where we made use of the slow

TABLE II. Quantitative anatomical analysis of the individual root types of maize. (Modified from Luxová and Kozinka, 1970)

Root type	Area of the transection through the stele in mm^2 $\bar{x} \pm s$	Number of large metaxylem vessels $\bar{x} \pm s$	Number of protoxylem poles $\bar{x} \pm s$
Primary seminal	0·265 ± 0·002	6·00 ± 0·74	14·5 ± 0·84
1st node	0·256 ± 0·003	10·10 ± 1·41	21·4 ± 2·39
2nd node	0·428 ± 0·003	12·00 ± 1·36	25·1 ± 2·16
3rd node	1·570 ± 0·058	18·40 ± 2·20	40·7 ± 4·90
4th node	3·833 ± 0·036	32·80 ± 5·24	59·2 ± 6·84
5th node	8·337 ± 0·091	54·30 ± 7·44	88·2 ± 9·69

process of tissue determination during embryogenesis in maize. Maize embryogenesis is completed in 45 days approximately. In our conditions, the histological determination of the root region of the embryo occurs within 15 to 25 days after pollination. This is a very slow process in comparison with the rate of tissue determination in the root tip after germination.

Excised, immature embryos, when transferred to artificial conditions, underwent little further embryonic development, but germinated precociously as small plantlets. The roots of these plantlets developed differently from embryos which had grown to maturity whilst attached in the mother plant. The results of this experiment are shown in Table III. Embryos of kernels matured under

TABLE III. Quantitative analysis of the primary root of *Zea mays*

Variant		Stele		Cortex
Embryos	Sucrose	Metaxylem $\bar{x} \pm s$	Number of poles $\bar{x} \pm s$	Number of layers $\bar{x} \pm s$
	2%	3·0 ± 0·887[a]	8·6 ± 0·812[a]	4·1 ± 0·812[a]
14 days	5%	4·1 ± 0·812[a]	10·2 ± 0·711[a]	4·7 ± 0·658[a]
	10%	4·0 ± 0·561[a]	10·2 ± 0·716[a]	5·0 ± 0·000[a]
	15%	4·1 ± 0·812[a]	10·0 ± 1·050[a]	4·9 ± 0·582[a]
	2%	5·0 ± 0·725	12·8 ± 1·105	7·5 ± 1·884[a]
20 days	5%	5·0 ± 0·788	12·8 ± 0·833	7·6 ± 0·904[a]
	10%	4·7 ± 0·638	12·7 ± 1·019	7·4 ± 1·209[a]
	15%	5·4 ± 0·680	13·1 ± 0·578	7·9 ± 0·831[a]
	2%	5·1 ± 0·552	13·2 ± 0·716	10·5 ± 0·412
30 days	5%	5·4 ± 0·680	13·3 ± 0·801	10·3 ± 0·865
	15%	5·1 ± 0·745	13·0 ± 1·168	10·0 ± 1·007
Control		5·5 ± 0·761	13·5 ± 0·998	10·0 ± 0·293

[a] determined by cultivating immature embryos *in vitro*.

natural conditions had, on average, 5·5 large metaxylem vessels and 13·5 alternating xylem and phloem poles preformed in the embryonic root region; there were ten cortical cell layers. With embryos excised 14 days after pollination there were changes induced by the artificial medium in the determination of all root tissues; in the central cylinder the numbers of prospective large metaxylem vessels and of xylem and phloem poles were reduced; in the cortical area the number of cell layers was decreased. With embryos excised after 20 days the effect of artificial conditions was evident only in the cortical layers, while with

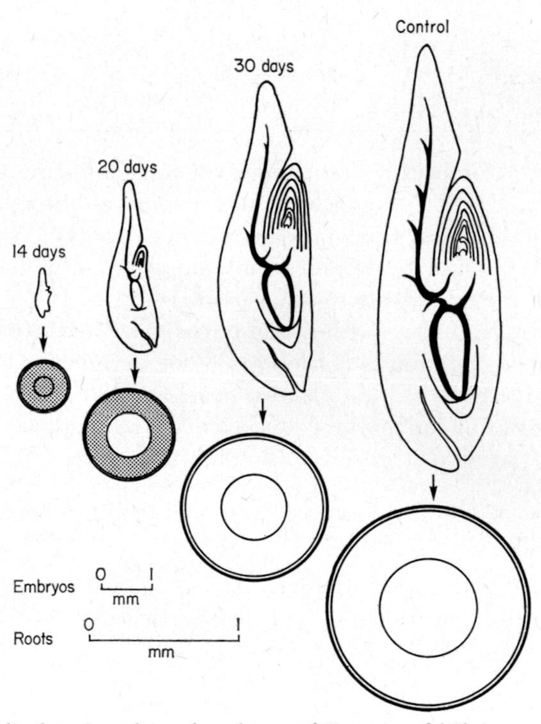

FIG. 6. Longitudinal sections through embryos of *Zea mays* of different age and transections of the corresponding roots fully or partially determined during embryogenesis *in vitro*.

embryos excised after 30 days the differentiation pattern did not differ quantitatively from the pattern found in root primordia of embryos developed under natural conditions.

The imperfect embryonic development seen under the conditions of cultivation *in vitro* was due to the inhibition of cell division and to continuous tissue differentiation without the interruption which usually occurs during seed dormancy. The inhibition of cell division occurred in conformity with the centrifugal course of tissue determination of the root region (Fig. 6); in embryos excised after 14 days the root primordium was formed by a group of meristematic cells which were not recognizably organized into a pattern and the reduced amount of cell division thus affected the entire root region. With embryos excised after 20 days the central cylinder was distinguishable in the root primordium. Since this was already determined histologically at the point of excision, *in vitro* cultivation influenced only the cortical region. With embryos excised after 30 days the root primordium was surrounded by coleorhiza and the whole primordium was histologically determined.

It was shown in a complementary experiment that the central stele did not determine the radial pattern in the remainder of the root. The control had an average of 5 central metaxylem vessels and 11 to 15 xylem and phloem poles in the primordium. In excised embryos with 5 central vessels determined in the central stele, a reduced number of poles 8 to 12 ($\bar{x} \pm s = 10{\cdot}44 \pm 1{\cdot}008$; $n = 65$)—was formed on the periphery of the stele in the course of development *in vitro*. The determination of peripheral stelar parts was thus accomplished independently of the central stele.

It is not yet clear how far the results obtained from the polyarch roots of *Zea* may be applied to other root types. From a study of the vascular pattern in diarch roots of *Sinapis alba*, Bünning (1951) observed that the radial polarity of the central metaxylem determined the subsequent development of the stele. It is possible that the increase in pole number in roots of this species described in Reinhard (1960), may have been preceded by a change in the structure of the central stele. The quantitative changes in the pattern caused by cultivating embryos of *Zea mays in vitro* far exceeded the natural variability due to genetic, maternal and ecological factors (Luxová, 1967, 1971b; Luxová and Polerecký, in press). These extreme, artificially induced changes of the pattern, as well as those which may be called natural changes, are the structural manifestation of the altered physiological conditions in which the root meristem is determined. This was most marked in embryos excised after 14 days (see Table III), in which cell division was strongly depressed at low sucrose concentrations in the incubation medium. In these, but not in embryos excised at later stages of their development, an increase in sucrose concentration from 2% to 5% elicited increased numbers of cells in both stele and cortex. In his paper dealing with the effects of auxin on the vascular pattern formation in triarch pea roots, Torrey (1957) expressed a similar opinion when stating: "A physiological homeostasis in intact pea roots produces a tissue system wherein the triarch pattern is normally produced. Upsetting this balance produces new physiological conditions in which a different pattern may be expressed."

C. Asynchronous Pattern Formation

The root is usually thought of as a radially symmetrical organ, but frequently asynchronous differentiation can be observed. This phenomenon is especially striking with geotropically curved roots. On analysing a series of sections prepared from curved roots of various species (*Hordeum*, *Sorghum*, *Zea*), we found that, on the shorter concave side, tissue determination advanced faster than on the longer convex side. This was associated with an enhanced mitotic activity in the root

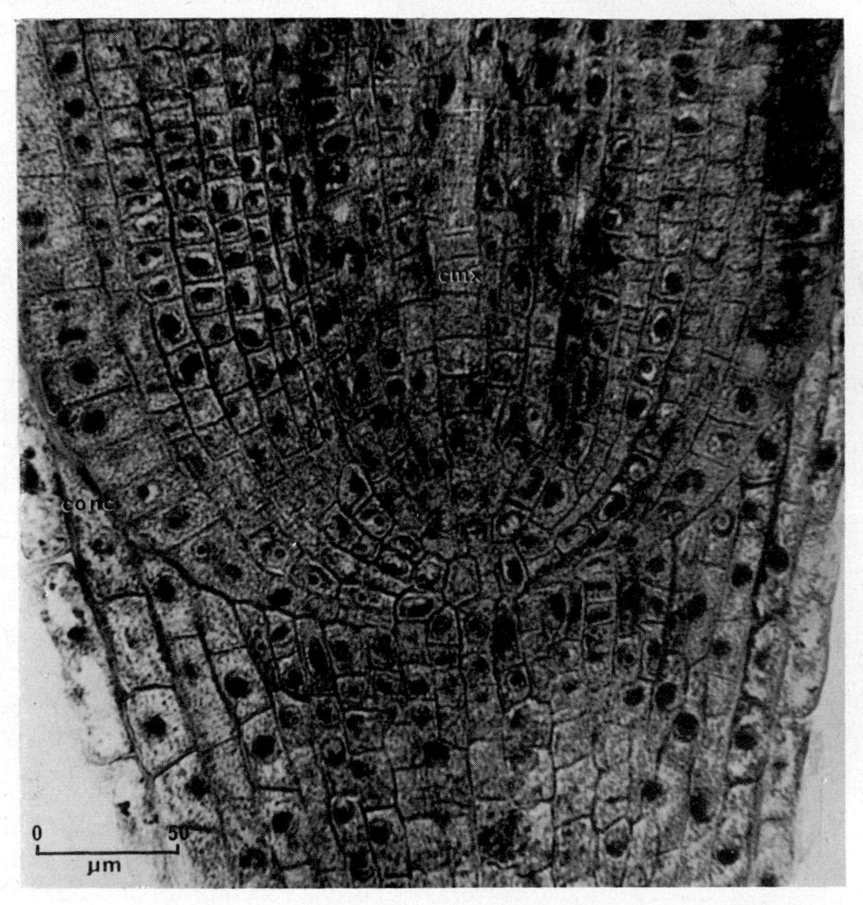

FIG. 7. Longitudinal section from an asymmetrical root tip of *Hordeum vulgare* showing more cell layers on the concave side (conc). cmv—the central metaxylem vessel.

apex on the concave side. This can be illustrated in the barley root which has a simple differentiation pattern with a single metaxylem vessel in the centre of the stele (Fig. 7). Histological determination and further differentiation were both more rapid on the concave side. The latter fact may be inferred macroscopically from the root hairs which grew closer to the root tip on the concave side (Fig. 8A). Histological analysis revealed that the first protophloem elements matured earlier on the concave side of the curved barley root; the first mature sieve tube of the protophloem was observed at a distance of 640 μm from the apex, while the first mature sieve tubes was seen at 710 μm from the apex on the convex side

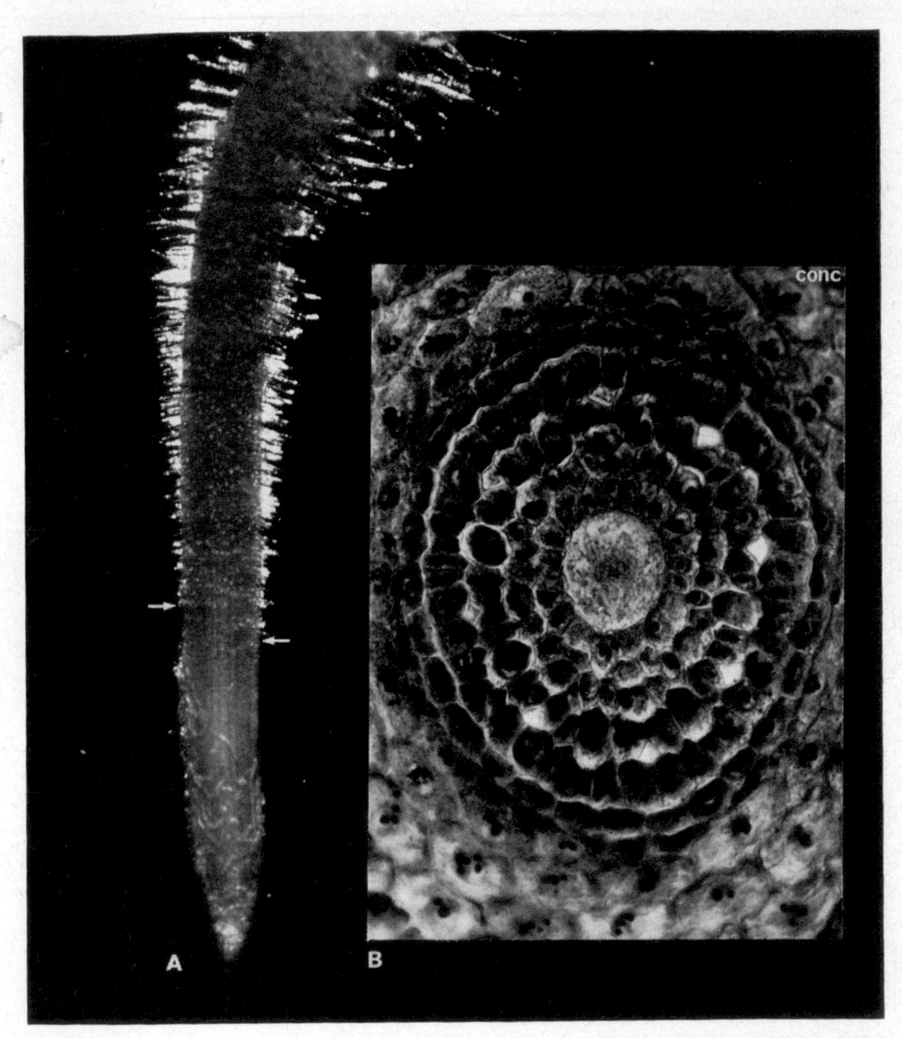

FIG. 8. Root tip of the curved root of *Hordeum vulgare*. Root hairs are formed more distally on the concave side (A). Transection through the similar root with the first mature protophloem sieve tubes on the concave (conc) side (B).

(Figure 8B). Similar asynchrony in the maturation of sieve tubes in barley roots was found by Hagemann (1957), and Esau (1943) observed it in pea roots. Peterson (1967) described the asynchronous maturation of both protophloem sieve tubes and protoxylem elements in the diarch root of *Sinapis alba*. It is well known that the inhibition of elongation growth in roots generally results in early differentiation of the primary conductive tissues, and the earlier maturation of vascular elements on the shorter concave side of the curved root is evidently related to this inhibition.

IV. Conclusions

The apical meristem is the morphogenetic centre from which all root tissues arise. The pattern of differentiation is determined by the frequency of cell division as well as by the orientation of the new cell walls. It depends on the plant species and on the morphological type of root. Periclinal and anticlinal cell divisions take place only within a relatively short distance of the meristematic initials so that the total length of the meristematic region is determined largely by transverse cell divisions. Meristematic activity in different tissues stops at varying differences from the root apex.

Meristematic cells of the derivative region are morphologically non-uniform. The first changes can be observed above the initials of the central stele. The sequence in which tissues are determined in the central cylinder is centrifugal; metaxylem is determined first followed by the peripherally situated protoxylem. The regions of the prospective xylem alternate with those of the prospective phloem. The sieve tubes of the protophloem are the first vascular elements to mature and they do so at a distance from the root apex where other cells are still dividing. Parts of the pericycle, endodermis and inner cortical cells adjacent to the phloem poles mature more rapidly than other regions of the same tissue, suggesting some relationship between the immediacy of supplies of carbohydrate from the sieve tubes and cell maturation.

By changing the frequency of cell division and orientation of the division walls, structural changes can occur in a given root. The general pattern may continue, but there may be changes in the number of the cell layers or in the number of cells in an individual layer. Such changes of pattern reflect the physiological conditions experienced by the meristematic region. The development of each tissue in the root proceeds with a certain degree of autonomy. Experiments with immature embryos cultured *in vitro* showed that the central cells in the stele did not determine the radial pattern of other cells in the periphery of the stele or in the cortex.

Asynchronous formation of the differentiation pattern can be frequently observed and is especially striking with roots which are geotropically curved. On the shorter, concave, side the tissue determination is usually found to be more advanced than on the longer convex side. The root meristem represents a complex system which responds sensitively to the changes in the physiological balance. This may result not only in the inhibition or stimulation of root growth, but in quantitative changes in the pattern of differentiation and also in the formation of an asynchronous pattern of development. The changes indicate a considerable functional independence of the individual regions of the meristematic complex.

Acknowledgements

The author wishes to thank Dr. A. Lux, C.Sc. for being so kind as to make photographs. Thanks are also due to Miss J. Glasová for her excellent technical assistance.

References

Alfieri, I. R. and Evert, R. F. (1968). Analysis of meristematic activity in the root tip of *Melilotus alba*. *New Phytol.* **67**; 641–647.

Balodis, V. A. (1968). Regularity of mitosis distribution in the root tip. (In Russ.) *Tsitologiya* **10**; 1374–1382.

Bünning, E. (1951) Über die Differenzierungsvorgänge in der Cruciferenwurzel. *Planta (Berl.)* **39**; 126–153.

Clowes, F. A. L. (1958). Development of quiescent centres in root meristems. *New Phytol.* **57**; 85–88.

Clowes, F. A. L. (1961). "Apical meristems". Blackwell Scientific Publications, Oxford.

Esau, K. (1941). Phloem anatomy of tobacco affected with curly top and mosaic. *Hilgardia* **13**; 437–490.

Esau, K. (1943). Vascular differentiation in the pear root. *Hilgardia* **15**; 299–311.

Goodwin, R. H. and Stepka, W. (1945). Growth and differentiation in the root tip of *Phleum pratense*. *Am. J. Bot.* **32**; 36–46.

Guttenberg, H. von (1947). Studien über die Entwicklung des Wurzelvegetationspunktes der Dikotyledonen. *Planta* (*Berl.*) **35**; 360–396.

Guttenberg, H. von (1960) "Grundzüge der Histogenese höheren Pflanzen I. Die Angiospermen." Gebrüder Borntraeger, Berlin.

Hagemann, R. (1957). Anatomische Untersuchungen an Gerstenwurzeln. *Die Kulturpflanze* **5**; 75–107.

Heimsch, Ch. (1951). Development of vascular tissues in barley roots. *Am. J. Bot.* **38**; 523–538.

Jackson, V. G. (1922). Anatomical structure of the roots of barley. *Ann. Bot.* **36**; 21–39.

Luxová, M. (1967). Determination of radial arrangement of vascular tissues in the radicle and the primary root of *Zea mays* L. *Biológia* (Bratislava) **22**; 161–165.

Luxová, M. (1971a). Histogenesis of the primary root primordium in excised embryos of *Zea mays* cultured *in vitro*. Presented at the International Symposium "Morphogenesis in plant cell, tissue and organ cultures," Delhi.

Luxová, M. (1971b). Concerning variability in the delimitation of the vascular system in root primordia of maize embryos. *Biol. Plant.* (Praha) **13**; 79–87.

Luxová, M. and Kozinka, V. (1970). Structure and conductivity of the corn root system. *Biol. Plant.* (Praha) **12**; 47–57.

Luxová, M. and Lux, A. (1974). Notes on the origin and development of primary root tissues. *In* "Structure and function of primary root tissues," Symp. Proc., Czechoslovakia, Sept. 7–10, 1971 (J. Kolek, ed.). pp. 37–40, Bratislava.

LUXOVÁ, M. and MURÍN, A. (1973). The extent and differences in mitotic activity of the root tip of *Vicia faba* L. *Biol. Plant.* (Praha) **15**; 37–43.

LUXOVÁ, M. and POLERECKÝ, O. Heterogeneity of the root primordia in maize embryos. (in press).

NĚMEC, B. (1930). "Cytology and Anatomy of Plants." (In Czech). Aventinum, Praha.

PETERSON, R. L. (1967). Differentiation and maturation of primary tissues in white mustard root tips. *Can. J. Bot.* **45**; 319–331.

REINHARD, E. (1960) Über die Rückregulierung des Gefassbündelmusters von *Sinapis alba*. *Ber. Deutsch. Bot. Ges.* **73**; 19–23.

TORREY, J. G. (1957). Auxin control of vascular pattern formation in regenerating pea root meristems grown *in vitro*. *Am. J. Bot.* **44**; 859–870.

Chapter 5

Further Studies on Primary Vascular Tissue Pattern Formation in Roots

JOHN G. TORREY and WILLIAM D. WALLACE

Cabot Foundation, Harvard University, Petersham, Massachusetts, U.S.A.

I. Introduction

From studies on the development of 0·5 mm root tips of the garden pea grown in excised root culture and on the regeneration of decapitated pea roots after 0·5 mm decapitation (Torrey, 1955, 1957), it was concluded that the apical meristem of the root is the pattern-controlling center for primary vascular tissue formation in the root and that the dimensions of the root apex, at the level of the blocking-out of the tissue pattern, were correlated directly with the complexity of the primary vascular tissue pattern. In pea roots, the normal vascular pattern of the primary root is triarch; with auxin treatment, regenerating root apices were induced which formed vascular patterns varying from triarch to hexarch. The responsiveness of the system to hormones of the auxin and cytokinin type (Torrey, 1963) led to the further conclusions that these hormones, as endogenous agents influencing cell division activity and orientation in the apex, were probably involved in the control of pattern formation at the apex.

These studies on the control of vascular pattern formation were made on excised roots grown in sterile nutrient culture. The question can be raised as to the general applicability of these conclusions to intact systems, including seedling roots, mature, branched root systems, or roots of adventitious origin. Recent studies by Dyanat-Nejad and Neville (1973) pointed to an important determining role by the cotyledons in the vascular pattern formation of seedling roots of *Theobroma cacao* L. Samantarai and Sinha (1957) reported the influence of nutrients, especially sucrose and ammonium sulfate, on the complexity of the vascular pattern formed in adventitious roots induced on isolated leaves. Lux and Luxová (1967) and Lux (1971) found that the age of the cutting and the environmental conditions influenced the number of vascular strands formed in adventitious roots from woody stem cuttings of *Populus deltoides*.

These concerns have led to a re-examination of the problem of vascular pattern formation in seedling roots of the garden pea, *Pisum sativum* L. and of seedling roots and cultured roots of *Convolvulus arvensis* L.

In the course of studies on the formation of lateral roots and endogenous buds by cultured roots of *Convolvulus arvensis* L. (Bonnett and Torrey, 1966), it was observed that the primary vascular tissue pattern varied considerably in different root explants all derived from the same original root clone. A systematic study was undertaken of the variation in pattern of the primary vascular tissues of seedling and excised roots grown in culture. The present paper reports these variations, ranging from diarch to hexarch, the changing patterns which occur in the main axis and in lateral roots and the influence of culture and explant length on pattern formation. The suitability of the cultured root system for further experimental work on pattern formation is emphasized.

II. Materials and Methods

Seeds of *Pisum sativum* L., c.v. Little Marvel were germinated at 25°C in sterile distilled water in 10 cm petri plates in the dark for 2–3 days and the roots examined directly by hand-sectioning. Other seedlings were transferred to a greenhouse growing box where the roots developed in a mist of nutrient solution and the shoots were exposed to normal greenhouse conditions. Root patterns were observed from free-hand sections.

Seeds from wild collections of *Convolvulus arvensis* L. were germinated, after cutting the hard seed coat with a scalpel, in "perlite" in plastic pots. Seedlings were grown at 25°C (day temperature) and 22·7°C (night temperature) in controlled environment chambers illuminated with warm white fluorescent lamps supplemented with incandescent bulbs giving about 1500 f.c. using either short days (8 h light per 24 h) or long days (16 h light per 24 h).

All cultured roots were derived from a single root started in culture in 1952 and grown continuously with periods of subculture varying from monthly to as long as half-yearly. During the experiments reported here, roots were regularly subcultured in modified Bonner liquid medium as described earlier (Torrey, 1958). Several root segments (2–5), 15–30 mm in length, were propagated in each 125 ml Erlenmeyer flask. Flasks were placed as standing cultures in the dark at 25°C with occasional fluorescent light at the time of examination or transfer. In standing cultures, only segments which floated formed lateral roots and could be further subcultured. Elongation rates of 15–30 mm/day were usual once the

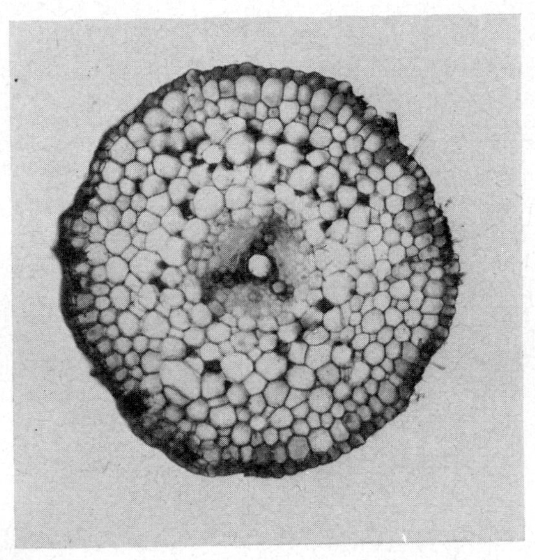

FIG. 1. Free-hand transverse section of a cultured root of *Convolvulus arvensis* L., grown in culture, showing the primary tissues of the root including the distinctive triarch arrangement of the primary xylem. Note the numerous intercellular spaces in the root cortex. Section was stained briefly with toluidine blue. Approx. ×75.

lateral root had established itself. Short root segments grown in liquid culture (10 mm or less) floated or sank unpredictably depending upon the air remaining in inter-cellular spaces of the root cortex. For these shorter segments it was found that agitation of the flasks on a rotary shaker at 80 r.p.m. gave more uniform results in establishing lateral root development and good root elongation.

Histological examination of the roots involved two methods. Seedling and cultured roots of *Convolvulus* are of 1–2 mm diameter and when turgid can be sectioned easily by hand with a sharp single-edged razor blade. Even small diameter lateral roots can be sectioned and the primary vascular tissue pattern determined (Fig. 1). Sections were floated in a drop of water, and examined without staining or with polarized light or were stained briefly with 0·05%

toluidine blue in phosphate buffer at pH 6·8, rinsed and mounted on a slide with a cover slip. Such temporary mounts provided a rapid method of analysing long roots and all their laterals quite quickly and accurately. For such studies a "root map" was made on graph paper on which were recorded diagrammatically the root length and the vascular pattern along the length of the main axis of the root as well as each lateral. When it became evident that changes in vascular pattern were to be expected, frequent observations along the length of the root were made to ensure that a complete record was obtained.

For detailed histological analysis, especially of the apical region of the root, it was necessary to fix the roots, embed them in paraffin and section them using standard methods. The roots were fixed in freshly prepared formalin-acetic acid-alcohol fixative, dehydrated using a t-butyl alcohol series, embedded in Tissuemat and sectioned at 10 μm. Stained serial sections were prepared using Heidenhein's haematoxylin and safranin. Photomicrographs were made of these sectioned roots.

The cultured roots of *Convolvulus* used in this study are from a clone, derived from the root of a single plant. The original primary root axis in culture was cut into segments each of which formed a new root. For convenience, the root formed by an excised root segment will be termed throughout this paper a lateral root. Branches on the lateral root will be termed secondary roots. Each segment (derived by excision from a lateral root) in turn will form a lateral root which elongates rapidly forming the dominant root axis in culture. Secondary roots in this context may be compared to "short" roots described in some other natural root systems.

III. Observations and Experimental Results

A. The Primary Vascular Pattern in Seedling Roots

1. *Pisum sativum* L.

The primary vascular pattern in the seedling radicle of *Pisum sativum* is triarch. In hundreds of seedling radicles examined, no exception to this rule was observed. What was striking was the variation in the vascular pattern of the lateral roots formed by the seedling radicle (Fig. 2a–c). Laterals initiated along the main axis showed diarch, triarch, tetrarch and pentarch patterns almost at random. Thus it is clear, as was evident in experiments with cultured pea roots, that pea has a genetic capacity for a wide range of pattern formation. The further striking observation was that individual lateral roots showed pattern changes during their elongation, increasing or decreasing by one or even two strand numbers over relatively small distances along the length of the lateral root. Figure 2a–c summarizes observations made on a number of seedling root systems examined at different ages.

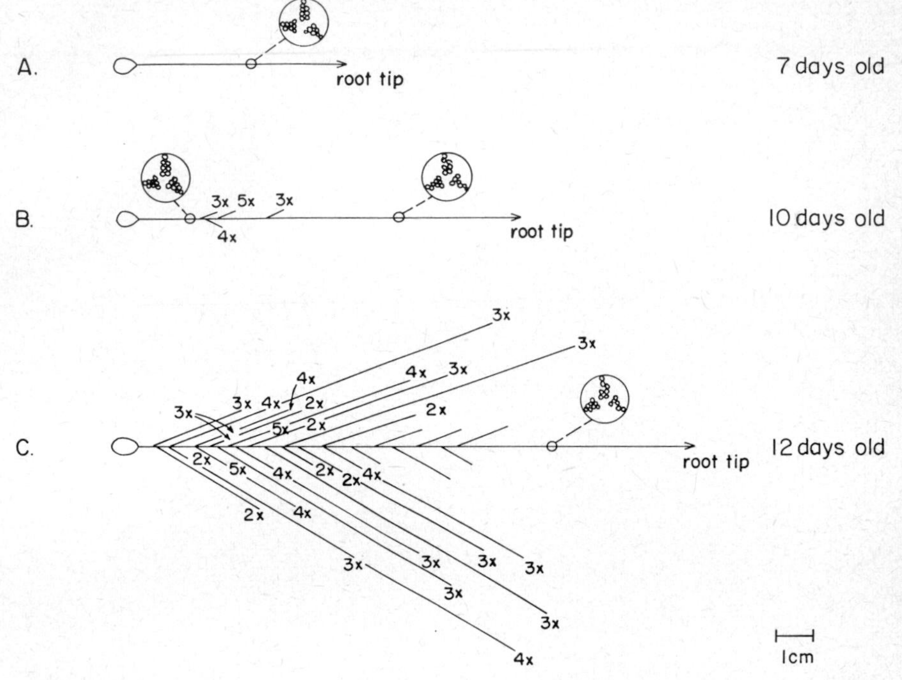

FIG. 2. Diagrammatic representation of the root systems of 7, 10- and 12-day old seedling roots of *Pisum sativum* L., showing the primary vascular tissue pattern in the main radicle axis and in the lateral roots. 2x-diarch; 3x-triarch; 4x-tetrarch; 5x-pentarch.

2. *Convolvulus arvensis* L.

In the seedling radicle of *Convolvulus*, the primary vascular tissue pattern was typically tetrarch. In ninety seedlings examined, all showed a tetrarch arrangement in the main root axis (Fig. 3a). One could say that in *Convolvulus* the genetically determined primary root vascular pattern is tetrarch. Even when lateral root formation begins on the third day, the primary axis pattern remains tetrarch. Newly formed lateral root branches are typically diarch (2×) or triarch (3×). A typical 3-day-old seedling is shown in Fig. 3b.

Thereafter, root elongation is rapid and root branching is frequent. By the eighth day after germination, the root is several decimeters in length and has formed a large number of lateral root branches. A typical root system is illustrated in Fig. 3c. Several striking features of the root vascular system can be observed. First is the change in the primary vascular tissue pattern of the primary root itself which, although initiated as tetrarch, increases in complexity to pentarch as the root elongates (5×). The transition occurs rather abruptly as a new radial strand of primary xylem appears and the pattern is adjusted to a symmetrical arrangement involving equal spacing of primary xylem and primary phloem at alternating radii equidistant from each other.

This same transition toward increased vascular complexity is observed in the first-formed lateral roots initiated as branches from the main axis root. As is seen in Fig. 3c, lateral roots initiated as triarch change to tetrarch as they elongate. In

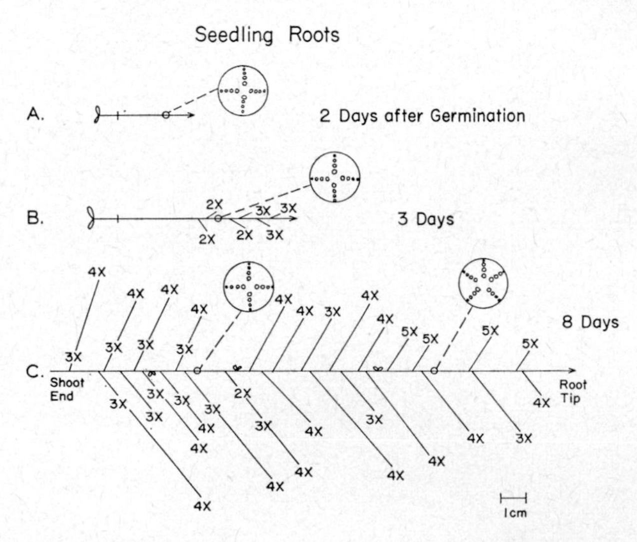

FIG. 3. Diagrammatic representations of the root systems of 2-, 3-, and 8-day old seedling roots of *Convolvulus arvensis* L., showing the primary vascular tissue patterns in the main radicle axis and in the laterals. Abbreviations as in Fig. 2.

one case a lateral root initiated with a diarch pattern, changed to triarch and then to tetrarch at the tip. So the same phenomenon of a change toward increased vascular complexity is observed in both the main axis and in lateral branches as the root system develops.

Another distinct phenomenon was also observed. Lateral root branches initiated along the primary root axis are initiated with more complex patterns the more distal the position of initiation. Thus, although the first-formed lateral roots are typically diarch and triarch, later formed lateral roots are initiated with tetrarch and pentarch vascular patterns. The primary vascular pattern of the lateral roots bears no precise relation to the vascular pattern of the primary axis. Thus a primary root which is tetrarch can give rise to diarch, triarch, or tetrarch roots and a pentarch root can initiate lateral roots which are triarch, tetrarch or pentarch. Essentially similar root systems with respect to these vascular patterns were observed in seedlings grown in perlite in growth chambers in long days (16 h light/24 h day) and in short days (8 h light/24 h day).

B. The Primary Vascular Pattern in Excised Roots grown in Culture

Isolated 15 mm-long root segments excised from roots of *Convolvulus* which had been grown for about two months in culture readily formed one lateral root at the distal end of the segment within 1–2 mm of the cut end. Usually only one lateral root was initiated, which elongated and became the new main axis of the root in culture. This process of subculturing has been carried on for many years without significant diminution in the capacity of the roots to form lateral roots. Typically such an elongating *Convolvulus* root, formed in culture from an excised segment, branched sparsely and these secondary roots in turn elongated very little.

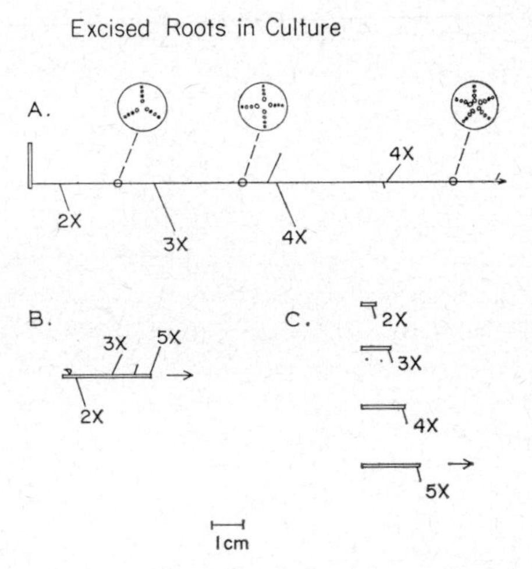

FIG. 4. Diagrammatic representations of the roots developed from excised root segments of *Convolvulus arvensis* L., grown in sterile nutrient medium, showing the primary vascular tissue patterns formed. Abbreviations as in Fig. 2.

Thus there appeared to be a marked apical dominance of each newly established lateral root on an excised segment. This behavior in culture is unchanged from that described by Torrey (1958).

The primary vascular tissue pattern of an elongating lateral root, formed on a 15-mm excised segment, is illustrated diagrammatically in Fig. 4a. Although considerable variation was seen in the many samples studied, the root shown in Fig. 4a represents the average situation and illustrates its essential features. The newly-initiated lateral root was typically triarch or tetrarch. If triarch on initiation, it became tetrarch as it elongated in culture and then still later became pentarch. Thus, segments excised for sub-culture from such a root might possess either a triarch, tetrarch or pentarch pattern. The vascular pattern of the segment itself appeared to have no influence on the dimensions or pattern of any newly initiated lateral root.

Short secondary roots were formed at infrequent intervals along the axis of the elongating lateral root. These short secondary roots showed the progressive change in vascular pattern at initiation that is illustrated in Fig. 4a. Thus, at the proximal end of the main axis, diarch or triarch secondary branches developed while at the more distal end, secondaries which had formed later in culture were typically tetrarch. Thus, with longer periods in culture, newly initiated secondary roots developed with more complex vascular patterns.

Lateral roots formed most readily on long excised root segments, that is, 15 mm or longer, up to 30 mm. In shorter segments, lateral root initiation was less regular. Thus 5 mm segments did not all initiate a lateral branch promptly when subcultured and 2·5 mm segments seldom formed a lateral in the control medium. It was found that, in short segments, less than 5 mm in length, lateral root initiation could be assured by treating the segments with 10^{-5} M indoleacetic acid for the first week in culture. Thereafter, once initiated, the lateral root would elongate in culture.

Experiments were performed in which excised segments of different lengths were used as the inoculum and the vascular patterns of lateral roots from such segments were analyzed. The results of such an experiment, summarized in Fig. 4c, show in an idealized way, the relation between segment length and the complexity of the primary vascular pattern in the lateral root produced. It was found that the longer the segment the more complex the primary vascular pattern formed in the lateral root. Segments 5 mm long typically formed triarch laterals while segments 30 mm long typically formed pentarch lateral roots. Even laterals 1·5 mm long treated with 10^{-5} M indoleacetic acid typically formed diarch or triarch lateral roots. The auxin treatment in short segments was essential for lateral root initiation as was shown by Bonnett and Torrey (1965) but the auxin treatment had no effect on the number of vascular strands formed. Occasionally, a long segment formed more than one secondary root along its length (Fig. 4b). The vascular patterns in these branches showed the same relationship, i.e. reduced vascular strand numbers in proximal branches and higher numbers in secondaries initiated more distally.

IV. Discussion and Conclusions

A. Natural Diversity in Vascular Patterns

There exist in the literature few detailed accounts of the vascular patterns to be found in root systems taken as a whole. When DeBary (1884) characterized roots of the Cruciferae as diarch, those of the Cucurbitaceae as tetrarch, *Pisum sativum* as triarch and *Phaseolus vulgaris* as tetrarch, he was referring to the typical situation in the primary root or radicle of the seedling roots of dicots. Higher numbers of vascular strands characterize vascular patterns in the main root axis of monocotyledonous plants. Few systematic studies of vascular patterns of lateral root branches have been published. It is usually conceded that, in lateral roots, the vascular pattern remains the same as, or shows fewer numbers of vascular strands than, the main axis root to which they are attached and that diarch lateral roots are probably of most frequent occurrence. The vascular connection of diarch lateral roots to the main root axis involves vertical orientation of the xylem poles so that attachment occurs to protoxylem above and below the lateral root attachment in the plane of the mother root (Dodel, 1872).

Following the development of the concept by Bower (1930) of internal form changing in relation to size, greater interest developed in the relationship between root size and the relative complexity of vascular systems. Wardlaw (1928) summarized the data showing that, in general, the larger the diameter of the mature root, whether in monocots, dicots or in the lower vascular plants, the greater the complexity of the vascular pattern. Subsequently, Wardlaw (1947) emphasized that, whatever the final advantage of increased internal vascular complexity, larger functional systems showing a size-form relationship, must be interpreted in dynamic developmental terms which take into account the events in and immediately adjacent to the apical meristem.

The diversity of vascular pattern to be found within the root system of a single plant seems not to have been generally appreciated. Careful systematic studies of vascular patterns throughout the root system, such as we have reported for *Pisum sativum* and *Convolvulus arvensis* seedling roots seem not to have been made. Rather, changes along the main axis have been reported in some cases, as for example, by F. J. Meyer (1930) in the roots of *Asphodelus*, *Hyacinthus* and *Maranta*, where the number of strands increased in a swollen region of the root and decreased along the root past the swollen zone. Bond (1932), in refuting earlier claims by Flaskämper (1910) that malnutrition influenced vascular strand number in the roots of *Vicia faba* seedlings, demonstrated that the seedling roots frequently showed a reduction from hexarch to pentarch or pentarch to tetrarch as a normal developmental event. Preston (1943) showed that in the seedling radicle of lodge pole pine (*Pinus contorta*) the root was first tetrarch, then during elongation became triarch and finally diarch. Lateral long-roots and short-roots were all diarch. Wilcox (1962a, 1962b) reported variation in vascular pattern in seedling roots of the incense cedar *Libocedrus decurrens* and found a low correlation to exist between root diameter and the number of protoxylem poles.

B. Experimental Manipulation of Root Vascular Patterns

Experiments designed deliberately to modify vascular patterns in roots go back to about the time of Jost (1931–32) who set out to answer the question: What determined the number of vascular strands in roots? Experimental work up until about 1965 was reviewed by Torrey (1965) and relatively little new work directed towards the problem has been published.

Luxová (1971 and in this volume, p. 73) studied vascular pattern determination during embryonic development in *Zea mays*. She found that embryos excised and placed in nutrient culture at different stages in their embryonic development showed modifications in root vascular patterns. The earlier the excision of the developing embryo, the smaller the number of protoxylem poles; whereas fully mature embryo radicles had an average of 13–14 protoxylem poles, embryos excised at 14 days after anthesis and grown in culture had only about 10 protoxylem poles. She concluded that vascular pattern determination occurred in the embryo during the first half of embryogenesis.

In a recent study, Dyanat-Nejad and Neville (1973) reported that in the seedling radicle of *Theobroma cacao* L. the vascular pattern is typically hexarch with a central cylinder of very large diameter at the base of the seedling root. As the seedling radicle elongated, its diameter decreased rapidly, including a marked reduction in the diameter of the central cylinder. This reduction was accompanied, at the 4th to 7th day, by a marked increase in the number of vascular strands in the central cylinder, frequently reaching 12 to 14 bundles by branching of existing bundles or by initiation of wholly new strands. From the 7th day on the seedling root diameter and central cylinder diameter continued to decrease, accompanied by a gradual reduction in strand number by fusions or disappearance until the original hexarch pattern was re-established. These authors showed, by excision of cotyledons or by horizontal cuts in the seedling root, that the sudden increase in vascular bundle number was associated with the active phase of cotyledon development. They believed the vascular pattern change was produced by materials moving from the cotyledons into the radicle. No characterization of the cotyledonary materials transported to the roots was attempted.

Our studies reported here show that in the seedling main axis there may exist fairly stable determination of the vascular pattern which seems to be a genetic characteristic. In *Pisum sativum* there is a strong determination in the main axis for a triarch vascular pattern which persists for the continued growth of the primary axis. We know, however, from experimental studies (Torrey, 1955, 1957) that this same axis can be changed after regeneration and hormone treatment to form a hexarch, pentarch, tetrarch or even diarch pattern. The work of Reinhard (1960) further supported the idea of a stable vascular pattern in *Sinapis alba* subject to experimental manipulation. In pea, newly initiated lateral root primordia may be initiated naturally with any of these patterns and, unlike the main axis, may undergo transition either to a reduced strand number or an increased number. Similarly, in the seedling root of *Convolvulus* the radicle is typically tetrarch. However, during development, the pattern of the main axis may change. Lateral root branches vary from diarch to pentarch and lateral roots may change their vascular pattern by adding or deleting an extra strand. In all of this vascular pattern instability one may deduce that pattern formation occurs at the root apex and that conditions to which the apex is exposed at any given time, including perhaps the growth rate of the root itself, determine the pattern which is formed.

C. Conclusions

The evidence from past work and recent studies, summarized above, points to the validity of the idea that vascular patterns in the root are determined at and by the apical meristem of the root. Pattern formation is under genetic control but can be modified within a root system by changes, either external or internal, which modify or change the activity of the pattern-forming center in the apex. There is strong evidence to suggest that hormones are involved in controlling the cellular activity in the apex.

Two types of pattern initiation can be distinguished in the systems described above. The first case involves the initiation of a new meristem when a lateral root develops. Each new lateral root primordium formed is subject to the influences acting on it at the time of initiation. As these change, the vascular pattern of the new root is determined and may vary within the genetic limits of the system. Thus, pattern formation here is determined at the time the lateral root emerges from the cortex of the mother root. Influences could come from the root apex or more distal parts of the root system or from the proximal end including older lateral roots, cotyledons or the shoot itself.

The second case of new pattern formation involves the transition from an existing pattern to either a reduction or an increase in pattern complexity by the loss or gain of a new vascular strand. This type of change also is most easily understood in terms of apical meristem activity, i.e. changes in hormone production in the tip itself or flow of materials into the tip region, either nutritional or hormonal, including both stimulators and inhibitors of cellular activity.

The nature of the specific substances acting on the root apex in the pattern forming process is not well understood. From the earlier work on pea roots, auxin of the indoleacetic acid type seems to be involved. All the evidence suggests that it is translocated from the root base towards the apex (see this volume pp. 299, 325). Attempts to demonstrate a role of auxin in pattern formation in cultured roots of *Convolvulus* have failed although polar transport of auxin to the root tip is well demonstrated and its role in lateral root initiation seems incontrovertible.

Cytokinins are present in the root apex and there is strong evidence to suggest that they are synthesized there, but a role for them in pattern determination remains to be demonstrated. The generally established idea that auxins and cytokinins interact intimately in controlling cell division may well apply to the root apex but needs further confirmation from direct experimentation.

The cultured root system of *Convolvulus arvensis* L. seems to offer an ideal experimental system with which to study further the general problem of root tip organization and vascular tissue pattern formation. What is needed is greater experimental expertise and ingenuity to unravel the intricacies of these complicated but highly organized intercellular events.

Acknowledgements

This work was supported in part by funds from the Maria Moors Cabot Foundation for Botanical Research of Harvard University. W. D. Wallace was supported by a pre-doctoral research fellowship from the National Institute of General Medical Sciences of the Public Health Service.

References

BOND, G. (1932). The effect of malnutrition on root structure. *Proc. Roy. Soc. Edinburgh* **52**; 159–173.

BONNETT, H. T., JR. and TORREY, J. G. (1965). Chemical control of organ formation in root segments of *Convolvulus* cultured in vitro. *Plant Physiol.* **40**; 1228–1236.

BONNETT, H. T., JR. and TORREY, J. G. (1966). Comparative anatomy of endogenous bud and lateral root formation in *Convolvulus arvensis* roots cultured in vitro. *Am. J. Bot.* **53**; 496–507.

BOWER, F. O. (1930) "Size and Form in Plants." Macmillan and Co., Ltd. London. 232 pp.

DEBARY, A. (1884) "Comparative anatomy of the vegetative organs of the Phanerogams and ferns" Transl. F. O. Bower and D. H. Scott. Clarendon Press, Oxford.

DODEL, A. (1872). Der Uebergang des Dicotyledonen-Stengels in die Pfahl-Wurzel. *Jahb. wiss. Botanik.* **8**; 149–193.

DYANAT-NEJAD, H. and NEVILLE, P. (1973). Variation du nombre de faisceaux dans la racine principale du Cacaoyer (*Theobroma cacao* L.) *Rev. Gen. Bot.* **80**; 41–74.

FLASKÄMPER, P. (1910). Untersuchungen über die Abhängigkeit der Gefäsz- und Schlerenchymbildung von äusseren Faktoren nebst einige Bermerkungen über die angebliche Heterorhizie bei Dikotylen. *Flora* **101**; 181–219.

JOST, L. (1931–1932). Die Determinierung der Wurzelstruktur. *Z. Botanik* (Jena) **25**; 481–522.

LUX, A. and LUXOVÁ, M. (1967). Development of vascular bundles in poplar roots growing on shoot cuttings of varying quality. *Biologia* (Bratislava) **22**; 166–171.

LUX, A. (1971). Experimentelle Modifikationen des Leitungssystems bei beginnender Wurzelbildung an Sprosstecklingen von Pappeln. *Biologia Plantarum* (Praha) **13**; 313–319.

LUXOVÁ, M. (1971). Concerning variability in the delimitation of the vascular system in root primordia of maize embryos. *Biologia Plantarum* (Praha) **13**; 79–87.

MEYER, F. JÜRGEN (1930). Die Leitbündel der Radices filipendulae (Wurzelanschwellungen) von *Maranta Kerchoveana* Morr. *Ber. Deutsch. bot. Gesell.* **48**; 51–57.

PRESTON, R. J., JR. (1943). Anatomical studies of the roots of juvenile lodge-pole pine. *Bot. Gaz.* **104**; 443–448.

REINHARD, E. (1960). Über die Rückregulierung des Gefässbündel-musters von *Sinapis alba*. *Ber. Deutsch. bot. Gesell.* **73**; 19–23.

SAMANTARAI, B. and SINHA, S. K. (1957). Factors concerned in the control of vascular pattern in the induced roots of isolated leaves. *J. Indian Bot. Soc.* **36**; 1–11.

TORREY, J. G. (1955). On the determination of vascular patterns during tissue differentiation in excised pea roots. *Am. J. Bot.* **42**; 183–198.

TORREY, J. G. (1957). Auxin control of vascular pattern formation in regenerating pea root meristems grown in vitro. *Am. J. Bot.* **44**; 859–870.

TORREY, J. G. (1958). Endogenous bud and root formation by isolated roots of Convolvulus grown in vitro. *Plant Physiol.* **33**; 258–263.

TORREY, J. G. (1963). Cellular patterns in developing roots. *Symp. Soc. Exp. Biol.* **XVII**; 285–314.

TORREY, J. G. (1965). Physiological bases of organization and development in the root. *Handb. Pflanzenphysiol.* **XV**(1): 1256–1327.

WARDLAW, C. W. (1928). Size in relation to internal morphology. 3. The vascular system of roots. *Trans. Roy. Soc. Edinb.* **56**; 19–55.

WARDLAW, C. W. (1947). Experimental and analytical studies of pteridophytes. X. The size-structure correlation in the Filicinean vascular system. *Ann. Bot.* **11**; 203–217.

WILCOX, H. (1962a). Growth studies of the root of incense cedar, *Libocedrus decurrens*. I. The origin and development of primary tissues. *Am. J. Bot.* **49**; 221–236.

WILCOX, H. (1962b). Growth studies of the root of incense cedar, *Libocedrus decurrens*. II. Morphological features of the root system and growth behavior. *Am. J. Bot.* **49**; 237–245.

Chapter 6

The Development of Lateral Roots

MARGARET E. McCULLY
Biology Department, Carleton University, Ottawa, Canada

I. Introduction

Lateral roots generally form some distance behind main root apices and thus in partially or fully differentiated root tissues. The new meristems originate endogenously and therefore must traverse living tissues as they emerge from the parent root. Lateral roots arise with some predictability as to position, in general their order of appearance is acropetal and they are placed with some relation to the internal vascular pattern of the main root.

Consideration of these features of lateral roots gives rise to three main questions which this paper will attempt to deal with:

1. How is the new meristem organized, especially in highly differential tissues?
2. How do lateral roots penetrate and emerge from the cortex of the parent root?
3. What regulates the positions at which lateral roots arise?

II. The Organization of the New Meristem

A. Changes in the Cells of the Parent Root Involved in Lateral Root Initiation

Except for roots of some of the pteridophytes, which branch dichotomously, and a few aquatic angiosperms (e.g. *Hydrocharis*, *Eichornia*, *Pistia*) where the new meristem can form within the apical meristem of the parent root, lateral root meristems are initiated from tissues lying outside the apical meristem. From the earliest publications on root structure (see Von Guttenberg, 1968) it was known that the initial cell divisions forming these new meristems occur in the pericycle and/or the endodermis of the parent root. Later work confirmed that in most ferns the first divisions are in the endodermis, while in angiosperms and gymnosperms the earliest divisions are seen in the pericycle, followed inevitably by some division in the endodermis.

In exceptional cases such as the aquatic plants mentioned above and occasionally in embryonic roots (e.g. see Fig. 192 Huber, 1961, redrawn from original by E. Heinricher; see also O'Dell and Foard, 1969) the new meristems are organized in relatively undifferentiated tissue, but in most instances lateral meristems are formed at some distance basipetal to the youngest root hairs. Thus the initial divisions occur in pericycle and endodermal cells that are at least partially differentiated (see Popham, 1955a; Esau, 1965; Fahn, 1967 re sequence of differentiation of primary root tissues). Pericycle cells at this level in the root are usually elongated along the root axis and highly vacuolated. In the case of the grasses at least, they may have thick secondary walls which are lignified. The endodermal cells in the region are also elongated periclinally and are highly vacuolated and have differentiated the Casparian strip, a structure always associated with this cell type (see Clarkson and Robards, pp. 415–436 this volume) and found elsewhere only infrequently. In some dicotyledons, the endodermis at this level may also possess a suberized lamella. This is a thin layer stained by lipophilic dyes which is deposited between the primary cellulosic wall and the cell membrane, and which completely encloses the endodermal cells except where it is perforated by the plasmodesmata (e.g. see Fig. 7—Karas and McCully, 1973). In grasses not only are the endodermal cells suberized, but they also have secondary fibrillar walls, thickened especially on the inner tangential side of the cells; in the case of *Zea* these appear to be lignified. The process by which these differentiated pericycle and endodermal cells become meristematic has only recently been looked at in any depth, and then only in two species, i.e. *Zea mays* (Bell and McCully, 1970; Karas and McCully, 1973) and *Convolvulus arvensis* (Bonnett, 1969). Vonhöne (1880) described a thinning of walls in pericycle and endodermal cells of the orchid *Laelia barkeri* in the region of lateral root initiation and Bloch (1935) observed in another orchid, *Cattleya*, that the endodermal cells involved in lateral formation showed "all stages of dedifferentiation such as separation of its secondary lamellae from one another or very thin collapsing suberin lamellae".

The study of *Convolvulus* by Bonnett and Torrey (1966) and Bonnett (1969) showed that following initial divisions and radial enlargement of cells in the pericycle, the endodermal cells surrounding the bulge divide in radial, transverse and tangential planes, and form the outer layer of the young primordium. Bonnett showed beautifully that once activated during the primordial development, the endodermal cells lose the capacity to form new Casparian strips and fail to maintain the distinctive region of plasma membrane and the intimate association of the membrane with the primary wall in the region of the strip which is characteristic of all endodermal cells lacking suberized lamella or secondary walls (Bonnett, 1968; Ledbetter and Porter, 1970). In *Convolvulus*, as the original endodermal cells involved in formation of a lateral meristem divide, cell wall material is deposited not only to form the new cross walls but all around the cytoplasm. The portion of the plasma membrane adhering to the Casparian strip is then severed and externalized by newly formed plasma membrane. The additional cell wall material embeds the isolated membrane and Casparian strip. This new wall has no special features in this region indicating that the information dictating Casparian strip formation is no longer functional.

In corn, the fate of the parent root cells participating in lateral meristem formation has been followed closely (Bell and McCully, 1970; Karas and McCully, 1973). In these roots cells of the stelar parenchyma, the pericycle and the endodermis involved in the formation of a lateral meristem all have thick, lignified, secondary walls. The endodermis at this level in the corn root is particularly specialized, having centripetal to the primary wall and Casparian strip (isolated at this stage) a suberized lamella on all sides (perforated only by the plasmodesmata) and a secondary fibrillar wall which is much thicker on the inner tangential side and the centripetal ends of the anticlinal walls. This secondary wall is lignified.

The first indication of the initiation of a lateral root primordium in *Zea* are changes in the cytoplasm and walls of a few pericycle and stelar parenchyma cells close to a xylem pole. These changes are accompanied or followed closely by (we have been unable to determine which) changes in the structure of cytoplasm and cell walls of the endodermal cells tangential to the activated pericycle cells. In each tissue the structural changes precede cell division, although in the case of the pericycle cell they may not be complete before cell division. Pericycle and stelar parenchyma cells involved in the initiation of the lateral meristem develop very basophilic cytoplasm, an increase in cytoplasmic volume and enlarged nuclei at the same time as wall changes begin. The thick secondary walls are thinned, apparently by hydrolysis, and their positive reaction for lignin disappears. This is particularly noticeable in the pericycle cells (e.g. see Figs. 2 and 3—Bell and McCully, 1970; and Figs. 5–9—Karas and McCully, 1973). In corn, the first cell divisions initiating the lateral meristem are periclinal in a few adjacent pericycle cells which divide asymmetrically, each forming a small outer and a much larger inner daughter cell. This wave of division is followed closely by asymmetric anticlinal divisions in the daughter cells. These divisions, and elongation of some of the daughter cells in the anticlinal direction result in a cellular configuration as shown in Fig. 1 (see also Figs. 6 and 7—Bell and McCully, 1970). This stage of lateral initiation is not difficult to find in corn and can be seen in hand sections of fresh material (e.g. Fig. 3—Ashford and McCully, 1973). The endodermal cells overlying the bulging pericycle become more basophilic and cytoplasmic volume increases; at the same time the secondary fibrillar wall disappears as does also the prominent suberized lamella (see Figs. 5 to 10—Karas and McCully, 1973). As in *Convolvulus*, the original embedded Casparian strip regions do not appear to change and remnants of them can occasionally be seen on some of the anticlinal walls of the cells derived from the endodermis which cover the young primordium.

The changes in the walls of the stelar parenchyma, pericycle and endodermal cells in corn must involve enzymatic hydrolysis. Controlled autolysis of non-lignified walls is common during higher plant development especially in xylem differentiation (e.g. see O'Brien, 1970), but autolysis of lignified or suberized walls is not. Vonhöne (1880) described the disappearance of phloroglucinol-HCL-stained material from the thick walls of exodermal cells in the pathway of emerging lateral roots in *Laelia* although Bloch (1935) suggested that Vonhöne was mistaken and was really describing cells which had not yet formed secondary walls. Bloch (1941) has described delignification following wounding in higher plants. The only reference to any possibility of suberin autolysis that I have been able to find is that of Bloch (1935) quoted above. Roelofsen (1959) in fact pointed out that he found no microorganisms capable of degrading suberin.

Regarding the apparent hydrolysis of lignin and suberin during lateral root development in corn, it must be emphasized that identification of these compounds is based only on staining reactions and their exact chemical nature is not known. In corn the positive reaction for lignin disappears from the walls of the stelar parenchyma, pericycle and endodermal cells involved in lateral root initiation but it is unchanged in the adjacent xylem walls suggesting either a very localized enzyme action or a difference in the materials giving the lignin reaction.

Dramatic changes in wall structure of stelar parenchyma, pericycle and endodermal cells involved in lateral root initiation could be expected to occur frequently in monocotyledons especially the grasses where secondary walls are laid down early. Except in *Zea*, this has not been looked into in detail but preliminary examination of hand sections of roots of bamboo, *Coix lacryma-jobii* and sugar cane show that, depending on the growth condition, laterals may be initiated in a position either acropetal or basipetal to the region where all pericycle walls are thickened and where all endodermal cells have a suberized lamella and a secondary wall. In the former case laterals presumably are initiated before the secondary walls develop. In the variety of corn examined (Seneca Chief), secondary walls always develop in all pericycle cells, and suberized lamellae form in all endodermal cells in a region acropetal to the zone of lateral root initiation regardless of growth conditions.

The most remarkable feature of the changes in the cytoplasm and walls of the parent root cells at the site of lateral root initiation in corn is that they apparently can precede any of the related cell divisions. This suggests that a considerable reprogramming of a differentiated cell can occur without an intervening mitosis. Two studies show that activation of individual cells of the pericycle and endodermis participating in early stages of formation of a lateral can include changes in cell shape and size without cell division. Foard *et al.* (1965) found that in wheat roots in which mitosis is blocked with colchicine, structures which they called primordiomorphs develop in the region where lateral root formation would be expected. These primordiomorphs consist of pericycle cells asymmetrically elongated in the radial direction producing a bulge into the cortex which is accommodated by enlargement of the overlying endodermal cells. The cell configuration of these primordiomorphs as seen in longitudinal section resembles that of early lateral root primordia except that in the latter case a number of cell divisions have occurred (e.g. compare Figs. 6 and 7 of Bell and McCully, 1970, with Fig. 11 of Foard *et al.*, 1965). Bayer *et al.* (1967) have described somewhat similar swellings in positions where laterals would normally form in roots of cotton treated with the antimitotic agent, trifluralin. In this case also the bulge in the pericycle is accompanied by enlargement of the overlying endodermal cells. Unfortunately details of cytoplasm and wall structure were not examined in either of these studies with mitotic inhibitors, so it is not known whether such premitotic changes as occur at the site of lateral root formation in corn also are found in the same regions if mitosis is inhibited.

B. Contribution of the Progeny of Main Root Tissues to the Lateral Meristem

In all the angiosperm roots that have been described, derivatives of both the parent pericycle and parent endodermis contribute to the tissues of the new meristem although, as will be seen, in many cases the derivatives of the endodermis are short-lived. In addition, a number of studies have shown that in some plants derivatives of parenchyma of the parent stele contribute to the basal tissues

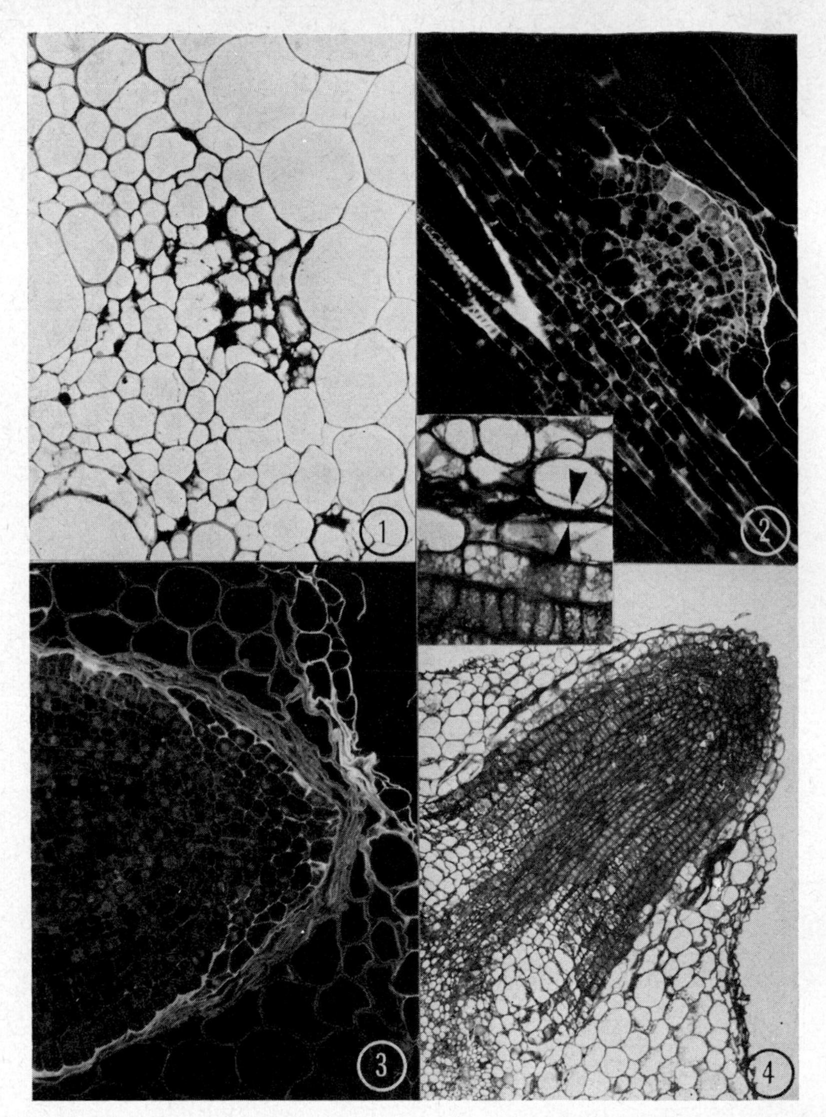

FIG. 1. Transverse section (TS) of the main root of *Zea mays* L. showing early stage of development of a lateral primordium. The stimulated pericycle cells become thin-walled and several have divided periclinally. An increase in cytoplasm is seen in these cells, in the stelar parenchyma adjacent to the protoxylem and in the as yet undivided cells of the endodermis. Tissue embedded in glycol methacrylate, stained with toluidine blue 0. ×320.

FIG. 2. Longitudinal section (LS) of main root of *Zea mays* L. showing median LS of young lateral. Continuity can be seen between the parent root endodermis and the intensely cytoplasmic epidermal cells of the lateral. Also apparent is the underlying protoxylem and the new cells formed from the stimulated xylem parenchyma. Cell walls of destroyed cortical cells are accumulating in front of the growing primordium. Tissue embedded in glycol methacrylate. Stained with an acriflavin pseudo-Schiff's reagent and viewed with a fluorescence microscope. The fluorescence in the cytoplasm and nuclei is induced by the fixative. ×240. (Reprinted from Karas and McCully (1973) with permission of Springer-Verlag.)

of the lateral and the connecting bridge between the primary and secondary or secondary and tertiary roots. In a few cases there is evidence that parent root cortical cells contribute to the new meristem.

In corn, the parent endodermis clearly gives rise to the epidermis of the lateral (Bell and McCully, 1970; Karas and McCully, 1973). The cells of the root epidermis in corn are distinctive, being columnar, densely cytoplasmic and secreting a thick coat of periodic acid–Schiff's positive mucilage to the outer surface of their outer tangential wall. These unmistakable characteristics appear in the derivatives of the endodermis at a very early stage in the development of a lateral primordium (Fig. 2, and see Fig. 14—Karas and McCully, 1973). During the early development of the primordium these epidermal cells divide only anticlinally, producing similar daughter cells; however, when the young primordium is about one-third of the way across the cortex a few epidermal cells at its tip divide periclinally and thus produce the root cap initials (see Fig. 17—Bell and McCully, 1970). The daughter cells of these periclinal divisions lack the intense basophilia of the parent epidermal cells, their polarity is changed and they elongate periclinally. Unlike the epidermal cells these first rootcap initials have amyloplasts containing statolith-like starch accumulation (Fig. 17, Karas and McCully, 1973) and they do not secrete the strongly periodic acid–Schiff's positive mucilage characteristic of epidermal cells. Derivatives of these initials produce normal rootcap cells by periclinal and anticlinal division (Fig. 3) and these in turn secrete all around their periphery large amounts of the weakly periodic acid–Schiff's positive mucilage characteristic of root caps of corn. Before the lateral emerges it has a well-formed root cap with central statolith-containing cells which have been formed by periclinal divisions of the initials, and a few periclinally elongated flank cells apparently derived from anticlinal divisions of some of the central cells (Fig. 3, and see Fig. 29—Bell and McGully, 1970). Thus, in corn a distinct lineage can be traced between parent endodermal cells, epidermis of the lateral and the root cap.

FIG. 3. LS of main root of *Zea mays* L. showing median LS of lateral root just before emergence. The young rootcap has formed and the columnar epidermal cells with their mucilaginous outer surface can be seen. Note the accumulated walls of dead cortical cells ahead of the rootcap and back along the flanks over the lateral epidermis. Tissue embedded in glycol methacrylate. Fluorescence microscopy of acriflavin pseudo-Schiff's stained section. ×400.

FIG. 4. LS main root of *Phaseolus vulgaris* L. showing median LS of emerging lateral. Note the difference in organization between this root and that of corn. The outer two layers of vacuolated cells are derived from the endodermis but are later stretched and discarded. The inset, a portion of the figure, shows details of the strongly basophilic epidermal cells (derived from the pericycle derivatives), the vacuolated endodermal derivatives, and accumulated cell walls of crushed cortical cells (arrows) at the interface between lateral and parent cortex. Cells of parent cortex lie toward the top of the inset figure. Tissue embedded in glycol methacrylate, stained with toluidine blue 0 and the Periodic acid–Schiff's reaction. Main Fig. ×280; Inset ×580.

Quite a different situation exists in many of the dicotyledonous roots that have been described in detail (e.g. carrot: Esau, 1940; *Fagus*: Clowes, 1950; *Convolvulus*: Bonnett and Torrey, 1966; *Malva*: Byrne, 1973; *Ipomoea*: Seago, 1973). Here derivatives of the pericycle form the primordial epidermis and root cap as well as the interior tissue of the lateral. The parent endodermis proliferates and forms a covering over the young primordium. This covering, or "Tasche" as it has been called in the German literature, consists of one or more layers of cells depending on the species. These endodermal derivatives are easily distinguished from the underlying epidermis of the lateral, the latter being somewhat columnar and intensely basophilic (Fig. 4) while the endodermal derivatives are usually highly vacuolated even in young primordia (e.g. Fig. 4, and Figs. 1–3—Bonnett, 1969). In roots of this type divisions in the endodermal derivatives at first keep pace with the developing primordium but by the time the lateral emerges divisions have stopped and the cell layer becomes discontinuous. The closest examination that has been made of the cells derived from the endodermis has been in *Malva silvestris* (Byrne, 1973). In this plant most of these cells continue to incorporate ^{3}H-thymidine until the lateral roots are up to 0·5 cm long. Beyond this stage those cells lying over the top of the root are stretched and broken.

In some aquatic angiosperms the "Tasche" originating from the endodermis is particularly well-developed (Schade and von Guttenberg, 1951). Conventional root caps are not formed in these plants and in some cases (e.g. *Lemna*) the "Tasche" persists and forms a substitute root cap, in others (e.g. *Pistia*), the epidermis of the lateral is derived from the inner layer of the endodermal derivatives and persists, while the remainder of the "Tasche" is shed.

In contrast, Popham (1955b) has shown that in pea the endodermal derivatives contribute even more to the cell population of the lateral meristem than they do in corn. In pea, these cells form the region Popham calls a transversal meristem which in turn produces the cortex and epidermis of the new root, the pericycle derivatives contributing only to the embryonic stele.

The studies mentioned above clearly show a wide range of contribution by main root endodermis and pericycle to new root meristems and it appears to be species specific. There has been controversy in the early literature regarding this point. Von Guttenberg (1968) reviewed this literature and concluded that claims that the endodermis can form part of the lateral meristem in some plants are incorrect. Berthon (1943) considered that the role played by the parent endodermis in formation of the lateral root depends upon its maturity at the level of lateral initiation (i.e., if the endodermis already has Casparian strips there is little or no involvement in the lateral development). These conclusions clearly must be evaluated again in the light of the more recent studies.

In some roots, some of the parenchyma cells of the stele also contribute to the basal tissue of the lateral roots and certainly in most cases they must be the source of the bridging vascular tissue. In corn the participation of these cells is particularly marked (Rywosch, 1909; Bell and McCully, 1970; Ashford and McCully, 1973). During lateral initiation, increased basophilia and cell wall changes occur in parenchyma adjacent to the xylem pole as soon as they appear in the overlying pericycle (Fig. 1). Later these cells divide although their exact contribution to the lateral has not been followed closely. The induced activity in these stelar parenchyma cells includes wall changes, cell division and increase in acid phosphatase activity. By the time the developing lateral emerges the activity in the stelar parenchyma has spread to include those surrounding five or six adjacent xylem poles (e.g. Fig. 2—Ashford and McCully, 1970). Early activation of stelar parenchyma and undifferentiated protoxylem at the site of lateral root initiation is apparently not confined to grasses and has also been described in *Ipomoea* (Seago, 1973). Most other studies of lateral development do not describe involvement of these cells.

The remaining parent root tissue which could contribute to the tissues of the lateral is the cortex outside the endodermis. Studies of corn (Bell and McCully, 1970), *Convolvulus* (Bonnett and Torrey, 1966), *Ipomoema* (Seago, 1973), and *Malva* (Byrne, 1973) have shown no involvement of these cortical cells in lateral formation. However, it has been claimed that they proliferate along with endodermis to form the outer covering of lateral primordia in a number of terrestrial angiosperms (Janczewski, 1874), and to contribute to the "Tasche" in some water plants (see von Guttenberg, 1968).

C. The Course of Differentiation of Tissues in Lateral Primordia

There is not much information available on the course of differentiation of tissues in lateral roots (Esau, 1965) and studies, similar to those of Bünning (1952) and Popham (1955a) on primary roots, are needed on them. The most extensive investigation of the arrangement of bridging xylem in the parent stele and the base of the laterals is that of Fourcroy (1942) who showed that the arrangement of the parent xylem system is changed for some distance around the point of insertion of the lateral. However, this study only dealt with xylem, and did not clarify the differentiation sequence of these elements nor did it discuss vascularization of the lateral itself. It has been considered that the bridging vascular tissue differentiates from cells derived from the parent root pericycle (Esau, 1965; Fahn, 1967) but the studies of Bell and McCully (1970) and Seago (1973) show that in *Zea* and *Ipomoea* this tissue originates from provascular tissue of main root and stelar parenchyma cells or their derivatives.

The timing and direction of endodermis differentiation in the young lateral roots is of particular interest. For some time at least following lateral initiation the barrier in the apoplast of the main root formed by the Casparian strips of the endodermis must be broken since the cells proliferated from the parent endodermis do not form these structures. Studies by Esau (1940) and Dumbroff and Peirson

(1971) have shown that, as the lateral primordium matures, cells lying in the cortical region of the base of the lateral differentiate to form a uniseriate endodermis which links the parent endodermis with the newly differentiated one in the lateral (Fig. 5). The details and course of the differentiation have not been studied although this information could be of particular significance to studies of ion uptake in the region of development of the lateral root primordia.

D. Cell Proliferation in Lateral Meristems

Various aspects of cell proliferation have been studied extensively in lateral root primordia of the species *Vicia faba* (Socher and Davidson, 1970; Friedberg and Davidson, 1971; MacLeod, 1972, 1973, 1974). These studies show that there are two population of cells, fast cycling cells and slowly cycling cells in the lateral primordia. These are distinguishable in very young primordia of less than 1500 cells and their proportions remain more or less constant in fully emerged meristems. The time difference in the cell cycles of both populations occur in the G2 period of interphase. Two other populations of cells can be observed in primordia from an early stage. These are differentiated on the basis of differences in mitotic activity. They are also spatially distinct, the larger group forming the core of the primordium and the smaller group the peripheral shell. These two populations are distinct from the two groups of cells distinguished by difference in cell cycle time and these latter cells are distributed throughout the meristem.

It is particularly interesting that at about 24 h before emergence the central cells, which prior to this time have a higher mitotic activity than the peripheral cells, become almost completely quiescent, arrested in the G1 stage of interphase. Mitotic activity does not decrease in the peripheral cells at this time. With emergence, mitotic activity resumes in the central cells and immediately before this becomes low in the peripheral cells. It is not known if such a quiescence of a large proportion of the cells of a lateral meristem prior to emergence is a widespread phenomenon. Unfortunately, there is not available a detailed anatomical study of lateral root development in *Vicia faba* so that the two spatially distinct cell populations cannot be linked to differences in their origin from parental tissues.

The formation in lateral primordia of the quiescent centre characteristic of root apices (see Clowes, pp. 3–19 this volume) has only been followed in a few species. In *Pistia* and *Eichornia* all the cells of the young lateral primordia are meristematic but before emergence a small group of cells at the apex of the stele stop dividing, and a fully developed quiescent centre is present by the time the laterals have emerged (Clowes, 1958). The laterals of these two species are, however, not typical since division in all cells stops soon after emergence and subsequent growth is by elongation. In *Malva silvestris*, a species in which lateral development is presumably more like that in most non-aquatic higher plants, a quiescent centre is not detectable until the new laterals have grown out at least 0·5 cm (Byrne, 1973). Friedberg and Davidson (1971) suggest that the core of quiescent cells which they observed in the lateral primordia of *Vicia faba* at about the time of emergence later become meristematic and thus do not constitute an early developmental stage of a quiescent centre. MacLeod and McLachlan (1974) find no evidence of a quiescent centre in *V. faba* lateral roots until they have emerged by ca. 2 mm. The newly organized quiescent centre increases in volume until the lateral is ca. 4 cm long.

III. Passage of Lateral Primordia through Parent Root Tissues

Lateral root primordia must grow through parent root cortex and epidermis either by mechanical means, by controlled enzymatic hydrolysis or a combination of both factors. All the early workers who described lateral roots speculated on the method of passage and some involvement of enzymatic hydrolysis was the most popular but by no means unanimous view (see references in Bell and McCully, 1970). More recent work (Bonnett, 1969; Bell and McCully, 1970; Friedberg and Davidson, 1971; Karas and McCully, 1973) has suggested purely on morphological evidence that death and hydrolysis of cortical tissues occurs ahead of the advancing lateral and that mechanical pressure may play little or no role in the breakthrough. The first attempt to demonstrate the role of enzymes in lateral penetration was that of Sutcliffe and Sexton (1968). Using histochemical methods they found high β-glycerophosphatase activity in cortical cells adjacent to lateral primordia and concluded that this enzyme was assisting in breakdown of the cortical cells and had been induced in these cells by mechanical pressure from the growing lateral. In contrast, using different histochemical methods including a different substrate, Ashford and McCully (1970) found high acid phosphatase activity in only a few of the crushed cortical cells lying against the young primordium in corn. In contrast to the results of Sutcliffe and Sexton who reported no acid phosphatase activity in epidermal or cortical cells of the lateral, Ashford and McCully found high levels of activity in this region in both corn and peas. At first glance this high enzyme activity in the epidermis suggested that these cells were secreting the enzyme and hydrolysing cortical cells but further work showed that high levels of acid phosphatase are characteristic of meristematic cells in both main and lateral meristems where it is obviously not involved with cell degradation. A similar high level of acid phosphatase activity has been observed in the main root epidermis of a number of species and in the lateral root epidermis of *Regnellidium* (Shaykh and Roberts, 1974).

Ashford and McCully (1973) have also examined the activity of β-glycosidases in regions of lateral root penetration in corn. A very low level of activity was observed among the crushed cell walls of cortical cells along the lateral primordium but it could not be determined if this represented enzyme involved in final breakdown of cellulose walls. The low enzyme activity in this location was contrasted with very high activity in young epidermal and root cap cells of both the lateral meristem and the main root apex. There is no evidence that this latter activity is involved in cell wall break-down. Thus the question of whether or not

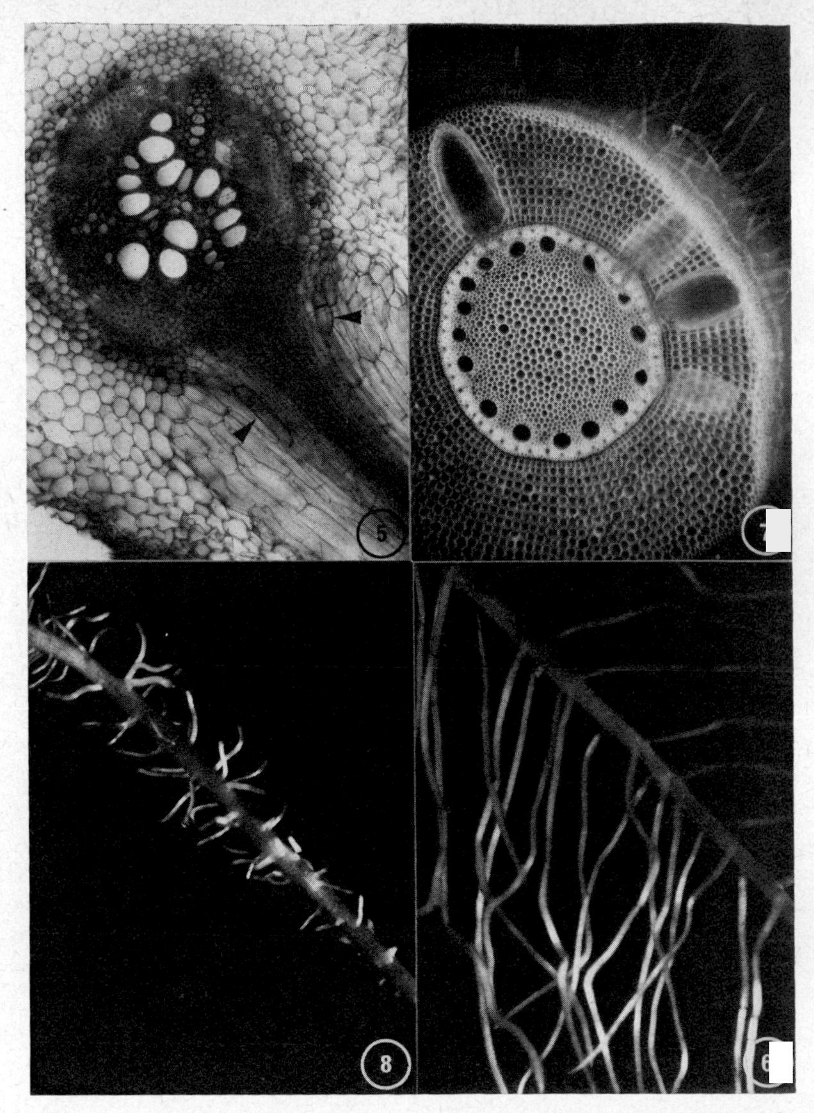

FIG. 5. TS of main root of *Pisum sativum* L. showing median LS through base of lateral emerged ca. 1 cm. The connection between the endodermis of the parent root and that of the lateral can be seen because at this stage of development many of the endodermal cells are forming suberized lamellae. Arrows show cortical cells of the lateral which have formed this suberin layer. Hand section stained with Sudan IV. ×68.

FIG. 6. Main root of *Pisum sativum* grown in vermiculite showing arrangement of laterals. ×3.

FIG. 7. TS of a straight piece of main root of *Croix lacryma-jobii* L. showing the tendency of this plant to produce laterals in clumps. Hand section "stained" with the optical brightener Calcofluor M2R New and viewed with a fluorescence microscope. ×48.

FIG. 8. Adventitious root of *Saccharum officinarum* L. rooted in tap water showing bunched laterals. ×3.

enzymes play a role in degradation of parent tissues during lateral root growth has not been settled.

A number of early workers made the interesting observation that renewed cell divisions sometimes occur in parent cortex in the vicinity of emerging lateral primordia. This phenomenon has been examined closely by Tschermak-Woess and Dolezal (1953). They found that in 53 of 71 species mitotic figures could be observed in parent cortical cells near lateral primordia and in 31 of these species spontaneous mitosis did not occur in cortical cells outside the lateral initiating zone. The mitotic figures could be seen in cells several cell diameters removed from the surface of the primordium which apparently were not under mechanical stress. Mitosis in these cells was followed by cross wall formation but not accompanied by much cell enlargement. More recently Gramberg (1971) has observed induced synthesis of RNA in nuclei of cortical cells lying ahead of emerging adventitious root primordia of *Phaseolus*. These observations suggest that a substance diffusing from the young primordia can influence cortical cells in its path. It is not clear, however, how these changes in the cortical cells could facilitate passage of the lateral through the cortex.

IV. Branching Patterns of Roots

A. Distribution of Laterals and Variation in their Growth Potential

The most obvious generalization about branch root position is that they are initiated some distance behind the parent root apex in an acropetal sequence (a few exceptions are noted elsewhere in this paper). Furthermore, their location is related to the vascular pattern of the parent root. Most studies of dicotyledons have shown that laterals are initiated in the pericycle adjacent to the protoxylem pole (Esau, 1965) except in some diarch roots (e.g. tomato) where they form opposite the phloem (see Fig. 113—Fahn, 1967), although Knobloch (1954) reported laterals forming opposite protoxylem in diarch roots of chicory and Russian dandelion. It is generally considered (Esau, 1965; Fahn, 1967) that laterals form opposite the phloem poles in many monocotyledons, but this is not true in corn (Bell and McCully, 1970) where initiation is always close to protoxylem (see Fig. 1). At later stages of development, however, the primordia in corn cover a number of vascular poles and sometimes appear centred over phloem. Thus in plants with many vascular poles it is only possible to determine the site of lateral meristem initiation by examining the earliest developmental stages.

The relationship of initiating site to protoxylem position results in linear arrays of laterals along the length of the root. These are particularly apparent in dicotyledons with small numbers of xylem poles (Fig. 6) but the most spectacular case described is in the fern *Ceratopteris thalictroides* (Mallory *et al.*, 1970) where the lateral roots form along two spirals reflecting the twisting of the diarch parent root during growth.

Except for their relationship to parent root vascular tissue the orientation of laterals relative to each other has received little attention until recently. This is surprising since casual observation of roots of many plants shows a tendency toward lateral clumping. In dicotyledons they are often in pairs, less frequently in groups of 3 to 5 at approximately the same level on the parent root (Fig. 6). In a number of monocotyledons the tendency is to large groups (Figs. 7 and 8). Riopel (1966, 1969) studied lateral placement along the main axis in several monocotyledons all with many protoxylem poles. Using two different methods of pattern analysis he concluded that the branch roots are non-randomly dispersed and the mean distance between nearest laterals exceeds that expected if distribution were random. Mallory *et al.* (1970) observed that in the plants studied with a small number of vascular poles laterals tend to occur in groups of two or more at the same level, separated from adjacent groups by relatively large distances. In each group the laterals appear to have a specific arrangement repeated in successive groups along the parent root. Yorke and Sagar (1970) have also noted a tendency to clumping of branch roots in *Pisum sativum*.

Perhaps the most spectacular example of clumping of laterals are the proteoid roots in the family Proteaceae (Lamont, 1972). There are short sections of main root bearing a very large number of short, unbranched, laterals. The laterals form only over protoxylem (or in two files on either side of each pole) but they are packed as close as physically possible in each longitudinal file.

Quite a different type of aggregation of laterals occurs on curved portions of main roots in many plants. When roots are curved (means of curvature unimportant) in the region of lateral formation, all the laterals develop on the convex side and where primordia have been initiated just before curvature they will grow toward this side. This phenomenon was first observed by Noll in 1900 (described in Torrey, 1965) and later by de Haan (1936). In many cases the laterals appear close together in one or two files along the convex side.

Another aspect of root branching which has been examined recently is variation in the growth potential of lateral meristems initiated at different positions on a root system. Yorke and Sagar (1970) showed that in *Pisum sativum* branch roots initiated in clumps midway along lateral bearing parts of main roots elongate the fastest and grow longest. Laterals in other clumps elongate more than single branch roots. Casual observations of roots of a large number of species confirms this tendency towards increased length of branch roots which are clumped.

An interesting variation in form of lateral roots on an individual plant occurs in the Cyperaceae (Davies *et al.*, 1973). In many species, in addition to normal branch roots, markedly swollen, very hairy laterals develop at intervals along the main roots. These are anatomically quite different from normal laterals, having more cortical cells and a columnar epidermis which secretes a thick layer of mucilage. Similar diversity of branch root anatomy has not been described in other species.

B. Factors Influencing Root Branching Patterns

A variety of exogenous factors are known to influence the branching pattern of roots. These have been reviewed by Torrey (1965) and Street (1969) and will not be considered here except to mention the recent studies of Hackett (1972) and Drew *et al.* (1973). These show clear effects of the local application of nitrates, with a stimulation of lateral initiation and growth only in the region of application. These results have obvious significance for agriculturalists.

There have been some very good studies showing endogenous control of root branching by various portions of the plant. Again, many of these have been reviewed by Torrey (1965) and Street (1969) and I only intend to emphasize a few points and introduce some of the recent literature.

It is not disputed that the apical meristem of main roots inhibit the formation of laterals for some distance behind them. This distance is quite constant for a given species under standard conditions and it varies directly with the growth rate of the main apex (Geissbühler, 1953). In some plants, notably ferns and aquatic angiosperms, the distance between the main apex and the youngest lateral can be very short. Generally it is considered that the apex produces an inhibitor of lateral formation although it is not clear if it is always initiation that is inhibited or if it can be just later stages of development. It is also possible that the influence of the main apex is indirect acting through its control of maturation of tissues in which laterals are initiated. The removal of the main root apex usually stimulates lateral development up to the wound and those produced nearest to the cut end are frequently longer (Dyanat-Nejad and Neville, 1972). It was noted, however, very early by Thimann (1936) and subsequently by others that while the removal of the main apex greatly stimulates lateral production in some species it may have little effect in others. In general lateral primordia develop in an acropetal sequence behind the main apex but there have not been a sufficient number of studies which looked for early stages of primordial development or followed the sequence of lateral emergence in intact roots to determine how strict this sequence is. There are two notable exceptions recorded in the literature. Rippel (1937) induced lateral initiation in a basipetal direction in *Vicia faba* by the removal of the cotyledons although Geissbühler (1953) failed to repeat this observation. Recently Dyanat-Nejad and Neville (1972) described the post-germination development of laterals in basipetal sequence on the portion of the radicle of *Theobroma cacao* L. which is present before germination.

A number of workers have shown that the cotyledons produce a stimulation of lateral formation which is influential in early stages of development and Dyanat-Nejad and Neville (1972) have evidence that this factor travels to the site of lateral initiation in the vascular tissue directly connected to the cotyledon involved. On the other hand, Wightman and Pohl (unpublished data) find no convincing evidence that the cotyledons produce a stimulator of lateral root development during the early stages of development of *Pisum sativum*. At the same time these latter workers have found a significant inhibition of lateral initiation by the removal of the shoot of very young seedlings (60 → 84 h). Removal of the shoot after this time had little or no effect. Most other workers have not found an effect of shoot removal on lateral root initiation but they have usually dealt with somewhat older seedlings. Dyanat-Nejad and Neville (1972), however, report a subtle but significant inhibition of lateral initiation following shoot removal in *Theobroma cacao*.

The effect of older root tissue on lateral initiation distal to it has been clearly established by Pecket (1957) and a number of other workers (see reviews by Torrey, 1965, and Street, 1969). It seems clear that in root pieces in culture there is an acropetal movement of a factor from the more mature tissue which stimulates lateral root initiation. Whether this same factor also controls the subsequent development of the primordia has not been established although the studies of Wightman (1954) on *Pisum sativum* suggest that this is so.

The effect of existing lateral primordia on the formation of new ones is not really clear. Bünning (1953) pointed out that in many cases there are more initial stages of lateral development present than emerged laterals and that the latter inhibit the former. How general this phenomenon is can only be clarified by studies using techniques such as those of Wightman (1954), and Hackett and Stewart (1969) which allow visualization of quite early stages of root primordium formation in cleared specimens. The pattern analysis studies of Riopel (1966 and 1969) and Mallory *et al.* (1970) suggest that there is mutual inhibition between lateral root primordia, but the results of Mallory *et al.* at least suggest it is more effective in the longitudinal than the radial direction. On the other hand the results of Goldacre (1959) indicate that lateral primordia may produce a factor stimulating the initiation of more primordia in the vicinity.

Torrey (1962) and others (see reviews by Torrey, 1965, and Street, 1969) have clearly demonstrated that auxins and cytokinins in the medium of cultured roots can markedly influence lateral initiation and development. But it is not possible from this type of study to relate these hormones to the factors which mediate correlations within the intact plant. The recent study of *Raphanus sativus* by Webster and Radin (1972) is a move in the right direction. These workers applied auxins and a cytokinin only to the basal end of cultured roots thus much more closely approximating the possible gradients in an intact plant. They found that the auxins greatly stimulate lateral development while the cytokinin completely inhibits it.

A detailed knowledge of the time course of lateral initiation and development in intact plants under controlled conditions is essential if the various correlative phenomena observed in experimental situations are to be interpreted correctly. Such information has been absent until the recent work of Klasová *et al.* (1971 and 1972). More studies of this type are necessary.

Obviously the most important control of root branching pattern is genetic and this is discussed by Zobel in this symposium (see pp. 261–275 this volume). However, it is worth mentioning here the diageotropic tomato mutant reported by Zobel because this plant offers the only good direct evidence to date of the nature of an inherent factor controlling lateral initiation—in this case ethylene. This report immediately suggests an explanation for two of the branching patterns commonly observed in roots. The tendency to clumping of laterals could be an ethylene effect, levels of ethylene stimulating initiation of additional primordia being produced by the wounding of cortical cells during exit of the original primordium. Ethylene produced by wounding of cortical or pericycle cells on the convex side of bent roots could also account for the resulting lateral distribution.

Acknowledgements

I would like to thank Patrick Michiel for Fig. 4 and Roger Britton for making up the plates.

References

ASHFORD, A. E. and MCCULLY, M. E. (1970). Localization of Naphthol AS-B1 phosphatase activity in lateral and main root meristems of corn. *Protoplasma* **70**; 441–456.

ASHFORD, A. E. and MCCULLY, M. E. (1973). Histochemical localization of β-glycosidases in roots of *Zea mays* III. β-glucosidase activity in the meristems of lateral roots. *Protoplasma* **77**; 411–425.

BAYER, D. E., FOY, C. L., MALLORY, T. E. and CUTTER, E. G. (1967). Morphological and histological effects of trifluralin on root development. *Am. J. Bot.* **54**; 945–952.

BELL, J. K. and MCCULLY, M. E. (1970). A histological study of lateral root initiation and development in *Zea mays*. *Protoplasma* **70**; 179–205.

BERTHON, R. (1943). Sur l'origin des radicelles chez les Angiospermes. *C.R. Acad. Sci.* **216**; 308–309.

BLOCH, R. (1935). Observations on the relation of adventitious root formation to structure of air roots of orchids. *Proc. Leeds Phil. Lit. Soc.* **3**; 92–101.

BLOCH, R. (1941). Wound healing in higher plants. *Bot. Rev.* **7**; 110–146.

BONNETT, H. T. JR. (1968). The root endodermis: fine structure and function. *J. Cell. Biol.* **37**; 199–205.

BONNETT, H. T. Jr. (1969). Cortical cell death during lateral root formation. *J. Cell. Biol.* **40**; 144–159.

BONNETT, H. T. JR. and TORREY, J. G. (1966). Comparative anatomy of endogenous bud and lateral root formation in *Convolvulus arvensis* roots cultured in vitro. *Am. J. Bot.* **53**; 496–507.

BÜNNING, E. (1952). Weitere Untersuchungen über die Differenzierungsvorgänge in Wurzeln. *Z. Bot.* **40**; 385–406.

BÜNNING, E. (1953). "Entwicklungs- und Bewegungsphysiologie der Pflanze" Springer-Verlag, Berlin.

BYRNE, J. M. (1973). The root apex of *Malva silvestris*. III Lateral root development and the quiescent centre. *Am. J. Bot.* **60**; 657–662.

CLOWES, F. A. L. (1950). Root apical meristems of *Fagus sylvatica*. *New Phytol.* **49**; 248–268.

CLOWES, F. A. L. (1958). Development of quiescent centre in root meristems. *New Phytol.* **57**; 85–88.

DAVIES, J., BRIARTY, L. G. and RIELEY, J. O. (1973). Observations on the swollen lateral roots of the *Cyperaceae*. *New Phytol.* **72**; 167–174.

DREW, M. C., SAKER, L. R. and ASHLEY, T. W. (1973). Nutrient supply and the growth of the seminal root system in barley. *J. exp. Bot.* **24**; 1189–1202.

DUMBROFF, E. B. and PEIRSON, D. R. (1971). Possible sites for a passive movement of ions across the endodermis. *Can. J. Bot.* **49**; 35–38.

DYANAT-NEJAD, H. and NEVILLE, P. (1972). Étude expérimentale de l'initiation et de la croissance des racines latérales précoces du cacaoyer (*Theobroma cacoa* L.). *Ann. Sci. nat. Bot., Paris Sér.* 12, **13**; 211–246.

ESAU, K. (1940). Developmental anatomy of the fleshy storage organ of *Daucus carota*. *Hilgardia* **13**; 175–226.

ESAU, K. (1965). "Plant Anatomy" John Wiley and Sons, Inc., New York.

FAHN, A. (1967). "Plant Anatomy" Pergamon Press, Toronto.

FOARD, D. E., HABER, A. H. and FISHMAN, T. N. (1965). Initiation of lateral root primordia without completion of mitosis and without cytokinesis in uniseriate pericycle. *Am. J. Bot.* **52**; 580–590.

FOURCROY, M. (1942). Perturbations anatomiques interessant le faiceau vasculaire de la racine au voisinage des radicelles. *Ann. des Sci. Nat., Bot., Ser.* 11, **3**; 177–198.

FRIEDBERG, S. H. and DAVIDSON, D. (1971). Cell population studies in developing root primordia. *Ann. Bot.* **35**; 523–533.

GEISSBÜHLER, H. (1953). Untersuchungen über die korrelative und hormonale Steuerung der Seitenwurzelbildung. *Ber schweiz. bot Ges.* **63**; 27–89.

GOLDACRE, P. L. (1959). Potentiation of lateral root induction by root initials in isolated flax roots. *Aust. J. Biol. Sci.* **12**; 388–394.

GRAMBERG, J. J. (1971). The first stages of the formation of adventitious roots in petioles of *Phaseolus vulgaris*. *Koninkl. Nederl. Akad. Wetenshappen Amsterdam* **74**; 42–45.

GUTTENBERG, H. VON (1968). "Der primäre Bau der Angiospermenwurzel" Gebrüder Borntraeger, Berlin.

HACKETT, C. (1972). A method of applying nutrients locally to roots under controlled conditions, and some morphological effects of locally applied nitrate on the branching of wheat roots. *Aust. J. Biol. Sci.* **25**; 1169–1180.

HACKETT, C. and STEWART, H. E. (1969). A method for determining the position and size of lateral primordia in the axes of roots without sectioning. *Ann. Bot.* **33**; 679–683.

DE HAAN, I. (1936). Polar root formation. *Rec. Trav. Bot. Neerl.* **33**; 292–309.

HUBER, B. (1961). "Grundzüge der Pflanzenanatomie" Springer-Verlag, Berlin.

JANCZEWSKI, E. (1874). Recherches sur le developpement des radicelles dans les phanerogames. *Ann. Sci. Nat. Bot.* **20**; 208–233.

KARAS, I. and MCCULLY, M. E. (1973). Further studies of the histology of lateral root development in *Zea mays*. *Protoplasma* **77**; 243–269.

KLASOVÁ, A., KOLEK, J. and KLAS, J. (1971). A statistical study of the formation of lateral roots in *Pisum sativum* L. under constant conditions. *Biol. Plant.* **13**; 209–215.

KLASOVÁ, A., KOLEK, J. and KLAS, J. (1972). Time dynamics of primary root branching in *Pisum sativum* L. *Biol. Plant.* **14**; 249–253.

KNOBLOCH, I. W. (1954). Development anatomy of chicory—the root. *Phytomorphology* **4**; 47–54.

LAMONT, B. (1972). The morphology and anatomy of proteoid roots in the genus *Hakea*. *Aust. J. Bot.* **20**; 155–174.

LEDBETTER, M. C. and PORTER, K. R. (1970). "Introduction to the Fine Structure of Plant Cells" Springer-Verlag, New York.

MACLEOD, R. D. (1972). Lateral root formation in *Vicia faba* L. 1. The development of large primordia. *Chromosoma* **39**; 341–350.

MACLEOD, R. D. (1973). The emergence and early growth of the lateral root in *Vicia faba* L. *Ann. Bot.* **37**; 69–75.

MACLEOD, R. D. (1974). Some observations on the growth of lateral roots of *Vicia faba* L. and their response to colchicine treatment. *New Phytol.* **73**; 147–155.

MACLEOD, R. D. and MCLACHLAN, S. M. (1974). The development of a quiescent centre in lateral roots of *Vicia faba* L. *Ann. Bot.* **38;** 535–544.

MALLORY, T. E., CHIANG, S., CUTTER, E. G. and GIFFORD, E. M. JR. (1970). Sequence and pattern of lateral root formation in five selected species. *Am. J. Bot.* **57**; 800–809.

O'BRIEN, T. P. (1970). Further observations on hydrolysis of the cell wall in the xylem. *Protoplasma* **69**; 1–14.

O'DELL, D. H. and FOARD, D. E. (1969). The presence of lateral root primordia in the radicle of buckwheat embryos. *Bull. Torrey Bot. Club.* **96**; 1–3.

PECKET, R. C. (1957). The initiation and development of lateral meristems in the pea root. I. The effect of young and mature tissue. *J. expt. Bot.* **8**; 172–180.

POPHAM, R. A. (1955a). Levels of tissue differentiation in primary roots of *Pisum sativum*. *Am. J. Bot.* **42**; 529–540.

POPHAM, R. A. (1955b). Zonation of primary and lateral root apices of *Pisum sativum*. *Am. J. Bot.* **42**; 267–273.

RIOPEL, J. L. (1966). The distribution of lateral roots in *Musa acuminata* "Gros Michel". *Am. J. Bot.* **53**; 403–407.

RIOPEL, J. L. (1969). Regulation of lateral root positions. *Bot. Gaz.* **130**; 80–83.

RIPPEL, K. (1937). Umkehr der Seitenwurzelgenese bei Leguminosen als korrelative Störung. *Ber. dtsch. bot. Ges.* **55**; 288–292.

ROELOFSEN, P. A. (1959). *In* "Handbuch der Pflanzenanatomie III" (K. Linsbauer, ed.); Gebrüder Borntraeger, Berlin.

RYWOSCH, S. (1909). Untersuchungen über die Entwicklungsgeschichte der Seitenwurzeln der Monocotyledonen. *Z. Bot.* **1**; 253–283.

SCHADE, C. and VON GUTTENBERG, H. (1951). Über die Entwicklung des Wurzelvegetations Punktes der Monokotyledonen. *Planta* **31**; 170–198.

SEAGO, J. L. (1973). Developmental anatomy in roots of *Ipomoea purpurea* II. Initiation and development of secondary roots. *Am. J. Bot.* **60**; 607–618.

SHAYKH, M. M. and ROBERTS, L. W. (1974). A histochemical study of phosphatases in root apical meristems. *Ann. Bot.* **38**; 165–174.

SOCHER, S. H. and DAVIDSON, D. (1970). Heterogeneity in G_2 duration during lateral root development. *Chromosoma* **31**; 478–484.

STREET, H. E. (1969). *In* "Root Growth" (W. J. Whittington, ed.); pp. 20–41. Butterworths, London.

SUTCLIFFE, J. F. and SEXTON, R. (1968). β-glycerophosphatase and lateral root development. *Nature, Lond.* **217**; 1285.

THIMANN, K. V. (1936). Auxins and the growth of roots. *Am. J. Bot.* **23**; 561–569.

TORREY, J. G. (1962). Auxin and purine interactions in lateral root initiation in isolated pea root segments. *Physiol. Plant.* **15**; 177–185.

TORREY, J. G. (1965). *In* "Encyclopedia of Plant Physiology" (A. Lang, ed.); pp. 1256–1327. Springer-Verlag, Berlin.

TSCHERMAK-WOESS, E. and DOLEZAL, R. (1953). Durch Seitenwurzelbildung induzierte und spontane Mitosen in den Dauergeweben der Wurzel. *Ost. bot. Zeit.* **100**; 358–402.

VONHÖNE, H. (1880). Über das Hervorbrechen endogener Organe aus dem Mutterorgane. *Flora* **83**; 268–274.

WEBSTER, B. D. and RADIN, J. W. (1972). Growth and development of cultured radish roots. *Am. J. Bot.* **59**; 744–751.

WIGHTMAN, F. (1954). "Studies on factors affecting cell division in pea root meristems". Ph.D. thesis, University of Leeds.

YORKE, J. S. and SAGAR, G. R. (1970). Distribution of secondary root growth potential in the root system of *Pisum sativum*. *Can. J. Bot.* **48**; 699–704.

Chapter 7

The Initiation and Development of Root Buds

R. L. PETERSON

Department of Botany and Genetics, University of Guelph, Guelph, Ontario, Canada.

I. Introduction

The ability of roots of many species of plants to form buds which develop into new shoots has been recognized and described in the botanical literature over a long period of time and extensive lists of species capable of forming "root buds" have been compiled (Wittrock, 1884; Beijerinck, 1887; Priestley and Swingle, 1929; Holm, 1925; Rauh, 1937; McVeigh, 1937; Raju *et al.*, 1966). In those species forming root buds there is a wide variation in the degree to which the root system is involved in vegetative reproduction. In some species buds occur sporadically on roots often only after the root has been excised or injured (Wittrock, 1884) while in other species, for example in many of the aquatic members of the Podostemonaceae, one of the main functions of the root system appears to be the production of root buds (Goebel, 1905; Rauh, 1937). It has been suggested (Raju *et al.*, 1966; Steeves *et al.*, 1966) that some of the most persistent dicotyledonous weedy species on the North American prairies have been able to survive rather adverse conditions partly through the ability to reproduce by root buds. Certainly some of the most difficult weedy species to eradicate are capable of reproducing by root buds. Therefore, much of the recent literature on root buds concerns methods of cultivation and chemical treatment to prevent bud initiation and growth. On the other hand, root buds are important in the propagation of some commercially important species. For example, raspberries have been successfully propagated from root cuttings (Hudson, 1954, 1955) and various techniques of improving bud initiation have been published (Heydecker and Marston, 1968; Marston and Village, 1972). Apple varieties (Stoutemeyer, 1937), cabbage (Isbell, 1945) and brussel sprouts (North, 1953) as well as several ornamental trees (Fowells, 1965) can be propagated from their roots.

Aside from the interest in vegetative propagation by roots from a practical point of view, the occurrence of buds on roots has stimulated considerable interest among anatomists and physiologists. It is the purpose of this paper to consider some of the anatomical aspects of bud initiation and development and some of the factors involved in the control of root bud formation.

II. Sites of Root Bud Initiation

A. Pericycle

A variety of root tissues may be involved in bud formation and therefore the developmental patterns vary considerably depending on the region of the root in which bud initiation occurs. One such tissue is the pericycle, the region in the root which is normally the site of lateral root initiation. Descriptions of buds arising from the pericycle (Irmisch, 1857; Beijerinck, 1887; Van Tieghem, 1887; Van Tieghem and Duliot, 1888; Priestley and Swingle, 1929) established that the sites of initiation are usually similar to those of lateral root primordia within the same species. For example, in species like *Euphorbia exigia* and *Epilobium angustifolium* which have tetrarch root steles, buds originate opposite the protoxylem poles and therefore can be seen to occupy four longitudinal rows along the root (Van Tieghem, 1887). If two lateral root primordia usually arise in relation to a single protoxylem pole as, for example, in *Alliaria officinalis* then bud initials form in a similar pattern (Van Tieghem, 1887; Van Tieghem and Duliot, 1888). These characteristics of bud initiation plus the fact that early stages of bud and lateral root initiation are similar anatomically led Goebel (1902) to the conclusion that endogenous buds are simply modified lateral root primordia.

More recently, additional species with pericyclic-derived buds have been described (Rauh, 1937; Myers *et al.*, 1964; Wilkinson, 1966; Kormanik and Brown, 1967) but no analysis of the initiation and development of the buds was included. Torrey (1958) studied pericyclic buds in cultured roots of *Convolvulus* and drew attention to the advantages of such a root system in which both bud and root primordia originate from the pericycle in studies of organogenic processes. In a subsequent paper, Bonnett and Torrey (1966) showed that very early stages in the initiation of both bud and lateral root primordia were identical as seen in the light microscope. Soon after their initiation, however, marked differences in rate of development, the behaviour of the endodermis, cellular patterns within each type of primordium, and the participation of central cylinder cells around the site of organ initiation occurred. Cleared root segments showed that rather well-developed bud initials could be distinguished from lateral root primordia because they were broader, the first xylem differentiated obliquely in relation to the parent root stele as opposed to at right angles for the root primordia and, as a bud formed, a terminal protoxylem element in the parent root stele became separated from the remainder of the xylem. More work along these lines with other species should contribute to a better understanding of differences between endogenous bud and lateral root development.

Root buds of pericyclic origin have been described in the Polypodiaceous fern, *Amphoradenium* (White, 1969) and although the initial divisions involved pericyclic cells, the endodermis and thin-walled cells of the cortex soon divided and took part in the formation of a meristematic mass of cells within which the bud apical meristem, the first leaf, and the first root differentiated. This process of organogenesis is reminiscent of events taking place within the embryo of such ferns.

Since buds forming from the pericycle are endogenous they have to traverse the cortex and epidermis before emerging. There is some indication that the endodermis may play a part in the formation of the bud meristem in some species (Beijerinck, 1887; White, 1969) or the suggestion has been made that the endodermis along with cortical cells may be digested by the enlarging bud (Van

Tieghem and Duliot, 1888; Dore, 1955). There is, however, no good evidence for this view and in fact Bonnett and Torrey (1966) found that in *Convolvulus* roots the endodermis associated with bud initiation underwent radial divisions followed by cell enlargement so that it was difficult to distinguish the endodermal cells from the cells of the adjacent cortex or from cells in the apical region of the bud primordium. Clearly, more detailed studies are needed before any conclusions can be reached concerning the mode of emergence of endogenous buds to the surface of roots.

B. Phellogen or Other Tissues Derived from the Pericycle

A few species have root buds arising from the phellogen or related tissues (Priestley and Swingle, 1929). Even though some of these tissues are superficial, the buds initiated within them should be considered as endogenous in origin according to Priestley and Swingle (1929) because these tissues are ultimately derived from the pericycle. On this basis these authors corrected an earlier claim made by Wilson (1927) that buds arising from the cork tissue in *Roripa* are exogenous in origin. In poplar roots, buds are initiated from the phellogen (Brown, 1935; Schier, 1973a) and in apple from the phelloderm (Stoutemeyer, 1937).

In creeping-rooted alfalfa, divisions in various regions within the phellogen and recent derivatives of the phellogen lead to the formation of a dome of meristematic cells (primordial dome) within which one or more shoot buds are initiated (Murray, 1957). Non-creeping varieties of alfalfa may develop meristematic domes but subsequent development results in the formation of whorls of vascular tissue without the initiation of bud primordia.

In *Euphorbia esula* roots, the pericycle forms a pericyclic cambium (Bakshi and Coupland, 1959) within which lateral organs are initiated at any position and are not necessarily associated with the protoxylem poles. These authors could not determine whether the primordia were roots or buds until they emerged from the parent root. Although this may well be the case, a re-examination of this species (Myers *et al.*, 1964) indicated that lateral roots and endogenous buds are distinguished before they rupture the periphery of the root in several samples examined.

In some woody species (Brown and Kormanik, 1967; Schier, 1973a) buds derived from the phellogen or its derivatives may remain as suppressed buds for a considerable period of time until conditions are favorable for their development.

C. Bud Initiation in Relation to Lateral Roots

The initiation of buds associated in some manner with the emergence of lateral roots has been discussed in length by Beijerinck (1887) and Priestley and Swingle (1929) and these authors observed that when buds occur normally on root systems they form most commonly in relation to lateral roots. Beijerinck (1887) illustrated many examples of buds forming in an axillary position ("unterachsel") in reference to the lateral root and usually in a callus-like proliferation at the base of the lateral. Bud meristems initiated in the proximity of lateral roots may arise from various tissues including the phellogen of the main root (Beijerinck, 1887; Priestley and Swingle, 1929; Emery, 1955; Dore, 1955), the cortex of the main root (Priestley and Swingle, 1929; Bakshi and Coupland, 1960) or in outer tissues of the lateral root itself when it is still within the parent root (Beijerinck, 1887; Van Tieghem, 1887; Priestley and Swingle, 1929; Charlton, 1965, 1966).

There has been much discussion concerning the frequency of species showing buds associated with lateral roots and some explanations have been suggested to account for this position of bud initiation. Goebel (1905) claimed that an advantage of bud meristems initiating close to lateral roots is to provide a direct access by the shortest route to the soil water. Both Goebel (1905) and Dore (1955) drew attention to the fact that the root bud with its associated lateral root is very similar to an axillary bud of the shoot with its subtending leaf. The suggestion has been made (Priestley and Swingle, 1929) that buds are initiated close to lateral roots due to a leakage of "sap" from the parent root stele when the lateral root primordium breaks through the endodermis. The "sap" was visualized as filling the intercellular spaces in the cortex adjacent to the base of lateral roots thereby modifying the environment in such a way as to induce cell divisions in this region. Examples of roots which do not form root buds and in which lateral root primordia apparently do not break the endodermal layer were cited as support for this concept. The authors also discussed at some length the potentialities of the pericyclic cells in most roots in terms of their approximation to the conducting system of the stele. Although no critical evidence was presented for the view that bud initiation is stimulated by substances originating in the stele and no indication was given concerning the nature of the "sap", this concept could be tested experimentally.

Dore (1955) stressed the importance of the lateral root trace in the initiation of buds in the phellogen of horseradish roots and suggested that either substances from the lateral root trace somehow lower the surface tension in the region close to the lateral and thus create conditions conducive to cell divisions or that special bud-forming substances may diffuse from the root traces. In a later paper (Dore and Williams, 1956) plugs of horseradish root tissue without a lateral root trace failed to form primordia while tissue plugs which included root trace tissue formed buds, indicating to the authors the importance of substances in the trace for bud formation. Similar observations had been made earlier by Lindner (1940) for the same species. In *Chamaenerion* it was pointed out (Emery, 1955) that buds do not arise near the lateral root until after the vascular connection with the parent root has been severed by lateral and interxylary periderms, and therefore is was concluded that stimulating substances could not be emanating from the lateral root trace. However, the possibility remains that substances are retained in the elements of the lateral root trace even after it is no longer functional in transport.

Recent observations on bud initiation adjacent to lateral roots in the hawkweed, *Hieracium florentinum*, may have some bearing on the question of the importance of lateral root traces to bud initiation. Bud formation on roots of this species was first described by Peterson and Thomas (1971) as a normal phenomenon occurring in both field and *in vitro* conditions. Figure 1 illustrates buds forming on roots of a plant excavated from the field. Normally, few buds form on the roots in any one season. Observations of bud development on roots of this species were made from plants grown in full strength Hoagland's solution in aerated hydroponic cultures. Root tissue was fixed in 4% glutaraldehyde and processed according to the techniques outlined in Feder and O'Brien (1968). Buds are initiated adjacent to lateral roots usually after some secondary growth has occurred in the main root (Fig. 2). Two or more bud meristems may be initiated in relation to a single lateral root (Fig. 2) each being apparent initially as a locus of densely staining cells in the cortex (Figs. 2 and 3).

As these meristems are being initiated, considerable change occurs in the surrounding cortical cells of the main root. Many cells undergo divisions resulting in the subdivision of the larger cortical cells into groups of smaller cells (Figs. 2 and 3). Cortical cells immediately adjacent to the bud meristem also divide (Figs 2 and 3). Eventually the bud meristem initiates leaf primordia in which procambial tissue differentiates basipetally towards the stele of the main root (Fig. 5). The leaf primordia are initiated in a spiral phyllotaxis as they are from the main shoot apex of this species. Observations of the parenchyma cells both within the lateral root trace and adjacent to the protoxylem pole of the main root from which the lateral has been initiated, suggest that a mechanism exists for the extraction of substances from the xylem stream at these positions in the root system. The darkly staining cells (indicated by arrows in Fig. 2) adjacent to the protoxylem of the diarch root are parenchymatous and possess wall ingrowths apparent at higher magnifications. A similar region in another root bearing root buds shows the extensive development of wall ingrowths in those cells abutting on tracheary elements (Fig. 4). A median longitudinal section through the stele of the lateral root shown in Fig. 2 shows tracheary elements bordered by darkly staining cells (Fig. 6). These cells also have wall ingrowths on the wall adjacent to the tracheary element (Fig. 7). Parenchymatous cells with these structural characteristics are known as transfer cells (Gunning and Pate, 1969) and are thought to be involved in various transport phenomena within the plant. It is reasonable to suggest that they are serving in a similar capacity in the present situation. It is not known whether these transfer cells are present before buds are initiated and therefore could be involved with the inception of buds or whether

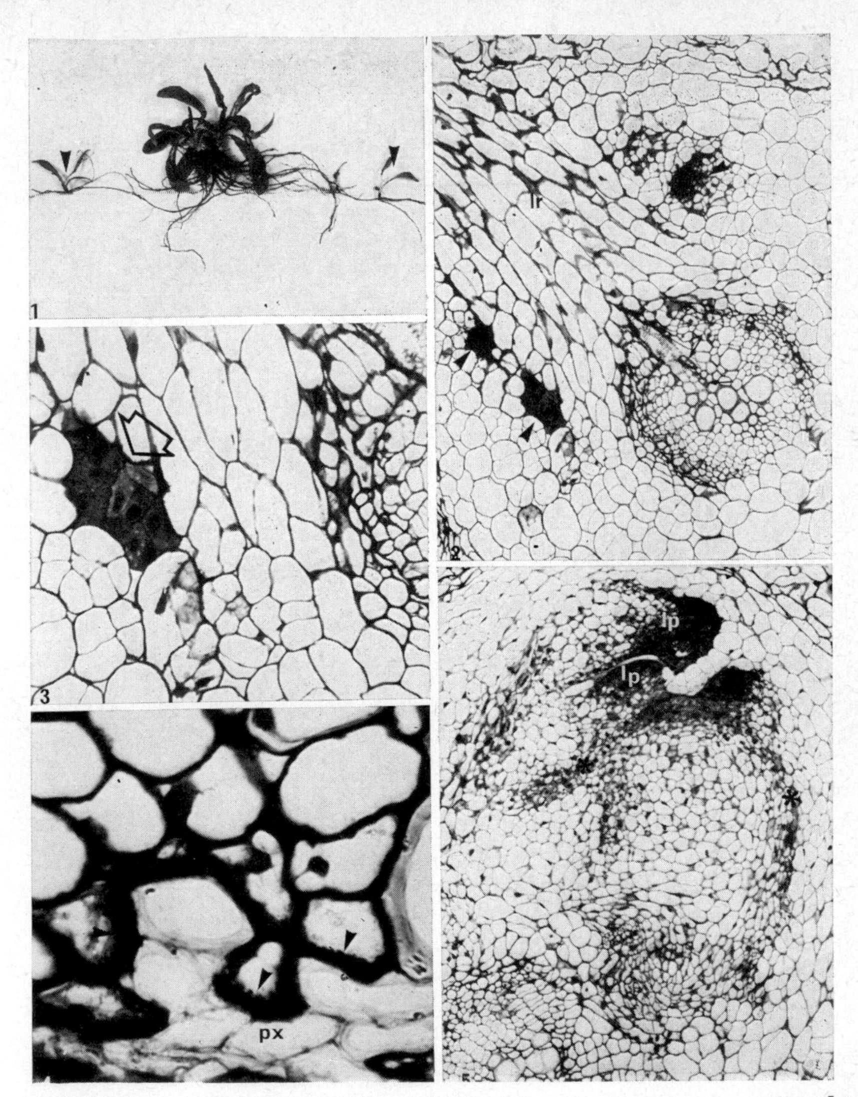

FIGS. 1–5. Bud initiation on roots of *Hieracium florentinum*. FIG. 1. Plant excavated from the field showing bud development on the roots (arrows) (from Peterson and Thomas, 1971. ×1/6. FIG. 2. Transverse section of root showing bud meristems (short arrows) in the cortex adjacent to a lateral root (lr). Secondary growth has occurred in the main root stele and numerous divisions are apparent in the cortex adjacent to the initiated buds. Darkly staining cells (long arrows) are present next to one of the protoxylem poles. ×175. FIG. 3. Early stage in bud initiation (arrow) showing the organization of the apical meristem. Cortical cells immediately above the bud meristem have divided. ×225. FIG. 4. Cells adjacent to the protoxylem pole (px) at which the lateral root trace joins as shown in Fig. 2 with numerous wall ingrowths (arrows). Wall ingrowths are formed on the sides of the parenchyma cells adjacent to the tracheary elements. ×800.

FIG. 5. An older bud with spirally arranged leaf primordia (lp) and differentiating procambial traces (*). A portion of the parent root stele is visible in the bottom left corner of the photomicrograph. ×175.

these cells develop wall ingrowths as a response to a gradient between the growing bud and the stele of the lateral root. Preliminary observations suggest that transfer cells are not present at the junction of the lateral root trace and the main root stele in the absence of buds but further work is required to clarify the involvement of transfer cells in bud initiation.

D. Buds from the Cortex not Related to Lateral Roots

In a very early paper (Trécul, 1847) the origin of root buds from the cortex was described for a number of species. In two species of the fern *Ophioglossum* which have been studied in some detail (Wardlaw, 1953a, b; Peterson, 1968) root segments or roots with the apex damaged initiate buds from the cortex. These species lack lateral roots and secondary growth so that stages in bud ontogeny are rather easy to follow, and there is no possibility that the initiated buds are modified lateral root primordia, a view expressed by Goebel (1902). As shown by Wardlaw (1953a, b) for *O. vulgatum* and Peterson (1968) for *O. petiolatum* the buds arise randomly along the length of root segments (Fig. 8). The ontogeny of these buds has been followed in paraffin embedded tissue. Initially a small group of cortical cells divide (Fig. 9) and the divisions progress until a meristematic mass of cells is formed (Fig. 10). Within these dedifferentiated cortical cells an apical meristem with an associated cavity develops (Fig. 11). Immediately after the organization of the apex, leaf primordia are formed (Fig. 11) which eventually grow and protrude through the cortex and epidermis of the root. Procambial tissue differentiates from the leaf primordia basipetally eventually joining to the stele of the root. Outer cortical cells form structures which are similar to leaf sheaths which normally develop in association with leaf primordia from the main apex of the plant (Fig. 12). Eventually root primordia are initiated in relation to the developing leaves and these plants can become independent. The induction of divisions in certain cortical cells leading to bud formation is difficult to explain in terms of either bud-forming factors or physical factors since the sites of initiation occur randomly along the length of the root (Wardlaw, 1953b). These species are of interest because of their unique characteristics of lacking lateral roots and secondary growth, and therefore provide a rather simple system in which to study bud initiation.

In *O. vulgatum* the induced buds have a different sequence of organ formation than do embryos of the same species (Wardlaw, 1954) in that the shoot apical meristem and first leaf primordium differentiate first, while in the embryo a root meristem is the first organ to be organized. Wardlaw explained this difference on the basis of the markedly different nutritional resources of the tissues supporting the bud as compared to the embryo.

In a recent paper (Zenkteler, 1971) buds of cortical origin were described in cultured roots of *Atropa belladonna* but little information was given concerning their ontogeny.

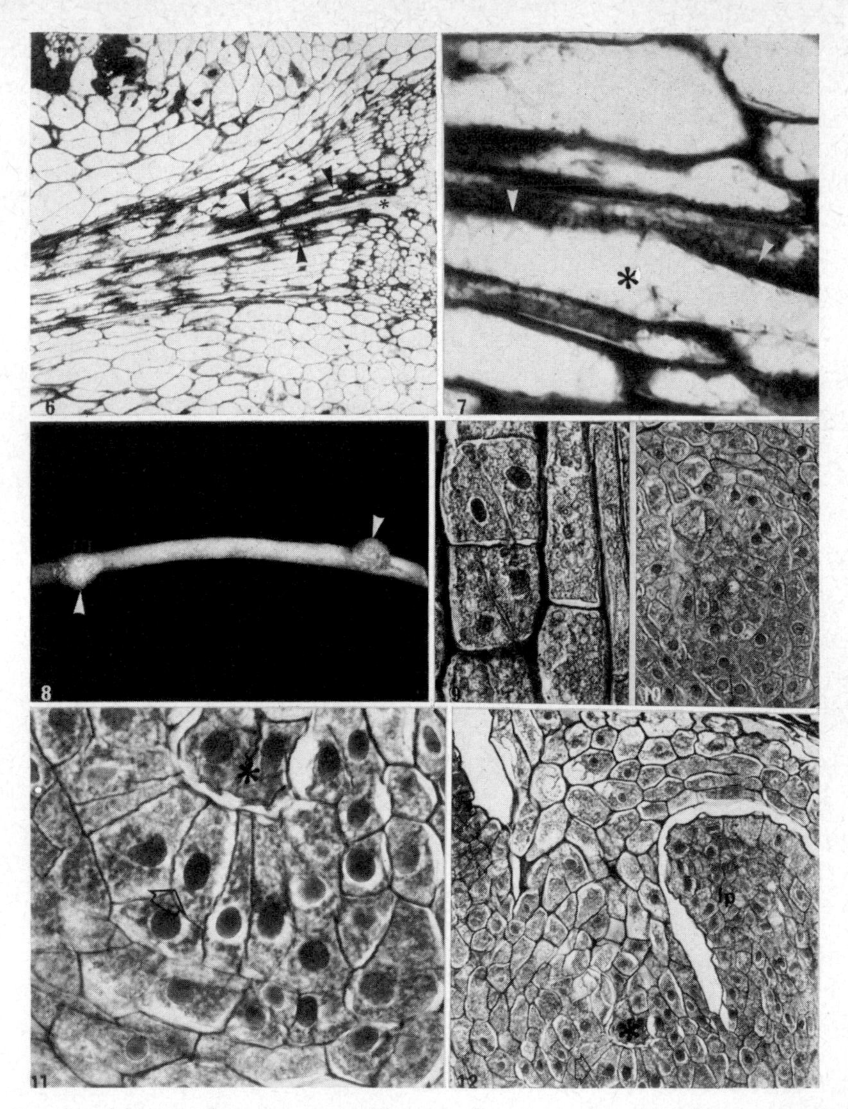

FIGS 6 and 7. Lateral root traces in *Hieracium florentinum*. FIG. 6. Median longitudinal section through the lateral root indicated in Fig. 2 showing tracheary elements (*) bordered by darkly staining cells (arrows). × 175. FIG. 7. Some of the cells indicated by arrows in Fig. 6 showing the development of wall ingrowths (arrows) on the walls adjacent to tracheary elements (*). × 800.

FIGS 8–12. Bud development in the cortex of *Ophioglossum petiolatum* root segments. All sectioned material is from FAA-fixed, paraffin-embedded root tissue. FIG. 8. Root segment with two initiated buds (arrows) in which leaf primordia have not yet emerged. × 2. FIG. 9. Initial divisions in the cortex indicating the inception of a bud. × 500. FIG. 10. Meristematic mass of cells formed by divisions of cortical cells. × 75. FIG. 11. Organization of a bud meristem (arrow) with an associated breakdown of adjacent cortical cells(*) × 240. FIG. 12. Initiated bud showing the apical meristem (arrow), cavity formation (*), and leaf primordia (lp). Note the sheathing layers of cortical cells around the leaf primordium on the right. × 100.

E. Buds Developed from the Root Apex

For well over a century there has been a fascination with the idea that a bud meristem could arise from a transformed root meristem and conversely that a bud meristem could be converted into a root meristem. Rostowzew (1890) summarized the previous evidence for the transformation of a root apex into a shoot apex and also described in some detail the events leading to shoot formation from root apices in two ferns, *Asplenium* and *Platycerium*. Reviewing many earlier papers he concluded that the only two cases of a true transformation from a root apical meristem to a bud apical meristem in the seed plants is in the orchid, *Neottia nidus-avis* and in *Anthurium longifolium*. However, if one evaluates the evidence for either of these cases it is found to lack critical anatomical details. Rostowzew (1890) seemed willing to accept as definitive proof of this transformation the change in vascular pattern from that typical of roots to that characteristic of shoots. A careful study of the fate of the root apical meristem itself during the transformation to a bud was not made. The report of root transformation in *Anthurium* (Goebel, 1878) especially includes no precise anatomical data. In the case of *Neottia* it has been more recently shown (Champagnat and Champagnat, 1965; Champagnat, 1971) that before shoot buds are formed at the root apex, the root meristem dedifferentiates and forms a structure referred to as a protocorm in which shoot primordia arise. Therefore in *Neottia nidus-avis* the root apical meristem does not change directly into a shoot apical meristem.

Rostowzew (1890) described the formation of a bud at the root apex of *Asplenium esculentum* and various *Platycerium* species and concluded that at the time of bud initiation, the tetrahedral apical cell of the root ceases to divide toward the rootcap and the root apical cell from this point becomes the apical cell of the shoot which then divides to form segments on three sides only. The root cap remains over the developing shoot for a time but is eventually sloughed. In *Platycerium*, leaf and scale primordia originate from derivatives of the shoot apical meristem but in *Asplenium* only leaf primordia are formed. In *Asplenium* lateral root primordia within the cortex of the parent root can also be transformed into buds so that regular rows of shoot primordia can be observed. In both genera there is a transition from root to shoot vasculature similar to what one finds in the hypocotyl of higher vascular plants.

In a discussion of the implications of his findings, Rostowzew (1890) concluded that because of the ability of a root apex to transform itself into a bud meristem it becomes difficult to be definitive about the morphological distinction between root and shoot. It may well be that in the ferns and other lower vascular plants there is a certain degree of plasticity in organ determination (see Bierhorst, 1971) and therefore these conversions of one type of organ into another should perhaps not be too surprising. In a recent text (Bierhorst, 1971) an example of this phenomenon is presented for the fern *Actinostachys*, in which a root tip may change

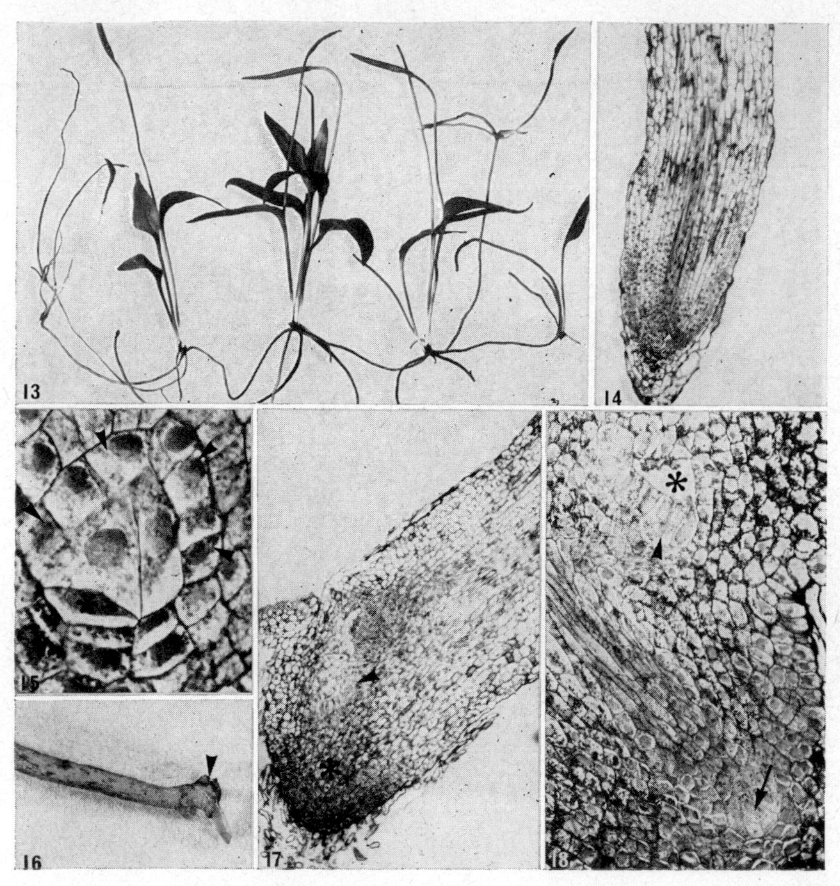

FIGS 13–18. Bud formation at the root apex of *Ophioglossum petiolatum*. Sectioned material is from paraffin-embedded root tissue. FIG. 13. A clone of plants formed by successive bud formation at root tips. $\times \frac{1}{4}$. FIG. 14. Longitudinal section of a root apex not forming a bud showing its narrow, pointed structure with a shallow root cap covering a tetrahedral apical cell (arrow) (from Peterson, 1970). $\times 65$. FIG. 15. The root apical cell of the root shown in Fig. 14 showing a derivative cell being formed towards the root cap(*) and several derivatives forming on the flanks of the apical cell (arrows) which divide to form the normal tissue pattern of the root (from Peterson, 1970). $\times 250$. FIG. 16. A root tip with an initiated bud (arrow). The root apex has become deflected from its usual direction of growth. $\times 1{\cdot}5$. FIG. 17. Longitudinal section through a root tip with a well developed bud showing the position of the bud meristem (arrow) and the continued growth of the root apex (*). $\times 75$. FIG. 18. An initiated bud at the root apex showing the bud apical meristem (short arrow), the cavity formed by the breakdown of root cap cells (*) and the apex of the root (long arrow). $\times 150$.

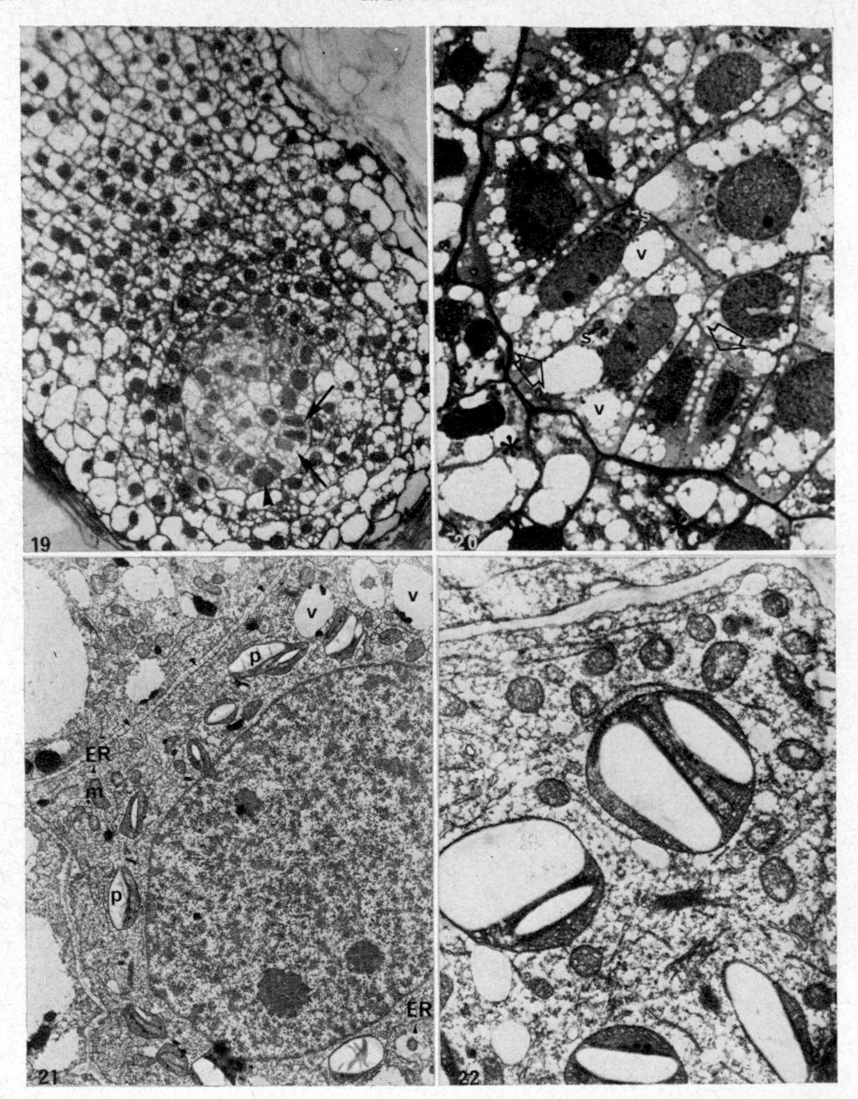

FIGS 19–20. Early stages of bud initiation at the root apex of *Ophioglossum petiolatum*. All tissue was fixed in 4% glutaraldehyde in Pipe's buffer (Salema and Brandão, 1973) post fixed in 2% osmium tetroxide in the same buffer, embedded in Spurr's embedding medium and sectioned with glass knives. Thick sections for light microscopy were stained with 0·05% Toluidine Blue 0 in 1% borax while sections for electron microscopy were stained with uranyl acetate followed by lead citrate. FIG. 19. L.S. of a root apex with an early stage of bud initiation. The root apex has become rounded and changes have occurred in the apical meristem. The root apical cell (short arrow) does not form derivatives towards the root cap but several large derivatives are formed on one flank of the apical cell (long arrows), some of which become cells of the bud meristem. × 150. FIG. 20. The root apical cell (filled arrow) and deriva-

its orientation of growth, cease forming derivatives towards the root cap, and form a shoot apical meristem directly from the root apex.

Therefore, it does seem probable that root apical meristem transformations can occur, particularly in the ferns and these species should provide excellent materials for experimental studies on the potentialities of meristems.

Two publications since the review of Rostowzew (1890) indicating the change from root apical meristem to shoot apical meristem are those of Brundin (1895) and Carlson (1938). Unfortunately, insufficient detailed anatomical information was presented to prove this type of transformation.

Although not involving the transformation of the root apical meristem, there are some interesting cases of bud development in close proximity to the root apical meristem while the latter retains its identity. Rostowzew (1891) corrected earlier workers who listed the fern *Ophioglossum vulgatum* among those capable of forming a bud by the transformation of the root apical meristem, and supplied a rather detailed analysis by line drawings of the initiation of a bud at the root apex. The bud is initiated from recent derivatives of the root apical cell, which retains its identity throughout bud formation and becomes active again after the bud is established. Poirault (1893) confirmed these findings for the same species. In another species in the same genus, *O. petiolatum*, essentially the same series of events in the ontogeny of a bud from the root apex takes place (Peterson, 1970). Since this last publication, additional observations have been made on bud formation at the root apex of *O. petiolatum*. In nature or under greenhouse conditions this species reproduces predominantly by means of buds initiated at the root apex and frequently a number of attached plants forming a clone can be seen in a population because of this means of propagation (Fig. 13). When the root apex is not involved in forming a bud it is rather slender with a shallow root cap covering a large tetrahedral apical cell (Fig. 14). The apical cell forms derivatives towards the root cap and on its flanks which divide to initiate series of cells determining the tissue pattern of the normal root (Fig. 15). The events leading to the formation of a bud at the apex has been described from paraffin-embedded material (Peterson, 1970) and it was observed that as the bud is initiated the root apex is deflected from its normal direction of growth (Fig. 16) and longitudinal sections through root tips with obvious buds showed that the bud meristem

---

tives (open arrows) have large nuclei with pronounced nucleoli, vacuoles (v) with small osmiophilic bodies and some starch grains (s) visible. Adjacent root cap cells are separated from the root apical cell and its derivatives by a distinct boundary. Adjacent root cap cells (*) have smaller nuclei with no obvious nucleoli. ×375. FIG. 21. A portion of the root apical cell showing the large nucleus with several nucleoli, plastids with starch grains (p), mitochondria (m), endoplasmic reticulum (ER) and vacuoles (v) with associated osmiophilic bodies. The walls are traversed by numerous plasmodesmata (arrows). ×4200. FIG. 22. Portion of a derivative cell showing plastids with starch grains, mitochondria, ER and dictyosomes. ×11,500.

assumed a lateral position after the continued elongation of the main root (Figs 17 and 18). Another feature commented on previously was the influence the bud meristem has on the surrounding root cap cells. As the bud meristem is established, a cavity, formed by the lysis of the adjacent root cap cells, is formed (Fig. 18) and the leaf primordia initiated by the bud meristem grow out through the surrounding root tissue (Fig. 18).

Although the early stages of bud initiation have also been considered from paraffin-embedded material (Peterson, 1970), additional observations at the light and electron microscope level from plastic-embedded materials are included. As the apex switches into the initiation of a bud, the root tip becomes rounded and the organization of the apex changes (Fig. 19). The derivatives of the root apical cell fail to divide in the usual pattern but become enlarged with pronounced nuclei and nucleoli (Fig. 20). The large derivatives formed on the flank of the root apical cell (Figs 19 and 20) become cells of the shoot meristem (Peterson, 1970). The root apical cell retains its identity but ceases to form derivatives towards the root cap during this phase of bud development (Figs 19 and 20). It has a large nucleus with several nucleoli, many mitochondria, plastids with starch grains, osmiophilic bodies and considerable endoplasmic reticulum (Figs 20 and 21). Numerous plasmodesmata are present between the apical cell and adjacent cells (Fig. 21). The derivatives of the apical cell have organelles similar to those of the apical cell but have large vacuoles (Fig. 20). Figure 22 shows a portion of a derivative cell with cytoplasm rich in mitochondria, plastids and starch grains, dictyosomes and endoplasmic reticulum. These cells also have numerous plasmodesmata in their primary walls (Fig. 23) and osmiophilic deposits many of which seem to be associated with vacuoles (Fig. 24). Small osmiophilic bodies of a different nature but definitely within vacuoles have also been observed (Fig. 25). The nature of these bodies is at present not clear.

As has been shown previously (Peterson, 1970) the root cap cells adjacent to the bud meristem undergo lysis forming a cavity (see also Fig. 18). An early stage in the initiation of a bud at the root apex indicates considerable differences between the bud meristem cells and the adjacent root cap cells (Fig. 20). The nucleus in these cap cells is rather dense with no obvious nucleoli and the cells generally appear more vacuolated than the bud meristem cells (Fig. 20). The distinct boundary between the bud meristem and the root cap (Fig. 20) is seen with the electron microscope as an amorphous osmiophilic material in the intercellular matrix (Fig. 26). No histochemical studies have been made on this layer at the ultra-structural level but it has been noted that this region stains rather intensely in PAS-stained paraffin sections indicating that some of this material may be of a polysaccharide nature. Besides being rather vacuolated, these root cap cells have a cytoplasm quite different from that of the root apical cell or its immediate derivatives. The endoplasmic reticulum becomes somewhat vesiculated (Fig. 27); osmiophilic deposits are evident, some of which seem to be associated either with

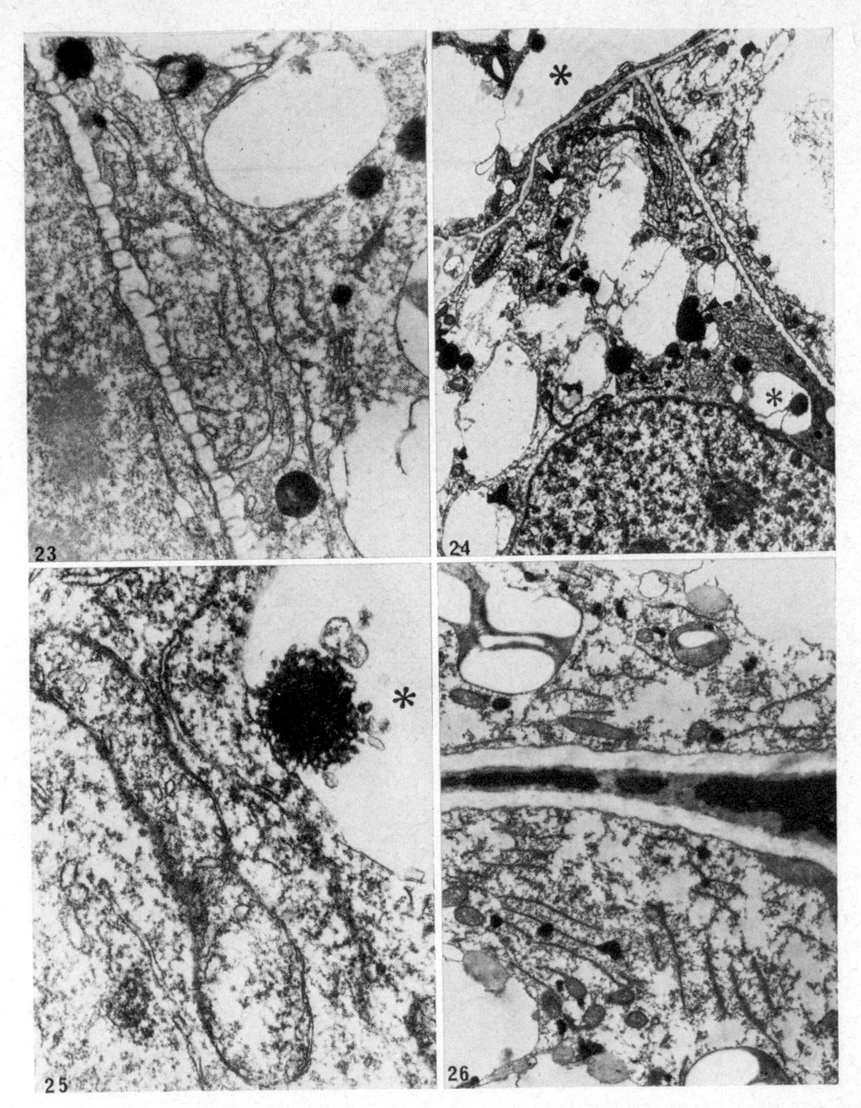

Figs 23–26. Early stages of bud initiation at the root apex of *Ophioglossum petiolatum*. All tissue was fixed in 4% glutaraldehyde in Pipe's buffer (Salema and Brandão, 1973) post fixed in 2% osmium tetroxide in the same buffer, embedded in Spurr's embedding medium and sectioned with glass knives. Thick sections for light microscopy were stained with 0·05% Toluidine Blue 0 in 1% borax while sections for electron microscopy were stained with uranyl acetate followed by lead citrate. Fig. 23. Portion of a derivative cell with numerous plasmodesmata between adjacent cells. ER, dictyosomes and osmiophilic bodies. ×12,900. Fig. 24. Derivative cell showing dense osmiophilic bodies in the cytoplasm (arrows) and in the vacuoles (*). ×4,250. Fig. 25. Osmiophilic body in the vacuole (*) of a derivative cell showing the loose organization at the periphery of this body. ×30,000. Fig. 26. Amorphous osmiophilic material in the intercellular matrix apparent as the distinct boundary between the root cap and the developing bud in Fig. 20. ×8,750.

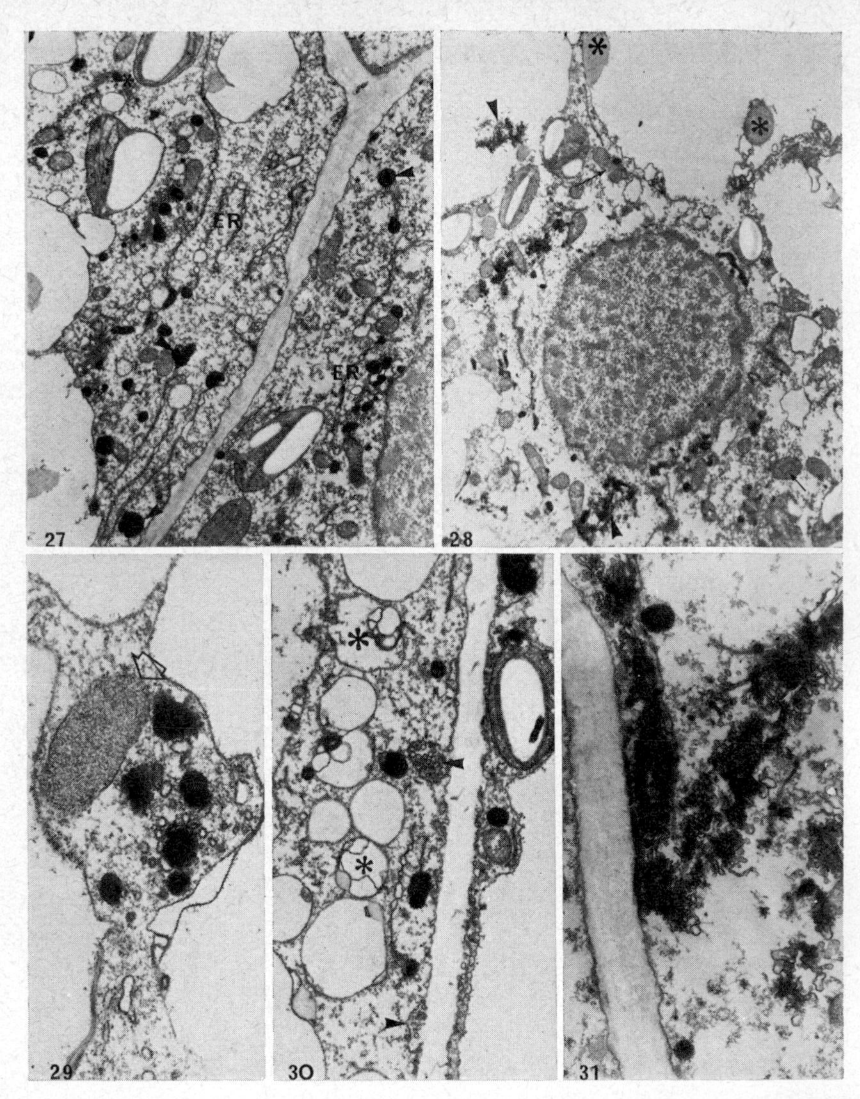

FIGS 27–31. Early stages of bud initiation at the root apex of *Ophioglossum petiolatum*. All tissue was fixed in 4% glutaraldehyde in Pipe's buffer (Salema and Brandão, 1973) post fixed in 2% osmium tetroxide in the same buffer, embedded in Spurr's embedding medium and sectioned with glass knives. Thick sections for light microscopy were stained with 0·05% Toluidine Blue 0 in 1% borax while sections for electron microscopy were stained with uranyl acetate followed by lead citrate. Root cap cells adjacent to the initiated bud meristem. FIG. 27. Cytoplasm of root cap cells adjacent to the newly-initiated bud meristem cells showing vesiculated endoplasmic reticulum (ER), dictyosomes(*), osmiophilic bodies (arrows), some of which appear to be associated with ER and dictyosomes. Plasmodesmata were scarce in walls of these cells. × 10,000. FIG. 28. Root cap cells at an older stage of bud development showing the vacuolation of the cytoplasm with disorganized membrane systems (short arrows), microbodies (long arrows) and deposits (*) in the vacuoles showing little osmiophilia.

the endoplasmic reticulum or dictyosomes (Fig. 27) and others showing less osmiophilia and associated with vacuoles (Figs 26, 28 and 30). Few plasmodesmata were observed in any of the walls of these cells (Figs 27, 30 and 31). With time, the cells adjacent to the bud meristem become very vacuolated (Figs 28–31) and pronounced changes occur within the cytoplasm. Small membrane-bound bodies resembling microbodies (see arrows in Figs 28 and 29) become quite numerous, multi-vesiculated bodies appear in many places next to cell walls (Fig. 30) and the system of membranes including the endoplasmic reticulum and dictyosomes becomes completely disorganized (Figs 28 and 31). The final stages in the breakdown of these cells to form a cavity similar to that shown in Fig. 18 has not been seen at the ultrastructural level so it is not clear what events occur in the degradation of the walls of these cells.

These observations on the initiation of a bud at the root apex of *Ophioglossum petiolatum* raise a number of questions, some of which have already been discussed (Peterson, 1970). One of the most intriguing but difficult problems is related to the factors involved in the triggering of a root apical cell into forming derivatives which initiate a shoot meristem instead of undergoing the usual pattern of cell divisions leading to root tissue formation. At the present time it is difficult to explain this triggering in terms of a change in hormonal balance at the root apex because of the close proximity of the newly formed bud meristem to the functional root apical cell and the periodic changes the apex goes through to produce a succession of buds. An understanding of the factors involved in the breakdown of root cap cells in the vicinity of the bud meristem requires further ultrastructural studies including enzyme histochemistry, and this work is in progress. Since these cells do not break down until the bud meristem is established and since they are not crushed by bud growth, it is evident that they must be destroyed by autolysis, triggered in some way by the presence of the bud meristem. There are many problems which remain concerning cellular relationships during bud initiation at the apex of this and other species of ferns (Rostowzew, 1891; Bierhorst, 1971), some of which are being investigated currently.

There are a few recent observations concerning bud initiation at the apex of adventitious roots after experimental treatment of the developing roots. For example, experiments using isolated nodes of *Nasturtium officinale* with axillary root primordia ("radicular plate") showed the plasticity of organogensis in these primordia (Champagnat *et al.*, 1967; Ballade, 1968, 1970, 1971, 1972). Application of kinetin in the vicinity of the petiolar axis caused the primordia to develop into

---

×11,250. FIG. 29. Microbody (arrow) and several dense osmiophilic bodies in degenerating root cap cells similar to that shown in Fig. 28. ×21,250. FIG. 30. Multivesiculated bodies (arrows), vacuoles (*) with various deposits and dense osmiophilic bodies in a degenerating root cap cell. ×10,600. FIG. 31. Degenerating root cap cell showing disorganization of membrane systems. ×15,000.

bud meristems (Champagnat *et al.*, 1967; Ballade, 1968) when the primordia were within a certain size range (Ballade, 1970). Essentially the same results were obtained with cultured isolated nodes (Ballade, 1971) and cultured radicular plates (Ballade, 1972). A histological study (Ballade, 1968) showed that the root meristem does not change into a shoot meristem directly but that the entire meristem differentiates into vacuolated parenchyma cells with vascular tissue differentiating close to the apex. Following this, cell divisions are initiated in the superficial layer covering the modified root apex and a shoot meristem capable of initiating leaf primordia is organized. A similar case of bud formation at the apex of adventitious roots in a species of poplar has been reported (Brand and Venverloo, 1973).

F. Buds from Root Callus

Many species of plants form root buds only after the root system has been damaged or otherwise treated so as to induce callus proliferations in which bud meristems are initiated (Priestley and Swingle, 1929; Naylor, 1941; White, 1943; North, 1953; Norton and Boll, 1954). Also, in some species capable of forming root buds normally, additional buds may occur in callus proliferations under certain conditions (Torrey, 1958; Bonnett and Torrey, 1965; Peterson, 1969b; Cuthbertson, 1972). Although the formation of buds in root-derived callus is an important phenomenon in itself, since very precise organization of cells must occur to initiate bud meristems, the factors controlling the inception of these organized structures are probably more similar to those involved in organogenesis in callus systems in general than to factors involved in the initiation of buds on undamaged roots. For this reason this topic will not be considered here.

III. Buds Formed on Cultured Roots

Although isolated roots of many species have been cultured (see for example Butcher and Street, 1964) relatively few species producing root buds have been grown *in vitro*. Shoot initiation has been obtained from callus tissue in cultured roots of dandelion (White, 1943), tomato (Norton and Boll, 1954) *Isatis tinctoria* (Danckwardt-Lillieström, 1957), *Atropa belladona* (Zenkteler, 1971; Thomas and Street, 1972) and *Hieracium florentinum* (Peterson, unpublished observations). Endogenous buds not related to callus formation have been induced in cultured excised roots of *Robinia* (Seeliger, 1956), *Convolvulus* (Torrey, 1958; Bonnett and Torrey, 1965, 1966; Bonnett, 1972) and *Linaria vulgaris* (Charlton, 1965). Cultured roots bearing both lateral roots and endogenous buds are valuable in studies of factors controlling the initiation and development of both types of organs (Torrey, 1958) but have been used infrequently in experimental studies. Some results using cultured roots are considered under other headings.

Tissue plugs of large storage roots of *Cichorium intybus* have been used successfully in producing flower buds *in vitro* (e.g. Paulet and Nitsch, 1964a, b; Nitsch and Nitsch, 1964; Margara, 1970, 1973a, b; Margara and Bouniols, 1967) and a large number of experiments concerning the regenerative process and factors most conducive to floral bud initiation have been described. This system has been used extensively to study many physiological aspects of flowering (see papers cited in Margara, 1973a, b).

IV. Effect of Growth Regulators on Bud Initiation and Development

A. Auxin and other Phenoxy Compounds

Early in this century most workers interested in regeneration from root cuttings were concerned with the polarity exhibited in organ initiation from fleshy roots. With the discovery of auxin as a natural plant hormone, this polarity was usually explained in terms of gradients from the proximal to distal* ends in root cuttings. High auxin concentrations were conducive to root initiation while low concentrations were conducive to bud initiation (see Dore, 1965 and literature cited). Applications of rather high concentrations of exogenous auxin generally resulted in upsetting the polarity in root segments by suppressing bud formation at the proximal end with or without a concomitant production of root primordia in this region (Lindner, 1938, 1940; Plant, 1940; Stoughton and Plant, 1938; Warmke and Warmke, 1950). Various other treatments designed to decrease the level of auxin at the distal end thereby providing an environment more conducive to bud initiation had some success in that buds were induced at the distal end in a limited number of cases (e.g. Plant, 1940; Stoughton and Plant, 1938; Warmke and Warmke, 1950). In addition to experiments on polarity of regeneration from roots, an effect of auxins on bud number on root segments has been noted. In general, exogenous applications of auxin and related phenoxy compounds have been found to suppress bud development on isolated root segments (Lindner, 1938, 1940; Stoughton and Plant, 1938; Went, 1939; Emery, 1955; Dore and Williams, 1956; Eliasson, 1961; Marcavillaca *et al.*, 1971; Kefford and Caso, 1972; Caso and Kefford, 1973), but in most cases the concentration of auxin used was very high. In some experiments lower concentrations of auxins were found to have no effect on bud development (Torrey, 1958; Peterson, 1969a) or to be stimulatory to bud initiation (Bonnett and Torrey, 1965; Kefford and Caso, 1972). In the latter cases, however, higher concentrations of auxin were inhibitory to bud initiation. On cultured root segments of *Convolvulus*, auxin can have different effects on the same tissue depending on the concentration used and the duration of treatment (Bonnett and Torrey, 1965). In very short segments

* proximal refers to that portion of the root closest to the shoot and distal, that portion of the root closest to the original root apex.

(1·5 mm) low concentrations of IAA were more effective in stimulating bud production than either the synthetic cytokinin, phenylamino purine or combinations of this cytokinin and IAA. In 15-mm segments bud initiation was suppressed with higher concentrations of IAA if the root segments were exposed to the auxin for eight weeks but if the segments were exposed to auxin for 4–6 days and then transferred to a medium without auxin for the balance of the 8-week period then there was a striking increase in the number of buds formed. The greatest number of lateral roots were formed, however, in segments that had been exposed to a relatively high auxin concentration throughout the experiment.

Attempts have been made to correlate auxin distribution within root segments and regenerative phenomena. Warmke and Warmke (1950) extracted auxins from various regions of chicory roots and found that the proximal ends of root fragments had less free and neutral auxin at particular times after root segment excision and suggested that this may be associated with the initiation of buds in this region. In a later study of the same species (Vardjan and Nitsch, 1961) several auxin-like substances were isolated and it was shown that the level of these substances, particularly substance B which may have been tryptophane and substance C which may be a complex of indolyl-3-acetic acid dropped sharply at the proximal end of the root segment within the first few days after excision. A low auxin status presumably provided conditions favorable to bud initiation.

Correlations between shoot growth and a number of developmental processes in roots including root bud initiation have been described in the literature. In alfalfa, regeneration from roots occurs easily from plants either in winter dormancy or in full bloom but not from actively growing plants (Smith, 1950). Surgically removing parts of the shoot has stimulated bud initiation on roots in a number of species (Carlson *et al.*, 1964a; Eliasson, 1971b; Marston and Village, 1972) as has girdling the phloem in the upper part of the root or lower part of the stem (Farmer, 1962; Kormanik and Brown, 1967). The stimulation in root bud initiation resulting from these treatments has been generally explained by the influence on the auxin balance in the plant with a reduction of auxin content in the root system favoring bud initiation (Eliasson, 1971b).

B. Cytokinins

With the exception of the publication of Montaldi (1972) and the results included in the present study which show that cytokinins can stimulate bud initiation on roots of intact plants, all other reported effects of cytokinins on root bud initiation have involved excised root segments or adventitious roots on stem segments. In aseptically grown *Hieracium florentinum* plants, buds can be stimulated to form on young plants by the application of cytokinins to the culture medium (Fig. 33) as compared to the controls (Fig. 32) in which a well-developed root system has been produced but in which buds have not formed at this stage of growth. At concentrations of 1·0 ppm or higher, the cytokinins inhibit plant growth and induce

FIGS 32–34. Root buds in *Hieracium florentinum*. Plants grown from seedlings under aseptic conditions on nutrient media. For photographs, plants were removed from the culture flasks without disturbing the root systems. FIG. 32. Plant grown on medium without cytokinin showing a well developed root system with no root buds. ×1·2. FIG. 33. Plant grown on nutrient medium containing 0·1 ppm benzylaminopurine. Several root buds (arrows) have formed. ×1·2. FIG. 34. Plant grown in nutrient medium with 1·0 ppm benzylaminopurine showing callus proliferations from the shoot and root within which numerous small buds (arrows) formed. ×1·2.

FIGS 35–37. FIG. 35. *Hieracium florentinum* plants grown in soil and treated with either water (control), low nitrogen or high nitrogen as explained in the text. Plants receiving high nitrogen treatment had more numerous and larger emerged buds than in plants of the two other treatments. ×1/3. FIG. 36. *H. florentinum* plant grown in aerated hydroponic nutrient solution containing high nitrogen level showing numerous initiated buds on the root system. Plants grown in low nitrogen or water had few root buds at this stage. ×1/5. FIG. 37. Well developed plants on roots of *H. florentinum* grown in high nitrogen nutrient solution. ×2.

considerable proliferation of root tissues within which numerous plantlets develop (Fig. 34). Sections through these callus proliferations showed numerous bud meristems in addition to those that were visible externally.

Application of cytokinins to isolated root segments have been reported to have various effects on bud formation. Commonly, the number of buds initiated per segment was increased (Danckwardt-Lilliestrōm, 1957; Torrey, 1958; Bonnett and Torrey, 1965; Marcavillaca *et al.*, 1971; Kefford and Caso, 1972). However, in some species there was no effect of cytokinins on root bud number (Charlton, 1965). In many instances the increase in bud number was associated with the stimulation of callus tissue at the cut proximal end of the root segment with a concomitant initiation of buds in this callus. Torrey (1958) and Bonnett and Torrey (1965) referred to these as "kinin-induced buds" in cultured *Convolvulus* roots in order to distinguish them from the pericyclic-derived buds in this species which seemed little affected by cytokinins. In *Isatis tinctoria* root segments, buds formed spontaneously from callus at the cut proximal end in freshly excised segments and in root segments which had been transferred as many as three times to fresh culture medium (Danckwardt-Lilliestrōm, 1957). However, in the subsequent transfers, shoot initiation did not occur unless cytokinin was added to the culture medium indicating to the author that cytokinins are essential for bud initiation. Since no mention was made of the ability of successively transferred root segments to form callus at the proximal end it is possible that cytokinin is limiting for callus initiation which is prerequisite for bud initiation in this species.

Applications of cytokinins can alter the polarity of bud formation along the root segment. In root segments of *Chondrilla juncea* treatment with cytokinins extended the zone along which buds would form, i.e. only 20% of initiated buds occurred at the proximal end as compared to 80% in water controls (Kefford and Caso, 1972). In *Ophioglossum* root segments the application site of cytokinins had a determining effect on the positioning of new buds in that application to the proximal end resulted in most of the buds being initiated immediately adjacent to this region while application to the distal end resulted in the majority of buds forming some distance from this end (Peterson, 1969a). Perhaps in these roots the applied cytokinin moved preferentially in a basipetal direction but this has not been verified. In *Convolvulus* roots, Bonnett and Torrey (1965) found that cytokinin treatment inhibited lateral root initiation at the distal end. Since there was no concomitant increase in bud number at the distal end it was apparent that cytokinin application did not cause primordia which would have developed into roots to develop into buds.

The inherent polarity in most root segments, which manifests itself by the regeneration of buds at the proximal end and roots at the distal end, may be due to the polar distribution of more than one endogenous hormone. As already indicated, Vardjan and Nitsch (1961) isolated a number of auxin-like substances from root segments and showed a polar distribution of at least two of them. In addition, the same authors isolated a substance which gave positive responses in two cytokinin bioassays and had its highest concentration at the proximal end of root segments up to three days after excision of the segment. After three days there was a drastic reduction in cytokinin content at the proximal end, and this was probably correlated with the initiation of root buds in this region. Following the initiation of root buds the cytokinin level again increased at the proximal end. The cytokinin extracted from these segments was later identified as ribosyl-zeatin (Bui-Dang-Ha and Nitsch, 1970).

One of the most interesting effects of cytokinins in relation to the initiation of root buds is the effect they may have on the root apical meristem prior to bud initiation. For example, root primordia located on "radicular plates" in axils of leaves of *Nasturtium officinale* can be inhibited in their development as roots if they are treated with kinetin before they reach a certain critical stage in development (Champagnat *et al.*, 1967; Ballade, 1968, 1970, 1971, 1972). The cells of the root meristem become enlarged and vacuolated and a bud is initiated in the superficial layer which has remained less differentiated as described in a previous section of this paper. In stem segments of poplar bearing adventitious root primordia cytokinin applications also inhibit normal root apical meristematic activity with the subsequent initiation of bud meristems (Brand and Venverloo, 1973). In this case, however, a considerable proliferation of root tissues precedes bud initiation.

A similar development of buds associated with proliferations at the apex of *Ophioglossum* roots treated with cytokinins had previously been published (Peterson, 1969b). With cytokinin applications greater than 3·0 ppm, the root apex proliferated and formed numerous bud apical meristems in close proximity to each other. None of these buds emerged from the swollen region of the root indicating that either they were incompletely developed or that they were inhibiting each other. The evidence for the role of cytokinins in the initiation of buds in normal positions in roots is very limited. However, there is evidence that cytokinins can alter the normal pattern of root development usually by the induction of callus-like proliferations in which bud meristems may become organized. In cases of this nature it is difficult to be sure that the cytokinins are exerting an effect on bud initiation directly or only through the stimulation of callus masses in which the conditions are favorable for the initiation of bud meristems.

C. Gibberellins

There have been relatively few studies on the effects of gibberellins on root bud initiation. In one study (Charlton, 1965) no effect of gibberellic acid on bud number in regenerating root segments of *Linaria* was found, but in another study (Kefford and Caso, 1972) high concentrations of gibberellic acid applied as an overnight immersion treatment stimulated the number of buds on root segments of *Chondrilla juncea*. In the latter study it was mentioned that gibberellic acid applied along with naphthaleneacetic acid counteracted the inhibitory effect of this auxin on root bud initiation. In creeping-rooted alfalfa in which bud formation is stimulated on plants grown under short day conditions, gibberellic acid application to the shoot reduced bud initiation on the roots (Carlson *et al.*, 1964b). Since the application of gibberellic acid changed the habit of shoot growth in these plants to that typical of plants grown in long days, and the application of the same hormone to plants grown in long days had no effect on root bud number, the authors concluded that the suppressing effect of gibberellic acid on bud initiation in short day plants was caused by an indirect effect on shoot growth. Differences in rates of shoot growth were thought to alter the hormone balance in the plants, therefore, affecting bud regeneration. Application of gibberellic acid to root segments of trembling aspen (*Populus tremuloides*) had different effects on root buds depending on the stage of bud initiation at the time of treatment (Schier, 1973b). Root segments without bud primordia or with primordia in early stages of development had fewer buds after gibberellic acid treatment compared to controls, while those root segments with well-formed bud primordia at the time of excision and treatment with gibberellic acid had more emerged buds after six weeks compared to the controls. Application of EL-531, a substituted pyrimidine known to inhibit gibberellin-controlled internode elongation, decreased the number of emergent buds. These results suggested that in poplar, gibberellic acid stimulates shoot elongation of developed bud primordia but inhibits the early stages of bud initiation. Buds which did form on gibberellin-treated root segments of poplar were spindly with narrow, chlorotic leaves (Schier, 1973b), an effect also reported for buds formed on cultured, gibberellin-treated root segments of *Convolvulus* (Bonnett, 1972).

Root segments of chicory cultured in medium containing gibberellic acid form buds which initiate floral organs directly, unlike the controls in which a few leaves precede the development of floral organs (Paulet and Nitsch, 1964b).

At present there is no evidence that gibberellic acid is necessary for the initiation of root buds but it is evident that this hormone enhances the subsequent growth of these initiated buds.

D. Other substances

A wide range of substances has been tested as to their possible role in organ initiation and development on roots. As early as 1940, Lindner was concerned with the control of root bud initiation but one wonders at his choice of test substances which numbered in the forties and ranged from ammonium sulfate to zinc sulfate and included such unusual substances as saliva! All substances tested were without effect on regeneration from horseradish root segments.

Warmke and Warmke (1950), interested in polarity in regenerating root pieces, found that certain substances such as ethylene chlorohydrin and chloral hydrate could induce buds at the distal end of root cuttings in a small number of cases while other substances such as iodoacetate, 2,3,5-triiodobenzoic acid (TIBA) and iso-leucine, reported to affect the availability of auxin, had no such effect on altering the polarity of regeneration. More recently, Bonnett and Torrey (1965) found TIBA to have no effect on bud number or position in regenerating segments of *Convolvulus* roots. On the other hand, TIBA was found to be stimulatory to bud initiation and another antiauxin, p-chlorophenoxyisobutyric acid (PCPIB), was found to counteract the inhibitory effect of 2,4-D on bud initiation on root segments of *Chondrilla juncea* (Kefford and Caso, 1972).

Although adenine has been found to be stimulatory to bud initiation in a number of callus systems this substance has little effect on bud initiation from root segments (Dore and Williams, 1956; Torrey, 1958; Charlton, 1965; Kefford and Caso, 1972). Carbohydrate level also seems to play a minor role in regeneration from roots (Tew, 1970; Schier and Zasada, 1973). Torrey (1958) investigated a variety of substances known to be growth promoters in other systems but found them all to be without effect on regeneration from root segments of *Convolvulus*. A number of growth retardants showed a general inhibition of bud initiation from root segments of *Chondrilla* (Kefford and Caso, 1972).

With the more recent interest in ethylene, abscisic acid (ABA) and morphactins as regulators of plant morphogenesis these substances have also been tested in a limited number of experiments on regenerating root segments. Schier (1973c) found that ABA was inhibitory to root bud development in *Populus*, particularly if segments used had early stages of bud primordia as opposed to later stages of bud initiation. At the same time callus tissue was promoted at the cut ends of root segments. There was a partial reversal of the inhibitory effect of ABA with applications of GA. In one report (Ilyas, 1973), morphactin (chlorfluorenol IT 3456) applied to the shoot apex of cauliflower greatly stimulated bud production on the roots. Ethylene treatment of cultured root fragments of *Cicorium* stimulated the total number of buds per root fragment (Bouriquet, 1972) but the results were not striking and this effect needs to be re-investigated.

V. Effect of Nutrient Level on Bud Formation

An early report (Ossenbeck, 1927) indicated that nutrient levels may be important in root bud formation. In intact plants or root segments of *Bryophyllum proliferum* regeneration of roots was affected by whether or not mineral nutrients were provided. Although limited data were given it appeared that the best conditions for regeneration were the presence of light and a good nutrient supply. Varying the nitrogen level affected the development of root buds in *Euphorbia esula* in that twice as many root buds developed on plants receiving high nitrogen levels as compared to those on low nitrogen regimes (McIntyre and Raju, 1967). Most of the additional buds on roots of plants receiving high nitrogen treatment were formed on lateral roots, which increased both in number and size under these conditions. It was also found that the regenerative capacity of root segments from plants receiving high nitrogen levels was far greater than on roots of plants receiving low nitrogen.

In a study of the correlative inhibition of the shoot on root bud development in *Euphorbia esula*, nitrogen level was found to modify the pattern of shoot growth which in turn influenced the development of root buds (McIntyre, 1972). Shoot axillary buds were released from apical dominance in plants receiving high levels of nitrogen and the rapid development of these buds inhibited root bud outgrowth. Decapitation of the axillary branches induced root bud elongation. Under low nitrogen levels, the axillary shoot buds remained inhibited but the elongation of the root buds was minimal even after removal of the shoot system from the plants. Thus, even though the influence of the shoot system had been removed, plants under low nitrogen had poor regenerative capacity from their roots. Measurements did show however that the root buds on plants receiving low nitrogen maintained slow but continuous development while the axillary shoots were completely inhibited. It was suggested that because of the endogenous origin of root buds they were able to compete favorably for the limited nitrogen supply available.

In a recent study (Kefford and Caso, 1972), root segments of *Chondrilla* were grown under different nutritional levels without an effect on bud number but bud growth was best on segments treated with weak nutrient solutions.

Recent experiments on the rosette plant *Hieracium florentinum* indicate the importance of nitrogen supply on root bud initiation in this species. Applications of full strength Hoagland's nutrient solution containing 210 ppm nitrogen (high nitrogen) to plants grown in soil resulted in more and larger emergent root buds than if plants were treated with Hoagland's solution with the nitrogen level reduced to 10·5 ppm by the equimolar substitution of K_2SO_4 and $CaCl_2$ for KNO_3 and $Ca(NO_3)_2$ respectively (low nitrogen) or with distilled water (Fig. 35). Removing the soil from the roots showed that not only were there more emergent buds on the plants receiving high nitrogen but there were considerably more buds initiated. Even more striking results were obtained, however, if plants were grown in aerated hydroponic solutions using the same three treatment solutions as in the above experiment. Again there was a marked increase in the number of buds on plants grown in high nitrogen solutions (Fig. 36) compared to those grown in solutions containing low nitrogen or in water which had few or no root buds. With time nearly every lateral root had one or more buds associated with it (Fig. 37) and therefore hundreds of root buds developed on each plant. The results are so consistent from one experiment to another that *Hieracium florentinum* plants are routinely grown in full strength Hoagland's solution to produce buds for developmental studies. The role of nitrogen level in bud initiation in this species is being explored in detail since an understanding of nutrient effects on root bud initiation may play a significant role in controlling weedy species reproducing by this means on agricultural land.

VI. Effect of Physical Factors

A. Seasonal Variation in Root Bud Formation

It has been demonstrated for a number of species that bud formation from roots varies, sometimes markedly, with the season of the year (see Dore, 1965). Hudson (1953, 1954) first noted definite "on" and "off" seasons for bud initiation in roots of raspberries with the "off" season being from May–July. In a recent paper (Marston and Village, 1972) it was shown that regeneration could be achieved in raspberry root cuttings grown in summer conditions, which normally lead to poor regeneration, by applying various treatments to the shoot prior to root removal The greatest regeneration occurred on roots from plants that had one-half the shoot and all axillary buds removed several weeks before root excision.

Seasonal differences in bud regeneration from roots have been recorded for a number of other species (Hudson, 1955; Emery, 1955; Raju *et al.*, 1964; Rosenthal *et al.*, 1968; Sterrett *et al.*, 1968; Eliasson, 1971a; Cuthbertson, 1972; Schier, 1973d). Several authors (Dore, 1953; Raju *et al.*, 1964; Rosenthal *et al.*, 1968; Cuthbertson, 1972) suggest that there may be a relationship between flowering and bud regeneration on roots since in the species studied the poorest regeneration occurred at a time when flowering occurred. Emery (1955) also claims that in *Chamaenerion* the best regeneration occurred after the flowering season but no data were given. The only experimental proof of a relationship between the flowering state and regeneration from roots is in a study of *Chondrilla* (Cuthbertson, 1972) in which root segments of vernalized plants in the early flower bud stage produced significantly fewer buds than root segments from non-vernalized vegetative plants.

The seasonal differences in regeneration may be due to differences in hormonal concentrations (Sterrett *et al.*, 1968; Schier, 1973d). In trembling aspen (Schier, 1973d), it was found that there was a peak in auxin content in the roots in June which was correlated with a flush of shoot growth and a marked decline in the regenerative capacity of roots. It is quite possible that changes in hormone levels might account for the other seasonal variations in bud formation from roots described, especially since the experiments of Marston and Village (1972) showed that removal of various parts of the shoot during the off season could increase the regenerative ability of roots. Certainly more experimental work needs to be carried out particularly in relation to the effect of the stage of shoot development on bud regeneration from roots.

B. Light

Light has little or no effect on the initiation of buds from root segments (Seeliger, 1956; Torrey, 1958; Bouriquet, 1966). However, in *Bryophyllum proliferum*, it appears to be essential for bud initiation on root pieces if the nutrient conditions are poor (Ossenbeck, 1927). Bud development, subsequent to initiation, is, however, markedly affected by light in that dark-grown segments have shoots which fail to develop normal leaves (Seeliger, 1956; Torrey, 1958; Bouriquet, 1966). In *Convolvulus*, light of different wavelengths affects the development of buds on isolated cultured roots (Bonnett, 1972). Cultures grown in continuous darkness have buds which fail to elongate shoots with normal expanded leaves. Segments grown in darkness but with short exposures to red light have elongated buds which appear similar to etiolated seedling shoots, an effect not usually attributed to red light. Exposure of root segments to short periods of far red light, either alone or after exposure to red light, inhibits bud elongation to the level of those growing in darkness. Further experiments showed that the bud primordia, although developing within the first two weeks in culture, are unable to respond to the red light stimulus during this period of growth. Only older primordia responded to red light. Bonnett (1972) suggests that root buds initiated on plants in the field will not elongate unless the roots grow close enough to the surface of the ground to perceive some light. Therefore, on any one *Convolvulus* plant there will be a number of buds potentially capable of elongating if light conditions are favorable.

Daylength has also been shown to influence bud development from roots. In creeping-rooted alfalfa (Carlson, 1965; Carlson *et al.*, 1964a, b) and in *Coronilla* and *Rumex* (Carlson, 1965), short days were found to be more conducive to root bud initiation than long days. It was suggested (Carlson, 1965) that the inhibitory effect of long days on bud initiation in these species is related to a strong apical dominance exerted by the rapidly developing shoot apex. Under short days shoot growth is reduced and therefore there is less influence of this part of the plant on the roots. In *Cichorium*, whether or not vegetative buds or floral buds form from cultured root pieces depends in part on the light conditions under which the cultures are maintained (Paulet and Nitsch, 1964a, b). The most critical period of exposure of the regenerating root segments to long days is between the 7th and 21st day, the time at which bud primordia are being formed within the callus proliferations.

C. Temperature

A number of workers have noted optimum temperatures for regeneration from root fragments, the optimum depending on the species being investigated. Maini and Horton (1966) found that in aspen roots bud regeneration was inhibited at either very low or high temperatures. Zasada and Schier (1973) confirmed that bud regeneration in aspen was inhibited at low temperatures and suggested that this might explain the lack of aspen stands on cooler sites in Alaska. In a number of species warm temperatures were found to be more conducive to bud formation than cool temperatures (Carlson *et al.*, 1964a; Sterrett *et al.*, 1968; Kefford and Caso, 1972) while in horseradish the initiation of buds was found to be rather temperature insensitive but the subsequent development was more temperature dependent (Lindner, 1940; Williams *et al.*, 1957). In *Cirsium arvense*, alternating temperatures of either cold–warm or warm–hot reduced the number of fragments regenerating; however, holding roots at 0°C for 1–5 weeks and then placing them in warm temperatures did not result in bud inhibition (Hamdoun, 1972).

In most of these studies it is not clear what effect temperature has on bud initiation since the results are always based on the number of emergent buds visible after treatment and therefore the effect in each case could simply be on bud growth.

D. Depth of Planting

When species regenerate from roots in the field it is normally the roots closest to the surface of the soil that are responsible for bud initiation. However, buds have been found on roots at considerable depths (Coupland and Alex, 1955; Coupland *et al.*, 1955). A consideration of the ability of root fragments to regenerate at various depths in the soil is of considerable interest from the standpoint of cultivation practices to eradicate weedy species known to propagate by root buds. In some species, depth of planting of root segments may have little effect on the regenerative capacity of the segment (Häkansson, 1969; Cuthbertson, 1972) or it may have a significant effect under certain growth conditions (Heydecker and Marston, 1968). In some cases the depth of planting of root fragments affects the number of emerging shoots (Häkansson and Wallgren, 1972; Hamdoun, 1972) but it is not clear from the experiments whether the number of buds initiated is influenced since the authors did not examine the roots for early stages of bud development. In leafy spurge, fragments derived from roots at different depths were tested for their regenerative capacity and it was found that all fragments, regardless of the depth of origin, retained the capacity for bud regeneration (Raju *et al.*, 1964). Cultivation practices which disturb the root system therefore often increase the amount of regeneration from the plant.

E. Segment Length and Diameter

Various effects of length and diameter on the regenerative capacity of root segments of a variety of species have been noted. In general, for each species there is a minimum length below which regeneration of buds is limited or does not occur (Prentiss, 1888; Lindner, 1940; Warmke and Warmke, 1950; Bonnett and Torrey, 1965; Heydecker and Marston, 1968). In some cases very short segments can be induced to regenerate by the application of hormones (Bonnett and Torrey,

1965). A positive correlation has been shown either between length of root segments and number of regenerated buds (Bonnett and Torrey, 1965) or root segment diameter and number of buds (Way, 1954). In some species with fleshy roots, the critical size of the root fragment capable of regenerating buds is dependent more on the tissues included in the fragment than on the size of the fragment itself. For example, in horseradish roots, tissue plugs must include a root scar and its associated lateral root trace in order for regeneration to occur (Dore and Williams, 1956).

VII. Conclusions

The production of root buds must be recognized as an important and integral part of the development of many plant species and more studies should be approached keeping the interrelationships within the entire plant in mind.

A consideration of the previous anatomical studies on root bud initiation and development indicates several areas requiring further study. For example, in those species in which root and bud primordia originate in the same tissue region it has been suggested that the early primordia are undetermined and may develop into either organ depending on the conditions present at the time or shortly after their initiation. In horseradish root segments the anatomical similarity in early stages of root and bud primordia and the modification of the ratio of buds to roots without a change in the total number of organs regenerated after experimental treatments was interpreted as evidence for the existence of undetermined primordia in early stages of development (Dore, 1955). In contrast, the identical appearance of early root and bud primordia in *Convolvulus* roots was not considered as conclusive evidence that undetermined primordia exist (Bonnett and Torrey, 1966). Furthermore, in a study of the chemical control of organ formation in isolated roots of the same species (Bonnett and Torrey, 1965) there was no evidence that potential root primordia could be induced to develop into shoots. In an earlier paper on endogenous organs in the root of *Euphorbia esula* (Bakshi and Coupland, 1959) it was claimed that both lateral root and bud primordia remain undetermined until they emerge from the epidermis at which time the environment influences the direction in which the primordia differentiate. In a later paper on the same species (Myers *et al.*, 1964), buds with leaf primordia could be observed when the bud was still within the cortex of the parent root. Additional species with bud and lateral root primordia arising from the same tissue region need to be analysed to gain a better understanding of the characteristics of early primordia in such situations.

Since most root buds are initiated endogenously within the parent root, tissues of the latter are greatly influenced by the inception and outgrowth of the bud. When buds are initiated within the pericycle, the involvement of the endodermis and other adjacent tissues is in question. In some species there is an indication that the endodermis may be involved in forming certain peripheral layers of the bud (Beijerinck, 1887; Bonnett and Torrey, 1966) but in other species the developing bud apparently destroys the endodermis (Van Tieghem and Duliot, 1888). Associated with the outgrowth of endogenous buds regardless of their site of formation, is the question of the mechanism by which these organs penetrate the adjacent tissues to reach the surface of the root. Some authors claim that a digestive process occurs in front of the advancing primordium (Van Tieghem and Duliot, 1888; Dore, 1955) while others (Bonnett and Torrey, 1966) find no evidence for this view. The majority of investigators however have not considered this aspect of bud development and only through ultrastructural and histochemical studies similar to those concerning the outgrowth of lateral root primordia (see article by M. McCully, pp. 105–124, in this volume) will this problem be resolved.

Bud initiation in relation to lateral roots or root traces should be studied using a variety of techniques. The suggestion in the earlier literature (see Priestley and Swingle, 1929) that substances emanating from the root stele may be stimulatory to bud initiation needs to be critically evaluated particularly in relation to the influence the developing lateral root primordium has on the endodermis which is generally thought to be a barrier to the movement of substances from the stele. Once the anatomical relationships between the lateral root and the surrounding parent root tissues are understood for species with root buds initiated in the proximity of the lateral root then more precise experiments may indicate what influences substances transported in the stele have on root bud initiation.

The species of ferns in which buds are initiated either by a transformation of the root apex (Rostowzew, 1890; Bierhorst, 1971) or from recent derivatives of the root apical cell (Rostowzew, 1891; Peterson, 1970) provide a unique opportunity to study factors determining the expression of shoot or root characteristics. To date, very few experimental studies have involved species showing this type of bud initiation (Peterson, 1968, 1969b); further studies, particularly of cultured roots, should increase our understanding of the potentialities of root meristems under controlled conditions.

Therefore, although a considerable amount of anatomical information on bud initiation and development exists, many aspects of these processes deserve more critical study. With recent advances in techniques of tissue preparation for microscopy new approaches to problems posed in the early literature are possible.

Of the various factors affecting the initiation and development of root buds, auxins and cytokinins have received the most attention. Generally, auxins are inhibitory to bud initiation but with some exceptions (Torrey, 1958). Cytokinins, on the other hand, are stimulatory to bud initiation in a number of species and in one study (Vardjan and Nitsch, 1961) a high level of endogenous cytokinins preceded bud initiation on root segments of chicory. However, in this and many other cases in which cytokinins have stimulated bud initiation on root segments, their initiation has been preceded by the development of a callus proliferation in which the bud meristems are initiated. It is difficult, therefore, to be certain that cytokinins influence root bud initiation directly or only through the induction of callus tissue in which buds form. The role played by other growth regulators such as gibberellic acid, ethylene and abscisic acid and by nutrient conditions is even less clear. It is suggested that before an understanding of the factors controlling root bud initiation can be attained the ontogeny of bud initiation and development of each species being used in experimental studies must be understood.

There remain many challenging problems concerning what must be considered a normal developmental feature of a wide variety of plant species—the ability to initiate root buds which subsequently develop into new shoots. Since shoots developing on root segments of the same plant are genetically uniform, clonal material for critical physiological experiments can be obtained (see Budd, 1973).

Acknowledgements

Financial support was provided for part of this work by the National Research Council of Canada. I thank Mrs. Laura Dobrindt for excellent technical assistance, Dr. Carol Peterson for help with translations and for her comments on the manuscript, and Mrs. May Bangham for typing the manuscript.

References

Bakshi, T. S. and Coupland, R. T. (1959). An anatomical study of the subterranean organs of *Euphorbia esula* in relation to its control. *Can. J. Bot.* **37**; 613–620.

Bakshi, T. S. and Coupland, R. T. (1960). Vegetative propagation in *Linaria vulgaris*. *Can. J. Bot.* **38**; 243–249.

Ballade, P. (1968). Caulogénèse apicale sur les jeunes racines axillaires du cresson (*Nasturtium officinale* R.Br.). *Soc. Bot. Fr. Mem.* **115**; 250–258.

Ballade, P. (1970). Precisions nouvelles sur la caulogénèse apicale des racines axillaires du cresson (*Nasturtium officinale* R.Br.). *Planta* **92**; 138–145.

Ballade, P. (1971). Etude expérimentale de l'organogenèse axillaire chez *Nasturtium officinale* R.Br. Comportement de noeuds isolés cultivés in vitro. *C. R. hebd. Séanc. Acad. Sci., Paris* **273**; 2079–2082.

Ballade, P. (1972). Etude expérimentale de l'organogenèse axillaire chez *Nasturtium officinale* R.Br.: Comportement d'aisselles isolées ou de parties d'aisselles cultivées in vitro. *C.R. hebd. Séanc. Acad. Sci., Paris* **274**; (9); 1282–1285.

Beijerinck, M. W. (1887). Beobachtungen und Betrachtungen Über Wurzelknospen und Nebenwurzeln. *Natuurk. Verh. der Kon. Akad. Wet. C., Amsterdam* **25**; 1–150.

Bierhorst, D. W. (1971). "Morphology of Vascular Plants." pp. 560. The Macmillan Company, New York.

Bonnett, H. T. Jr. (1972). Phytochrome regulation of endogenous bud development in root cultures of *Convolvulus arvensis*. *Planta* **1–6**; (4) 325–330.

Bonnett, H. T. Jr. and Torrey, J. G. (1965). Chemical control of organ formation in root segments of *Convolvulus* cultured in vitro. *Pl. Physiol., Lancaster* **40**; 1228–1236.

BONNETT, H. T. JR. and TORREY, J. G. (1966). Comparative anatomy of endogenous bud and lateral root formation in *Convolvulus arvensis* roots cultured in vitro. *Am. J. Bot.* **53**; 496–507.

BOURIQUET, R. (1966). Action de la lumière sur le dévelopment des tissus de feuilles d'Endive cultivées *in vitro*. *Photochem. Photobiol.* **5**; 391–395.

BOURIQUET, R. (1972). Action de l'éthylène sur le bourgeonnement et la floraison in vitro de fragments de racines d'endive. *C.R. hebd. Seanc. Acad. Sci. Paris* **275** (1); 33–34.

BRAND, R. and VENVERLOO, C. J. (1973). The formation of adventitious organs. II. The origin of buds formed on young adventitious roots of *Populus nigra* L. "Italica". *Acta bot. neerl.* **22** (4); 399–406.

BROWN, A. B. (1935). Cambial activity, root habit and sucker shoot development in two species of poplar. *New Phytol.* **34**; 163–179.

BROWN, C. L. and KORMANIK, P. P. (1967). Suppressed buds on lateral roots of *Liquidambar styraciflua*. *Bot. Gaz.* **128** (3–4); 208–211.

BRUNDIN, I. A. Z. (1895). Über Wurzelsprosse bei *Listera cordata* L. *Bihang. K. Su. Vet.-Akad. Halgr.* **21**; 3–11.

BUDD, T. W. (1973). An excellent source of vegetative buds for use in plant hormone studies on apical dominance. *Pl. Physiol., Lancaster* **52**; 82–83.

BUI-DANG-HA, D. and NITSCH, J. P. (1970). Isolation of zeatin riboside from the chicory root. *Planta* **95**; 119–126.

BUTCHER, D. N. and STREET, H. E. (1964). Excised root culture. *Bot. Rev.* **30** (4); 513–586.

CARLSON, G. E. (1965). Photoperiodic control of adventitious stem initiation on roots. *Crop. Sci.* **5**; 248–250.

CARLSON, G. E., SPRAGUE, V. G. and WASHKO, J. W. (1964a). Effects of temperature, day-length and defoliation on the creeping rooted habit of alfalfa. *Crop. Sci.* **4**; 284–286.

CARLSON, G. E., SPRAGUE, V. G. and WASHKO, J. B. (1964b). Effects of daylength and gibberellic acid on the creeping-rooted habit of alfalfa. *Crop. Sci.* **4**; 397–399.

CARLSON, M. C. (1938). Origin and development of shoots from the tips of roots of *Pogonia ophioglossoides*. *Bot. Gaz.* **100**; 215–225.

CASO, O. H. and KEFFORD, N. P. (1973). Control of regeneration in roots of the deep rooted weed *Chondrilla juncea* L. *Weed Res.* **13**; 148–157.

CHAMPAGNAT, M. (1971). Recherches sur la multiplication végétative de *Neottia nidus-avis* Rich. *Ann. Sci. nat. Bot. Ser.* **12**; 209–248.

CHAMPAGNAT, M., BALLADE, P. and PORTE, J. (1967). Introduction à l'étude expérimentale de l'orangogenèse axillaire du Cresson (*Nasturtium officinale* R.Br.). *Mem. Soc. Bot. Fr.* **114**; 109–122.

CHAMPAGNAT, M. and CHAMPAGNAT, P. (1965). Vers une étude expérimentale des problèmes de morphogenèse posés par *Neottia nidus-avis* Rich. *Ann. Fac. Sci. Clermont* **26**; 115–124.

CHARLTON, W. A. (1965). Bud initiation in excised roots of *Linaria vulgaris*. *Nature, Lond.* **207**; 781–782.

CHARLTON, W. A. (1966). The root system of *Linaria vulgaris* Mill. I. Morphology and anatomy. *Can. J. Bot.* **44**; 1111–1116.

COUPLAND, R. T. and ALEX, J. F. (1955). Distribution of vegetative buds on the underground parts of leafy spurge (*Euphorbia esula* L.). *Can. J. agric. Sci.* **35**; 76–82.

COUPLAND, R. T., SELLECK, G. W. and ALEX, J. F. (1955). The reproductive capacity of vegetative buds on the underground parts of leafy spurge (*Euphorbia esula* L.). *Can. J. agric. Sci.* **35**; 477–484.

CUTHBERTSON, E. G. (1972). *Chondrilla juncea* in Australia. 4. Root morphology and re-generation from root fragments. *Aust. J. exp. Agric. Anim. Husb.* **12** (58); 528–534.

DANCKWARDT-LILLIESTRÖM, C. (1957). Kinetin induced shoot formation from isolated roots of *Isatis tinctoria*. *Physiologia Pl.* **10**; 794–797.

DORE, J. (1953). Seasonal variation in the regeneration of root cuttings. *Nature, Lond.* **172**; 1189.

DORE, J. (1955)l Studies in the regeneration of horseradish. I. A re-examination of the morphology and anatomy of regeneration. *Ann. Bot.* **19**; 127–137.

DORE, J. (1965). "Physiology of Regeneration in Cormophytes. Handbuch der Pflanzenphysiologie," Vol XV 2, pp. 1–91. Springer-Verlag, Berlin.

DORE, J. and WILLIAMS, W. T. (1956). Studies in the regeneration of horseradish. II. Correlation phenomena. *Ann. Bot.* **20**; 231–249.

ELIASSON, L. (1961). The influence of growth substances on the formation of shoots from aspen roots. *Physiologia Pl.* **14**; 150–156.

ELIASSON, L. (1971a). Growth regulators in *Populus tremula.* III. Variation of auxin and inhibitor level in roots in relation to root sucker formation. *Physiologia Pl.* **25**; 118–121.

ELIASSON, L. (1971b). Growth regulators in *Populus tremula.* IV. Apical dominance and suckering in young plants. *Physiologia Pl.* **25**; 263–267.

EMERY, A. E. H. (1955). The formation of buds on the roots of *Chamaenerion angustifolium* (L.) Scop. *Phytomophology* **5**; 139–145.

FARMER, R. E. (1962). Aspen root sucker formation and apical dominance. *Forest Sci.* **8**; 403–410.

FEDER, N. and O'BRIEN, T. P. (1968). Plant microtechnique: some principles and new methods. *Am. J. Bot.* **55**; 123–144.

FOWELLS, H. A. (1965). Silvics of forest trees of the United States. *U.S.D.A. Agric. Handb.* **271**; 1–762.

GOEBEL, K. (1878). Ueber Wurzelsprosse von *Anthurium longifolium. Bot. Ztg.* **36**; 645–648.

GOEBEL, K. (1902). Ueber Regeneration in Pflanzenreich. *Biol. Zbl.* **22**; 385–397, 417–438, 481–505.

GOEBEL, K. (1905). "Organography of Plants". Part II. Clarendon Press, Oxford.

GUNNING, B. E. S. and PATE, J. S. (1969). "Transfer cells". Plant cells with wall ingrowths, specialized in relation to short distance transport of solutes—their occurrence, structure and development. *Protoplasma* **68**; 107–133.

HÅKANSSON, S. (1969). Experiments with *Sonchus arvensis* L. I. Development and growth, and response to burial and defoliation in different developmental stages. *Lantkr. Annlr.* **35**; 989–1030.

HÅKANSSON, S. and WALLGREN, B. (1972). Experiments with *Sonchus arvensis* L. III. The development from reproductive roots cut into different lengths and planted at different depths, with and without competition from barley. *Swedish J. agric. Res.* **2**; 15–26.

HAMDOUN, A. M. (1972). Regenerative capacity of root fragments of *Cirsium arvense* (L). *Scop. Weed Res.* **12** (2); 128–136.

HEYDECKER, W. and MARSTON, MARGARET E. (1968). Quantitative studies on the regeneration of raspberries from root cuttings. *Hort. Res.* **8**; 142–146.

HOLM, T. (1925). On the development of buds upon roots and leaves. *Ann. Bot.* **39**; 867–881.

HUDSON, J. P. (1953). Factors affecting the regeneration of root-cuttings. *Nature, Lond.* **172**; 411–412.

HUDSON, J. P. (1954). Propagation of plants by root cuttings. I. Regeneration of raspberry root cuttings. *J. hort. Sci.* **29**; 27–43.

HUDSON, J. P. (1955). Propagation of plants by root cuttings. II. Seasonal fluctuation of capacity to regenerate from roots. *J. hort. Sci.* **30**; 242–251.

ILYAS, M. (1973). Shoot initiation on cauliflower roots caused by morphactin. *Experientia* **29**; 130.

IRMISCH, T. (1857). Über die Keimung und die Erneuerungsweise von *Convolvulus sepium* und *Convolvulus arvensis* so wie uber hypocotylische Adventivknospen bei krautartigen phanerogamen Pflanzen. *Bot. Ztg.* **15**; 433–443, 449–472, 481–484, 489–497.

ISBELL, C. L. (1945). Propagating cabbage by root cuttings. *Proc. Am. Soc. hort. Sci.* **45**; 341–344.

KEFFORD, N. P. and CASO, O. H. (1972). Organ regeneration on excised roots of *Chondrilla juncea* and its chemical regulation. *Aust. J. biol. Sci.* **25**; 691–706.

KORMANIK, P. P. and BROWN, C. L. (1967). Root buds and the development of root suckers in sweetgum. *Forest Sci.* **13**; 338–345.

LINDNER, R. C. (1938). Effects of indoleacetic and naphthylacetic acids on development of buds and roots in horseradish. *Bot. Gaz.* **100**; 500–527.

LINDNER, R. C. (1940). Factors affecting regeneration of the horse-radish root. *Pl. Physiol., Lancaster* **15**; 161–181.

MAINI, J. S. and HORTON, K. W. (1966). Vegetative propagation of *Populus* spp. I. Influence of temperature on the formation and initial growth of aspen suckers. *Can. J. Bot.* **44**; 1183–1189.

MARCAVILLACA, M. C., CASO, O. H. and MONTALDI, E. R. (1971). Regeneración en segmentos de raices de *Medicago sativa* L. cv. Rambler. *Rev. Invest. Agropecu. Ser. 2, Biol. Prod. Veg.* **8**; (5); 179–199.

MARGARA, J. (1970). Sur la réversibilité de l'aptitude a l'induction florale photopériodique des bourgeons de *Cichorium intybus* L. néoformés in vitro. *C. R. hedb. Séanc. Acad. Sci., Paris* **270**; 1104–1107.

MARGARA, J. (1973a). Influence de l'isolement des bourgeons néoformés de *Cichorium intybus* L. sur leur évolution, végétative ou inflorescentielle. *C.R. hedb. Séanc. Acad. Sci., Paris* **277**; 497–500.

MARGARA, J. (1973b). Interaction de facteurs nutrionnels sur l'initiation florale des bourgeons; de *Cichorium intybus* L. néoformés in vitro. *C.R. hebd. Seanc. Acad. Sci., Paris* **277**; 2673–2676.

MARGARA, J. and BOUNIOLS, A. (1967). Comparaison in vitro de l'influence du milieu liquide ou gélosé, sur l'initiation florale, chez *Cichorium intybus* L. *C.R. hebd. Séanc. Acad. Sci., Paris* **264**; 1166–1168.

MARSTON, M. E. and VILLAGE, P. J. (1972). Regeneration of raspberries from root cuttings in response to physical treatment of the shoots. *Hort. Res.* **12** (3); 177–182.

MCINTYRE, G. (1972). Developmental studies on *Euphorbia esula*. The influence of the nitrogen supply on the correlative inhibition of root bud activity. *Can. J. Bot.* **50**; 949–956.

MCINTYRE, G. I. and RAJU, M. V. S. (1967). Developmental studies on *Euphorbia esula* L. Some effects of the nitrogen supply on the growth and development of the seedling. *Can. J. Bot.* **45**; 975–984.

MCVEIGH, I. (1937). Vegetative reproduction of the fern sporophyte. *Bot. Rev.* **3**; 457–497.

MONTALDI, E. R. (1972). Kinetin induction of bud differentiation on roots of entire plants. *Z. Pflphysiol.* **67**; 43–44.

MURRAY, B. E. (1957). The ontogeny of adventitious stems on roots of creeping-rooted alfalfa. *Can. J. Bot.* **35**; 463–475.

MYERS, G. A., BEASLEY, C. A. and DERSCHEID, L. A. (1964). Anatomical studies of *Euphorbia esula* L. *Weeds* **12**; 291–295.

NAYLOR, E. (1941). The proliferation of dandelions from roots. *Bull. Torrey bot. Club* **68**; 351–358.

NITSCH, J. P. and NITSCH, C. (1964). Néoformation de boutons floraux sur cultures *in vitro* de feuilles et de racines de *Cichorium intybus* L.—Existence d'un état vernalisé en l'absence de bourgeons. *Bull. Soc. bot. Fr.* **111**; 299–304.

NORTH, C. (1953). Experiments with root cuttings of Brussels sprout. *Ann. appl. Biol.* **40**; 250–261.

NORTON, J. P. and BOLL. W. G. (1954). Callus and shoot formation from tomato roots *in vitro*. *Science, N.Y.* **119**; 220–221.

OSSENBECK, C. (1927). Kritische und experimentelle Untersuchungen an *Bryophyllum*. *Flora, Jena* **122**; 342–387.

PAULET, P. and NITSCH, J. P. (1964a). Néoformation de fleurs in vitro sur des cultures de tissus de racines de *Cichorium intybus* L. *C.R.hebd. Seanc. Acad. Sci., Paris* **258**; 5952–5955.

PAULET, P. and NITSCH, J. P. (1964b). La néoformation de fleurs sur cultures in vitro de racines de *Cichorium intybus* L.: étude physiologique. *Annls. Physiol. vég., Paris* **6**; (4); 333–345.

PETERSON, R. L. (1968). "Developmental and morphogenetic studies on *Ophioglossum petiolatum* Hook". Ph.D. Thesis, University of California, Davis.

PETERSON, R. L. (1969a). Bud formation on root segments of *Ophioglossum petiolatum*: effect of application site of cytokinins and auxin. *Can. J. Bot.* **47** (8); 1285–1287.

PETERSON, R. L. (1969b). Observations on bud formation and proliferations on cytokinin-treated root segments of *Ophioglossum petiolatum*. *Can. J. Bot.* **47**; 1579–1583.

PETERSON, R. L. (1970). Bud formation at the root apex of *Ophioglossum petiolatum*. *Phytomorphology* **20**; 183–190.

PETERSON, R. L. and THOMAS, A. G. (1971). Buds on the roots of *Hieracium florentinum* (hawkweed). *Can. J. Bot.* **49**; 53–54.

PLANT, W. (1940). The role of growth substances in the regeneration of root cuttings. *Ann. Bot.* **4**; 607–615.

POIRAULT, G. (1893). Recherches anatomiques sur les cryptogames vasculaires. *Annls. Sci. nat. Bot. Ser.* 7 **18**; 113–256.

PRENTIS, A. M. (1888). On root propagation of Canada thistle. *Bull. Cornell Univ. agric. Exp. Stn.* **15**; 190–192.

PRIESTLEY, J. H. and SWINGLE, C. F. (1929). Vegetative propagation from the standpoint of plant anatomy. *USDA Tech. Bull.* **151**; 1–98.

RAJU, M. V. S., COUPLAND, R. T. and STEEVES, T. A. (1966). On the occurrence of root buds on perennial plants in Saskatchewan. *Can. J. Bot.* **44**; 33–37.

RAJU, M. V. S., STEEVES, T. A. and COUPLAND, R. T. (1964). On the regeneration of root fragments of leafy spurge (*Euphorbia esula* Ll). *Weed Res.* **4**; 2–11.

RAUH, W. (1937). Die Bildung von Hypokotyl-und Wurzelsprossen und ihre Bedeutung für die Wuchsformen der Pflanzen. *Nova Acta Leopoldina* **4**; 395–553.

ROSENTHAL, R. N., SCHIRMAN, R. and ROBOCKER, W. C. (1968). Root development of rush skeletonweed. *Weed Sci.* **16**; 213–217.

ROSTOWZEW, S. (1890). Beiträge zur kenntniss der Gefasskryptogamen. I. Umbildung von Wurzeln in Sprosse. *Flora, Jena* **48**; 155–168.

ROSTOWZEW, S. (1891). Recherches sur l'*Ophioglossum vulgatum* L. *Overs. K. danske Vidensk. Selsk. Forh.* **2**; 54–83.

SCHIER, G. A. (1973a). Origin and development of Aspen root suckers. *Can. J. For. Res.* **3**; 45–53.

SCHIER, G. A. (1973b). Effects of gibberellic acid and an inhibitor of gibberellin action on suckering from Aspen root cuttings. *Can. J. For. Res.* **3**; 39–44.

SCHIER, G. A. (1973c). Effect of abscisic acid on sucker development and callus formation on excised roots of *Populus tremuloides*. *Physiologia Pl.* **28**; 143–145.

SCHIER, G. A. (1973d). Seasonal variation in sucker production from excised roots of *Populus tremuloides* and the role of endogenous auxin. *Can. J. For. Res.* **3**(3); 459–461.

SCHIER, G. A. and ZASADA, J. C. (1973). Role of carbohydrate reserves in the development of root suckers in *Populus tremuloides*. *Can. J. For. Res.* **3**; 243–250.

SEELIGER, I. (1956). Über die Kultur isolierter Wurzeln de Robinie (*Robinia pseudoacacia* L.). *Flora, Jena* **144**; 47–83.

SALEMA, R. and BRANDÃO, I. (1973). The use of Pipe's buffer in the fixation of plant cells for electron microscopy. *J. Submicr. Cytol.* **5**; 79–96.

SMITH, D. (1950). The occurrence of adventitious shoots on severed alfalfa roots. *Agron. J.* **42**; 398–401.

STEEVES, T. A., COUPLAND, R. T. and RAJU, M. V. S. (1966). Vegetative propagation in relation to the aggressiveness of species. *In* "The Evolution of Canada's Flora" (Roy L. Taylor and R. A. Ludwig, eds.) pp. 121–137. University of Toronto Press, Toronto.

STERRETT, J. P., CHAPPELL, W. E. and SHEAR, C. M. (1968). Temperature and annual growth cycle effects on root suckering in black locust. *Weed Sci.* **16**; 250–251.

STOUGHTON, R. H. and PLANT, W. (1938). Regeneration of root cuttings as influenced by plant hormones. *Nature, Lond.* **142**; 293–294.

STOUTEMEYER, V. T. (1937). Regeneration in various types of apple wood. *Res. Bull. Iowa agric. Exp. Stn.* **220**; 308–352.

TEW, R. K. (1970). Root carbohydrate reserves in vegetative reproduction of aspen. *Forest Sci.* **16**; 318–320.

THOMAS, E. and STREET, H. E. (1972). Factors influencing morphogenesis in excised roots and suspension cultures of *Atropa belladonna*. *Ann. Bot.* **36**; 239–247.

TORREY, J. G. (1958). Endogenous bud and root formation by isolated roots of *Convolvulus* grown in vitro. *Pl. Physiol., Lancaster* **33**; 258–263.

TRÉCUL, A. (1847). Récherches sur l'origine des bourgeons adventifs. *Ann. Sci. nat. Bot. Ser.* 3, **8**; 268–295.

VAN TIEGHEM, P. (1887). Récherches sur la disposition des radicelles et des bourgeons. *Ann. Sci. nat. Bot. Ser.* 7, **5**; 130–151.

VAN TIEGHEM, P. and DULIOT, H. (1888). Récherches comparatives sut l'origine des members endogènes. *Ann. Sci. nat. Bot. Ser.* 7, **8**; 1–660.

VARDJAN, M. and NITSCH, J. P. (1961). La régénération chez *Cichorum endivia* L.: étude des auxines et des ((kinines)) endogènes. *Bull. Soc. bot., Fr.* **108**; 363–374.

WARDLAW, C. W. (1953a). Endogenous buds in *Ophioglossum vulgatum* L. *Nature, Lond.* **171**; 88.

WARDLAW, C. W. (1953b). Experimental and analytical studies of Pteriodophytes. XXIII. The induction of buds in *Ophioglossum vulgatum* L. *Ann. Bot.* **17**; 513–527.

WARDLAW, C. W. (1954). Experimental and analytical studies of Pteridophytes. XXVI. *Ophioglossum vulgatum*: Comparative morphogenesis in embryos and induced buds. *Ann. Bot.* **18**; 397–406.

WARMKE, H. E. and WARMKE, G. L. (1950). The role of auxin in the differentiation of root and shoot primordia from root cuttings of *Taraxacum* and *Cichorium*. *Am. J. Bot.* **37**; 272–280.

WAY, D. W. (1954). The relationship of diameter to regenerative organ differentiation in apple roots. *Proc. kon. ned. Akad. Wet C.* **57**; 601–605.

WENT. F. W. (1939). Some experiments on bud growth. *Am. J. Bot.* **26**; 109–117.

WHITE, P. K. (1943). "A Handbook of Plant Tissue Culture". Jacques Cattel Press, Lancaster. pp. 277.

WHITE, R. A. (1969). Vegetative reproduction in the ferns. II. Root buds in *Amphoradenium*. *Bull. Torrey bot. Club* **96**; 10–19.

WILKINSON, R. E. (1966). Adventitious shoots on saltcedar roots. *Bot. Gaz.* **127**; 103–104.

WILLIAMS, W. T., DORE, J. and PATTERSON, D. G. (1957). Studies in the regeneration of horseradish. III. External factors. *Ann. Bot.* **21**; 627–632.

WILSON, C. L. (1927). Adventitious roots and shoots in an introduced weed. *Bull. Torrey bot. Club* **54**; 35–38.

WITTROCK, V. B. (1884). Ueber Wurzelsprossen bei krautartigen Gewächsen, mit besonderer Rüchsicht auf ihre verschiedene biologische Bedeutung. *Bot. Zbl.* **17**; 227–232; 258–264.

ZASADA, J. C. and SCHIER, G. A. (1973). Aspen root suckering in Alaska. Effect of clone, collection date and temperature. *NW Sci.* **47**; 100–104.

ZENKTELER, M. A. (1971). Development of new plants from leaves and roots of *Atropa belladonna* L. in the *in vitro* culture. *Acta. Soc. Bot. Pol.* **60**; 305–315.

Chapter 8

Tree Rootlets and their Distribution

E. R. C. REYNOLDS

Department of Forestry, University of Oxford, England

I. Introduction

It is rarely appropriate to discuss the root system without any reference to the whole plant. This interrelationship within the plant has long been the subject of allometry and growth analysis; the commonest parameter studied has been the ratio between the weights of the root and the shoot. The underlying interest in the relationship is that in the specialization of the different parts, each becomes, in a sense, heterotrophic with respect to obtaining certain necessities from the environment. The comparative size and growth of the parts are, therefore, measures of the degree of dependence of the shoot on the absorption and export of water and mineral nutrients by the root, and of the root in turn on the import of photosynthates from the shoot. The hormonal mechanisms controlling the relative growth of the parts and the resulting efficiency of the whole plant are of equal interest. Here I concentrate on the efficiency.

We can express this efficiency in another way, using the terminology of plant adaptation to the environment. Quasi-teleological phraseology has been defended by Parkhurst and Loucks (1972) and, in particular, the ideas of strategy and tactics have been profitably employed by Harper and his associates (1967, etc.) in studying the establishment of weeds. We may say that roots have as their primary strategy that of obtaining water and mineral nutrients to enable the plant to function. Tactical adaptations have probably evolved such that the foraging roots drain the plant of photosynthates only within the limits of efficiency.

I have selected for discussion only the tactics of spatial distribution of roots, the longevity of rootlets, dormancy, and more cursorily, the tactic of the development of root hairs. For each in turn, I wish to erect hypotheses for the tactical advantages conferred and, using both some of my own data and that published by others, I will try to produce evidence that these tactics are in fact employed.

II. Site, Materials and Methods

My own work which I report is based exclusively on a plantation of Douglas fir [*Pseudotsuga menziesii* (Mirb.) Franco] in Bagley Wood, near Oxford, on a sandy facies of Pleistocene Plateau Gravel. The trees were planted on a south-facing slope in 1929, and have been thinned frequently since. In March 1974 the stocking was 389 stems ha^{-1}, the mean girth 99 cm, and height 22·7 m. A crown projection map is in Reynolds (1970). There is very little undergrowth. The soil is a slightly podsolized sandy loam with the water table deeper than 3 m. Rooting is sparse below 2·6 m.

Volumetric core sampling (details Reynolds, 1970) down to 1·3 m at random (or stratified random) positions was performed on untreated parts of the plantation, and in a watering experiment where small plots between trees received (a) natural precipitation, (b) no water, or irrigation at a rate of (c) half the potential evapotranspiration, or (d) one and a half times. Other soil samples were obtained as contiguous 7·5 cm cubes from the surface mineral soil in a line between two tree stems (see Fig. 1). The roots and other organic debris were sorted from the mineral soil in Cahoon and Morton's (1961) apparatus. The roots were sorted into size classes ($<$0·5 mm, 0·5–1 mm, 1–2 mm, and $>$2 mm diameter) and their lengths measured. Latterly the sorting and measuring were accomplished using Newman's grids (1966). Some attempt was made to assess the vitality of the roots as judged by the persistence of the xylem revealed by dissection.

On three occasions in 1973 roots in the upper soil were lifted, killed in 70% acetic alcohol, and examined morphologically and, on microscope sections, anatomically. Sections of root fragments from the volumetric soil samples were also examined microscopically.

Currently a root observation chamber inserted in the plantation is being used to observe extension rates, areas of activity, rootlet longevity, root hair distribution, and the dormancy of apices. The glass windows extend to a depth of about 2·1 m below the soil surface.

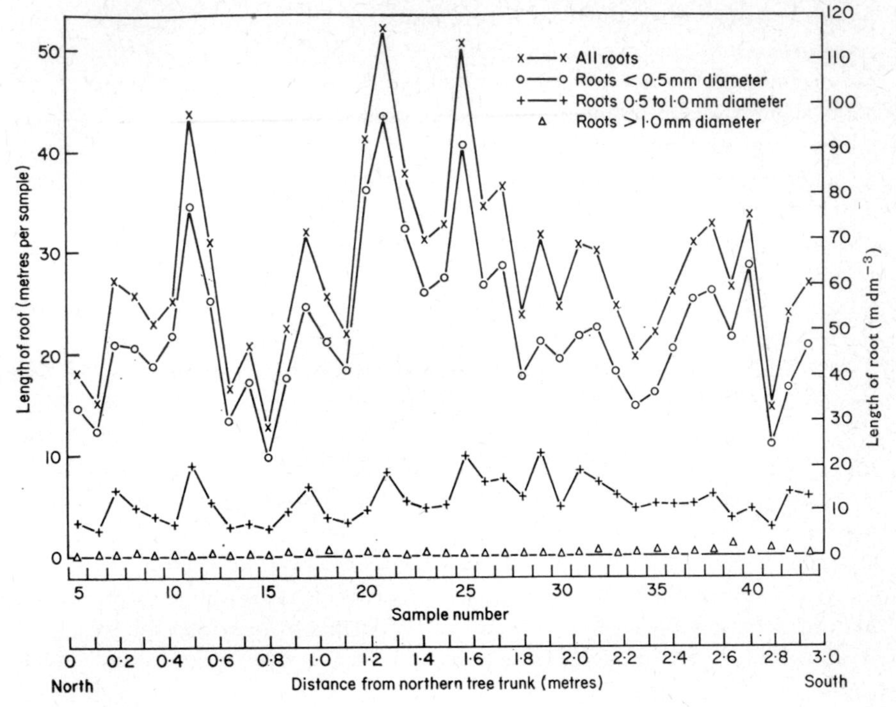

FIG. 1. Lengths of root by diameter classes in surface soil in contiguous 7·5 cm cubic samples along a line between two Douglas fir trees. Data from Bagley Wood, Oxford, November 1968.

III. Adaptive Tactics of Roots

A. Rootlet Death

There may be several good reasons for classifying roots according to their diameters. However, it is inadvisable to think that the classes are more than crudely related to function. To use tendentious names for the classes as "feeding" or "conducting" has frequently been castigated (e.g. Leshem, 1965), but some workers persist in this uncritical practice. Mycorrhizal roots are morphologically distinguishable, but otherwise there is a continuous spectrum of root size. In Douglas fir the finest roots were 0·15 mm in diameter. Except in a dry season, when many of the fine roots appeared to die, approximately 50% of the root length was less than 0·5 mm, and 45% between 0·5 and 1 mm (Reynolds, 1970). The peak of the root length/diameter distribution was 0·3 mm.

While it may not be possible to identify function with root diameter, I suggest that the thinner laterals are characterized by determinate growth and are ephemeral if not deciduous. There are several lines of deductive reasoning which would lead us to expect this tactic to be adopted. Available water, in the soil where roots are found, is a renewable resource. It may be depleted by absorption and only refurbished by precipitation after an interval of up to several months. The roots would not be achieving their strategic function while the soil was dry, yet they would consume photosynthates in maintaining their existence. Taking a reasonable estimate of their rate of respiration, 5 ml 0_2 hr^{-1} g dry $weight^{-1}$ (Grable, 1966), roots would consume energy reserves in one week equivalent to their own dry weight. Apart from tactics involving reduced metabolic rates, it would be advantageous to the tree to reject many of the lateral branches and retain only a skeleton of living roots from which new laterals could emerge after the soil has become rewetted. It may be productive to consider the analogy with deciduous leaves. The tactic is common and undoubtedly advantageous above ground and should not be unexpected in the root system. When Kozlowski's (1973) new book is available for lengthy perusal, we may be able to appreciate more fully the significance of the deciduous habit above and below ground.

Again, on rewetting a soil must be re-exploited. Since most absorption takes place immediately behind the apex (but, see p. 433 for a different view), and absorption is probably enhanced if not dependent upon, on-growing root tips, then replacement of lateral roots at least seems a necessity. Without discarding the majority of roots, absorption of water and nutrients from within the root mass would produce an impossibly intricate and long system of roots and lateral roots. Much of the system would be composed of a reasonably leak-proof contorted transport network. Useful contrasts can be drawn between an efficient shoot system fixing solar energy, and the ideal root system absorbing renewable soil moisture.

In the root observation chamber, there was some evidence for distinguishing determinate from indeterminate growth among laterals. The frequency distribution of their lengths had a maximum around 1·2 cm, but this was not associated with any obvious acropetal trend of shorter rootlets. Instead, one or two roots at intervals of perhaps 9 cm were materially longer. The distal lateral(s) were only seen to be actively extending where the tip of the parent root appeared inactive.

Morphological examination of Douglas fir roots reveals numerous scars, often approximately alternate, along the length of roots of apparently indeterminate growth. These correspond to short lateral roots which probably disappeared before excavation. As many as perhaps 1·1 cm^{-1} occurred on one root examined, in contrast to 0·8 cm^{-1} lateral roots which were still attached. Few if any of the attached laterals could be adjudged dead though this is difficult to assess objectively. Microscope sections through the scars failed to reveal any abscission layer; the lateral merely parts at the level of the outer xylem tracheids of the parent root. This absence of a distinct mechanism is implicit in Kozlowski's (1973) comments.

These observations accord with those of Fayle (1965) who observed in *Acer saccharum* that there is progressive reduction of the numbers of laterals borne by a root—from 3·3 cm^{-1} to 1·3 cm^{-1} after 5 yr and 0·2 cm^{-1} by 20 yr. This suggests dehiscence of laterals up to a considerable age and size.

It is to be expected that soil samples will contain some dead roots. Since it was difficult to distinguish objectively between dead and living roots, the figures I have reported elsewhere (Reynolds, 1970) should be received cautiously. At least they suggest that the numbers of dead rootlets extracted are probably considerable, in many samples equal in dry weight to the living fine roots. Microscope sections of roots from soil samples generally seemed to lack cell contents. The first tissues to disintegrate seemed to be the phloem and the pericycle, so that the xylem was loose in a cortical cylinder. Next the xylem disappeared to leave a persistent hollow cylinder of cortical tissues. These observations need to be repeated in greater detail and with more care to prevent changes after the moment of sampling.

The loss of weight by decay is likely to be considerable. Provisional figures from a mixed deciduous woodland in N.W. England (Hibberd and Sykes, personal communication) for roots up to 30 mm diameter, suggest dry weight losses of 20 to 37% in the first year, depending on species. The dead root biomass was 22 ± 5% of the live root biomass. In root observation chambers in the same woodland, 30 to 40% of roots less than 5 mm in diameter which were produced in the first year of observation, died in that year (Hibberd and Sykes, 1973).

The chamber in Bagley Wood in its first year of operation indicates that for Douglas fir, whereas many of the lateral roots are dark in colour and are not growing, it is not possible to say if they are dead, and none have become detached.

B. Dynamic Distribution

Deductive reasoning would lead us to suspect that it would be more efficient for plant roots to exploit local reserves intensively and then to pass into a quiescent period while another portion of the soil supplies the required water and nutrients. A high density of absorbing tips would ensure rapid depletion through short diffusion paths. Long-lived roots more widely spaced would probably be a greater drain on carbohydrate reserves.

I have argued previously (Reynolds, 1970), on the basis of variability in the amounts of root in soil samples from the Douglas fir plantation and from differences in the proportion of roots which were presumed dead, that there are cycles of root extension and death in discreet "cells" of the soil.

If I am correct in believing that root systems function in this cellular fashion, then the variability of root amounts should be patterned. The hypothesis can be tested by finding if such a pattern exists, and estimating the size of the postulated "cells". However, it would be important to prove that the pattern did not simply reflect static variability of the soil environment. Mathematical analyses devised by ecologists enable us to make these tests. The methods require data from contiguous samples, each of such a size as to be relatively small compared with the probable size of the pattern. In the present investigation, the samples adjoined to form transects or vertical files, and the data from the samples were soil or root parameters. The analysis proceeds by combining the data from adjacent samples into blocks of samples and calculating the variance attributable to each block size. If pattern exists, the relation between block size and variance shows one or more maxima. The size of a block which corresponds with a maximum variance, physically represents the average width of a pattern which, if the pattern is composed of isodiametric elements, is the mean length of chords across the elements. There is a variety of analyses available, and instead of the insensitive method used previously (Reynolds, 1974), I have adopted Hill's (1973) "2-term local variance" method of analysis. This method diminishes the effect of start point in the linear series of samples and allows variance to be apportioned to block sizes of every whole number of samples, up to half the total number in a transect, instead of only factorial numbers of samples which are required in the older methods based on analysis of variance. The use of Hill's pattern intensity was inappropriate with data transformations producing mean values of zero.

As a measure of the environmental variability among the small cubical samples along a transect between two trees, soil pH was measured, although the range was only from 3·6 to 4·1 units. When these data were subjected to Hill's pattern analysis, there was no indication whatever of any pattern in the variation. The legnths of roots (without discriminating between dead and living ones) of different size classes extracted from the samples are shown in Fig. 1, and were similarly analysed. The class less than 0·5 mm diameter (nearly 80% of the total root length) had a pattern corresponding to 84 cm. Roots 0·5 to 1·0 mm had a still larger pattern size, 107 cm across, and accounted for about 19% of the total root length. The larger roots comprising just over 1% of the total seemed to have a pattern of around 60 cm.

Trends interfere with pattern analysis and, although they are, of course, large elements of pattern, they are commonly removed before analysis. It is evident from the data in Fig. 1 that, especially in the quantities of fine roots, there are trends along the transect which are probably related to increasing rain throughfall away from the trunks (Reynolds, 1970). Radial transects would be expected to exaggerate annular patterns around each tree. However, for the sum of all root classes, the variability (standard error 5·1% mean) was reduced only by 25% through fitting a pair of linear regressions of opposite sign, by 24% with a quadratic equation for each half of the transect, and by 67% upon regarding the two halves as mirror images and calculating differences from the mean of equivalent pairs. The residual variability after removing trend in any of these three ways was subjected again to pattern analysis. Differences from linear regressions show a single peak at blocks of 53 cm. This pattern dimension can be observed in Fig. 1, and corresponds to the inner, outer, and intermingled edges of the projected crowns of the two trees. Trend corrections by quadratic curves, which fitted the data rather poorly, produced a complex block size/variance relationship with four peaks. The rather drastic mirror image treatment produced peaks at 46 cm and 114 cm which confirmed irregularities in the other curves.

Data from the vertical series of core samples taken from the watering experiment were treated similarly, removing trend before analysis by subtracting from each figure the relevant mean for that treatment and stratum. I suspect that most of the data from vertical soil profiles which I have subjected to pattern analysis should be further transformed to remove trends of decreasing variance with depth. Due to the sample size and the shallowness of the profiles, only detection of pattern to a maximum size of 61 cm was possible.

Again, pH data were used to examine the root environment for relatively static patterns. Agreement of pattern between the replicate profiles in each treatment was poor: generally, variance of pH differences increased up to 61 cm but no peak could be identified with certainty. Although stones in the core samples reduced the accuracy of estimates of bulk densities, these were similarly analysed. As with pH, no clear indication of pattern emerged. Attributes of the moisture characteristic curves were also analysed after trend removal to distinguish any static patterns in these soils. No peak variance was reached; generally variance increased with block size up to 61 cm. The variance of the differences in moisture content between one third and 15 atmospheres showed little effect on block size beyond a slight tendency to peak at 38 cm. As a crude measure of penetrability, the number of hammer blows needed to insert the core sampler for each sample were recorded. This would be related to the moisture status at the time of sampling which resulted from the watering regime and the absorption by the tree roots. Pattern analysis of these data showed a rather more consistent increase in variance with block size up to 46 cm (the limit of the data) for both May and September results. Moisture contents of core samples collected in May 1971, 12 months after the watering regimes were imposed, might be expected to demonstrate any pattern of root activity. Analysis of this parameter showed, as with the other analyses, that any pattern there might have been exceeded the size imposed by the limitations of the data, but that there was considerable inconsistency between the variance/block size relationships for the replicate profiles.

Weller (1971) has given data on the numbers of root tips of apple trees for soil profiles under a cover crop or sod-mulched. In order to remove trend from these data which was due to the effect of depth on the numbers of root tips, the mean from each depth was subtracted from the observed value and subsequently pattern analysis was performed. The ten profiles around each tree are not very consistent among themselves, and the mean variance shows a regular increase with block size for profiles under a cover crop. The sod-mulched plot showed peak variance at blocks of 90 cm thickness, with slight irregularities at 20 and 60 cm.

The data of Watson (1955) for the numbers of live roots of Douglas fir and *Ulmus* sp. are rather more comparable with my own. The data were obtained from counts from a grid on the face of soil profile trenches midway between trees in pure or mixed stands in Bagley Wood. By treating vertical files of the grid as separate transects and by analysing differences from the strata means to remove trend, evidence for the presence of pattern was sought. Interpretation is equivocal, but where either species grew in a pure stand, variance was usually greatest for a block size of 15 cm; no pattern was discernable when the species grew together. The profile was analysed only to a depth of 61 cm; large scales of pattern were, therefore, excluded. The individual horizontal rows were likewise treated as replicate transects and were analysed for pattern but without any attempt to remove trend. The sample size in this dimension was 15 cm and there were only six samples at each level on each face. The analysis did not detect any clear pattern in the data.

My previous data (Reynolds, 1970) were unsuitable for normal pattern analysis since horizontal distribution was by stratified randomization and not by transect. However, I have re-examined these figures by looking at differences of live root lengths between pairs of samples on each plot, classified by their horizontal separation. Unfortunately, the method is subject to plot edge effects of uncertain significance (P. S. Lloyd, personal communication); also the same sample may appear several times in different pairs. In each separation class, the variances of the differences and ratios of the pairs were calculated and compared with variances derived from random pairing of the data. On two of the plots, variances of differences between pairs 45 cm apart were very large (2 to 4 times that of random pairing) whereas smaller and larger separations had variances similar or much smaller than random pairs. The third plot had few live rootlets, having been sampled in a dry summer (1964), and did not show variances much larger than that found with random pairing. On this plot, comparison of mean differences within random and classified pairs, not their variances, suggested a peak difference at 60 cm separation. If these maxima correspond to mean horizontal widths of "cells" of root density, then 45 cm would be equivalent to a diameter of 57 cm and 60 cm to 76 cm (see Reynolds, 1974).

The considerable variability in the root system at an instant is not confined to root density and moisture absorption, but samples collected at the same time near to one another, but not necessarily from the same tree, had apices of very different shape and colour, probably indicative of differences in root extension. On the root window extension rates are far from uniform in space, many roots are alive but not extending, while others nearby are extending as much as 2 mm day^{-1}. This surely cannot be a simple response to local soil environment, but may reflect tactical organization.

I wish to illustrate this lack of synchrony by commenting on the starch contents of root samples collected on 2 March, 14 June, and 11 July 1973. Starch in Douglas fir seedlings is probably a minor form of energy reserve in the root system, and is not positively correlated with other energy materials throughout the year (Kruegar and Trappe, 1967) though Rutter (1957) showed that dry weight changes paralleled changes in starch content of Scots pine seedling needles. Starch content alone will only give a partial picture, and unless the other reserves are assessed, warrants no more than arbitrary grading systems such as Wight (1933) and Rutter (1957) have applied to observations on microscope sections. Laing (1932) discussed the disposition of starch grains in Norway spruce needles in terms of dynamic deposition and utilization, i.e. that gradients are interpretable in terms of mobilization. My observations corroborate such an explanation: where there was little starch, it was to be found in the outer pericycle cells most distant from the problem. Rutter's classes (1957) thus appear to incorporate gradients in different directions. Starch grains varied in size, number per cell and distribution between the tissues in my sections. The amount decreased towards apical meristems and it was absent within 1·7 cm of active apices (cf. Bogar and Smith, 1965, who found that starch disappeared from the vicinity of the earliest stages of root initials in Douglas fir). The bulk root system shows an annual trend of starch content in Douglas fir seedlings (Kruegar and Trappe, 1967): a maximum occurs in late winter and early spring with regular reduction thereafter until October. My observations rather emphasized local variations in the pattern. On the first occasion, two out of five root samples had no detectable starch, whereas on each of the other two dates three had only very small grains yet two samples had the parenchymatous cells full of large starch grains. The starch contents seemed to be unrelated to any morphological or ecological features of the samples. Some short and apparently inactive laterals had more starch than the parent root, and some much less. Laing (1932) noted that young, active roots of Norway spruce had a seasonal variation of starch content which differed from that in older roots. The work of Wight (1933) and Laing (1932) strongly suggests that, in conifers, the starch in the root follows a different seasonal cycle from that in the shoot: Kozlowski and Winget (1964) found that very little of the stored reserves are returned to the shoot in *Pinus resinosa.* All the evidence seems to suggest starch reserves are of local significance, not only within major divisions of the plant body, but perhaps even to individual rootlets.

C. Root Hair Production

It might be surmised that root hairs, for the quantities of structural carbohydrates involved, are an economic method of increasing the radial diffusivity of water and solutes to the roots under certain conditions. These conditions may be defined as situations when conduction of water through pores and films in the soil is relatively poor. This could occur when the general soil water potential is not high, and is not too low for root extension. When the root is passing through a soil void wider than the diameter of the root, the production of root hairs could be advantageous, as also to overcome "internal self-mulching" (Penman, 1956) produced in the soil immediately behind the root apex by absorption. Obviously root hairs are not merely produced as responses to these situations: in certain conditions, they may be abundant in water cultures. But in natural conditions tactical advantages of root hairs may be deduced.

The root hairs of Douglas fir are persistent and can be observed, though they are atrophied and are undoubtedly non-functional, many months after they are produced. Samples of roots showed that the production of root hairs in this species is extremely versatile: perhaps more than four fifths of the surface of the roots are bare, and elsewhere the density of root hairs and their lengths are quite variable. This suggests that root hair production is an adaptive response to certain stimuli. They are usually symmetrically disposed around the root, are more common on the straighter, lighter coloured parts of roots where laterals are further apart, and distribution often correlates between main roots and their laterals. I interpret these observations as indicative evidence in favour of my deductive reasoning: the root hairs are produced in response to conditions of intermediate soil moisture supply when roots are growing rapidly. There may be three or four "generations" of root hairs along a 30 cm length of root; tentatively I interpret these as epochs of rather more favourable conditions for growth. Conversely, structures which I hope to demonstrate are indicative of arrested growth are free of root hairs.

The root observation chamber provides an opportunity for relating the production of root hairs to local soil conditions, although the environment is somewhat artificial: voids are probably more frequent and water condensing on the glass provides an unnatural source of water. Thus, in contrast to the samples described above, more than half the root surface could bear hairs. The density of root hairs was positively correlated with their length: where most dense they could exceed 1·5 mm and this condition was only seen in roots traversing voids. Perhaps significantly, as roots entered or left voids there were usually a few millimetres devoid of hairs. It appeared that in the rather artificial situation of the chamber, the proximity of the soil particles largely governed the presence of root hairs: they sometimes occurred sparsely where the soil was in close contact with the root. Where a new and an old root occurred side by side, the densities of root hairs were not necessarily equivalent. This suggests that a time variable is also involved.

D. Dormancy: Metacutization

If the supply of photosynthates from the shoot is at a premium, an alternative tactic to discarding parts would be to restrict their respiration through a dormancy system. This would only be appropriate for the select roots of indeterminate growth. Dormancy seems virtually synonymous with protective coatings. These have long been known to occur in plant roots. Leshem (1965) has described the anatomy of the "metacutized" apices in *Pinus halepensis*: a layer of suberized cells (the metacutis), joining the endodermis behind the apex, isolates the meristem from the root cap in front and from the primary cortex laterally.

My observations extend the list of species to Douglas fir growing in the mild oceanic climate of England. Presuming Plaunt's (1909) "intercuticle" is absent, the situation in this species apparently conforms to his type II of metacutization, as the metacutis joins to the endodermis. I have observed on microscope sections the sequence of events leading to apical dormancy, with tissue differentiation drawing progressively closer to the meristem. Upon the breaking of dormancy, differentiation of the endodermis progresses forwards from the junction with the metacutis, tissue further out suberizes, and a frill of cortical scales is produced as extension proceeds. This frill marks the remains of the metacutized "protection" of the dormant apex. The process sometimes occurred several times within a very short distance. Sometimes the tip had emerged eccentrically, resulting in a characteristic kink in the root. The root was often lighter coloured distally to the swelling.

The dormant apex is swollen and club-shaped. The process of cell enlargement producing it is illustrated in Fig. 2, where the radial and longitudinal dimensions of the cells are given for the section drawn in outline. Longitudinal extension evidently diminished as the apex approached dormancy. The cortical tissue was also reduced in radial extension so much that a typical neck was formed behind the dormant apex. In the apex, radial enlargement of the lateral cells was considerable: it was apparently independent of longitudinal extension which was still reduced. Plaut's (1909) sections do not show this lateral swelling whereas Leshem's (1965) probably do. Cortical cells in the swollen region collapsed, whereas those in the neck region were rucked and thrown up into wrinkles. Suberization of the cortical cells was progressively more pronounced towards the dormant apex; cells just distal to the neck became virtually disorganized at the base of the metacutis which here formed a hollow cone. The tissue distal to the swelling when growth recommenced was characterized by narrow, long, fusiform parenchymatous cells with intrusive growth in distinction to the short abutting cells of the swollen region and just proximal to the swelling.

Some attempt has been made to classify the root tips in the observation chamber in early spring in order to gain some idea of the proportions of roots which morphologically appear to be metacutized. The numbers seem to vary widely, from as small a proportion as 10% to as many as 90% of the apices borne by a portion of root system. At the same season the roots which broke out of this dormancy were seen to be not confined to the apices of the longer roots: some were recently formed laterals.

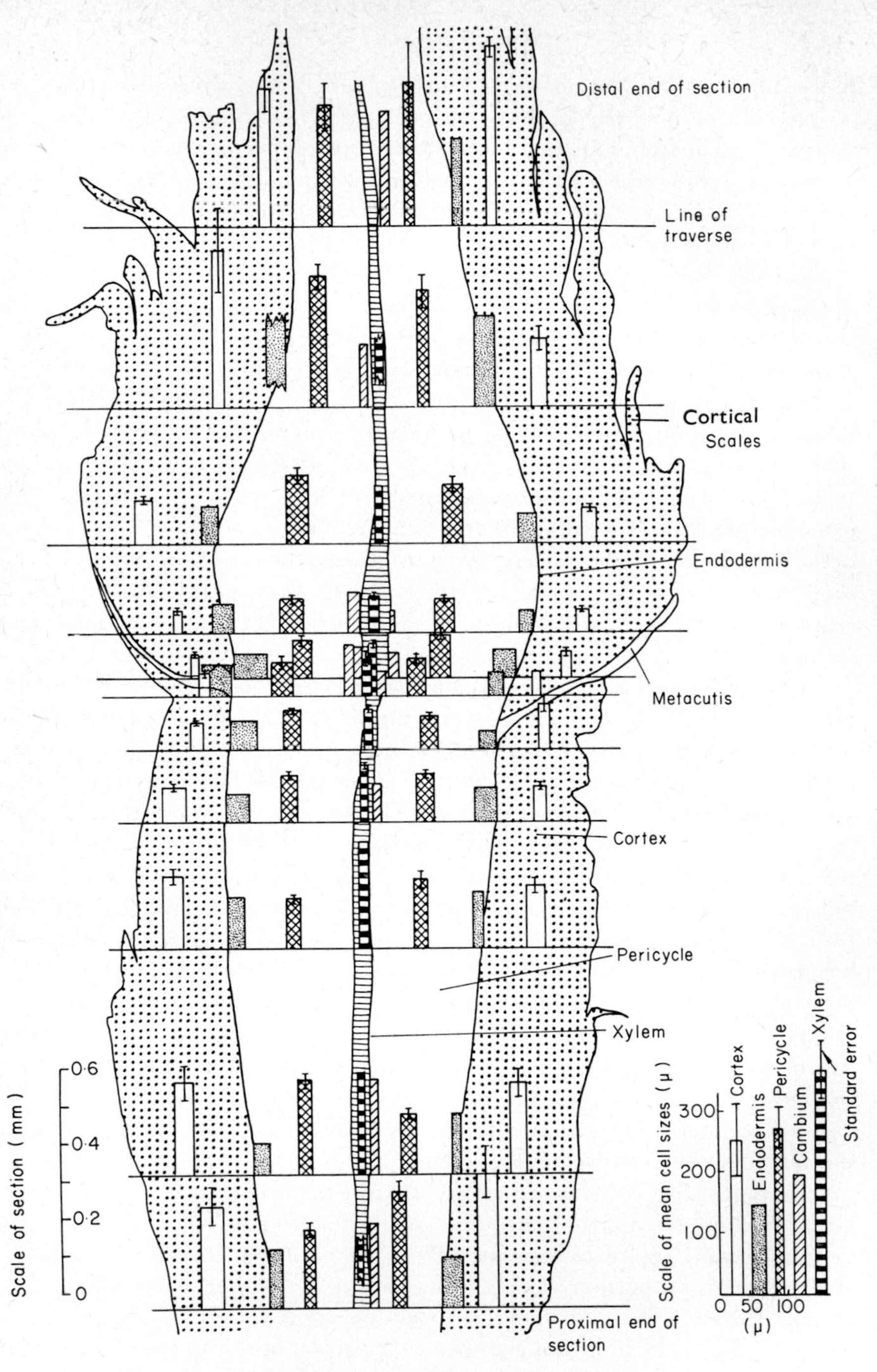

FIG. 2. Median longitudinal section of a root of Douglas fir collected 14 June, 1973, with mean lengths and widths of cells of various tissues at selected distances along its length. Standard errors of the mean widths were about 12%.

IV. Discussion and Conclusions

Among the many variables of root systems, some may be genetically fixed, others may be phenotypic responses to the environment. Several could be adaptations which are tactically advantageous in the highly competitive edaphic environment. I have only considered four, but other tactics, such as mycorrhizal associations, variations of root branching, geotropism and the depth of rooting, and the toleration of poor soil aeration, are amongst those which potentially might be at least as important.

All I have managed to do is to look for evidence from a specific site that, within a species, these four tactics have been adopted. I may say that the evidence for apical dormancy and root shedding (see Kozlowski, 1973, for examples) is quite well established both in my investigations on Douglas fir and by other workers on a wide variety of plant species. The plasticity of Douglas fir as regards the production of root hairs is evident even though I have made so few observations. Dynamic variations in the density of roots of Douglas fir I still regard as plausible but unproven. Observations are confounded by the heterogeneity of the soil, which probably produces stable spatial differences in root density, and by the difficulties of distinguishing living or active roots from inactive or dead roots. Pattern analysis techniques are still evolving: it may be premature to apply them in this area. Appropriate tests of significance of the patterns elucidated have yet to be devised, although I have tried to give due attention to the reproducibility of pattern where transects or profiles were replicated. There is no doubt, however, that the roots in the Douglas fir plantation are synchronous in neither growth nor starch mobilization.

The more challenging investigation still remains—to show what quantitative advantage each adaptation confers over the absence of the tactic or the employment of an alternative.

Leshem (1970) has attempted to assess the adaptive advantage of a metacutis in *Pinus halepensis* but unfortunately he used rather inappropriate treatments, i.e. exposure to 80% R.H. air, warm water or saline solutions. Rather than assessing survival and growth recovery after metacutization, he employed the tetrazolium chloride test which gave negative results. He seemed to show convincingly that dormant tips were produced as a response to increasing drought after irrigation of potted trees was stopped. Rather less convincingly, after irrigation of older plantation trees the numbers of dormant root tips were thought to decrease. In fact, if, as may be inferred, his figures roughly equate with densities of root tips, there were 13·3 dm^{-3} dormant tips before irrigation and 15·0 dm^{-3} ten days after: but the real difference was in the density of growing tips, being 2·1 dm^{-3} before and 26·9 dm^{-3} after watering, perhaps suggesting that growing tips proliferated rather than that dormancy was broken. Once again this may point to tactical lack of synchrony within root systems.

As we begin to understand what are the advantageous tactics of root systems, what degree of advantage each confers, and the role of these tactics in the ecosystem, we may hope to combine this with knowledge of the biochemical mechanisms controlling them to approach a holistic understanding of the underground parts of plants.

Acknowledgements

My thanks are due to Mr. B. J. Kemp for making many of the observations in the chamber. Thanks are also due to my colleagues Mr. I. A. Andrew for statistical advice and computer programming and Mr. H. L. Wright for providing computational facilities.

References

Bogar, G. D. and Smith, F. H. (1965). Anatomy of seedling roots of *Pseudotsuga menziesii*. *Am. J. Bot.* **52**; 720–729.

Cahoon, G. A. and Morton, E. S. (1961). An apparatus for the quantitative separation of plant roots from soil. *Proc. Am. Soc. hort. Sci.* **78**; 593–596.

Fayle, D. C. F. (1965). Rooting habit of Sugar Maple and Yellow Birch. *Publ. Dep. For. Can.*, No. 1120.

Grable, A. R. (1966). Soil aeration and plant growth. *Adv. Agron.* **18**; 58–106.

Harper, J. L. (1967). A Darwinian approach to plant ecology. *J. Ecol.* **55**; 247–270.

Hibberd, J. K. and Sykes, J. M. (1973). Root production by trees and shrubs (Abstract only). *Bull. Brit. ecol. Soc.* **4**(3); 3–4.

Hill, M. O. (1973). The intensity of spatial pattern in plant communities. *J. Ecol.* **61**; 225–235.

Kozlowski, T. T. (ed.) (1973). "Shedding of Plant Parts" Academic Press, New York, London and San Francisco.

Kozlowski, T. T. and Winget, C. H. (1964). The role of reserves in leaves, branches, stems and roots on shoot growth of Red Pine. *Am. J. Bot.* **51**; 522–529.

Kruegar, K. W. and Trappe, J. M. (1967). Food reserves and seasonal growth of Douglas Fir seedlings. *Forest Sci.* **13**; 192–202.

Laing, E. V. (1932). Studies on tree roots. *Bull. For. Commn., Lond.* No. 13.

Leshem, B. (1965). The annual activity of intermediary roots of the Aleppe Pine. *Forest Sci.* **11**; 291–298.

Leshem, B. (1970). Resting roots of *Pinus halepensis*: structure, function, and reaction to water stress. *Bot. Gaz.* **131**; 99–104.

Newman, E. I. (1966). A method for estimating the total length of root in a sample. *J. appl. Ecol.* **3**; 139–145.

Parkhurst, D. F. and Loucks, O. L. (1972). Optimal leaf size in relation to environment. *J. Ecol.* **60**; 505–537.

Penman, H. L. (1956). The movement and availability of soil water. *Soils Fertil.* **19**; 211–225.

Plaut, M. (1909). Untersuchungenz ur Kenntnis der physiologischen Scheiden bei den Gymnospermen, Equiseten und Bryophyten. *Jb. wiss. Bot.* **47**; 121–185.

Reynolds, E. R. C. (1970). Root distribution and the cause of its spatial variability in *Pseudotsuga taxifolia* (Poir.) Britt. *Pl. Soil* **32**; 501–517.

Reynolds, E. R. C. (1974). The distribution pattern of fine roots of trees. *In: Internat. Symp. Ecology and Physiology of root growth, Potsdam*, pp. 101–112, Akademie-Verlag Berlin.

RUTTER, A. J. (1957). Studies in the growth of young plants of *Pinus sylvestris* L. *Ann. Bot.* **21**; 399–426.

WATSON, J. W. (1955). An Investigation of the Rooting Depths of Elm and Douglas Fir in Pure and Mixed Stands. Unpublished Thesis, Commonwealth Forestry Institute, Oxford University.

WELLER, F. (1971). A method for studying the distribution of absorbing roots of fruit trees. *Exp. Agric.* **7**; 351–361.

WIGHT, W. (1933). Radial growth of the xylem and starch reserves of *Pinus sylvestris*: a preliminary survey. *New Phytol.* **32**; 77–96.

Chapter 9

Rhizography of Non-woody Roots of Trees in the Forest Floor

W. H. LYFORD

Harvard Forest, Harvard University, Petersham, Massachusetts, U.S.A.

I. Introduction

Much information is available about root systems of individual forest trees (Sutton, 1969; Lyr and Hoffmann, 1967; Köstler *et al.*, 1968), but information about forest stand root systems is scarce. Many studies have shown that forest tree roots are especially prevalent in upper soil horizons; however, the in-place relationships of roots to each other are less well known.

Organic debris accumulates as a forest floor to depths of several cm on the surface of the mineral soil in many forests, particularly in cool-temperate and boreal regions. This forest floor is the habitat of a wide variety of fauna and flora and in many places supports a large population of underground stems and small roots.

Most roots of forest trees in the forest floor are less than 0·5 mm in diameter and often are overlooked because they are so small and fragile. The importance of the forest floor as a suitable habitat for these small-diameter roots becomes evident whenever a complete root system is examined.

This report deals primarily with the rhizography of small-diameter, non-woody roots in the forest floor under mature mixed hardwood stands.

II. Materials, Methods and Terminology

A. Materials

Most field and laboratory observations were made on roots in the forest floor under natural 50 to 70-year-old mixed hardwood stands at the Harvard Forest in central Massachusetts. These stands are composed principally of red oak (*Quercus rubra* L.), red maple (*Acer rubrum* L.), paper birch (*Betula papyrifera* Marsh) and black birch (*B. lenta* L.) and are growing on strongly acid, well drained Entic Haplorthods of the Gloucester series. These soils developed from stony, sandy glacial till derived from granite and schist.

The forest floor under mixed hardwood stands at the Harvard Forest has an average thickness of 4–5 cm and is a mor consisting of L, F and H horizons. Immediately underlying the H horizon, and sharply separate from it, is the nearly black mineral A1 horizon consisting of a mixture of mineral soil particles, disintegrated organic matter, micro and mesorganisms, and plant roots. Although the 4–6 cm thick A1 horizon results largely from mixing by burrowing and tunnelling small mammals (particularly moles, shrews and voles), the mixing in any one place occurs at intervals of several years so roots of forest trees are not continually broken or disturbed and are able to populate the overlying organic layers of the forest floor. By contrast in warm-temperate or tropical forests, or anywhere that earthworms are prevalent, the organic debris that reaches the forest floor is intermixed, consumed or otherwise disintegrated at such a high rate that forest tree roots are necessarily confined to the mineral soil.

B. Methods

Root systems of individual trees were examined first by exposing the 3 to 10 large framework woody laterals at the base of the stem and then working outward along one of these, using small pointed tools to avoid cutting the numerous small-diameter, non-woody, first order laterals. Direct examination of the small roots was possible, but the roots are so small and numerous that most details were worked out from 10 cm square blocks cut from the forest floor. These blocks are a convenient size for examining under a stereomicroscope and provide root segments of suitable length for determining the spacing between laterals, number of root tips per unit length and other parameters.

Many unsuccessful attempts were made with detergents and cleaning agents to remove the matrix between roots to determine their arrangement in the soil without injuring the individual root segments. Removal of disintegrating leaves and other organic matter bit by bit with forceps was the only satisfactory means found.

Thin sections were tried as an alternative procedure for in-place root examination. Horizontal one-cm-thick slices through the 10 cm square blocks allowed removal of the main axes of the small roots but only after considerable injury to the root tips and laterals. Four cm square columns of the forest floor impregnated with paraffin, and two mm square columns impregnated with resin, were also sectioned serially in the usual manner and provided suitable thin sections. These were useful for examination of spacing between root tips and the soil matrix, but of limited usefulness for study of root axes.

C. Terminology

Rhizography (synonymous with geography) is a term used in this paper for the description and distribution of roots.

Tree root systems, as with other plant root systems, commonly are classed into taproots and laterals. Laterals are further sub-classed by orders of branching—first, second, third, fourth order and so on. In the case of plants with secondary xylem thickening it is convenient to have a term for those portions of roots with sufficient secondary xylem thickening to make the root stiff (woody), another term for those portions that have so little secondary xylem thickening the roots remain flexible (non-woody), and still another term for the apices of roots (root tips) that have only primary tissue. There is no standard terminology for tree root laterals and both Sutton (1969) and Lyr and Hoffmann (1967) note that a rather large number of names have been applied, some based on function, some on anatomy, and some on morphology. Long and short are terms frequently used to distinguish "woody" and "non-woody" roots yet in deciduous tree root systems many "short" roots are a metre or more in length. While these are short relative to roots 15–20 m long the classification of root sections removed from small samples of the forest floor is necessarily on the basis of diameter, degree of flexibility, and presence or absence of live epidermal and cortical tissue rather than length. There seem to be no brief, completely satisfactory morphologic terms, and in this report the writer chooses to use woody and non-woody, believing they are more connotative than long and short for distinguishing laterals of hardwood trees. Perhaps more suitable from a strict morphologic standpoint would be stiff, cork-covered roots versus flexible, cork-covered roots. But these names are unnecessarily long.

As used in this report a root tip is that portion of the root extending from the terminus of the root apex back to the first emergent lateral or, if no lateral has developed, back to the parent root. This morphological definition facilitates counting and yet encompasses those portions of dicotyledonous forest tree roots that contain only primary tissue. Using this definition the first emergent acropetal lateral is entirely root tip until such time as it elongates and develops its own first lateral or until it becomes suberized. Thus, on a rapidly elongating root, there is a root tip at the terminus of the parent root and often several others along the axis of the parent root. There is no minimum length limit for a root tip—only that the apex be clearly emergent. Neither is there a maximum length limit; in the absence of laterals the root tip encompasses only that portion of the axis consisting of primary tissue.

III. Results and Discussion

A. Gross Rhizography of Mature Forest Trees

Root systems of forest trees are widespread and roots of many forest trees extend as much as 15–20 m from the base of the tree. An extensive lateral root system originates from the single taproot, and within about 2 m of the base of the tree laterals are numerous, close together and nearly fill the soil. This portion of the root system is known as the central root zone (Lyr and Hoffmann, 1967) and it is here that woody roots achieve their deepest penetration into the soil. At distances greater than 2 m—the outer (peripheral) root zone—most woody roots tend to be horizontal and within 40 cm of the soil surface.

1. *Woody Roots*

Some tree species have large, strong, woody taproots that penetrate vertically downward for several metres in suitable soils to provide support as well as a source of laterals. Other species have short, weak, taproots that serve primarily as a source of laterals.

All trees have 3 to 10 or more first order woody laterals spaced rather evenly around the base of the stem. Zobel (pp. 261–275, this volume) proposes the name "basal" roots for these laterals that develop at or near the root–shoot junction inasmuch as they may be under different genetical control than the taproot itself, or other laterals. These basal roots form a strong framework as well as a support and conduction system. They originate near the surface of the soil and descend obliquely for several cm before becoming horizontal. Woody basal laterals may be as large as 20–40 cm in diameter. They have eccentric radial growth and some tendency for buttressing. Most taper rapidly within a metre or two to a diameter of 2–4 cm and then very gradually become smaller in diameter. Many woody laterals of spruce, fir and some other forest trees remain close to the surface and may grow within the forest floor itself (Fig. 1A), whereas woody laterals of other trees tend to be stratified at various depths within the mineral soil (Stout, 1956).

Laterals arise acropetally on first order woody roots just behind the elongating portion of the root tips and are spaced 2–3 mm apart on average. The majority of these laterals remain as small-diameter, first order, non-woody roots, but near the base of the stem a few may enlarge and become woody framework roots. These large second order woody laterals are spaced 25–100 cm apart near the base of the tree but are much less frequent in the outer root zone if, indeed, they occur at all. In the outer root zone, branching into woody laterals occurs primarily after injury to the rapidly elongating, fleshy, root tips at the terminus of woody roots. Replacement roots form just behind the injury and woody forks result (Wilson, 1970). These forks are generally bi- or tri-membered and may occur at intervals of 1–5 m along the framework woody laterals. They are an important means by which the tree root system becomes widespread.

Because of the spreading character of the woody root systems of forest trees, the roots of adjacent trees intermingle and there are few areas in a mixed hardwood stand where woody roots of different species are separated by more than 10–20 cm.

2. *Non-woody Roots*

The numerous non-woody laterals that arise just behind the large-diameter apices of woody roots soon lose their outer live epidermal and cortical primary tissues and subsequently have a cork-covered surface. A small amount of secondary xylem develops in the central cylinder but not enough to cause appreciable thickening and stiffening. Diameters of most non-woody laterals tend to be one mm or less. With age or prolific branching diameters may increase to 2–3 mm or more.

Most non-woody hardwood roots are at least 20 cm long and many extend a metre or two (Fig. 1B). They branch and rebranch abundantly until they fill the space between woody roots. If there is a suitable forest floor environment many grow upwards into this and branch into several higher orders. If the parent woody roots are near the surface of the mineral soil, or in the forest floor itself, this upward growth is not particularly noticeable. If, as in the case of red oak, the woody laterals are at depths of 30–40 cm the oblique upward growth of the non-woody laterals is such a contrasting feature that attention is called to it and to the proliferation of roots that takes place when the roots reach the forest floor.

3. *Root Tips*

Root tips are at the terminus of every live lateral and also along the younger portions of many laterals. Tips of woody roots on a mature forest tree root system number in the hundreds whereas those of non-woody roots number in the hundreds of thousands. Although root tips of non-woody laterals are small in diameter and length as compared with those of woody roots their enormous number overshadows the larger root tips in surface area.

4. *Spacing of Laterals*

The number of laterals per cm along the parent axis of non-woody roots in the forest floor can be low or high depending on species, soil conditions (moisture, temperature, nutrient availability, friability) and biota, especially the fauna that feed upon roots, and infectious microorganisms. Spacing of laterals on replacement roots of mature trees grown in a rhizotron (Lyford and Wilson, 1966) are listed in Table I. These roots grew in water, soil or gravel under near-ideal

TABLE I. Spacing of non-woody laterals on woody replacement roots growing in a rhizotron

Species	Spacing of non-woody laterals		
	1st order on woody replacement axes	2nd order on 1st	3rd order on 2nd
		No. per cm	
Red oak			
Water	4	9	11
Soil	4	6	8
Gravel	5	8	5
Aspen	4	8	—
Yellow birch	3	5	—
Red maple	3	3	—
White ash	5	—	—

conditions and seem to be non-mycorrhizal. On the basis of these observations values of 4, 6 and 8 laterals per cm are close approximations for the spacing of non-mycorrhizal first, second and third order laterals respectively.

An estimate of the number of root tips produced on a single lateral is obtained from counts made on a red oak first order, non-woody lateral that grew entirely in water on a woody replacement root. This 14·8 cm long lateral by actual count had 1066 root tips—one on the first order axis, 135 on the second order axes and 930 third order root tips—an average of 72 root tips per cm of first order axis.

From these figures a very rough idea can be obtained of the number of root tips on a mature red oak root system. Using an estimate of 70 live root tips per cm for first order non-woody laterals (as the laterals increase in length there will be branching to fourth and fifth orders but this increase in numbers of root tips will be balanced by dying-off) and assuming each first order, non-woody lateral is only 25 cm long with four of these per cm of woody root; and further assuming only 5 basal woody roots per tree, each 10 m long and each with woody branch roots that total another 10 m; the live root tip total is 70 million. These are exceedingly conservative assumptions especially with respect to the total length of woody roots and the true value could range anywhere from half of this up to a value 10 times greater. 500 million live root tips is a reasonable first estimate of the number on a mature red oak root system.

B. Rhizography of Non-woody Roots in the Forest Floor

Having examined the overall rhizography of the tree root system, some details of non-woody roots in the forest floor will be described.

Several rather strong impressions are obtained when roots in the forest floors of mixed hardwood stands are first observed. Especially noted are the small diameters of roots, their closeness to the surface, their horizontal growth habit, the intimate intermingling of root axes, the firmness by which the roots are held in place and the large number of root tips.

1. *Diameters*

Most roots in the forest floors of hardwood stands are non-woody and range from 0·5–0·2 mm in diameter and are easily broken when the forest floor is pulled apart. For this reason one first notices the few roots a mm or greater in diameter as these are strong enough to resist breaking when ordinary hand tools are used. The much more numerous smaller diameter roots tend to be overlooked unless they are unusually strong or branched. Certainly the thread-like size of these roots seems incompatible with the large size of the shoot system and they are likely to be considered roots of grasses or other ground cover plants.

2. *Closeness to Surface*

If the forest floor is examined from the top downward the first roots are generally encountered either in the two-year-old litter layer or at the junction of the L and F horizons. (If mosses occur on the surface of the forest floor roots may be found within the living portion of the moss clump or carpet). No roots enter the loose litter resulting from the last increment of leaves or needles but as soon as this is covered by a new fall of organic debris it becomes a suitable medium and roots elongate into it. Root growth into this thin layer is dependent on satisfactory moisture and temperature conditions and the leaves and needles must be essentially immovable. This means that root invasion occurs only after the leaves and needles are well tied together by mycelium. For the most part this occurs in the autumn after leaf fall or early in the next growing season.

Careful removal of the upper leaves of the newly stabilized thin layer exposes the small-diameter advancing roots. For quick demonstration of these roots the 1–2 cm apical portions can be exposed by repeatedly rubbing the forest floor with the tips of the fingers and blowing away the loosened material. Older root parts are gradually exposed as the process is continued.

3. *Horizontal Growth Habits*

Most roots in the forest floor are in a horizontal position, partly the result of a trophic response and partly the result of the character of the forest floor. New extension of roots into the upper forest floor layers is mostly in the recently stabilized layer of leaves or needles; and they are necessarily limited to this thin layer. Moreover, deciduous leaves are generally flattened and closely pressed to one another like shingles on a roof and the small-diameter roots tend to grow between the horizontally arranged leaves. Roots commonly enter between leaves by way of the overlapped edges although occasionally they penetrate through a leaf by dissolution of the parenchyma. Diarch roots, like those of red maple, readily grow between two closely appressed leaves and the laterals extend on either side of the root axis in a typical fan shape with no evidence of being forced into this shape. Triarch and other polyarch roots also elongate between closely appressed leaves but the axes of these roots are likely to show curves and contortions suggesting that the laterals emerging from three or more sectors of the root axis are less readily confined to a single plane. As the polyarch roots become older and thicker the fan shape becomes even less evident.

A root tip apparently dissolves rather than pushes its way through the leaf parenchyma; however, there is no evidence of etching of the leaf surface when the root elongates on the surface of closely appressed leaves. Nor is there any adhesiveness along the root. If loose fungal hyphae are numerous on ectomycorrhizae, some of these may penetrate the leaf enough to hold the root in place. Most root tips in the mixed hardwood forest floor are ectomycorrhizal and as a consequence root hairs are not at all common. In fact most ectomycorrhizae are completely sheathed with fungal tissue by the time they emerge through the parent root epidermis. In any event root tips, whether mycorrhizal or non-mycorrhizal, on the small-diameter deciduous tree root axes are too small to be differentiated clearly with the naked eye and the first impression is of root axes and root branches rather than root tips.

4. *Intimate Intermingling*

Roots in the forest floor intermingle closely; in a forest stand with a closed canopy there are virtually no portions of the forest floor that lack non-woody roots. Root axes of the same and different species cross each other, and although they may be close to one another, they seldom actually touch inasmuch as they are separated during elongation by leaves or needles. When the leaves or needles disintegrate, their place is taken by mycelia and the roots are still efficiently separated. Elongating root tips seldom touch even when they happen to grow between the same two leaves.

The arrangement of root axes in a 10 cm square block of forest floor is shown in Fig. 1C. The crisscross arrangement of the larger root axes (which survived the vigorous washing and kneading used to remove the soil matrix) indicates that the forest floor is well occupied by roots. Arrangement of roots in one-cm-thick horizontal slices through a continuous 10 cm square block are also shown in Fig. 1C. The small diameter of the roots in the upper part of the forest floor is evident. Some of the root axes in the lower portion are larger in diameter and are the parent roots for many laterals in the upper part.

5. *Firmness in Place*

Considerable difficulty was encountered in attempts to remove the matrix between roots. Detergents and clearing agents were of little value. Finally it became inescapable that the roots are too tightly held in place within a felty mass of mycelium for removal of roots without considerable injury to the finer branches and root tips. Even 5 mm thick sections of the forest floor are still too thick for successful removal of root segments. Small individual root segments can be removed with most laterals intact but many root tips on the laterals are broken. Roots which elongate into the upper portion of the forest floor tend to be isolated from each other and are between leaves or needles. These can be removed by careful use of forceps whereas extraction of uninjured root segments from a felty matrix is nearly impossible.

While this is true for roots in the felty mor under deciduous tree stands at the Harvard Forest it is not necessarily true for all mors. The matrix of some is composed primarily of finely granular fecal pellets and these, at times, are rather easily shaken out leaving the roots themselves well exposed.

6. *Root Tips*

Root tips on the thread-like, highly branched, non-woody roots are tiny (0·1–0·2 mm diameter) and not easily seen with the naked eye. They are very numerous, however, particularly on those roots that have mycorrhizal clusters. To obtain some idea of the number of root tips and laterals, several one cm cubes were cut from the F and H horizons of the 70-year-old mixed hardwood forest floor. From these small cubes root segments were removed as carefully as possible by forceps. The large number of root tips, an average of over 1100 per cubic centimeter (Table II), may represent about the maximum for root tips in forest

TABLE II. Number of root tips and length of root segments in one cm cubes cut from F and H forest floor horizons in a 70-year-old mixed hardwood stand.

	No. root tips	Length of roots Metres
	1073	2·43
	1418	2·85
	1281	—
	979	0·93
	954	0·97
Average	1141	1·80

FIG. 1. Rhizography of tree roots in the forest floor. A: Woody white spruce root (painted white) at forest floor-mineral soil junction, extending 20 m into an adjacent white pine stand. Each branch is the result of injury. B: Much branched first order, non-woody red

floors although Meyer and Göttsche (1971) reported 990 tips in one forest floor sample.

C. Characteristics of Individual Non-woody Roots

As a means of showing relationships among root tips, laterals, and the parent axis of individual roots, a few short segments were dissected as carefully as possible from 10-cm-square blocks of forest floor. Some of the segments had linear configuration and others were clustered. On linear portions the laterals tend to be spaced irregularly and root tips and laterals seldom bunch together close enough to touch. Clustered portions have laterals emerging in such close proximity that root tips of adjacent laterals or root tips nearly or actually touch.

1. *Spacing of Laterals*

Spacing of laterals on linear root segments from the upper part of the forest floor averages about 5 per cm (Table III) and is less regular than on those grown under controlled conditions in the rhizotron (Table I). Horsley (1971) listed several reasons for the irregularity, among which are injury, slow elongation, dormancy and curving. The most obvious causes for the irregular spacing of laterals in the forest floor are injuries by fauna, infection by pathogens, increased elaboration following infection by mycorrhizal fungi, and lack of laterals on the concave side of strongly curved roots (Fig. 1F).

Average spacing of laterals on the linear non-woody roots in the forest floor is comparable to that on the second and third order roots grown in the rhizotron (Table I) in spite of the irregularity of spacing. Forking after injury and clustering compensate for some loss of laterals. This is shown by the maximum number of laterals or root tips observed on a one cm long portion of the short segments. The maximum of 41 and 42 on two of the clustered roots gives an indication of the large number of root tips that result after some kinds of infection by mycorrhizal fungi (Fig. 1E). This degree of elaboration is comparable to that in nodules on roots of leguminous and non-leguminous plants.

---

maple lateral and its parent woody root showing how spaces among woody roots soon become occupied by a non-woody network. C: Root distribution within the forest floor. Upper left, 10 cm square block of forest floor. Lower left, all roots within a contiguous block. Right, top to bottom, roots in successive one cm thick horizontal slices through a third contiguous block. D: Paper birch mycorrhizal cluster—1786 root tips on a 44 mm long root axis with many fourth and a few fifth order laterals. E: Red oak mycorrhizal cluster—588 root tips on a 19 mm long axis with many third and no fourth order laterals. F: Red oak showing both intercalary root tip proliferation and characteristic development of laterals on outside of a sharp curve. Scale in mm.

TABLE III. Number of root tips and spacing of laterals on root segments dissected from the forest floor.

Species	No. root tips per cm parent axis	Max. No. laterals or root tips on any one cm long portion	Aver. spacing of derived laterals 1st parent axis	2nd on 1st
Linear roots with few or no clusters			No. per cm	
Birch	46	20	4	7
Birch	62	21	4	7
Birch	60	8	8	2
Birch	64	8	2	5
Birch	16	10	1	7
Birch	58	14	6	6
Birch	26	12	3	3
Oak	54	14	3	6
Oak	56	21	9	11
Oak	29	28	10	4
Average	47			
Roots with clusters				
Birch	100	41	6	13
Birch	150	—	7	—
Birch	405	10	6	10
Oak	310	42	20	22
Oak	150	18	13	18
Average	227			

2. *Orders of Branching*

All laterals on root segments listed in Table III had at least three orders of branching, some had four, and two of the segments had five or more orders of branching. It should be pointed out, however, that determination of orders of branching depends on ability to distinguish between forking due to injury and normal unequal growth of laterals. This is difficult to do consistently on the higher order laterals. Furthermore, the exact order of branching of laterals on non-woody segments remains unknown unless one is fortunate enough to find the root segment attached to the parent woody root. Probably the largest diameter non-woody roots extracted from small blocks of forest floor are first or second order and so it is probable that the non-woody roots commonly branch into at least three or four higher orders and in clusters perhaps into as many as six or seven orders.

3. *Root Tips*

The average number of root tips per cm of parent axis on linear roots (Table III) is 47, with a range of 16–64. On clustered roots the average is 227 with a range of 100–405. The large number of root tips on some clusters is illustrated by the white birch mycorrhizal cluster shown in Fig. 1D which has 1786 root tips on an axis 4·4 cm long. On this particular cluster some laterals have as many as five orders of branching.

Most root tips are ectomycorrhizal and have no root hairs. Red maple roots are endomycorrhizal and root hairs are common on some axes although rarely on portions of the root that are beaded. Some ectomycorrhizae are bristly and black, others are sheathed with loose wefts of white or yellow mycelium and this extends out into the surrounding soil matrix for several and sometimes many mm. By far the majority of the mycorrhizae are smooth and cream-colored and the root tips have a slightly swollen appearance.

Diameters of root tips on non-woody hardwood roots are generally 0·1–0·2 mm and the root tips average 0·2–0·5 mm in length.

D. Longevity, Injury and Health

Because forest tree root systems are larger and older than those of many plants the effects of longevity, injury and overall health are evident over a period of years. For example, there is a wide-spread belief that the small-diameter non-woody laterals of trees are short-lived (ephemeral); and there is the implication that their death results solely from age and that roots are shed periodically, much as leaves or needles. That a high proportion of small-diameter laterals on tree roots disappear is indisputable because there are many bare spaces or stumps of former laterals along woody roots in places where laterals must have been numerous just after they developed in acropetal sequence. Nevertheless there have been few direct studies of individual roots to determine exactly what happens to them over time and whether, in fact, ageing alone is involved or whether disappearance is due primarily to injury by pathogens, to root consumers, or to starvation. Death of roots by starvation doubtless is a major factor in suppressed trees but this is not the same as ageing. Head (1973) made direct observations on roots of perennial woody plants and noted the complete disappearance of some roots within a matter of months as a result of decay, yet he noted that some roots lasted for several years. Head also reviewed previous work dealing with death of small roots. Kalela (1955) and Heikurainen (1955) sampled pine (*Pinus sylvestris*) forests periodically during the growing season and found a much greater number of root tips and greater length of small-diameter laterals in the early part of the growing period than at other times. Orlov (1957) made direct observations on spruce roots in a 100-year-old forest and found that small-diameter tips did not die off in the two year period he had them under observation. Kolesnikov (1971) suggests that root death and renewal is a natural process for all trees and that this bears a resemblance to leaf shedding in evergreen plants. He stops short, however, of saying that death of roots is the result of ageing and notes that the process is irregular and depends on environmental conditions. Lyr and Hoffmann (1967) consider longevity of tree roots and conclude it would be difficult to formulate any generally valid statement on the longevity of fine roots.

Several causes for the death of small laterals of forest trees are observable on roots growing in rhizotrons. For example, exposure to dry air of the small-diameter third and fourth order laterals for only a few minutes is sufficient to make them shrivel and die. If roots are exposed on the soil surface and moisture and temperature remains high the exposed tips decay within a day or two. If root tips are not well protected they are eaten avidly by millipedes, slugs and crickets. There is no dearth of organisms causing root death and Sutton (1969) called attention to many of them. Recently Edwards *et al.* (1970) reviewed causes of root breakdown by soil invertebrates, Ruehle (1973) discussed the injurious action of nematodes on tree roots, and Campbell and Hendrix (1974) described feeder root necrosis, particularly that caused by pathogens. In the forest floor continual destruction of small roots by small mammals and dipterous and lepidopterous larvae is readily evident and the after-effects of previous injuries are evidenced by the prevalence of forked roots.

Formation of replacement roots and forking after injury of the large-diameter root tips at the terminus of woody roots, is common (if not universal) in forest tree root systems (Wilson and Horsley, 1970) (Fig. 1A). Injury is common also on non-woody laterals and formation of replacement roots and forks occurs even on second and third order laterals. On most third and fourth order non-woody roots the formation of replacement roots does not follow injury and only the stump persists. If these higher order laterals are mycorrhizal a great elaboration of intercalary root tips often occurs (Fig. 1F) and this more than balances the inability to form replacement roots.

In the absence of direct evidence it may be best to reserve judgement about the actual cause of the disappearance of small non-woody laterals. Age by itself as a cause may be unusual. Some small non-woody tree root laterals are known to be old. For example, a study of a woody red oak root 17 m long was made at the Harvard Forest and all persistent laterals and all stumps of laterals were charted. There were 300 persistent laterals per metre on the youngest part of the root whereas there were only about 15 stumps and persistent laterals on the older part. Even so, a few first order, small-diameter, non-woody laterals persisted in the zone of rapid taper at the very base of the stem and these were as old as the woody root itself—at least 40–50 years. These non-woody laterals arose acropetally but were soon encased in the ray tissue of the woody parent root and in large part protected over the years from physical injury.

Stumps of laterals persist for a long time on non-woody roots because the parent axis does not increase appreciably in diameter with age. A study of both stumps and persistent laterals might provide worthwhile information about the cause and rate of dying-off of small-diameter roots.

A noteworthy feature of non-woody roots in the forest floor is that dead root tissue does not persist for long. In fact no evidence—or, at least, no conspicuous evidence—of former roots remains for long in the soil; decay or complete consumption of dead small-diameter roots is probably accomplished in only a few days. Judging from the unmistakable presence of black, bristly mycorrhizal fragments in insect feces seen in thin sections of the forest floor, some root tips may be consumed almost immediately after death or, indeed, before death.

Ideally it should be possible to examine the root system of a tree and evaluate its degree of health. Continual death of so-called ephermeral roots seems to take place on all tree root systems but whether a moderate amount of death or injury to non-woody laterals is harmful or beneficial is not well known. Forest pathologists probably have evaluated the health of root systems more than other scientists. Leaphart and Copeland (1957) found an average of 4·4% mortality of roots less than one-mm-diameter in a healthy stand of western white pine, with a range of 1·5–9·6%. Copeland (1952) studied little leaf disease on loblolly and shortleaf pine and decided there is a critical percentage of dead roots, somewhere between 18 and 34%, beyond which a tree cannot continue normal growth. Spaulding and MacAloney (1931) recorded 3·8% root mortality on a healthy yellow birch tree and from 8·4 to 16·3% in trees that showed evidence of dieback.

Hawboldt and Skolko (1948) found 30% mortality in the roots of a healthy yellow birch tree but this tree was in an area of rapid dieback and may have been in early stages of dieback. Although Leaphart and Copeland (1957) were concerned primarily with the mortality of non-woody laterals and the tips on these small roots, it is difficult to be sure just what portions of the root system were studied in the earlier research—probably only woody roots and the tips of these roots.

E. Minimum Diameter and Maximum Length

After non-woody roots branch and rebranch to three, four or five orders the vascular cylinder reaches a very small diameter. For red oak, red maple, and paper birch this diameter is 60–70 microns. Each higher order seems to have a slightly smaller diameter than the parent so eventually there comes a time when laterals no longer can form. But an interesting and significant feature is that when laterals of near-minimum diameter become mycorrhizal they are sometimes stimulated to form more laterals (Marx and Zak, 1965). An elaboration of intercalary tips is common on small-diameter, non-woody roots in the forest floor (Fig. 1F).

Even though repeated elaboration occurs after a root reaches minimum diameter there must be some ultimate limit to this process and, in the absence of appreciable injury, it is conceivable that minimum diameter and maximum length exert fundamental control over the extent of tree root systems.

When a root reaches minimum diameter it may still have the ability to elongate and conceivably could become very long, unbranched and thread-like. If so, there should come a time when the root tip would be so far from the stem that distance would play a major role. Many forest tree roots extend 15–20 m from the stem and in such long roots the availability of photosynthate may diminish gradually with distance because other portions of the root intercept most of the supply. In this way the root tip becomes subject to starvation. There is, of course, no specific requirement that photosynthate should arrive at the root tip the same day it is produced in the leaves and there may be several intermediate links between leaf and root tip. Noteworthy in this connection are red maple roots, growing in the rhizotron, which elongated all winter even though the tree to which they were attached was dormant. One of these red maple roots also continued to elongate for about two weeks after it had been completely severed about 70 cm behind the apex. In this connection one might speculate that if a root tip in the forest floor does not readily obtain energy-supplying substances from the tree itself it could obtain some from the soil medium by way of its fungal symbionts. Garrett (1970) notes that some fungi can obtain all necessary growth substances from decomposing debris in the forest floor or from other organic material and Went (1971) indicates that photosynthates fixed in cell materials of the litter may also be transformed to sugars and transported to roots by way of rhizomorphs and mycorrhizae. Reid (1971) has demonstrated that carbohydrates can be transported from the root system of one pine seedling to another by mycelial strands. Root tips exude many substances, including carbohydrates. If root tips are close together a large volume of soil could be a single mass of rhizosphere (Bowen and Rovira, 1969). Thus, the transfer of necessary energy substances from one root tip to another may be possible. In fact, Akhromeiko (1965) makes a strong case for the transfer of substances from one tree to another as a result of root excretions. It may not be unrealistic, therefore, to speculate that if excised roots *in vitro* can obtain necessary energy when bathed in proper substances the same could be true for intact roots in the soil.

F. Future Needs

More detailed knowledge about the rhizography of non-woody laterals in mature forest stands is needed before the character of roots in small soil cores can be evaluated satisfactorily. Up to the present woody roots and root tips have been given most attention while the small-diameter branching network of fine roots in between has been neglected.

Soil itself is sampled and analysed for evaluating most root-soil conditions; roots in the soil are not given much attention unless there is a question of root health. Yet there are no good standards for root health; a few missing laterals, for example, do not necessarily mean poor health. As a means of establishing root health benchmarks, it would be valuable to have some detailed morphological analyses on entire non-woody laterals growing in natural forest stands, paralleled by similar analyses of roots grown under controlled conditions in the same stands. This would be possible because inexpensive controlled experiments on roots can be carried out in natural forest stands, as replacement roots are so easily and quickly developed, and these are representative of a large part (perhaps a major part) of the root system. Newly regenerated roots of mature trees grow satisfactorily on or under plastic sheeting, or in simple wooden boxes or trays lined with plastic sheeting and filled with water, nutrients or various soil mixtures.

Sampling forest soil is fairly simple because soil is a continuous mantle and reasonably uniform. By contrast, the root system of an individual tree is not randomly distributed and some soil samples may contain no roots of a given tree at all. Still, the problem lies not in sampling but in evaluating a myriad of tips and laterals. Present techniques are time consuming and questionable. Counting root tips seems somewhat analogous to counting all sand grains in a soil sample. For routine use there is need for quick laboratory methods—something like dye absorption.

Radioisotopes are extremely useful for some purposes but are unlikely to be used for routine work. Nevertheless, at present, counting root tips and root stumps and obtaining the length, spacing and branchiness of laterals is the best means available for making comparisons between forest stands and between soil horizons. Perhaps until detailed studies of whole systems are available it may be useful when sampling forest soils to collect a few samples and make detailed morphological root analyses on representative root segments extracted from these samples.

References

AKHROMEIKO, A. I. (1965). "Physiological Basis for the Establishment of Hardy Forests" (Translation from Russian). Israel Program for Scientific Translations, Jerusalem. 1968.

BOWEN, G. D. and ROVIRA, A. D. (1969). *In*: "Root Growth" (W. J. Whittington, ed.); pp. 170–201. Butterworth, London.

CAMPBELL, W. A. and HENDRIX, F. F. JR. (1974) *In*: "The Plant Root and its Environment" E. W. Carson, ed.): pp. 219–243. Univ. Press of Virginia, Charlottesville.

COPELAND, O. L. JR. (1952). Root mortality of shortleaf and loblolly pine in relation to soils and littleleaf disease. *J. For.* **50**; 21–25.

EDWARDS, C. E., REICHLE, D. E. and CROSSLEY, D. A. (1970). *In*: "Analysis of Temperate Forest Ecosystems. Ecological Studies I" (D. A. Reichle, ed.): pp. 147–171. Springer-Verlag. Berlin–Heidelberg–New York.

GARRETT, S. D. (1970). "Pathogenic Root-infecting Fungi." Cambridge Univ. Press. Cambridge.

HAWBOLDT, L. S. and SKOLKO, A. J. (1948). Investigation of yellow birch dieback in Nova Scotia. *J. For.* **46**; 659–671.

HEAD, G. C. (1973). In: "Shedding of Plant Parts" (T. T. Kozlowski, ed.) pp. 237–293. Academic Press, New York, London and San Francisco.

HEIKURAINEN, L. (1955). Über Veränderungen in den Wurzelverhältnissen der Kiefernstände auf Moorboden im Laufe des Jahres. *Acta Forestalia Fennica* **65**; No. 2, 1–70.

HORSLEY, S. B. (1971). Root tip injury and development of the paper birch root system. *For. Sci.* **17**; 341–348.

KALELA, E. K. (1955). Über Veränderungen in den Wurzelverhältnissen der Kiefernbestände in Laufe der Vegetationsperiode. *Acta Forestalia Fennica* **65**; No. 1, 1–42.

KOLESNIKOV, V. A. (1971). "The Root System of Fruit Plants" (Translated from the Russian). Mir Publishers, Moscow.

KÖSTLER, J. N., BRÜCKNER, E. and BIBELRIETHEN, H. (1968). "Die Wurzeln der Waldbäume" Paul Parey, Hamburg.

LEAPHART, C. D. and COPELAND, O. L. JR. (1957). Root and soil relationships associated with the pole blight disease of western white pine. *Soil Sci. Soc. Am. Proc.* **21**; 551–554.

LYFORD, W. H. and WILSON, B. F. (1966). Controlled growth of forest tree roots: technique and application. Harvard Forest Paper No. 16.

LYR, H. and HOFFMANN, G. (1967). Growth rates and growth periodicity of tree roots. *Int. Rev. For. Res.* **1**; 181–236.

MARX, D. H. and ZAK, B. (1965). The effect of pH on mycorrhizal formation of slash pine in aseptic culture. *For. Sci.* **11**; 66–75.

MEYER, F. H. and GÖTTSCHE, D. (1971) "Integrated Experimental Ecology" (H. Ellenberg, ed.) pp. 48–52. Springer-Verlag, New York–Heidelberg–Berlin.

ORLOV, A. Y. (1957). Observations on absorbing roots of spruce (*Picea excelsa* Link) in natural conditions. *Botanicheskii zhurnal, USSR* **42**; 1172–1181. (English translation, Israel Program for Scientific Translations, Jerusalem).

REID, C. P. P. (1971). *In*: "Mycorrhizae" (E. Hacskaylo, ed.) pp. 222–227. *U.S. Dept. Agric.—Forest Service Misc. Pub.* 1189.

RUEHLE, J. L. (1973). Nematodes and Forest trees—types of damage to tree roots. *A. Rev. Phytopathol.* **11**; 99–118.

SPAULDING, P. and MACALONEY, H. J. (1931). A study of organic factors concerned in the decadence of birch on cut-over lands in northern New England. *J. For.* **29**; 1134–1149.

STOUT, B. B. (1956). Studies of the root systems of deciduous trees. *Black Rock Forest Bull. No.* 15.

SUTTON, R. F. (1969). "Form and Development of Conifer Root Systems." *Commonwealth Forestry Bureau, Tech. Comm. No.* 7, *London.*

WENT, F. W. (1971). *In*: "Mycorrhizae" (E. Hacskaylo, ed.) pp. 230–232. *U.S. Dept. Agric.—Forest Service Misc. Pub. No.* 1189.

WILSON, B. F. (1970). Evidence for injury as a cause of tree root branching. *Can. J. Bot.* **48**; 1497–1498.

WILSON, B. F. and HORSLEY, S. B. (1970). Ontogenetic analysis of tree roots in *Acer rubrum* and *Betula papyrifera*. *Am. J. Bot.* **57**(2); 161–164.

Chapter 10

Distribution of Secondary Thickening in Tree Root Systems

B. F. WILSON

*Department of Forestry and Wildlife Management, University of Massachusetts, Amherst, Massachusetts, U.S.A.**

I. Introduction

In a closed forest, each year approximately 2 tonnes of new tree root material is produced per hectare; this represents about one-fifth of the total annual production of organic matter by the trees (Bray, 1963). Although more information is needed about tree roots, research is slowed by the physical problems of studying

* Manuscript prepared while on sabbatical leave at the Department of Forestry, Australian National University, Canberra.

roots, especially under natural conditions, and by the complexity of the root systems. Most research on tree root systems has been related to gross morphology, or elongation of root tips (Lyr and Hoffmann, 1967; Sutton, 1969). Since the classic paper by von Mohl (1862), few papers have considered secondary thickening of tree roots (see refs. in Fayle, 1968; Riedl, 1937).

This paper is concerned with the distribution of secondary thickening in tree roots and considers the distribution both of dry weight accumulation between the root and shoot systems and of thickening within and between roots in a root

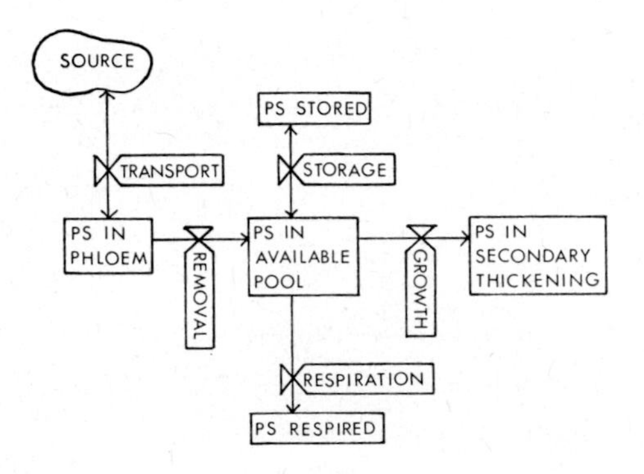

FIG. 1. Diagram of the flow of photosynthate (PS). Following the conventions of Forrester (1968), boxes are levels, or amounts, valve symbols are rates of flow, the cloud is an external source or sink. Feedbacks that determine the rates are omitted.

system. These problems are discussed in relation to the flow of photosynthate (PS) from the shoot system to the root system (Fig. 1). The amount of PS flowing into secondary thickening is regulated by the rates of transport, storage, respiration and growth. To account for variations in amount of secondary thickening, information is needed on the interactions between the factors determining these rates. Throughout this paper an attempt is made to develop the flow chart in Fig. 1 into a useful descriptive model of the process of distributing secondary thickening.

II. Distribution Between the Root and Shoot Systems

A. The Root: Shoot Ratio

The root:shoot ratio (R:S) is a measure of the distribution of dry weight between the root and shoot systems. Due to the difficulty of accurately determining the root weights of large trees, most available measurements are from seedling material. This is unfortunate because the positive and immediate relationship between root and shoot growth seen in seedlings (Richardson, 1956) and in herbaceous plants (Hatrick and Bowling, 1973) is lacking in older trees (Richardson, 1956; Lyford and Wilson, 1966). Much of the relevant data have been summarized by Ledig *et al.* (1970).

Where root dry weight is determined, often no attempt is made to distinguish between dry weight in secondary thickening and in primary tissue. Ovington (1957) found for Scots pine that the percent of total root dry weight of roots greater than 0·5 mm in diameter increased from 0% at age 3 years to a fairly stable 60–70% from 20–55 years. Although these figures probably represent an under-estimate of the percent dry weight in secondary thickening, they provide a good indication of the changing proportions of secondary tissue with increasing age.

Through the use of allometric analysis (linear regression of log shoot weight against log root weight) Ledig *et al.* (1970) showed that many variations in R:S reported under experimental conditions are due to normal changes in R:S with increasing plant size. For instance, in many cases the apparent reduction in R:S due to shading is merely due to the reduced plant size in the shade. Data from Lyr *et al.* (1964), however, suggest that at extremely low light intensities (1% of incident sunlight) PS is preferentially distributed to the shoot system. In two-year-old spruce the root system even lost weight while the shoot system gained. Ledig *et al.* (1970) stated, "conclusions based on the allometric approach . . . indicate a developmental pattern which—within a wide range of environments—operates to maintain a generally balanced growth of root and shoot". Apparently this balanced growth is disturbed only under conditions of marginal PS production. The R:S allometric relations can be shifted experimentally for perennial ryegrass (Hunt and Burnett, 1973), but no similar experiments have been reported for tree seedlings (and may never be reported for large trees).

The studies of R:S suggest mechanisms exist that regulate the distribution of PS between the shoot and root systems. Brouwer and deWitt (1969) give a general model for the role of roots in crop growth. One possible mechanism, the action of root- and shoot-produced hormones on the growth of the plant, is discussed later because it applies to many other aspects of the distribution of secondary growth. Another mechanism, modelled in detail by Borchert (1973), is the balance between the surface for water uptake in the root system and the surface for transpiration in the shoot system. If the root system is too small, water stress develops and stops leaf growth and production until enough roots have been produced to achieve a balance and reduce water stress. Another mechanism which could be particularly important in determining thickening in the zone of rapid taper (ZRT) of roots is the mechanical stimulation by the swaying of the stem (Jacobs, 1939, 1954; Fayle, 1968). Basically, the bigger the stem, the more it sways; the more it sways, the more mechanical stimulation the roots receive so the ZRT becomes thicker. This mechanism will be discussed in detail in Section III, B5.

B. Hormonal Feedback Control

The idea of a hormonal feedback control between the root and shoot systems is not new. In 1938 Went proposed "rhizocalines" and "caulocalines" to explain why, for growth, the shoot system needs hormones from the root system and vice versa for the root system.

There is little research on the hormonal control of secondary thickening in roots. Torrey and Loomis (1967) found that isolated radish roots grown in culture never show secondary thickening when growth regulators are put in the culture medium around the tip. Secondary thickening could be induced, however, by adding auxin and cytokinin through a vial feeding the base of the root, as though the compounds were moving into the root from the shoot system. Fayle and Farrar (1965) have demonstrated that auxin transport is polarized toward the tip in woody tree roots. These results support the conclusions of Brown (1935) and Wight (1933) that growing root tips do not cause cambial activation in roots. Numerous experiments on cambial activity in stems (e.g. Digby and Wareing, 1966) have shown that initiation of normal cambial activity in dormant, disbudded twigs requires the addition of both auxin and gibberellin to the top of the twig. Hejnowicz and Tomaszewski (1969) found that both cytokinin and gibberellin increase auxin-induced cambial activity in intact pines. Wareing *et al.* (1964) noted that experiments with conifer twigs are unsuccessful if the twigs are detached from the root system.

The general conclusions from these various experiments are that auxin, gibberellin and probably cytokinin are involved in the regulation of cambial activity. Although auxin presumably comes from the shoot, both gibberellin and cytokinin could be supplied from the roots. More investigators are finding that cytokinins and gibberellins are present in physiologically active concentrations in the xylem sap moving up the plant from the root system (see chapters by Feldman, pp. 55–72, and by Skene, pp. 365–395, in this volume). At present there is a contradiction between the normal occurrence of cytokinin in xylem sap and the finding of Torrey and Loomis (1967) that for optimal cambial activity in excised radish roots cytokinin had to be added to the base of the root. They did not test whether cytokinin could be added to the tip of the root and only auxin to the base (J. G. Torrey, personal communication). Although yet to be established, these hormones in the xylem sap may be the endogenous source required for normal cambial activity. Thus, cambial activity (in both root and shoot) could require hormones from both roots and shoots.

A simple model of the operation of such a feedback system in the control of secondary growth is given in Fig. 2. If growth regulators are mostly produced by growing cells of the primary tissues, the amount of hormones is largely determined by the amount of growing tissue. The rate of secondary growth is in turn determined by the amount of the hormones. The rate of primary growth in the shoot depends on the rate of primary growth in the root, both through hormonal feedbacks and through nutritional and water stress feedbacks as suggested earlier. Obviously, if photosynthesis and growth slow down in the shoot due to water stress or some other factor, then root growth will slow because of low PS and hormone production by the shoot. In extreme conditions of suppression, in fact,

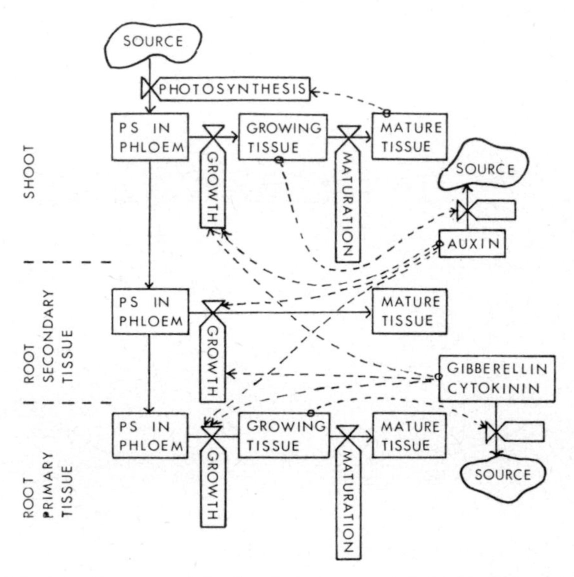

FIG. 2. Model of hormonal feedbacks between the root and shoot systems.

secondary growth only occurs towards the tops of white pine trees (Bormann, 1965), presumably because auxin is exhausted before it can stimulate root thickening. It is the interaction of all these various feedback systems that maintains a relatively constant R:S.

III. Distribution Within the Root System

A. General Morphology and Terminology of Tree Root Systems

Despite differences in morphology within a species growing on one or different sites (Bannan, 1940), and despite changes with age and size, some characteristics are common to most tree root systems (Fig. 3). The most obvious feature of the root system of a tree is the woody framework of roots that have undergone extensive secondary thickening. Seedlings and young trees may have a thickened taproot, but in older trees, as horizontal roots develop, the taproot usually stops growing and may be occluded by growth of other roots. Horizontal roots are usually found in the upper 30 cm of soil. They are relatively straight, branch at irregular intervals of often more than one metre and may be more than 30 metres long. Horizontal roots usually have a zone of rapid taper within a metre or two of the trunk, beyond which they are characteristically "rope-like" with little taper between branch points and sudden decreases in diameter following branching. Sinker roots grow vertically down from horizontal roots, but are usually found only within a few metres of the stem. When vertical roots, sinkers or the taproot reach hard layers of soil they branch frequently, to become brush-like at the end. The woody framework of horizontal and vertical roots bears many short

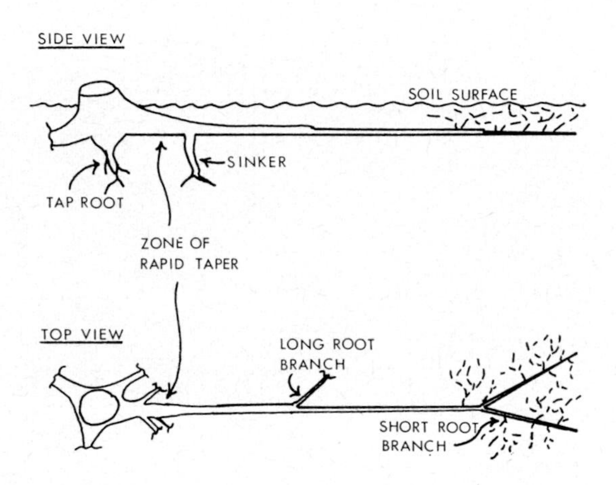

FIG. 3. Diagram of a portion of a tree root system. Long roots are drawn with solid lines and short roots with dashed lines.

root branches that have reduced cambial activity and may only live for a few weeks or years. In many tree root systems the two different types of roots (long and short roots) are clearly separable but in some species there is a general gradation in root size and type (Wilcox, 1964).

B. Distribution Along Roots and Between Branches

Growth ring widths measured at intervals along the main axis of a long root usually follow the somewhat hyperbolic pattern shown in Fig. 4 (Fayle, 1968). With greater distance from the stem, ring width decreases fairly rapidly in the ZRT and less rapidly beyond. Differences between the top and the bottom of the root will be discussed in Section III, D. As the tree grows, the ZRT becomes longer although rarely exceeding 2 m.

Two main features associated with decreasing ring width are (1) the decrease in ring width and root diameter commonly found following branching (Wilson, 1964) and (2) the progressive decrease in ring width that occurs between branching points, i.e. the "natural" taper of the root. Because these two features are independent, they will be discussed separately. As an example of the effects of the two factors, the longest, thinnest roots with the least branching are usually found on suppressed trees, while the roots on dominant trees are usually thicker, shorter and more branched. A third feature, that will not be discussed in detail, is that with increasing distance from the stem the root is younger and, therefore, thinner. This is only significant, for ring width, in the growth ring next to the pith (Wilson and Horsley, 1970).

In a tree root system three types of lateral roots branch from long roots (Fig. 3). In the first type the new lateral is itself a long root which undergoes considerable

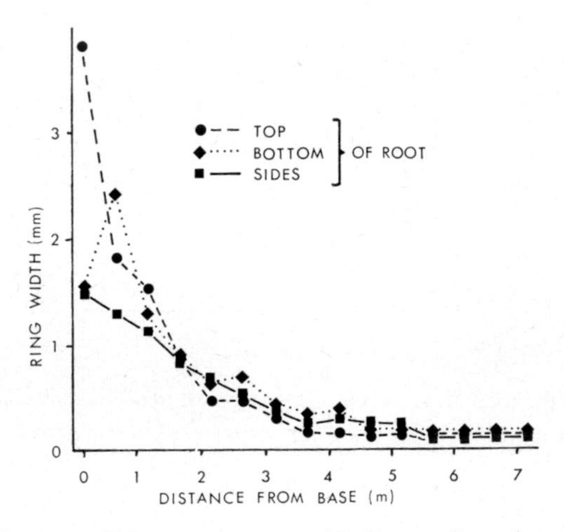

FIG. 4. Changes in ring width around a root with distance from the stem in *Pinus strobus* (mean of 10 roots).

secondary thickening and forms a permanent part of the woody framework of the root system. Long root branches are usually formed by replacement roots following injury to a long root tip (Wilson, 1970; Horsley, 1971), the replacement tips usually having about the same diameter as the parent, injured tip (Horsley, 1971). The second, and most common type, are short root branches. The lateral root has a small diameter tip, is short-lived and undergoes limited secondary thickening. The third root type develops when a short root lateral is converted to a long root through radial enlargement of the root tip (Horsley, 1971; Wilson and Horsley, 1970). This type of branch may become indistinguishable, externally, from the first type.

1. *Long Root Branches*

In long root branches the salient fact about distribution of PS is that the cross-sectional area of the parent root before branching roughly equals the sum of the areas of the branches just after the branching point, even though the total circumference of the branches is greater than of the parent (Fig. 5). A similar kind of relationship occurs in the branches of the shoot system (Zimmermann and Brown, 1971, p. 121).

The fact of relatively constant area despite widely fluctuating cambial circumference has interesting implications for the distribution of secondary thickening. In order to maintain the same rate of growth in cross-sectional area when the cambial circumference has increased, the rate of radial growth must decrease. This behaviour can be understood in terms of the flow diagram shown in Fig. 1.

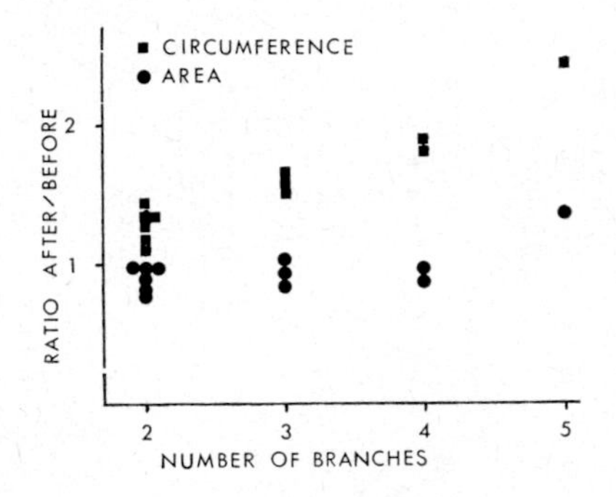

FIG. 5. Ratios of the total areas and circumferences of *Pinus strobus* long root branches just after the branching point to the areas and circumferences of the parent root just before the branching point.

At a long root branch the PS transport system works as though at a given time period there is a unit of PS in the phloem before the branch point. A small percentage of the PS is removed and, after an appropriate time lag and with respiratory and other losses, is converted into secondary thickening. In the next time period the unit of PS, now slightly diminished, is split into smaller units that move into each branch. The mechanism for dividing the unit could be, in the simplest case, the relative transport capacity of each branch. Because the thickness of the phloem appears to be the same in long root branches as in the parent root, the relative circumference of each root would be a reasonable measure of the relative transport capacity. For instance, if one root branch had 60% of the total circumference of all the root branches then it would receive 60% of the PS unit.

Once into the root branches, the unit of PS is spread over a greater phloem circumference than before so the amount per unit circumference would decrease. Given the same phloem thickness before and after branching, then the concentration of PS should diminish. If, for each unit of circumference, the same percentage of PS is removed in the root branches as in the parent root, then the total amount of PS removed will be roughly equal before and after branching. Thus, the total growth in cross-sectional area would also be roughly the same.

2. *Short Root Branches*

The initial difference between short root branches and long root branches is that the short root tip has a much smaller diameter than its parent long root. In short roots the circumference and thickness of the phloem is small relative to the parent root (Table I). This initial low phloem transport capacity may be the cause of both the small amount of secondary growth in short roots and their short life.

TABLE I. Measurements from long roots and their short root laterals in *Acer rubrum*[a]

Root age (yrs)	Parent long root xylem diam. (mm) Primary	S3condary	Phloem Thickness (mm)	Lateral short root xylem diam. (mm) Primary	Secondary	Phloem thickness (mm)	Relative phloem area LR/SR
1	0·29	1·11	0·12	0·13	0·18	0·03	24
1	0·31	1·28	0·21	0·17	0·21	0·01	121
5	0·21	2s34	0·12	0·05	0·56	0·06	8
8	0·18	5·9	0·14	0·06	0·36	0·04	58
10	0·18	5·9	0·14	0·06	0·36	0·04	58
10	0·15	2·76	0·18	0·08	0·28	0·04	45
10	0·24	4·8	0·18	0·15	1·71	0·24	2
10	0·27	4·9	0·21	0·08	0·51	0·08	25

[a] Primary xylem diameters are proportional to root tip diameters when the tip was at the point of measurement. Phloem areas of long roots (LR) and short roots (SR) were calculated by multiplying the secondary xylem circumference by the phloem thickness.

Short roots elongate more slowly than long roots (Lyford and Wilson, 1964) but they may reach lengths of over 30 cm and become much branched. All the PS for this mass of roots must pass through the base of the root which, because of limited secondary thickening, has a relatively constant transport capacity. As the root tips elongate and branches increase, there is a greater demand for PS for growth and respiration. It can be postulated that eventually the demand will approach the supply capacity and the root may be killed if it is unsuccessful in satisfying the respiratory requirement in periods of environmental stress.

This model of short root development is consistent with observations, but does not explain why short roots do *not* undergo secondary thickening while long roots do. Aside from the possible adaptive value of a population of short-lived roots (see chapter by Reynolds, pp. 163–177, in this volume), the answer may lie in the relationship of root tip activity to cambial activity proximal to the tip. Obviously PS moves along the short roots, because it reaches the tips for elongation, but only enough is removed along the root to meet respiratory requirements and little is used in growth. There cannot be any direct "competition" between the tip and the cambial zone for PS because the PS passes the cambial zone before it reaches the tip. Possibly the percentage removed is too small to provide for further growth because a lack of growth hormones limits either the rate of growth or the rate of removal.

Sometimes short roots are converted to long roots. Most lateral short root tips enlarge slightly over time, but few become long roots. If the tip reaches a sufficiently large diameter there is an associated thickening at the base of the root so that it becomes in all respects a long root. One possible interpretation is that

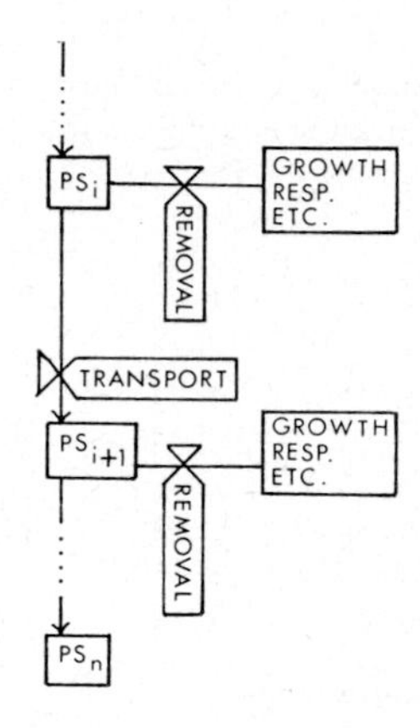

FIG. 6. Model for PS distribution along a root between long root branches.

as the short root tip enlarges it produces more of the growth regulators required for secondary thickening proximal to the tip and the increased phloem transport capacity permits more movement of hormones and PS from the shoot. An interesting parallel is the observation by Torrey (1951) that pea roots growing in culture, which normally do not show cambial activity, develop a cambium and secondary tissues in the main root following injury to the main root tip. This secondary thickening is associated with the development of multiple replacement roots. Possibly an increase in the number of root tips increases the amount of hormones enough to support cambial activity, just as the increase in root tip size may act in the conversion of short roots to long roots.

3. *Distribution Among Long Root Branches*

A segment of root between long root branches can be modelled as a series of short, connected segments each of which removes some PS from the flow in the phloem (Fig. 6). Several basic questions about the functioning of this system can be answered by determining the pattern of secondary thickening along a root, undisturbed by long root branches.

The width of a growth ring generally decreases with increasing distance from the stem between successive long root branches. This decrease is somewhat curvilinear so that the regression of the log of ring width (log y) against distance along the root is consistently a better fit to the data than the regression of ring width (y) on distance along the root in two-year-old maple trees (Table II). Additionally the slopes for regressions of log y on distance are virtually independent of mean ring width ($r^2 = 0{\cdot}10$) while the regressions of y on distance are quite dependent on mean ring width ($r^2 = 0{\cdot}56$). This means that when modelling longitudinal distribution of a growing ring the same slope of log y against distance may be used in the ZRT as in the rest of the root, even though in the ZRT the rings are much wider and taper more rapidly.

Within any one root system there is considerable variation in the slope of log y on distance (standard deviation = 26% of the mean slope). Judging from the limited available data, neither ring width, root tip diameter nor annual elongation of the tip contributes significantly to the variation. Comparing trees of quite different sizes grown under different light intensities, there is no consistent difference between the slope of log y on distance, so there is no apparent effect of tree size (Table II). Knowledge of the major source(s) of variation would greatly assist in modelling the longitudinal distribution of secondary thickening.

Pinus radiata roots have the same type of distribution among root branches as in maple trees. The longest distance between root branches in the maples was 70 cm, but in the pine roots one segment was 270 cm long. Over this distance the outermost xylem growth ring decreased from 0·18 mm to 0·09 mm, while the width of uncollapsed phloem remained at 0·17–0·20 mm (r^2 for log y = 0·856, for y = 0·826). The slope of log y on distance was −0·0011, whereas in 55 maple root segments the slope averaged −0·0245 ± 0·0062 (minimum slope −0·0047). Therefore, the only valid generalization between species, until more data are available, is that there is an exponential decrease in ring width with increasing distance from the stem and this decrease is separate from that attributable to the effect of root branching.

The model developed for distribution of secondary thickening at branching points can be extended to cover the exponential decrease in ring width (Fig. 6). At any specific time a particular segment of a root has a certain amount of PS in the phloem. Over a short period of time a small percentage of the PS is removed and eventually converted to secondary growth. The actual percentage is determined by the amounts used in growth, respiration, etc. In the next time period the PS in the phloem, minus the percentage removed, is transported to the next segment and the whole process is repeated. Thus, as the unit of PS moves along the root, it is gradually depleted. If the percentage removed stays fairly stable, then the amount of secondary thickening will show an exponential decrease as observed. If an *absolute* amount of PS were removed at each point along the root, then the decrease in PS would be linear so that PS and secondary thickening

TABLE II. Regression parameters for the relationship of ring width (y, in 0·01 mm) on distance from the stem (cm) for 2-year-old *Acer rubrum* root systems.

% Sunlight	Root wet wt. (g)	Number of segments	y on distance		log y on distance	
			slope	r^2	slope	r^2
100	63	18	− 2·432 ± 0·596	0·812 ± 0·035	0·0285 ± 0·0050	0·801 ± 0·035
94	82	12	− 1·435 ± 0·307	0·823 ± 0·023	0·0195 ± 0·0073	0·864 ± 0·026
50	20	7	− 2·265 ± 0·856	0·857 ± 0·050	0·0247 ± 0·0083	0·871 ± 0·059
50	29	13	− 2·520 ± 0·354	0·871 ± 0·036	0·0365 ± 0·0079	0·913 ± 0·040
30	14–38[a]	6	− 2·162 ± 0·406	0·833 ± 0·025	0·0235 ± 0·0072	0·912 ± 0·017

[a] 6 trees are combined for the 30% sunlight data.

might cease towards the tips of long roots. In fact, despite discontinuous rings in the root, thickening occurs up to the root tips. The decrease in PS at the end of long roots could, however, be a factor in determining the maximum length of long roots by affecting root elongation.

4. *Changes in Longitudinal Distribution*

Longitudinal distribution of secondary growth can best be changed by restraining the stem from swaying. In restrained trees thickening is reduced both in the lower

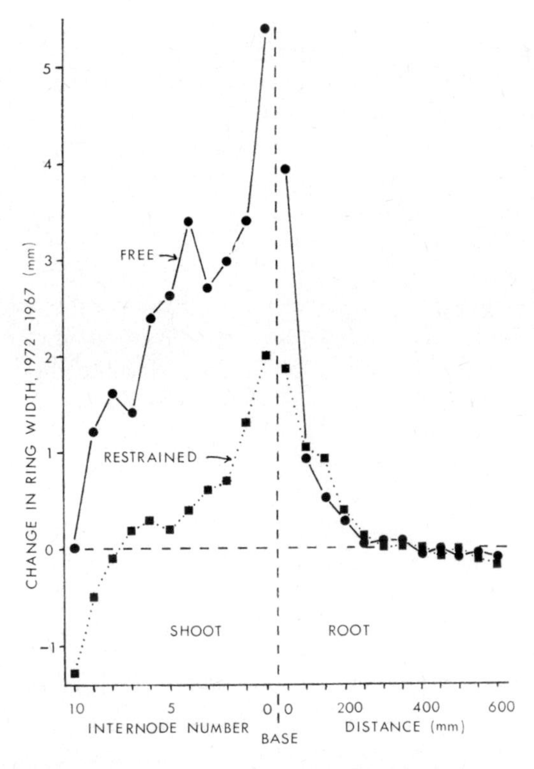

FIG. 7. Change in ring width on top of root 5 years after thinning in restrained and free-to-way *Pinus strobus* trees. Each point is the mean for 5 trees, one stem per tree and two roots.

stem and in the ZRT (Jacobs, 1939, 1954; Fayle, 1968). In a similar experiment at Amherst, Massachusetts a dense stand of *Pinus strobus* with trees 10 m tall was thinned heavily in April 1969. Five trees were restrained by building a rigid, triangular framework around the stem and then guying the framework so that it was immobile. The towers were constructed in 1969 and the 5 restrained trees and 5 free-to-sway trees were cut in June 1973.

Two major points of interest about roots from this study were (1) thinning only increased secondary thickening within the ZRT, (2) stimulation of thickening by stem sway occurred only at the very base of the roots (Fig. 7). The thinned, free-to-sway trees increased dramatically in ring width in the lower stem and in the ZRT with the increase tapering away from the base of the tree in both directions. Beyond 350 cm from the stem there was a slight, but consistent, decrease in root ring width after thinning. Restraining the tree greatly reduced the stimulation of thickening down the stem and at the very base of the root but more than 50 cm from the base the root ring width of the restrained trees was the same as in the free-to-sway trees.

The results are basically in agreement with those of Jacobs (1939, 1954) and Fayle (1968), except that they did not take measurements far enough away from the stem to show a decrease in the effect of sway stimulation. Fayle (1968) also reported some success in stimulating thickening in the ZRT by removing the soil from underneath the ZRT to permit more root movement and illustrated the changes in growth ring width occurring during a drought. Growth decreased in the ZRT but not beyond it, and Fayle states that drought causes growth to shift out along the root.

Since both thinning and drought affect the growth of the whole tree, it is not possible to separate changes in shoot activity from those in root activity. Thus, changes in distribution along a root can only be related to changes in "vigor" of the whole tree. Thinning increases vigor, increases thickening in the ZRT and decreases thickening in the distal portion of the roots. Drought decreases vigor, decreases thickening in the ZRT and has no effect or increases thickening in the distal portions.

These data can be interpreted in terms of the model in Fig. 6. Increased tree vigor could increase both the amount of PS available from the stem and also the rate of removal along the root through a general increase in shoot- and root-produced hormones. If this is the case, thickening would be greatly accelerated near the base of the root, the ZRT, but could be decreased beyond it because of the increased rate of removal. The effect would be reversed when tree vigor was decreased by drought, for the amount of PS and hormones would diminish and the rate of removal would decrease. The ZRT would thicken less while the distal portions of the root might thicken more (Fig. 8).

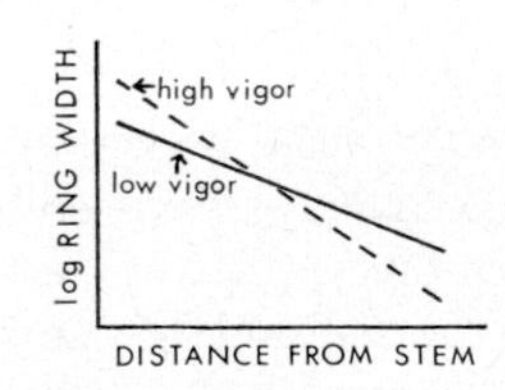

FIG. 8. Model for changes in the distribution of secondary thickening along a root with short term changes in tree vigor.

Although stem sway stimulates thickening at the very base of the ZRT, sway is not, by itself, responsible for the ZRT. The interpretation of restraining experiments is somewhat complicated since restraining reduces thickening over the whole lower stem and ultimately reduces tree vigor (Jacobs, 1954). Although sway stimulation clearly exists, the mechanism cannot be explained fully.

C. Local Stimulation of Thickening

Woody roots do not always decrease in diameter with increasing distance from the stem. There are several reports of root diameter increasing, at least for a while, with increasing distance from the stem (e.g. Stout, 1956). Local stimulation of thickening has been observed after exposure of buried roots to the light (Fayle, 1968), on the concave side of vertically undulating horizontal roots (Rigg and Harrar, 1931) and I have measured increases in diameter just before long root branching points. Any of these local stimulations could account for measured increases in diameter with increasing distance from the stem, and there are probably other factors involved.

The behavior of woody roots exposed to the light is particularly interesting and has a bearing on other topics considered in this paper. If a segment of a woody root is exposed, the exposed segment usually shows increased thickening and the new wood is "shoot-type" wood (Westing, 1965; Fayle, 1968). The growth rings are wider, reaction wood may form in horizontal roots and angiosperm wood has fewer, smaller vessels, more fibres and less parenchyma than in the buried root. When an exposed root is buried, it reverts to producing root-type wood. The reversibility of the developmental patterns is shown by the fact that buried shoots often show decreased thickening and produce root-type wood. Fayle's (1968) experiments suggest that light is the major factor in affecting these developmental changes in cambial activity.

These local effects of light on secondary thickening indicate that (1) there appears to be no intrinsic difference between cambial activity in the root and the shoot and (2) the whole process can be highly modified in a local area. The mechanism for modification is unknown, but probably the differences are related to growth hormones, which in turn affect the rates of growth and of removal of PS. Possibly light stimulates the production of more growth regulator(s) but this seems unlikely, because translocation of the hormone would produce a more general effect. Another possible explanation is that light changes the sensitivity of the system to the hormones, this would allow for the complex set of changes that occur, while keeping the effect local. Almost certainly, light somehow induces the lateral transport of auxin or some other mechanism to cause the unequal auxin distribution associated with reaction wood formation (Wardrop, 1964; Westing, 1965). Clearly, experimental data are needed before the mechanism for this light effect can be understood.

D. Distribution Around a Root

1. *Thickening on the Upper and Lower Sides*

The distribution of secondary thickening between the top and the bottom of a root changes with increasing distance from the stem (Fig. 4). Within the ZRT, growth is predominantly in the vertical plane, often forming plate-like, oval roots. The extreme expression of this tendency is in the buttress or plank roots of many tropical trees. Beyond the ZRT the roots are usually round, although they may be flattened opposite protoxylem poles of diarch roots (Wilson, 1964; Fayle, 1965).

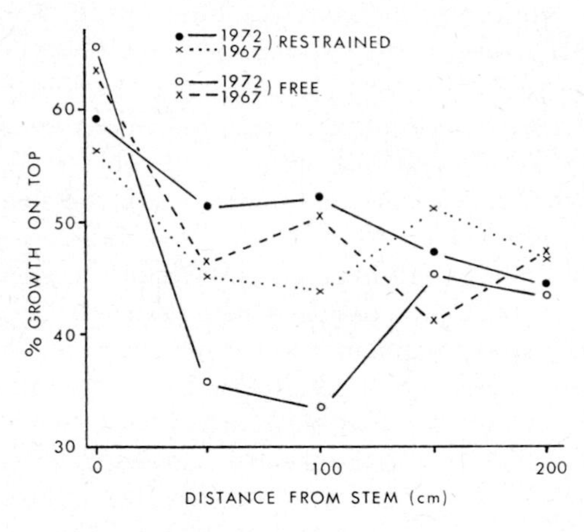

FIG. 9. Changes in distribution of growth between the top and bottom of *Pinus strobus* roots 5 years after thinning in restrained and free-to-sway trees (each point the mean of 10 roots).

In pines, growth is predominantly on top of roots at the base of the ZRT. Within the ZRT the growth maximum switches to the underside. Beyond the ZRT the position of maximum growth is not related to any particular side of the root (Fig. 9, Fayle, 1968). In angiosperm trees that I have worked with (birch and maple) and in root buttresses (Richards, 1964) growth is on the upper side at the base of the ZRT and within the buttress. Beyond the ZRT or the buttress, growth is evenly distributed. I do not know of any measurements for hardwoods showing a shift of the growth maximum to the underside as in pines.

Growth on the upper side is not merely the growth increase associated with reaction wood formation. If this were so, the increased growth would be on top in angiosperms and on the underside in gymnosperms. Conifer and angiosperm roots do have a common structural feature that may be related to the growth on the upper side. As most tree root systems develop, the horizontal roots thicken, fuse at the base with adjacent horizontal roots, and growth of the taproot and other central roots usually stops (Büsgen and Münch, 1929). Apparently the central roots are being "starved" by diversion of PS to the horizontal roots, probably because the vascular tissue at the junction of two adjacent horizontal roots becomes distorted and convoluted, just as in the axil of a branch on a stem. This distorted tissue is probably inefficient for transport and so acts as a partial structural block to movement of PS to the underside of the root. Further along the root, lateral transport would eventually bring PS to the lower side.

In the thinning and restraining experiments described previously (III B4), the treatments markedly changed the distribution of growth between the top and bottom at 50 and 100 cm from the base of the root (Fig. 9). After thinning, growth in the restrained trees shifted towards the top of the root, while growth in trees free-to-sway shifted to the bottom of the root. Statistical analysis of the mean changes from 1967 to 1972 in percent of growth on the top of the root showed significant differences between restrained and free-to-sway trees at both 50 and 100 cm from the base at $p = 0{\cdot}05$. There were no other significant differences in the shifts of percentage growth on the top of the root. Thus, stem sway appears to stimulate growth on the bottom of the root rather than the top. This apparent sway-induced strain stimulus may be related to the shift in growth maximum normally observed in pines.

Excessive growth on the upper side of roots is best developed in buttress and plank roots. Root buttresses are most common on trees growing in the lowland tropics, although their development is genetically determined and subject to some modification by ecological conditions (Richards, 1964; Kozlowski, 1971; Smith, 1972).

The physiological basis for buttress formation is unknown, but may be different only in degree, rather than in kind, from the causes of ZRT formation in temperate trees. It seems unlikely that stress or strain induced by stem sway plays a significant role in buttress formation. Johnson (1972) found no significant correlation between wind direction and the orientation of buttresses. He did find that buttresses were best developed on the "tension" side of leaning stems. This latter observation could result from an extension of the increased growth associated with tension wood formation on the upper side of leaning stems. Another argument against any stress–strain stimulation is that some trees have ribbon-like buttresses that extend many metres from the stem, presumably far beyond the extent of any stress–strain stimulation (Kozlowski, 1971; and see Gill, pp. 237–260, in this volume). In addition, judging from the results in pine, stress–strain stimulation is limited to the very base of the root.

The increased growth of exposed roots (Section III, C) may be another factor in ZRT and buttress formation. The upper part of the ZRT, and of buttresses, is exposed to the light. The wood in the ZRT is more stem-like than root-like (Wilson, 1964), but exposure to light cannot be the only factor, because the ZRT often goes into the buried portion of the root. Also, Gill (pp. 237–260 in this volume) shows the example of *Xylocarpus*, where buttress formation starts underground. It is unclear whether the top of the ZRT is exposed because of growth on the top of the root, or whether exposure has caused growth on top of the root.

2. *Discontinuous Rings*

In the root distal to the ZRT, growth rings are frequently discontinuous around the root (Wilson, 1964; Fayle, 1965) but it is not known if these discontinuous rings are annual. One problem posed by discontinuous rings is that it is difficult, often impossible, to determine the age of a root segment by counting the growth rings. From the point of view of secondary thickening, it seems that a portion of the cambial circumference can go through a cycle of activity with the rest of the cambium inactive.

The inactive portion of the cambium is often, but not always, opposite the protoxylem poles where the lateral short roots emerge. Because these lateral roots are usually in longitudinal rows along the parent root, the parent root may become flattened or even develop longitudinal grooves (Wilson, 1964). The role of growth hormones in cambial activity has been discussed (Section II, B). Cambial activity is assumed to be initiated in roots by auxin moving down from the stem, if cytokinin and gibberellins are present. The absence of cambial activity in many short roots and in some parent roots, near the base of the short roots in the parent root, could be connected. If cambial activity does not occur in short roots due to the lack of root-produced hormones, then this effect should carry along into the parent root through the vascular tissue forming the direct connection between the parent and lateral. When the short root dies, the associated grooves fill in, perhaps because that portion of the cambium is now supplied by the main xylem stream.

IV. Conclusions

Although root systems are tremendously variable, the features common to secondary thickening in different systems allow the creation of a model for the distribution of secondary thickening (Fig. 10). This model is based on the flow diagram in Fig. 1 with some factors governing the rates of flow added. The rates of transport, removal from the phloem and growth have been considered in detail but respiration only briefly. The modelling of these rates will be considered later. The rate of storage was not considered. Clearly, the rate of storage and subsequent release of PS to the available pool could drastically affect the distribution of secondary thickening because storage can be independent of growth. Virtually nothing is known about storage in relation to the distribution of secondary thickening. Ziegler (1964) has reviewed aspects of storage and mobilization.

I have assumed that phloem transport operates basically as a mass flow system with PS thought of as moving through the phloem along the root in relatively discrete packets. The amount of PS may be reduced through removal into the available pool, or it may be split into smaller packets at branching points as some of the phloem goes into the branch. Both PS supply and phloem transport capacity appear to limit the amount of PS that is moved. Distribution of PS at a branching point appears to be determined by the relative transport capacity of the branches. Relative transport capacity is proportional to the cross sectional

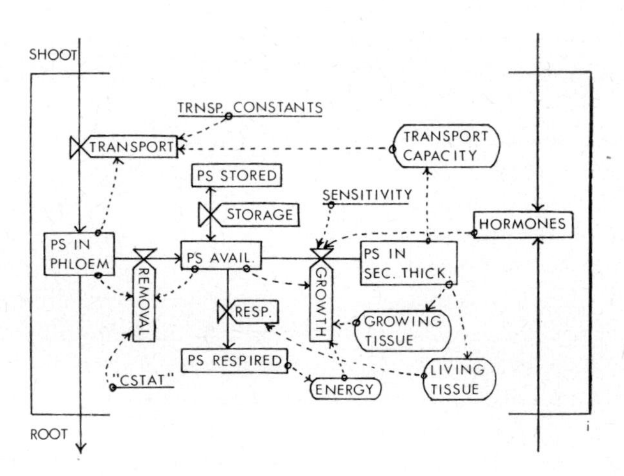

FIG. 10. Model for the flow of PS within a small segment of root. The flows of PS and hormones are solid arrows, the feedbacks determining rates are dashed arrows. Constants are underlined and intermediate values calculated from levels are in rounded boxes. ("C stat" = "concentration stat").

area of transporting phloem and varies with the circumference and thickness of the phloem. Distortion of vascular tissue at the base of horizontal roots may be able to limit transport capacity to the underside of the root. Thus, there is an important positive feedback between the amount of PS diverted into the formation of secondary phloem, which determines transport capacities, and the rate of PS transport (Fig. 10). A long root with rapid secondary thickening has more transport capacity than a slow growing short root. Therefore the long root has more PS in the phloem and is capable of even more growth, while the short root may starve.

The rate of removal from the phloem can be assumed to operate like the "concentrationstat" in the phloem transport model of Zimmermann and Brown (1971, p. 271). The concentrations of PS in the phloem and in the available pool are maintained at an equilibrium. When the equilibrium is disturbed, usually by removal of PS from the available pool, PS is removed from the phloem with the rate of removal determined by the difference in concentrations and by rate constants associated with the removal process (Fig. 10).

The main significance of the rate of respiration in the model is as a sink for PS. Energy produced by respiration is required for all the other rates in the model, perhaps particularly in growth. In this sense, respiration connected with growth is just an inevitable cost in PS associated with any growth. But, respiration must also continue in the absence of growth to maintain the living root tissue. Therefore, there is another feedback between the amount of secondary thickening in living tissue and the rate of maintenance respiration. If, as in short roots, the amount of PS required for maintenance respiration approaches the amount in the available pool, little will be left for growth and, if the available pool decreases, the root may die.

The major factors determining the growth rate can be specified for the model, but their manner of operation is largely unknown. Respiration is required for energy, PS is required for structural material, other growth factors such as minerals and vitamins are needed, but the most complicated and sensitive controlling factors are the growth hormones. I have suggested in this paper that interaction of both root- and shoot- produced hormones is required for secondary thickening. Thus, not only the amount of each hormone is important, but also the relative amounts of each. An additional complicating factor is that the response, or "sensitivity" of a particular root segment to a given level of hormones may be changed by local conditions, for instance by exposure. No quantitative data are available on the amounts of hormone required for root cambial activity in trees. Further development of the growth model will depend on such data.

The ZRT, with its characteristic "excessive" thickening, vertical orientation of thickening and predominant growth on the upper side, seems to be formed by a number of different factors. (1) The exponential decrease in ring width and the high frequency of long root branching near the stem both contribute to a fast rate of taper at the base of the root near the stem. Branching itself accounts for a 50%, or more, reduction in ring width within 50 cm in the ZRT of some pines we have measured. (2) Stem sway stimulates root thickening next to the stem. (3) The exposed portion of the root near the base may be stimulated by light to produce thicker growth rings and "stem-like" wood. (4) The fusion of adjacent horizontal roots may tend to block PS transport to the bottom of the root. At present we do not known if additional factors are involved, particularly in the formation of some of the spectacular types of buttressed roots in the tropics. An interesting suggestion by Wight (1933) is that the ZRT can be thought of as really stem, not root. In this sense, the ZRT and buttresses are merely extensions of shoot wood out over the root system.

A question raised throughout this paper has been: What is the role of the root tip in the secondary thickening of roots? There are several structural relationships between root tips and secondary thickening. Thickening can only occur on the framework provided by the elongation of root tips. Thus, the activity of the root tips is the major determinant of the form of the root system. Probably the initial size of a lateral root determines the phloem transport capacity which, in turn, determines whether the root becomes a long root, or a short root. Therefore, the initial size of a root tip determines whether the root becomes thickened and part of the woody framework, or whether it soon dies.

It has been suggested that the root tips produce growth hormones, cytokinins and gibberellins, and these move in the xylem sap and are required for root cambial activity. Because of the lack of intrinsic differences between root and shoot cambial activity, root-produced hormones are presumably also required for secondary thickening of the shoot. Currently, there is only circumstantial evidence to support this hypothesis. The growing interest and knowledge of root-produced hormones represented by other papers in this volume suggest that the actual role of root-produced hormones in secondary thickening may eventually be described.

Acknowledgements

Original data reported in this paper have been collected over a number of years with the assistance of, among others, Charles Boland, Jane Difley, Ronald Klotz, David Levy and Alan Page. The work was supported by Massachusetts Experiment Station Projects McIntire-Stennis 9 and 10 and National Science Foundation Grants GB-8063 and GK-31490.

I thank the Australian National University, Department of Forestry for providing facilities and assistance and my special thanks go to Professor J. D. Ovington and Drs. E. P. Bachelard and I. A. Wardlaw for their comments on the manuscript.

References

BANNAN, M. W. (1940). The root systems of northern Ontario conifers growing in sand. *Am. J. Bot.* **27**; 108–114.

BORCHERT, R. (1973). Simulation of rhythmic tree growth under constant conditions. *Phys. Plant.* **29**; 173–181.

BORMANN, F. H. (1965). Changes in the growth pattern of white pine trees undergoing suppression. *Ecology* **46**; 269–277.

BRAY, J. R. (1963). Root production and the estimation of net productivity. *Can. J. Bot.* **41**; 65–72.

BROUWER, R. and DEWIT, C. T. (1969). *In*: "Root Growth" (W. J. Whittington, Ed.); pp. 242–242. Butterworth's, London.

BROWN, A. B. (1935). Cambial activity, root habit and sucker shoot development in two species of poplar. *New Phyt.* **34**; 163–179.

BÜSGEN, M. and MÜNCH, E. (1929). "The Structure and Life of Forest Trees" (3rd ed. T. Thomson Trans.) Chapman and Hall, London.

DIGBY, J. and WAREING, P. F. (1966). The effect of applied growth hormones on cambial division and the differentiation of cambial derivatives. *Ann. Bot.* **30**; 539–548.

FAYLE, D. C. F. (1965). Rooting habit of sugar maple and yellow birch. *Can. Dept. For. Publ.* 1120: 1–31.

FAYLE, D. C. F. (1968). Radial growth in tree roots. *Fac. of Forestry Univ. of Toronto Tech. Rep.* 9: 1–183.

FAYLE, D. C. F. and FARRAR, J. L. (1965). A note on the polar transport of exogenous auxin in woody root cuttings. *Can. J. Bot.* **43**; 1004–1007.

FORRESTER, J. W. (1968). "Principles of Systems" (2nd prelim. ed.) Wright-Allen Press, Cambridge, Massachusetts.

HATRICK, A. A. and BOWLING, D. J. F. (1973). A study of the relationship between root and shoot metabolism. *J. Exp. Bot.* **24**; 607–613.

HEJNOWICZ, A. and TOMASZEWSKI, M. (1969). Growth regulators and wood formation in *Pinus sylvestris*. *Phys. Plant* **22**; 984–992.

HORSLEY, S. B. (1971). Root tip injury and development of the paper birch root system. *Forest Sci.* **17**; 341–348.

HUNT, R. and BURNETT, J. (1973). The effects of light intensity and external potassium level on root/shoot ratio and rates of potassium uptake in perennial ryegrass (*Lolium perenne* L.). *Ann. Bot.* **37**; 519–537.

JACOBS, M. R. (1939). A study of the effect of sway on trees. *Comm. For. Bur. Aust. Bull.* **26**; 1–19.

JACOBS, M. R. (1954). The effect of wind sway on the form and development of *Pinus radiata* D.Don. Aust. J. Bot. **2**; 35–51.

JOHNSON, P. W. (1972). Factors affecting buttressing in *Triplochiton scleroxylon* K. Schum. Ghana J. Agri. Sci. **5**; 13–21 (Seen in *For. Abst.* **34**; 520).

KOZLOWSKI, T. T. (1971). "Growth and Development of Trees, Volume II". Academic Press, New York, London and San Francisco.

LEDIG, F. T., BORMANN, F. H. and WENGER, K. F. (1970). The distribution of dry matter growth between shoot and roots in loblolly pine. *Bot. Gaz.* **131**; 349–359.

LYFORD, W. H. and WILSON, B. F. (1964). Development of the root system of *Acer rubrum* L. *Harvard Forest Paper* 10: 1–17.

LYFORD, W. H. and WILSON, B. F. (1966). Controlled growth of forest tree roots: technique and application. *Harvard Forest Paper* 16: 1–21.

LYR, H. and HOFFMANN, G. (1967). Growth rates and growth periodicity of tree roots. *Int. Rev. Forestry Res* **2**; 181–236.

LYR, H., HOFFMANN, G. and ENGEL, W. (1964). Über den Einfluss unterschiedlicher Beschattung auf die Stoffproduktion von Jungpflanzen einiger Waldbaume. *Flora* **155**; 305–330.

MOHL, H. VON (1862). Einige anatomische und physiologische Untersuchungen uber das Holz der Baumwurzeln. *Bot. Zeit.* **20**; 225–230, 233–239, 269–278, 279–287, 289–295, 313–319, 321–327.

OVINGTON, J. D. (1957). Dry matter production by *Pinus sylvestris* L. *Ann. Bot.* **21**; 287–314.

RICHARDS, P. W. (1964). "The Tropical Rain Forest". Cambridge University Press.

RICHARDSON, S. D. (1956). Studies of root growth in *Acer saccharinum* L. III. The influence of seedling age on the short term relation between photosynthesis and root growth. *Proc. Ned. Akad. Wet.* **59C**; 416–427.

RIEDL, H. (1937). Bau und Leistungen des Wurzelholzes. *Jahr. Wiss. Bot.* **85**; 1–72.

RIGG. G. B. and HARRAR, E. S. (1931). The root systems of trees growing in sphagnum. *Am. J. Bot.* **18**; 391–397.

SMITH, A. P. (1972). Buttressing of tropical trees: a descriptive model and new hypothesis. *Am. Nat.* **106**; 32–46.

Stout, B. B. (1956). Studies of the root systems of deciduous trees. *Black Rock Forest Bull.* **15**; 1–45.

Sutton, J. F. (1969). Form and development of conifer root systems. *Comm. Forestry Bur. Oxford, Tech. Comm.* **7**; 1–131.

Torrey, J. G. (1951). Cambial formation in isolated pea roots following decapitation. *Am. J. Bot.* **38**; 596–604.

Torrey, J. G. and Loomis, R. S. (1967). Auxin-cytokinin control of secondary vascular tissue formation in isolated roots of *Raphanus*. *Am. J. Bot.* **54**; 1098–1106.

Wardrop, A. B. (1964). *In*: "The Formation of Wood in Forest Trees" (M. H. Zimmermann ed.) pp. 405–456. Academic Press, New York, London and San Francisco.

Wareing, P. F., Hanney, C. E. A. and Digby, J. (1964). *In*: "The Formation of Wood in Forest Trees" (M. H. Zimmermann ed.) pp. 323–344. Academic Press, New York, London and San Francisco.

Went, F. W. (1938). Specific factors other than auxin affecting growth and root formation. *Plant Physiol.* **13**; 55–80.

Westing, A. H. (1965). Formation and function of compression wood in gymnosperms. *Bot. Rev.* **31**; 381–480.

Wight, W. (1933). Radial growth of the xylem and the starch reserves of *Pinus sylvestris*: a preliminary survey. *New Phyt.* **32**; 77–96.

Wilcox, H. (1964). *In*: "The Formation of Wood in Forest Trees" (M. H. Zimmermann, ed.) pp. 459–479. Academic Press, New York, London and San Francisco.

Wilson, B. F. (1964). Structure and growth of woody roots of *Acer rubrum* L. *Harvard Forest Paper* **11**; 1–14.

Wilson, B. F. (1970). Evidence for injury as a cause of tree root branching. *Can. J. Bot.* **48**; 1497–1498.

Wilson, B. F. and Horsley, S. R. (1970). Ontogenetic analysis of tree roots in *Acer rubrum* and *Betula papyrifera*. *Am. J. Bot.* **57**; 161–164.

Ziegler, H. (1964). *In*: "The Formation of Wood in Forest Trees" (M. H. Zimmermann, ed.) pp. 303–320. Academic Press, New York, London and San Francisco.

Zimmermann, M. H. and Brown, C. L. (1971). "Trees, Structure and Function". Springer, New York.

Chapter 11

The Structure and Function of Roots in Aquatic Vascular Plants

J. M. BRISTOW

Biology Department, Queen's University, Kingston, Ontario.

In aquatic vascular plants, roots may be totally absent. Even when present they constitute a relatively small proportion of the plant biomass compared to those of land plants, and their vascular tissue is often vestigial. These features might suggest that in these plants roots are unimportant and serve merely as anchoring devices. However, there is increasing evidence that hydrophyte roots are functionally similar to those of terrestrial plants. It is the purpose of this review to discuss this evidence and to stress the unique structural and physiological adaptations which roots of aquatic plants exhibit.

I. Structural and Physiological Adaptations to the Aquatic Environment

In the Lentibulariaceae, Ceratophyllaceae, Salviniaceae, and some Lemnaceae roots are never formed. But most hydrophytes, whether floating or attached to the substrate, develop a root system which makes up 1–10% of the plant biomass, (and even as much as 50% in some Nymphaeaceae). The roots are generally adventitious from stems or rhizomes (Fig. 1). Root hairs are, or can be, formed by the majority of species (Shannon, 1953).

In the evolution of aquatic vascular plants from terrestrial ancestors, many structural modifications no doubt have taken place. The marked differences in the root anatomy of typical land and water plants may be regarded as genetic adaptations to certain factors in the aquatic environment, notably the smaller physical stresses to which hydrophytes are generally exposed, and the much slower rates of oxygen and carbon dioxide diffusion in aqueous solution. The anatomy of aquatic vascular plant roots has been reviewed at length by Arber (1920), Guttenberg (1968), and Sculthorpe (1967), and therefore is not treated in detail

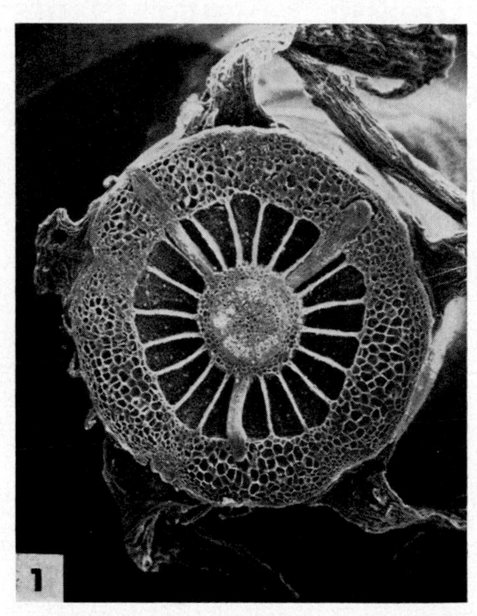

FIG. 1. Scanning electron micrograph illustrating the root structure of hydrophytes. *Myriophyllum brasilense* stem (land form), cut at a node. Three adventitious roots have developed from the central stele close to the diaphragm. $\times 32$.

here. The vascular system is much reduced and poorly lignified especially in Monocotyledons, and in some species of *Elodea*, *Najas*, *Potamogeton*, *Sagittaria* and *Vallisneria* the conducting elements of the xylem degenerate, leaving a central lacuna which probably functions similarly (Figs 2–5). Well-developed sieve tubes and companion cells are present however.

In most species, there are large air spaces continuous with those in the stem and leaves. Depending on species, and age and size of root, these air spaces constitute between 30 and 60% of the total cross-sectional area (Figs 2, 4–7). In many plants the air spaces are intercepted at short intervals by diaphragms (Figs 8–9). These probably serve both for strengthening and to prevent water from entering the air space system in the event of the organ being wounded (Williams and Barber, 1961). Plants lacking an air space system, such as members of the Podostemaceae, are restricted to environments such as fast-flowing streams and waterfalls where oxygen is not a limiting factor (Pannier, 1960).

Several environmental factors affect the normal development of roots, and partially determine species distribution. In species which are anchored, the type of substrate is perhaps the most important factor. Thus in cultures grown in aquarium tanks, roots of *Bidens beckii*, *Elodea canadensis*, *Myriophyllum brasilense* and *Najas flexilis* grew better in gyttja (organic mud) than in gravel while those of *Myriophyllum exalbescens* grew equally well in both types of substrate (J. M. Bristow, unpublished observations). The latter species is the only one which is normally found in both mud and gravel. Rooting depth has not been carefully investigated but in most cases is considerably less than 0·5 metres and is related to the stability of the substrate, being greater in mud than in gravel (Rudescu, 1965; Bjork, 1967). In *Thalassia*, a tropical marine grass, rooting depth depends on E_h, being greater when the substrate surface is well-aerated (Patriquin, 1972). Many emergents such as *Phragmites* and *Typha* form two types of roots, soil- and water-roots, which are morphologically distinct. When the substrate is anaerobic, water-roots are developed to a much greater extent and the soil-roots are sometimes more superficial (Weaver and Himmel, 1930; Dean, 1933; Rudescu, 1965; Bjork, 1967). Removal of water-roots in *Phragmites* reduces plant vigour (Liubich and Arbuzova, 1964), and these roots are perhaps formed in response to anaerobic conditions in the substrate (Weaver and Himmel, 1930). Some free-floating species such as *Ludwigia peruviana* form aerenchymatous air roots for the same reason (Guttenberg, 1968).

In those plants in which a well-developed air space system is formed, there is a ready diffusion of oxygen from the stems or leaves down to the roots, and root growth is not thought to be normally oxygen-limited (Vallance and Coult, 1951; Barber *et al.*, 1962; Teal and Kanwisher, 1966). Some aquatic and bog species such as *Isoëtes lacustris*, *Littorella uniflora*, *Lobelia dortmanna*, *Menyanthes trifoliata*, *Nyssa aquatica*, *Oryza sativa*, and *Spartina alterniflora* excrete oxygen into the substrate. The oxidized substrate in most cases is a more favourable environment for root growth (van Raalte, 1944; Teal and Kanwisher, 1966; Armstrong and Boatman, 1967; Armstrong, 1969; Hook *et al.*, 1972; Wium-Andersen and Andersen, 1972b). Also it has been reported that in British marshes the distribution of *Glyceria maxima* is limited by high redox potentials in the mud (Buttery *et al.*, 1965). However, some species grow better with their roots in an anaerobic substrate (Bergman, 1920; Laing, 1941), and *Thalassia testudinum* actually requires a reduced substrate (Patriquin, 1972).

In *Elodea densa* and several other species, root hairs do not develop unless the roots penetrate the substrate (D'Almeida, 1942; Guttenberg, 1968). This is probably an indirect effect of increased carbon dioxide concentration which reduces the cuticularization of the root epidermis (Dale, 1951). Increased root hair development and the consequent marked increase in the surface area of roots

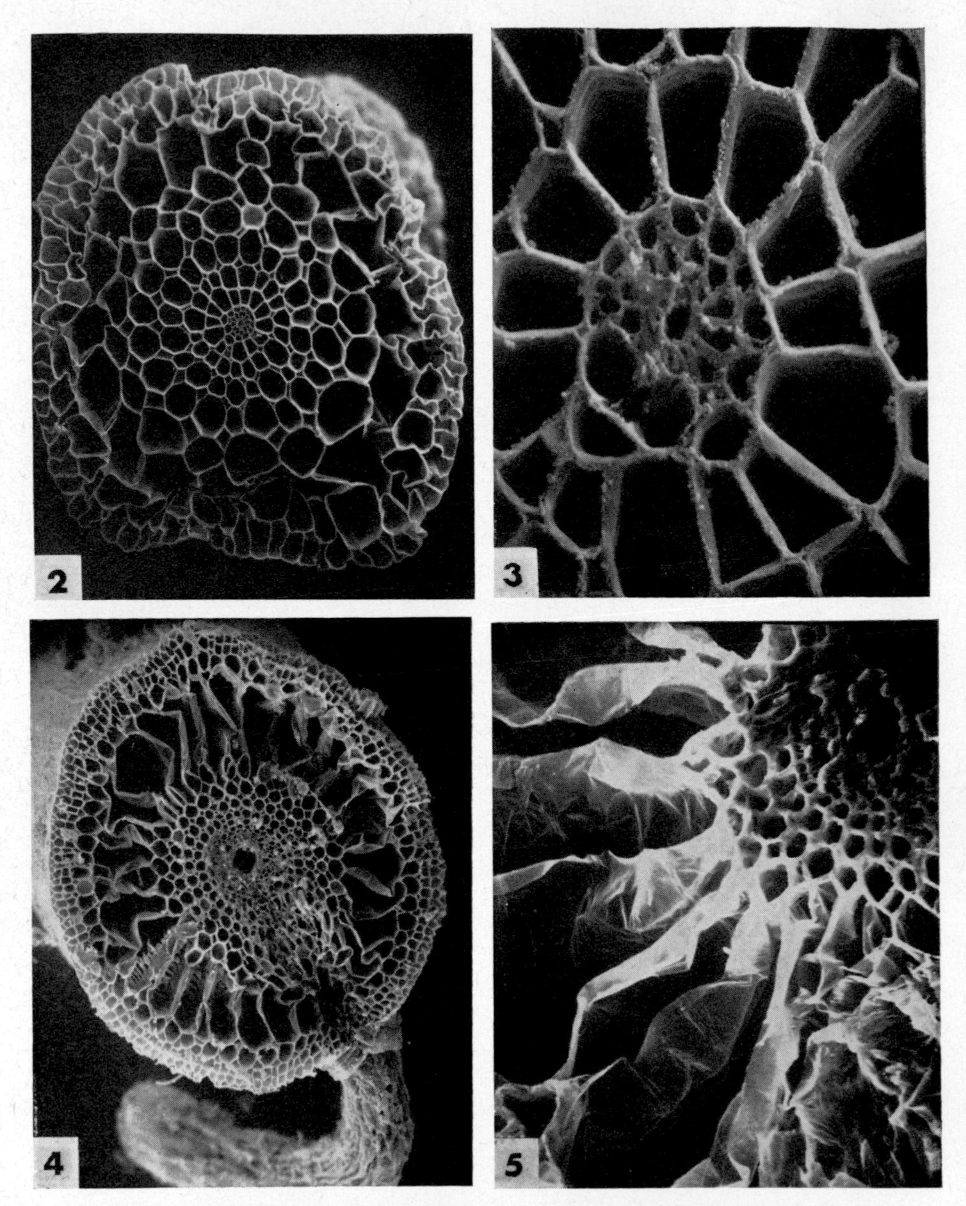

FIGS. 2–5. Scanning electron micrographs illustrating the root structure of hydrophytes. FIG. 2. *Vallisneria* sp. root. ×70. FIG. 3. *Vallisneria* sp. root: Central stele and part of cortex. ×350. FIG. 4. *Sagittaria latifolia* root. ×32. FIG. 5. *S. latifolia* root: Central stele and part of cortical air space tissue. ×70.

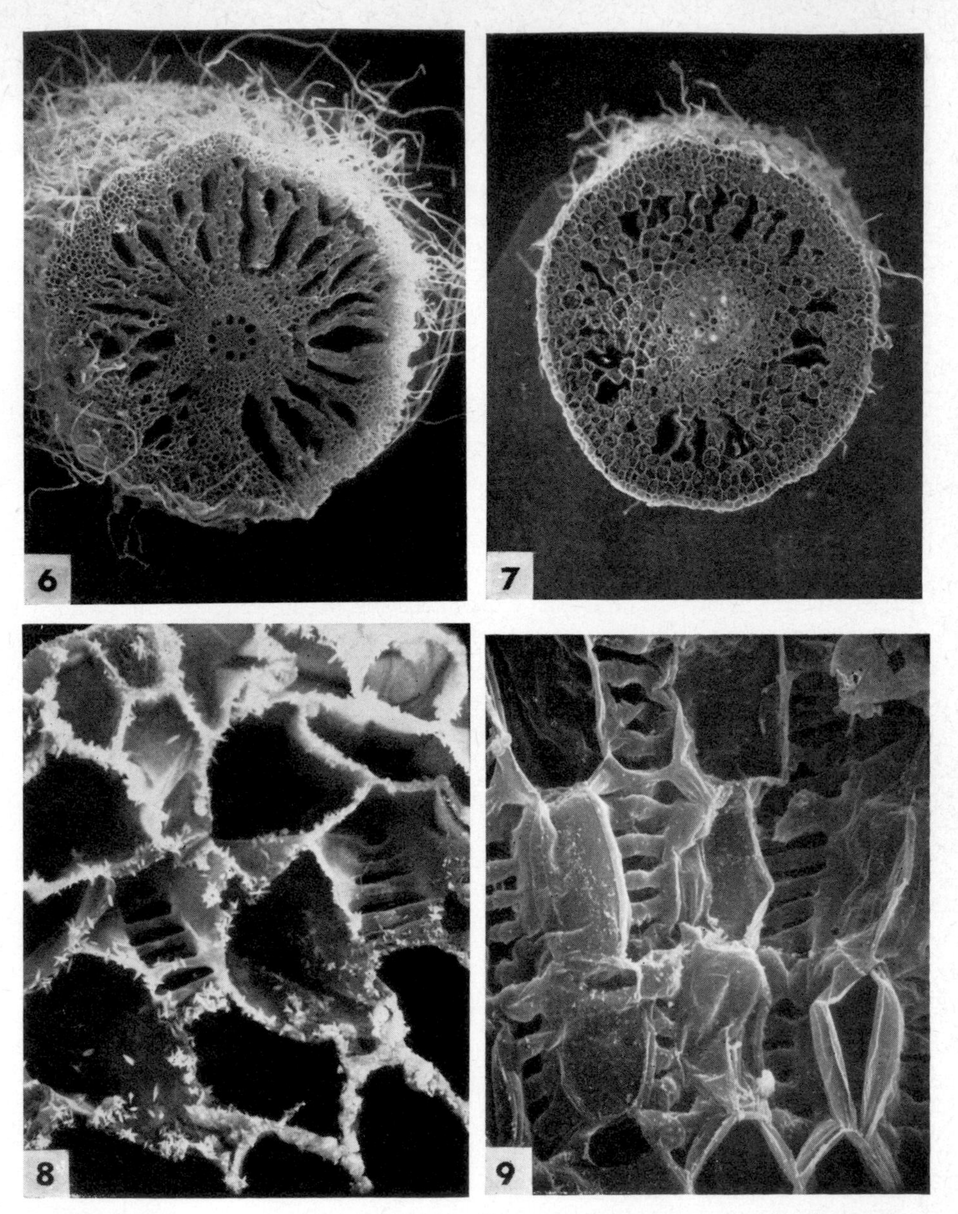

FIGS. 6–9. Scanning electron micrographs illustrating the root structure of hydrophytes. FIG. 6. *Typha* sp. root. ×32. FIG. 7. *Myriophyllum brasilense* root. ×37. FIG. 8. *Vallisneria* sp. root: Part of cortex with diaphragms. ×315. FIG. 9. *S. latifolia* root: Part of cortex with diaphragms. ×315.

The methods of preparation for scanning electron micrographs was as follows: Fig. 1: Fixation in FAA followed by critical point drying. All other sections prepared in the following manner: (1) 2 mm pieces of freshly-cut root were placed in thiourea: t-butyl alcohol (4·4 g 1^{-1}) for 24 h at room temperature, then (2) dried near a radiator for several hours, (3) mounted on stubs with silver paint, and (4) coated with carbon and gold. This formula was originally devised by Mr. Jack Webb of Queen's University Biology Dept. as a means of preserving the natural form and colour of flowers.

probably explains why many species grow better when their roots penetrate the substrate than when suspended in water. Promotion of root growth by darkness is perhaps another factor (Iltis, 1903). In two free-floating species the length of the roots (or their equivalent structures in *Salvinia*) is increased by high light intensity and/or nitrogen starvation (White, 1936, 1938; Gaudet, 1973).

Besides anchoring the plant, roots act as sites of mineral nutrient absorption (see next section) and as sites of synthesis of cytokinins (Waisel and Shapira, 1971) and other unidentified growth substances (Sircar and Ray, 1961) (cf. also Skene, pp. 365–395, in this volume); plastids are often present (Kawamatu, 1965). In free-floating hydrophytes, roots help maintain plant equilibrium and, in species which are attached to the substrate, may exhibit modifications such as spiral coiling which help the plant to anchor itself to what is often an unconsolidated and unstable substrate.

One remarkable characteristic is probably unique to certain species, including *Lobelia dortmanna*, *Isoëtes lacustris* and *Littorella uniflora*, which may grow in lakes where there is relatively little total CO_2 in the water and a rich ground-water supply of free CO_2 in the substrate. In these plants, the root system is relatively massive, and, as has been clearly demonstrated with one of them, *Lobelia*, absorbs CO_2 from the substrate. The gas then diffuses through the air space system to the leaves where it is photosynthetically fixed (Wium-Andersen, 1971; Wium-Andersen and Andersen, 1972a). The leaves of these plants are covered with a thick cuticle and no stomata are developed, so that virtually all gas exchange takes place through the roots.

Hydrophytes growing in waterlogged soil do not form mycorrhizal associations, but the latter may develop in the same plants under drier conditions (Maeda, 1954; Mejstřík, 1965).

II. Mineral Ion Uptake and Translocation

There is good evidence for ammonium, calcium, chloride, iron, phosphate, rubidium and sodium uptake and translocation by roots of several freshwater and marine hydrophytes (Waisel and Shapira, 1971; McRoy and Barsdate, 1970; McRoy *et al.*, 1972; Reimold, 1972; Toetz, 1974; DeMarte and Hartmann, 1974). Most, but not all, authors hold that roots play a significant role in nutrient uptake by aquatic plants (Arber, 1920; Butcher, 1933; Hartog and Segal, 1964; Buttery *et al.*, 1965; Sculthorpe, 1967; Spence, 1967; Martin *et al.*, 1969; Peltier and Welch, 1969; Tomlinson, 1969; Boyd, 1971; Waisel and Shapira, 1971; Denny, 1972; Keeney, 1972; Patriquin, 1972; Wali *et al.*, 1972; Dykyjová and Hradecká, 1973). However, the relative magnitude of ion uptake from the substrate by roots and from the water by leaves of submergent or floating-leaved species remains a matter of debate. It seems reasonable to think that the relative magnitude will vary in any one species according to nutrient availability (McRoy and

Barsdate, 1970; Denny, 1972; Patriquin, 1972), but there is a dearth of experimental evidence on the subject. This is an important question since many sediments are very rich in available nutrients which do not readily diffuse across an oxidized soil–water interface (Boyd, 1971). They could be utilized by rooted plants and thus indirectly contribute to the productivity of aquatic food chains. An attempt has been made in the author's laboratory to gain information on this question, using a two-chamber apparatus, in which the lower portion of the plant including roots may be kept under different conditions from the upper leafy part (Bristow and Whitcombe, 1971).

In the simplest type of experiment, the ability of either roots or leaves to serve as the sole site of mineral nutrient uptake was investigated. *Egeria densa*, *Bidens beckii* and *Myriophyllum exalbescens* were grown for two weeks in the two-compartment apparatus. One of the compartments contained only distilled water, and the other a complete nutrient solution. The lower compartment was painted black to exclude light, and 5% CO_2 was bubbled through both parts. Media were renewed daily. When nutrients were provided in the lower compartment, there was some growth of the leafy shoot though considerably less than when nutrients were provided directly (Table I). The development of the root system

TABLE I. Growth of plants in a two-compartment apparatus over a 2-week growth period when all mineral nutrients were supplied to one of the compartments only (J. M. Bristow, unpublished observations). 1/10 strength Gaudet medium (Gaudet 1963).

Species	Compartment containing nutrients	No. replicates	Av. DW(g)		Length (cm)	
			shoots (upper) comp.	roots (lower) comp.	shoots	roots
Egeria (Elodea) densa	top	2	78·5	5·7	5·3	21
	bottom	3	31·8	8·2	3·8	35
Bidens beckii	top	2	47·1	20·1	6·5	85
	bottom	2	35·5	10·4	4·5	68
Myriophyllum exalbescens	top	3	44·6	4·7	9·3	38
	bottom	6	24·6	11·5	6·6	129

was also more vigorous when roots were bathed in nutrient solution, except in *B. beckii*. Previously, shoots of *Myriophyllum brasilense* developing from axillary buds were found to grow as vigorously when the upper portion of the plant was bathed in nutrient solution as when nutrients were only available to the roots

(Bristow and Whitcombe, 1971). In all four of these species the leafy shoots were normal in morphology and grew considerably in length during the course of the experiment. It may be concluded that all mineral nutrients *can* be obtained via the roots of these plants for a limited period of growth, or alternatively (in the first three species mentioned), via the leaves.

In another series of experiments the uptake of radioactive phosphate was studied. All but one node of the stem was cut off in the top compartment, and an axillary shoot allowed to grow from this node over a ten-day experimental period. The bud was initially very small so that nutrients must have been obtained from the surrounding medium, or from another part of the plant. The nutrient solution was the same in both compartments. Using $^{32}PO_4$ and calculating the specific activity of the phosphate in the newly formed axillary shoot, it was possible to show that a large proportion of the phosphate was derived from the lower compartment medium via the root system (Table II). Different levels of

TABLE II. Estimate of phosphate absorbed by roots and then translocated into axillary shoots over a ten-day growth period. ^{32}P added to the top compartment (Bristow and Whitcombe, 1971). (1085 cpm ml^{-1} in medium supplied; 0·1 μC ^{32}P μMP^{-1}.)

Species	Av. cpm μMP^{-1} in axillary shoot	% total P in axillary shoot derived via roots ± S.E.
E. densa	2857	73·9 ± 3·3
M. exalbescens	4437	59·1 ± 8·0
M. brasilense	169	98·5 ± 1·7

phosphate in the medium of the upper compartment did not significantly alter the pattern of ^{32}P-translocation. Likewise ^{32}P-uptake from the medium of the upper compartment was little affected by the nutrient status of the rooted base, or by the removal of roots (Table III). The amount of ^{32}P taken up by leaves of *M. brasilense* after 48 h was not significantly different from that taken up in 24 h, even though the radioactive medium was renewed after the first day. This result strongly suggests that leaves of this amphibious species are not active in nutrient uptake. Uptake of ^{32}P by leaves of *Zostera* also seems to be limited (McRoy and Barsdate, 1970, Fig. 2).

The results of the ^{32}P experiments lead one to the conclusion that uptake by leaves of submergent or amphibious species is considerably less than that by roots when nutrients are made available to both, and that uptake by the one is not readily influenced by the nutrient environment surrounding the other. In *M. brasilense* (and probably in amphibious and emergent species generally) most

TABLE III. Uptake of ^{32}P from the upper compartment medium, when roots and/or stem base were kept under various conditions. (345 and 742 cpm ml^{-1} in medium supplied for *Myriophyllum brasilense* and *Egeria densa* respectively).

Nutrient status of lower compartment	Duration of expt.	Species	No. replicates	Av. cpm mg^{-1} DW in upper compartment stem + leaves
1. Minus PO^4	24 h	*M. brasilense*	5	251[a]
		E. densa	5	732
2. Minus PO^4	48 h	*M. brasilense*	3	311[a]
		E. densa	5	1286
3. Plus PO^4	24 h	*M. brasilense*	5	240
		E. densa	5	783
4. Plus PO^4; roots removed	24 h	*M. brasilense*	5	185
		E. densa	5	772

[a] Not significantly different at the 5% level.

nutrient uptake occurs via the root system. It seems likely in fact that if nutrients in the substrate become limiting to the plants, the development of water-roots is stimulated. In the experiments already described, roots often developed in the upper compartments when the original root system in the lower compartment was kept in distilled water for more than a few days. From the results shown in Table I, it seems likely that in some submergent species, certain essential nutrient ions are more readily absorbed by leaves, in contrast to the uptake of phosphate. Waisel and Shapira (1971) concluded that uptake and translocation of phosphate and other ions by roots of *Myriophyllum spicatum* (a submergent) are not significant, and that most ion uptake occurs directly through the leaf surface. Differences in methodology are probably sufficient to explain the discrepancy between their results and those of the author. These workers measured uptake of ^{32}P over one or two days into pre-existing shoots, and therefore any ^{32}P translocated from the roots would have been greatly diluted by the non-radioactive phosphate already present in the shoot and roots, as was found by Bristow and Whitcombe (1971).

Potassium absorption in *Lemna minor* may take place at least as readily through the ventral surface of the frond as through the roots themselves (M. Young, personal communication), and the latter may be removed without decreasing the frond multiplication rate (Blackman and Robertson-Cuninghame, 1955).

III. Interaction Between Roots and Microorganisms in the Rhizosphere

Little is known about the rhizosphere of hydrophytes, but it is probable that similar interactions occur there between the root and associated bacteria as in land plants. Considerable quantities of organic metabolites are excreted by aquatic plants (Wetzel and Manny, 1972) and, as in *Oryza* (Macrae and Castro, 1967), it is likely that some of this material is excreted by the roots. In the substrate immediately adjacent to the root, it would have a marked effect on both the types and numbers of bacteria present. Thus in *Thalassia*, numbers of N_2-fixing bateria were found to be 50–300 times higher in rhizospheral than in non-rhizospheral sediments (Patriquin and Knowles, 1972).

The importance of nitrogen fixation by bacteria in the rhizosphere of non-nodulated terrestrial and aquatic species is becoming increasingly evident (Knowles, 1974). Thus *Thalassia testudinum*, a tropical marine grass with a very high productivity, often grows in water containing very little combined nitrogen. Patriquin and Knowles (1972) have concluded that in the Caribbean this species obtains most or all of its nitrogen through the activities of N_2-fixing bacteria in the rhizosphere. Rates of N_2-fixation depend very much on the vigour of plant growth, and in Florida during the winter, much lower N_2-fixing activity was found (McRoy *at al.*, 1974). Other marine grasses and rice probably rely on the activities of these bacteria to varying extents depending on the availability of sources of combined nitrogen in the environment. (Yoshida and Ancajas, 1971; Patriquin and Knowles, 1972; Dommergues *et al.*, 1973). Other aquatic species have also been examined, including *Carex aquatilis*, *Cyperus tetragonus*, *Eichhornia crassipes*, *Hydrocotyle umbellata*, *Hydrilla verticillata*, and mangrove (species unknown). In the absence of an added energy source, it is doubtful whether any of these show evidence of significant rhizospheral N_2-fixation (McRoy and Alexander, 1973; Silver and Jump, 1974).

As in soybeans (Bach *at al.*, 1958), metabolites excreted by roots no doubt provide an energy source for the N_2-fixing bacteria, since N_2-fixation in non-rhizospheral sediments is usually much lower. N_2-fixation in the rhizosphere of rice is also closely correlated with the photosynthetic activity of the plant, presumably due to the rapid excretion of photosynthetic products through the roots (Dommergues *et al.*, 1973). N_2-fixing bacteria are also reported to occur in large numbers on the roots of two free-floating species, *Lemna minor* and *Azolla caroliniana* (Reinke, 1904), and on the leaves of the water hyacinth, *Eichhornia crassipes* (Iswaran *et al.*, 1973). The latter species often occurs in conditions where there are only low levels of combined nitrogen, and the ammonia formed by the bacteria is thought to be either absorbed directly or washed down the stem and absorbed by the roots.

The conditions favouring N_2-fixation in the sediments surrounding the roots of rice and seagrasses also pertain in natural freshwater environments. This fact prompted the author to undertake a study of N_2-fixation associated with two freshwater plants, *Typha* sp. and *Glyceria borealis* (Bristow, 1974). The acetylene reduction method was used as an assay. Plants were collected from the field and incubated the same day, in the presence or absence of rhizosphere soil, and under various experimental conditions. Ethylene production showed a 12 h lag period in *Glyceria*, but no lag in *Typha*; after this time the rate remained essentially constant for three days. The rates of ethylene production averaged over three days are shown in Tables IV and V.

TABLE IV. Rates of acetylene reduction in *Glyceria* (Bristow, 1974)

System	nmoles C_2H_4 g^{-1} root + rhizomes day^{-1} or g^{-1} rhizosphere soil day^{-1} [a]	
	Expt. I	Expt. II
Unamended		
1. Anaerobic. Plant and rhizosphere soil.	51,200[a]	24,700[A]
2. Aerobic. Plant + rhizosphere soil.	28,300[b]	—
3. Anaerobic. Washed plant.	19,000[d]	6,800[C]
4. Aerobic. Washed plant.	8,100[e]	—
5. Anaerobic. Rhizosphere soil.	20[f]	20[D]
6. Aerobic. Rhizosphere soil.	10[f]	—
Glucose amended		
7. Anaerobic. Plant and rhizosphere soil.	193,900[c]	106,300[B]

[a] Treatments in the same experiment with different superscripts are significantly different at the 5% level of significance.

High rates of acetylene reduction were found in all systems which included the plant rootstocks, especially in *Glyceria*. In both species the highest rates (except when glucose was added as an energy source) were found when the plants were incubated anaerobically with the sediment still adhering to them (treatment 1). The effect of washing roots was to decrease acetylene reduction with *Glyceria* but not *Typha*. Aerobic conditions in most cases decreased the rates of reduction. No ethylene was detected in controls which were incubated in the absence of acetylene, so that the ethylene production was not a result of normal metabolism or damage to the plant.

TABLE V. Rates of acetylene reduction in *Typha* (Bristow, 1974).

System	nmoles C_2H_4 g^{-1} root + rhizome day^{-1} or g^{-1} rhizosphere soil day^{-1} [a]	
	Expt. I	Expt. II
1. Anaerobic .Plant + rhizosphere soil.	3400[a]	3500[A]
2. Aerobic. Plant + rhizosphere soil.	400[b]	1950[A]
3. Anaerobic. Washed plant.	2900[a]	2200[A]
4. Anaerobic. Rhizosphere soil.	1[c]	20[B]

[a] Treatments in the same experiment with different superscripts are significantly different at the 5% level.

Assuming a ratio of 3 moles of acetylene to 1 mole of N_2 reduced (Hardy *et al.*, 1973; Patriquin and Knowles, 1972), the rates of N_2 fixation in plant plus sediment systems (treatment 1) can be estimated, from the mean of the two experiments, to be 1·2 μ M $N_2(C_2H_2)$ g^{-1} roots and rhizomes day^{-1} and 12·6 μ M $N_2(C_2H_2)$ g^{-1} day^{-1} for *Typha* and *Glyceria* respectively. Taking into account various factors, including the biomass of roots and shoots, the length of the growing season, and the Q_{10} of the N_2-fixing system (4·5 for *Glyceria*), a rough estimate can be made of the percentage of the standing crop nitrogen requirements which might be supplied by N_2-fixation. In *Typha*, this would not exceed 10%, while in *Glyceria* it might reach 50% or more, depending on the availability of combined nitrogen. The rate of $N_2[C_2H_2]$ fixation in *Glyceria* may be estimated to be about 60 kg N_2 ha^{-1} yr^{-1}. This rate of fixation may be compared to a range of 100–220 kg N_2 ha^{-1} yr^{-1} in legumes (Stewart, 1966), 100–1700 kg N_2 ha^{-1} yr^{-1} in *Thalassia* (Patriquin and Knowles, 1972), and 115 kg N_2 ha^{-1} yr^{-1} in rice (Yoshida and Ancajas, 1973). A knowledge of the extent of N_2-fixation associated with aquatic roots under a variety of natural conditions is important, and this aspect obviously deserves much more detailed study.

It seems clear even from the relatively few studies discussed in this paper that the roots of aquatic vascular plants perform a variety of functions, which are important and often essential for the normal growth of the plant. Most of the physiological experiments demonstrating this fact have been carried out only in recent years and, for the most part, the remarkable structural and physiological adaptive features of hydrophytes are still rather poorly understood. The study of these adaptations including those relating to root function should provide a fertile field for future investigations.

Acknowledgements

The author wishes to thank Drs. Adele Crowder and Hugh Dale for their critical reading of the manuscript, and also Jack Webb for his help in the preparation of plant material for the scanning electron microscope and with the photography.

References

ARBER, A. (1920). "Water Plants: A Study of Aquatic Angiosperms". University Press, Cambridge.

ARMSTRONG, W. (1969). Rhizosphere oxidation in Rice: an analysis of intervarietal differences in oxygen flux from roots. *Physiol. Plant.* **22**; 296–303.

ARMSTRONG, W. and BOATMAN, D. J. (1967). Some field observations relating the growth of bog plants to conditions of soil aeration. *J. Ecol.* **55**; 101–110.

BACH. M. K., MAGEE, W. E. and BURRIS, R. H. (1958). Translocation of photosynthetic products to soybean nodules and their role in nitrogen fixation. *Plant Physiol.* **33**; 118–124.

BARBER, D. A., EBERT, M. and EVANS, N. T. S. (1962). The movement of $^{15}O_2$ through barley and rice plants. *J. Exp. Bot.* **13**; 397–403.

BERGMAN, H. F. (1920). The relation of aeration to the growth and activity of roots and its influence on the ecesis of plants in swamps. *Ann. Bot.* **34**; 13–33.

BJORK, S. (1967). Ecologic investigations of *Phragmites communis*. *Folia Limnol. Scand.* **14**; 1–248.

BLACKMAN, G. E. and ROBERTSON-CUNINGHAME, R. C. (1955). Inter-relationships between light intensity, temperature, and the physiological effects of 2,4-dichlorophenoxyacetic acid on the growth of *Lemna minor*. *J. Exp. Bot.* **6**; 156–176.

BOYD, C. E. (1971). The limnological role of aquatic macrophytes and their relationship to reservoir management. pp. 153–166. *In*: "Reservoir Fisheries and Limnology", (Ed. G. E. Hall) Special Publ. No. 8. *Amer. Fisheries Society, Washington, D.C.*

BRISTOW, J. M. (1974). Nitrogen fixation in the rhizosphere of freshwater angiosperms. *Can. J. Bot.* **52**; 217–221.

BRISTOW, J. M. and WHITCOMBE, M. (1971). The role of roots in the aquatic vascular plants. *Amer. J. Botany* **58**; 8–13.

BUTCHER, R. W. (1933). Studies on the ecology of rivers. I. On the distribution of macrophytic vegetation in the rivers of Britain. *J. Ecol.* **21**; 58–91.

BUTTERY, B. R., WILLIAMS, W. T. and LAMBERT, J. M. (1965). Competition between *Glyceria maxima* and *Phragmites communis* in the region of Surlingham Broad. I. The competition mechanism. *J. Ecol.* **53**; 163–81.

DALE, H. M. (1951). Carbon dioxide and root hair development in *Anacharis* (*Elodea*). *Science, N.Y.* **114**; 438–39.

D'ALMEIDA, J. F. R. (1942). A contribution to the study of the biology and physiological anatomy of Indian marsh and aquatic plants. *J. Bombay nat. Hist. Soc.* **43**; 92–96.

DEAN, E. B. (1933). Effect of soil type and aeration upon root systems of certain aquatic plants. *Pl. Physiol., Lancaster*, **8**; 203–22.

DEMARTE, J. A. and HARTMANN, R. T. (1974). Studies on absorption of ^{32}P, ^{59}Fe, and ^{45}Ca by water milfoil (*Myriophyllum exalbescens* Fernald). *Ecology* **55**; 188–194.

DENNY, P. (1972). Sites of nutrient absorption in aquatic macrophytes. *J. Ecol.* **60**; 819–829.

DOMMERGUES, Y., BALANDREAU, J., RINAUDO, G., and WEINHARD, P. (1973). Non-symbiotic nitrogen fixation in the rhizospheres of rice, maize, and different tropical grasses. *Soil Biol. Biochem.* **5**; 83–89.

Dykyjová, D. and Hradecká, D. (1973). Productivity of reed-bed stands in relation to the ecotype, microclimate and trophic conditions of the habitat. *Pol. Arch. Hydrobiol.* **20**; 111–119.

Gaudet, J. (1963). *Marsilea vestita*: conversion of the water form to the land form by darkness and by far-red light. *Science, N.Y.* **140**; 975–976.

Gaudet, J. (1973). Growth of a floating aquatic weed, *Salvinia*, under standard conditions. *Hydrobiologia* **41**; 77–106.

Guttenberg, H. von (1968). Die Wurzel der Hydro- und Hygrophyten. pp. 260–303. *In*: "Handbuch der Pflanzenanatomie" VIII 5. Gebruder Borntraeger, Berlin.

Hardy, R. W. F., Burns, R. C. and Holsten, R. D. (1973). Applications of the acetylene-ethylene assay for measurement of nitrogen fixation. *Soil Biol. Biochem.* **5**; 47–81.

Hartog, C. den and Segal, S. (1964). A new classification of water-plant communities. *Acta bot. Neerl.* **13**; 367–393.

Hook, D. D., Brown, C. L. and Wetmore, R. H. (1972). Aeration in trees. *Bot. Gaz.* **133**; 443–454.

Iltis, H. (1903). Über den Einfluss von Licht und Dunkel auf das Langenwachstum der Adventivwurzeln bei Wasserpflanzen. *Ber. deutsch. bot. Ges.* **21**; 508–517.

Iswaran, V., Sen, A. and Rajne Apte. (1973). *Azotobacter chroococcum* in the phyllosphere of water hyacinth (*Eichhornia crassipes* Mort. Solms). *Plant and Soil* **39**; 461–463.

Kawamatu, S. (1965). Electron microscope observations on the leaf of *Azolla imbricata* Nakai. *Cytologia* **30**; 80–87.

Keeney, D. R. (1972). "The Fate of Nitrogen in Aquatic Ecosystems". Literature Review No. 3. 60 pp. The University of Wisconsin, Water Resources Center, Madison, Wisconsin.

Knowles, R. (1975). The significance of asymbiotic dinitrogen fixation by bacteria. *In*: "Dinitrogen Fixation" ed. by R. W. F. Hardy *et al.* Wiley Interscience, N.Y.

Laing, H. E. (1941). Effect of concentration of oxygen and pressure of water upon growth of rhizomes of semi-submerged water plants. *Bot. Gaz.* **102**; 712–24.

Liubich, P. P. and Arbuzova, L. R. (1964). Biologicheskoye znacheniye vodnik pridatochnik korni ou *Phragmites communis* Trin. (The biological significance of aquatic adventitious roots in *Phragmites communis* Trin.). *Bot. Zhurnal.* **49**; 1299–1301.

Macrae, I. C. and Castro, T. F. (1967). Root exudates of the rice plant in relation to Akagare, a physiological disorder of rice. *Plant. and Soil* **26**; 317–323.

Maeda, M. (1954). The meaning of mycorrhiza in regard to systematic botany. *Kumamoto J. Sci., Ser. B.*, **3**; 57–84.

Martin, J. B. Bradford, B. N. and Kennedy, H. B. (1969). "Factors Affecting the Growth of *Najas* in Pickwick Reservoir (Tennessee)." 47 pp. National Fertilizer Development Centre, Tennessee Valley Authority, Muscle Shoals, Ala.

McRoy, C. P. and Alexander, V. (1973). Nitrogen kinetics in emergent aquatic plants in arctic Alaska. *Abstr. Amer. Soc. Limnol. Oceanogr. 36th Ann. meeting, Salt Lake City, Utah.*

McRoy, C. P. and Barsdate, R. J. (1970). Phosphate absorption in eelgrass. *Limnol. Oceanogr.* **15**; 6–13.

McRoy, C. P., Barsdate, R. J. and Nebert, M. (1972). Phosphorus cycling in an eelgrass (*Zostera marina* L.) ecosystem. *Limnol. Oceanogr.* **17**; 58–67.

McRoy, C. P., Goering, J. J. and Chaney, B. (1974). Nitrogen fixation associated with seagrasses. *Limnol. Oceanogr.* **18**; 998–1002.

Mejstřík, V. (1965). Study on the development of endotrophic mycorrhiza in the association of *Cladietum marisci*. *In*: "Plant Microbe Relationships" pp. 283–290. *Czech. Acad. Sci. Prague.*

Pannier, F. (1960). Physiological responses of Podostemaceae in their natural habitat. *Int. Revue ges. Hydrobiol.* **45**; 347–354.

PATRIQUIN, D. G. (1972). The origin of nitrogen and phosphorus for growth of the marine angiosperm, *Thalassia testudinum*. *Mar. Biol.* **15**; 35–46.

PATRIQUIN, D. G. and KNOWLES, R. (1972). Nitrogen fixation in the rhizosphere of marine angiosperms. *Mar. Biol.* **16**; 49–58.

PELTIER, W. H. and WELCH, E. B. (1969). Factors affecting growth of rooted aquatics in a river. *Weed Science* **17**; 412–416.

RAALTE, M. H. VAN. (1944). On the oxidation of the environment by the roots of rice (*Oryza sativa* L.). *Ann. bot. Gdn. Buitenz., Hors Série* 15–33.

REIMOLD, R. J. (1972). The movement of phosphorus through the salt marsh cord grass, *Spartina alterniflora* Loisel. *Limnol. Oceanogr.* **17**; 606–611.

REINKE, J. (1904). Zur kenntnis der Lebensbedingungen von *Azotobacter*. *Ber. deutsch. bot. Ges.* **22**; 95–100.

RUDESCU, L. (1965). Neue biologische Probleme bei den Phragmiteskulturarbeiten im Donandelta. *Arch. Hydrobiol Suppl.* **30**; 80–111.

SCULTHORPE, C. D. (1967). "The Biology of Aquatic Vascular Plants". Edward Arnold, London.

SHANNON, E. L. (1953). The production of root hairs by aquatic plants. *Am. Midland Nat.* **50**; 474–479.

SILVER, W. S. and JUMP, A. (1974). Nitrogen fixation associated with vascular aquatic macrophytes. Paper presented at IBP-PP Synthesis Meeting entitled "Nitrogen fixation and the biosphere" Edinburgh, September 1973.

SIRCAR, S. M. and RAY, A. (1961). Growth substances separated from the root of water hyacinth by paper chromatography. *Nature, Lond.* **190**; 1213–1214.

SPENCE, D. H. N. (1967). Factors controlling the distribution of freshwater macrophytes with particular reference to the lochs of Scotland. *J. Ecol.* **55**; 147–170.

STEWART, W. D. P. (1966). "Nitrogen Fixation in Plants". Athlone Press, London, 168 pp.

TEAL, J. M. and KANWISHER, J. W. (1966). Gas transport in the marsh grass *Spartina alterniflora*. *J. Exp. Bot.* **17**; 355–361.

TOETZ, D. W. (1974). Uptake and translocation of ammonia by freshwater hydrophytes. *Ecology* **55**; 199–201.

TOMLINSON, P. B. (1969). On the morphology and anatomy of turtle grass, *Thalassia testudinum* (Hydrocharitaceae). II. Anatomy and development of the root in relation to function. *Bull. Mar. Sci.* **19**; 57–91.

VALLANCE, K. B. and COULT, D. A. (1951). Observations on the gaseous exchanges which take place between *Menyanthes trifoliata* L. and its environment. *J. exp. Bot.* **2**; 212–222.

WAISEL, Y. and SHAPIRA, Z. (1971). Functions performed by roots of some submerged hydrophytes. *Israel J. Bot.* **20**; 69–77.

WALI, M. R., GRUENDLING, G. K. and BLINN, D. W. (1972). Observations on the nutrient composition of a freshwater lake ecosystem. *Archiv. Hydrobiol.* **69**; 452–464.

WEAVER, J. E. and HIMMEL, W. J. (1930). The relation of increased water content and decreased aeration to root development in hydrophytes. *Pl. Physiol., Lancaster*, **5**; 69–92.

WETZEL, R. G. and MANNY, B. A. (1972). Secretion of dissolved organic carbon and nitrogen by aquatic macrophytes. *Verh. int. verein. Limnol.* **18**; 162–170.

WHITE, H. L. (1936). The interaction of factors in the growth of *Lemna*. VIII. The effect of nitrogen on growth and multiplication. *Ann. Bot.* **50**; 403–417.

WHITE, H. L. (1938). The interaction of factors in the growth of *Lemna*. XIII. The interaction of potassium and light intensity in relation to root length. *Ann. Bot. N.S.* **2**; 911–917.

WILLIAMS, W. T. and BARBER, D. A. (1961). The functional significance of aerenchyma in plants. *Symp. Soc. exp. Biol.* **15**; 132–144.

WIUM-ANDERSEN, J. (1971). Photosynthetic uptake of free CO_2 by the roots of *Lobelia dortmanna*. *Physiol. Plant.* **25**; 245–248.

WIUM-ANDERSEN, J. and ANDERSEN, J. M. (1972a). Carbon dioxide content of the interstitial water in the sediment of Grane Langsø, a Danish *Lobelia* lake. *Limnol. Oceanogr.* **17**; 943–947.

WIUM-ANDERSEN, J. and ANDERSEN, J. M. (1972b). The influence of vegetation on the redox profile of the sediment of Grane Langsø, a Danish *Lobelia* Lake. *Limnol. Oceanogr.* **17**; 948–952.

YOSHIDA, T. and ANCAJAS, R. (1971). Nitrogen fixation by bacteria in the root zone of rice. *Proc. Soil Sci. Soc. Amer.* **35**; 156–158.

YOSHIDA, T. and ANCAJAS, R. (1973). Nitrogen-fixing activity in upland and flooded rice fields. *Proc. Soil Sci. Soc. Amer.* **37**; 42–46.

Chapter 12

Aerial Roots: An Array of Forms and Functions

A. M. GILL

CSIRO Division of Plant Industry, P.O. Box 1600, Canberra, City, A.C.T. Australia.

and

P. B. TOMLINSON

Harvard University, Harvard Forest, Petersham, Massachusetts, U.S.A.
(and Research Collaborator, Fairchild Tropical Garden, Miami, Florida, U.S.A.)

I. Introduction

Roots are often regarded as uniform in contrast to the greater diversity of shoots. In general this is probably true but this attitude is largely the result of our ignorance of roots. Aerial roots, which are quite common in tropical plants, provide the opportunity to witness a wide spectrum of form and function in roots which shows that they are capable of more diversity than is generally appreciated. It is our object to draw attention to this diversity.

The term "aerial root" has been used in three rather arbitrarily differing senses, which may be considered separately:

1. Roots arising from an aerial stem, but still embedded within, or enclosed by other plant tissues or organs. Examples include the aerial roots of some tree ferns (e.g. *Cyathea australis*) which remain enclosed within a thick mantle of old, hairy leaf bases. In Southeastern Australia where this tree-fern grows the roots rarely appear to reach the ground, but this may provide some protection against occasional fires. A number of monocotyledons, notably in the families Bromeliaceae, have "intracauline" roots, i.e. adventitious roots which originate at the periphery of the central cylinder, grow downwards through the cortical tissues of the parent stem usually breaking through the surface layers to become exposed. In the Bromeliaceae this feature may be correlated with epiphytism or succulence, but it occurs also in terrestrial species. In the erect stems of larger Velloziaceae, the surface roots may form a rigid mantle of some mechanical significance (Weber, 1953, 1954). Similar roots occur in *Kingia australis* (Xanthorrhoeaceae) (I. Staff, personal communication).
2. Roots which arise from an aerial (or even a subterranean) organ, become exposed, but remain appressed to some adjacent surface, e.g. the surface of a host plant if an epiphyte, or of a rock if a lithophyte. The roots of many woody lianes are in this category where they may function as "clasping roots", providing anchorage, or as "feeding roots", on humus accumulated in the crotch between tree branches. The dimorphism of this kind of aerial root within a single individual is often striking and is characteristic of certain families or genera, notably

climbers in the Araceae and *Freycinetia* in the Pandanaceae. The topic was discussed in detail as long ago as 1895 by F. A. F. C. Went. Temperate climbers may show similar clasping roots, e.g. *Toxicodendron* spp. (Anacardiaceae); *Hedera* (Aquifoliaceae); some *Metrosideros* spp. (Myrtaceae) such as *M. fulgens* Sol. ex Gaertn., *M. scandens* J.R. & G. Forst. of New Zealand; *Pieris phillyreifolius* Hook. [=*Ampelothamnus phillyreifolius* (Hook.). Small] an unusual climber in the Ericaceae.

An extremely specialized example of this kind of root is provided by those genera of epiphytic orchids (e.g. *Angraecum*, *Doritis*, *Harrisiella*, *Polyrhiza*, *Sacrochilus*, *Taeniophyllum*, *Thrixspermum*) which have flattened, photosynthetic roots, the shoot system being virtually leafless and largely an inflorescence. Energy from root photosynthesis has been claimed to support the plant entirely (Withner, 1959) but a partially saprophytic habit is also possible. Another group of epiphytes, this time woody, includes those which are colloquially described as "stranglers". These become established high in a host tree and send down roots which creep over the surface of the support, to thicken when they reach the ground. They often fuse to form a solid mass of tissue around the host plant which dies. The mechanical process of "strangling" has probably been much exaggerated. *Ficus* (Moraceae) provides the best known examples (e.g. *F. aurea*, *F. citrifolia* in South Florida) but species of the following genera also have the same habit (Walter, 1971): *Metrosideros* (Myrtaceae), *Griselinia* (Cornaceae), *Nothopanax* (Araliaceae) and *Clusia* (Guttiferae). Most of these examples will grow as normal trees if they germinate in the soil; aerial roots are often uncommon on plants of terrestrial habit.

3. Roots exposed to air for at least 50% of each day and free of any support or substrate; they may originate either above or below ground. Aerial roots of mangrove and swamp plants belong here and they may be immersed in salt or fresh water part of the day.

This last category represents the most specialized kind of aerial roots and ranges from the free-hanging aerial roots of tropical lianes to the erect pneumatophores of swamp plants and the spiny roots of some palms. The diversity of these aerial roots is best illustrated by a number of specific case histories. Development is emphasized, with items of special biological history highlighted. Contrasts and similarities between aerial and subterranean roots will be drawn in some instances.

II. Rhizophora mangle L. (Rhizophoraceae)

A. Environment

This mangrove species is a tree or shrub of the warmer coasts of the Americas and West Africa with an outlier in the Fiji–Samoa–Tonga area. The climate in which this plant grows varies from the warm but dry type of the Galapagos Islands to the warm, moist, non-seasonal Singapore type. Frost limits distribution in some areas (e.g. central Florida). Soil environments are typically anaerobic but may be of mud, sand or even coral rock. Substrate salinities vary enormously. The tree grows well in the open sea, provided it can become established, but it seems equally at home in fresh water, with only occasional intrusion of salt water. Consequently flooding of the environment may be daily, seasonal or continuous. The mangrove substrate is subject to much change; erosion or sedimentation are common; *Rhizophora* root systems may give rise to peat.

B. Structure and Development

The overall form of the mature plant varies from low shrubs scarcely 1 m high to tall trees exceeding 30 m. Aerial roots may form an interlacing network, either supporting smaller plants (Fig. 1c) or as an intricate arching system from the trunks of larger plants (Fig. 1a, b). Aerial roots may comprise at least 24% of the above-ground biomass (Golley *et al.*, 1962).

The following general description is taken from the papers of Gill and Tomlinson (1969, 1971) together with unpublished observations. Aerial roots about 5–10 mm in diameter appear first on the hypocotyl or lower internodes of seedlings and subsequently in a generally acropetal order up the trunk. Aerial roots can grow from high branches of tall trees. Branching of existing aerial roots can occur, according to rules discussed below. The initiation of aerial roots has not been investigated.

Aerial roots tend to emerge perpendicular to the parent organ (Fig. 1d). They elongate at rates of up to 9 mm per day and have a very long zone of elongation (up to 23 cm) behind the apex. This long zone of extension accounts, in part, for some unusual features of anatomy, as emphasized by Gill and Tomlinson (1971). The root is polyarch, with a wide medulla. Protoxylem and protophloem are initiated on alternate radii, quite close to the apical meristem, with the protoxylem exarch, as in a normal root. The earliest protoxylem is obliterated and newer elements continue to be differentiated, first in a mesarch and finally in an endarch direction. At this time phloem is also initiated opposite the xylem, giving the appearance of a collateral bundle. These features which result from the continued differentiation of vascular elements within the elongating zone, and their continued disruption and dissolution, give the root some anatomical features apparently of a stem. There is abundant and continued development of H-shaped trichosclereids within the ground parenchyma of medulla and cortex. The arms of these cells grow by apical intrusive growth, seemingly independent of overall root elongation. They probably have an important function in maintaining the rigidity of this very long meristem. Behind the zone of elongation, a continuous cambial cylinder is established, forming secondary xylem which establishes the rigidity of the root. A further pecularity of this root is that the ground parenchyma cells remain short i.e. extension is largely by new cell formation and not primarily

FIG. 1. Examples of mangrove root systems. (a) Mangrove forest, Matheson Hammock, Miami, Florida, showing a mass of aerial arched roots of *Rhizophora mangle*. (b) Old, secondarily-thickened aerial roots of *R. mangle* and the base of the trunk which tapers below. (c) Scrub habit of *R. mangle* in the Everglades National Park, Florida, with many aerial roots arising in the crown of a depauperate tree. (d) Aerial roots (arrows) of *R. mangle* emergent from the stem. Note the black root cap. (e) *Avicennia germinans*, vertical pneumatophores and foliage; vicinity of Matheson Hammock, Miami, Florida. (f) *Ceriops* sp. (cf. *C. tagal*), pneumatophores of the *Bruguiera*-type; Philippines.

cell elongation. Yet these dividing cells are densely tanniniferous. The process of growth in these roots remains quite mysterious.

The root cap is conspicuous, since it is black, but thin and quite short (Fig. 2a). The apex which it covers shows virtually no geotropic or phototropic response; roots often emerge at right angles to their parent axes and seem to bend by virtue of their own weight. We can continue with the peculiarities of this root system. In the absence of injury aerial roots remain unbranched; they produce no lateral root primordia. In the event of injury such that the apex is destroyed or ceases to grow, several replacement roots may develop close to the injury. In Miami, Florida, emergence of these lateral roots takes an average of six months after injury in winter, but only three months after injury in summer.

Eventually, by virtue of curving growth, the root apex meets and penetrates the substrate. In this new milieu, the root apex undergoes a pronounced morphogenetic change, which is indicated by the adoption of a range of new physiological responses and marked anatomical changes (Fig. 2a, b). The external color of the root changes from tan to white, as the thin surface layers lose their chlorophyll and thickened walls; trichosclereids are no longer formed and the ground parenchyma becomes lacunose (Fig. 2c, d). The most dramatic change is that lateral roots are formed without stimulation from injury. Several branch orders are produced which become smaller in diameter with depth and with each higher order of branching. The ultimate branches ("capillary rootlets") are about 0·1 mm in diameter and with a very reduced vascular system (Attims and Cremers, 1967). The main axis itself tapers in diameter and ceases growth at depth. It is reasonable to suppose that the underground root has only a very short zone of elongation, in contrast to that of the aerial root.

Changes can still occur in the proximal, aerial, portion of a root after its apex has become buried. Most notable is the formation of further branch roots, usually on the outer side of the arched aerial segment. These lateral roots grow out to repeat the sequence of development already described and a series of root loops develops (Fig. 1a). Secondary thickening proceeds preferentially along the arches at the expense of the supportive columns which remain less woody. The columns show most extensive development of aerenchyma just above the substrate, in association with abundant lenticels, and this seems to account for the aeration of the subterranean root system. In effect one has a series of vertical aerating organs close to the absorptive system as in *Avicennia* described later, but the whole is connected via an above-ground rather than an underground horizontal system.

C. Function

Anaerobic soil conditions develop rapidly upon flooding (Ponneramperuma, 1972) so that roots growing in this anaerobic medium require a ready source of oxygen which must find its way into the plant, with some method of gas conduction. In *Rhizophora* the atmosphere is the source, the lenticular root column

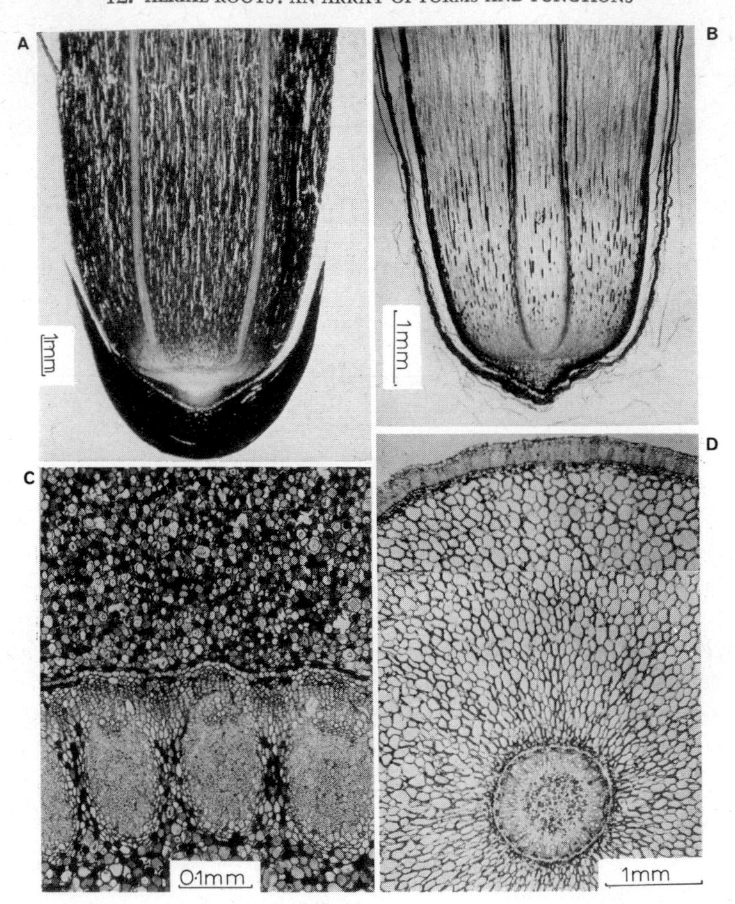

FIG. 2. *Rhizophora mangle*, anatomy of aerial and underground roots contrasted. (a) Longitudinal section of apex of aerial root showing the root cap and densely tanniniferous ground tissue of the cortex and medulla. (b) Longitudinal section of the apex of an underground root illustrating the change in organization after the meristem becomes submerged. (c) Transverse section of an aerial root in the region of the vascular cylinder. Original protoxylem poles are represented by fibrous strands at this stage. The cortex and medulla include abundant trichoblasts. (d) Transverse section of an underground root to contrast with (c); the stele is narrow and more typical of a root; the cortex is lacunose and lacks trichoblasts.

is the entry zone and conduction is facilitated by aerenchyma. The aerial root itself normally has only about 5% gas space before penetration into the soil, compared with about 50% gas space in the large subterranean roots and branches after soil penetration. Scholander *et al.* (1955) have studied the mechanism for gas transport into *Rhizophora* in tidal situations. (It should be noted that *Rhizophora* also frequently grows in non-tidal situations.) As the water level rises around the roots, the lenticels are covered and oxygen supply to the subterranean roots is cut off. Pressure within the system drops as oxygen consumption continues—but without equivalent release of carbon dioxide. As the tide recedes and lenticels are uncovered, air is drawn into the root system along a negative pressure gradient.

D. Summary

Physiological considerations apart, the number of distinctive morphogenetic responses of the *Rhizophora* root system is remarkable and it offers itself as an accessible experimental system. The descriptive work of Gill and Tomlinson (1971 and unpublished) serves simply to draw attention to this system.

III. Woody Swamp Plants Generally

Rhizophora is but one example of a variety of aerial root structures, in woody swamp plants, of which further examples are illustrated in Fig. 1e, f and 3 (see also Jenik, 1967). All of these systems are assumed to have an aerating function but experiments designed to demonstrate this function are largely inconclusive. The frequently-described "knees" of swamp cypress (*Taxodium*) provide an example; there is a lengthy bibliography on the subject but no precise evaluation of the significance of these roots (Kramer *et al.*, 1952). However, circumstantial evidence that aerial roots of swamp plants are involved in the oxygenation of subterranean roots growing in anaerobic soils is strong and goes back to the publications of Goebel (1886) and Westermaier (1900). Evidence of the following kind has been provided for various species:

(i) aerial roots develop on plants growing in water-logged conditions, but not when conditions are aerobic (Ernould, 1922);
(ii) the lengths of the aerial roots (*pneumatophores*) may be in proportion to the depth of flooding (Scholander *et al.*, 1955);
(iii) most obvious is the high frequency of species with aerial roots growing in swampy situations and the anaerobic conditions of waterlogged soils.

The mangroves, of course, provide the most striking example, but many woody plants in unrelated families in the tropics growing in wet situations develop aerial roots (e.g. in the families Avicenniaceae (Fig. 1e), Combretaceae, Palmae, Pandanaceae, Rhizophoraceae (Fig. 1f), Rubiaceae, and Sonneratiaceae). In some species aerial root production is genetically fixed while in others it is environmentally induced. In the latter case, anaerobic conditions in the subterranean roots may stimulate the formation of aerating roots which, in turn, are the agency for the alleviation of an oxygen deficiency.

An important aspect of the development of aerial "breathing roots" is that the root system also is adaptive in accommodating changing substrate levels, the result of either erosion or sedimentation. Such changes are common in aquatic environments (Davis, 1940; Thom, 1967; Craighead, 1971; Bird, 1971). Aerial roots may allow rapid exploitation of new sediment and allow roots to develop in newly deposited soil layers which are both nutrient-rich and better aerated. That aerial roots can accommodate to accretion of sediments was shown by Troll and Dragendorff (1926), McCarthy (1962), and Jenik (1967).

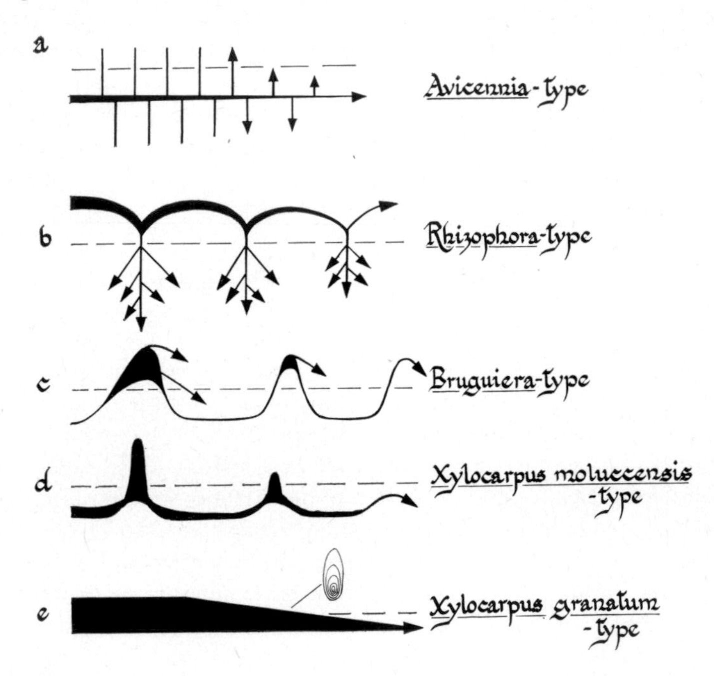

FIG. 3. Diagrammatic illustration for comparative purposes of growth and development of aerial root systems in a number of swamp plants, especially mangroves. Only the major features of the root systems are shown; diversification of branch systems is ignored. The dashed lines represent the soil surface; thinner lines represent roots with little or no secondary thickening; arrows represent actively growing apices. (a) *Avicennia*-type: pneumatophores are negatively geotropic lateral roots from horizontal, wide-spreading parent roots; other branch roots are positively geotropic. (b) *Rhizophora*-type: arching aerial roots grow sympodially. (c) *Bruguiera*-type: horizontal roots produce emergent pneumatophores by looping out of the soil; these subsequently thicken and produce branch roots (cf. Fig. 1f). (d) *Xylocarpus moluccensis*-type: pneumatophores are regions of localized secondary thickening. *Taxodium* is of this type. (e) *Xylocarpus granatum*-type: horizontal roots become plank-like sinuous structures with eccentric secondary thickening.

There has been virtually no work on the morphogenetic factors controlling differentiation and growth of aerial roots, or the influence of environmental factors on the growth of pneumatophores. Control by air humidity in promoting continued upgrowth of roots of *Mitragyna* or in promoting outgrowth of new roots from above the water surface was suggested by McCarthy (1962). Jenik (1967) suggested that aeration controls root orientation by an "aerotropic" response in *Mitragyna ciliata* Aubrev. & Pellegr. (Rubiaceae). Tropistic behaviour offers scope for experimental work. Lateral roots in *Avicennia* appear to be positive or negatively geotropic simply according to their position of origin around the circumference of the parent horizontal root—erect roots from the upper side (Fig. 1e), descending roots from the lower side (Fig. 3a). Adventitious roots on the trunk are often positively geotropic. In *Bruguiera* the same root apex (Fig. 3c) shows positive and negative geotropism, and ageotropism at different times in its growth; a horizontally-growing root turns upwards to emerge from the substrate, then downward with apparent negative geotropism to return to the substrate before continuing its horizontal growth. This behaviour produces the "knees" of this species (Fig. 3c); *Ceriops* is similar (Fig. 1f).

IV. Pandanus spp. (Pandanaceae)

A. Distribution

This genus is widespread in the Old World tropics. Arborescent *Pandanus* species occupy a wide variety of environments, from lowland swamps to coastal beaches, volcanic craters and montane slopes (van Steenis, 1948). A few smaller species are epiphytic. In *Pandanus* aerial roots are a generic feature, their development as "stilt-" or "prop-roots" is related, in part, to the method of establishment growth whereby plants without secondary thickening growth produce an obconical axis which can be, in the absence of such roots, mechanically and physiologically unstable. *Pandanus* in fact represents in some ways an extreme of development common in monocotyledons e.g. *Zea*, where adventitious roots are formed at progressively higher nodes. In *Pandanus* species where aerial roots arise from low branches (e.g. *P. candelabrum*) they undoubtedly "short- circuit" the translocating pathway. Frequently, species occupying swampy situations also develop pneumatophores from underground roots.

B. Structure and Development

The extent and size of aerial roots and whether they are restricted to the base of the trunk (Fig. 4a) or also arise on branches (Fig. 4b) varies appreciably from species to species and may provide diagnostic features. Most of our observations refer to undetermined species cultivated in botanical gardens in Florida and Hawaii and wild populations in New Guinea. The largest apices we have seen are those of an undetermined *Pandanus* species in New Guinea, up to 11 cm in diameter. It is interesting to speculate about the size of possible quiescent centers in meristems which generate roots of this size (cf. Fig. 5b). The forces developed by such roots when they penetrate soils must be considerable in view of the fact that studies to determine mechanical forces have so far been restricted to tapered root tips up to 1·5 mm in diameter (Stolzy and Barley, 1968). In these large *Pandanus* roots the apex itself is about 1 cm across, but the sub-apical region is obviously

FIG. 4. Diversity in aerial roots, from specimens cultivated at Fairchild Tropical Garden, Miami, Florida. (a) *Pandanus utilis*. Aerial roots are restricted to a cluster at the stem base. (b) *Pandanus* sp. (cf. *P. baptistii*). Well-developed aerial roots grow from stem and lower branches. (c) *Pandanus* sp. An aerial root with massive root cap and files of inhibited lateral roots. (d) *Philodendron* sp. (Araceae). Pendulous unbranched aerial roots originate from lianescent stems. (e) *Ficus platyphylla*. Aerial roots have been induced by ringing a branch; in many *Ficus* species similar aerial roots develop without injury. (f) *Cryosophila* (syn. *Acanthorhiza*) *warscewiczii*. Branched spinous aerial roots are shown. Immature root to left (arrow) still retains its root cap.

larger. Such roots occur on plants growing on sandy soils near Lae (personal observations), but elsewhere occur on finer-textured soils (A. N. Gillison, personal communication).

The further observations reported below refer to a less spectacular species (c.f. *P. baptistii*), growing at Fairchild Tropical Garden, Miami, Florida, with roots up to 4 cm in diameter (Fig. 4b, c). Root primordia on the main trunk are spirally-arranged, being evident as large swellings on the trunk. This regularity in root initiation has not been studied further, but it is useful to point out the very precise arrangement in *Pandanus tectorius* growing in the Marquesas, as described by F. Hallé (personal communication), where the root primordia bear a very fixed relation to the leaf spiral. In *P. baptistii*, roots emerge in a more or less acropetal sequence and are usually about 2–4 cm in diameter (measured 20 cm behind their apices), but roots as narrow as 6–7 mm in diameter have been seen to emerge from within the crown. The topic of root initiation itself offers much further scope for investigation.

Aerial roots lack secondary thickening. They are capped by a many-layered root cap several centimeters long, with the individual layers eventually quite papery (Figs 4c, 5b). Compression of these layers opposite the apex itself is in strong contrast to their texture laterally. In the wild, these root caps may provide an important microhabitat for epiphytes and insects (van Steenis, 1948).

Our measurement of aerial root growth in this species growing in the seasonal climate of Miami show an S-shaped curve (plotting length against time). Growth is very slow (rate of elongation 1 cm per month or less) or even ceases in winter (November–December), but begins again at any time between January and May, with a maximum rate of elongation of about 1 cm per day. The zone of elongation is short. The position of winter dormancy is indicated on each root by a distinct articulation in the surface texture of the root together with a slight constriction and lack of lateral rootlets. Consequently, annual increments in length can be seen although few roots take more than two years to reach the soil. The beginning of each new season's growth is marked by several prominent lateral rootlets around the root circumference. These rootlets however, still remain inhibited.

C. Lateral Rootlets

Between each pair of articulations the root axis supports 5 to 12 longitudinal files of lateral "rootlets", i.e. branch roots of small diameter (Fig. 4c). These are no more than 1–2 mm in diameter and always abort before they reach a length of 2 mm. They persist as sclerotic and often quite sharp spines (Fig. 4c).

The distribution of these lateral rootlets is not at random. There are usually between 40 and 60 rootlets along each 10 cm of root, a frequency unrelated to the overall rate of elongation of the parent root. They occur in regular longitudinal files, the total number of which varies from year to year and may even change in a single year, especially during the early period of active growth. There seems to be a simple linear relationship between the number of rootlet files and the diameter of the parent root. Files are fairly evenly spaced, while rootlets within an individual file vary in their arrangements (cf. Riopel, 1969, who found non-random distributions in lateral roots of subterranean roots in other monocotyledons). When a root grows into water, rootlets occur in the same files, but are more numerous. Rootlets are initiated acropetally within the sub-apical meristem of the parent, i.e. they are never "secondary" in origin. These general observations suggest that *Pandanus* species may provide useful material for studying the control of lateral root initiation, with a minimum of disturbance (cf. McCully, pp. 105–124 this volume).

The anatomy of large roots in relation to rootlet insertion deserves mention. Protoxylem poles are numerous (e.g., Fig. 5c; their number seems to be determined by root diameter); in addition to this polyarch system, which corresponds to the normal vascular system of a monocotyledonous root, there is an extensive medullary system of separate xylem and phloem strands, such that the whole of the stele is uniformly vasculated (Fig. 5c, d). Cinematographic analysis of a large root (a length of about 40 cm) by the method of Zimmermann and Tomlinson (1966) shows that these conducting strands are continuously differentiated, without any deviation from a strict longitudinal course and without any interconnection. The only interchange which is possible between these strands is at the very periphery of the stele where a rootlet makes connection to more than one protoxylem or peripheral phloem strand, thus providing an indirect link between adjacent or nearby strands (Fig. 5e). The physiological significance of these links for indirect interchange remains to be explored. Rootlets may be equally important as lenticel-like structures in aeration of bulky root tissues.

D. Branch Roots

Under a number of circumstances laterals of large diameter (of the order of 1 cm diameter) may form on the aerial root axis. These circumstances are:

(i) injury to the apical meristem;
(ii) after a long period of rest or dormancy of the apex (i.e. when it still retains the capacity to resume subsequent growth);
(iii) after the apex reaches and becomes anchored in the soil.

These circumstances are similar to those producing replacement roots in *Rhizophora*. As with other root systems, lateral replacement roots formed as a result of injury are of larger diameter than those formed without it. In *Pandanus* this is striking since rootlets and lateral roots differ in diameter by about an order of magnitude. In this respect, lateral roots formed just beyond articulations (i.e.

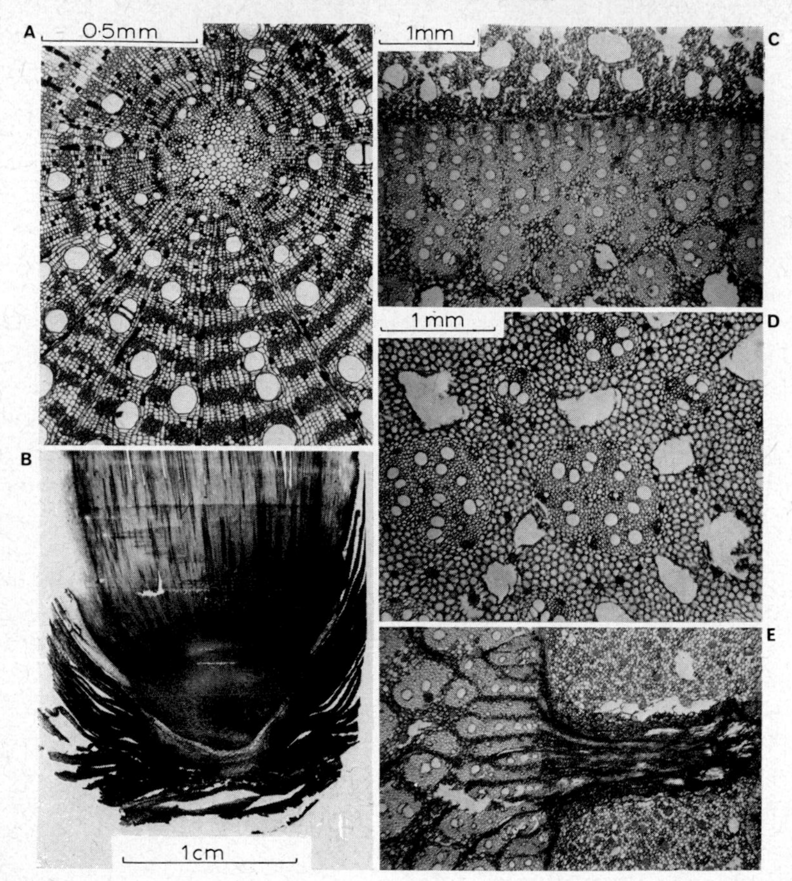

FIG. 5. Anatomy of aerial roots in *Ficus* (a) and *Pandanus* (b)–(e). (a) *Ficus benjamina*. Transverse section of aerial root 5 mm in diameter, after it has become anchored in the soil. The original pentarch primary xylem can be recognized. Secondary xylem is characterized by bands of fibers showing the wall structure typical of reaction wood fibers. (b) *Pandanus* sp. Longitudinal section of apex of aerial root showing the lamellate root cap (cf. Fig. 4c). (c) Transverse section of the periphery of the stele showing part of the polyarch peripheral tissue and the medullary vascular tissue. (d) Transverse section in the region of the medulla with vascular strands including both xylem and phloem. (e) Transverse section of the periphery of the stele showing the insertion of a lateral rootlet (cf. Fig. 4c). Air-spaces in (c)–(e) are not artefacts; the ground tissue also includes strands of fibers and crystal cells.

after a period of inactivity of the apical meristem) resemble lateral roots induced by injury, though they are rarely as large.

Recovery from injury is very slow in aerial roots of *Pandanus*, though comparable with rates for *Rhizophora*. In January 1969 a 20-cm apical length was removed from each of six roots and a 5-cm apical length from each of a further six roots. The first lateral to appear on the cut roots was in July, the last to respond had done so by mid-October. There was no obvious difference in the effect of the length of the apical portion removed. This response points to interesting morphogenetic events, because it must be again recalled that there is no secondary thickening meristem in *Pandanus* roots. Furthermore, the zone of apical elongation is quite short in *Pandanus* and corresponds approximately to the region enclosed by the root cap. Experimental treatments would thus have removed all meristematic tissue. This process of replacement undoubtedly deserves further anatomical investigation.

V. Cissus sicyoides L. (Vitaceae)

Cissus is a pantropical genus of woody vines supported by tendrils; *C. sicyoides* is widespread in the New World tropics (Elias, 1968). We have made observations on this species cultivated in Miami, Florida, where it is somewhat weedy. The rope-like stems which scramble through the crowns of supporting trees send down long festoons of delicate pink aerial roots arising from the nodes (Zimmerman and Hitchcock, 1935). The plant is difficult to eradicate because cutting of the stem merely promotes the development of more aerial roots which can rapidly re-establish contact with the soil. *Cissus* seems typical of many woody or herbaceous dicotyledonous and monocotyledonous vines in the tropics (e.g. Fig. 3d) but is distinguished by the exceptional vigor of its growth.

Aerial roots of *Cissus* remain unbranched unless injured. They are very slender in the free-hanging state; roots have been observed with a diameter of about 1 mm throughout their total extent of 8 m. This narrowness is related to remarkable rates of elongation (as much as 24 cm per day has been measured) and very low rates of dry matter accretion (of the order of 0·04 g per day). As in *Rhizophora*, the zone of elongation is long; lengths up to 100 cm have been measured in *C. pubifolia* by Blaauw (1912). Unlike *Rhizophora*, however, this length seems independent of the rate of elongation i.e. both slowly and rapidly growing roots have an equally long zone of extension.

Aerial roots, on penetrating the soil or entering water, form lateral roots. Zimmerman and Hitchcock (1935) treated aerial roots with growth substances of the auxin type and found that elongation was slowed while lateral root formation was enhanced. When the elongation rate of the main axis resumed its pre-treatment level, lateral root development (i.e., initiation) was again completely inhibited. If the induced zone of lateral root formation was kept in water, however, formerly initiated lateral roots were released from the inhibition of the main axis which continued to grow in air.

The type of aerial root development found in *Cissus* is probably common in woody vines of the tropics. *Cissus*, however, is a particularly suitable experimental subject, since short lengths of stem, kept in water will continue to produce aerial roots for periods of several weeks, even in a dry atmosphere.

VI. Coralloid Roots of Cycads

Although less spectacular than the aerial roots of other tropical plants, the coralloid roots which emerge above the soil surface close to the trunk are a characteristic feature of cycads (Chamberlain, 1919). Their association with blue–green algae is assumed to be a symbiotic one (Bergersen *et al.*, 1965). Little morphogenetic work has been done on these aerial roots. They vary somewhat in form, but shapes are not diagnostic for larger taxa. We restrict our discussion mainly to *Macrozamia communis* L. Johnson, an acaulescent or at most short-trunked plant which grows in the forested country of eastern New South Wales (Johnson, 1959).

Seedlings develop "normal" plagiotropic roots as laterals from the deeper parts of the fleshy, contractile root-hypocotyl axis. Negatively geotropic roots arise from the same axis near the soil surface (McLuckie, 1922; Wittmann *et al.*, 1965). The coralloid appearance of these aerial roots is largely due to their erect habit and frequent forking. This branching is pseudo-dichotomous, and results from abortion of the apex and its substitution by two adjacent laterals (Life, 1901). Older plants may continue to develop coralloid roots at some depth in the soil (Bergersen *et al.*, 1965).

Coralloid roots are of two types, depending on whether they contain an algal endophyte or not. Those with the algae may fix nitrogen, those without do not (Bergersen *et al.*, 1965). After algal infection coralloid roots become thicker and lose their strict negative geotropism (Wittmann *et al.*, 1965). In nature the apices of the algae-free coralloid roots were characteristically near and even above the soil surface, while the algae-containing roots were found down to depths of 30 cm (Wittmann *et al.*, 1965), which seems anomalous if the endophyte is photosynthetic. Coralloid roots with the endophyte, on plants grown in a glasshouse, were found to develop near the soil surface and fixed more nitrogen when illuminated than in the dark; endophyte-containing roots at some depth in the soil in field conditions fixed nitrogen but did not respond to illumination (Bergersen *et al.*, 1965). Subsequent work with the endophyte of *Macrozamia lucida* by Hoard *et al.* (1971) has shown that it will grow heterotrophically on glucose in the dark, with or without oxygen, but the ability to fix atmospheric nitrogen under these cultural conditions was not determined.

The mechanism of induction of coralloid roots is obscure. McLuckie (1922) found them to be absent from plants grown under sterile conditions, but present when grown under septic conditions. McLuckie claimed that coralloid root formation is dependent on bacterial infection, but other investigators have been unable to isolate bacteria from the roots (see Wittmann *et al.*, 1965).

From this brief survey it is clear that there is no consistency in the two main functions which have been ascribed to the coralloid roots of cycads i.e. aeration and N_2 fixation. Coralloid roots occur on plants grown in well-aerated soil; not all coralloid roots contain a nitrogen-fixing endophyte. It seems reasonable to conclude, however, that coralloid roots with an endophyte can fix nitrogen and that this is likely to contribute to the growth of the cycad.

VII. Ficus (Moraceae)

Ficus is a pantropical genus of about 800 species. It includes a wide variety of growth forms and occupies a diversity of habitats. Among the epiphytes and trees, buttressing, root-climbing and the so-called "strangling" habit have been much discussed (e.g. Corner, 1967) but there is little detailed or quantitative information. We concern ourselves here mainly with *Ficus benjamina* L. representative of those species with free-hanging aerial roots which arise from the branches and which subsequent to their becoming rooted, undergo extensive secondary thickening to form massive root pillars. This species has been studied by Zimmermann *et al.* (1968) with particular attention to anatomical features.

The growth of aerial roots may be conveniently divided into three stages:

(i) free-hanging phase, with largely primary growth;
(ii) an initial phase of attachment with some growth in diameter during which typical tension-wood fibers develop in the secondary xylem (Fig. 5a) leading to overall root contraction;
(iii) continued growth in diameter with production of normal wood (i.e. without tension-wood fibers).

During this later phase the root may come under compressive forces, as when it supports the weight of the branch from which it has arisen.

The origin of aerial roots in *Ficus benjamina* has not been studied in detail. Their initiation may be promoted by high humidity and by injuries which sever the phloem; thus, horizontal cuts promote the initiation of aerial roots (Fig. 4c), while vertical cuts have little effect. New roots will develop on either stems or existing roots with well-developed secondary tissues. Aerial roots in the free-hanging phase are of the order of 2 mm in diameter and are rarely branched, unless injured. Tropisms in these roots seem slight; they tend to be weakly negatively phototropic, or at most weakly positively geotropic, so that they hang vertically downward largely by their own weight. Elongation rates of up to 1 cm per day have been measured in Miami; elongation is confined to a short zone behind the apex. Natural grafts between aerial roots have been recorded in *Ficus globosa* (Rao, 1966). One report suggests that aerial roots in *Ficus benghalensis* form lateral roots without injury, but mainly in spring or early summer (Kapil and Rustagi, 1966), suggesting that this is a response to resumption of growth after a period of dormancy.

Secondary xylem, which is little developed in free-hanging roots, is produced in quantity when the root is anchored and the root contracts to the extent that it becomes quite taut. The earliest layers of the newly formed xylem (up to a width of about 5 mm) are typical tension wood, with the wood fibers having the appropriate secondary wall structure (Fig. 5a) as demonstrated by Zimmermann *et al.* (1968). As these authors emphasized, tension wood is the cause and not the consequence of root contraction. Roots planted in loose pots were able to lift the pots from the ground. Some attempts to manipulate the type of secondary xylem, by application of auxin, were unsuccessful.

As the root continues its thickening growth, normal xylem is formed and the root becomes thick enough to support the weight of its parent branch; under these circumstances it is under compressive and not tensile forces. The adaptive significance of the initial tensile state is now appreciated as a mechanism whereby the originally loose root becomes straightened so that the ultimately supporting root pillar is mechanically most efficient. Furthermore, the translocating pathway is shortened.

Aerial roots of epiphytic "strangling" figs (e.g., *F. aurea, F. citrifolia* in South Florida) may also develop tension wood in early stages of establishment. Here roots are not normally free-hanging but follow the contours of the supporting host tree. The initial phase of contraction is likely to be useful in providing a rigid supporting framework. Such contraction undoubtedly contributes to a degree of mechanical compression of the host. However, the "strangling" process in such species has been much overdramatized and competition with the host tree for light and soil moisture is probably the most significant cause of decline of the host. The fig can scarcely "strangle" a palm tree, which has no cambium; palm trees are common hosts for strangling figs.

Tension wood occurs in the aerial roots of some other genera of the Moraceae, e.g. *Cecropia* species.

VIII. Root Thorns

Modified aerial roots with determinate growth may become prominent spines. Initially such roots have well-developed root caps but the root apex eventually aborts and becomes sclerotic (Fig. 4f). Subterranean thorny root systems can also occur, as in *Dioscorea prehensilis* and *Moraea* (Scott and Wager, 1897). Recently described aerial examples are those in the Central American palm *Cryosophila guagara* Allen presented by McArthur and Steeves (1969), and those described by Jenik and Harris (1969) for a number of African woody dicotyledons from the families Euphorbiaceae, Irvingiaceae, Burseraceae, Guttiferae, Loganiaceae and Sterculiaceae. It should also be mentioned that the short, aborted lateral rootlets on the stilt roots of many *Pandanus* spp. and on stilt-palms like *Socratea* and *Iriartea* are sclerotic and sharp. Such roots have been used as natural graters by aboriginal peoples.

In the palm *Cryosophila guagara*, adventitious roots of several kinds are produced from the trunk. Roots of relatively large diameter emerge near the base of the plant; they do not themselves become spines directly since they reach the soil, but some of their sparsely produced laterals may become spines. Distally, narrower roots become spines directly (to what extent diameter is significant in this differential behaviour is not made clear by McArthur and Steeves). A third kind of aerial root is distinguished by these authors as a "crown root" which emerges from the trunk via the leaf base and initially grows vertically upward, to a length of as much as 1 m, becoming pendulous as it lengthens. Two orders of branch roots may be produced and all axes become spinous. These roots seem to be lost as the plant matures. No comparison between aerial and subterranean roots of *Cryosophila guagara* was made by McArthur and Steeves and the stimulus for the initiation of root spines remains uninvestigated.

Similar spinous roots occur in other species of *Cryosophila*, e.g. *C.* (=*Acanthorhiza*) *nana* and *C. warscewiczii* (H. Wendl.) Bartlett which species, however, seem to lack the "crown roots" described above. *Mauritiella* species [e.g. *M. aculeata* (H.B.K.) Burrett] develop short, unbranched adventitious roots from the aerial trunk which persist as short, wide spines. Several palms, typically of wet habitats, develop erect roots within the persistent leaf bases (e.g. *Mauritia*, *Metroxylon*, *Raphia*) but these do not become sclerotic like the "crown roots" of *Cryosophila guagara*. They are seemingly induced by the water which accumulates between leaf base and trunk. This indicates the strong potential for adventitious roots to develop throughout the trunk of a palm, under appropriate circumstances.

Spiny roots of palms are but one of a large number of protective mechanisms developed by palms to discourage predators (Uhl and Moore, 1973).

IX. Discussion and Conclusions

A. Functions

We may conclude our brief description with a summary of some of the functions aerial roots may have in tropical plants. They may provide the plant with:

(i) protection from large animals (e.g. *Cryosophila* and other spiny-rooted palms);
(ii) channels to subterranean roots for gas exchange in anaerobic soils (e.g. *Rhizophora* and other swamp or mangrove plants);
(iii) support for horizontal branches of trees (e.g. *Ficus*);
(iv) mechanical support for axes without secondary thickening (e.g. *Pandanus* and most Iriarteoid palms);
(v) short cuts for translocation between the soil and high-climbing shoot systems (e.g. *Cissus* and many other tropical lianes);

(vi) hosts for symbiotic micro-organisms (coralloid nitrogen-fixing roots of cycads);
(vii) anchoring roots for woody climbers (e.g. members of the Cyclanthaceae, *Freycinetia*, *Hedera*, *Toxicodendron* and many other root-climbing lianes of the tropics).
(viii) a photosynthetic area (epiphytic orchids).
This list could certainly be amplified. Of interest are situations where more than one of these functions are combined within a single plant; for example, in *Rhizophora*, aerial roots to some extent combine functions (ii), (iii), and (v).

B. Physiological Processes

Less obvious biochemical functions of aerial roots should be mentioned. For example, Kursanov (1960) introduced the idea that the aerial roots of *Ficus* in the free-hanging phase function as glands. He speculated that materials moved down from the shoot system to be transmuted within the root apices whence they were returned to the shoot system in the modified state. This process, of course, may differ little from that which occurs in any root system, but the biochemical reactions might well be different since aerial roots are exposed to light.

An obvious biochemical specialization of some aerial roots is frequently evident to travellers in the tropical rain forest. The apex of free-hanging aerial roots is sometimes coated with a thick, gelatinous mucilage, presumably excreted by the root [e.g. *Hedyosmum arborescens* described by Gill (1969)]. This coat may simply prevent the root from drying out, but it may be a protection against predators. It provides a substrate for a variety of microscopic algae (Gill, 1969).

C. Morphogenetic Responses

The most striking morphological change which we have repeatedly commented upon is that which occurs when the environment of the aerial root is modified—usually when the aerial root makes contact with the soil. Typically, development of large lateral roots is more or less completely inhibited in the air (e.g. *Cissus*, *Ficus*, *Pandanus*, *Rhizophora*) but in the soil root development is more "normal" in that laterals are produced regularly. Our incomplete observations suggest that the environmental stimuli promoting these changes may vary. In *Cissus* it is sufficient to place the root in water to induce the formation of laterals but in *Rhizophora* the chief stimulus is darkness. Other environmental effects on root morphology and anatomy have been described earlier for *Rhizophora* (p. 242).

In *Cissus* and *Rhizophora* it is likely that a morphogenetic effect of the transition of a root from the air to the soil is a decrease in the length of the zone of elongation, since, as noted by Barley (1967) a short elongation zone is mechanically adaptive in permitting the root apex to circumvent frictional barriers in the soil. It is scarcely conceivable, for example, that a *Rhizophora* root could penetrate the soil while retaining its aerial elongation zone of 10–20 cm.

Of particular interest are the belated changes occurring in the aerial portion of roots of *Rhizophora* after the apex has reached the soil. Branching occurs and this seems associated with another response, the marked development of secondary tissues. To what extent these processes (branching without injury and producing well-developed secondary thickening) are correlated remains uninvestigated. In the example of *Ficus*, development of tension wood in the aerial portion of the root after it becomes anchored distally is possibly due to the movement of growth substances from the buried to the aerial part of the root.

Tropic responses are evidently varied in aerial roots but the importance of environment in affecting root behaviour in this way is worthy of further emphasis. We have already commented on the opposed directions of growth in *Avicennia* (Fig. 3a) of lateral roots arising from underground horizontal roots, which may be negatively or positively geotropic. The changing tropic responses of *Bruguiera* (Fig. 3c) have been mentioned and these seem to vary as the environment into which the root apex develops is also changed.

With some species secondary thickening seems sensitive to environmental stimuli. In *Bruguiera* there is very little secondary thickening in subterranean roots, but in aerial roots secondary thickening is marked on the upper or convex side of the root. Parallels are found in *Xylocarpus moluccensis* (unpublished observations of A.M.G.), in *Nyssa* (Penfound, 1934), and in *Taxodium* (Wilson, 1889) where the local exaggeration of secondary thickening begins underground. Inhibition of secondary thickening by continuous water-logging is apparent in *Taxodium* (Kurz and Demaree, 1934) and *Bruguiera* while upper limits to secondary thickening seem to be set by the maximum heights of fluctuating water levels (see Kurz and Demaree, 1934). The upwardly curving arch of the primary roots may provide a local site for increased secondary activity because of better aeration and high hydration but an internal effect associated with root curvature should not be dismissed.

D. Research Potential

The reader at this point may be dissatisfied with the qualitative scope of our resumé and the repeated reference to the need for detailed investigation. This unsatisfactory state does, however, truly represent present knowledge of this interesting field. Aerial roots can be studied *in situ* without disturbance, can be observed directly on mature plants in the field, and offer examples of roots with extreme properties, e.g. the absence of lateral roots. They suggest a wide variety of biological adaptations and of morphogenetic response. The opportunity that aerial roots may offer to the study of more general structural phenomena is indicated by the work of Zimmermann *et al.* (1968) on tension wood formation in *Ficus* and of Berquam (1972) on size–structure correlations in *Cissus*. These studies illustrate the fact that the specialized organs we have mentioned, if studied in comparative and experimental ways, can throw light on general features of plant growth and development. We urge that investigators take up this challenge and reduce our present limited contribution to that of a historical document.

Acknowledgements

One of us (A.M.G.) would like to thank the National Geographic Society for a travel grant which enabled him to study mangrove root systems at first hand in the western Pacific. At that time (1970) both authors were employed by Fairchild Tropical Garden, Miami, Florida, under whose auspices many of the miscellaneous previously unpublished observations mentioned in this paper were collected.

References

Attims, Y. and Cremers, G. (1967). Les radicelles capillaires des palétuviers dans une mangrove de Côte d'Ivoire. *Adansonia* ser. 2, **7**(4); 547–551.

Barley, K. P. (1967). Mechanical resistance as a soil factor influencing the growth of roots and underground shoots. *Adv. Agron.* **19**; 1–43.

Bergersen, F. J., Kennedy, G. S. and Wittmann, W. (1965). Nitrogen fixation in the coralloid roots of *Macrozamia communis* L. Johnson. *Aust. J. biol. Sci.* **18;** 1135–1142.

Berquam, D. L. (1972). Size-structure correlation in developing roots of *Cissus* and *Syngonium. Minn. Acad. Sci. J.* **38**(1); 42–45.

Bird, E. C. F. (1971). Mangroves as land builders. *Vict. Natur.* **88**(7); 189–197.

Blaauw, A. H. (1912). Das Wachstum der Luftwurzeln einer *Cissus*- Art. *Anns Jard. bot. Buitenz.* **26** (n.s.11); 266–293.

Chamberlain, C. J. (1919). "The Living Cycads" University of Chicago Press, Chicago.

Corner, E. J. H. (1967). *Ficus* in the Solomon Islands and its bearing on the Post-Jurassic history of Melanesia. *Phil. Trans. R. Soc. Lond.* B **253**(783); 23–159.

Craighead, F. C. (1971). "The Trees of South Florida, Vol. I. The Natural Environments and Succession". University of Miami Press, Coral Gables.

Davis, J. H. (1940). The ecology and geologic role of mangroves in Florida. *Carnegie Instit. Wash.* (*Publ.* 517) **32;** 307–409.

Elias, T. S. (1968). Flora of Panama VI. Fam. 112 Vitaceae. *Ann. Missouri Bot. Gard.* **55**(2); 81–92.

Ernould, M. (1922). Recherches anatomiques et physiologiques sur les racines respiratoires. *Mem. Acad. r. Belg. Cl.* **2**(6); 3–50.

Gill, A. M. (1969). The ecology of an elfin forest in Puerto Rico, 6. Aerial roots. *J. Arn. Arb.* **50**(2); 197–209.

Gill, A. M. and Tomlinson, P. B. (1969). Studies on the growth of red mangrove (*Rhizophora mangle L.*) I. Habit and general morphology. *Biotropica* **1**; 1–9.

Gill, A. M. and Tomlinson, P. B. (1971). Studies on the growth of red mangrove (*Rhizophora mangle L.*) II. Growth and differentiation of aerial roots. *Biotropica* **3**; 63–77.

Goebel, K. (1886). Ueber die Luftwurzeln von *Sonneratia. Ber. Deut. Bot. Ges.* **4**; 249–255.

Golley, F., Odum, H. T. and Wilson, R. F. (1962). The structure and metabolism of a Puerto Rican red mangrove forest in May. *Ecology* **43**; 9–19.

Hoard, D. S., Ingram, L. O., Thurston, E. L. and Walkup, R. (1971). Dark heterotrophic growth of an endophytic blue-green alga. *Arch. Mikrobiol.* **78**(4); 310–321.

JENIK, J. (1967). Root adaptations in west African trees. *J. Linn. Soc.* (*Bot.*) **60**; 25–29.

JENIK, J. and HARRIS, B. J. (1969). Root-spines and spine-roots in dicotyledonous trees of tropical Africa. *Osterr. Bot. Z.* **117**; 128–138.

JOHNSON, L. A. S. (1959). The families of cycads and the Zamiaceae of Australia. *Proc. Linn. Soc. N.S.W.* **84**(1); 64–117.

KAPIL, R. N. and RUSTAGI, P. N. (1966). Anatomy of aerial and terrestrial roots of *Ficus benghalensis* L. *Phytomorphology* **16**(3); 382–386.

KRAMER, P. J., RILEY, W. S., BANNISTER, T. T. (1952). Gas exchange of cypress knees. *Ecology* **33**; 117–121.

KURSANOV, A. L. (1960). "The Physiology of the Whole Plant." Occ. Publ. No. 12 Wye College, Univ. London.

KURZ, H. and DEMAREE, D. (1934). Cypress buttresses and knees in relation to water and air. *Ecology* **15**; 36–41.

LIFE, A. C. (1901). The tuber-like rootlets of *Cycas revoluta*. *Bot. Gaz.* **31**; 265–271.

MCARTHUR, I. C. S. and STEEVES, T. A. (1969). On the occurrence of root thorns on a Central American palm. *Can. J. Bot.* **47**(9); 1377–1382.

MCCARTHY, J. (1962). The form and development of knee roots in *Mitragyna stipulosa*. *Phytomorphology* **12**; 20–30.

MCLUCKIE, J. (1922). Studies in symbiosis II: The ageotropic roots of *Macrozamia spiralis* and their physiological significance. *Proc. Linn. Soc. N.S.W.* **22**; 319–322.

PENFOUND, W. T. (1934). Comparative structure of the wood in the "knees", swollen bases, and normal trunks of tulepo gum (*Nyssa aquatica* L.) *Am. J. Bot.* **21**; 623–631.

PONNERAMPERUMA, F. N. (1972). The chemistry of submerged soils. *Adv. Agron.* **24**; 29–96.

RAO, A. N. (1966). Developmental anatomy of natural root grafts in *Ficus globosa*. *Aust. J. Bot.* **14**; 269–276.

RIOPEL, J. L. (1969). Regulation of lateral root positions. *Bot. Gaz.* **130**; 80–83.

SCHOLANDER, P. F., VAN DAM, L. and SCHOLANDER, S. I. (1955). Gas exchange in the roots of mangroves. *Am. J. Bot.* **42**; 92–98.

SCOTT, D. H. and WAGER, H. (1897). On two new instances of spinous roots. *Ann. Bot.* **11**; 327–332.

STEENIS, C. G. G. J. VAN (1948). *Pandanus* in Malaysian vegetation types. *Flora Malesiana* **1**(4); 3–12.

STOLZY, L. H. and BARLEY, K. P. (1968). Mechanical resistance encountered by roots entering compact soils. *Soil Sci.* **105**(5); 297–301.

THOM, B. G. (1967). Mangrove ecology and deltaic geomorphology: Tabasco, Mexico. *J. Ecol.* **55**; 301–343.

TROLL, C. W. and DRAGENDORFF, O. (1926). Ueber die Luftwurzeln von *Sonneratia* Linn. f. und ihre biologische Bedeutung (mit einem rechnerischen Anhang von H. Fromherz). *Planta Archiv. wiss. Bot.* **13**; 311–473.

UHL. N. W. and MOORE, H. E. (1973). The protection of pollen and ovules in palms. *Principes* **17**; 111–149.

WALTER, H. (1971). "Ecology of Tropical and Subtropical Vegetation" *ed.* J. H. Burnett. Oliver and Boyd, Edinburgh.

WEBER, H. (1953). Las raices internes de *Navia* y *Vellozia*. *Mutisia* **13**; 1–4.

WEBER, H. (1954). Wurzelstudien an tropischen Pflanzen I. *Adh. math. naturw. Kl. Akad. Wiss. Mainz* **4**; 211–249.

WENT, F. A. F. C. (1895). Ueber Haft- und Nahrwurzel bei Kletterpflanzen und Epiphyten. *Ann. Jard. Bot. Buitenz* **12**; p. 1 *et seq*.

WESTERMAIER, M. (1900). Zur Kenntniss der Pneumatophoren. Botanische Untersuchungen in Anschluss an eine Tropenreise. Heft 1. Freiburg.

WILSON, W. P. (1889). The production of aerating organs in the roots of swamp and other plants. *Proc. Acad. Nat. Sci.* Philadelphia, 67–69.

WITHNER, C. L. (1959). Orchid physiology. *In*: "The Orchids: A Scientific Survey" *ed.* C. L. Withner. Ronald Press, N.Y.

WITTMANN, W., BERGERSEN, P. J. and KENNEDY, G. S. (1965). The coralloid roots of *Macrozamia communis* L. Johnson. *Aust. J. biol. Sci.* **18**; 1129–1134.

ZIMMERMAN, P. W. and HITCHCOCK, A. E. (1935). The response of roots to "root-forming" substances. *Contrib. Boyce Thompson Instit.* **7**; 439–445.

ZIMMERMANN, M. H. and TOMLINSON, P. B. (1966). Analysis of complex vascular systems in plants: optical shuttle method. *Science, N.Y.* **152**; 72–73.

ZIMMERMANN, M. H., WARDROP, A. B. and TOMLINSON, P. B. (1968). Tension wood in aerial roots of *Ficus benjamina* L. *Wood Sci. Technol.* **2**; 95–104.

Chapter 13

The Genetics of Root Development

R. W. ZOBEL*

Cabot Foundation, Harvard University, Petersham, Massachusetts, U.S.A.

I. Introduction

Genes are ultimately responsible for all physiological, developmental, and morphological characteristics of plants. In this volume many facets of root development will be discussed, and I would like to underscore the point that all of these characters and responses are under some form of genetic control. The exact nature of this control is, unfortunately, not well understood. Studies on the genetic control of root development have suffered from a distinct paucity of both researchers and experimental tools. Of course the principal blame for this lack is the natural habitat of the roots. Studies of root systems involve either time consuming excavation or growth in the limited spaces provided by hydroponics and aeroponics. Since genetic studies rely heavily upon relatively large numbers of plants to arrive at statistically significant results, these are major disadvantages.

Because of these disadvantages, little concrete information is available about

* Current address: Monsanto Commercial Products Company, Agricultural Division, 800 N. Lindbergh Blvd., St. Louis, Missouri, U.S.A.

the amount and type of genetic variability existent in current crop varieties and wild species of all plants. The classic works of Cannon (1949), Weaver (1958), and Krasilnikov (1968), attempting to establish systems for the classification of root systems, point out clearly the variability among genera and families of plants. Also clear from their studies is that much of this variability parallels the differences in the ecological niche which the plants inhabit. On the other hand, anatomical studies of roots from widely differing species of plants demonstrate a phenomenal amount of similarity in developmental processes. One would like to assume that because of this underlying similarity, conclusions based upon results with one plant may be directly applicable to many if not all others.

II. Existent Variability

A cursory look at root systems of plants which differ only by the few genes which differentiate sibling varieties, gives the careful observer an indication of the existent variability. The use of a spade to excavate roots in a commercial planting of any horticultural crop turns up much variability and is a simple and quick method of surveying for possible problems in horticultural practice.

A. Tomato

Figures 1 and 2 demonstrate some of the genetic and genetic–environmental variability to be found in existing cultivars of closely related parentage. Each of the groups of root systems shown was excavated in a single shovel full of soil, and washed off for observation. The two figures represent different field plots, but both have in common the fact that the soil type graded from light sandy loam at one end to moderate clay at the other. Varieties numbered 2, 3, and 4 in Fig. 1 demonstrate some of the genetic differences commonly found; number 2 has a strong tap and secondary roots; number 3 has a strong tap root but the secondaries are much smaller and weaker by comparison, and number 4 has a strong tap root with few secondaries and many fibrous roots. These differences were observed under four different field conditions in three subsequent years, and are thought to be characteristic of the different varieties. Variety 1 on the other hand demonstrates a genetic–environmental interaction. Where numbers 2, 3, and 4 maintained the same morphology throughout the field, number 1 showed a change as roots were successively sampled from the south to the north end of the field. One will note that rep. 1 (replication 1, south end of the field where the soil type was light sandy loam) demonstrates a root type similar to variety number 2 while reps. 2, 3 are very much like variety number 3, and rep. 4 (north end of the field, moderate clay type soil) is very similar to variety number 4 in its development. The normal morphological type for variety number 1 is demonstrated by rep. 1. The ability of variety number 1 to modify its morphology in response to the differing soil conditions is of much interest.

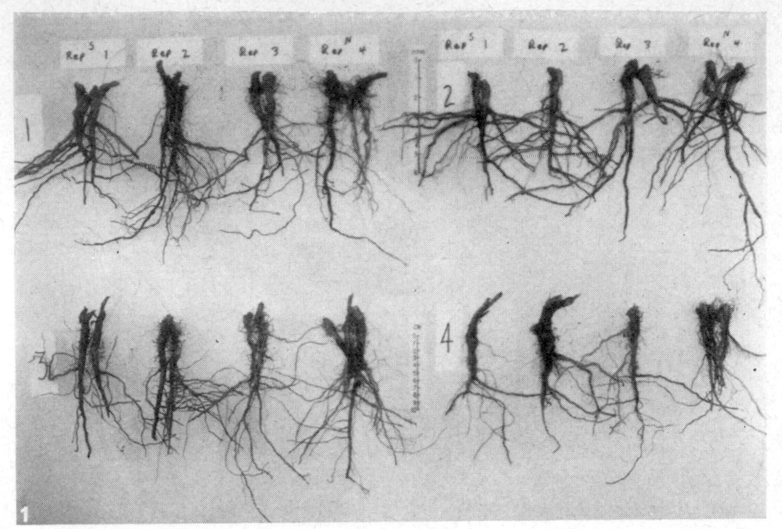

FIG. 1. Roots of tomato plants grown in sandy loam (south end of field-s) grading into light clay towards the north end of the field (n). Four replications: rep 1-south end; rep 2-south middle; rep 3-north middle; rep 4-north quarter of the field. Varieties represented are: number 1-cv. VF-10; 2-cv. VF-109; 3-cv. Pickrite; 4-UC-31-6-4-3. Each cluster of 2–3 roots represents a clump of plants growing together in the row. Individual clumps were excavated in a single spade-full of soil at the center of the replication. Replications were laid out in a randomized complete block design with five (5) other varieties. (Plant material and yield data generously provided by Ray King, San Joaquin County Agricultural Extension Service, Stockton, California).

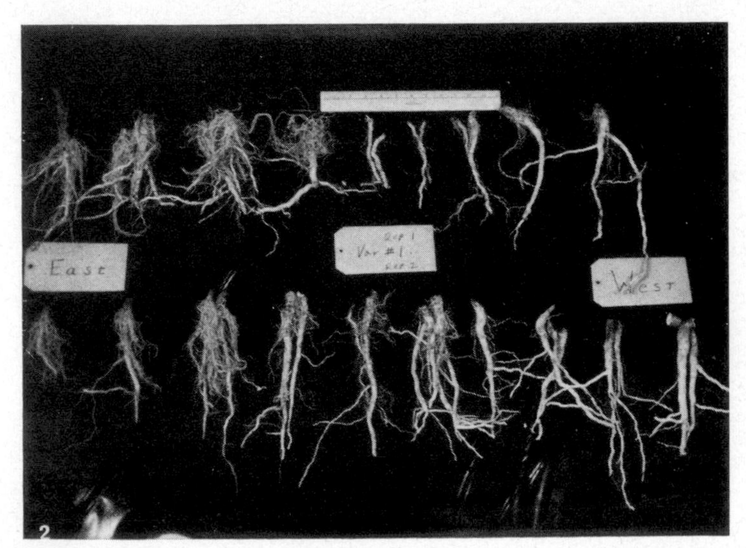

FIG. 2. Roots of the tomato variety VF145-78-79 grown in one-half-mile rows running east and west. Replicate 1 above; replicate 2 below. (Plant material and yield data generously provided by Mel Zobel, Yolo County Agricultural Extension Service, Woodland, California).

The roots shown in Fig. 2 are from a fifth variety, cv. VF145-78-79 (the predominant mechanical harvest variety in California). This variety demonstrates the same modification in response to environment as was shown by variety number 1 in Fig. 1. The roots are from two parallel replications taken from half-mile-long rows running east to west. The soil at the east end was clay and at the west end, light sandy loam. The standard root morphology of this variety is that of the roots in rep. 2 at the west end, i.e., a strong tap root with 1–2 major secondaries. Although there appears to be some change in overall morphology similar to that seen in variety 1, Fig. 1, the character to be noticed here is the increase in relatively thin fibrous roots towards the east end of the field. Although these fibrous roots may be mycorrhizal (Zobel, unpublished observations), they were only observed when the root system was under a stress situation. Yield studies with this variety and others showing similar characteristics established that a stress-induced decrease in yield was accompanied by the development of fibrous roots. Varieties with well-developed fibrous root systems were unable to produce further roots and therefore suffered extreme losses in yield under conditions where cv. VF145-78-79 suffered only slight to moderate reductions. The indication is that the development of fibrous roots is an attempt by the plant to maintain the nutritional balance necessary for the highest possible yields under conditions which are less than optimum. This reaction appears to be a developmental response to the environment which is under different genetic control in different cultivars.

B. Beans

Figure 3 illustrates the root system of two pole bean varieties excavated from a summer garden. Although both are pole beans, the root types are quite different. Those in Fig. 3a have a tap root with strong secondaries originating from the base of the stem and tap root, while those in 3b have only a reduced tap root with few secondary roots. The difference in ability of these plants to utilize differing regions of the soil and different strata seems obvious. Since these plants were grown in the same part of the garden, the differences are probably genetic rather than environmental.

Tomato, pea, brussels sprout and bean varieties differing only by several generations of selection for characters other than root characteristics often have widely differing root systems. The existence of this variability in closely related lines is indicative of a rather complex system of genetic interactions, resulting in the final form of the root system for any one variety, or species of plant.

The study of these complex interactions should ultimately lead to an understanding of the role the root system plays in overall growth and development of the plant. However, a genetic basis for root growth and development must be clearly established using simply inherited characteristics before complex interactions may be approached with any reasonable chance of success. The genes

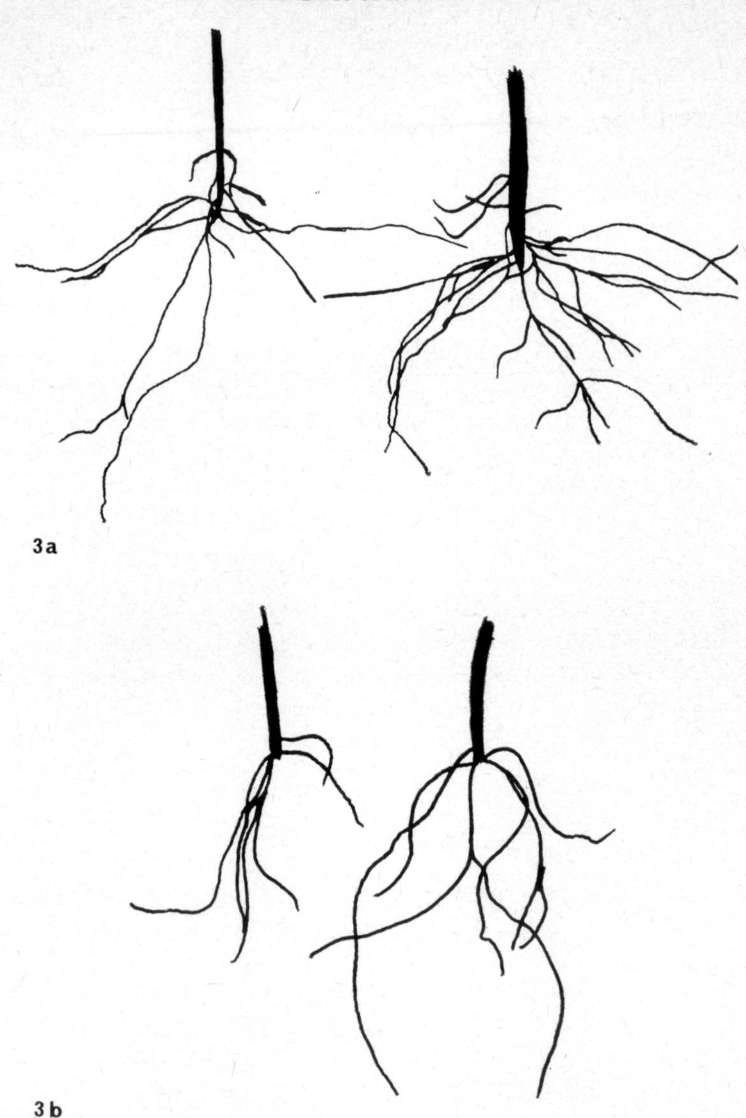

FIG. 3. Silhouettes of roots of two pole bean varieties [(a) cv. Blue Lake; (b) cv. Romano showing differences in their overall habit. × 1/8.

involved in the determination of natural variability are, almost without exception, genes with small phenotypic effects, and therefore extremely difficult to identify even when in an isogenic condition. To study adequately the genetics of root growth and development, a source of single gene variants with large phenotypic effects or otherwise easily measurable characteristics is required,

III. Induced Variability

There are very few single gene root variants available (Zobel, 1972b; see also Troughton and Whittington, 1968). One very successful method of attacking this problem of a paucity of root mutants is the use of mutagenesis (Zobel, 1973a). A statistical analysis of the progeny of a mutagenic program also gives some insight into the proportion of genes in the genome which code for processes which ultimately control root growth and development.

A. Mutagenesis

In this laboratory we have studied the effects of mutagenesis on the root systems of two plants: *Lycopersicon esculentum* Mill. (tomato, experimental line LeA-24) and *Pisum sativum* L. (garden pea, experimental line PsA-1, cv. "Little Marvel"). All variants were scored by visual identification of characteristics which diverge from normal. For statistical purposes the non-root-mutant categories, such as chlorophyll modifications, leaf shape, growth habit, and plant size (to mention a few) were lumped together. From 56 M_2 lines in tomato, 13 distinct root variants were noted, with a total of 39 variants of all classes. Thirty percent of these variants were root variants indicating that 30% of the genome conditions some aspect of root growth and development. Due to the exceptionally poor fertility of even the normal untreated pea line, there was no attempt to derive any statistical data from the pea mutagenesis treatments. Out of 96 M_2 seedlings screened, 75 were modified in their nodulation characteristics and 5 had shoot modifications. This gives an extremely high incidence of variability for plant control over the symbiotic relationship.

B. Nodulation Genetics

The symbiotic relationship between pea and rhizobium is a complex interaction of two metabolisms. Many rhizobium mutants are known which condition nodulation variants such as size of nodule, number of nodules, presence or absence of leghaemoglobin, lack of nitrogen fixation in otherwise apparently normal nodules, and lack of infection. Only a few plant mutants are known which effect a change in the symbiotic relationship; amongst the best known are those in clover studied by Nutman (1956).

In our mutagenic program most of the rhizobial variants were mimicked in their nodulation pattern by differing pea variants. A few examples of the more interesting ones are: peas with normal nodules, but without leghaemoglobin; nodules with leghaemoglobin but unable to fix nitrogen (as determined by the acetylene reduction assay); smaller or larger nodules, greater or fewer nodules, and no nodules. Overall, the variability induced with mutagenesis is very great and highly diverse, possibly indicating that the symbiotic relationship is controlled more by the metabolic balance of the plant (something easily upset by even the most minute mutation) than by one or two major genes. Because of difficulties with the growing conditions in this laboratory, we have been unable, to date, to establish clearly the genetic bases of any of our more promising variants. All of the most likely candidates have expired during attempts at increasing seed. We have now turned to a more suitable strain of pea (L110, experimental line PsA-2), and hope to isolate some single gene variants.

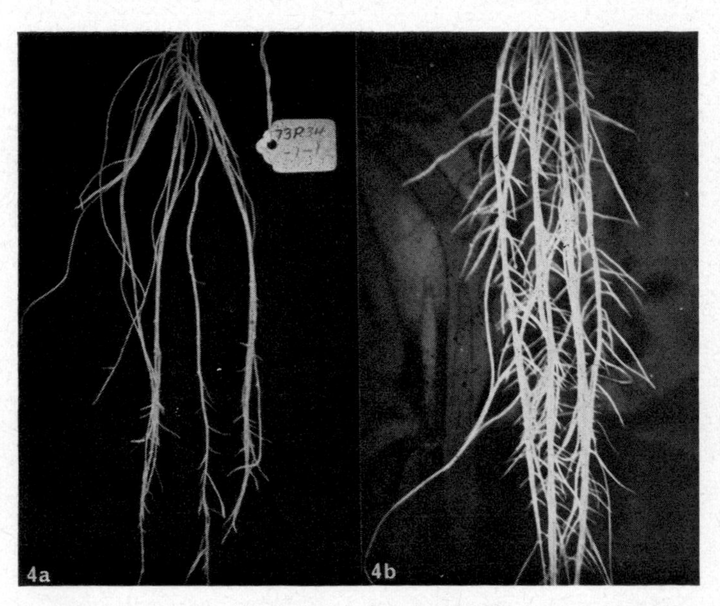

FIG. 4. Two pea root systems: (a) a variant infected with effective rhizobium but showing a characteristic reduction of nodule formation. Note the absence of lateral roots in the region of initial infection and the few small nodules and the presence of laterals on the remaining parts of the roots. ×1·25 (b) normal root systems not infected with rhizobium. Note the high frequency of tertiary roots. ×1·25

One very intriguing aspect of the legume symbiosis is the relationship between nodules and lateral roots, and between existing nodules and later infection and nodulation. Although the mutagenic data do not point to any clear genetic picture, they do demonstrate that these relationships are very easily modified. Normal pea plants which have been grown in aeroponics without infection with rhizobia show a proliferation of lateral (tertiary) roots from the primary and secondary roots (Fig. 4b). On the other hand, a nodulated pea plant has far fewer tertiary roots in the region of nodulation. This observation recalls the proposal of Nutman (1948) concerning a mutual interrelationship between nodules and lateral roots.

Figure 4a shows a mutant pea plant which has few nodules and few tertiary roots in the region where nodulation normally takes place; however, below this region there appear to be normal numbers of tertiary roots. The presence of several nodules and the relative absence of tertiary roots suggest that rhizobial infection was normal but further development of the nodules was inhibited. The infection appears to have prevented initiation of tertiary roots, thus conditionally supporting Nutman's (1948) thesis. The presence of tertiary roots below the region of infection demonstrates the ability of the plant to produce lateral primordia under non-infection conditions. However, variants with normal numbers of nodules and normal numbers of tertiary roots (similar in quantity to non-nodulated controls) have been found in the region of nodulation, arguing against a strict relationship between nodulation and lateral root initiation. Variants which have nodulation restricted to a narrow zone of the roots, perhaps 2–4 cm wide, and variants with nodulation throughout the root system have also been observed. Several conclusions may be logically derived from this information: (1) nodulation does not necessarily preclude lateral initiation—this is genetically controlled by the plant; (2) initial nodulation does not necessarily preclude further infection and nodulation in contiguous or distant regions of the root system; this is also genetically controlled by the plant; (3) it appears that the nodulation–lateral primordia interaction is separate from the nodulation–later-infection interaction.

C. Developmental Genetics of Tomato Roots

Developmental genetics as a term has been used in many ways but here is defined as the study of the genes involved in control of development *and* their interaction with each other.

1. *Mutants*

The exceptionally high mutability of the genome of the tomato line used in this study renders it exceptionally useful for the production of mutants. Although 30% appears to be an extremely high proportion of variants as root variants, two-thirds of these are also modified in their shoot characters, thus reducing the total number of strictly root variants to 10% of the total; perhaps more accurately, 6% of the screened lines contained root mutants which were not coupled to any mutant shoot character. These data can be interpreted to mean that 30% of the tomato genome conditions normal root growth, and that 10% conditions only root growth and development. These ratios would seem to compare favorably with those for any one of the other major plant organs (Monaco, 1967).

A number of interesting and potentially very useful root mutants have been isolated from these and earlier mutagenic treatments (Zobel, 1971, 1972b, 1973a) (Fig. 5–7). The first mutant, dwarf root (*drt*) (Fig. 5), has not been studied in detail, but, has been shown to condition a dwarf shoot and root when grown in soil or one-quarter strength Hoagland's solution. If *drt* is placed in full-strength Hoagland's solution, the shoot grows normally and only the roots are modified.

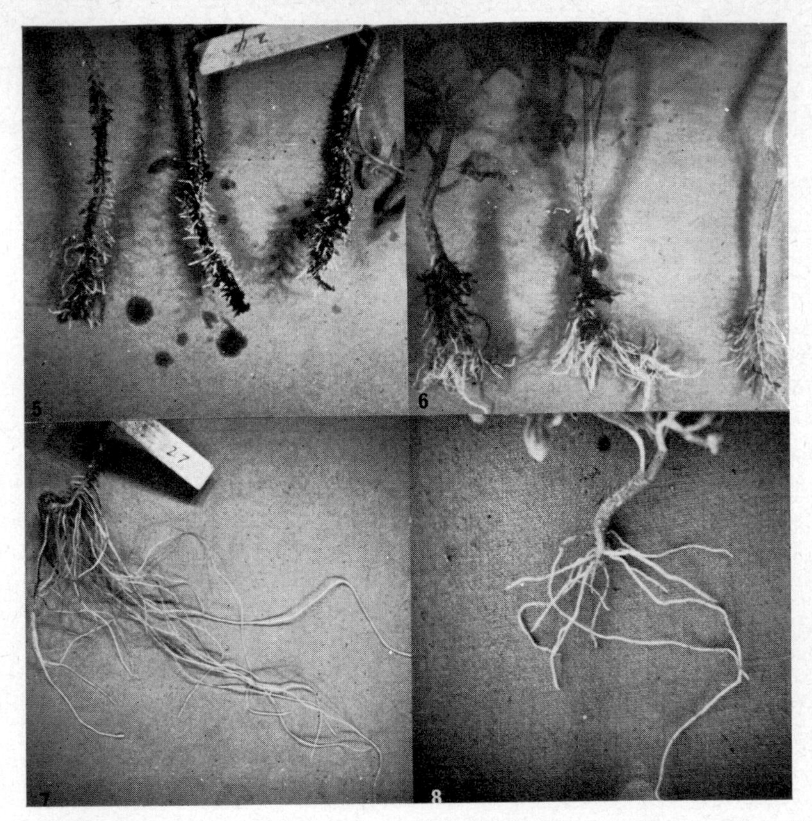

FIG. 5. Adventitious roots of the dwarf root (*drt*) mutant of tomato. Germinated in soil and transplanted to hydroponics. × 1/2.

FIG. 6. Adventitious roots of the bushy root (*brt*) mutant of tomato. Cultured as in Fig. 5. × 1/4.

FIG. 7. Adventitious roots of the diageotropica (*dgt*) mutant demonstrating the characteristic lack of lateral roots. × 1/8.

FIG. 8. Roots and hypocotyl of the *dgt/ro* double homozygote. Note the localization of the roots in the basal portion of the hypocotyl and radicle. × 3/4.

Clearly the primary effect of the mutation is on the root system, while the reduced size of the root system conditions a nutritional deficit thereby stunting the shoot.

The second mutant, *brt* (Fig. 6), has been studied in some detail (Zobel, 1972b). This mutant gene has been shown to be localized on the long arm of chromosome 12 and to condition an accumulation of starch granules in the basal cortical tissues of the root-hypocotyl region. Coupled to this is a dramatic proliferation of roots in this basal region of the plant. Starch accumulation is apparently induced by the production of a substance in the root apex and its subsequent transport to the basal portion of the root where it modifies normal sugar transport causing the characteristic development of starch. Grafting and excised root studies clearly support this hypothesis by showing that the mutant shoot has little or no effect on normal roots and that normal amounts of sucrose condition starch accumulation and root proliferation in excised roots cultured with basal feeding according to the method of Raggio and Raggio (1956). It should also be noted that when the apices of the majority of intact roots are 15 cm or more distance from the basal region of the plant, the shoot and root systems assume a normal habit (Zobel, 1972b).

Culture of this mutant in excised root culture has raised some interesting questions about the most common methods of root culture. It has long been assumed, though not with absolute conviction, that root systems in a total immersion type of excised root culture are able to establish transport gradients for sucrose and other metabolites to the extent necessary for normal developmental and physiological responses. The mutant, *brt*, fails to conform to the mutant phenotype when the excised root is grown in total immersion culture; it expresses, rather, a normal phenotype. When the root is placed with the cut basal end in a vial containing sucrose and the organic nutrients of White's medium, and the apex is placed in a solution of inorganic salts (White, 1954), it shows complete expression of the mutant phenotype. Experiments such as those of Raggio and Raggio (1956), Torrey (1963) and Zobel (1972b) indicate the need for a very careful study of the differences in physiological and developmental characters in roots cultured under the two differing conditions.

The third mutant, diageotropica (*dgt*), has been the most extensively studied root mutant, though primarily for its striking shoot characteristics rather than the root characters. The mutant gene has been localized on the long arm of chromosome 1 and shown to be a point mutation (Zobel, 1972a). The primary effect of the gene is considered to be the production of a non-functional enzyme which normally is responsible for auxin-induced ethylene production (Zobel, 1973b). Whole *dgt* plants treated with as little as 0·7 nl/l ethylene show a complete conversion to normal phenotpye within 24 hours. This response is seen only in those portions of the plant which are expanding or developing at the time of treatment, thus indicating a morphogenetic role for ethylene. The root characters of the mutant are diageotropism and a lack of lateral roots (Fig. 7). The mutant, however, has normal numbers of adventitious roots, indicating a clear developmental difference between those two types of root.

Anatomical studies of the root system of *dgt* show a lack of lateral root primordia in the untreated plant. It must be assumed, therefore, that ethylene has a role in the localized proliferation of the pericycle which gives rise to lateral roots. As was noted by Zobel (1974), ethylene or auxin treatment of excised roots does not result in the initiation of lateral roots, whereas treatment of intact *dgt* plants with ethylene induces lateral root initiation. Treatment of excised *dgt* roots basally with IAA or with ethylene, however, does induce a *generalized* proliferation of the pericycle rather than a localized one. Grafting experiments clearly point to a polarly transported substance other than IAA which is synthesized in the shoot in response to ethylene and is transported to the roots to produce or stimulate the localized proliferation of the pericycle which leads to lateral roots.

A fourth mutant rosette (*ro*) discovered and studied for its rosette-type shoot character is now considered to be an adventitious rootless mutant (Zobel, unpublished). This mutant, described and studied genetically by Butler (1954), involves a single gene which is localized on the second chromosome. We have, therefore, three root mutants which are well known genetically and which involve several different though related aspects of root development and morphogenesis.

2. *Interactions of the Mutants* (*Developmental Genetics*)

If *brt*, the bushy root mutant, is hybridized with *dgt*, the lateral-less mutant, the hybrid should be normal and its progeny expected to segregate one-fourth *dgt*, and one-fourth *brt*. One-fourth of the *dgt* or *brt* populations should be homozygous for the other gene, resulting in one-sixteenth of the total population being double homozygotes. The results from the F_2 segregation confirm this; the ratio was 9 = +, 3 = *brt*, 3 = *dgt*, 1 = *brt*/*dgt*. The double homozygote can be classified by a proliferation of adventitious roots and lack of lateral roots, whereas *brt* has proliferation of both types. The dwarf-like habit of *brt* and hyponastic curling of *dgt* cotyledons also allow scoring of the double homozygote. The proliferation of roots in the *dgt*/*brt* double homozygote is limited to the upper region of the radicle and lower region of the hypocotyl. This is the region of the plant where the vascular pattern of the shoot changes over to the vascular pattern of the root. For convenience it will be called the basal region and will include the upper 1 cm of radicle and lower 1 cm of hypocotyl in fully expanded tomato seedlings with 5 cm in hypocotyl length. It should be noted that this region of tomato seedlings has the greatest number of isoenzyme bands of any plant tissue (Zobel, unpublished; Rick *et al.*, 1974).

Based upon these results, it would be expected that a double recombinant between *dgt* (lateral-less) and *ro* (adventitious-less) should have a single long unbranched primary root and no other roots. When the cross was made, the F_1 was normal in appearance, and the F_2 had a slightly distorted ratio based upon shoot and root characters. The segregation ratio for shoot characters was: 9 = +, 3 = *dgt*, 4 = *ro* and for root characters 12 = +, 4 = *dgt*. The double homozygote, rather than having a single long root, had a *dgt*-type root system and *ro*-type shoot. These plants had roots, in addition to the elongated radicle, which originated in the basal region of the plant (Fig. 8). They also appeared to be initiated in rows, as if they developed from proliferations of the pericycle from between the phloem and xylem poles. The appearance of these roots on a plant which genetically should have been unable to form adventitious or lateral roots provides an example of the information to be derived from developmental genetics.

3. *Basal Roots*

Before considering the matter of these roots in the basal region of the *dgt/ro* double homozygote (*dgt/dgt*, *ro/ro*), the definition of the differing types of roots to be found in most plants should be discussed. Esau (1960) discusses three types of root: (1) the radicle or primary root which forms the tap root; (2) adventitious roots; and (3) lateral roots. The radicle originates in the embryo and on germination develops into the primary root. Adventitious roots arise from cambial and non-cambial tissues in all parts of the plant except roots; here Esau is unclear because she later states that they occasionally arise from very old roots. Lateral roots on the other hand are limited to those arising from roots and are derived from pericyclic divisions opposite a vascular pole or between them, depending on the species.

Since the parents of the double homozygote *dgt/ro* are genetically and developmentally unable to form roots of their respective types, the double homozygote should be genetically unable to form either adventitious or lateral roots. Of 20 double homozygous plants only one had adventitious roots in areas other than the basal region and they were only 1 cm away. These plants also exhibited a very low number of lateral roots on each plant. All mutants appear to be leaky when grown in soil; *dgt* and *ro* developed occasional roots at about the same frequency as were observed in the double homozygote. The phenotype of the double homozygote confirms that it is effectively lateral-less and adventitious-less.

What then are these roots in the basal region? Since the double homozygote regularly develops 4–5 roots in the hypocotyl portion of the basal region, these normally would be termed adventitious roots. As mentioned earlier, they develop in parallel rows characteristic of laterals developing opposite a vascular strand. Cross sections of the basal region demonstrate that root primordia are initiated in the pericycle from between the phloem and xylem poles. These roots which arise in the basal region can then be defined as either adventitious roots (because of their region of origin) or lateral roots (because of their tissue and position of origin). Since they fit both definitions and are genetically, and presumably developmentally, distinct from either type it would be appropriate to term these "basal roots". A further requisite to classifying these as distinct from other previously accepted root types is to demonstrate their existence in other plants or genetic backgrounds. Recent research by Byrne and Aung (1974) described the anatomical and morphological uniqueness of the basal region and the roots derived therefrom in tomato. They, therefore, provide a further basis of support for acknowledging basal roots as distinct from other roots and for giving them a characteristic name.

Close observation of normal tomatoes at the early stages of growth shows the presence of basal roots in nearly all plants. The *dgt/brt* double homozygote previously described had its greatest amount of root proliferation in this region. Reference to Figs. 1 through 3 which show root systems of mature plants excavated from soil makes clear that many of the main secondaries of tomato and bean appear to originate in this particular region of the plant. The question may be legitimately raised whether basal roots are the predominant source of the strong secondaries found in most dicotyledonous plants, as for example, in deciduous trees.

IV. Conclusions

Though this discussion of the genetics of root development has been wide ranging, it has emphasized the existence of considerable genetic variability with which one may study in detail any aspect of root development. The initial discussion pointed to the lack of available knowledge about naturally existing variability and by discussing genetic and genetic–environmental differences between related varieties of plants, demonstrated the existence of considerable amounts of information to be uncovered. While the conclusions derived at this time are hypotheses rather than facts, they serve to emphasize the need for a continuation of that type of study and the likelihood of obtaining significant results.

The results of mutagenic treatments point to the wealth of information possible with this approach. As mentioned earlier the pea-nodulation results established the need for further studies of this type with more suitable strains. The generally accepted ideas about nodules vs. lateral primordia, and nodule control over further nodulation are brought into question with these results. This work has indicated that the host pea plant exerts a very strong and complex control over many phases of the symbiotic relationship leading to nitrogen fixation.

The tomato research described gives the best indication of the unlimited potential of a genetic approach to studies of root growth and development. Of the four root mutants discussed, three should make major contributions to our knowledge of the genetic and developmental control over root growth. One mutant demonstrated that more knowledge is needed about the normality of total immersion excised root culture, especially in relation to the role of transport of metabolites and of concentration gradients on growth and development. Another mutant established the presence of a substance(s) responsible for the localized proliferations of the root pericycle leading to lateral roots, and that this substance is (1) produced in response to ethylene, (2) produced in the shoot and transported to the roots, and (3) that it is neither IAA nor ethylene. These data also clearly separate lateral and adventitious roots in terms of the biochemistry of their initiation. This same mutant used in conjunction with another which does not normally exhibit adventitious rooting demonstrates clearly that there exist four main types of root based upon the genetic, biochemical and developmental control of their initiation. These four types are: (1) radicle or primary root (the embryological root), (2) lateral roots (the branch roots arising from proliferation of the root pericycle), (3) adventitious roots (arising from plant parts other than roots), and (4) basal roots (developing in the basal region of the plant, that region in tomato comprising the lower 1 cm of the hypocotyl and upper 1 cm of radicle).

Clearly, since approximately 30% of the plant genome is involved in root growth and development, there is a wealth of information available. With the technique of excised root culture, biochemical, developmental and physiological studies with root mutants should be and can be easily undertaken where such studies with other plant parts suffer severe handicaps. During a period when the use of somatic cell genetics with plants is growing, it must be mentioned that only with studies of whole plants or at least plant organs like the root can adequate information be gained about the biochemistry, physiology and development of tissues which, through their interactions with each other, continually modify their own environment in the intact plant. The coupled use of mutagenesis and excised root culture offers such an approach.

References

Butler, L. (1954). Rosette, *ro*. Report Tomato Genetics Cooperative. **4**; 9.

Byrne, J. and Aung, L. (1974). Adventitious root development in the hypocotyl of *Lycopersicon esculentum* Mill. var. Fire Ball. *Am. J. Botany*. **61**; suppl. p. 54 (abstract).

Cannon, W. A. (1949). A tentative classification of root systems. *Ecology* **30**; 542–548.

Esau, K. (1960) "Anatomy of Seed Plants". John Wiley and Sons, New York (first printing) pp. 186 and 200.

Krasilnikov, P. K. (1968). On the classification of the root system of trees and shrubs. *In*: "Methods of Productivity Studies in Root Systems and Rhizosphere Organisms." (an International Symposium) NAUKA, Leningrad. pp. 106–114.

Monaco, L. C. (1967). "Relative Effectiveness of Fast Neutrons, X-rays and Ethylmethanesulfonate as Mutagens for Tomato Pollen." Ph.D. dissertation. Univ. Calif., Davis.

Nutman, P. S. (1948). Physiological studies on nodule formation I. The relation between nodulation and lateral root formation in red clover. *Ann. Bot.* N.S. Vol. **XII**; 81–96.

Nutman, P. S. (1956). The influence of the legume in root nodule symbiosis—A comparative study of host determinants and functions. *Biol. Rev.* **31**; 109–151.

Raggio, M. and Raggio, N. (1956). A new method for the cultivation of isolated roots. *Plant. Physiol.* **9**; 466–469.

Rick, C. M., Zobel, R. W. and Fobes, J. (1974). Four peroxidase loci in red-fruited tomato species: genetics and geographic distribution. *Proc. Natl. Acad. Sci. U.S.A.* **71**; 835–839.

Torrey, J. G. (1963). Cellular patterns in developing roots. *Symp. Soc. Exp. Biol.* **17**; 285–317.

Troughton, A. and Whittington, W. J. (1968). Significance of genetic variation in root systems. *In*: "Root Growth" (W. J. Whittington, Ed.) Plenum Press, New York pp. 296–314.

Weaver, J. E. (1958). Classification of root systems of forbes of grassland and a consideration of their significance. *Ecology*, **39**; 393–401.

WHITE, P. R. (1954). "The Cultivation of Animal and Plant Cells." The Ronald Press Co., New York, New York.

ZOBEL, R. W. (1971). Root mutants of tomato. Report Tomato Genetics Cooperative. **21**; 42.

ZOBEL, R. W. (1972a). Genetics of the diageotropica mutant in the tomato. *J. Heredity* **63**; 94–97.

ZOBEL, R. W. (1972b). "Genetics and Physiology of Two Root Mutants in Tomato, *Lycopersicon esculentum* Mill." Ph.D. Dissertation, Univ. Calif., Davis.

ZOBEL, R. W. (1973a). Use of mutagenesis in physiological genetic studies with plant root systems. *Genetics* **74**; 5306–abstract.

ZOBEL, R. W. (1973b). Some physiological characteristics of the ethylene-requiring tomato mutant diageotropica. *Plant. Physiol.* **52**; 385–389.

ZOBEL, R. W. (1974). Control of morphogenesis in the ethylene-requiring tomato mutant, diageotropica. *Can. J. Bot.* **52**; 735–741.

Part II

PHYSIOLOGICAL ASPECTS OF ROOT FUNCTION

Chapter 14

Physiology of Growing Root Cells

NATALIE V. OBROUCHEVA

Institute of Plant Physiology of the U.S.S.R. Academy of Sciences, Moscow, U.S.S.R.

I. Introduction

To investigate successfully such complicated processes as growth, the metabolic changes in growing cells should be investigated side by side with the physiology of the intact growing organs. The idea of this method is as follows. The growing root tip is divided into separate zones on the basis of anatomical data, distribution of mitoses and cell length. Each root zone thus contains cells most of which are at the same growth stage. Thus, the root tip is divided into meristem (cell division zone), elongation zone and zone of mature cells which no longer increase in length. In these zones the content of cell constituents, the rate of respiration and other processes, and the activities of various enzymes are investigated. At the same time the number of cells in each zone is estimated and the content of any substance is calculated per cell. This technique permits one to investigate the content of various cell components and the activities of enzymes responsible for their metabolism in the course of cell growth. Thus one can compare the biochemical and cytological data with success. In general, this method allows one to characterize the metabolism of root cells at each growth stage and to reveal ways in which the metabolism changes as a root cell transforms from meristematic into elongating and then into the mature state.

The calculation of experimental data per average cell of growing organs was suggested and applied to roots by Reid (1941). R. Brown was the first who managed to work out a simple and convenient technique for root cell counting (Brown and Rickless, 1949), which permitted him and his colleagues to reveal the main features of growing root cells in a large series of experiments (Brown and Broadbent, 1950; Brown and Sutcliffe, 1950; Robinson and Brown, 1952, 1954; Brown and Cartwright, 1953; Brown and Robinson, 1955; Robinson, 1956; Robinson and Cartwright, 1958; Heyes, 1959, 1960, 1963; Brown, 1963). This approach was then used by numerous investigators in a number of countries and some reviews have appeared (Heyes and Brown, 1965; Torrey, 1965; Obroucheva, 1965; Street, 1966; Khavkin, 1973). The present paper may be considered an attempt to summarize and interpret these results.

II. Metabolism of Carbohydrates by Roots

The formation of new cells in the growing root tips and their further modification in the course of elongation and maturation are based on the metabolic utilization of photosynthates translocated from the lower leaves to the roots. The transport form of photosynthates is sucrose, or its polymers, with galactose, as was shown in the experiments with leaves fed $^{14}CO_2$ (Pristupa, 1959; Shiroya *et al.*, 1962). The fact that sucrose is practically the only starting material for root metabolism was confirmed by the sterile culture of isolated roots, sucrose being the only carbohydrate capable of providing their continuous growth (Smirnov, 1970). Sucrose is metabolically inert during translocation (Arnold, 1968) as the sieve tubes lack invertase. Sucrose is metabolized only in roots, the most active invertase being present in the elongation zone (Brown and Robinson, 1955; Hellebust and Forward, 1962; Sutcliffe and Sexton, 1969) where the phloem sieve tubes terminate. Alkaline invertase is localized in the stele while acid invertase is active predominantly in the cortex (Lyne and Ap Rees, 1971).

The metabolism of monosaccharides formed from sucrose by invertase occurs in meristematic and elongating cells at a fast rate. In growing cells their level is very low (Ramshorn and Koenig, 1959; Rogozinska, 1965; Rogozinska *et al.*, 1965; Göring, 1970; Ste-Marie and Weinberger, 1971). Their derivatives, i.e. free amino acids, organic acids and nucleotides (Dudchenko and Sytnik, 1966;

Khavkin and Varakina, 1967; Potapov and Sumanova, 1968; Ste-Marie and Weinberger, 1970; Mamedova and Rassulov, 1970; Sytnik and Dudchenko, 1972; Naumova and Khavkin, 1974) are present in minute amounts as well. The low value of their content does not mean that growing cells are deficient in the syntheses of these compounds. On the contrary, these compounds are formed to be used immediately for the syntheses of polymer end products, i.e. polysaccharides, proteins, nucleic acids and nucleoproteins, accompanied by active respiration.

Proteins make up some 50% of the dry substances (dry weight) in the meristematic cells, and a little less in elongating cells. The ratio between protein and total nitrogen is the following: protein nitrogen represents 70–80% in meristematic cells and 60–75% in elongating cells (Erickson and Goddard, 1951; Obroucheva, 1965). Hence the predominance of proteins over low molecular nitrogen compounds is obvious. The same tendency is observed in phosphorus metabolism, viz. the ratio P(nucleic) + P(lipid): P(acid-soluble organic) is equal to 2–5:1 in meristematic cells of pumpkin and pea roots and 1·5–2:1 in elongating cells (Dudchenko and Sytnik, 1966; Sytnik and Dudchenko, 1972). Another part of the dry substances in growing cells is represented mainly by polysaccharides of the cell walls. The most rapid synthesis of uronic acids takes place in the meristem, while that of pentosans and cellulose occurs in cells, which have begun to elongate (Roberts and Butt, 1969; Gournay and Péaud-LeNoel, 1961).

Thus, rapid transformation of sucrose leads to the rapid accumulation of cell constituents forming cell wall, cytoplasm and organelles. For example, the number of mitochondria and their heterogeneity increase in elongating cells (Potapov and Salamatova, 1963; Yoo, 1970), the number of ribosomes also increases (Hsiao, 1970), the ratio between DNA, RNA and proteins changes in nuclei during elongation (Lyndon, 1963) and so on. If a portion of translocated sucrose is not used immediately for the metabolic needs, it is stored in the meristem as starch and can then enter the metabolic pathways during the elongation (Rogozinska, 1965; Rogozinska *et al.*, 1965; Ste-Marie and Weinberger, 1971).

III. Differences between Meristematic and Elongating Cells

The general trends of metabolism are similar in meristematic and elongating cells. Moreover, the relative rates of protein synthesis in these cells are almost the same (Khavkin *et al.*, 1967). These cells differ mainly in polymer composition and the level of their cell constituents.

A. Cell Constituents

The content of various compounds per cell turned out to be much higher in elongating cells when compared with meristematic ones. This difference in cell content concerns both compounds of primary metabolism and the end products, i.e. different polysaccharides (Jensen and Ashton, 1960; Jensen, 1961); proteins (Table I); RNA (Potapov *et al.*, 1959; Sunderland and McLeish, 1961);

TABLE I. Average content of protein nitrogen in root cells (g × 10^{-10}/cell)

	Meristem	Zone of elongation	Zone of mature cells	References
Vicia faba	1·8	1·9	2·9	Ramshorn, 1957
	1·4	1·5	2·6	Morgan and Reith, 1954
Pisum	1·0	2·0	2·5	Brown and Broadbent, 1950
sativum	0·5	0·7	1·4	Heyes, 1959
	0·5	0·8	1·2	Lyndon, 1963
	0·7	1·5	1·9	Sytnik and Dudchenko, 1970
Lupinus	0·9	1·2	0·9	Obroucheva, 1965
angustifolius	0·6	1·2	2·3	Potapov and Sumanova, 1966
	0·2	0·8	0·6	Potapov and Salamatova, 1964
Zea mays	0·9	1·3	2·0	Brown and Cartwright, 1953
	1·0	1·4	1·6	Cook, 1959
	1·2	1·5	1·3	Erickson and Goddard, 1951

DNA (Holmes *et al.*, 1955; Heyes, 1959; McLeish and Sunderland, 1961; Lyndon, 1963; Woodstock and Skoog, 1962); phosphatides (Dudchenko and Sytnik, 1966; Sytnik and Dudchenko, 1972); monosaccharides and sucrose (Baldovinos, 1953; Ramshorn and Koenig, 1959; Rogozinska, 1965; Rogozinska *et al.*, 1965; Göring, 1970; Ste-Marie and Weinberger, 1971; Lyne and Ap Rees, 1971); organic acids (Naumova and Khavkin, 1974); amino acids (Morgan and Reith, 1954; Khavkin and Varakina, 1967; Potapov and Sumanova, 1968; Ramshorn and Blohm, 1968; Ste-Marie and Weinberger, 1970); phenols (Khavkin and Pereljaeva, 1970); ascorbic acid and SH-group containing compounds (Trezzi *et al.*, 1959; Kossey-Fejér, 1961); and nucleotides (Dudchenko and Sytnik, 1966; Mamedova and Rassulov, 1970; Sytnik and Dudchenko, 1972). Most compounds of primary metabolism are osmotically active and localized in vacuoles. The increase in their level in elongating cells favors increased osmotic potential and water uptake.

B. Enzymes

The increase in the content in elongating cells of all substances mentioned above compared with meristematic ones closely parallels the elevation of the activities of the enzymes taking part in their synthesis and transformation. In the elongating cells the following enzymes are known to be more active: invertase (Brown and

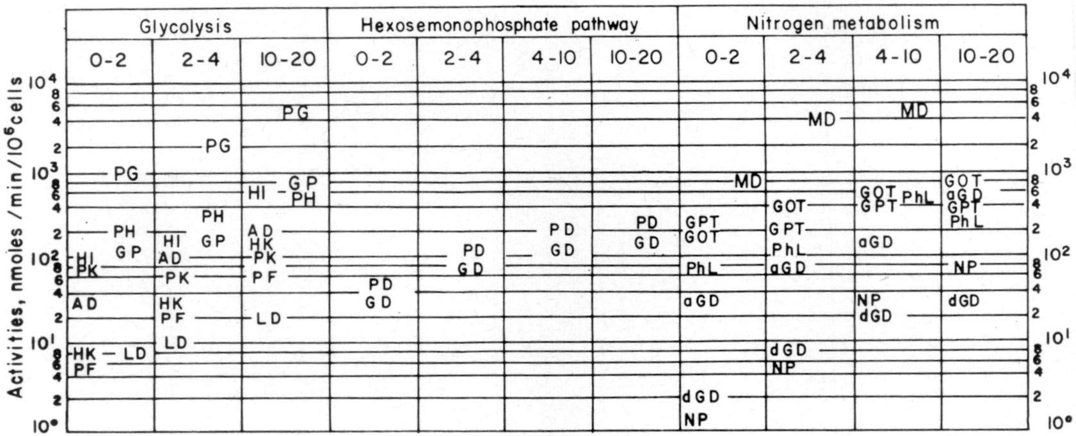

FIG. 1. The enzyme systems of growing and mature cells of maize root. The distance from the root tip is given in mm. 0–2 mm corresponds to meristem, 2–4 mm and 4–10 mm—two portions of elongation zone, 10–20 mm—zone of mature cells. The activities of enzymes are plotted on a logarithmic scale for convenience. Abbreviations: AD—alcohol dehydrogenase (EC 1.1.1.1); GP—glyceraldehydephosphate dehydrogenase (EC 1.2.1.12); HK—hexokinase (EC 2.7.1.1); HI—hexosephosphate isomerase (EC 5.3.1.9); LD—lactate dehydrogenase (EC 1.1.1.27); PF—phosphofructokinase (EC 2.7.1.11); PK—pyruvate kinase (EC 2.7.1.40); PH—phosphopyruvate hydratase (EC 4.2.1.11); PG—phosphoglycerate kinase (EC 2.7.2.3); PD—6-phosphogluconate dehydrogenase (EC 1.1.1.44); GD—glucose-6-phosphate dehydrogenase (EC 1.1.1.49); aGD—aminating glutamate dehydrogenase (EC 1.4.1.2–4); dGD—deaminating glutamate dehydrogenase (EC 1.4.1.2–4); GOT—glutamate-oxaloacetate-transaminase (EC 2.6.1.1); GPT—glutamate-pyruvate-transaminase (EC 2.6.1.2); NR—nitrate reductase (EC 1.6.6.1–3); PhL—phenylalanine-ammonium-lyase (EC 4.3.1.5); MD—malate dehydrogenase (EC 1.1.1.37). Reproduced with permission from Dr. E. Khavkin.

Robinson, 1955; Hellebust and Forward, 1962; Sutcliffe and Sexton, 1969; Lyne and Ap Rees, 1971); β-glucosidase and galactosidase (Zeleneva *et al.*, 1972); β-glycerophosphatase (Sutcliffe and Sexton, 1969; Sexton and Sutcliffe, 1969a); acid phosphatase (Potapov *et al.*, 1970; Zeleneva *et al.*, 1972); ATPase (Salamatova and Balajan, 1966; Sexton and Sutcliffe, 1969b); peroxidase (Potapov *et al.*, 1959); nucleases (Robinson and Cartwright, 1958); and proteolytic enzymes (Robinson and Brown, 1952; Robinson, 1956). Furthermore the same tendency was observed in the case of enzymes taking part in nitrogen metabolism, i.e., hydroxylamine reductase (Potapov and Sumanova, 1966; Khavkin *et al.*, 1968), transaminase (Cook, 1959; Khavkin *et al.*, 1968), and others (see Fig. 1). In addition the rate of respiration is much higher in elongating cells (Table II); at

TABLE II. Rate of respiration of growing root cells (μl $O_2 \times 10^{-6}$ per cell per hour)

	Meristem	Zone of elongation	Zone of mature cells	References
Pisum				
sativum	10	30	45	Brown and Broadbent, 1950
Zea mays	10	38	45	Goddard and Bonner, 1960
	9	54	42	Khavkin and Varakina, 1969
Allium cepa	15	50	20	Wanner, 1950
	20	68	53	Norris *et al.*, 1959
Lupinus	3	18	37	Obroucheva, 1965
angustifolius	4	10	64	Potapov and Salamatova, 1964

the same time, activities of glycolytic enzymes, and enzymes of the hexosemonophosphate pathway and respiratory chain rise as well (Fig. 1). Hence the elongating cells represent the site of greater metabolic activity than meristematic cells. Thus elongation apparently cannot be restricted on the one hand to modifications of cell walls and on the other hand to osmotic water uptake. The transformation of a meristematic cell to an elongating one, and the elongation itself, are accompanied by the acceleration of all metabolic processes.

The elongating cells always try to achieve the normal pattern of metabolism specific for mature cells of this plant species. If the root grows under unfavorable conditions (for instance, at oxygen deficiency), it would be reasonable to assume that the level of cell constituents would be lower in mature cells. However, the elongation process itself is decelerated, but the normal pattern of metabolism develops by the end of elongation phase (Table III).

Comparison of various metabolic processes at the cell level has shown that the difference between meristematic and elongating cells appears to be mainly

TABLE III. The content of some cell constituents in an average elongating root cell of lupine seedling under various oxygen condition (Obroucheva, 1965).

	Dry substances, g × 10^{-11}/ cell	Protein g × 10^{-11}/ cell	RNA g × 10^{-11}/ cell	Protein nitrogen as percentage of total nitrogen	Duration of elongation, hrs
Strong aeration	140 ± 17	72 ± 6	3·16 ± 0·2	73	10
Weak aeration	170 ± 24	60 ± 6	3·16 ± 0·3	58	18

quantitative. In no case was there observed the first appearance or the complete loss of any enzyme activity in elongating cells (Khavkin, 1973). It is still an open question as to what metabolic process may be considered as providing a controlling mechanism in the transformation of meristematic cells into elongating cells. The use of cellular levels of activity or content for comparison has not yet proved successful. To settle this question attention should be paid to qualitative differences of metabolic patterns in the growing cells.

IV. Growth in Relation to Metabolism

A. Aerobic Respiration and Fermentation

The relationship between growth and metabolism was investigated from different points of view. By means of respiratory studies, so-called "aerobic fermentation" (aerobe Gärung) was found in root meristems of some plants. This phenomenon implies that under aerobic conditions when seedlings grow in the open air, typical alcoholic fermentation occurs in meristems; it is distinguished by increased CO_2 output, i.e. high respiratory quotient (RQ), and ethanol production (Ruhland and Ramshorn, 1938; Ramshorn, 1957; Betz, 1955, 1957, 1958, 1960). However, aerobic fermentation cannot be considered as a peculiar property of root meristems because the RQ falls to 1 and ethanol does not accumulate if the roots are transferred into an atmosphere of pure oxygen (Berry and Norris, 1949; Kandler, 1950; Norris, 1951; Ramshorn, 1957). Root growth in strongly aerated mineral solutions also results in the decrease of RQ to 1 (Obroucheva, 1965). Aerobic fermentation turned out to be the usual alcoholic fermentation occurring simultaneously with respiration. It takes place only in thick roots in which meristems are badly ventilated (Obroucheva, 1965; Buder, 1967). Indeed, under normal conditions the RQ does not exceed 1 in thin roots of wheat, rye, barley and mustard (Fig. 2); in thick roots (corn, lupin, horsebean) it is much higher. Hence meristems respond to partial anaerobiosis by the appearance of fermentation as also occurs in other parts of roots, and other plant tissues.

In the course of respiratory studies attention was drawn to the ratio between glycolysis and the hexosemonophosphate (HMP) pathway of oxidation. The HMP-pathway is connected with the syntheses of cell wall pentosans, phenolic compounds and lipid constituents of membranes. The activity of the HMP-pathway increases in elongating cells as compared with meristematic ones (Hadačova, 1968, Khavkin and Varakina, 1969; Fowler and Ap Rees, 1970; Khavkin, 1973), but its percentage of total respiration shows almost no change. Hence the respiratory pathway through glycolysis, the Krebs cycle and the respiratory chain remains the principal one during both growth phases.

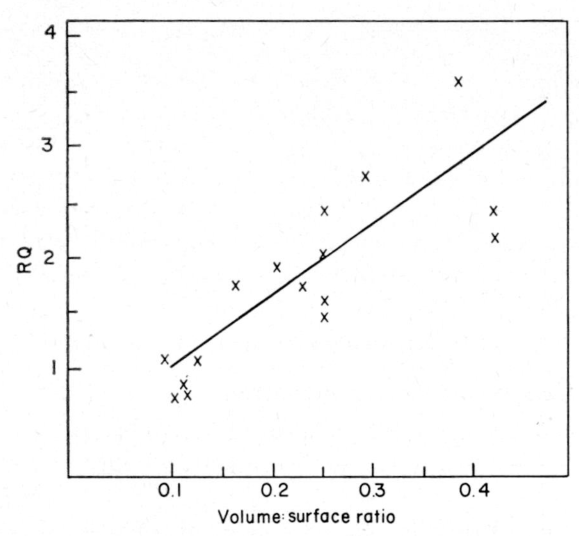

FIG. 2. The respiratory quotient (RQ) in root meristems of different plants (Obroucheva, 1965).

The investigations of metabolic specificity in cells transforming from the meristematic into the elongating phase have shown that both the ratios among cell wall constituents and their actual composition change (Jensen and Ashton, 1960; Jensen, 1961; Dever *et al.*, 1968; Roberts *et al.*, 1968). They appear to be closely related to cell wall modification at the beginning of elongation, i.e. its elasticity and extensibility (Burström *et al.*, 1970). However, it is rather difficult to assume that the synthesis of cell wall constituents would control the change of growth pattern. It would be more reasonable to suppose that nucleic acids and(or) proteins play a key role in it. We have already mentioned that the content of DNA per cell increased during growth and maturation of cells. It is now evident that DNA synthesis is not restricted to the S-period in meristematic cells but proceeds in some tissues during cell elongation and later in mature cells.

DNA synthesis without concomitant cell division is an extraordinary phenomenon and results in polyploidy in some tissues, but it is too early to conclude that such DNA plays a role in any growth-controlling mechanism. This mechanism must provide the qualitative peculiarities at every growth stage, which probably are due to some difference in DNA. But the data in our possession concerning the amount of DNA, localization of polyploid nuclei in tissues, ^{32}P-incorporation into RNA fractions, the nucleotide composition etc., do not permit us to decide whether DNA exerts the controlling influence on the transformation of meristematic cells to elongating and mature ones.

B. Changes in Protein Composition

We can consider whether DNA participates in growth regulation in an indirect way, i.e., by considering changes in the protein composition during growth. This hypothesis was put forward by R. Brown (Brown, 1963; Heyes and Brown, 1965). He supposed that specific sets of proteins (enzymes) are characteristic of every phase of root cell growth. The protein composition was investigated in growing root cells by a number of methods. Disk electrophoresis has not revealed any difference in band distribution of proteins extracted from meristem and elongation zone (Hadačova and Sahulka, 1967; Dudchenko and Sytnik, 1967; Polter, 1967; Besemer and Clauss, 1968; Polter and Müller-Stoll, 1969a, b; Sytnik and Dudchenko, 1970). These results cannot be considered as final, because bands with similar Rf can contain different proteins (Steward *et al.*, 1965). If a decrease in band number takes place, it can be interpreted as a decrease of protein amounts but not their omission (Steward *et al.*, 1965). Zymographic analysis of proteins also is open to criticism (Khavkin, 1973). Chromatographic separation of proteins from different zones of corn roots (Khavkin and Varakina 1967) has shown that in elongating cells the fraction of high molecular RNA-rich proteins decreased, but the fraction of low molecular proteins and their heterogeneity increased. The best results were obtained by immunochemical methods on corn root tips (Reimers and Khavkin, 1970; Khavkin *et al.*, 1971). An antigen specific for meristematic cells, two antigens specific for elongating cells and an antigen common for both types of cells were detected. Their enzymic nature, intracellular localization, molecular weights and charge values were investigated (Khavkin *et al.*, 1972b). The proteins specific for the elongation zone turned out to be mitochondrial, and their "appearance" in this zone can be explained by their accumulation in the course of enhanced mitochondrial genesis but not by their synthesis *de novo*.

Hence the detailed analysis of proteins of the same material by means of chromatography and electrophoresis separately and in combination with immunochemical methods of identification permits one to conclude that marked quantitative changes in protein composition take place during cell growth while qualitative changes in the sets of proteins are not yet evident (Khavkin, 1973).

To summarize the results just cited the pure metabolic approaches have not yet allowed us to offer a satisfactory explanation of growth regularities in root cells. It is always implied that growth substances exercise a controlling influence on cell growth. However, the discussion of their role in the growth regulation of root cells is beyond the scope of the present article for two reasons. Firstly, these investigations are based on treatments with added amounts of growth substances (as compared with endogenous amounts) while this paper deals with natural levels of cell constituents; secondly, there are still no reliable data concerning the content of endogenous growth substances at the cell level.

V. Mitosis versus Cell Elongation

Cytological observations indicate that elongation of cells does not necessarily follow the completion of divisions in the meristem (Balodis and Ivanov, 1970). Mitoses can proceed in the pericycle during elongation or may, in some tissues, stop in the meristem long before the beginning of cell elongation. The transformation of meristematic cells into elongating ones was observed in the corn roots irradiated with high doses of X-rays (50–100 kr), i.e. in roots with completely suppressed divisions (Ivanov, 1968). After some time meristematic cells began to elongate in both non-irradiated and irradiated roots simultaneously. Hence the transformation into elongating cells does not depend upon whether meristematic cells divide in the meristem or not. This period was named by V. Ivanov as the "life span of a cell in the meristem" (1970a). In corn roots this life span is about 60–70 hrs. During this period in non-irradiated roots a meristematic cell will divide 6 times after leaving the quiescent centre before beginning to elongate. In irradiated roots such a cell "lives" in the meristem without divisions for 60–70 hrs and then begins to elongate in the usual way. This life span of cells in the meristem is not influenced by chloramphenicol (50 mg/l), puromycin (100 mg/l), cycloheximide (0·5–2 mg/l) or other treatments. It seems reasonable to assume that the mechanism controlling the start of elongation, after termination of the life span, is almost insensitive to inhibitors of protein and nucleic acid synthesis (Ivanov, 1973, 1974), although it does depend on temperature. Thus the process of transformation of meristematic cells into elongating ones is quite different from division and elongation which depend on the normal course of protein and nucleic acid metabolism (Ivanov *et al.*, 1967; Ivanov, 1970b). This assumption is in accordance with the above mentioned failure to detect new proteins in cells beginning elongation, and with data concerning the inhibition of synthesis of numerous enzymes in meristematic and elongating cells by chloramphenicol and other inhibitors (Khavkin, 1973). It is tempting to suppose that during the life span of cells in the meristem a kind of auxin activation (Burström, 1968, 1969; Burström *et al.*, 1970) occurs. Auxin exerts an effect on cells beginning elongation, increasing the extensibility of cell walls and inducing elongation.

VI. Cell Maturation

We have considered the metabolic processes in root tip cells, comparing the meristematic and elongating cells and trying to analyse them from the point of view of the growth mechanism. However, we need to look at the matter the other way round. During completion of growth root cells become mature and serve the main function of roots, i.e. the uptake of nutritional elements and water. Cell maturation begins during growth and proceeds some time after its completion. For this reason the comparison of metabolic patterns of elongating and mature cells could characterize maturation, i.e. the formation of the metabolism typical for actively functioning root cells.

A. Metabolic Patterns—Organic Components

The metabolic pattern in mature cells markedly differs from that in growing cells. Photosynthates translocated from shoots are used here mainly for respiration, for further modifications of cell walls (the formation of cellulose, lignin and slime, covering the root surface), for root hair growth, and for syntheses of numerous compounds of the so-called primary metabolism. The label from ^{14}C-photosynthates translocated to roots or from ^{14}C-glucose in root incubation medium was revealed after short exposures to be mainly in sucrose, glucose and fructose, the other portion of ^{14}C-compounds being used for respiration and detected in glucose-1-P, glucose-6-P, organic acids, amino acids and UDP-glucose (Pristupa and Kursanov, 1957; Ivanov and Jacobson, 1966; Mertz and Nordin, 1971). After 3 h exposure, the label was found also in respiratory $^{14}CO_2$ and alcohol-insoluble compounds, viz. proteins, nucleic acids and polysaccharides (Nelson *et al.*, 1961; Anisimov *et al.*, 1964; Fowler and Ap Rees, 1970; Greenway, 1970). Later the output of $^{14}CO_2$ and label incorporation into lipids and polysaccharides increased but the labelling remained higher in alcohol-soluble compounds as compared with alcohol-insoluble ones.

Unlike growing cells with their fast synthesis of polymeric compounds and low level of compounds of primary metabolism, mature cells synthesize the polymers rather more slowly, but accumulate the compounds of primary metabolism and have a high rate of respiration. In mature cells of the root hair zone the content of proteins and nucleic acids shows almost no change with the exception of polyploid nuclei in some tissues. Protein nitrogen constitutes only 30–50% of total nitrogen, hence the low molecular forms of nitrogen compounds predominate (Erickson and Goddard, 1951; Obroucheva, 1965). Mature cells accumulate free amino acids, mono- and disaccharides, organic acids, nucleotides and so on (Khavkin and Varakina, 1967; Potapov and Sumanova, 1968; Ramshorn and Blohm, 1968; Ste-Marie and Weinberger, 1970; Dudchenko and Sytnik, 1966; Mamedova and Rassulov, 1970; Sytnik and Dudchenko, 1972; Naumova and Khavkin, 1974). Their synthesis occurs in the course of respiratory metabolism; the rate of respiration of mature cells is high (Table II), as it provides the energy for ion uptake. Thus in the course of respiratory metabolism photosynthates furnish the mature cells with diverse cell constituents which provide for the functional activity of roots, in particular, the primary stages of ion assimilation. Maximum ion uptake occurs in mature cells, as a calculation of the amount of absorbed ions per cell indicates. The relationship between ion uptake and metabolic features of mature cells should be considered.

Organic acids are formed through the Krebs cycle and related ornithine and glyoxalate cycles; the other pathway in roots is dark CO_2 fixation. Cation uptake is accompanied by an equivalent increase in the level of organic acids in roots and their accumulation in vacuoles as salts, that in turn promotes the uptake of new portions of cations. This relationship is known for potassium and malic acid (Jacoby and Laties, 1971) and for iron and citric acid (Brown, 1966; Brown and Chaney, 1971), Fe-chelate being formed in the latter case.

The level of amino acids in mature cells is closely related to the incorporation of ammonium ions into the molecules of α-ketoglutaric acid and subsequent transfer of amino groups to other keto acids. Glutamic acid and glutamine not only detoxify free ammonium but serve as nitrogen reserve, the amino group of the former and the amido group of the latter being easily utilized in synthesis and turnover of proteins, in synthesis of purine bases, nucleotides and nucleic acids and other N-organic compounds.

The level of nucleotides and hexosephosphates in roots is related to the uptake of inorganic phosphates because the first stage of phosphate metabolism is the formation of these compounds during the first minute of ^{32}P uptake (Loughman and Russel, 1957; Miettinen and Savioja, 1959; Tyszkiewicz, 1959; Kursanov and Viskrebentzeva, 1960; Jackson and Hagen, 1960).

The mature cells of roots carry out numerous reactions of secondary metabolism. For instance, they contain many phenols and very active phenylalanine-ammonium-lyase (Fig. 1), i.e. the enzyme diverting a portion of free amino acids to the synthesis of polyphenols and their derivatives (Khavkin and Pereljaeva, 1970). At the same time another portion of free amino acids in mature cells is the starting material for the synthesis of alkaloids (Obroucheva, 1974). Analysis of data mentioned above leads to the conclusion that the pattern of metabolism of growing root cells is similar in plants of different species but mature cells are characterized by a metabolic pattern typical for every plant species.

B. Metabolic Patterns—Enzymes

The metabolic pattern of mature cells is carried out with the aid of numerous enzyme systems formed earlier in elongating cells. The specificity of mature cell metabolism is established in the course of elongation (Khavkin, 1973). For instance, the enzyme system of assimilation and transformation of nitrogen entering the root is formed (Fig. 1), viz. (a) nitrate reductase and hydroxylamine reductase, reducing NO_3 to NH_4 (Potapov and Sumanova, 1966; Khavkin and Varakina, 1967); (b) glutamate dehydrogenase, providing NH_4 acceptance by α-ketoglutaric acid of the Krebs cycle (Khavkin *et al.*, 1968; Khavkin, 1973); and (c) transaminases, transferring NH_4 from glutamic acid to other organic acids and forming a number of amino acids (Cook, 1959; Khavkin, 1973). The activities of these enzymes proceed to increase also after growth completion though their level at the end of elongation was already high. This accounts for the maximum uptake of nitrates by mature root cells. It should be noted that the formation of enzyme systems of nitrogen metabolism in the course of elongation does not depend upon whether nitrates are present in the medium or amino nitrogen is translocated to the root from the cotyledons (Potapov and Sumanova, 1966).

The respiratory system can offer another example of the formation of "mature" enzyme systems in elongating cells (Fowler and Ap Rees, 1970; Khavkin and Varakina, 1970; Khavkin *et al.*, 1971; Zeleneva and Khavkin, 1971a, b; 1973; Khavkin and Zeleneva, 1972; Polikarpochkina *et al.*, 1973; Varakina *et al.*, 1974; Khavkin *et al.*, 1974). The enzymes carrying out glycolysis, the hexosemonophosphate pathway, the Krebs cycle and terminal oxidation are synthesized in the course of elongation (Fig. 1). It is quite obvious in two cases:

(1) Formation of the "constant proportion group" of glycolytic enzymes, i.e. group of enzymes the activities of which are in rather constant ratio (Zeleneva and Khavkin, 1971a, b; Khavkin and Zeleneva, 1972). The rates of accumulation of three individual enzyme proteins taking part in this "constant proportion group", i.e. glyceraldehydephosphate dehydrogenase (GP), phosphoglycerate kinase (PG) and phosphopyruvate hydratase (PH), are different, as are also their rates of turnover. As a result, in the course of elongation the typical glycolytic pattern (Fig. 1) is provided.

(2) The formation of the mitochondrial enzyme pattern. During the elongation period the content of mitochondrial proteins increased 3 or 4-fold and the rate of ^{14}C-amino acid incorporation into proteins increased too. However, the matrix proteins accumulated at more than double the rate of protein synthesis of inner membranes of the mitochondria. In elongating cells the differential accumulation of cytochromes, i.e. an increase of cytochrome "b" relative to cytochrome "a", takes place. At the same time the rise of activity of cytochrome

"c" oxidase leaves behind an accumulation of cytochrome "a". Thus, specialized synthesis of mitochondrial enzymes and cytochromes in the course of cell elongation provide the rearrangement of the mitochondrial respiratory chain; in its turn, it increases the power of mitochondrial respiration and the high rate of respiration in mature cells in general.

In conclusion it should be noted also that mature cells are the cells in which a portion of photosynthate translocated to roots is utilized for root exudates (Pearson and Parkinson, 1961).

References

ANISIMOV, A. A., DUBOVSKAJA, I. S. and DOBRJAKOVA, L. A. (1964). The action of phosphorus and nitrogen nutrition on ^{14}C-incorporation into assimilates and their translocation in wheat. *Fiziologia rastenii*, **11**; 5, 793–796 (In Russian).

ARNOLD, W. N. (1968). The selection of sucrose as the translocate of higher plants. *J. Theor. Biol.* **21**; 1, 13–20.

BALDOVINOS, DE LA PENA, G. (1953). Growth of the root tip. *In*: "Growth and Differentiation in Plants". (W. Loomis ed.). The Iowa State College Press, Ames, Iowa.

BALODIS, V. A. and IVANOV, V. B. (1970).The investigation of multiplication of root cells transforming from meristematic into elongating phase. *Cytologia*, **12**; 8, 983–992. (In Russian).

BERRY, L. Y. and NORRIS, W. E. (1949). Studies of onion root respiration. I. Velocity of oxygen consumption in different segments of root at different temperatures as a function of partial pressure of oxygen. *Bioch. biophys. acta.* **3**; 5–6, 593–608.

BESEMER, J. and CLAUSS, H. (1968). Disk-Elektrophorese von löslichen Pflanzenproteinen. *Z.Naturforsch.*, **23**; 5, 707–716.

BETZ, A. (1955). Zur Atmung wachsender Wurzelspitzen. *Planta*, **46**; 4, 381–402.

BETZ, A. (1957). Zur Atmung wachsender Wurzelspitzen. III. Das Verhalten in Stickstoff- und hochprozentiger Sauerstoffatmosphäre, die Pasteurische Reaktion. *Planta*, **50**; 2, 122–143.

BETZ, A. (1958). Der Äthanolumsatz in meristematischen Wurzelspitzen von Pisum sativum. *Flora*, **146**; 4, 532–545.

BETZ, A. (1960). Aerobe Gärung in aktiven Meristemen höherer Pflanzen. *In*: "Encycl. Plant physiol." (W. Ruhland ed.), **12/2**; 88–114. Springer Verlag, Berlin.

BROWN, J. C. (1966). Fe and Ca uptake as related to root-sap and stem exudate citrate in soybeans. *Physiol. Plant.*, **19**; 968–976.

BROWN, J. C. and CHANEY, R. L. (1971). Effect of iron on the transport of citrate into the xylem of soybeans and tomatoes. *Plant. physiol.* **47**; 6, 836–840.

BROWN, R. (1963). Cellular differentiation in the root. *Symp. Soc. exp. biol.* **17**; 1–17.

BROWN, R. and BROADBENT, D. (1950). The development of cells in the growing zones of the root. *J. exp. bot.*, **1**; 3, 249–263.

BROWN, R. and CARTWRIGHT, P. M. (1953). The absorption of potassium by cells in the apex of the root. *J. exp. bot.* **4**; 11, 197–211.

BROWN, R. and RICKLESS, P. (1949). A new method for the study of cell division and cell extension with some preliminary observations on the effect of temperature and of nutrients. *Proc. R. Soc.*, ser. B., **136,** 882, 110–125.

BROWN, R. and ROBINSON, E. (1955). Cellular differentiation and the development of enzyme proteins in plants. *In*: "Biological Specificity and Growth", 93–118. New Jersey, Princeton.

BROWN, R. and SUTCLIFFE, J. F. (1950). The effects of sugar and potassium in extension growth of the root. *J. exp. bot.* **1**; 1, 88–113.

BUDER, E. (1967). Der anaerobe Anteil der meristematischen Gärung im Wurzelmeristem von Vicia faba *Flora.* **158A**; N3, 298–324.

BURSTRÖM, H. G. (1968). Root growth activity of barban in relation to auxin and other growth factors. *Physiol. plant.* **21**; N6, 1137–1155.

BURSTRÖM, H. G. (1969). Influence of the tonic effect of gravitation and auxin on cell elongation and polarity in roots. *Am. J. Bot.* **56**; N7, 679–684.

BURSTRÖM, H. G., UHRSTRÖM, I. and OLAUSSON, B. (1970). Influence of auxin on Young's modulus in stems and roots of Pisum and the theory of changing the modulus in tissues. *Physiol. plant.* **23**; 1223–1233.

COOK, F. S. (1959). Generative cycles of protein and transaminase in the growing corn radicle. *Can. J. Bot.* **37**; N4, 621–640.

DEVER, J. E., BANDURSKI, R. S. and KIVILAAN, A. (1968). Partial chemical characterization of corn root cell walls. *Plant. physiol.* **43**; N1, 50–56.

DUDCHENKO, L. G. and SYTNIK, K. M. (1966). Phosphorus metabolism in growth zones of pumpkin roots. *Ukrainskii Botanichnii jurnal*, **23**; N6, 14–17. (In Ukranian).

DUDCHENKO, L. G. and SYTNIK, K. M. (1967). On qualitative composition of proteins in root growth zones. *Dokladi AN UkrSSR*, **ser. B**; N11, 1024–1027 (In Ukranian).

ERICKSON, R. O. and GODDARD, D. R. (1951). An analysis of root growth in cellular and biochemical terms. *Growth, Symposium*, **X**; 89–116.

FOWLER, M. W. and AP REES T. (1970). Carbohydrate oxidation during differentiation in roots of Pisum sativum. *Bioch. Bioph. Acta.* **201**; N1, 33–44.

GODDARD, D. R. and BONNER, W. D. (1960). Cellular respiration. *In*: "Plant Physiology". (A.F.C. Steward, Ed.) **1**; Academic Press, New York, London and San Francisco.

GÖRING, H. (1970). Zur Regulation des Zuckerspiegels in pflanzlichen Geweben. *Biol. Zentralbl.* **89**; N3, 343–358.

GOURNAY-MARGERIE, C. DE, and PÉAUD-LENOEL, C. (1961). Cinetique de la biosynthèse de la cellulose dans les racines de blé. *Bioch. Bioph. Acta*, **47**; N2, 275–287.

GREENWAY, H. (1970). Effects of slowly permeating osmotica on metabolism of vacuolated and non-vacuolated tissues. *Plant. physiol.* **46**; N2, 254–258.

HADAČOVA, V. (1968). Der Einfluß einiger Atmungshemmstoffe auf die Atmung der Wurzelzonen von Vicia faba L. *Biol. plantarum*, **10**; N6, 385–397.

HADAČOVA, V. and SAHULKA, J. (1967). Electrophoretic investigation of proteins in different root zones of Vicia faba L. *Biol. plantarum.* **9**; N5, 396–400.

HELLEBUST, J. A. and FORWARD, D. F. (1962). The invertase of the corn radicle and its activity in successive stages of growth. *Can. J. Bot.* **40**; N1, 113–126.

HEYES, J. K. (1959). The nucleic acids and plant growth and development. *Symp. Soc. exp. biol.* **13**; 365–385.

HEYES, J. K. (1960). Nucleic acid changes during cell expansion in the root. *Proc. R. Soc.* ser. B., **152**; N947, 218–230.

HEYES, J. K. (1963). The role of nucleic acids in cell growth and differentiation. *Symp. Soc. exp. biol.* **17**; 40–56.

HEYES, J. K. and BROWN, R. (1965). Cytochemical changes in cell growth and differentiation in plants. *In*: Encycl. Plant Physiol. (W. Ruhland, Ed.), **15/1**; 189–212, Springer Verlag, Berlin.

HOLMES, B. E., MEE, L. K., HORNSEY, S. and GRAY, L. H. (1955). The nucleic acid content of cells in the meristematic elongating and fully elongated segments of roots of Vicia faba. *Exp. Cell. Res.* **8**; N1, 101–113.

HSIAO, T. C. (1970). Ribosomes during development of root cells of Zea mays. *Plant. physiol.* **45**; N1, 104–106.

IVANOV, V. B. (1968). Cell growth in maize seedling roots irradiated by high doses of X-rays. II. Cell growth under conditions of complete suppression of mitoses, *Cytologia*, **10**; N9, 1105–1117. (In Russian).

IVANOV, V. B. (1970a). Cell interactions in the growing root tip. *In*: "Intercellular Interactions in the Course of Differentiation and Growth," pp. 226–240, Nauka Publ. House, Moscow. (In Russian).

IVANOV, V. B. (1970b). The action of some inhibitors of protein synthesis on division and elongation processes in root cells. *Dokladi AN SSSR*, **121**; N1, 224–227. (In Russian).

IVANOV, V. B. (1973). Growth and multiplication of cells in roots. *In*: "Physiology of Plants", V.1, "Physiology of the root", pp. 7–57. *VINITI AN SSSR* Publ. House, Moscow. (In Russian).

IVANOV, V. B. (1974). The Cellular Bases of Plant Growth. Nauka Publ. House, Moscow. (In Russian).

IVANOV, V. B. and JACOBSON, G. A. (1966). The influence of rhizosphere microorganisms on mutual exchange of higher plants with root exudates. *In*: "Physiologo-biochemical Bases of Mutual Influence of Plants in Phytocenosis," pp. 280–285, Nauka Publ. House, Moscow. (In Russian).

IVANOV, V. B., OBROUCHEVA, N. V. and LITINSKAJA, T. K. (1967). The analysis of chloramphenicol action on growth of maize roots. *Fiziologia rastenii*, **14**; N5, 785–795. (In Russian).

JACKSON, P. C. and HAGEN, C. E. (1960). Products of orthophosphate absorption by barley roots. *Plant. physiol.* **35**; N3, 326–332.

JACOBY, B. and LATIES, G. G. (1971). Bicarbonate fixation and malate compartmentation in relation to salt-induced stoichiometric synthesis of organic acids. *Plant physiol.* **47**; N4, 525–531.

JENSEN, W. A. (1961). Relation of primary cell wall formation to cell development in plants. Society for the study of development and growth. 19th Symposium "Molecular and Cellular Synthesis," pp. 89–110. Ronald Press Co.

JENSEN, W. A. and ASHTON, M. (1960). The composition of developing primary wall in onion root tip cells. I. Quantitative analyses. *Plant physiol.* **35**; N3, 313–323.

KANDLER, O. (1950). Untersuchungen über den Zusammenhang zwischen Atmungsstoffwechsel und Wachstumsvorgängen bei in vitro kultivierten Maiswurzeln. *Z.Naturforsch.* **56**; N4, 203–210.

KHAVKIN, E. E. (1973). The metabolism of growing root cells. *In* "Physiology of Plants," V.1, "Physiology of the Root," pp. 58–106, VINITI AN SSSR Publ. House, Moscow. (In Russian).

KHAVKIN, E. E. and PERELJAEVA, A. I. (1970). Phenylalanine ammonium-lyase and accumulation of free phenolic compounds in growing and mature cells of roots and coleoptiles of maize. Dokladi AN SSSR, **193**; N1, 231–234. (In Russian).

KHAVKIN, E. E. and VARAKINA, N. N. (1967). The composition of free amino acids and soluble proteins in root growth zones of maize seedlings. *Agrochimia*, N12, 18–26. (In Russian).

KHAVKIN, E. E. and VARAKINA, N. N. (1969). Hexose monophosphate pathway of glucose utilization in growing cells of maize root and coleopile. *Fiziologia rastenii* **16**; N6, 1064–1073. (In Russian).

KHAVKIN, E. E. and VARAKINA, N. N. (1970). The role of protein synthesis in promotion of Na N_3—sensitive respiration during elongation of root cells. *Dokladi AN SSSR*, **193**; N3, 716–719. (In Russian).

KHAVKIN, E. E. and ZELENEVA, I. V. (1972). Glycolytic enzyme pattern and constant proportion group in plant cells as related to their developmental and functional state. *FEBS Letters*, **21**; N3, 269–272.

KHAVKIN, E. E., TOKAREVA, E. V. and OBROUCHEVA, N. V. (1967). The rate of protein synthesis in root growth zones of maize seedlings. *Fiziologia rastenii*, **14**; N6, 997–1005. (In Russian).

KHAVKIN, E. E., POLIKARPOCHKINA, R. T., BABURINA, O. M. and TOKAREVA, E. V. (1968). Nitrate reductase, glutamate dehydrogenase and aminotransferase activities in growth zones of maize roots. *Doklady AN SSSR*, **178**; N3, 737–739. (In Russian).

KHAVKIN, E. E., ANTIPINA, A. I. and MISHARIN, S. I. (1971). Immunochemical investigation of proteins from growth zones of maize roots. *Dokladi AN SSSR*, **199**; N4, 972–975. (In Russian).

KHAVKIN, E. E., VARAKINA, N. N. and PERELJAEVA, A. I. (1971). The relationship between protein synthesis in mitochondria and cytoplasm and respiratory enhancement in elongating cells of maize roots. *Fiziologia rastenii*, **18**; 1, N42–49. (In Russian).

KHAVKIN, E. E., KOHL, J.-G., MISHARIN, S. I. and IVANOV, W. N. (1972a). Enzymatische Identifikation der Antigene der wachsenden Wurzelzellen von Zea mays L. *Biochem. Physiol. Pflanzen.* **163**; N3, 308–315.

KHAVKIN, E. E., MISHARIN, S. I., IVANOV, W. N. and KOHL, J.-G. (1972b). The identification of mitochondrial antigens in growth zones of maize roots. *Fiziologia rastenii*, **19**; N1, 160–168. (In Russian).

KHAVKIN, E. E., ZELENEVA, I. V. and VARAKINA, N. N. (1974). The development of respiration in growing root cells. *In*: "Structure and Function of Primary Root Tissues", Symp. Proc. Czechoslovakia, Tatranska Lomniza, Sept. 7–10, 1971.

KOSSEY-FEJÉR, O. (1961). Anderung des gesamten titrierbaren Sulfhydrylgehaltes in den Wurzeln von Keimpflanzen. *Narurwissenschaften* **48**; N11, 434–435.

KURSANOV, A. L. and VISKREBENTZEVA, E. I. (1960). The primary step of phosphate incorporation in root metabolism. *Fiziologia rastenii*, **7**; N3, 276–286. (In Russian).

LOUGHMAN, B. C. and RUSSEL, R. S. (1957). The absorption and utilization of phosphate by young barley plants. IV The initial stages of phosphate metabolism in roots. *J. exp. bot.* **8**; 280–293.

LYNDON, R. F. (1963). Changes in the nucleus during cellular development in the pea seedlings. *J. exp. bot.* **14**; N42, 419–430.

LYNE, R. L. and AP REES, T. (1971). Invertase and sugar content during differentiation of roots of Pisum sativum. *Phytochem.* **10**; N11, 2593–2600.

MAMEDOVA, T. K. and RASSULOV, F. A. (1970). Free nucleotides in growth zones of pumpkin roots grown under calcium deficiency. *Dokladi AN AzSSR*, **26**; N10–11, 71–76. (In Russian).

MCLEISH, J. and SUNDERLAND, N. (1961). Measurements of desoxyribosonucleic acid in higher plants by Feulgen photometry and chemical methods. *Exp. Cell. Res.* **24**; N3, 527–540.

MERTZ, J. and NORDIN, P. (1971). Uptake of labeled glucose by root tips from etiolated wheat seedlings. *Phytochem.* **10**; N6, 1223–1228.

MIETTINEN, J. K. and SAVIOJA, T. (1959). Uptake of orthophosphate by the pea plant. II. Identification of the organic phosphate esters formed. *Acta Chem. Scand.* **13**; N10, 1693–1698.

MORGAN, C. and REITH, W. S. (1954). The composition and quantitative relations of protein and related fractions in developing root cells. *J. exp. bot.* **5**; N13, 119–135.

NAUMOVA, Z. I. and KHAVKIN, E. E. (1974). The regulation of glycolysis by substrates in growing and mature cells of maize seedlings. The concentration of metabolites and saturation of the key enzymes. *Fiziologia rastenii*, **21**; N3. (In Russian).

NELSON, C. D., CLAUSS, H., MORTIMER, D. C. and GORHAM, P. R. (1961). Selective translocation of products of photosynthesis in soybean. *Plant. physiol.* **36**; N5, 581–588.

NORRIS, W. E. (1951). Studies of onion root respiration. V. Effect of culturing temperature and seed sample on root respiration and diameter. *Bioch. Bioph. Acta.* **7**; N2, 225–237.

NORRIS, W. E., HARBER, E. J. and BUTLER, J. E. (1959). Cellular respiration in onion root tips. *Bot. Gaz.* **120**; N3, 131–137.

OBROUCHEVA, N. V. (1965). The Physiology of Growing Root Cells. Nauka Publ. House, Moscow. (In Russian).

OBROUCHEVA, N. V. (1973). The specificity of root metabolism. *In*: "Physiology of Plants", V.1 "Physiology of the Root," pp. 107–163. VINITI AN SSSR Publ. House, Moscow. (In Russian).

OBROUCHEVA, N. V. (1974). The relationship between growth and synthesis of secondary compounds in roots. *In* "Structure and Function of Primary Root Tissues", Symp. Proc. Czechoslovakia, Tatranska Lomniza, Sept. 7–10, 1971.

PEARSON, R. and PARKINSON, D. (1961). The sites of excretion of ninhydrin-positive substances by broad bean seedlings. *Plant. and Soil.* **13**; N4, 391–396.

POLIKARPOCHKINA, R. T., VARAKINA, N. N. and KHAVKIN, E. E. (1973). Cytochromes of mitochondrial respiratory chain in growing and mature cells of maize roots. *Dokladi AN SSSR*, **209**; N2, 492–495. (In Russian).

POLTER, C. (1967). Trennung von loslicher Plasmaproteinen der Erbsenwurzel durch Acrylamid-Scheibenelektrophorese. *Z.Naturforsch.* **22b**; N3, 340–347.

POLTER, C. and MÜLLER-STOLL, W. R. (1969a). Zur Frage der Identität und Homologie der durch Acrylamid-Scheibenelektrophorese getrennten Proteine aus Leguminosen Wurzel. *Z.Naturforsch.* **24b**; N3, 333–341.

POLTER, C. and MÜLLER-STOLL, W. R. (1969b). Über das Verhalten der in Acrylamid getrennten löslichen Proteine aus Erbsenwurzeln bei verlängerter Elektrophorese Dauer. *Z.Naturforsch.* **24b**; N9, 1180–1183.

POTAPOV, N. G. and SALAMATOVA, T. S. (1963). The amount of mitochondria in the cells of the growing zones of lupine root. *Acta biol. acad. sci. Hung.* **14**; N2, 155–159.

POTAPOV, N. G., and SALAMATOVA, T. S. (1964). The effect of some inhibitors on respiration of growth zones of lupine roots. *Fiziologia rastenii*, **11**; N5, 761–768. (In Russian).

POTAPOV, N. G. and SUMANOVA, V. E. (1966). The role of growth zones of lupine roots in uptake and metabolization of nitrates. *Fiziologia rastenii*, **13**; N2, 231–235. (In Russian).

POTAPOV, N. G. and SUMANOVA, V. E. (1968). Amino acid composition of cells in growth zones of lupine roots. *Nauchnie dokladi vysshei shkoli*, seria biologia, N2, 93–98. (In Russian).

POTAPOV, N. G., OBROUCHEVA, N. V. and MAROTI, M. (1959). On physiology of root system growth. *In*: "Plant Growth", Lvov university press, Lvov. (In Russian).

POTAPOV, N. G., KOSULINA, L. G. and LAPIKOVA, V. P. (1970). Some remarks to the problem of mechanism of nutrient uptake by roots. *Selskokhozaistvennaja biologia*, **5**; N2, 302–310. (In Russian).

PRISTUPA, N. A. (1959). The transport form of carbohydrates in pumpkin. *Fiziologia rastenii*, **6**; N1, 30–35. (In Russian).

PRISTUPA, N. A. and KURSANOV, A. L. (1957). Assimilate translocation to roots in relation to their nutrient uptake. *Fiziologia rastenii*, **4**; N5. (In Russian).

RAMSHORN, K. (1957). Zur partiellen "aeroben" Gärung in der Wurzeln von Vicia faba L. *Flora*, **145**; N1, 1–35.

RAMSHORN, K. and BLOHM, D. (1968). Über die freien Aminosäuren von Wurzelzellen sukzessiver Entwicklungsstadien bei Keimpflanzen von Zea mays L. *Biol. Zbl.* **87**; N2, 207–216.

RAMSHORN, K. and KOENIG, R. (1959). Die Bedeutung der Substratqualität und der Sauerstoffspannung für Gaswechsel und Wachstum isolierter Wurzelspitzen. *Flora*, **147**; N3, 358–380.

REID, M. E. (1941). A study of physical and chemical changes in the growing region of primary roots of cowpea seedlings. *Am. J. Bot.*, **28**; N1, 45–51.

REIMERS, F. E. and KHAVKIN, E. E. (1970). De novo protein synthesis in growing cells. *Fiziologia rastenii*, **17**; N2, 337–347. (In Russian).

ROBERTS, R. M. (1967). The incorporation of ^{14}C-labelled D-glucuronate and D-galactose into segments of the root tip of corn. *Phytochem.* **6**; N4, 525–533.

ROBERTS, R. M. and BUTT, V. S. (1967). Patterns of cellulose synthesis in maize root tips. A chemical and autoradiographic study. *Exp. Cell. Res.*, **46**; N3, 495–510.

ROBERTS, R. M. and BUTT, V. S. (1968). Patterns of incorporation of pentose and uronic acid into the cell walls of maize root tips. *Exp. Cell. Res.*, **51**; N213, 519–530.

ROBERTS, R. M. and BUTT, V. S. (1969). Patterns of incorporation of D-galactose into cell wall polysaccharides of growing maize roots. *Planta*, **84**; N3, 250–262.

ROBERTS, R. M. and BUTT, V. S. (1970). Incorporation of ^{14}C-L- Arabinose into polysaccharides of maize root-tips. *Planta.* **94**; N3, 175–183.

ROBERTS, R. M., DESHUSSES, J. and LOEWUS, F. (1968). Inositol metabolism in plants V. Conversion of myo-inositol to uronic acid and pentose unit of acidic polysaccharides in root-tips of Zea mays. *Plant. physiol.* **43**; N6, 979–989.

ROBINSON, E. (1956). Proteolytic enzymes in growing root cells. *J. exp. bot.* **7**; N20, 296–305.

ROBINSON, E. and BROWN, R. (1952). The development of the enzyme complement in growing root cells. *J. exp. bot.* **3**; N9, 356.

ROBINSON, E. and BROWN, R. (1954). Enzyme changes in relation to cell growth in excised root tissues. *J. exp. bot.* **5**; N13, 71–78.

ROBINSON, E. and CARTWRIGHT, P. M. (1958). Nucleolytic enzymes in growing root cells. *J. exp. bot.* **9**; 27.

ROGOZINSKA, J. (1965). The carbohydrate distribution in maize root apex in early growth stage. *Acta Soc. Bot. Polon.*, **34**; N4, 627–635.

ROGOZINSKA, J. H., BRYAN, P. A. and WHALEY, W. G. (1965). Developmental changes in the distribution of carbohydrates in the maize root apex. *Phytochem.* **4**; N6, 919–924.

RUHLAND, W. and RAMSHORN, K. (1938). Aerobe Gärung in aktiven pflanzlichen Meristemen. *Planta*, **28**; N3, 471–514.

SALAMATOVA, T. S. and BALAJAN, E. N. (1966). Adenosine triphosphatase of cell fractions from growth zones of lupine roots. *Fiziologia rastenii*, **13**; N1, 76–81. (In Russian).

SEXTON, R. and SUTCLIFFE, J. F. (1969a). The distribution of β-glycerophosphatase in young roots of Pisum sativum. *Ann. Bot.* **33**; N131, 407–419.

SEXTON, R. and SUTCLIFFE, J. F. (1969b). Some observatons on the characteristics and distribution of adenosine triphosphatases in young roots of Pisum sativum, cultivar Alaska. *Ann. Bot.* **33**, N132, 683–694.

SHIROYA, T., LISTER, G. R., SLANKIS, V., KROTKOV, G. and NELSON, C. D. (1962). Translocation of the products of photosynthesis to roots of pine seedlings. *Can. J. Bot.* **40**; N8, 1125–1137.

SMIRNOV, A. M. (1970). "Growth and Metabolism of Isolated Roots in Sterile Culture." Nauka Publ. House, Moscow. (In Russian).

STE-MARIE, G. and WEINBERGER, P. (1970). Changes in nitrogen metabolism in the wheat root tip following growth and vernalization. *Can. J. Bot.* **48**; N4, 671–681.

STE-MARIE, G. and WEINBERGER, P. (1971). Changes in carbohydrate metabolism in the wheat root tip after growth and vernalization. *Can. J. Bot.* **49**; N2, 195–200.

STEWARD, F. C., LYNDON R. F. and BARBER, J. T. (1965). Acrylamide gel electrophoresis of soluble plant proteins: a study on pea seedlings in relation to development. *Am. J. Bot.* **52**; N2, 155–164.

STREET, H. E. (1966). The physiology of root growth. *Ann. Rev. Plant. Physiol.* **17**; 315–344.

SUNDERLAND, N. and McLEISH, J. (1961). Nucleic acid content and concentration in root cells of higher plants. *Exp. Cell Res.* **24**; N3, 541–554.

SUTCLIFFE, J. F. and SEXTON, R. (1969). Cell differentiation in the root in relation to physiological function. *In*: "Root Growth". (W. J. Whittington, ed.), pp. 80–100. London, Butterworths.

SYTNIK, K. M. and DUDCHENKO, L. G. (1970). Some remarks about heterogeneity of proteins from different growth zones of roots. *Fiziologia i biochimia kulturnikh rastenii*, **2**; N6, 598–603. (In Russian).

SYTNIK, K. M. and DUDCHENKO, L. G. (1972). Metabolism of phosphorous compounds in growth zones of roots under stimulation. *Fiziologia i biochimia kulturnikh rastenii*, **4**; N5, 458–463. (In Russian).

TORREY, J. G. (1965). Physiological bases of organization and development in the root. *In* "Encycl. Plant Physiol." **15/1**, 1256–1327. (Ed. by W. Ruhland). Springer Verlag, Berlin.

TREZZI, F., PEGORARO, L. and VACCARI, E. (1959). Richerche sulla fisiologia della crescita della radice. II Caratteristiche della radice di pisello. *Atti Acad. naz. dei Lincei Rendiconti; Cl. sci. fis. matem. natur.* **26**; N1, 40.

TYSZKIEWICZ, E. (1959). Premiers produits du metabolisme des ion phosphates chez les jeunes plantules d'orge. *C.r.Acad. Sci.* **249**; N19, 1926–1928.

VARAKINA, N. N., ZELENEVA, I. V., POLIKARPOCHKINA, R. T. and KHAVKIN, E. E. (1974). The formation of enzymic pattern of respiration in growing cells. 2. The reorganization of mitochondrial respiratory chain in the maize root tip. *Ontogenez*, **5**; N1, 61–69. (In Russian).

WANNER, H. (1950). Histologische und Physiologische Gradienten in der Wurzelspitze. *Ber. Schweiz. Bot. Ges.* **60**; 404–425.

WOODSTOCK, L. W. and SKOOG, F. (1962). Distributions of growth, nucleic acids and nucleic acid synthesis in seedling roots of Zea mays. *Am. J. Bot.* **49**; N6, 623–633.

YOO, B. Y. (1970). Ultrastructural changes in cells of pea embryo radicles during germination. *J. cell biol.* **45**; N1, 158–171.

ZELENEVA, I. V. and KHAVKIN, E. E. (1971a). Glycolytic enzymes in growing cells of maize roots. *Dokladi AN SSSR*, **199**; N2, 481–484. (In Russian).

ZELENEVA, I. V. and KHAVKIN, E. E. (1971b). The formation of enzymic pattern of respiration in growing cells. I. The glycolytic enzymes in maize seedlings. *Ontogenez*, **2**; N3, 311–320. (In Russian).

ZELENEVA, I. V. and KHAVKIN, E. E. (1973). The regulation of the level of glycolytic enzymes in elongating root cells of maize. *Dokladi AN SSSR*, **208**; N4, 991–994. (In Russian).

ZELENEVA, I. V., REIMERS, F. E. and KHAVKIN, E. E. (1972). The organization of enzyme systems in the growing root cells. *Fiziologia rastenii*, **19**; N6, 1298–1305. (In Russian).

Chapter 15

Auxin Transport in Roots: Its Characteristics and Relationship to Growth

M. W. BATRA, K. L. EDWARDS* AND T. K. SCOTT

Department of Botany, University of North Carolina, Chapel Hill, North Carolina, 27514. *U.S.A*

* Present address: Department of Biology, Kline Tower, Yale University, New Haven, Connecticut, 06520, U.S.A.

I. Introduction

Numerous controversies exist in the literature regarding the presence of auxin in roots. Various compounds with auxin-like properties have been reported to occur in several species (see review by Scott, 1972). A recent and more definitive proof for the presence of auxin, clearly characterized as indole-3-acetic acid (IAA), has been shown in roots of corn seedlings by Elliott and Greenwood (1974). This auxin was found to be present both in the stele and the cortex, although considerably more was isolated from stelar tissue (Bridges *et al.*, 1973; Greenwood *et al.*, 1973).

Auxin is transported in a predominantly acropetal direction in roots, which is to say it moves more towards the tip of the root rather than away from it. This polarity of movement is now well-documented. Studies, over the past ten years, have shown that cultured roots (Bonnett and Torrey, 1965; Bonnett, 1972) or root segments utilized in the classical donor-receiver agar method (Pilet, 1964; Kirk and Jacobs, 1968; Scott and Wilkins, 1968, 1969; Wilkins and Scott, 1968a, b; Zaer, 1968; Cane and Wilkins, 1970; Hillman and Phillips, 1970; Iversen and Aasheim, 1970; Wilkins and Cane, 1970; Aasheim and Iversen, 1971; Wilkins *et al.*, 1972a, b; Scott and Batra, 1973) exhibit this property of auxin behavior. Auxin transport occurs in the same direction in roots of intact seedlings (Hejnowicz, 1968; Konings, 1969; Konings and Gayadin, 1971).

Some characteristics of acropetal polar transport, as shown by studies with root segments, are that it is dependent upon metabolic energy (Wilkins and Scott, 1968b), that it is enhanced under light conditions during the transport period (Scott and Wilkins, 1969), and that it is temperature-sensitive (Wilkins and Scott, 1968b; Wilkins and Cane, 1970). Further, the acropetal transport of auxin is most pronounced at the extreme tip of the root (Cane and Wilkins, 1970; Hillman and Phillips, 1970). Thus a transport gradient, like that demonstrated for stems (Scott and Briggs, 1960; Leopold and Lam, 1962), is also present in roots. The transport of auxin in roots apparently occurs both in the stele and the cortex (Bowen *et al.*, 1972). However, as predicted by Hejnowicz (1968) and Scott (1972), the transport is far greater in the stele (Bowen *et al.*, 1972).

A small basipetal component of auxin transport, which appears to take place with equal magnitude in both the stele and the cortex (Bowen *et al.*, 1972), is present along the entire length of the root (Cane and Wilkins, 1970; Hillman and Phillips, 1970).

The polar transport of auxin in roots may well be related to the acropetal development of cambial activity (Torrey, 1963), to the acropetal regeneration of xylem (Fayle and Farrar, 1965), to lateral root formation (Torrey, 1958), to the polarity of organ formation in root cuttings (Torrey, 1958), as well as to the growth of the root itself (Scott, 1972). The earlier literature, however, abounds with reports of inhibitory effects of auxin on root elongation (Thimann, 1936,

1937; see also review by Åberg, 1957) and thus the view that auxin inhibits root growth at concentrations which promote growth of shoots has had wide acceptance. Extremely low concentrations of auxin (10^{-9} to 10^{-11}M) have been reported to be stimulatory to root elongation (see Torrey, 1956). On the other hand, the findings of Lundegårdh (1950), Leopold and Guernsey (1953), Libbert (1964), Hejnowicz (1968), Burström (1969), and Edwards and Scott (1974b) suggest a stimulatory role of auxin on root growth at concentrations which promote growth of shoots. Evidence in favor of the stimulation of growth of roots under the influence of relatively high concentrations of auxin in "slow-growth" studies (elongation measured at the end of a 6 h period) and in "rapid-growth" studies (elongation recorded immediately and at 5 min intervals) are presented here in an effort to resolve this issue. In addition, the effect of pH on transport was investigated since pH influences both "fast" and "slow" growth responses. Results indicate that there is a correlation between pH, auxin transport, and growth, and arguments pertaining to cause and effect relationships are presented.

The source of the auxin which is moving towards the tip of the root is still unknown (Scott, 1972; Sheldrake, 1973). It could reside somewhere in the basal region of the root (Scott, 1972), or in shoot parts of seedlings (Eliasson, 1972; Eschrich, 1968; Iversen *et al.*, 1971; Morris *et al.*, 1969; McDavid *et al.*, 1972). A final source might, obviously, be the germinating seed itself (Kaldewey and Kraus, 1972). Another aspect of the present investigation was to determine the likely sources or origins of auxin in the root by applying IAA in different areas of the corn seedling and following distribution patterns and profiles in the root over a period of hours and days following application.

II. Materials and Methods

A. Plant Material

Seeds of *Zea mays* L. var. Burpee Snowcross were surface-sterilized with 2% Clorox for 10 min, rinsed and soaked for about 7 h in distilled water, were planted, embryos facing up, on 1% agar slabs in covered plastic boxes, and seeds were then germinated in total darkness at 25° ± 1C. Roots were harvested 72–73 h (3-day-old) after soaking and those selected for use were 50–60 mm in length. When the effect of pH on the transport of auxin was studied, citrate-phosphate buffer (0·001M) of pH ranging from 4·0–8·0 was used throughout, that is, during soaking of seeds, germination, and in the donor and receiver blocks. All pH measurements were made with a Fisher Accumet 520 digital pH meter.

B. Auxin

For most of the present study indole-3-acetic acid-1-^{14}C ammonium salt (1–^{14}C IAA) of specific activity of 26–57 mCi/mM was used. A comparison was, however, made of the transport of 1–^{14}C IAA with indole-3-acetic acid-2-^{14}C (2–^{14}C IAA) having a specific activity of 48 mCi/mM. The two acids were purchased from Radio-chemical Centre, Amersham Searle. The non-radioactive (unlabelled) indole-3-acetic acid was obtained from CalBiochem, California.

C. Experimental Procedure for Root Segments—Transport Studies

Unless otherwise noted, root segments 6 mm were excised 1 mm behind the apex of the primary root under light conditions prevailing in the laboratory. Twenty segments were mounted vertically in Plexiglas holders (see Scott and Wilkins, 1968). The IAA 1–^{14}C and 2–^{14}C were supplied at a concentration of $2{\cdot}5 \times 10^{-7}$ M to one end of segments in the lower donor block ($3 \times 20 \times 25$ mm) of 1·5% agar. The radioactivity transported through the segment was collected at the other end in an upper plain 1·5% agar receiver block ($1 \times 20 \times 25$ mm). The root segments were oriented either with their basal ends touching the donor block (acropetal transport = A) or their apical ends touching the donor block (basipetal transport = B). The holders were kept in moist chambers which were stored in the dark during the 6 h transport period.

At the end of the transport period, the receiver and the donor blocks were placed on planchets (2 inch diameter) and were melted and dried. In those experiments where radioactivity in tissue segments was counted, root segments were dried at 60°C for about 24 h. The dried tissue was finely ground in chloroform and the slurry was then transferred to planchets where the chloroform was allowed to evaporate. The radioactivity was counted on a Spectro/Shield Nuclear-Chicago, low background (2·8–4·0 cpm), automatic gas flow system. The data are presented as cpm or as the percentages of radioactivity moved from donor to either receiver block or the 20 tissue segments, following the procedure of Scott and Wilkins (1968). Data are in all cases corrected for background and corrected, where appropriate, for self-absorption of blocks and tissue. In each transport experiment, at least three replicate holders were set up per treatment and each was carried out on three or more separate occasions.

D. Experimental Procedure for Root Segments—Growth Studies

1. A similar set-up was used for investigating the relationship of IAA to the growth of root segments following a period of 6 h ("slow-growth"). The IAA supplied in the donor blocks was, however, *not* radioactive. Plain receiver blocks placed at the other end of the root segments served to prevent drying of cut ends. The IAA was initially tested at three different concentrations (10^{-7}, 10^{-8}, and 10^{-9}M) but for subsequent studies with buffers at pH 4·0 and 7·0 only 10^{-7}M was used. For each growth experiment two holders per treatment were set up. Each experiment was carried out on three or more separate occasions. At the end of 6 h the segments were shadowgraphed. Increase in length of root segments was

measured at a magnification of 5·1–5·3 × but was corrected for actual size in the data presented.

2. In a second series of experiments segments were floated in petri dishes for 6 h (and measured as above) in a more classical fashion for purposes of comparison.

3. Finally, a third method was employed to study "fast-growth" responses. Ten segments, 2 mm in length, were excised 1 mm behind the apex of the root and were stacked in a holder comprised of four thin steel rods set in a small square of Plexiglas, forming a rectangular restrainer without sides. The tissue holder was inserted into a Plexiglas sling, and the assemblage was then set into a specifically designed small Plexiglas chamber (Edwards and Scott, 1974a). The chamber consisted of two compartments separated by a perforated Plexiglas plate. During the experiment, the chamber was continuously irrigated. The liquid entering the chamber was aerated with oxygen via an air stone in one chamber half. The aerated liquid then passed through the perforated plate to the other half of the chamber containing the root segments. Solutions such as buffer, water and IAA were maintained in large reservoirs above the chamber and were connected to the inlet tubing by Y connections. Elongation of the segments was recorded using a travelling horizontal microscope focused on a small lead weight placed on top of the root segment. The rate of elongation (RE) was recorded generally at 5 min intervals.

E. Experimental Procedure for Intact Roots—Distribution Studies

1. *Mesocotyl Application*

Three-day-old seedlings were selected for uniformity (roots 50–60 mm long) and each root was placed singly through a perforation in a strip of filter paper and parafilm covering the mouth of a 100 ml conical flask containing a few ml of distilled water. The seed rested outside on top of the filter paper. The coleoptile and a portion of mesocotyl were removed leaving behind about 6–8 mm of mesocotyl. A small donor block of agar (3 × 5 × 5 mm) containing 1–^{14}C IAA at 2·5 × 10^{-5}M was balanced on the cut end of each mesocotyl. Each flask was then placed in a beaker (1,000 ml) lined with damp absorbent paper. The top of the beaker was covered with damp filter paper and absorbent paper which hung down over the edge of the beaker into a trough of water. These setups were kept either in the dark or in under 12 h of light at 25°C for 24–72 h, and the roots in the high humidity atmosphere appeared fresh and turgid throughout the duration of the experiment. New donor blocks were applied to a freshly cut surface every 24 h. Two seedlings were used for each treatment and each experiment was carried out on two different occasions. The radioactivity in control donor blocks varied from 33,403·9 to 52,439·5 with a mean of 46,609·85 cpm. At the termination of the experiment, the seedling was cut into segments 6 mm long, starting from the root tip end. Each segment was dried, ground, and counted separately.

2. *Root and Seed Application*

Auxin distribution was observed in experiments using intact roots as well. Seeds, after being surface-sterilized, were germinated in the dark in test tubes containing moist filter paper. After 72 h, seedlings with roots measuring 30–40 mm in length were transferred to tubes having two zones of agar. The lower zone had 10–15 ml of 1% agar mixed with activated charcoal (0·25%) for purposes of distinguishing the two zones. The upper zone consisted of 4 ml agar (1%) with 1–^{14}C IAA at a concentration of 5×10^{-7}M. The radioactivity in this zone varied from 38,606·8 to 30,650·3 with a mean of 34,641·0 cpm. The two agar preparations were autoclaved before pouring into the tubes and allowed to harden separately. The seedlings were inserted into these tubes in such a way that only the basal region of the root (the region adjacent to the seed) was in contact with the zone of agar mixed with 1–^{14}C IAA. The apical portion of the root was pushed through a hole in the radioactive agar layer into the cold layer and was done so rapidly so as to minimize exposure to 1–^{14}C IAA. Three to four seedlings were used for each treatment per experiment and each experiment was carried out two times. Seeds were also embedded in the agar zone containing 1–^{14}C IAA and were allowed to germinate so that when the root emerged it penetrated only the cold layer. At the end of the experiment (four days after planting the seed) seedlings were divided into 6 mm segments and were counted for radioactivity as mentioned earlier.

The "Students" t-test and analysis of variance using the method of Keuls were employed for statistical evaluation of the results (Snedecor, 1967).

III. Results

A. Amounts of 1-^{14}C IAA Transported through Segments Harvested various distances from the Primary Root Apex

Root segments, 6 mm long, were cut 1, 2, 3, and 7 mm behind the root apex and oriented for both A and B transports. In each case, as is evident from Fig. 1, more radioactivity moved acropetally than basipetally. The largest acropetal flux was, however, localized in the extreme tip of the root, in the segment taken 1 mm behind the tip. In this most apical segment, the number of counts moving acropetally and basipetally was approximately 18 and 14 times higher than in the segment 7 mm behind the apex. Segments 1 and 2 mm behind the apex transported significantly more auxin then segments 3 and 7 mm behind the apex. Further, the amount of radioactivity transported by segments 3 and 7 mm behind the apex was not significantly different. Thus, there is an acropetal polarity along the root, or at least up to 13 mm. However, the flux diminishes dramatically as the distance from the apex increases. Therefore, since segments 1 mm behind the apex transport by far the greatest amount of auxin, this region was used in subsequent studies.

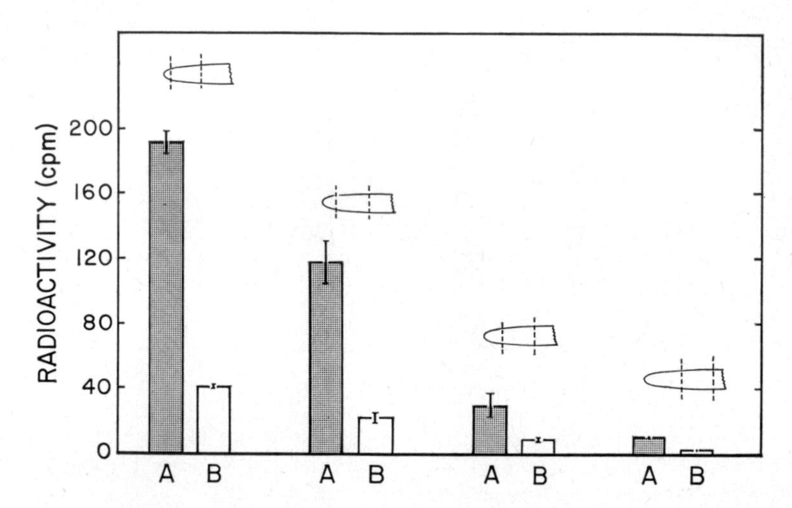

FIG. 1. The acropetal (A) and basipetal (B) transport of ^{14}C through 6 mm root segments excised 1, 2, 3, and 7 mm behind the apex. The vertical bars in the figure represent standard error of the mean.

B. Effect of Orientation of Roots During Germination on Polar Movement of 1-^{14}C IAA

Subapical 6 mm root segments excised from roots grown exclusively horizontally on agar slabs (Scott and Wilkins, 1968) or grown vertically for the last 12–20 h before harvesting were compared in 1–^{14}C transport studies. The transport of radioactivity as measured by the movement of counts into the receiver block is presented in Fig. 2. At the end of the 6 h transport period as much as 5 times more radioactivity was transported acropetally than basipetally through segments taken from roots grown horizontally. This difference is significant at the 0·05 level. However, if roots were oriented in a vertical position a considerable and significant increase in acropetal flux of ^{14}C was observed. Basipetal transport was not increased significantly. The ratio of acropetal to basipetal movement of ^{14}C was 12·9 : 1. Thus the vertical orientation of roots during germination of seeds favors greater movement of radioactivity. The aqueous environment does not appear to cause this difference since flooding the horizontal roots, 12–20 h prior to harvesting, does not change the recovery significantly. The ratio of acropetal to basipetal transport of radioactivity when roots were flooded was 5·5 :1.

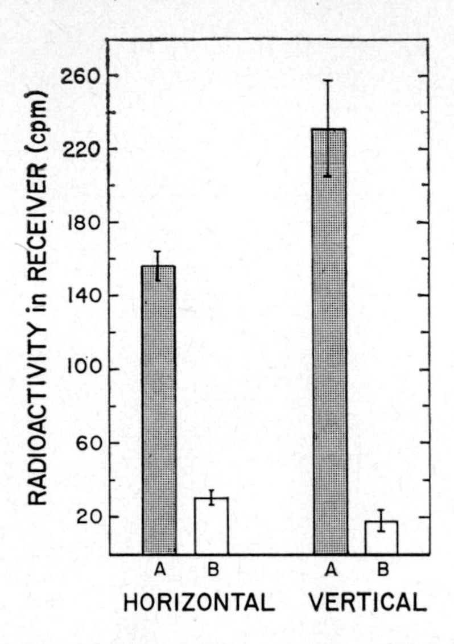

FIG. 2. The acropetal (A) and basipetal (B) movement of ^{14}C through 6 mm segments excised from roots grown horizontally and vertically (12–20 h. before harvesting). The vertical lines in the figure represent standard error of the mean.

C. Comparison of the Transport of 1-^{14}C and 2-^{14}C IAA through Sub-apical 6 mm Segments

The radioactivity in the receiver blocks, tissue segments and the original (OD) and used (D) donor blocks was counted at the end of 6 h. About 1·49% and 1·02% respectively of the radioactivity added in the original donors for 1–^{14}C and 2–^{14}C reached the respective apical receivers (Table I). The percent of radioactivity retained in the tissue segment and that lost from donor blocks were also comparable for the two IAA molecules. Furthermore, recoveries approaching 100% of the radioactivity lost from the donor blocks (OD–D) were obtained in the receiver blocks and tissue segments for both 1–^{14}C and 2–^{14}C IAA. Thus, the two IAA's labelled in the different positions were transported both acropetally and basipetally with equal effectiveness and destruction of 1–^{14}C IAA by decarboxylation, which is often a concern when using tissue with cut surfaces and which is commonly believed to take place in root tissue, was not a factor in our study.

D. Effect of pH on the Transport of 1-^{14}C IAA

Citrate–phosphate buffer (0·001 M) over a pH range of 4·0 to 8·0 was incorporated throughout the handling of the plant material—that is, during soaking of seeds, during germination, and in the agar of the donor and receiver blocks. Distilled water (DW) was also used as a basis of comparison.

TABLE I. Comparison of the transport of 1—^{14}C and 2—^{14}C IAA (A = acropetal transport; B = basipetal transport) through subapical 6 mm root segment of *Zea mays* after a 6 h transport period.[a]

IAA		Radioactivity in Receiver (R)		Radioactivity in Tissue (T)		Radioactivity lost from Donors[b]		Unaccounted Loss[c]
		cpm	R/OD × 100	cpm	T/OD × 100	cpm	OD-D/OD × 100	
1—^{14}C	A	168 ± 7	1·49	593 ± 65	5·27	1000 ± 199	8·89	− 2·13
	B	44·1 ± 3·6	0·39	940 ± 110	8·35	1799 ± 210	15·99	− 7·25
2—^{14}C	A	102 ± 12	1·02	661 ± 60	6·60	702 ± 115	7·00	+ 0·62
	B	32·3 ± 4·2	0·32	1016 ± 63	10·2	1203 ± 158	12·02	− 1·50

[a] All data corrected for self-absorption. [b] OD = Original Donor. D = Used Donor. Average counts for 1—^{14}C IAA = 11249 ± 529. Average counts for 2—^{14}C IAA = 9998 ± 93. [c]OD-D/OD × 100 − R/OD × 100 + T/OD × 100.

The amount of radioactivity in the receiver block, for different pH's and DW, expressed as a percentage of the radioactivity of original donor, is given in Table II. In each case, acropetal movement predominated over basipetal. The highest

TABLE II. Effect of pH and distilled water (DW) on the relative levels of radioactivity[a] in receiver blocks (A = acropetal transport; B = basipetal transport) after a 6 h transport period.[b]

Block	4·0	5·0	6·0	7·0	8·0	DW
A	0·63	0·74	0·96	0·79	1·00	1·04
B	0·03	0·10	0·25	0·07	0·24	0·18
A/B	21·0	7·4	3·8	11·3	4·2	5·8

[a] Radioactivity expressed as percent recovery of the original donor block in the receiver block.
[b] All data corrected for self-absorption.

percentage of the radioactivity in apical receivers was in the DW treatment and the lowest was in the pH 4·0 treatment. However, the ratio of recoveries in apical and basal receivers was maximum at pH 4·0 (21:1) and minimum at pH 6·0 (3·8:1). The other ratios were 7·4:1 (pH 5·0), 11·3:1 (pH 7·0), 4·2:1 (pH 8·0), and 5·8:1 (DW). The pH of distilled water was checked and was found to be close to pH 6·0 but considerable fluctuation was observed.

A detailed study of the recoveries of radioactivity in the receiver block, tissue segments, original and used donor blocks at pH 4·0, 7·0 and DW was undertaken. The results are presented in Table III. In each case, there was more net loss from donors (OD–D) during basipetal (B) than during acropetal (A) transport at pH 4·0, 7·0 and DW. Although the maximum loss of radioactivity from donors was at pH 4·0, only 0·63% of the radioactivity supplied in original donors reached apical receivers (A) as compared to 0·79% at pH 7·0 and 1·49% in DW. The corresponding values for basal (B) receivers were also higher at pH 7·0 and DW. The transport of 1–^{14}C IAA, therefore, was very much reduced at pH 4·0. The recoveries of ^{14}C in tissue segments, like donor loss, were greater during basipetal than during acropetal transport in these two cases. However, the highest retention of radioactivity during both acropetal and basipetal transport took place at pH 4·0. It is noteworthy that with DW almost 100% of the radioactivity supplied in the original donor was recovered in the receiver blocks, tissue segments and used donors; this is not the case at pH 7·0 and 4·0. At pH 7·0, the unaccounted loss was not great (5–8·5%). However, at pH 4·0 there was 19–23% loss from the donors which cannot be accounted for.

Thus, although the acropetal polarity of auxin transport is most pronounced at 4·0, the amount transported *per se* is reduced over that found at all other pH's tested (Table II). It is to be remembered, however, that the greatest amount of tissue recovery of ^{14}C is at pH 4·0.

E. Effect of IAA and pH on Root Growth

1. "*Slow-growth*" *Studies Employing Distilled Water*

IAA at three different concentrations (10^{-9}, 10^{-8}, 10^{-7} M) was tested for effects on root growth over a period of 6 h (Fig. 3). Subapical 6 mm segments were

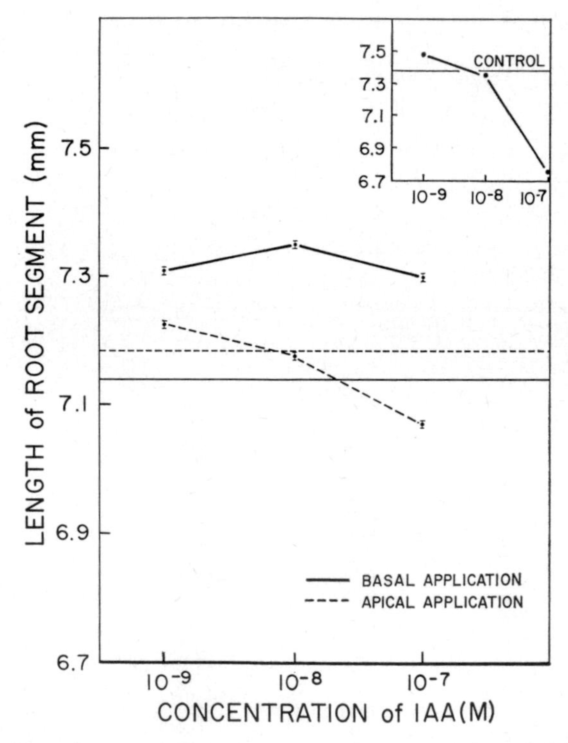

FIG. 3. The growth of 6 mm root segments as influenced by the apical (B) and basal (A) application of IAA. The vertical bars are the standard error of the mean. The insert represents the growth of 6 mm root segments when floated in petri dishes. Horizontal lines (solid and dashed) represent growth of distilled water controls.

placed in the holders with their ends touching the donor blocks. DW blocks without IAA served as controls. More growth of the root segments occurred when DW donors were placed apically (B) than basally (A). Highly significant growth promotion due to IAA (10^{-9}, 10^{-8}, and 10^{-7} M) over the controls was observed when auxin was provided basally (A) and in a fashion in which there was greatest acropetal transport. Apically applied (B) auxin inhibited growth at 10^{-7} and 10^{-8} M. Promotion of growth, however, was observed at 10^{-9} M under conditions of basipetal transport (B).

The insert in Fig. 3 represents the growth of root segments when floated in petri dishes for 6 h. Statistically significant inhibition of growth over the controls is seen at a concentration of 10^{-7} M. The slight promotion of growth at 10^{-9} M is, however, not significantly different from the control.

Thus, IAA at similar concentrations evokes different kinds of growth responses when supplied in different ways. The concentrations (10^{-7} and 10^{-8} M) which proved to be inhibitory under petri dish conditions promoted growth in roots which were oriented and had the IAA applied at a point source.

2. "*Slow-growth*" *Studies Employing Buffer*

Root segments, 6 mm long, were placed in holders with their ends touching the donor blocks. Only the donor blocks were buffered at pH 4·0 and 7·0 and half the donors included IAA at 10^{-7} M. At the end of the 6 h transport period, root segments were shadowgraphed and segment lengths were measured. The results of a representative experiment are presented in Fig. 4.

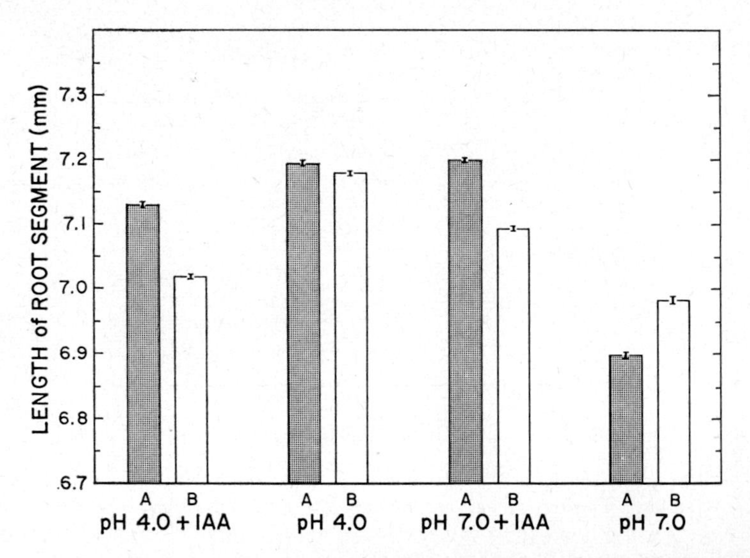

FIG. 4. Effect of apical (B) and basal (A) application of IAA (10^{-7}M) and pH 4·0 and 7·0 agar blocks on the growth of 6 mm root segments. The data presented are from a representative experiment which was repeated twice. The vertical bars in the figure are standard error of the mean.

Greatest growth promotion of segments was observed when a buffered donor at pH 4·0 was supplied either basally or apically. Addition of IAA at pH 4·0, however, inhibited elongation of segments below that of pH 4·0 alone. Inhibition was more when IAA was supplied apically (B) than basally (A). These results are correlated with the transport results where it was found that the IAA transported (recovery in receiver) at pH 4·0 was considerably reduced in both directions (Table III). A greater retention of ^{14}C in the tissue at pH 4·0 took place

TABLE III. Effect of pH 4·0, 7·0 and distilled water (DW) on the levels of radioactivity in receiver blocks, tissue segments and donor blocks (A = acropetal transport; B = basipetal transport) after a 6-h transport period.[a]

pH		Radioactivity of original Donor in Receiver (R) R/OD × 100	A/B	Radioactivity of original Donor in Tissue (T) T/OD × 100	A/B	Radioactivity lost from Donors OD-D/OD × 100	Unaccounted Loss[b]
4·0	A	0·63	21·0	9·39	0·69	29·00	18·98
	B	0·03		13·61		36·38	22·74
7·0	A	0·79	11·3	5·29	0·60	11·07	4·99
	B	0·07		8·74		17·30	8·49
DW	A	1·49	3·8	5·27	0·63	8·89	2·13
	B	0·39		8·35		15·99	7·25

[a] All data corrected for self-absorption. [b] OD-D/OD × 100 − R/OD × 100 + T/OD × 100.

during the basipetal transport (B) than during the acropetal transport (A), which may explain why inhibition of growth due to IAA when it is supplied apically (B) is greater than when it is supplied basally (A) (Table III). The tissue retention of ^{14}C at pH 4·0 is also greater than that at pH 7·0 and in DW (Table III). Thus, this inhibition of elongation appears to be correlated with the increase of IAA retained in the tissue rather than with the IAA transported through the tissue. Although the unaccounted loss of radioactivity is highest at pH 4·0, as compared to that at pH 7·0 and DW, it is more or less the same, whether IAA is transported acropetally or basipetally. Thus it may be surmised that high H^+-ion concentration may inhibit the transport of IAA which in turn would induce an increase in the IAA concentration in the tissue. This non-transported IAA would act as a partial "brake" on the H^+-ion-induced elongation. Finally, as may be seen in Fig. 4, the addition of IAA at pH 4·0 or pH 7·0 causes greater growth when the blocks are applied basally (A). These results are in agreement with those shown in Fig. 3.

Treatment with pH 7·0 alone inhibited the elongation of the segment when applied basally (A) as compared to pH 4·0 alone (Fig. 4). Growth with apical application was, however, not as markedly inhibited. The presence of IAA in the donor buffered at pH 7·0 caused the maximum growth response when the donor was applied basally (A). This growth promotion is indistinguishable from that induced by pH 4·0 alone. Since more IAA at pH 7·0 was transported both acropetally and basipetally (Table III) than that at pH 4·0 and since amounts retained in tissue were lower than those at pH 4·0, the growth promotion due to IAA at pH 7·0 is most closely correlated with an acropetal IAA transport.

3. "*Fast-growth*" *Studies Employing Buffer*

Ten 2 mm root segments, cut 1 mm behind the apex, showed a large initial rate of elongation (RE) in citrate–phosphate buffer at pH 7·0. By the end of 2 h, however, this RE of root segments was considerably reduced and is indicated by line E in Figs. 5 and 6. The RE of root segments was increased above the

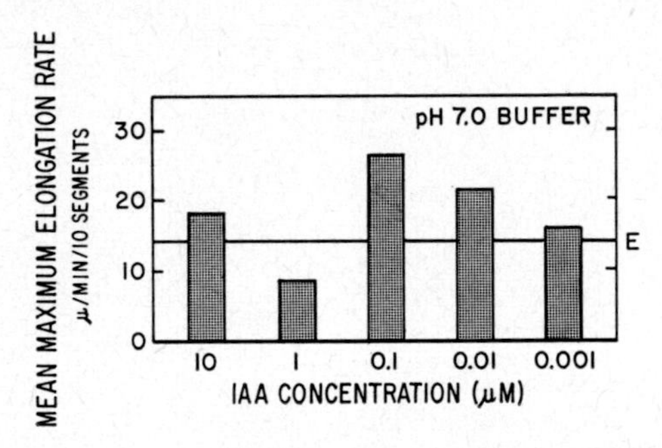

FIG. 5. Effect of IAA in buffer at pH 7·0 on the elongation rate of 2 mm root segments. E = equilibrium rate in buffer at pH 7·0.

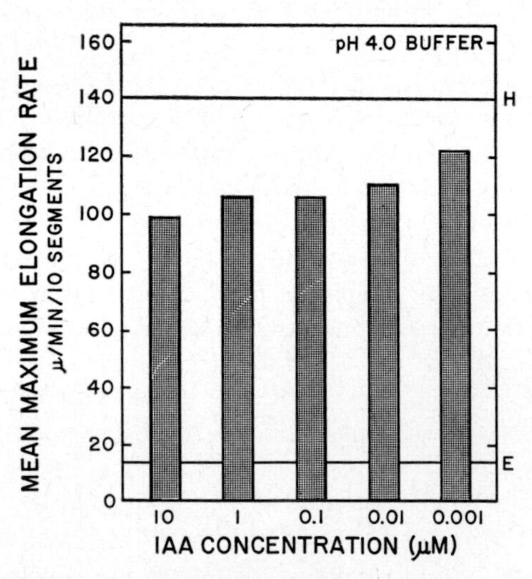

FIG. 6. Effect of IAA in buffer at pH 4·0 on the elongation rate of 2 mm root segments. H = elongation rate in pH 4·0 buffered solution; E = equilibrium rate in buffer at pH 7·0.

equilibrium RE (pH 7·0) when exposed to acid pH. Although the maximum RE was obtained at pH 3·4, pH 4·0 was generally used as the standard H^+-ion concentration. The increased RE induced by low pH was induced immediately and reached the maximum rate in approximately 13–14 min after the initial exposure.

With the addition of IAA in pH 7·0 buffer, the RE of root segments was increased above the pH 7·0 equilibrium rate with maximum stimulation occurring at a concentration of 10^{-7} M (Fig. 5). The maximum RE was reached within 40 min after exposure to IAA (Edwards and Scott, 1974b).

On the other hand, all concentraions of IAA tested in pH 4·0 decreased the pH 4·0 buffer-induced RE of root segments (Fig. 6). Upon removal of IAA, the acid-induced RE did not return to its normal maximum (the line designated H in Fig. 6). This may be due, in part, to the fact that considerable acid-induced elongation had already occurred in the presence of IAA (Edwards and Scott, 1974b).

F. Experiments with Intact Roots

1. *Distribution of Radioactivity in the Seedling when* 1–^{14}C *IAA was supplied to basal region of the root.*

As outlined in Materials and Methods, 1–^{14}C IAA (5×10^{-7} M) was supplied to the basal region (close to seed) of the root (which was 30–40 mm long) of a 3-day-old seedling. After 48 h (roots measured 55–70 mm) and 72 h (roots measured 70–84 mm) of treatment, 6 mm segments were cut from the seedling, dried and counted for radioactivity. The total number of counts moving down to the root after 48 and 72 h were considerably higher than those moving up to the mesocotyl and the coleoptile. The radioactivity at the root tip far exceeded that at the coleoptile tip (213 cpm vs. 8·7 cpm after 48 h; 202·3 cpm vs. 7·9 cpm after 72 h). The number of counts tended to accumulate in the extreme tip of the root in contrast to the subtending segments (Fig. 7). The distribution of ^{14}C in the roots, under these conditions, would appear to reflect an acropetal diffusion, as judged by the steep gradient, except for the tip where there is a marked increase in concentration and thus evidence for an active acropetal transport system. The extremely low radioactivity recovered in the shoot parts of the seedling strongly supports the presence of a weak acropetal transport system in mesocotyl and coleoptile tissues.

2. *Distribution of Radioactivity in the seedling when* 1–^{14}C *IAA was applied to seed*

During germination, seeds were exposed to radioactive IAA and the absorbed radioactivity was transported to all parts of the seedling (Fig. 8). Under these conditions of germination, the roots at the end of 4 days (the time in this case counted from the moment of planting the seeds on the agar containing 5×10^{-7}M 1–^{14}C IAA), were shorter (about 30–50 mm) compared to those grown on agar slabs for 3 days (50–60 mm) or in distilled water for 3 days (30–40 mm). Radioactivity moved upwards as far as the coleoptile where maximum recovery was

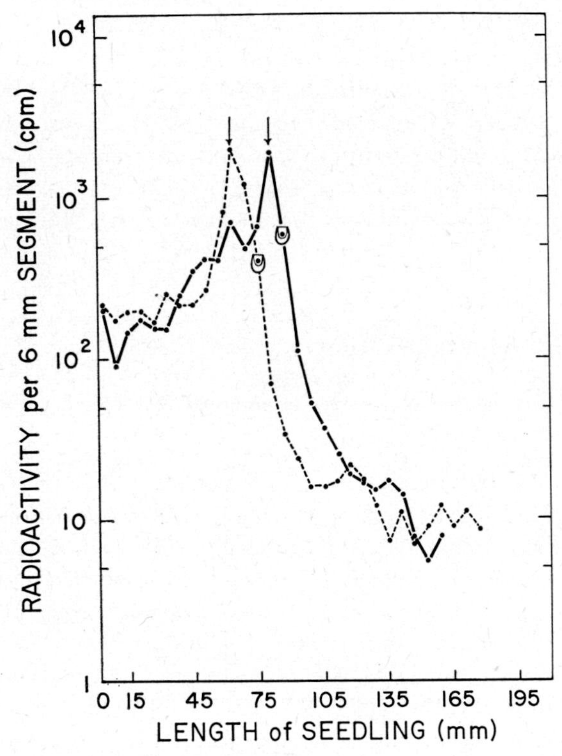

FIG. 7. Distribution of radioactivity in the seedling when 1–^{14}C IAA (5×10^{-7}M) was supplied to the basal region of the root (shown by arrow) 3 days following germination. The curve to the left of the seed represents the primary root and that to the right is mesocotyl and coleoptile. The data plotted at each treatment are from a representative seedling. (– – – –) after 48 h; (———) after 72 h.

7·0–21·0 cpm. It moved to the root tip as well and again it tended to accumulate in the apical 6 mm segment (189–209 cpm). Distribution of radioactivity along the root, but apart from the tip, suggests its movement by diffusion. Thus, auxin moving acropetally in roots *in vivo* could be derived from the fruits or the seeds during germination although profiles in Figs 7 and 8 are by no means direct evidence of this.

3. *Distribution of Radioactivity in the Seedling when* 1–^{14}C *IAA was supplied to the mesocotyl*

Small donor blocks containing 1–^{14}C IAA at $2{\cdot}5 \times 10^{-5}$ M were balanced on the cut end of the mesocotyl of 3-day-old seedlings (root length 50–60 mm). Although radioactivity was detected in all parts of the seedling (Fig. 9) after 24, 48, and 72 h treatment, the number of counts reaching the root tip were not

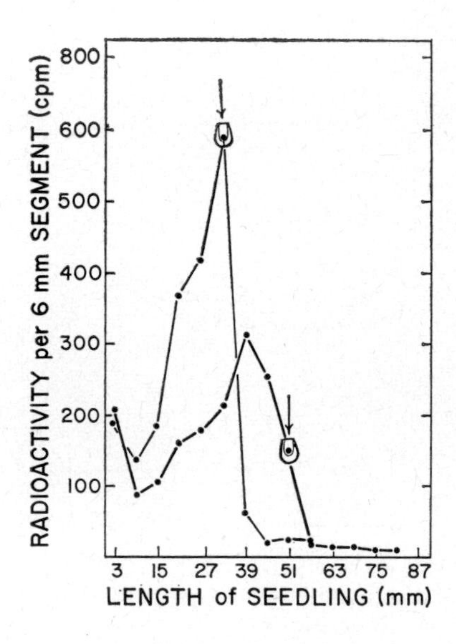

FIG. 8. Distribution pattern of radioactivity in the seedling when 1–^{14}C IAA (5×10^{-7}M) was supplied to the seed (shown by arrow) at the time of planting. The curve to the left of the seed represents the primary root and that to the right is mesocotyl and coleoptile. The data plotted are from two representative seedlings. (———) after 96 h.

appreciably different at the different time intervals. That is, 3–5 cpm reached the root tip at the end of 24 h, 4–6 cpm after 48 h and 6–24 cpm after 72 h. The total radioactivity supplied to the mesocotyl differed according to the duration of the experiment. Nevertheless, by 72 h radioactivity began to accumulate in the extreme tip of the root. The amount of radioactivity recovered at any place and at any time in the root was but a very small fraction of the high dosage applied.

Adventitious roots, which arose during the course of the experiment at the base of the mesocotyl very close to the seed, had appreciable amounts of radioactivity: 270 cpm after 24 h, 314 cpm after 48 h and 1598·5 cpm after 72 h. Radioactivity, moving down the mesocotyl, thus gets diverted to and accumulated in the newly arising adventitious roots.

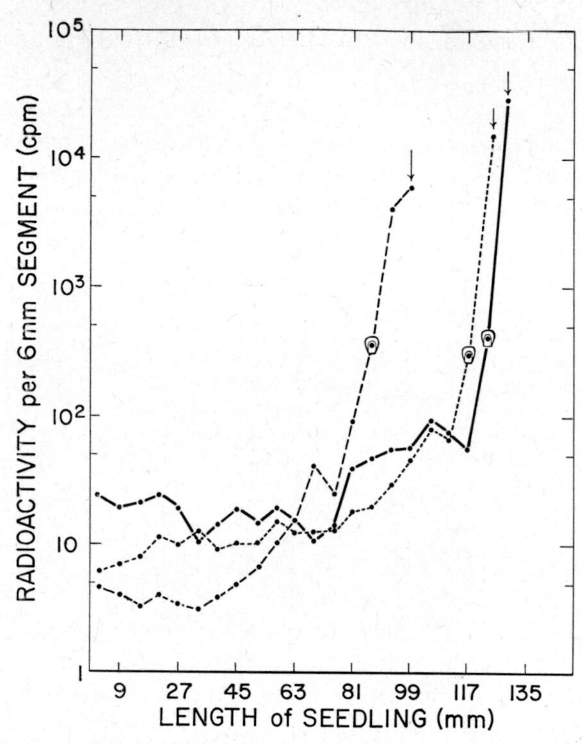

FIG. 9. Distribution of radioactivity in the seedling when 1–^{14}C IAA ($2{\cdot}5 \times 10^{-5}$M) was supplied to the cut end of the mesocotyl (shown by arrow) 3 days following germination. The curve to the left of the seed represents the primary root and that to the right is mesocotyl. The data plotted at each treatment are from a representative seedling. (— — —) after 24 h; (- - - - - -) after 48 h; (———) after 72 h.

IV. Discussion and Conclusions

The data presented in this paper show that auxin moves in a polar fashion in the primary root of *Zea mays*. There also exists an exponential transport gradient along the root up to 13 mm from the tip. The gradient terminates with the greatest amount of auxin moving acropetally and with the strongest polarity of movement in a region 1–3 mm from the apex. The results of Hillman and Phillips (1970) and Cane and Wilkins (1970) also suggest a similar kind of acropetal transport gradient in primary roots of pea and corn which appears to occur over a distance of 15 mm and 19 mm respectively.

Scott and Wilkins (1968) claimed that a 180° inversion from the normal orientation of the root segments does not affect the transport of auxin during the course of the transport period. However, Lyon (1965) reported a 2–5 fold increase of

2–^{14}C IAA in roots of mature bean and cabbage plants when the radioactivity was applied to the lower epidermis of fully expanded leaf blades of plants growing erect in relation to gravity as opposed to those maintained on a clinostat. In the present study, displacement of roots by 90°, during the period of their early growth and development, significantly affected the subsequent transport of 1–^{14}C IAA. More auxin was transported in segments cut from vertically grown roots (12–20 h before harvesting) than from those grown horizontally. The difference was both in magnitude (156 cpm vs. 231 cpm) and degree of polarity (5:1 vs. 12·9:1). Naqvi and Gordon (1966) demonstrated that inversion has no effect on auxin transport of the corn coleoptile of Burpee Snowcross, unless the tissue was inverted for a period of 5 h before the transport experiment. The result of such a pre-treatment was a reduction of both transport velocity and transport capacity of the coleoptiles. Similarly, Anker (1960) reported that a 90° displacement inhibited longitudinal growth in decapitated *Avena* coleoptile sections, even though the vertical and horizontal sections were supplied with the same amount of exogenous auxin. Thus there is an effect of orientation of tissue, whether in advance of harvesting the tissue or during experimentation, on both auxin transport characteristics and growth characteristics. Vertical and proper polar orientation favors more auxin transport as well as more growth.

The results presented in this paper show that 1–^{14}C and 2–^{14}C IAA are transported with equal effectiveness over a period of 6 h (Table I). The recoveries of both 1–^{14}C and 2–^{14}C in the receiver block, tissue segments, and the donor block are almost the same suggesting, therefore, that the enzymatic decarboxylation of 1–^{14}C IAA was not a factor in this study. The fact that decarboxylation of 1–^{14}C IAA starts after 6–8 h (Iversen and Aasheim, 1970; Wilkins *et al.*, 1972b) and of 2–^{14}C after 10–12 h (Wilkins *et al.*, 1972b) of transport in sunflower and corn root segments is consistent with our observation that there was no significant loss of radioactivity due to decarboxylation. Wilkins *et al.*, (1972a) also have shown the transport of 1–^{14}C and 2–^{14}C IAA through corn root segments to be closely identical. Further, Scott and Wilkins (1968, 1969) have shown that the radioactivity transported (found in receiver block) is confined to the IAA molecule. Thus the over-all loss of radioactivity found generally in this study, as small as it is, does not seem to be due to specific causes and may reflect experimental error.

On the other hand, there was a considerable amount of 1–^{14}C lost in experiments performed at pH 4·0 (in both basipetal and acropetal transport experiments), and it may be that this was the result of auxin destruction by an enzyme system whose pH optimum is low.

Since it has now been shown beyond doubt that auxin in roots is transported acropetally and since auxin transported *acropetally* in this study stimulates the growth of corn root segments, it may be inferred that there is a direct relationship between transport and growth. Others have reached a similar conclusion regarding roots using different methods of study (Hejnowicz, 1968; Konings, 1969). Hertel *et al* (1969), working with coleoptiles of corn, have drawn the strongest correlation between auxin-induced growth and auxin transport. They observed that the α-isomer of naphthaleneacetic acid (NAA), which is transported basipetally like IAA, and 2,4-dichlorophenoxyacetic acid (2,4-D) showed growth-promoting activity, whereas the β-isomer of NAA was neither transported nor did it stimulate the growth of the coleoptiles. Parallel results were obtained with (+) and (−) isomers of 3-indole-2-methylacetic acid and the antiauxin p-chlorophenoxyisobutyric acid. Therefore, the processes involved in the transport of auxin are believed by these workers to be linked to the primary action of auxin.

The effect of pH on the transport of auxin appears to be associated with an increased uptake by tissue (Smith and Jacobs, 1968; Mitchell and Davies, 1972; Rubery and Sheldrake, 1973) as well as with an accumulation of auxin within the tissue (Rubery and Sheldrake, 1973). We find that uptake (expressed as a function of radioactivity lost from donors) and retention of radioactivity within the tissue segments are greater at pH 4·0 than at pH 7·0. Although the ratio of acropetal to basipetal transport, and thus the degree of polarity, is maximum at pH 4·0, the transport of auxin *per se* is less than that found at pH 7·0 (Tables II and III). Mitchell and Davies (1972), however, reported increased transport of IAA-^{3}H (2×10^{-9} M) in intact roots of *Phaseolus*, when pH was decreased from 7·0 to 4·0. One might, in fact, interpret their observation as reflecting a pH-induced increase in auxin retention in the tissue.

Our results with "slow-growth" and "fast-growth" studies show that the growth of root segments is stimulated under acid conditions. The acid-induced "fast-growth" elongation responses of seedling shoot parts are numerous (for instance, see, Rayle and Cleland, 1970; Evans *et al*., 1971; Hager *et al*., 1971; Barkley and Leopold, 1973; Bridges and Wilkins, 1973; Rayle, 1973; Rehm and Cline, 1973). The only report known to us which deals with a rapid acid-induced elongation of roots is a paper of Lundegårdh (1950) dealing with wheat roots. He observed a stimulation of the growth of intact roots at pH 3·4. The response was immediate and reached a peak within 12 min. Two other pH's tested (4·7 and 2·7) were less effective but did promote growth. Of interest is the fact that acid stimulation of growth of wheat roots was roughly one half that which was observed with auxin-induced growth using IAA at 10^{-3} M. Of all concentrations tested (10^{-9}–10^{-3} M), 10^{-3} M was the highest and most promotive, again with an immediate response which peaked at about 8 min. Since the pH of 10^{-3} M IAA solution was 3·87, the growth reaction observed at this remarkably high IAA concentration may have been due to the variation of the dissociation of IAA and, in the end, to H^+-ions. To cite Lundegårdh (1950) himself, "The pH of the medium thus dominates the growth reaction, even if IAA is supplied from the outside" but the author leaves open the question of the direct effect of "auxin anions."

A comparison of the results of the transport experiments with those of "slow-growth" studies indicates that IAA (10^{-7} M) at pH 4·0 causes inhibition of elongation of root segments as compared to that of pH 4·0 alone (Fig. 4). This inhibition, however, is greater when IAA at pH 4·0 is applied at the apical end (B) than when it is applied at the basal end (A). This inhibition of elongation appears, on the one hand, to be correlated primarily with the increase of IAA retained in the tissue and to some extent IAA transported through the tissue segment. More IAA is retained in the tissue at pH 4·0 than when it is applied at pH 7·0. This increase in retention is greater when IAA, at pH 4·0, is applied apically (B) than when it is applied basally (A). Again comparing the two pH's, metabolic acropetal transport (A) at pH 4·0 is reduced less and passive basipetal transport (B) of IAA is reduced more comparable to pH 7·0 (Table III). The inhibition of elongation by IAA at pH 4·0 would not appear to be accounted for by the unaccountable loss of IAA from the donor block since the loss of IAA is roughly the same whether the IAA is applied apically (B) or basally (A). Indeed, if one subtracts loss at pH 7·0 from that of pH 4·0, the values are really identical for A and B. Thus, a workable hypothesis is that a high H^+-ion concentration induces transport inhibition which in turn causes an increase in the IAA concentration in the tissue. An accumulation in the tissue might thus result in greater degradation and the observed increase in over-all loss of radioactivity. This non-transportable IAA would act as a partial "brake" on the H^+-ion-induced elongation. Finally, polarity of the growth response results from more acropetal transport–less tissue retention as opposed to less basipetal transport–more tissue retention.

On the other hand, IAA at pH 7·0 promotes the elongation of root segments above those at pH 7·0 alone, and again more growth was observed when IAA was applied basally (A). IAA at pH 7·0 when applied apically (B) has less marked effect on elongation (Fig. 4). Furthermore, at pH 7·0, IAA transport (radioactivity in receiver) is increased over that of pH 4·0. The increase in the acropetal metabolic transport is less than the basipetal passive transport (Table III). In addition, at pH 7·0, less IAA is retained in the tissue segments as compared with those at pH 4·0. The amount of IAA retained is equally less in both apical (B) and basal (A) applications (4·1 vs. 4·8). Therefore, growth promotion induced by an apical application of IAA at pH 7·0 is equated to the promotion of metabolic acropetal transport. The increase in IAA transport may account for the decreased amount of IAA held in the tissue, which in turn, is at a promoting, rather than "braking" level. However, once again the polarity of transport is correlated with the maximum growth response. In fact, growth at pH 7·0 and growth studies with distilled water (Fig. 3) show a reversed polarity in the absence of IAA.

Slightly more growth of root segments occurs when distilled water–donor blocks are applied at the apical ends than when applied at the basal end (horizontal lines, Fig. 3). The presence of IAA in DW–donor blocks induces considerable elongation of the root segments when IAA is supplied basally (A) but causes inhibition when supplied apically (B) at a concentration of 10^{-7} M. The acropetal transport of IAA is greater using distilled water than buffer at any pH tested (Table III). Both acropetal and basipetal transport are considerably lower at pH 4·0 and 7·0. Furthermore, the amount of auxin retained in the tissue segment when IAA is transported acropetally (A) or basipetally (B) is less than that at pH 4·0 and equal to that at pH 7·0 (Table III). Therefore, although both metabolic and passive transport of auxin are promoted over pH 4·0 and 7·0, it is the increased metabolic acropetal auxin transport (A) which remains as the agent which induces an increase in the growth promotion of root segments.

The modes of auxin inhibitory and promotory action on elongation in root segments are considered to be different and occurring at different sites in the cell or cell wall. Auxin inhibition of growth is considered here to be stimulated by high concentrations of auxin in the cell, partially due to a reduction in auxin transport. Auxin promotion of growth is considered to be integrally related to increased acropetal metabolic transport only. These considerations are summarized in the two models below.

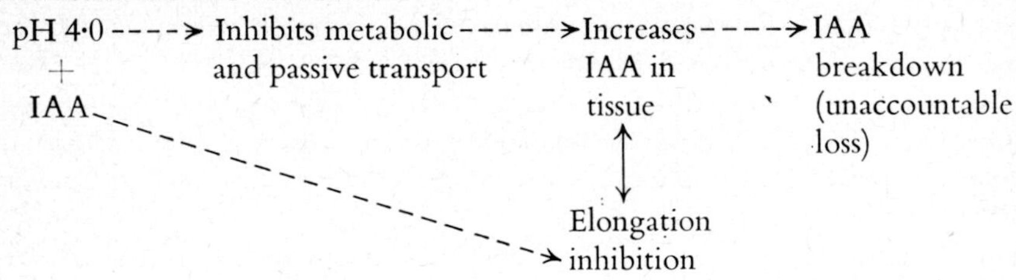

Model 1. IAA at pH 4·0: Growth Inhibition

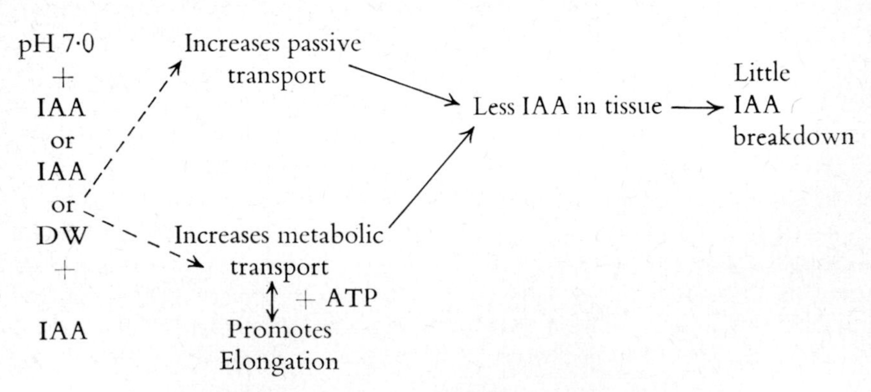

Model 2. IAA at pH 7·0: Growth Promotion
or
IAA in DW

By comparing "slow-growth" results (Figs. 3 and 4) to those of the "fast-growth" results (Figs. 5 and 6) one may see striking similarities. Figure 5 illustrates the low elongation rate of 2 mm segments at pH 7·0 and how it is increased by the addition of IAA. Figure 6 illustrates the acid-induced rate increase (line H vs. E) as well as the effect of IAA added at pH 4·0, which, as was the case for the "slow-growth" study, caused a growth inhibition but not to the point achieved with pH 7·0 alone.

Experiments done with intact roots in complete darkness or under 12 h of white light, in the present study, demonstrate that the label from 1–^{14}C IAA applied at the cut end of the mesocotyl of 3-day-old corn seedlings does not reach the root tip, 24–72 h after the application, in any appreciable amount (Fig. 9). Most of the radioactivity was recovered in the mesocotyl. Elliott (see Greenwood *et al.*, 1973), experimenting with corn seedlings, has also obtained similar results. The movement of a considerable amount of radioactivity into the adventitious roots, in the present study, suggests that the auxin which may arise from the shoot gets "side-tracked" into adventitious roots and does not move into the primary root to any great extent. Morris *et al.* (1969) have also shown the accumulation of label in the "developing adventitious" root primordia with no evidence that IAA reached the primary root. We have found little radioactivity in corn lateral roots. Morris *et al.*, (1969) used the terms "adventitious" and "lateral" interchangeably but it would appear that our results, in this case, are in disagreement with theirs since the tissue they studied was below the hypocotyl. Indirect evidence that auxin does not reach the root from the shoot in younger seedlings is provided in cases in which there does not appear to be endogenous auxin transport through the shoot–root transition zone (Thimann, 1934; Jacobs, 1950; Scott and Briggs, 1960).

Although Morris *et al.* (1969) found no radioactivity in the primary root as a result of shoot application as is reported here, they are not alone in finding movement of radioactivity to the developing root system. Other investigations report the movement of auxin into the root system when labelled IAA was applied to either leaf (Lyon, 1965; Eschrich, 1968; Eliasson, 1972) or to the decapitated plant (McDavid *et al.*, 1972).

Auxin in developing cereal grains and dicotyledonous seeds is in an esterified form, from which it is released when the seed germinates (see Sheldrake, 1973). Cartwright *et al.* (1956) presented evidence for the increase in free auxin content during the early stages of the germination of corn seeds. Much of this auxin was shown to be localized in the endosperm. Kaldewey and Kraus (1972) observed the transport of 2–^{14}C IAA into the primary root of cotton seedlings when auxin was applied to both the cotyledon and the hypocotyl, and the radiocarbon was shown to accumulate, although weakly, in the root tip in each case. Iversen *et al.* (1971) followed the movement of IAA–^{3}H into the root when radioactivity was injected into the basal root hypocotyl region of *Phaseolus*. Only a small portion of the IAA injected was translocated unchanged, and reached the tip of the root (where it piled up to some extent). Davies and Mitchell (1972) observed the movement of a small amount of radioactivity in the acropetal direction when IAA–^{3}H was applied to seedling roots of *Phaseolus*, 2 cm behind the apex. The presence of an exponential gradient of radioactivity along the entire length of the primary root of corn when 1–^{14}C IAA was applied to the seed or to the basal region of the root, in our experiments, suggests that the seed or the basal region of the root could be the source of auxin. Immediately following germination, the auxin would "find" its way readily to the transport system and the growing tip. However, later in development one must wonder how auxin produced by the seed or produced at the base of the root (or indeed the shoot) can reach the root tip some 50–80 mm distant when there is no apparent mechanism for delivering auxin to the active transport system and the growing tip. There is very indirect evidence for the presence of an acropetal auxin transport in older seedlings since in all experiments the terminal 6 mm segment showed an accumulation of radioactivity (Figs. 7 and 8). To be sure this may merely reflect a "sink" phenomenon since we do not know what molecule the radioactivity is associated with. However, there is a strong positional coincidence.

Acknowledgement

We wish to thank Miss Linda Overholser for critical reading of the manuscript.

References

AASHEIM, T. and IVERSEN, T.-H. (1971). Decarboxylation and transport of auxin in segments of sunflower and cabbage roots. II. A chromatographic study using IAA-1-^{14}C and IAA-5-^{3}H. *Physiol. Plant.* **24**; 325–329.

ÅBERG, B. (1957). Auxin relations in roots. *A. Rev. Plant Physiol.* **8**; 153–180.

ANKER, L. (1960). On a geo-growth reaction of the *Avena* coleoptile. *Acta Bot. Neerl.* **9**; 411–415.

BARKLEY, G. M. and LEOPOLD, A. C. (1973). Comparative effects of hydrogen ions, carbon dioxide and auxin on pea stem segment elongation. *Plant Physiol.* **52**; 76–78.

BONNETT, H. T. JR. (1972). Influence of growth centre on the transport of indoleacetic acid and sucrose. *Plant Physiol.* **49**; 55.

BONNETT, H. T. JR. and TORREY, J. G. (1965). Auxin transport in *Convolvulus* roots cultured in vitro. *Plant Physiol.* **40**; 813–818.

BOWEN, M. R., WILKINS, M. B., CANE, A. R. and MCCORQUODALE, I. (1972). Auxin transport in roots. VIII. The distribution of radioactivity in the tissues of *Zea* root segments. *Planta* **105**; 273–292.

BRIDGES, I. G., HILLMAN, J. R. and WILKINS, M. B. (1973). Identification and localisation of auxin in primary roots of *Zea mays* by mass spectrometry. *Planta* **115**; 189–192.

BRIDGES, I. G. and WILKINS, M. B. (1973). Acid-induced growth and the geotropic response of the wheat node. *Planta* **114**; 331–339.

BURSTRÖM, H. G. (1969). Influence of the tonic effect of gravitation and auxin on cell elongation and polarity in roots. *Am. J. Bot.* **56**; 679–684.

CANE, A. R. and WILKINS, M. B. (1970). Auxin transport in roots VI. Movement of IAA through different zones of *Zea* roots. *J. exp. Bot.* **21**; 212–218.

CARTWRIGHT, P. M., SYKES, J. T. and WAIN, R. L. (1956). The distribution of natural hormones in germinating seeds and seedling plants. *In*: "The Chemistry and Mode of action of Plant Growth Substances" (R. L. Wain and F. Wightman, eds.) pp. 32–39. Butterworths, London.

DAVIES, P. J. and MITCHELL, E. K. (1972). Transport of indoleacetic acid in intact roots of *Phaseolus coccineus*. *Planta* **105**; 139–154.

EDWARDS, K. L. and SCOTT, T. K. (1974a). Rapid-growth responses of corn root segments. Effect of pH on elongation. *Planta* **119**; 27–37.

EDWARDS, K. L. and SCOTT, T. K. (1974b). Rapid-growth responses of corn root segments. Effect of Indole-3-acetic acid at pH 7.0 and pH 4.0. *Planta* (in press).

ELIASSON, L. (1972). Translocation of shoot-applied indolyl-acetic acid into the roots of *Populus tremula*. *Physiol. Plant.* **27**; 412–416.

ELLIOT, M. C. and GREENWOOD, M. S. (1974). Indol-3yl-acetic acid in roots of *Zea mays*. *Phytochem.* **13**; 239–241.

ESCHRICH, W. (1968). Translokation radioaktiv markierter Indolyl-3-essigsäure in Siebröhren von *Vicia faba*. *Planta* **78**; 144–157.

EVANS, M. L., RAY, P. M. and Reinhold, L. (1971). Induction of coleoptile elongation by carbon dioxide. *Plant Physiol.* **47**; 335–341.

FAYLE, D. C. F. and FARRAR, J. L. (1965). A note on the polar transport of exogenous auxin in woody root cuttings. *Can. J. Bot.* **43**; 1004–1007.

GREENWOOD, M. S., HILLMAN, J. R., SHAW, S. and WILKINS, M. B. (1973). Localization and identification of auxin in roots of *Zea mays*. *Planta* **109**; 369–374.

HAGER, A., MENZEL, H. and KRAUSS, A. (1971). Versuche und Hypothese zur Primärwirkung des Auxins beim Streckungswachstum. *Planta* **100**; 47–75.

HEJNOWICZ, Z. (1968). Studies on the inhibitory action of auxin on root growth. *Acta Soc. Bot. Pol.* **37**; 451–460.

HERTEL, R., EVANS, M. L., LEOPOLD, A. C. and SELL, H. M. (1969). The specificity of the auxin transport system. *Planta* **85**; 238–249.

HILLMAN, S. K. and PHILLIPS, I. D. J. (1970). Transport and metabolism of indol-3yl-(acetic acid-2-^{14}C) in pea roots. *J. exp. Bot.* **21**; 959–967.

IVERSEN, T.-H. and AASHEIM, T. (1970). Decarboxylation and transport of auxin in segments of sunflower and cabbage roots. *Planta* **93**; 354–362.

IVERSEN, T.-H., AASHEIM, T. and PEDERSON, K. (1971). Transport and degradation of auxin in relation to geotropism in roots of *Phaseolus vulgaris*. *Physiol. Plant.* **25**; 417–424.

JACOBS, W. P. (1950). Control of elongation in the bean hypocotyl by the ability of the hypocotyl tip to transport auxin. *Am. J. Bot.* **37**; 551–555.

KALDEWEY, H. and KRAUS, H. (1972). Translocation and immobilization of radiocarbon in the hypocotyl and the primary root of *Gossypium hirsutum* L. after application of IAA-2-^{14}C to intact light-grown seedlings. *In*: "Hormonal Regulation in Plant Growth and Development" (H. Kaldewey and Y. Vardar, eds.) pp. 137–153. *Proc. Adv. Study Inst. Izmir* 1971,, Verlag Chemie, Weinheim.

KIRK, S. C. and JACOBS, W. P. (1968). Polar movement of indole-3-acetic acid-^{14}C in roots of *Lens* and *Phaseolus*. *Plant Physiol.* **43**; 675–682.

KONINGS, H. (1969). The influence of acropetally transported indoleacetic acid on the geotropism of intact pea roots and its modification by 2, 3, 5-triiodobenzoic acid. *Acta Bot. Neerl.* **18**; 528–537.

KONINGS, H. and GAYADIN, A. P. (1971). Transport, binding, and decarboxylation of carboxyl-labelled IAA-^{14}C in intact pea roots. *Acta, Bot. Neerl.* **20**; 646–654.

LEOPOLD, A. C. and GUERNSEY, F. S. (1953). Auxin polarity in the *Coleus* plant. *Bot. Gaz.* **115**; 147–154.

LEOPOLD, A. C. and LAM, S. L. (1962). The auxin transport gradient. *Physiol. Plant.* **15**; 631–638.

LIBBERT, E. (1964). Significance and mechanism of action of natural inhibitors. *In*: "Régulateurs Naturels de la Croissance Végétale" (J. P. Nitsch, ed.) pp. 387–405. Centre National de la Recherche Scientifique, Paris.

LUNDEGÅRDH, H. (1950). The influence of auxin anions on the growth of wheat roots. *Arkiv Botanik* **1**; 289–293.

LYON, C. J. (1965). Action of gravity on basipetal transport of auxin. *Plant. Physiol.* **40**; 953–961.

McDAVID, C. R., SAGAR, G. R. and MARSHALL, C. (1972). The effect of auxin from the shoot on root development in *Pisum sativum* L. *New Phytol.* **71**; 1027–1032.

MITCHELL, E. K. and DAVIES, P. J. (1972). The transport of indoleacetic acid in intact roots of *Phaseolus coccineus* L., *Plant Physiol.* **49**; 55.

MORRIS, D. A., BRIANT, R. E. and THOMPSON, P. G. (1969). The transport and metabolism of ^{14}C-labelled indoleacetic acid in intact pea seedlings. *Planta* **89**; 178–197.

NAQVI, S. M. and GORDON, S. A. (1966). Auxin transport in *Zea mays* L. coleoptiles. I. Influence of gravity on the transport of indoleacetic acid-2-^{14}C. *Plant. Physiol.* **41**; 1113–1118.

PILET, P. E. (1964). Auxin transport in roots. *Lens culinaris. Nature* (*Lond.*) **204**; 561–562.

RAYLE, D. L. (1973). Auxin-induced hydrogen-ion secretion in *Avena* coleoptiles and its implications. *Planta* **114**; 63–73.

RAYLE, D. L. and CLELAND, R. (1970). Enhancement of wall loosening and elongation by acid solutions. *Plant. Physiol.* **46**; 250–253.

REHM, M. M. and CLINE, M. G. (1973). Inhibition of low pH-induced elongation in *Avena* coleoptiles by abscisic acid. *Plant. Physiol.* **51**; 946–948.

RUBERY, P. H. and SHELDRAKE, A. R. (1973). Effect of pH and surface charge on cell uptake of auxin. *Nature* (*New Biol.*) **244**; 285–288.

SCOTT, T. K. (1972). Auxins and roots. A. Rev. Plant Physiol. **23**; 235–258.

SCOTT, T. K. and BATRA, M. (1973). Effect of pH on auxin transport in the roots of *Zea mays* L. *Plant Physiol.* **51**; 13.

SCOTT, T. K. and BRIGGS, W. R. (1960). Auxin relationships in the Alaska pea (*Pisum sativum*). *Am. J. Bot.* **47**; 492–499.

SCOTT, T. K. and WILKINS, M. B. (1968). Auxin transport in roots. II. Polar flux of IAA in *Zea* roots. Planta **83**; 323–334.

SCOTT, T. K. and WILKINS, M. B. (1969). Auxin transport in roots. IV. Effects of light on IAA movement and geotropic responsiveness in *Zea* roots. *Planta* **87**; 249–258.

SHELDRAKE, A. R. (1973). The production of hormones in higher plants. *Biol. Rev.* **48**; 509–599.

SMITH, C. W. and JACOBS, W. P. (1968). The movement of IAA-^{14}C in the hypocotyl of *Phaseolus vulgaris*. *In*: "The Transport of Plant Hormones" (Y. Vardar, ed.) pp. 48–64. *Proc. NATO/EGE Univ.* 1967. North-Holland Publ. Co., Amsterdam.

SNEDECOR, G. W. (1967). Statistical Methods, Iowa State Univ. Press, Ames, Iowa.

THIMANN, K. V. (1934). Studies on the growth hormone of plants. VI. The distribution of the growth substance in plant tissues. *J. Gen. Physiol.* **18**; 23–34.

THIMANN, K. V. (1936). Auxins and the growth of roots. *Am. J. Bot.* **23**; 561–569.

THIMANN, K. V. (1937). On the nature of inhibition caused by auxin. *Am. J. Bot.* **24**; 407–412.

TORREY, J. G. (1956). Physiology of root elongation. *A. Rev. Plant. Physiol.* **7**; 237–266.

TORREY, J G. (1958). Endogenous bud and root formation by isolated roots of *Convolvulus* grown in vitro. *Plant Physiol.* **33**; 258–263.

TORREY, J. G. (1963). Cellular patterns in developing roots. *Symp. Soc. Exptl. Biol.* **17**; 285–314.

WILKINS, M. B. and CANE, A. R. (1970). Auxin transport in roots. V. Effect of temperature on the movement of IAA in *Zea* roots. *J. exp. Bot.* **21**; 195–211.

WILKINS, M. B., CANE, A. R. and MCCORQUODALE, I. (1972a). Auxin transport in roots. VII. Uptake and movement of radioactivity from IAA-^{14}C by *Zea* roots. *Planta* **105**; 93–113.

WILKINS, M. B., CANE, A. R. and MCCORQUODALE, I. (1972b). Auxin transport in roots. IX. Movement, export, resorption, and loss of radioactivity from IAA by *Zea* root segments. *Planta* **106**; 291–310.

WILKINS, M. B. and SCOTT, T. K. (1968a). Auxin transport in roots. *Nature* (*Lond.*) **219**; 1388–1389.

WILKINS, M. B. and SCOTT, T. K. (1968b). Auxin transport in roots. III. Dependence of the polar flux of IAA in *Zea* roots upon metabolism. *Planta* **83**; 335–346.

ZAER, J. B. (1968). Transport gradient of indoleacetic acid in pine seedlings. *Physiol. Plant.* **21**; 1265–1269.

Chapter 16

Geotropism in Roots

L. J. AUDUS

Department of Botany, Bedford College, University of London, England.

I. Introduction

The study of the phenomena of geotropism poses three major problems. First, in the natural sequence of events which characterizes those phenomena, is the mechanism of gravity perception, involving the identification of the cell system on which the gravity force field operates to register the direction of the gravity vector in relation to some fixed cell or organ axis. Next, but third in order of operation, is the identification of the chemical regulator (hormone) system which controls the differential growth of the responding tissue, and hence the geotropic growth curvature bringing about the re-orientation of the organ concerned. Lastly there is the intermediate stage, which is still very largely restricted to speculation (see Audus, 1971) whereby the changes involved in the detection (perception) processes are linked to hormone changes leading to differential growth.

In this review of the phenomena in roots, attention will be concentrated mainly on current ideas of the mechanism of detection and hormone control as they have emerged from research over the last decade. However, in attempting finally to construct a unifying theory of the phenomena in roots some further speculation on linking processes will be necessary.

II. Perception

A. General

Two major theories of the primary action of gravity have dominated the literature; they are the starch statolith theory of Haberlandt (1900) and of Nemeč (1900) in which the sedimentation of the dense starch-laden amyloplasts constituted the act of detection, and the geo-electric theory (Brauner, 1942) in which the induction of an electrical potential difference between upper and lower sides of horizontal organs was attributed to the differential action of gravity on ion diffusion potentials across variously orientated cell membranes. However, since the work of Hertz and Grahm and of Wilkins and his colleagues (see Wilkins and Woodcock, 1965; Audus, 1969; Hertz, 1971; Woodcock and Hertz, 1972) has conclusively shown that the geo-electric effect is a symptom of processes which come much later in the sequence than the initiating action of perception, attention has recently reverted to starch statoliths.

Up until 1966 an impressive amount of "circumstantial" evidence had been accumulated in support of the starch statolith theory in a range of above-ground organs (see Audus, 1962) and much of the meagre evidence contrary to the hypothesis was susceptible of alternative explanations. However in 1966, Pickard and Thimann published results which showed that wheat coleoptiles, completely depleted of their starch content by incubation with hormone solutions (gibberellic acid plus kinetin) were still capable of curving when placed horizontal, presumably in response to a geotropic stimulus. Such a claim, threatening the second of the two major theories, has stimulated the search for other graviperceptor organelles; mitochondria in particular have been indicated as theoretically capable of redistribution in the cell under the action of gravity (Audus, 1962; Gordon, 1964). Such a search in the tip of the *Avena* coleoptile (Shen-Miller and Miller, 1972a and b; Shen-Miller, 1972) revealed that, coincident with the rapid sedimentation of amyloplasts, there was a redistribution of Golgi bodies, the bottom half of the cell having significantly greater numbers than the top half after 15 minutes or so of stimulation in the horizontal position. Furthermore a greater proportion of the lowermost Golgi bodies were active, as judged by their distended vesicles. Mitochondria showed similar distributions but the change occurred later than those of the Golgi bodies. These rapid changes in the position of the Golgi bodies are not easily explained in terms of sedimentation under gravity since the few direct observations which have been made in centrifugation experiments, in roots for example (Bouck, 1963a and b), suggest that dictyosomes have an effective density not vastly different from vacuoles and neither sediment or "cream" in the cell in high gravity fields. Certainly activity changes are most likely to be indirect effects of gravity. It is more logical to suppose that mitochondria and Golgi movements are in some way coupled to amyloplast movements, and the obvious step now to take is to re-investigate these changes in organs which have been depleted of their amyloplast starch with treatments such as those applied by Pickard and Thimann (1966). The question of graviperceptor mechanisms in aerial organs, at least in coleoptiles, is thus back in the melting pot.

In contrast to this the situation in roots has clarified over the same period and there is now wholly convincing evidence from various sources that sedimenting amyloplasts in root-cap cells are the gravity-sensing organelles. The more recent evidence for this is reviewed in the following section.

B. Amyloplast Statoliths in Root Cap Cells

Ever since the classical experiments of Ciesielski (1872) in which the removal of the extreme tips of roots eliminated their response to gravity, the root cap has been regarded as the most likely site of gravity perception. The ingenious "skew-centrifuge" experiments of Piccard (see Larsen, 1962) supported this in intact roots. However, the more direct and convincing decapitation experiments were, in the early days, always regarded with some suspicion since the degree of damage to the meristem, with concomitant restrictions of growth and hence curvature, was not known and could have accounted for the loss of sensitivity to gravity. Indeed experiments such as those of Younis (1954) in which the removal of 0·5, 1·0 and 2·0 mm of the tips of *Vicia faba* roots progressively reduced geotropic responses and elongation growth in parallel, indicated that the loss of sensitivity was due only to the removal of the meristematic growth centre and not to the statolith-containing root cap.

However more recent root-tip excision experiments have not confirmed the results of Younis. In experiments on the roots of 2-day-old *Pisum sativum* seedlings Konings (1968), using a specially constructed micro-guillotine of high precision, removed different lengths of root tip from 0·2 to 1·0 mm. Loss of the distal 0·2, 0·3 and 0·4 mm of cap reduced the response progressively but at 0·5 mm, which represents the removal of the entire root cap, the response (see Fig. 1A) disappeared. At the same time extension growth, at least for a period of 6 h during which the geotropic curvatures were studied, was not affected by tip excision, even up to 1·0 mm, an amount which included the quiescent centre and some of the meristem. Almost identical results were obtained by Zinke (1968; see Fig. 1B).

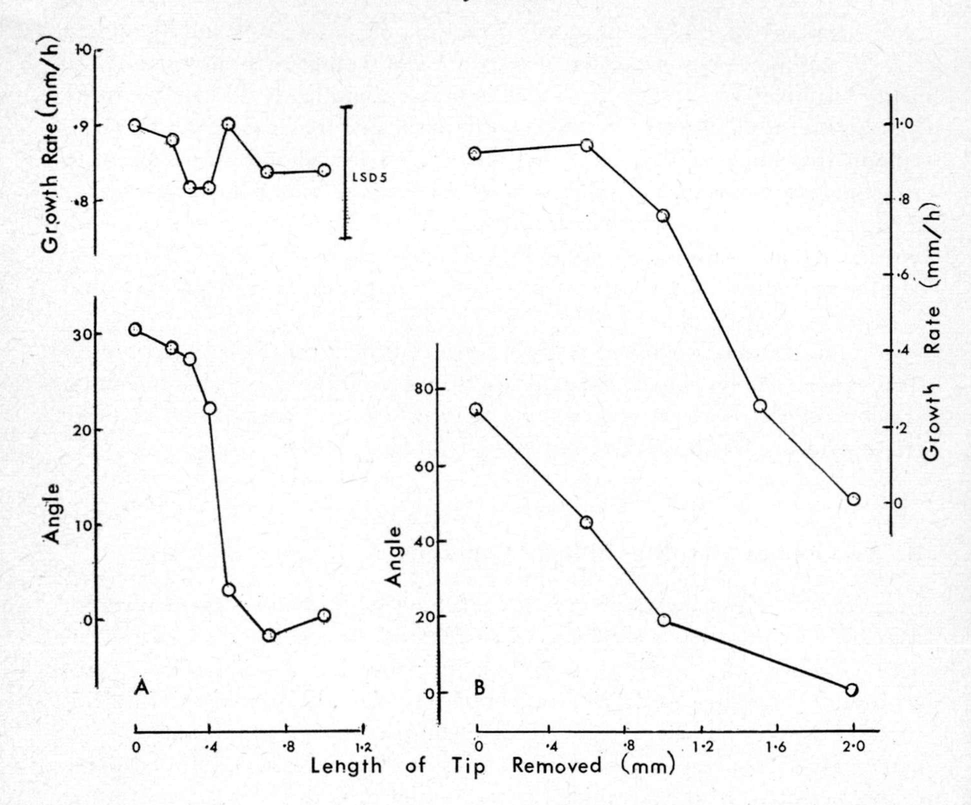

FIG. 1. The effect of tip excision on the growth rates and geotropic responses of roots. A. Data from 2-day-old pea seedlings (Konings, 1968). *Top:* Mean growth rate (mm/h) over the first 6 h after tip excision in relation to the amount of tip removed. Curves constructed from an analysis of variance of Table I. *Bottom*: Curvature after 6 h in relation to the amount of tip removed (data from Fig. 2). B. Data from 2-day-old pea roots. Relationships between rate of extension growth, curvature after 6 h and amount of tip removed (Zinke, 1968—replotted from Fig. 15).

The key role of the root cap was finally established as the result of the development by Dr. B. E. Juniper at Oxford of a new technique, whereby the root cap could be dissected from the tip of certain varieties of *Zea mays* without in any way damaging the meristem. Application of this technique to studies of geotropism demonstrated that these "decapped" roots lost *all* their sensitivity to gravity but continued to extend without any impairment of their extension-growth rate (Schachar, 1967; Juniper *et al.*, 1966). These findings have been subsequently confirmed by independent observers (Pilet, 1971a and b; 1972 a and b; Shaw and Wilkins, 1973; Dr. Peter Barlow, personal communication).

These observations by themselves are persuasive evidence enough, but what seemed to be even more convincing was the observation (Juniper *et al.*, 1966; Schachar, 1967) that root cap regeneration started some few hours after decapping and was claimed to be complete in about 36 h at 28°C. However the recovery of geotropic sensitivity occurred rather abruptly at about 16–24 h, depending on

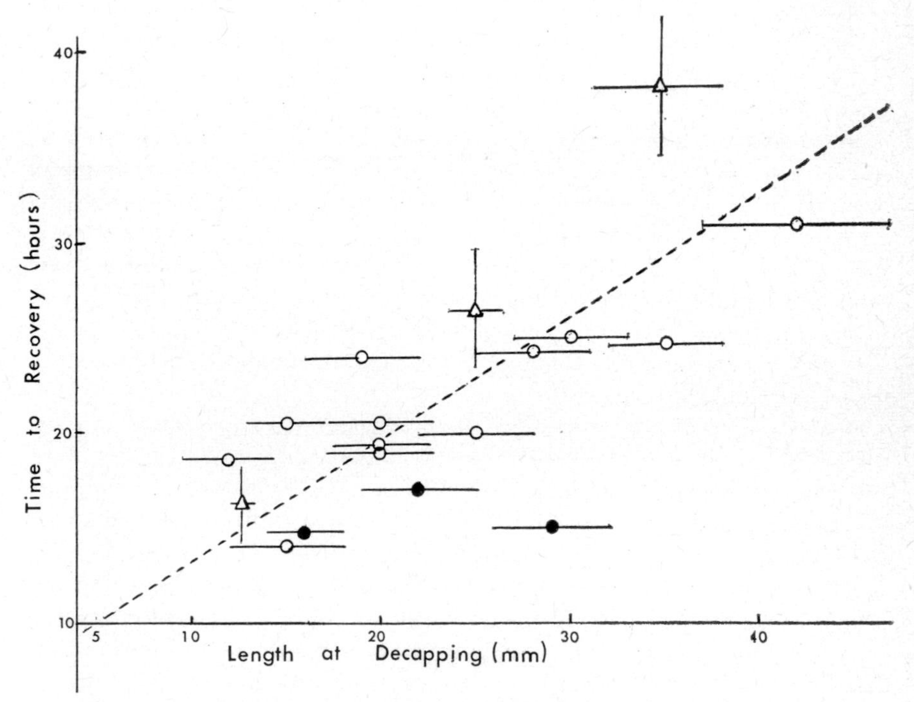

FIG. 2. Graph showing the relationship between the length of the root of maize at the time of decapping and the time taken to recover geotropic sensitivity. Data for roots decapped and left to regenerate; —○— means and range of root lengths in sample (Schachar, 1967); —△— means and standard deviations of root lengths and recovery times (Pilet, 1973). Data for roots decapped and then the cap replaced immediately —●— (Schachar, 1967). The regression line for the observations on decapped and not reheaded roots is given by: Recovery time (hours) = 9·20 + 13·04 × Root length (mm).

the initial root length, i.e. before cap regeneration was complete, but, as the electron microscope showed, at the same time as starch grains again appeared in the root tip (Schachar, 1967). The rate of recovery of sensitivity was a function of the age of the radicle at the time of decapping and was more rapid the younger the root (Schachar, 1967). This finding was confirmed by Pilet (1973) in completely independent experiments. The combined results of Schachar and Pilet are shown in Fig. 2. A second independent confirmation of a 14–22 h recovery time for geotropic sensitivity has come from studies by Barlow and Grundwag (1974).

However, the situation is not as simple as was first supposed by Juniper *et al.*, (1966), since the most recent work by Barlow and Grundwag (1974) has shown that in *Zea* the regeneration of the new cap is not complete until after 3 to 4 days and that at the time of re-appearance of gravi-sensitivity the cells of the quiescent centre have not yet started to divide. Nevertheless a careful time-sequence study of cytological events in the quiescent centre and meristem of *Zea* roots (Barlow and Grundwag, 1974) revealed that, whereas in intact roots only traces of starch grains can be seen in the plastids of such cells, decapping seems to trigger starch synthesis, and by 6 h the number of starch grains in plastids in the quiescent centre had increased by over 40 times. An examination of the electron micrographs of Schachar (1967), which were made 25 to 29 h after decapping, show essentially the same picture, except that by this time cell division in the quiescent centre was well under way. Of course the critical experiments, still to be performed, are those to determine if and when these plastids, newly stocked with starch grains, sediment in the cell in a way similar to those in the intact root cap and whether the onset and extent of this behaviour bears any relationship to the kinetics of the recovery of geotropic sensitivity.

Further support for the unique role of amyloplasts as graviperceptors in roots has come from the results of Iversen (1969) who applied the technique of Pickard and Thimann (1966) to the roots of *Lepidium sativum*. Incubation of such roots in solutions containing $4{\cdot}3 \times 10^{-5}$ M gibberellic acid and kinetin at 35°C for 29 h caused the complete disappearance of starch from the amyloplast of the root cap. These roots, when geotropically stimulated for 30 min and then rotated on a klinostat at 21°C, showed no trace of a curvature response after 60 min of rotation whereas growth in length proceeded normally at 0·48 mm/h. Control roots similarly pretreated at 35°C in water had curved to 22° after 60 min and had elongated at a similar rate (0·33 mm/h). Subsequent illumination of these hormone-treated seedlings induced the re-formation of starch in the amyloplasts after 20 to 24 h, and geotropic responsiveness re-appeared at the same time. Later Pilet and Nougarède (1971) and Nougarède and Pilet (1971) demonstrated similar behaviour in the roots of *Lens culinaris*.

Such results are reasonably conclusive evidence that sedimenting amyloplasts are the organelles *directly* acted on by gravity and that their induced movement, or a change in the pressure exerted by them on some cell component, initiates the sequence of processes which culminate in root curvature.

C. The Participation of other Organelles

Amyloplast movement *per se* is unlikely to trigger off the response sequence; it is much more likely that their movement *relative* to some other cell component gives rise to a mutual interaction and initiates the essential reaction chain. What these other cell components are has long been a topic for discussion and, in recent years, direct investigations. The early ideas (see Haberlandt, 1928) identify the component as the "physically lower portion of the ectoplast" which has in recent years been assumed to mean the plasmalemma, or its immediately contiguous layer of cytoplasm. Many observations by many people have shown that the migration of amyloplasts under gravity begins immediately on a change of organ orientation (e.g. Grifflths, 1963; see Fig. 3; see also Iversen *et al.*, 1968). Graviperception has always been regarded as a threshold phenomenon, a stimulus having to be applied to an organ for a minimum "presentation time", to trigger a response. Consequently perception involved the movement of a minimum number of amyloplasts to this "endoplasmic" sensitive surface and their application to it of a minimum pressure. This now seems unlikely. For example, observations under the electron microscope and also in the light microscope suggest that the amyloplasts never touch the plasmalemma but are arrested, before they reach it, by obstructing sheets of endoplasmic reticulum (Griffiths, 1965; Sievers and Volkmann, 1972; Iversen and Larsen, 1973). Furthermore the employment of more sensitive techniques for recording and measuring the geotropic response has indicated that "presentation time" for a 1 *g* field may be considerably lower than the classical observations claimed; thus in *Avena* coleoptiles it is less than 30 seconds (Johnson, 1971; Audus, unpublished data). In *Vicia faba* roots the lower limit of exposure to 1 *g* for a just perceptible response is about 18 seconds (Griffiths, 1955). Certainly (see Johnson, 1971) amyloplasts move only a very short distance across the cell during such short intervals and, unless already in close proximity to it, would not reach the plasmalemma in that time.

In the root cap of *Vicia* about 20% of the amyloplasts move from the upper to the lower half of the cell in the 18 seconds of the minimum presentation time (see Fig. 3). After this there is no appreciable movement for the next five minutes and then the amyloplasts remaining on the upper side migrate slowly to the lower side. Even after 15 minutes a few amyloplasts linger in the upper half of the cell.

Such facts have led to other theories of amyloplast/organelle interaction to explain the very rapid detection of the change in the direction of the gravity vector. The most recent is that of Sievers and Volkmanns (1972) who based their theory on the very regular and symmetrical structure of the root cap and its component cells in *Lepidium sativum*. In this theory the interacting organelle is the cup-shaped aggregation of endoplasmic reticulum found on the distal walls of the columella cells, the rim of the cup spreading partially up the lateral walls to an extent depending on the position of the cell concerned. The amyloplasts in normally orientated vertical roots sit at the bottom of the "cup". Sievers and Volkmanns regard the cells on the flanks of the central core of the columella as the receptor cells proper. Any change in the orientation of the root will *immediaately* alter the distribution of pressure of the amyloplasts in this endoplasmic

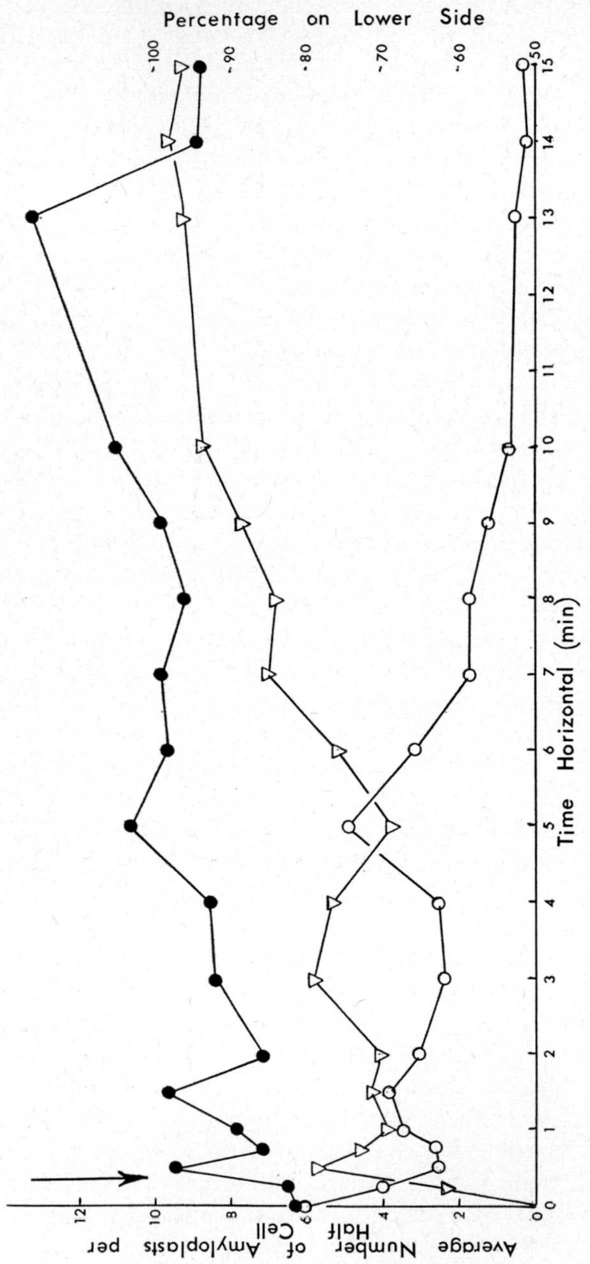

FIG. 3. The kinetics of amyloplast movement in the root cap cells of *Vicia faba*. The minimum stimulation time to obtain a just perceptible curvature during subsequent rotation on a klinostat is shown by the vertical arrow (from data of Griffiths, 1963). ●—● average number of amyloplasts in lower half of cell ○—○ average number of amyloplasts in upper half of cell. △—△ Percentage of total amyloplasts in lower half of cell.

reticular cup and, because the axis of symmetry of the cup is skew to the main axis of the root in these flank cells, the change of pressure in the cells on the lower side will be different from that on the upper side (see Fig. 4), and it would be this difference of pressure pattern between cells on the two sides of the cap which would constitute the initiating stimulus. Volkmann (1974), in a careful analysis of electronmicrographs of the two sets of six central columella cells on either side

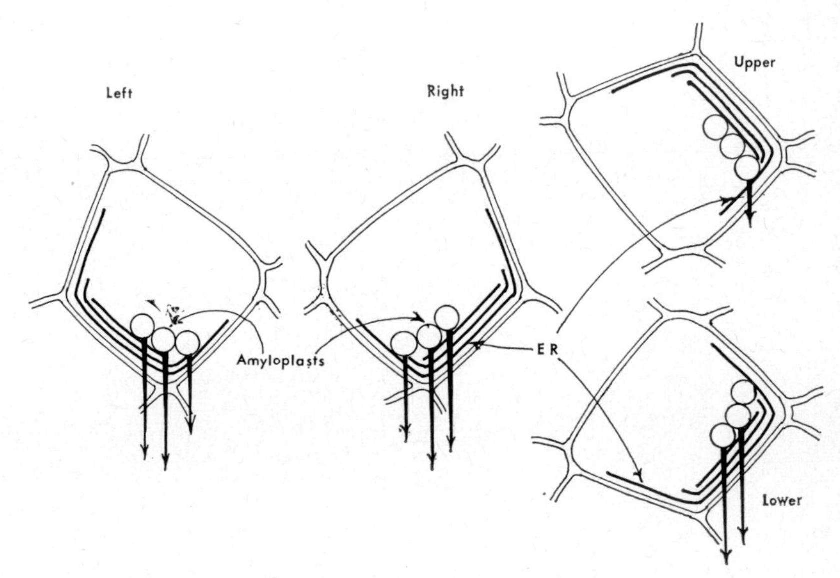

FIG. 4. Diagram of formalized cell from the lateral flank of the central core of the root cap of *Lepidium sativum*, showing the distribution of the endoplasmic reticulum and the amyloplasts. The thick arrows indicate the direction and intensities of the pressures exerted by the amyloplasts on the endoplasmic reticulum when the root is in the vertical (left pair of cells) or horizontal (right pair of cells) position. (Modified from Fig. 6 of Sievers and Volkmann, 1972.)

of the main axial line, in which the greater mobility of the amyloplasts is seen, has made an estimate of the pressure changes on the cisternae of the endoplasmic reticulum which might be expected from a change of orientation of 90°. He has dropped vectors from the central points of the amyloplasts in electron micrographs (as indicated for the diagram in Fig. 4) of vertically orientated roots. These have been parallel to the main axis and towards the tip ("normal orientation") or perpendicular to the axis and towards the "lower" side for the simulated situation for stimulation. The criterion of "pressure" was the average number of endoplasmic reticular cisternae cut on average per amyloplast. Figure 5 shows that wheras pressure is quite symmetrically distributed on either side of the main axis in normal vertical roots, pressure is lowered in the cells of "horizontal" roots and then much more on the "upper" than on the "lower" side.

This change of pressure distribution, causing a correlated differential change in the biochemical or biophysical properties of the endoplasmic reticulum (or perhaps the amyloplast membranes themselves) would put the response sequence in train. These changes could well be of the kind envisaged by Osborne (1968)

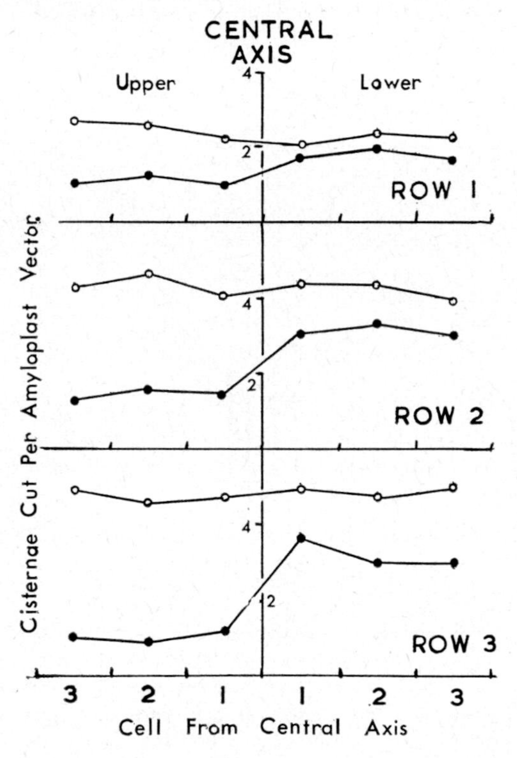

FIG. 5. Estimated changes in the pressure of amyloplasts on the endoplasmic reticulum in the eighteen central columella cells of the root of *Lepidium sativum*. (Redrawn from Volkmann, 1974). ○—○ Vectors drawn parallel to the main axis and towards the tip (as for left hand side of Fig. 4). ●—● Vectors drawn perpendicular to the main axis and towards the side prescribed as "lower" (as for right hand side of Fig. 4).

who suggested the endoplasmic reticulum as the site of synthesis of auxin trans-permeases, regulating the polar movement of auxin from cell to cell. Differential pressure of amyloplasts could well modulate transpermease synthesis in a similar differential manner and hence operate a differential polar flow of auxin from the cap (but see later for objections to this theory p. 343, and for discussions on other possible biochemical changes in membranes under pressure p. 358).

However there are difficulties with Sievers and Volkmann's theory, even when applied to *Lepidium* roots themselves. Thus Iversen and Larsen (1973) point out that it will not explain the geotropic response patterns of roots that have been pre-inverted before stimulation, when the amyloplasts start sedimenting from the proximal instead of the distal end of the calyptra cells when they were in contact with little or no endoplasmic reticulum, which remained at the distal end of the cell even after inversion. Furthermore, this particular and very regular pattern of distribution of endoplasmic reticulum is peculiar to *Lepidium*, although a similar but less regular arrangement is found in *Lens culinaris*, *Genista tinctoria*, *Daucus carota* and *Allium cepa* (Volkmann, 1974). In *Vicia faba*, although the distribution when the roots are vertical is roughly the same as in *Lepidium*, on geotropic stimulation it does not remain so but tends to move and accumulate in the upper distal corner of the cells as the amyloplasts sediment (Griffiths, 1963; Griffiths and Audus, 1964). In *Zea* roots the endoplasmic reticulum is arranged near and parallel to the lateral walls and is symmetrically distributed (Juniper and French, 1973). Placing a *Zea* root horizontal causes an aggregation of these membranes in the top distal corner as in *Vicia faba*. On the other hand *Pisum* resembles *Lepidium* in that the endoplasmic reticulum in root cap cells does not shift as a result of geotropic stimulation (Grundwag, M., 1971; quoted by Juniper and French, 1973). Thus unless we are going to propose different mechanisms for different genera of plants we must conclude that the amyloplast-endoplasmic reticulum interaction during stimulation may not be the initial step in the response chain.

Looking at other organelles it seems that, in contrast to the proposed situation in *Avena* coleoptile tips (Shen-Miller, 1972), Golgi bodies and mitochondria are not re-distributed directly under gravity, as was shown earlier by Griffiths (Griffiths, 1963, Griffiths and Audus, 1964). Slight movement of both organelles into the upper halves of cells could be explained entirely in terms of the bulk displacement of the protoplast brought about by amyloplast sedimentation, but these movements were so slight as to imply little physiological significance.

One structural feature of the core cells of root caps which cannot be displaced by gravity are the plasmodesmata which Juniper and Barlow (1969) have shown to predominate greatly on the transverse wall in *Zea*. Is it possible that the relative position of the amyloplasts in relation to these distributions of plasmodesmata in different cells may cause the differential changes we are seeking? If the plasmodesmata are responsible for the control of the major direction of information flow in the root cap cells, as suggested by Juniper and Barlow (1969), could the nature of this flow be differentially modified by the nearness or otherwise of the amyloplasts?

III. The Hormone Relationships

We have all been brought up on the classical Cholodny/Went theory of geotropism; although now widely accepted, at least in its basic concept, as valid for coleoptiles (and possibly other main aerial axes of seedlings) it was extrapolated to roots on the slenderest of evidence. The synthesis of auxin in the root tip, its "braking" control of extension growth at supra-optimal concentrations and its preferential movement to the lowermost cells of the elongation zone in horizontal roots were phenomena never established by rigorous experiments before they were accepted as lore. Although over the years voices have expressed disenchantment with this classical theory as applied to roots, particularly with the role of indol-3yl-acetic acid, only recently have improved techniques brought persuasive evidence that quite different hormones may be involved. The aim of this section will therefore be two-fold, first to assess in the modern context the role of indol-3yl-acetic acid in the geotropic response of roots and second to survey the evidence for the existence and participation of non-auxin hormones.

A. The Role of Indol-3yl-acetic acid

1. *Occurrence and Distribution in Roots*

The occurrence of auxins, i.e. substances giving growth responses in auxin assays, have been reported over and over again in roots. Furthermore there is a mass of indirect evidence showing that auxins can fundamentally alter extension growth processes in roots although the questions as to the relevance of these effects to *normal control* by endogenous auxins is still, after 30 years, very much a matter for discussion. A recent survey (Scott, 1972) has summarized the situation excellently and very little point would be served in reiteration. In the same review the uncertain situation concerning the occurrence of indol-3yl-acetic acid in roots is clearly set out. All results at that time indicated the presence of IAA-like substances which had Rf values in various chromatographic solvents and indole chromogenic reactions the same as IAA: attempts at specific identification however had so far failed. Thus Bennet-Clark *et al.*, (1959) showed that the indole auxin of root apices of *Vicia* did not co-chromatograph on starch columns with authentic IAA and that no trace of activity could be seen on chromatograms at the Rf of IAA. On the other hand Burnett *et al.*, (1965), from the same species, showed that the main auxin ran on paper chromatograms at the same Rf as IAA but that it had quite different fluorescent properties. Using the same purification techniques with DEAE cellulose chromatography columns, Woodruffe *et al.*, (1970) could find no trace of IAA among the four indole compounds isolated from wheat and tomato roots.

However we may well be at the end of this era of uncertainty. In recent years the advent of the sensitive and powerful techniques of gas chromatography combined with mass spectrometry, which have transformed the gibberellin scene, now promises to solve the enigma of IAA status in roots. Bridges *et al.*, (1973) have removed 5 mm tips of *Zea mays* roots and have found that the stele of the more mature parts can be easily withdrawn from the cortex after fracture at the endodermis. Tips, cortex and stele were separately extracted. After purification procedures a substance having chromatographic properties similar to that of IAA was reacted with bis-trimethylsilylacetamide and subsequent gas chromatography showed the trimethylsilyl derivative to have the same retention time as an authentic IAA derivative. The identity of the compound with IAA was then confirmed with mass spectrometry. Using single beam mass spectrometry they made quantitative determinations of the *N*-trimethylsilyl-3-methyleneindolyl radicle ($m/e = 202$). In mg/kg the concentrations found were: stele, 53·3; cortex, 4·8; tip, 29·0. The authors suggested that the small amounts detected in the cortex could easily be explained as coming from stele fragments incompletely removed from the cortical sleeve, and it seems likely also that most of that present in the tip could be located in the vascular tissues, as judged by the above figures. The extremely low values in the cortex may be explained by an active metabolism since the incubation of isolated cortices in IAA-(1-^{14}C) yielded at least two radioactive products, both unidentified, accompanied by a complete disappearance of labelled IAA. Steles similarly incubated showed only a trace of one of these products and no significant change in the level of labelled IAA (Greenwood *et al.*, 1973). Using virtually identical techniques, an equally positive identification of IAA has been made in *Zea* roots by Elliott and Greenwood (1974).

The questions posed by these findings, as they concern geotropic phenomena in roots, is where this stelar IAA is synthesized and how it gets into and moves from the stele, particularly into those regions of the root where active extension growth and hence geotropic curvature is going on. One suggestion is that it arises in the stele itself as a result of processes of cell differentiation in the xylem cells (Sheldrake, 1971) which involve cytoplasmic autolysis. On the other hand there is evidence (see Scott, 1972, p. 245) that auxins, including radioactive IAA, can move from the shoot into the root. In *Zea mays* it seems to move predominantly in the stele, as judged from micro-autoradiographic observations with labelled IAA and isolated subapical segments of roots (Bowen *et al.*, 1972) and it has been suggested (Hejnowicz, 1968) that phloem is specifically involved. The question now arises whether this stelar IAA is involved in geotropic response and if not whether there is any other source of IAA which is involved in roots. Light is thrown on these points by studies of the transport of exogenous IAA in root tissue.

2. *The Characteristics of IAA Transport in Roots in Relation to Gravity*

It is now about 10 years since doubts began to be thrown on the basic concepts of the Cholodny/Went theory as applied to roots, namely that the auxin regulator moved basipetally from the tip to the extension zone. Early work showed that movement of IAA was sluggish and what polarity there was seemed to be acropetal (Yeomans and Audus, 1964, Pilet, 1964). Since that time numerous independent studies have confirmed an acropetal polarity in isolated segments of the extending zones of roots of several plant species and have shown that, like the situation in coleoptile and hypocotyl segments, it is apparently a metabolically-powered, active process (for reference details see Scott, 1972). However, unlike the situation in coleoptiles, where polar movement seems to be in the parenchyma cells, the acropetal movement in root segments seems to be confined mainly, if not entirely, to the vascular system, as has been shown by the use of labelled IAA (Bowen *et al.*, 1972; Shaw and Wilkins, 1974). Since the non-living xylem is scarcely likely to be involved in any acropetal movement, these findings suggest that the phloem must be involved.

However the question which concerns us is whether this acropetal flow of auxin is concerned in geotropic response. Presumably for reasons of technical difficulty the movement and distribution of auxins in roots under the action of gravity have been little studied.

The most extensive recent work has been that of Konings, who has set out specifically to study the lateral distribution of applied IAA in geotropically stimulated roots of *Pisum sativum*. Earlier work with radioactive IAA by Ching and Fang (1958) had shown that such auxin, taken up from solutions by pea, lima bean and maize roots, was not measurably redistributed under gravity. Since this could possibly be explained by their having saturated any transport system by uniform application of high concentrations of the auxin, Konings (1967) refined the techniques by applying IAA in small blocks of agar to the extreme tip of intact roots. Various parts of the tip were subsequently split into upper and lower halves and counted for ^{14}C content. IAA was shown to move into the root from the tip of the cap and into the meristem and extending zones, although only a small percentage of that taken up could be detected beyond 4 mm from the extreme tip. In vertical roots the distribution was uniform but in horizontal roots much more moved to the lower side, reaching about two thirds of the total in half an hour and being reduced to about three fifths after 4 h. Furthermore, in spite of the marked decline in content with distance towards the base, this ratio was maintained virtually constant up to 4 mm from the tip. The most interesting feature of these experiments was the demonstration that this redistribution was dependent in some way on an intact root cap, since removal of 0·5 mm (or more), which eliminated virtually all the cap, also eliminated this gravity-induced lateral movement of IAA–^{14}C applied to the tip (Fig. 6). Since previous work (Konings, 1965) had indicated that a differential activity of IAA oxidase on upper and lower sides of pea roots might have a part to play in determining these concentration differentials of IAA, this possiblity was checked in the tip-application experiments. However methylene-labelled IAA, which was not degraded in root tips, showed the same lateral redistribution under gravity. Furthermore 2,4-dichlorophenol, which promotes IAA oxidase activity when applied to roots, could not abolish this redistribution although it did reduce the magnitude of the concentration differential (Fig. 6). Konings concluded that an induced difference in IAA oxidase activity between the two sides could not account for the redistribution of IAA under gravity.

Although these experiments are puzzling in view of the acropetal polarity of movement in roots, yet further experiment (Konings, 1969; Pilet, 1971) showed

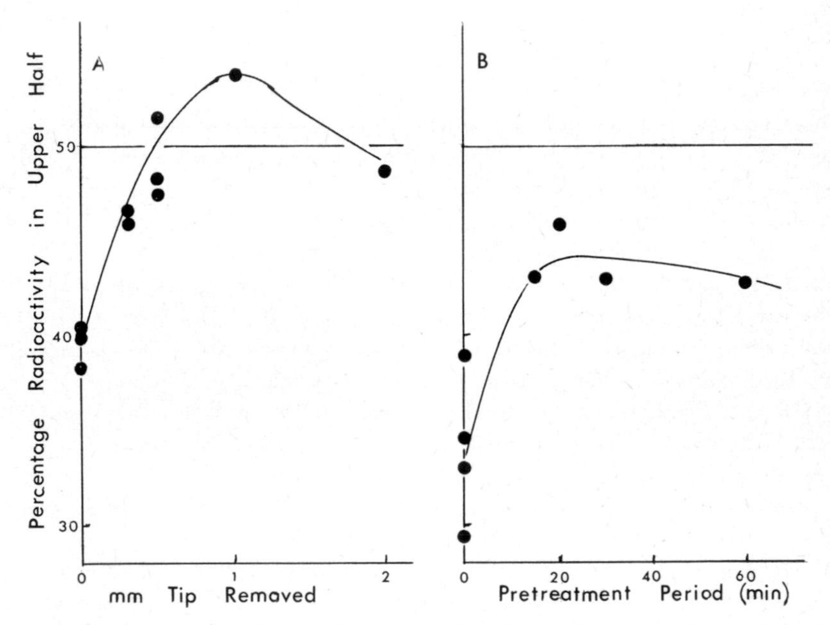

FIG. 6. The distribution of ^{14}C-IAA in the terminal 4 mm of pea roots after application to the tip in agar. The values represent percentages of the total radioactivity in the upper halves of geotropically stimulated roots after 1 to 3 h of stimulation (over this period the percentage did not change significantly, indicating that the differential accumulation was complete within one hour). A. The effect of different degrees of decapitation on the distribution. B. The effect of pretreatment of the intact roots with 10 μg ml 2,4-dichlorophenol for various times. (Data from Konings, 1967).

that acropetally transported IAA may be similarly redistributed. Thus Konings demonstrated that 0·05 μg of IAA, applied to the surface of intact roots of pea at 20 mm from the apex, significantly augmented the rate of geotropic response and that a ring of lanolin, containing 0·1% of 2,3,5-tri-iodobenzoic acid (TIBA) between the point of IAA application and the tip, not only eliminated the IAA effect but also greatly reduced or even stopped normal geotropic response. Since TIBA has long been known as an effective inhibitor of the polar transport of auxin in shoots (see McCready, 1966), the extrapolation was made that its blockage of the acropetal movement of IAA, either when applied or when moving naturally from the cotyledons, prevented accumulation in the root cap where gravity induces its unequal distribution and subsequent movement in unequal amounts to the upper and lower sides of the extending zone. Direct support for these suggestions subsequently came from Pilet (1971b) who applied methylene-labelled IAA to *Lens* roots 10 mm from the apex and after 4 h horizontal stimulation measured the distribution of radioactivity in the upper and lower halves of the subapical 4 mm tip. Pilet demonstrated that in such roots with intact caps the ratio upper side/lower side was 36·6/63·4, a value very close to those of Konings for tip application to pea roots. Removal of the cap however eliminated this difference, the ratio now being 49·2/50·8.

Further suggestions for the involvement of IAA transport in geotropic response comes from the work of Keitt (1960). Whereas high concentrations of auxins such as IAA, naphth-lyl-acetic acid, 2,4,-dichlorophenoxyacetic acid and 2,3,6-trichlorobenzoic acid all inhibit curvature in parallel with growth, two strong inhibitors of polar auxin transport in shoots, TIBA and *N*-1-naphthyl-phthalamic acid (NPA) (see McCready, 1968), reduce curvatures at concentrations lower than those reducing growth. This implies that these two compounds are reducing curvature by means other than a direct effect on growth, which is presumably the case with the synthetic auxins. They could logically be doing so by altering the lateral polarity of auxin movement under gravity. One is tempted to ask why then, if supra-optimal auxin concentrations are prevented from arising in the extending zone by this blockage of flow from the shoot, these curvature-preventing concentrations of TIBA and NPA do not *accelerate* extension growth. A similar reduction of geotropic response has been shown in *Lens* seedling roots after seed pre-treatment with NPA (Pilet, 1967a) but this was associated with an inversion of the differences induced by gravity on the activity of IAA oxidase on the two sides of the root. For control roots the ratio upper/lower was 32/21 and for NPA-treated the ratio was 44/57. What is more, the analogue *N*-(1-naphthoyl)-anthranilate, which has no action on geotropic response, similarly had no effect on either the *in vitro* or *in vivo* IAA oxidase activity, whereas NPA markedly inhibited them both. At that time Pilet regarded geotropic perturbations by NPA as definitely associated with IAA oxidase inhibition although he implies that IAA transport modification should not be ruled out. The results of Konings (1967) would definitely rule out any role of IAA oxidase in the gravity-induced lateral distribution of auxin and this would mean that Pilet's NPA effect on response can also be attributed to inhibition of lateral IAA transport. Although NPA has been shown to inhibit markedly the gravity-induced lateral movement of auxin in *Helianthus* hypocotyl segments (Abrol and Audus, 1973) such direct experiments have still to be done on root tissue.

Further pointers to a role of acropetally transported IAA in geotropic response come from the work of Scott and Wilkins (1969) who showed that the primary roots of *Zea mays* showed no geotropic response when grown in complete darkness; curvature occurred only when seedlings were exposed to light. Correlated with this effect was a promotion of the acropetal transport of IAA-(1-^{14}C) by light; this seems to be a direct action of light since it is effective only during the transport period. This light-dependence of geotropism in roots seems not to be a phenomenon specific to *Zea* since earlier studies (Lake and Slack, 1961) had indicated similar effects in *Callistephus*, *Matthiola*, *Lycopersicon* and *Cucumis* grown in transparent and in opaque pots.

However, the role of this acropetally moving auxin in geotropic response remains obscure. Iversen *et al.*, (1971) have shown that ^{3}H-IAA, injected basally into *Phaseolus* roots, is transported acropetally at a velocity of 7 mm/h and modifies the geotropic response. After 5 h the response is augmented (cf. Konings, 1969) while after 20 h it is decreased; root elongation remains unaffected. However much of the transported IAA is conjugated with aspartic acid, a situation reminiscent of the findings of Greenwood *et al.*, (1973) who indicated an active metabolism of IAA to unknown radioactive products in the cells of the cortex. It has still to be decided whether metabolism of this kind is to be regarded as detoxication machinery for rendering ineffective an excess of auxin applied to these experimental roots or whether metabolism itself is in any way associated with growth activity and hence geotropic response. This latter seems most unlikely.

3. *The Lateral Distribution of Endogenous Auxin*

It is quite astonishing how slender the evidence is for the redistribution of endogenous auxin in roots under the action of gravity. Hawker's (1932a and b) pioneering observations on decapitation, tip replacement etc. simply demonstrated that the tip was the source of an inhibitor which was redistributed under gravity. Only three other early observations identified this inhibitor as auxin. Boysen-Jensen (1933) collected auxin by diffusion into agar from the upper and lower cut surfaces of horizontal root tips of *Vicia faba*. Collection continued for 2 to 5 h since shorter times gave responses in the *Avena* coleoptile curvature assay which were too small for accurate determination. An upper/lower auxin ratio of 37/63 was claimed and my own statistical analysis of the primary data show this to be significantly different from unity. However since roots have developed a very substantial curvature by this time it is not impossible that the different auxin contents of the two sides may be the *result* of the different growth rates (i.e. different auxin consumptions) rather than their cause. Similar comments apply to similar redistributions revealed in chloroform extracts of the roots of the same species (Boysen-Jensen, 1936). Amlong (1939), using boiling agar to extract auxin (a very suspect method) and a very unconventional *Helianthus* hypocotyl curvature assay, obtained differences between extracts of upper and lower halves of geotropically stimulated *Vicia* roots which Larsen (1962b) calculated to represent equivalent IAA concentration ratios of upper/lower equal to 2/98, a rather startling result in view of ratios previously found in coleoptiles.

Subsequent results have not confirmed these claims. Thus Genkel (1960), on the strength of an indole colour reaction, claimed that after geotropic stimulation, "heteroauxin" was at a higher concentration on the convex (i.e. upper) sides of the roots. Much more sophisticated studies by Lahiri in my laboratory (1959, 1968) on *Vicia faba* roots involved the harvesting of upper and lower halves of 5 mm root tips after 1 h of stimulation. Such half tips were plunged immediately on harvesting into the extracting ethanol at –14°C and extraction proceeded for 20 h at that temperature in the dark. The acid fractions of such extracts were subsequently run on paper chromatograms and assayed by the *Avena* mesocotyl segment test. The auxin running at the position of IAA (corresponding to that which could not be identified as IAA by Burnett *et al.*, 1965) when estimated quantitatively from the chromatograms in terms of IAA equivalents, showed upper/lower ratios of 78/11 and 52/22, a result in line with those of Genkel. However a follow-up of these experiments by Most (1962) also in my laboratory with shorter stimulation times of from 10 to 40 minutes, corresponding more closely with the presentation time for these roots, could obtain no evidence whatsoever of any consistent differences in the concentrations of this particular auxin, or indeed any other growth-promoting substance detected on the chromatograms, between upper and lower halves of 7 mm *Vicia* root tips (Fig. 7). It will be seen that the period of horizontal exposure far exceeded the presentation time and covered the early stages of geotropic response (see Fig. 3). Of course the objection could be raised to the work of these last two authors that solvent-extracted auxin was studied and that any differences in the growth-effective fraction (diffusible auxin) could be swamped by large amounts of immobile (bound) auxin removed by the solvent from both halves. A repetition of Boysen–Jensen's classical diffusion experiments with modern, more refined techniques, is now well overdue.

B. The Special Inhibitor Theory

The idea that a special inhibitor, not indol-3yl-acetic acid, was produced by the root tip and was, following a Cholodny/Went type of theory, responsible for geotropic curvature response, was put forward by Audus and Brownbridge (1957a and b). These conclusions were based on two primary findings: (a) that during curvature the growth rates of *both sides* of the root were decreased, the lower more than the upper, and (b) that low concentrations of IAA *promoted* the rate of curvature by promoting the growth rate of both sides more or less in the same proportion. Such results are inexplicable if curvature is due to an inhibitory concentration of IAA on the lower side of the extending zone, but can easily be explained if applied IAA at these concentrations is *antagonizing* an inhibition of growth due to a special inhibitor. These results have been broadly confirmed by Konings (1965). Furthermore the results of an extension of these observations to response to applied, 2,4-dichlorophenol (DCP) and caffeic acid cast further doubts on the role of IAA as the geotropic inhibitor. Thus DCP promotes IAA oxidase and caffeic acid inhibits it and so one would expect, on the classical theory, that DCP would suppress geotropic response by reducing IAA levels and thus releasing the under side of the root from inhibition; by the

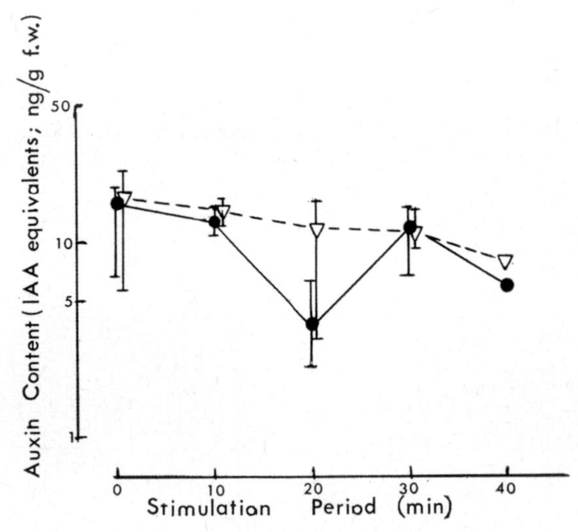

FIG. 7. The effect of different periods of geostimulation on the distribution of the major auxin in 7 mm tips of roots of *Vicia faba*. The auxin is an ether-soluble acid running at the same Rf as IAA (0·55 to 0·75 in isobutanol methanol water). Samples from 100 root tips were quick-frozen in methanol and extracted for 18 h at —18°C before fractionation and paper chromatography. Values represent the means and ranges of a number of replicates; ●—● upper half, △---△ lower half. (Data replotted from Most, 1962).

same token caffeic acid should reduce IAA oxidase activity, reduce IAA destruction and hence promote curvature. [Other expectations based on a gravity-induced redistribution of IAA oxidase activity on the two sides are ruled out by subsequent studies (Konings, 1967).] The relevant parts of his graphs have been replotted in Fig. 8. and reveal that during the stage of most rapid curvature in the first hour or so, DCP prevents root curvature by suppressing the growth rate of the upper side with no effect on the lower side while caffeic acid promotes curvature by increasing the growth rate of the upper side again with no significant effort on the lower. These results are completely inexplicable in terms of IAA inhibition on the lower side of the root.

Studies aimed at further characterization of this inhibitor have intensified in recent years and have been concerned in the first instance with experiments on the removal of its presumptive source, the root cap.

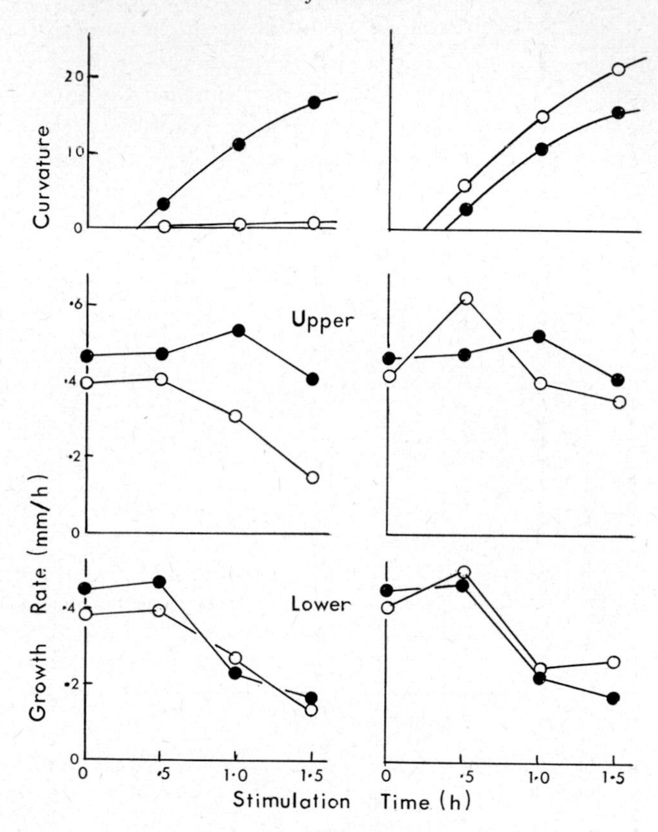

FIG. 8. Graphs showing the geotropic responses of roots of *Pisum sativum* under the action of 10^{-5} g/ml of 2,4-dichlorophenol (DCP) (left) and 10^{-8} to 10^{-9} g/ml caffeic acid (right). ●—● control roots in water. ○—○ treated roots in DCP or caffeic acid solutions. *Top*: geotropic curvatures; *centre*: growth rates of upper sides of roots (mm/h); *bottom*: growth rates of lower sides of roots (mm/h). Graphs redrawn in part from Konings (1965).

1. *The Indirect Evidence for a Special Tip Inhibitor*

The literature contains many scattered observations pointing to the production of an inhibitor by the root tip. In the 1920's and 1930's there were several claims that removal of the extreme 1 mm tips of the roots of several species caused an increase in the growth rates of those roots provided certain optimal environmental conditions obtained, and that "reheading" reimposed a slower growth rate. On the other hand an almost equal number claimed that they could detect either no change or a retardation of the growth rate [the literature up to 1958 is comprehensively summarized by Larsen (1962b).] Clearly these inconsistent results are difficult to interpret and responses are bound, to a large extent, to be complicated by damage done to the meristem by surgical operations. The success of recent "decapping" techniques with *Zea* roots in geotropic response experiments has prompted a further study.

Schachar (1967) in her geotropic sensitivity studies, also measured the growth rates of *Zea* roots at half-hourly intervals during the first three hours after decapping, and also after reheading with the caps. Although there was a progressive decline in growth rate over that period, decapped and reheaded roots showed growth rates which did not differ significantly from those of the control intact roots. Subsequent to this Pilet (1971a and b; 1972a) made a similar series of short-term observations on both *Lens* and *Zea* roots. In *Zea* (Pilet, 1971a) there was an overall decline in growth rate over the 14 h of the experiment. Although he claimed that decapping had no effect on longitudinal growth rates, a rough statistical check on his data indicates that it induced a general reduction in the growth rate over the 12 h period which was particularly marked and probably statistically significant in the 8–12 h period. Later studies of growth over shorter time intervals (Pilet, 1972a) revealed more consistent changes in which decapping produced a marked increase in growth rate over the first 2 to 3 h and then a decline to that of the intact root (Fig. 9). On the other hand in intact *Lens* roots tipped with a *Zea* root cap affixed with Ringer solution, the growth rate was reduced for the same period. Pilet concluded from these experiments that the root cap was the source of a growth inhibitor which was not specific, since the *Zea* cap could inhibit the growth of the intact *Lens* roots. These latter results are puzzling since they suggest that the *Zea* inhibitor is much more powerful than that from the cap of *Lens* itself. Unfortunately direct decapitation in *Lens* only added complication to the picture (Pilet, 197b). Surgical removal of 0·5 mm of the *Lens* root removes not only the cap but the quiescent centre and some of the meristem. In very young roots (5 to 15 mm) such a procedure significantly depressed the growth rate by approximately 50% during the subsequent 24 h (Fig. 9). What is surprising is that in somewhat older roots (*c* 20 mm long) such an operation *promotes* the growth rate for at least 24 h. Calculations from Pilet's Table 2 give rate ratios decapitated/control of 1·18, 3·92, and 2·62 for the 0–8, 8–16 and 16–24 h periods respectively. However very recent decapping experiments with *Zea* (var. Golden Bantam) could reveal no significant effect on growth rates at any time during the first 24 h after cap removal (Dr. P. Barlow, personal communication). Clearly the effects of cap removal on the rate of root extension need further careful study.

Fortunately more consistent, although less direct results have come from experiments on curvature resulting from removal of half tips and other related surgical operations. Thus Gibbons and Wilkins (1970) succeeded in dissecting the lateral halves of caps from roots of *Zea mays* by means of fine forceps and a binocular microscope. They found that such half-capped roots in the vertical position curved as actively as horizontal intact roots, the concave side being the side possessing the half cap. Furthermore, half-capped roots placed horizontal

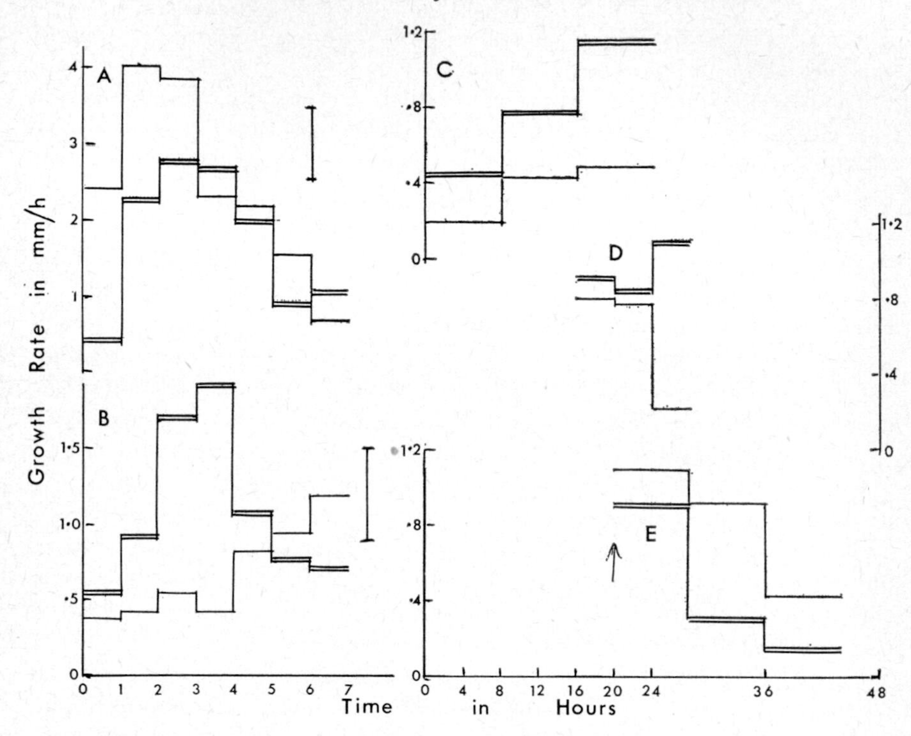

Fig. 9. Effects of cap removal and replacement on the rate of growth of *Zea* and *Lens* roots. (A) Growth rate progress curves for vertical roots of *Zea mays* initially 15 mm long showing the effects of cap removal. ═══ control roots; ——— roots decapped at start of observations. (B) Growth rate progress curves for vertical roots of *Lens culinaris* showing the effects of heading with root caps of *Zea mays*. ═══ normal roots; ——— roots with caps of *Zea* placed on tip. (A) and (B) redrawn from Pilet, 1972a. (C), (D) and (E). Growth rate progress curves of *Lens culinaris* roots, showing the effect of tip excision. ═══ intact roots; ——— detipped roots. Lengths of roots at the start of observations: (C) c. 6 mm; (D) c. 15 mm; (E) c. 20 min. The three graphs have been located on the time axis so that measurements on the same vertical axis are for roots of approximately comparable age. (Growth rates calculated from Tables 1 and 2 of Pilet 1971b).

with the half-cap side uppermost curved *upwards* "against" gravity as actively as similar vertical roots. This is a clear indication of inhibitor production from the cap tissue. Following this work Pilet (1971a, 1972b) checked the nature of this inhibitor by observing the effects on geotropic response of replacing the cap on the decapped tip, "sticking" it on with a film, either of Ringer solution or of "oleate" (i.e. an oleic oil). Pilet claimed that decapped roots which had lost their sensitivity to gravity regained it in the former case but not in the latter. This indicated to him that the inhibitor responsible for gravitational response was water- (but not lipid-) soluble. However inspection of his graphs suggest another interpretation. Thus as we have seen, recovery of sensitivity in decapped roots, (due presumably to the development of starch statoliths in the quiescent centre and distal parts of the meristem) started, as can be seen after extrapolation of his progress curves, at about 11·5 h. When reheaded using Ringer's solution the recovery took 4·0 h; could not he have been dealing with an accelerated recovery of the statolith apparatus? Reheading with "oleate" clearly prevented normal recovery and should be explained purely as a toxic action. These experiments therefore give little clear indication of the physical properties of the inhibitor.

Elaboration of the first half-cap experiments have further characterized the properties and behaviour of the inhibitor (Pilet, 1973b;Shaw and Wilkins, 1973). Pilet, using *Zea* roots, studied the effects of removing a lateral half of the tip, either with or without its half root cap. Half meristems only, in any of the three Gibbons and Wilkins' positions, produced completely non-significant curvatures (we do not known how the overall growth rate was affected), whereas half meristems *with* their half caps verified Gibbons and Wilkins' (1970) observations. This experiment demonstrated that the cap, not the meristem, was the source of the inhibitor. Furthermore the insertion of tiny mica barriers into the meristem on the half-cap side brought about reversals of the direction of curvature with all three orientations, indicating that the movement of the inhibitor from the cap could be deflected across the meristem to the opposite side, where inhibition of growth would cause the reversal of curvature. An ingenious variation of these experiments was the replacement of half tips (i.e. half meristems plus their half caps) from roots of one age by half tips of roots of another age. The difference of curvature of vertically orientated roots after these replacement operations suggested that inhibitor production decreases with root age. Shaw and Wilkins (1973) produced very closely similar results with half-tipped roots with or without half caps. The effect of the half cap in terms of curvature in horizontal roots with the half tip lowermost was almost three times that of similar vertical roots. Similarly a unilateral barrier (mica, polythene or metal) just behind the cap in intact roots produced more than five times the curvature when in the upper side of horizontal roots, than when in vertical roots. However when the barrier was on the lower side of horizontal roots or the half tip was on the upper side the curvature was upwards (i.e. a negative geotropic curvature), but was less than that of vertical roots similarly prepared (See Fig. 10).

These results are susceptible of two explanations. Firstly the production of the inhibitor by the cap is uniform and not affected by root orientation. In this case the curvature responses can be interpreted in terms of a blocking of longitudinal movement of this inhibitor by the barrier, coupled with its downward movement under gravity in the intact region of the root tip proximal to the barrier or, in half-tipped roots, the transverse cut. The second explanation is that there is no significant effect of gravity on lateral movement of this inhibitor in those regions but that the modifications of the barrier and half-cap responses brought about by root orientation is a reflection of a differential production of inhibitor by the cap under the action of gravity. If the production by the cells of the lower half is increased while that of the upper half is decreased as compared with the uniform level of production in vertical roots, then Shaw and Wilkins' results can be explained.

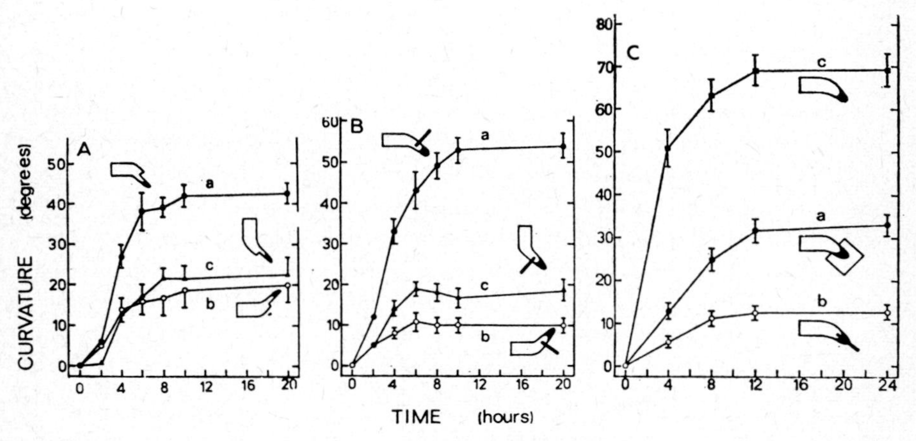

FIG. 10. Half-tip and barrier experiments of Shaw and Wilkins (1973). The graphs show the progress of curvature with time in *Zea* roots. A. Half-cap experiments (a) Roots horizontal, half tip lowermost. (b) Roots horizontal, half tip uppermost, (c) Vertical roots. B. Transverse barrier experiments. (a) Horizontal roots, barrier uppermost, (b) Horizontal roots, barrier lowermost, (c) Vertical roots. C. Longitudinal tip-barrier experiments on horizontal roots. (a) Barrier plane vertical, (b) barrier plane horizontal. (c) Intact roots.

When barriers were placed 4–5 mm from the tip of intact roots no interference with normal geotropic response could be observed and no curvature was induced in vertical roots in both *Zea* and *Pisum*. The great importance of these latter observations is that they suggest that auxin flowing acropetally from more mature regions of the root (see Konings, 1969; Pilet, 1971b) is *not* involved directly as the geotropic inhibitor but that the cap inhibitor alone is the prime mover.

However there are additional complications arising from experiments on the insertion of barriers axially through the tips of intact roots, i.e. the analogue of the classical Boysen-Jensen (1928) experiments on phototropism in *Avena* coleoptiles. These showed that such barriers with their planes vertical in horizontal roots reduced geotropic response by about 50% in *Zea* and by only 25% in *Pisum*. When the plane of the barrier was horizontal curvature was reduced by about 85% in *Zea* and 60% in *Pisum*. Thus, although the results from the horizontal barrier experiments could be explained in terms of differential inhibitor production on the upper and lower sides (see previous discussion) yet the greater curvatures in the vertical barrier experiments strongly indicate a lateral movement of the inhibitor across the tip. As Shaw and Wilkins (1973) point out these results suggest that in principle the Cholodny/Went hypothesis is still valid for roots although the nature of the inhibiting hormone is still unknown.

2. *The Nature of the Cap Inhibitor*

Dating from the pioneering work of Bennet-Clark and his colleagues, who applied in the early 1950's the newly established techniques of paper chromatography to plant hormone studies, natural plant growth inhibitors have continued to be demonstrated from all plant organs, including roots. The apparently widespread, and then uncharacterized, "inhibitor-β" was first demonstrated in *Vicia* roots by Bennet-Clark and Kefford (1953) and has since been demonstrated in the roots of wheat, pea, maize and others (see Hemberg, 1961). Owing perhaps to the domination of our thoughts by auxins in the past, no serious attempts have ever been made to explore the possible significance of this (and other) inhibitors in the control of root growth and behaviour. Except for a demonstration by Howell (1954) that an ether-soluble inhibitor of shoot elongation in pea roots occurred in greater amounts at the tip and declined in concentration with increasing distance from the tip, I can trace no systematic studies on the distribution of such inhibitors in roots or of changes induced in them by experimental conditions aiming to elucidate growth control phenomena. Now that modern physico-chemical techniques have shown that one of the major components of the "inhibitor-β" complex is abscisic acid (Cornforth *et al.*, 1965) which appears to be a major and ubiquitious component of the balanced hormone complex of plants (see Addicott and Lyon, 1969) it seemed to me that the time was ripe to re-open the study of inhibitor-β in roots and to look at the spatial and temporal concentration relationships in geotropic phenomena.

To this end Dr. Kundu and I redesigned the root inhibition bioassay of Audus and Thresh (1953) to increase its sensitivity and, using 2 mm segments of *Zea mays* roots cut 2 mm behind the extreme tip (i.e. from the zone just beginning to extend), an inhibition sensitivity to about 50 pg of IAA has been achieved (Kundu and Audus, 1974). Root caps have been dissected from the tips of roots of 6-day-old seedings of *Zea mays* var Golden Bantam, by the technique used for geotropic response studies. Segments without caps but including meristems and early extension zones were similarly harvested. Both were immediately frozen and extracted in methanol at —18°C. Acid fractions of such extracts were run on paper in an isopropanol: ammonia: water (10:1:1) solvent and were subsequently assayed by the above root-segment micro-assay.

Root caps showed predominantly one growth inhibitor (Fig. 11) which runs at Rf 0·5 to 0·7, corresponding with inhibitor-β (and also abscisic acid). Occasionally on some chromatograms a second inhibitor is seen at a higher Rf (0·9 to 1·0). In view of the doubts already expressed on the role of IAA, it is interesting that no trace of that hormone could be found in cap extracts. Since usually 100

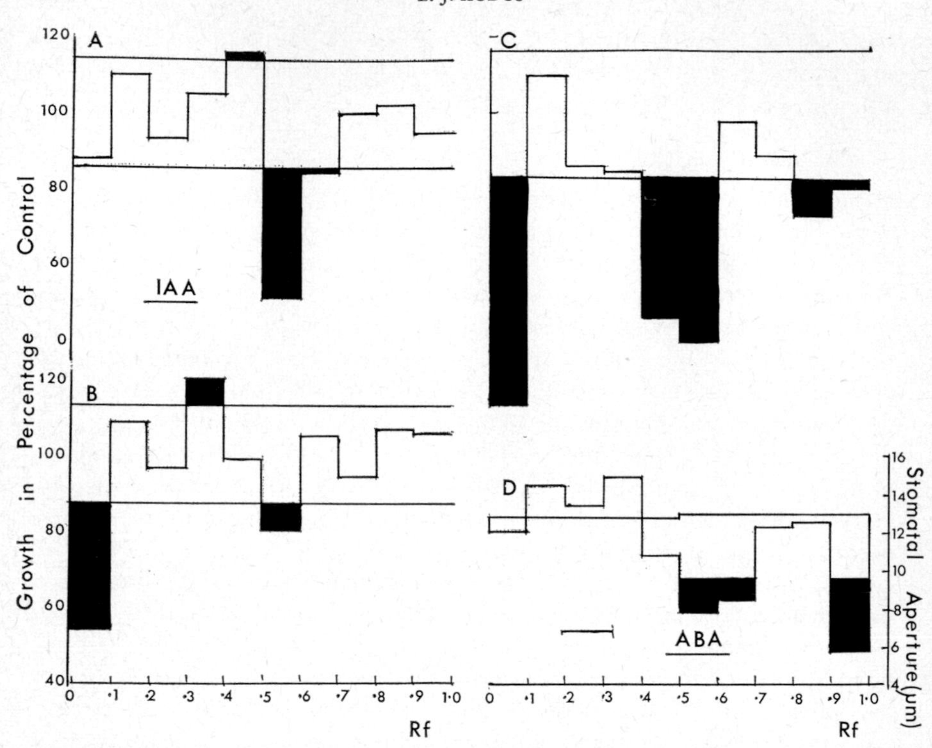

FIG. 11. Paper chromatograms of acid fractions of extracts of root tip tissue of *Zea mays* run in isopropanol/ammonia/water (10: 1: 1). A. Extract of 100 root caps assayed by the root-growth-inhibition micro-assay of Kundu and Audus (1974). B. Extracts of 100 5 mm root tips without caps, assayed as for (A.) C. Extracts of 100 5 mm entire root tips (i.e. with caps) assayed as for (A) and (B). D. Extracts of 100 5 mm entire root tips as for C but assayed by the stomata-closing test for abscisic acid-like substances (Tucker and Mansfield, 1971).

caps were used per chromatogram this means that the IAA content of such caps cannot be higher than about 0·5 pg per cap. Simple but rough calculations indicate that this upper limit would represent a content of about 1 ng/g (equivalent to about 5×10^{-9} M) a level which is below the threshold concentration for root growth inhibition (see Larsen, 1962b, Table 8) as judged by the results of external applications. This is further very strong evidence that IAA cannot be the geotropic response inhibitor produced by the cap.

Further evidence in support comes from the analysis of the extract of capless segments. Here the major inhibitor is not the same as that in the cap since it scarcely moves from the point of application on the chromatogram (Rf 0 to 0·1) (Fig. 11). Occasionally, traces of the main cap inhibitor have been seen in these extracts, but this could be explained by residual fragments of cap tissue having been included on the segments for extraction. In these extracts occasional small traces of activity (either stimulation or inhibition of segment growth) have been detected at the Rf of IAA and this is compatible with the results of Bridges *et al.* (1973) showing the presence of IAA in the vascular tissue; the activity at this spot on our chromatograms is probably due to IAA.

Extracts of whole tips have been assayed with the *Commelina* epidermal strip assay of Tucker and Mansfield (1971); this assay tests stomatal closing activity and is extremely sensitive to abscisic acid. Such assays have shown a very high activity on the chromatograms at Rf of the main cap inhibitor (Fig. 11) which corresponded precisely with the Rf of authentic samples of abscisic acid. Further stomata-closing activity was observed where a second root growth inhibitor was occasionally seen at Rf 0·9 to 1·0 in the capless segment extracts.

These results are a very strong indication that the "cap inhibitor" is abscisic acid and not indol-3yl-acetic acid.

The nature of the other stomata-closing substances found in the capless tip (Rf 0·9–1·0) seems fairly certain. It resembles xanthoxin, a naturally-occurring abscisic acid analogue with ABA-like properties (Taylor and Burden, 1970), in that it has a closely similar Rf on paper chromatograms in the solvent used. Furthermore Mrs Lesley Stanbury in my department has subjected extracts of whole seedling roots to a sequence of purification steps on Sephadex columns and on TLC. Subsequent GLC analysis showed the presence of a substance with the same retention time as xanthoxin and the methylation of this suspected xanthoxin fraction gave a mass spectrogram showing the undoubted presence of that compound (see Fig. 12).

Other work also points to the implication of ABA and/or xanthoxin in root growth control. Thus we have already indicated that in some species geotropic response in roots is conditional upon illumination (Scott and Wilkins, 1969; Tepfer and Bonnett, 1972). Furthermore in *Convolvulus arvensis* roots the light effect seems to be phytochrome-mediated since it is evoked by red light and reversed by far-red light (Tepfer and Bonnett, 1972). Pilet (1973c) has extended these observations by studying the effects of illumination on the curvature of vertical roots from which half-tips had been removed (see technique of Gibbons and Wilkins, 1970). Such roots do not curve in the dark but upon illumination a curvature is produced with the half-cap side of the root becoming concave. This suggests that the cap inhibitor is produced only in the light. Furthermore it has been known for a considerable time that light inhibits root growth (see Burström, 1960) and this could conceivably be due to the induction of inhibitor formation. Direct support for this comes from several sources. Thus there are two early reports that light puts up the inhibitor-β (abscisic acid?) content of *Helianthus* (Bayer, 1961) and wheat roots (Masuda, 1962). More recent studies have implicated xanthoxin in the light-induced inhibition. Thus studies by Wilkins *et al.*, (1974) have shown that a 10 μM solution of xanthoxin can inhibit root growth in seedling wheat by up to 83% in the dark but is less effective in the

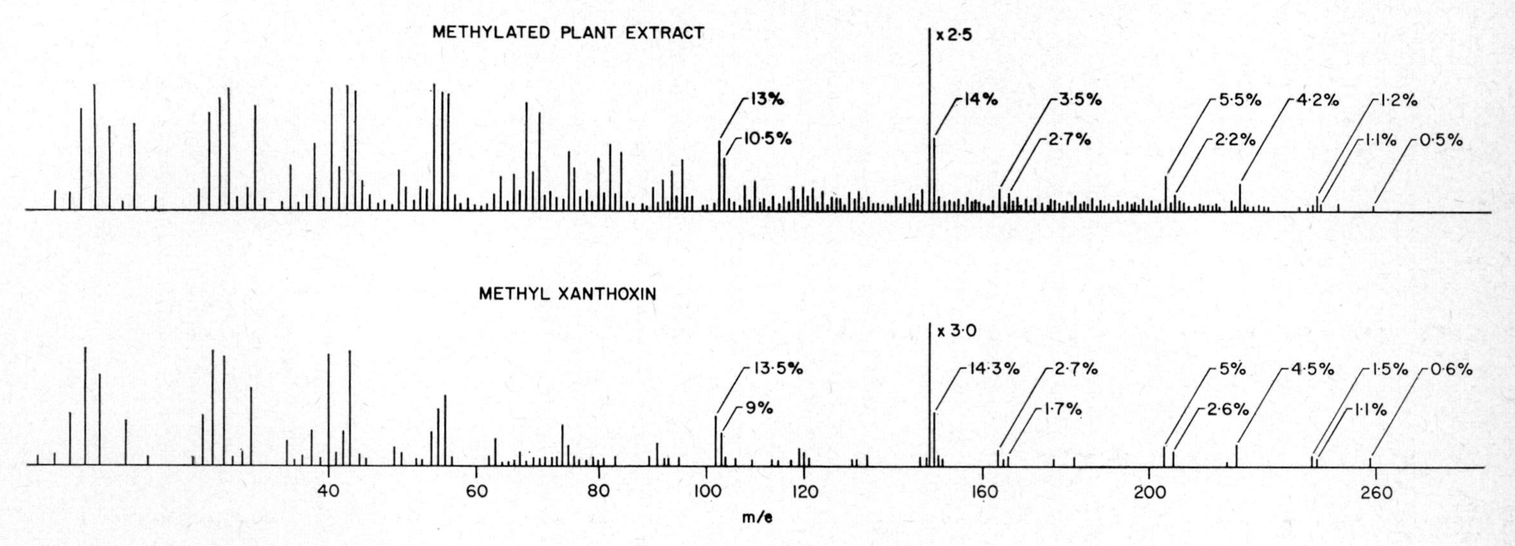

FIG. 12. Mass spectrograms of methylated extract of seedling roots of *Zea mays* c.v. Kelvedon 33, and of authentic xanthoxin. The acid fraction of the methanol extract was fractionated on Sephadex-LH20 and eluted with 90% ethanol plus 10^{-3}M HCl. The "xanthoxin" fraction was purified by three successive TLC runs (solvent hexane/acetic acid, 2/1). The fraction at Rf 0.29–0·36 was methylated with diazomethane and re-run on TLC [solvent light petroleum (BP 60–80°C) acetone, 4/1]. Purified methyl xanthoxin at fraction 0·02–0·12 contains both isomers. (By permission of Mrs. Lesley Stanbury).

light. Looked at another way the data show that in the presence of xanthoxin there is no significant effect of light on growth. Furthermore decapping of roots, which does not affect their growth in the dark, eliminates the light inhibition of growth. This is all consistent with the idea that the light inhibition of growth is due to the induction of the synthesis of xanthoxin which is the tip inhibitor. Unfortunately the authors were not able to isolate xanthoxin from 2-day-old *Zea* roots and suggest that this may be explained by rapid immobilization (during growth?).

Studies involving the illumination of *Convolvulus* roots with an 0·85 mm microbeam of light (Tepfer and Bonnett, 1972) have shown that the photosensitive region is located in the extreme tip, and the possibility exists that it is restricted to the root cap, Since this light-dependent geotropic mechanism is phytochrome-mediated (see above) one is strongly led to speculate whether the synthesis of the abscisic acid-like cap inhibitor, like the anthocyanins (Vince and Grill, 1966; Wagner and Mohr, 1966) is also regulated by phytochrome.

All this however leaves unexplained the baffling enigma as to why it is that root caps, being the source of the growth inhibitor, can be removed whole from roots without affecting the growth rate but when removed in part reveal an inhibition coming from the remaining part only. Can these discrepancies be explained in terms of the effects of light described above and the different lighting conditions of the several experiments? Or is the hormonal control mechanism much more complex than the above relatively simple considerations would suggest?

IV. An Attempted Synthesis

It is very hard from this survey to avoid the conclusion that root growth and hence geotropic response is under the control of at least two hormones or hormone systems.

In spite of the continuing difficulties of establishing indol-3yl-acetic acid as the regulator directly (and, as once thought, exclusively) in control, it now seems probable that this auxin has some part to play. Three main types of observation point to this; they are the apparent universal sensitivity of roots to very low concentrations of applied IAA, the recent clear demonstration that IAA occurs in the extension zone of *Zea* roots and the now inescapable evidence that it is transported there acropetally from more mature regions of the plant and is somehow deflected towards the lower side under the influence of gravity acting in the cap. However its apparent absence from the root cap itself, coupled with the observations that the one-sided interruption of its acropetal movement into the extension zone does not disturb the curvature mechanism, suggests very strongly that it is not the regulator in direct control of growth and curvature. It seems likely therefore that its presence in the extension zone is a *necessary condition* for growth and response but that it never acts as the *limiting factor*; such a "regulating" role would be assumed by the cap inhibitor, possibly abscisic acid or a related compound. This of course would mean the complete abandonment of the classical concept of root growth control by modulation of supra-optimal auxin concentrations; in fact it would imply the irrelevance, to the normal control of root growth, of the concentration/response curves so painstakingly acquired over the years from experiments with auxin applied to the culture medium. It would also rule out the theory of Chadwick and Burg (1967) that auxin-induced ethylene production is involved in root geotropism. The remarkable parallelism they showed to exist between ethylene generation and the inhibition of root growth by supra-optimal concentrations of exogenously applied auxin would thus be irrelevant to the natural hormone control of geotropic response and must be regarded as an artefact of high non-physiological auxin levels.

A further corollary of relinquishing this long-established concept and of accepting the idea of IAA as an "essential conditioner" but not a "direct modulator" of root growth, would be the irrelevance of the distribution of IAA under gravity in the extending zone. Could this not be a phenomenon which is relevant enough in shoots, but which is a kind of "evolutionary relic" in roots, playing no direct part in geotropic response, as seems also to be the case with the gravity-induced differential activity of IAA oxidase on the two sides of the root (Konings, 1967).

This would then leave the cap inhibitor in *direct* control and would require its deflection under gravity to the lower side of the extending zone so that the appropriate inhibition of growth could be exerted. As we have seen M. B. Wilkins and his colleagues have already obtained direct evidence of this in the cap itself although the possibility of a similar downward transport in meristem and extending zone is indicated although not yet unequivocally proved. However lack of such evidence should not preclude speculation and now that there are clear indications of the involvement of abscisic-acid-like compounds in the phenomenon, such speculation should soon have a firmer foundation in fact.

What is, and has long since been, the key question is the mechanism whereby gravity perception is linked to geotropic response. In the context of the root this means the coupling of amyloplast sedimentation to the differential movement of cap inhibitor to the extending zone. This is no new topic and I have already made one proposal (Audus, 1971) whereby a "message" (i.e. the cap inhibitor) is induced to arise in cells of the cap as a direct result of the mechanical pressure exerted by the amyloplasts on membranes along the *tangential longitudinal* walls of the cap cells. If there were differences in sensitivity between the abaxial and adaxial sides of the cell, then the two sides of the cap would produce a different level of "message" (in the current context this means transmissable cap inhibitor) which would be moved strictly basipetally in unequal amount to the extending zone and hence cause curvature. Alternatively, following Sievers and Volkmann (1972), if the sensitive membranes are the endoplasmic reticulum and are the source of the inhibitor, then the greater number on the abaxial side of the cell would give the required differential production and flow.

If this induced cap inhibitor is abscisic acid, then there is experimental support for this idea from its antagonistic action on the activity of IAA in root growth (Pilet, 1972c) where in *Lens* both growth inhibition by high, and promotion by low, IAA concentrations are counteracted by applied ABA.

However the fact that the geosensitivity of decapped roots is recovered before cap regeneration and is accompanied by the development of amyloplasts in the quiescent centre, has led to the suggestion (Dr. Peter Barlow—private discussions) that the amyloplasts themselves may be the source of the inhibitor and this would imply their possession of synthetic enzymes capable of producing ABA and/or its analogues. This seems a reasonable possibility. Thus chromoplasts, amyloplasts and chloroplasts must possess some basic organizational features in common since they arise from similar if not identical proplastids and, under certain conditions, can change one into the other (Clowes and Juniper, 1968). Furthermore etioplasts (and presumably chromoplasts) contain the full enzyme complement for the synthesis of their characteristic pigments (see Rebeiz and Castelfranco, 1973). These facts, coupled with the close chemical relationship between ABA and the carotenoid pigments, make it very plausible that amyloplasts possess the capability of synthesizing and presumably storing ABA and its chemical analogues. The most direct evidence however comes from the work of Milborrow (1974) who has demonstrated that cell-free preparations from chloroplasts of ripening Avocado fruits will synthesize abscisic acid from labelled mevalonic acid.

A very intriguing observation recently made is that the amyloplast membranes of the root caps of cabbage have a high "IAA-decarboxylating" activity, as judged by the precipitation of $SrCO_3$ from $SrCl_2$ solutions and viewed in sections under the electron microscope (Aasheim and Iversen, 1972). This indicates the occurrence of membrane-bound enzymes in the amyloplasts, a very interesting suggestion in view of the apparent absence of IAA from cap cells.

However, attractive as the suggestion is that ABA may be synthesized in or liberated from the amyloplast, we encounter certain difficulties when we try to reconcile it with the theory outlined above (Audus, 1971). Thus the production or liberation of ABA from the amyloplast would have to be differential, i.e. more intense on one side of the root than on the other. Any such modulation by gravity could act only by contact or pressure between the amyloplast and the other cell structures and would demand a radial asymmetry of some kind in the cell containing the amyloplast. This could be the case in Sievers and Volkmann's theory as applied to *Lepidium*: a reference to Fig. 4 will show that this differential effect could come about by the application of pressure to more amyloplasts in the cells on the lower side of the cap than in those on the upper. However in those roots where there is *no* such regular radial asymmetry, the number of amyloplasts subjected to pressure would, on average, be the same in the cells on the two sides of the cap and hence the production or release of inhibitor would also be the same on the two sides. In these roots at least the amyloplasts are unlikely to be the source of the inhibitor.

However, there is the other alternative that gravity acts on the lateral movement and not the production of the inhibitor. Once released from the amyloplast it would move to the lower side of the organ under the action of gravity; indeed, the results of Shaw and Wilkins (1973) have pointed to this possibility.

But why then does decapping *immediately* remove all capacity of the root to curve under gravity? Surely if gravity induced a lateral migration of the inhibitor in the extending zone then there should be enough residual inhibitor left after decapping to be deflected and give at least a transient curvature. This has never been observed and even if it had there would then be the additional problem of how this lateral gravity-induced movement could arise in the absence of sedimenting amyloplasts; clearly another gravity sensor would have to be invoked.

The strongest probability still seems to be a stimulated production of the inhibitor from some organelle put under mechanical stress by contact with the amyloplast. Again, if ABA is the inhibitor, there is supporting evidence for this suggestion from other phenomena. Thus one of the most dramatic results of water stress in plant cells is the marked rise in abscisic acid content (Wright, 1969; Wright and Hiron, 1969; Hiron and Wright, 1973), which is the result of a promoted synthesis from the mevalonic acid precursor (Milborrow and Robinson, 1973). The immediate effect of wilting on the cell is a loss of turgor and thus a drop in hydrostatic pressure of the cell contents. Pressures on internal membranes of one kind or another are bound to be changed as a result and a concomitant change in membrane permeability is the most logical outcome. Were abscisic acid contained in some membrane-bound compartment, such a change (presumably an increase) in permeability would result in release into the cell. But since increased *synthesis* is involved in these water-stress responses, and also that from externally-supplied precursors, some kind of enzyme activation must be the basis of the wilting response. If general pressure changes can induce these, presumably general, rises in ABA content, why should not *local* pressure differences caused by amyloplast contact (or lack of it) on some such membrane-bound enzymes cause *local* changes in ABA production?

So far there is little direct information on the effects of mechanical pressure on enzyme systems. Thus the activity of ATPase from potatoes seems to have an optimum at a hydrostatic pressure of about 5 atmospheres (Kuiper, 1971) but this involves a much larger force than is likely to be produced by the weight of an amyloplast on a membrane in the cell, which has been calculated to be of the order of 2 to 10 dyn cm^{-2} (Audus, 1962; Volkmann, 1974). However it is promising that pressures of this order will alter the activity of monomolecular films of acetylcholine esterase (Skou, 1959).

The next steps are clear. There must first be an unambiguous identification of the cap inhibitor. The site of its synthesis in the cap cell must be identified, a much more difficult matter. The modulation of its production by gravity must be checked and this is at present under way in my laboratory.

References

AASHEIM, T. and IVERSEN, T. H. (1972), Attempts towards a cytochemical detection of IAA-decarboxylating enzymes in cabbage roots. *In*: "Hormonal Regulation in Plant Growth and Development" (ed. Kaldeway, H. and Vardar, Y.) Verlag Chemie, Weinheim. pp. 171–173.

ABROL, B. K. and AUDUS, L. J. (1973). The effects of *N*-1-naphthylphthalamic acid and (2-chloroethyl)-phosphonic acid on the gravity-induced lateral transport of 2,4-dichlorophenoxyacetic acid. *J. exp. Bot.* **24**; 1224–1230.

ADDICOTT, F. T. and LYON, J. L. (1969). Physiology of abscisic acid and related substances. *A, Rev. Pl. Physiol.* **20**; 139–164.

AMLONG, H. U. (1939). Untersuchungen über Wirkung und Wanderung des Wuchsstoffes in der Wurzel. *Jb. wiss. Bot.* **88**; 421–469.

AUDUS, L. J. (1962). The mechanism of the perception of gravity by plants. *Symp. Soc. exp. Biol.* **16**; 197–226.

AUDUS, L. J. (1969). Geotropism. *In*: "Physiology of Plant Growth and Development" (ed. M. B. Wilkins) pp. 205–242. McGraw-Hill, London.

AUDUS, L. J. (1971). Linkage between detection and the mechanisms establishing differential growth factor concentrations. *In*: "Gravity and the Organism." (Eds. Solon A. Gordon and Melvin J. Cohen). pp. 137–151. Univ. Chicago Press.

AUDUS, L. J. and BROWNBRIDGE, M. E. (1957a). Studies on the geotropism of roots. I. Growth-rate distribution during response and the effects of applied auxins. *J. exp. Bot.* **8**; 105–124.

AUDUS, L. J. and BROWNBRIDGE, M. E. (1957b). Studies on the geotropism of roots. II. The effect of the auxin antagonist α (1-naphthylmethylsulphide)propionic acid (NMSP) and its interaction with applied auxin. *J. exp. Bot.* **8**; 235–249.

AUDUS, L. J. and THRESH, R. (1953). A method of plant growth substance assay for use in paper partition chromatography. *Physiologia Pl.* **6**; 451–465.

BARLOW, P. W. and GRUNDWAG, M. (1974). The development of amyloplasts in cells of the quiescent centre of *Zea* roots in response to removal of the root cap. *Z. PflPhysiol.* **73,** 56–64.

BAYER, M. (1961). Über die Aktivierung des Hemmstoffsystem von *Helianthus* durch kurtzfristige Belichtung. *Planta (Berl.)* **57**; 258–265.

BENNET-CLARK, T. A. and KEFFORD, N. P. (1953). Chromatography of the growth substance in plant extracts. *Nature, Lond.*, **171**; 645–647.

BENNET-CLARK, T. A., YOUNIS, A. F. and ESNAULT, R. (1959). Geotropic behaviour of roots. *J. exp. Bot.* **10**; 69–86.

BOUCK, G. B. (1963a). An examination of the effects of ultra-centrifugation on the organelles in living root tip cells. *Am. J. Bot.* **50**; 1046–1054.

BOUCK, G. B. (1963b). Stratification and subsequent behaviour of plant cell organelles. *J. Cell. Biol.* **18**; 441–457.

BOYSEN JENSEN, P. (1928). Die phototropischen Induktion in der Spitze der *Avena*-koleoptile. *Planta* (*Berl.*) **5**; 464–477.

BOYSEN JENSEN, P. (1933). Die Bedeutung des Wuchsstoffes für das Wachstum und die geotropische Krümmung der Wurzeln von *Vicia faba*. *Planta* (*Berl.*), **20**; 688–698.

BOYSEN JENSEN, P. (1936). Über die Verteilung des Wuchsstoffes in Keimstengeln und Wurzeln während der phototropischen und geotropischen Krümmung. *K. danske Vidensk. Selsk. Biol. Med.* **13**; pp.31.

BOWEN, M. R., WILKINS, M. B., CANE, A. R. and MCCORQUODALE, I. (1972). Auxin transport in roots VIII. The distribution of radioactivity in the tissues of *Zea* root segments. *Planta* (*Berl.*), **105**; 273–292.

BRAUNER, L. (1942). New experiments on the geo-electric effect in membranes. *Istanb. Univ. Fen. Fak. Mecm. Ser. B.* (*Sci. Nat.*) **7**; 46–102.

BRIDGES, I. G., HILLMAN, J. R. and WILKINS, M. B. (1973). Identification and localisation of auxin in primary roots of *Zea mays* by mass spectometry. *Planta* (*Berl.*) **115**; 189–192.

BURNETT, D., AUDUS, L. J. and ZINSMEISTER, H. D., (1965). Growth substances in the roots of *Vicia faba*. II. *Phytochemistry*, **4**; 891–904.

BURSTRÖM, H. (1960). Influence of iron and gibberellic acid on the light sensitivity of roots. *Physiologia Pl.* **13**; 597–615.

CHADWICK, A. V. and BURG, S. P. (1967). An explanation of the inhibition of root growth caused by indole-3-acetic acid. *Pl. Physiol.*, **42**; 415–20.

CHING, TE MAY, and FANG, S. C. (1958). The redistribution of radioactivity in geotropically stimulated plants pretreated with radioactive indoleacetic acid. *Physiologia Pl.* **11**; 722–727.

CIESIELSKI, T. (1872). Untersuchungen über die Abwärtskrümmung der Wurzel. *Beitr. Biol. Pfl.* **7**; 30.

CLOWES, F. A. L. and JUNIPER, B. E. (1968). "Plant Cells." Blackwell, Oxford and Edinburgh.

CORNFORTH, J. W., MILBORROW, B. V., RYBACK, G. and WAREING, P. F. (1965). Identity of sycamore "dormin" with abscisin II. *Nature, Lond.*, **205**; 1269–1270.

ELLIOTT, M. C. and GREENWOOD, M. S. (1974). Indol-3yl-acetic acid in roots of *Zea mays*. *Phytochemistry* **13**; 239–241.

GENKEL, P. A. (1960). The distribution of heteroauxin in the stems and roots of plants during geotropic bending. *Fiziol. Rastenii*, **7**(2); 167–172, (from *Biol. Abstr.* **36**(7), 21637, 1961).

GIBBONS, G. S. B. and WILKINS, M. B. (1970). Growth inhibitor production in root caps in relation to geotropic response. *Nature, Lond.*, **226**; 558–559.

GORDON, S. A. (1964). Gravity and plant development. Bases for experiment. *In*: "Proceedings of the 24th annual Biology Colloquium" (Ed. Gilfillan, F. A.) pp. 75–105. Oregon St. Univ. Press.

GREENWOOD, M. S., HILLMAN, J. R., SHAW, S. and WILKINS, M. B. (1973). Localisation and identification of auxin in roots of *Zea mays*. *Planta* (*Berl.*) **109**; 369–374.

GRIFFITHS, H. J. (1963). "Physiological and Cytological Studies of the Statolith Apparatus in Plants." Ph.D. Thesis, University of London, England.

GRIFFITHS, H. J. and AUDUS, L. J. (1964). Organelle distribution in the statocyte cells of the root tip of *Vicia faba* in relation to geotropic stimulation. *New Phytol.* **63**; 319–333.

HABERLANDT, G. (1900). Über die Perzeption des geotropischen Reizes. *Ber. dt. bot. Ges.* **18**; 261.

HABERLANDT, G. (1928). "Physiological Plant Anatomy." (Translated from the 4th edition by M. Drummond). pp. 597. MacMillan, London.

HAWKER, L. E. (1932a). Perception of gravity by roots of *Vicia faba*. *Nature, Lond.*, **129**; 364.

HAWKER, L. E. (1932b). Experiments on the perception of gravity by roots. *New Phytol.* **31**; 321–328.

HEJNOWICZ, Z. (1968). Studies on the inhibitory action of auxin in root growth. *Acta Soc. bot. pol.* **37**; 451–460.

HEMBERG, T. (1961). Biogenous Inhibitors. *In*: "Encycl. Pl. Physiol." (Ed. Rhuland, W.) **Vol. 14**; p. 1162–1184. Springer, Berlin, Göttingen, Heidelberg.

HERTZ, C. H. (1971). Bioelectric phenomena in graviperception. *In*: "Gravity and the Organism". (Eds Solon A. Gordon and Melvin J. Cohen) pp. 151–158. Univ. Chicago Press.

HIRON, R. W. P. and WRIGHT, S. T. C. (1973). The role of endogenous abscisic acid in the response of plants to stress. *J. exp. Bot.* **24**; 769–781.

HOWELL, R. W. (1954). The inhibiting effect of root tips on the elongation of excised *Pisum* epicotyls cultivated *in vitro*. *Pl. Physiol.*, **29**; 100–102.

IVERSEN, T.-H. (1969). Elimination of geotropic responsiveness in roots of cress (*Lepidium sativum*) by removal of statolith starch. *Physiologia Pl.* **22**; 1251–1262.

IVERSEN, T.-H., AASHEIM, T. and PEDERSEN, K. (1971). Transport and degradation of auxin in relation to geotropism in roots of *Phaseolus vulgaris*. *Physiologia Pl.* **25**; 417–424.

IVERSEN, T.-H. and LARSEN, P. (1973). Movement of amyloplasts in the statocytes of geotropically stimulated roots. The preinversion effect. *Physiologia Pl.* **28**; 172–181.

IVERSEN, T.-H., PEDERSEN, K. and LARSEN, P. (1968). Movement of amyloplasts in the root cap cells of geotropically sensitive roots. *Physiologia Pl.* **21**; 811–819.

JOHNSON, A. (1971). Investigations of the geotropic curvature of the *Avena* coleoptile I. The geotropic response curve. *Physiologia Pl.* **25**; 35–42.

JUNIPER, B. E. and BARLOW. P. W. (1969). The distribution of plasmodesmata in the root tip of maize. *Planta (Berl.)* **89**; 352–360.

JUNIPER, B. E. and FRENCH, A. (1973). The distribution and redistribution of endoplasmic reticulum (ER) in geoperceptive cells. *Planta (Berl.)* **109**; 211–224.

JUNIPER, B. E., GROVES, S., LANDAU-SCHACHAR, B. and AUDUS, L. J. (1966). Root cap and the perception of gravity. *Nature, Lond.*, **209**; 93–94.

KEITT, G. W. JR. (1960). Effect of certain growth substances on elongation and geotropic curvature of wheat roots. *Bot. Gaz.* **122**; 51–62.

KONINGS, H. (1965). On the indoleacetic acid converting enzyme of pea roots and its relation to geotropism, straight growth and cell wall properties. *Acta bot. neerl.* **13**; 566–622.

KONINGS, H. (1967). On the mechanism of the transverse distribution of auxin in geotropically exposed pea roots. *Acta bot. neerl.* **16**; 161–176.

KONINGS, H. (1968). The significance of the root cap for geotropism. *Acta bot. neerl.* **17**; 203–211.

KONINGS, H. (1969). The influence of acropetally transported indoleacetic acid on the geotropism of intact pea roots and its modification by 2,3,5-tri-iodobenzoic acid. *Acta bot. neerl.* **18**; 528–537.

KUIPER, P. J. C. (1971). ATPase; Sensitivity to hydrostatic pressure of a cold-labile form. *Biochim. biophys. Acta* **250**; 443–445.

KUNDU, K. K. and AUDUS, L. J. (1974). Root growth inhibitors from root cap and root meristem of *Zea mays* L. *J. exp. Bot.* **25**; 479–489.

LAKE, J. V. and SLACK, G. (1961). Dependence on light of geotropism in plant roots. *Nature, Lond.*, **191**; 300–301.

LAHIRI, A. N. (1959). "Studies in the Hormone Relations of Root Growth." Ph.D. Thesis, University of London, England.

LAHIRI, A. N. (1968). Distribution of endogenous growth regulators within the geotropically stimulated root tips. *Proc. natn. Inst. Sci. India, Part B (Biol. Sci.)* **34**; 21–26.

LARSEN, P. (1962a). Geotropism. An Introduction. *In*: "Encycl. Pl. Physiol." (Ed. Rhuland, W) **Vol. 17**(2), pp. 34–73. Springer, Berlin, Göttingen, Heidelberg.

LARSEN, P. (1962b). Orthogeotropism in roots. *In*: "Encycl. Pl. Physiol." (Ed. Rhuland, W.) **Vol. 17**(2), pp. 153–199. Springer, Berlin, Göttingen, Heidelberg.

Masuda, Y. (1962). Effect of light on a growth inhibitor in wheat roots. *Physiologia Pl.* **15**; 780–790.

McCready, C. C. (1966). Translocation of growth regulators. *A. Rev. Pl. Physiol.* **17**; 283–294.

McCready, C. C. (1968). The acropetal movement of auxin through segments excised from petioles of *Phaseolus vulgaris* L. *In*: "The Transport of Plant Hormones." (Ed. Vardar, Y.). pp. 108–129. North-Holland Publ. Co. Amsterdam.

Milborrow, B. V. (1974). Biosynthesis of abscisic acid by a cell-free system. *Phytochemistry* **13**; 131–136.

Milborrow, B. V. and Robinson, D. R. (1973). Factors affecting the biosynthesis of abscisic acid. *J. exp. Bot.* **24**; 537–548.

Most, B. H. R. (1962). "Studies on the Auxin Relationship of Geotropically Stimulated Roots". Ph.D. Thesis, University of London, England.

Němec, B. (1900) Über die Art der Wahrnehmung des Schwerkraftreizes bei den Pflanzen. *Jb. wiss. Bot.* **36**; 80.

Nougarède, A. and Pilet, P.-E. (1971). Action de l'acide gibbérellique (GA_3) sur l'ultrastructure des amyloplastes du statenchyme de la racine du *Lens culinaris* L. *C. r. hebd. Séanc. Acad. Sci. Paris. Ser. D.* **273**; 348–351.

Osborne, D. J. (1968). A theoretical model for polar auxin transport. *In*: "The Transport of Plant Hormones." (Ed. Vardar, Y.) pp. 97–107. North-Holland Publ. Co., Amsterdam.

Pickard, B. G. and Thimann, K. V. (1966). Geotropic response of wheat coleoptiles in absence of amyloplast starch. *J. gen. Physiol.* **49**; 1065–1086.

Pilet, P.-E. (1964). Auxin transport in roots. *Lens culinaris. Nature, Lond.*, **204**; 560–561.

Pilet, P.-E. (1967a). Effet de l'α-naphtylphtalamate sur le géotropisme et le catabolisme auxinique des racines de *Lens. C. r. hebd. Séanc. Acad. Sci. Paris.* **265**; 745–747.

Pilet, P.-E. (1967b). Action *in vivo* et *in vitro* de l'α-naphtylphtalamate sur la destruction *in vitro* de l'acide β-indolacétique. *C. r. hebd. Séanc. Acad. Sci. Paris.* **265**; 610–612.

Pilet, P.-E. (1971a). Root cap and geoperception. *Nature, New Biol.* **233**; 115–116.

Pilet, P.-E. (1971b). Role de l'apex radiculaire dans le croissance, le géotropisme et le transport des auxines. *Bull. Soc. bot. suisse.* **81**; 51–65.

Pilet, P.-E. (1972a). Root cap and root growth. *Planta (Berl.)* **106**; 169–171.

Pilet, P.-E. (1972b). Géotropisme et géoréaction racinaires. *Physiol. vég.* **10**; 347–367.

Pilet, P.-E. (1972c). ABA effects on growth in relation to auxin, RNA and ultrastructure. *In*: "Hormonal Regulation in Plant Growth and Development." (Eds. Kaldewey, H. and Vardar, Y.). pp. 297–315. Verlag Chemie, Weinheim.

Pilet, P.-E. (1973a). Georeaction of decapped roots. *Pl. Sci. Lett.* **1**; 137–140.

Pilet, P.-E. (1973b). Growth inhibitor from the root cap of *Zea mays. Planta (Berl.)* **111**; 275–278.

Pilet, P.-E. (1973c). Inhibiteur de croissance racinaire et énergie lumineuse. *C. r. hebd. Séanc. Acad. Sci. Paris.* **276**; 2529–2531.

Pilet, P.-E. and Nougarède, A. (1971). Action de l'acids gibbérellique sur la croissance et le géotropisme racinaires. *C. r. hedb. Séanc. Acad. Sci. Paris. Ser. D.* **272**; 418.

Rebeiz, C. A. and Castelfranco, P. A. (1973). Protochlorophyll and chlorophyll biosynthesis in cell-free systems from higher plants. *A. Rev. Pl. Physiol.* **24**; 129–172.

Scott, T. K. (1972). Auxins and Roots. *A. Rev. Pl. Physiol.* **23**; 235–258.

Scott, T. K. and Wilkins, M. B. (1969). Auxin transport in roots. IV. Effect of light on IAA movement and geotropic responsiveness in *Zea* roots. *Planta (Berl.)* **87**; 249–258.

Schachar, B. (1967). "The Root Cap and its Significance in Geoperception." Ph.D. Thesis University of London, England.

Shaw, S. and Wilkins, M. B. (1973). The source and lateral transport of growth inhibitors in geotropically stimulated roots of *Zea mays* and *Pisum sativum. Planta (Berl.)*, **109**; 11–26.

Shaw, S. and Wilkins, M. B. (1974). Auxin transport in roots X. Relative movement of radioactivity from IAA in the stele and cortex of *Zea* root segments. *J. exp. Bot.* **25**; 199–207.

SHELDRAKE, A. R. (1971). The occurrence and significance of auxin in the substrate of bryophytes. *New Phytol.* **70**; 519–526.

SHEN-MILLER, J. (1972). The golgi apparatus and geotropism. *In*: "Hormone Regulation of Plant Growth and Development." (Eds Kaldewey, H. and Vardar, Y.) pp. 365–376. Verlag Chemie, Weinheim.

SHEN-MILLER, J. and MILLER, C. (1972a). Distribution and activation of the golgi apparatus in geotropism. *Pl. Physiol.* **49**; 634–639.

SHEN-MILLER, J. and MILLER, C. (1972b). Intracellular distribution of mitochondria after geotropic stimulation of the oat coleoptile. *Pl. Physiol.* **50**; 51–54.

SIEVERS, A. and VOLKMANN, D. (1972). Verursacht differentieller Druck der Amyloplasten auf ein Komplexes Endomembran-system die Geoperzeption in Wurzeln? *Planta* (*Berl.*), **102**; 160–172.

SKOU, J. C. (1959). Studies on the influence of degree of unfolding and orientation of the side chains on the activity of a surface-spread enzyme. *Biochim. biophys. Acta.* **31**; 1–10.

TAYLOR, H. F. (1969). "Studies on the Regulation of Plant Growth with Chemicals." Ph.D. Thesis, University of London, England.

TAYLOR, H. F. and BURDEN, R. S. (1970). Xanthoxin, a new naturally occurring growth inhibitor. *Nature, Lond.*, **227**; 302–304.

TEPFER, D. A. and BONNETT, H. T. (1972). The role of phytochrome in the geotropic behaviour of roots of *Convolvulus arvensis*. *Planta* (*Berl.*), **106**; 311–324.

TUCKER, D. J. and MANSFIELD, T. A. (1971). A simple bioassay for detecting "Antitranspirant" Activity of naturally occurring compounds such as Abscisic Acid. *Planta* (*Berl.*), **98**; 157–163.

VINCE, D. and GRILL, R. (1966). The photoreceptors involved in anthocyanin synthesis. *Photochem. Photobiol.* **5**; 407–411.

VOLKMANN, D. (1974). Amyloplasten und Endomembranen; Das Geoperzeptionssystem der Primärwurzel. *Protoplasma* **79**; 159–183.

WAGNER, E. and MOHR, H. (1966). Kinetische Studien zur Interpretation der Wirkung von Sukzedanbestrahlungen mit Helbrot und Dunkelrot bei der Photomorphogenese (Anthocyaninsynthese bei *Sinapis alba* L.) *Planta* (*Berl.*), **70**; 34–41.

WILKINS, H., LARQUÉ-SAAVEDRA, A. and WAIN, R. L. (1974). Control of *Zea* root elongation by light and the action of 3,5-di-iodo-4-hydroxy-benzoic acid. *Nature, Lond.*, **248**; 449–450.

WILKINS, M. B. and WOODCOCK, A. E. R. (1965). Origin of the geo-electric effect in plants. *Nature, Lond.*, **208**; 990–992.

WOODCOCK, A. E. R. and HERTZ, C. H. (1972). The geo-electric effect in plant shoots V. A discussion. *J. exp. Bot.* **23**; 953–957.

WOODRUFFE, P., ANTHONY, A. and STREET, H. E. (1970). Ether-soluble indoles from seedling roots of wheat. *New Phytol.* **69**; 51–63.

WRIGHT, S. T. C. (1969). An increase in the "inhibitor-β" content of detached wheat leaves following a period of wilting. *Planta* (*Berl.*), **86**; 10–20.

WRIGHT, S. T. C. and HIRON, R. W. P. (1969). (+)-Abscisic acid, the growth inhibitor induced in detached leaves by a period of wilting. *Nature, Lond.*, **224**; 719–720.

YEOMANS, L. M. and AUDUS, L. J. (1964). Auxin transport in roots; *Vicia faba*. *Nature, Lond.*, **204**; 559–561.

YOUNIS, A. F. (1954). Experiments on the growth and geotropism of roots. *J. exp. Bot.* **5**; 357–372.

ZINKE, H. (1968). Versuche zur Analyse der positiven und der negativen geotropischen Reaktionen von Keimwurzeln. *Planta* (*Berl.*), **82**; 50–72.

Chapter 17

Cytokinin Production by Roots as a Factor in the Control of Plant Growth

K. G. M. SKENE

CSIRO Division of Horticultural Research, Adelaide, South Australia.

I. Introduction

In recent years a substantial body of evidence has accumulated in support of the concept that growth and development of the aerial parts of plants are dependent on hormones synthesized by the root systems. The present paper considers evidence for the participation of root-derived cytokinins in this hormonal control. Although other hormones such as gibberellins (Phillips and Jones, 1964; Carr *et al.*, 1964) and abscisic acid (Lenton *et al.*, 1968) also may be produced by roots, the discussion is restricted mainly to cytokinins. The scope of the paper does not include a detailed discussion of the interactions between cytokinins and other hormones in the shoot, or of the specialized role that bacteria (Phillips and Torrey, 1970) and mycorrhizal fungi (Miller, 1967) might play in supplying cytokinins to the plant.

A. Effects of Roots on Shoot Development

The concept of root-derived shoot factors originates from observations of Went (1938) on the growth of rootless pea seedlings and from Chibnall's studies (1939, 1954) on the protein metabolism of detached leaves. Went first proposed in 1938 that a factor necessary for shoot growth originated in the root. At about the same time Chibnall (1939) suggested that the regulation of protein levels in leaves might be under the influence of a hormone-like factor from the roots. Chibnall's views were reinforced when he eventually showed that the development of adventitious roots on the petioles of detached leaves did in fact prevent the rapid senescence that usually followed excision (Chibnall, 1954). Loss of leaf protein was arrested, and under appropriate conditions rooted leaves remained green and in a healthy condition for many weeks.

Mothes and Englebrecht (1956) subsequently reported that the formation of roots on detached leaves actually reversed the process of ageing. Protein levels increased and the leaf began to grow again. In related experiments on rejuvenation it was observed that if a tobacco plant were reduced to a single senescing leaf, thus putting the whole root system at the disposal of this leaf, it regreened quickly and resumed growth (Mothes and Baudisch, 1958). On the other hand, regular removal of lateral roots from pea seedlings accelerated senescence of older leaves and reduced their rate of $^{14}CO_2$ fixation (McDavid *et al.*, 1973). Mothes' review (1960) gives a detailed account of the early experiments on rejuvenation.

The influence of roots on shoot growth extends beyond the regulation of protein metabolism in leaves. Continued growth of cultured apices of *Carex flacca* is associated with the development of root primordia on the explant (Smith, 1968). Whilst roots promote, but are not essential for, initiation of inflorescences on cultured explants of *Carex*, their presence appears to be necessary for normal branching of the inflorescence (Smith, 1969). Growth factors from the roots have also been invoked to explain the adverse effects of root pruning on shoot elongation and leaf number of grape vines grown in nutrient culture solutions (Buttrose and Mullins, 1968). Similar conclusions on the role of root factors in the grape vine have been drawn from studies on root formation in relation to the retention of inflorescences by single node cuttings (Mullins, 1967, 1968), and from observations on the stimulatory effects of elevated root temperature on shoot growth and fruit set (Woodham and Alexander, 1966).

The control of lateral shoot growth may also be under the influence of root-produced growth factors. Wareing and Nasr (1961) observed that the outgrowth of lateral buds on variously oriented woody shoots was dependent on their proximity to the roots. Later work with *Salix viminalis* (Smith and Wareing, 1964) led to the proposal that roots synthesize substances necessary for shoot growth. This conclusion was based on the fact that the bending of *Salix* stems into loops inhibited lateral shoot growth from the apical region, and the inhibition could be relieved by the induction of roots at the base of the loop. The stimulus produced by the roots could be transmitted past a bark girdle, and did not appear to be merely nutritive.

The influence of roots also appears to be manifested in the form of the outgrowing lateral shoot. Single lateral buds on cuttings of *Solanum andigena* will develop either as leafy orthotropic shoots or as diageotropic stolons, depending on the presence or absence of roots (Kumar and Wareing, 1972); cultured cotyledonary buds of *Scrofularia arguta* develop into vegetative shoots in the presence of "non-absorbing" roots, whereas in their absence flowers are produced (Miginiac, 1971).

B. Cytokinins and Root Removal

There is good evidence to implicate cytokinins in the phenomena described above. Cytokinins, first defined by Skoog *et al.*, (1965) as compounds that promote cell division in cultured plant cells, are now known to influence a wide range of physiological and biochemical processes. As well as being involved in the control of cell division and cell enlargement, exogenous cytokinins affect the activities of enzymes and the translocation of metabolites. They modify the plant's response to stress, and delay the senescence of detached organs; they stimulate the growth of lateral buds, the development of inflorescences and the setting of fruits. These and other properties of cytokinins are thoroughly covered by several reviews that have appeared over the last few years (Letham, 1967; Skoog and Armstrong, 1970; Kende, 1971; Mothes, 1973; Hall, 1973).

The common synthetic cytokinins kinetin (6-furfurylaminopurine) and BAP (6-benzylaminopurine), and naturally-occurring cytokinins such as zeatin (6-(4-hydroxy-3-methylbut-*trans*-2-enyl)aminopurine) are N^6-(substituted) adenine derivatives. Zeatin together with its riboside and ribotide were originally identified in extracts of immature corn kernels (Miller, 1961; Letham *et al.*, 1964; Miller, 1965; Letham, 1966a, b, c). These or closely related compounds have subsequently been detected in a diverse range of plant genera; cytokinins are clearly ubiquitous plant hormones (Kende, 1971).

The initial observation by Richmond and Lang (1957) that kinetin retarded the senescence of detached leaves in a manner similar to Chibnall's postulated root factor has now been followed by numerous examples in which the effects of roots on varied aspects of shoot behaviour are duplicated by exogenous cytokinins.

Synthetic cytokinins stimulate the growth of isolated apple shoots in culture (Jones, 1967), and simulate the effects of roots in promoting normal branching of inflorescences on cultured explants of *Carex flacca* (Smith, 1969). Cytokinins, like roots, favour the production of vegetative rather than floral shoots by cultured cotyledonary buds of *Scrofularia arguta* (Miginiac, 1971). Cytokinins applied to unrooted single node cuttings of grape vines promote the retention and development of inflorescences (Mullins, 1967, 1968). In the absence of roots, these inflorescences fail to develop and eventually atrophy.

Removal of the roots of pea seedlings causes a reduction in stem growth. Although Kende and Sitton (1967) were unable to counteract the inhibition of stem growth with exogenous cytokinin, or with combinations of cytokinin and gibberellin, the results of Holm and Key (1969) with rootless soybean seedlings do indicate a role of cytokinins in controlling stem growth. Combinations of cytokinin and gibberellin restored growth of the apical section of the hypocotyl, whereas only gibberellin was needed to restore growth of the lower or elongating section. Whilst neither hormone affected DNA synthesis in the elongating section, DNA synthesis in the apical section could be restored by cytokinin alone.

The formation in rye seedlings of the photosynthetic enzymes carboxydismutase and NADP-dependent glyceraldehydephosphate dehydrogenase is strongly reduced by excision of the roots early in their development (Feierabend, 1969). Feeding kinetin to rootless seedlings restored a high rate of enzyme formation. Similarly, the accelerated senescence and reduced capacity for $^{14}CO_2$ fixation of older leaves of pea seedlings following regular removal of lateral roots could be largely reversed by applications of synthetic cytokinins to the shoots (McDavid *et al.*, 1973).

The effects of exogenous hormones in stimulating lateral bud growth on rootless cuttings of *Solanum andigena* apparently are obscured by residual root effects (Woolley and Wareing, 1972b). Woolley and Wareing prepared cuttings in which residual root effects had been eliminated by treatments presumed to exhaust the stem of stored cytokinins. In such pre-treated rootless cuttings, the outgrowth of lateral buds had an absolute dependency on exogenous hormones. Cytokinins, like roots, promoted the lateral bud to grow as a leafy shoot; gibberellins favoured development as a stolon. These workers also concluded that buds released from apical dominance do not synthesize their own cytokinin. They appear to require a continued external supply.

II. Occurrence of Cytokinins in Xylem Sap

The first evidence for the movement of cytokinin-like substances from roots to shoots was provided by Kulaeva's (1962) discovery that crude xylem exudate from decapitated tobacco plants had a weak effect in delaying the yellowing of isolated leaves from the same species. Soon after, by the use of refined chlorophyll retention tests (Loeffler and van Overbeek, 1964; Kende, 1964) or the more selective cell division assays (Kende, 1964, 1965; Nitsch and Nitsch 1965a) little doubt remained that xylem sap contained cytokinins. Substances with cytokinin activity have now been detected in xylem sap from a wide range of species, both perennial and annual, monocotyledenous and dicotyledenous (Table I).

Xylem sap has been collected by a variety of methods. It includes sap sucked under reduced pressure from the xylem of isolated stems (see Bollard, 1960) as

TABLE I. Occurrence of cytokinins in xylem sap

Plant	Cytokinin[a]	Reference
Acer saccharum	4	Nitsch and Nitsch (1965b)
Acer pseudoplatanus	1,2,3,4	Reid and Burrows(1968)
	2[b]	Horgan *et al.* (1973a)
Betula pubescens	1,2,3	Reid and Burrows (1968)
Coffea arabica	2,4	Browning (1973)
Helianthus annuus	1,2,3	Kende (1964, 1965); Kende and Sitton (1967)
	1	Klämbt (1968)
Impatiens glandulifera	1,2,3	Carr and Burrows (1966)
Lycoperiscon esculentum	4	Tal *et al.* (1970)
Lupinus angustifolius	2,3	Carr and Burrows (1966)
Malus sylvestris	4	Luckwill and Whyte (1968)
	2,3,	Jones (1973)
Musa sapientum	4	Vaadia and Itai (1969)
Nicotiana rustica	4	Kulaeva (1962); Itai and Vaadia (1971)
Oryza sativa	1,2,3	Yoshida *et al.* (1971)
Perilla frutescens	4	Beever and Woolhouse (1973)
Phaseolus vulgaris	1 and/or 2	Engelbrecht (1972)
Pisum arvense	1,2,3	Carr and Burrows (1966)
Populus x robusta	1,2	Hewett and Wareing (1973a)
Vitis vinifera	4	Loeffler and van Overbeek (1964); Nitsch and Nitsch (1965b)
	1,2,3	Skene (1972a, 1972b)
Vitis hybrids	1,2	Skene and Antcliff (1972)
Xanthium pensylvanicum	1,2,3	Carr and Burrows (1966)
X. strumarium	1 and/or 2,3,4	van Staden and Wareing (1972)
Zea mays	4	Andreenko *et al.* (1964)

[a] 1,2,3 = substances with chromatographic properties resembling zeatin, zeatin riboside and zeatin ribotide, respectively. 4 = insufficient information available for tentative identification.
[b] Positive identification by gas chromatography-mass spectrometry.

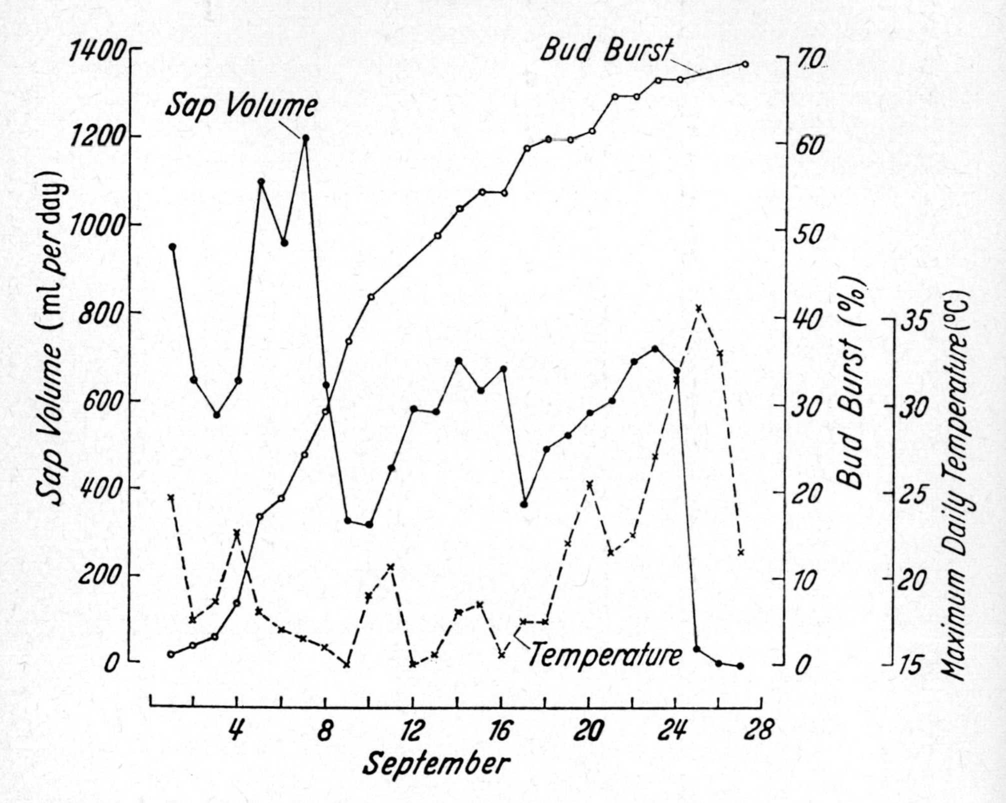

FIG. 1. Volumes of bleeding sap from fieldgrown sultana grape vines in the Southern Hemisphere. Percentage bud burst and maximum temperatures each day also are given. (From Skene, 1967; by permission of Springer-Verlag, Berlin, Heidelberg and New York).

well as the sap that bleeds (or exudes) from the xylem as a consequence of the plant's root pressure. Bleeding generally is induced by removal of the shoot (Kende, 1964), but in woody species sap also exudes from pruning cuts (Skene, 1967) (Fig. 1) or from holes bored in the trunk (Skene and Antcliff, 1972) as the plants emerge from winter rest. In all cases, if care is taken, direct contamination from the phloem is negligible (Bollard, 1960; Kende, 1965).

Cell division assays specific for cytokinins (see Letham, 1967) have been the main methods for detecting activity in sap. Tentative identifications have primarily been based on chromatographic behaviour, and sometimes physical and biochemical properties. Whilst chromatographic evidence alone is insufficient for conclusive identification, it does seem that cytokinins in xylem sap resemble the common cytokinins detected in other plant parts. Recently *trans*-zeatin riboside was positively identified in sycamore sap by combined gas liquid chromatography- mass spectrometry (Horgan *et al.*, 1973a); by less definitive methods substances suggestive of zeatin riboside, the free base and also the nucleotide have all been detected to varying extents in xylem sap. For instance, bleeding sap of grape vines grown in nutrient culture solutions contains three fractions active in the soybean callus assay that are suggestive of zeatin, its ribonucleoside and ribonucleotide (Skene, 1972a) (Fig. 2).

The qualitative pattern of activity in vine sap is not constant. The proportion of the presumed cytokinin nucleotide increases with time after decapitation (see Skene, 1972a, Table I), and its actual presence seems to depend on root temperature (Skene and Kerridge, 1967), and the nature of the medium supporting root growth. It is rarely detected in bleeding sap from vines growing in soil or other solid media (Skene, 1972a). Plant age also may affect the types of cytokinin detected in xylem sap, as Woolley and Wareing (1972c) obtained some evidence that the content of water-soluble cytokinins in sap from *Solanum andigena* increased with plant age, with a corresponding decrease in butanol-soluble cytokinins.

In contrast to bleeding sap of grape vines, no free base was detected in xylem sap collected under vacuum (Fig. 3), the riboside being the principal component of sap from stems supporting active growth (Skene, 1972b). Substances resembling zeatin riboside also appear to be the main cytokinins in vacuum-extracted sap of poplar (Hewett and Wareing, 1973a) and apple (Jones, 1973). In the case of apple, activity was qualitatively similar to that in bleeding sap; the free base was absent from both samples.

Hewett and Wareing (1973a) suggest that the riboside is the usual form in which cytokinins are transported in the xylem of woody species and this may also be the case for certain other species. After supplying (^{3}H)zeatin to the roots of radish seedlings, a range of metabolites eventually was detected throughout the plant, but zeatin riboside accounted for most of the radioactivity recovered from bleeding sap (Gordon *et al.*, 1974). Nevertheless, with our limited knowledge it would be unwise at this stage to attach undue significance to the particular forms in which cytokinins have been detected in sap. Nucleotides are often present (Table I) and it is possible that partial or complete hydrolysis of the more complex forms occurs during purification of extracts (e.g. Kende, 1965; Tegley *et al.*, 1971), or even during sap collection. The main point is that xylem sap contains cytokinins related to zeatin. In general, levels of activity are quite high, suggesting that substantial quantities of cytokinins pass in the transpiration stream to the above-ground parts of the plant each day.

III. Origin of Cytokinins in Xylem Sap

Are the cytokinins in xylem sap actually synthesized by roots, or are they merely recirculating from the root system after synthesis elsewhere in the plant? The latter alternative cannot be an exclusive explanation of the observed effects of

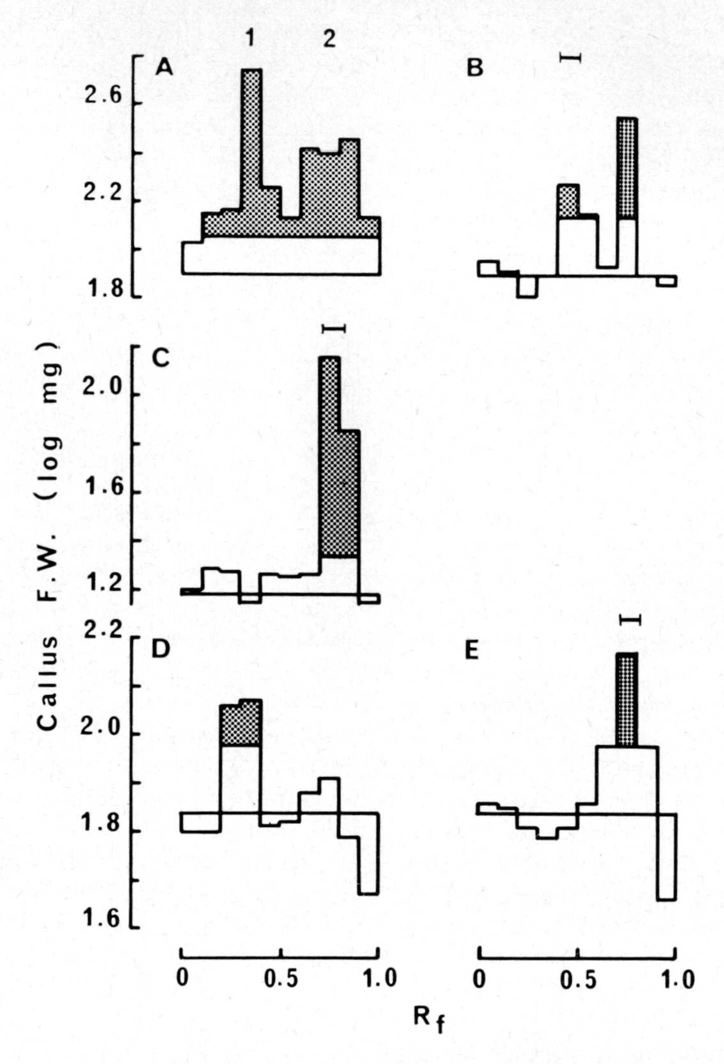

FIG. 2. Response of soybean callus to chromatogrammed extracts of bleeding sap from grape vines grown in nutrient culture solutions. (A) Unfractionated alcohol extracts developed in n-butanol/acetic acid/water (4:1:1); (B) n-Butanol fraction developed in 0·03 M borate at pH 8·4; (C), (D) and (E) Fractions separated from alcohol extract with barium acetate and chromatogrammed in sec. butanol/acetic acid/water (70:2:28). (C) is from the supernatant, (D) from the barium precipitate and (E) from the barium precipitate after alkaline phosphatase treatment. Significant responses (P = 0·05) to the extracts are indicated by shading of the histograms. Horizontal bars above the figures show the mobility of synthetic zeatin. (From Skene, 1972a; by permission of Springer-Verlag, Berlin, Heidelberg and New York.)

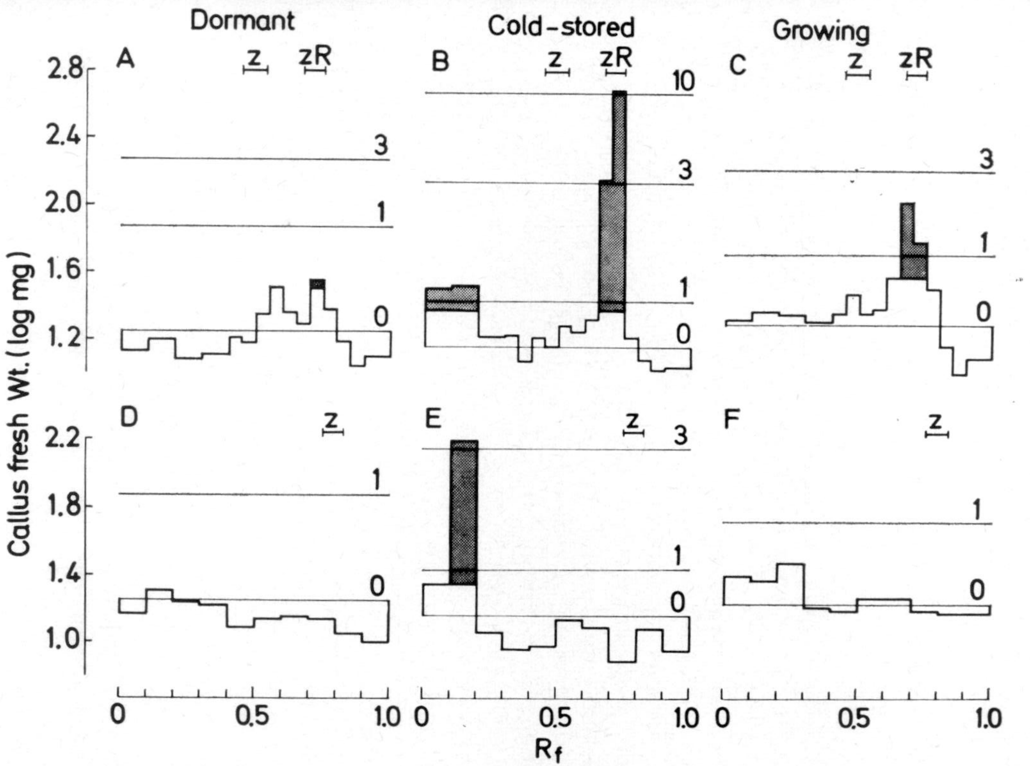

FIG. 3. Response of soybean callus to chromatogrammed extracts of xylem sap collected under reduced pressure from dormant (A, D), cold-stored (B, E) and growing (C, F) (grapevine canes. (A–C) n-butanol-soluble fractions, chromatogrammed in borate buffer D–F) water-soluble fractions, chromatogrammed in sec. butanol/acetic acid/water. Z and ZR :mobilities of zeatin and zeatin riboside, respectively. Significant responses ($P = 0{\cdot}05$) are indicated by shading the histograms. Horizontal lines show the response to kinetin standards (μg/1). (From Skene, 1972b; by permission of Springer-Verlag, Berlin, Heidelberg and New York.)

roots on leaf or shoot behaviour, although recirculation of other hormones sometimes may be involved (e.g. Phillips, 1964a, b). Recirculation of cytokinins is not supported by work with grape vines, where Skene (1972c) failed to detect a rise in the level of extractable cytokinins in shoots distal to a bark girdle. However, one cannot overlook the possibility that precursors or biologically-inactive derivatives of cytokinins from shoots are modified by roots before being returned to the shoot. A shoot–root–shoot recycling scheme has been proposed for gibberellin biosynthesis in *Phaseolus coccineus* seedlings (Crozier and Reid, 1972).

Kende's (1965) observation that the quantity of cytokinins in xylem sap of decapitated sunflowers did not decrease over a four-day period is in agreement with the concept that cytokinins in sap are synthesized in the roots. However, other workers (Carr and Reid, 1968; Skene, 1970; Wagner and Michael, 1971) report an eventual decline in activity over a similar period. It is difficult to interpret these results without information on (i) the extent to which cytokinins or their precursors are stored in the root, and (ii) how rapidly the nutritional status of the root is affected by the absence of the shoot.

More direct evidence for the production of cytokinins by roots comes from Engelbrecht's (1972) time-course studies on cytokinins in rooting leaf-cuttings of *Phaseolus vulgaris* (Fig. 4). Initially no cytokinins could be detected. Activity first appeared in a combined extract of roots and petioles shortly after the formation of roots, and did not appear in extracts of the leaf blades until several days later. One fraction, with the properties of a free base or riboside (Fraction Z), increased in "roots + petioles" during the first few days of root formation, but soon reached a constant level in these organs, and then began to increase in the leaf. Eventually a constant level was reached in the leaf as well. On the other hand, a second fraction (Fraction N), that Engelbracht suggested was a storage form of Fraction Z, appeared in the leaf at the same time as Fraction Z and showed a steady increase thereafter. It is significant that the appearance of cytokinins in the leaf blade coincided with the period when rooted leaves began to turn dark green and increased their capacity for protein synthesis (Parthier, 1964).

Not all experiments lead to similar conclusions. Since the formation of a cambium in cultured radish roots is dependent on an exogenous supply of auxin and cytokinin, Torrey and Loomis (1967) concluded that the shoot system of intact radish plants is the normal source of these hormones. Furthermore, Radin and Loomis (1971) found increasing amounts of three cytokinin fractions in developing roots of radishes. Two of these were chromatographically similar to zeatin ribotide and to zeatin or its riboside; the third was not identified but it did not seem to be a derivative of zeatin. The results of time-course and anatomical studies suggest that zeatin and its derivatives originate in the shoots of radish and regulate cambial activity, whereas the unidentified cytokinin may be

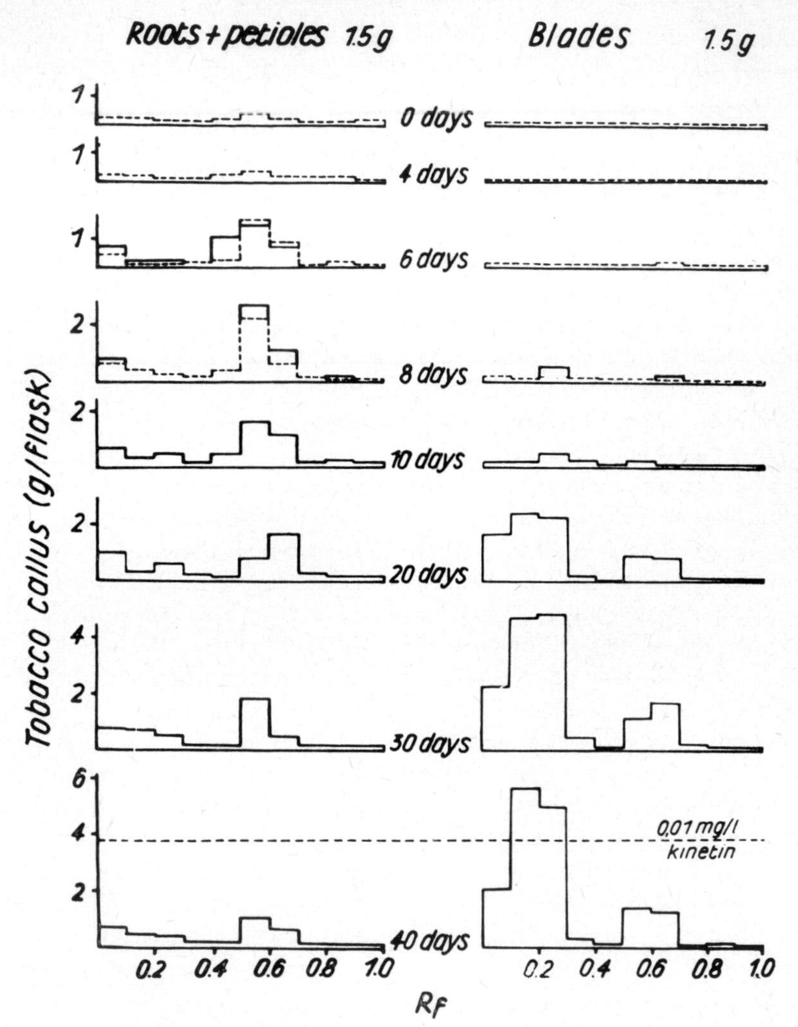

FIG. 4. Histograms representing cytokinin levels in *Phaseolus* leaf-cuttings during their development, using 1·5 g equivalents for each chromatogram. – – – – – – Cuttings in tap water, ——— in nutrient solution. Each column represents the average of 4 bioassays with a total of 16 pieces of tobacco callus. (From Engelbrecht, 1972; by permission of the author and Fischer-Verlag, Jena.)

synthesized in the root and transported to the shoot. It remains to be seen whether this interpretation of the results can be extended to other species. Radishes and similar plants with fleshy axes derived from both hypocotyl and root may not be typical examples.

Substances with cytokinin activity have also been detected in extracts of roots of sunflowers (Weiss and Vaadia, 1965), grape vines (Skene, 1970), peas

(Short and Torrey, 1972a), rice (Yoshida and Oritani, 1972) and *Solanum andigena* (Woolley and Wareing, 1972c). Most of the activity appears to be confined to the meristematic regions of the roots (Weiss and Vaadia, 1965; Short and Torrey, 1972a).

Chromatographic evidence suggests the presence of zeatin, its riboside and ribotide in roots. *Trans*-zeatin riboside has been positively identified in chicory roots (Bui-Dang-Ha and Nitsch, 1970). The behaviour of a cytokinin extracted from rice roots is suggestive of a glucoside of zeatin (Yoshida and Oritani, 1972). Its chromatographic properties are similar to those of a presumed cytokinin glucoside in extracts of mature poplar leaves (Hewett and Wareing, 1973b, d) as well as to those of an unidentified cytokinin in extracts of corn kernels (Letham, 1968), apple fruits (Letham and Williams, 1969), seeds of pumpkin (Gupta and Maheshwari, 1970) and watermelon (Prakash and Maheshwari, 1970), tips of pea roots (Short and Torrey, 1972a), pea root callus tissue (Short and Torrey, 1972b) and cultured sycamore cells (MacKenzie and Street, 1972). A glycoside of zeatin has recently been detected in corn kernels (Letham, 1973). Cytokinin glucosides have also been identified as major metabolites after supplying labelled 6-benzylaminopurine to potato tuber slices (Deleuze *et al.*, 1972) and various other tissues (Fox *et al.*, 1973). Glucosides also appear after application of (^{3}H) zeatin to seedlings of rootless (Parker *et al.*, 1972; Parker and Letham, 1973) and intact (Gordon *et al.*, 1974) radish and to seedlings of corn (Parker and Letham, 1974).

The physiological significance of naturally-occurring cytokinin–glucose complexes in roots and other tissues is yet to be elucidated, but it seems more than likely that they are storage compounds. Cytokinin glucosides possess only low activity in cell division tests (Yoshida and Oritani, 1972), and are remarkably stable, persisting for long periods in tissues (Deleuze *et al.*, 1972; Fox *et al.*, 1973; Parker and Letham, 1973). Unlike 6-benzyladenosine, 6-benzylamino-9-glucosylpurine is resistant to the action of nucleoside phosphorylases, and undergoes very little metabolic transformation in etiolated *Cicer* seedlings (Guern *et al.*, 1968).

Available evidence strongly favours the meristematic regions of the root as sites of cytokinin synthesis. Goldacre's (1959) observations on the preferential formation of lateral roots adjacent to preformed root initials in isolated flax roots led him to propose that root meristems synthesized cytokinins, and that cytokinin production may be a normal accompaniment of cell division. This view is supported by correlations between metabolic activity of root apices and cytokinin levels in bleeding sap of sunflowers (Burrows and Carr, 1969). Further support is provided by measurements of cytokinin levels in root tips.

Weiss and Vaadia (1965) found high levels of two cytokinin fractions, chromatographically similar to those of xylem sap, in extracts from the root apex of sunflowers. Very little activity was detected in physiologically older tissue.

Short and Torrey (1972a) examined cytokinins in both the free form and as constituents of transfer RNA in serial segments of young seedling roots of pea. Interest in tRNA centres around its hypothetical role as a source of cytokinins during nucleic acid catabolism (see Hall, 1973, especially Figure 4). As in the case of sunflowers, the cytokinin content of the meristematic region of the root apex was greater than in the more mature regions. A substance resembling zeatin was the most active free cytokinin detected, with evidence of its riboside, a ribotide and an unidentified cytokinin. No attempt was made to identify the cytokinin released after acid hydrolysis of tRNA; it too was restricted to the apex, but in much smaller quantities than the free forms. Short and Torrey suggest the quiescent zone and surrounding meristematic tissues as centres of cytokinin production in the root tips of pea seedlings. (For further discussion of cytokinins in the root apex see L. Feldman, pp. 55–72 this volume). Furthermore, unless there is an unusually rapid turnover of cytokinin-containing tRNA species in the root apex, the preponderance of free cytokinin over that in tRNA suggests that cytokinins in the root tip are produced by biosynthetic pathways separate from the catabolism of tRNA. Hall (1973) points out that 1-methyladenine, a modified component of tRNA in the starfish, is formed by pathways independent of tRNA (Shirai *et al.*, 1972).

Sheldrake and Northcote (1968) are of the opinion that cytokinins in xylem sap might be derived from the breakdown of nucleic acids during the autolysis of differentiating xylem elements. Insufficient information is available to decide whether a significant proportion of total cytokinin activity in sap could arise in this way.

Whilst it is probable that cytokinins are synthesized by root tips, the possibility cannot be entirely discounted that cytokinins are transported to the meristematic regions of the root, where they accumulate. There are similar difficulties in interpreting the high levels of cytokinins in other meristems, such as cambial tissue (Nitsch and Nitsch, 1965a) and young fruits and seeds (Zwar *et al.*, 1963; Letham, 1966a; Letham and Williams, 1969). However, indirect evidence suggests that the seed does produce a significant proportion of its own cytokinin (e.g. Blumenfeld and Gazit, 1971) and it is quite clear that a range of cultured tissues are capable of synthesizing cytokinins. Cytokinin activity has been detected in a cytokinin-autotrophic strain of soybean callus (Miura and Miller, 1969), in suspension cultures of cells from sycamore cambium (MacKenzie and Street, 1972), in pea root callus tissue (Short and Torrey, 1972b), and in cotyledon callus from the avocado (Blumenfeld and Gazit, 1971). Labelled compounds with the chromatographic properties of zeatin and N^6-($\triangle^2$-isopentenyl) adenine (i^6Ade) were isolated from a cytokinin-autotrophic tobacco callus after supplying (^{14}C) adenine (Einset and Skoog, 1973). The riboside of i^6Ade is regarded as a key compound in the biosynthesis of zeatin and its derivatives (Hall, 1973); it is converted to *trans*-ribosyl zeatin after incubation with corn endosperm or slices of the kernel (Miura and Hall, 1973). Miura and Hall suggest that the latent capacity to synthesize cytokinins is carried by all tissues, and the site of synthesis might change during the course of development. It is even possible that circumstances might alter the relative contributions of various potential sites of synthesis.

The relative importance of sources of cytokinins other than the root during budburst of woody species has been discussed by Skene (1972b). One year-old

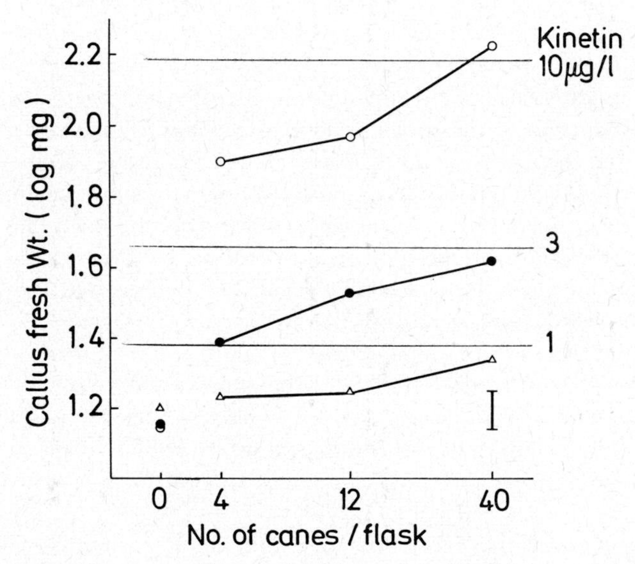

FIG. 5. Response of soybean callus to serial dilutions of xylem sap extracted under reduced pressure from dormant (△), cold-stored (○) and growing (●) grape vine canes. Assay flasks contained sap extracted from the equivalent of 4, 12 or 40 canes. Vertical bar is the LSD (P = 0·05) between treatments for each dilution. Horizontal lines show the responses to kinetin standards (μg/1). (From Skene, 1972b; by permission of Springer-Verlag, Berlin, Heidelberg and New York.)

woody stems of grape vine, collected during winter rest and stored for six months at 1°C accumulated high concentrations of cytokinin in their xylem sap, independently of a root system (Fig. 5). Essentially similar results have been obtained for sap from artificially-chilled poplar twigs as well as for twigs remaining on the plant during winter (Hewett and Wareing, 1973a). Whether these cytokinins are synthesized by the cambium or are released from an inactive storage form is not known, but it is clear that xylem sap expressed under vacuum need not necessarily be derived directly from roots. The results raise the possibility that stems of woody species may be significant sources of cytokinins, especially during periods when contributions from the roots are low. Such a situation might be expected to arise towards the end of winter rest prior to budburst.

IV. Cytokinins in Sap and Roots: Relationship to Plant Development, and Factors Affecting Production

Studies on the cytokinin content of xylem sap in relation to developmental aspects of plant behaviour are comparatively few in number. The results that are available provide further insights into the varied ways in which cytokinin production by roots may influence overall development of the plant. In particular, synthesis of cytokinins by roots seems to be influenced by factors of the environment that either directly or indirectly affect root physiology. These factors are discussed below.

A. pH; Nutrition

Low pH of the root medium adversely affects the growth of many plants. Cytokinins could be detected in xylem sap of maize plants grown in medium of pH 7·0 but not pH 4·0 (Andreenko *et al.*, 1964). The nutritional status of the plant also affects cytokinin production by roots. Reduced quantities of cytokinins were found in xylem sap and root extracts of sunflowers (Wagner and Michael, 1969, 1971) and in root extracts of *Solanum andigena* (Woolley and Wareing, 1972c) when grown in media containing low levels of nitrogen. The changes in sunflowers could be correlated with growth of the root tips. Wagner and Michael cautiously suggested that cytokinin synthesis at the tip might depend on protein synthesis rather than cell division. Colchicine inhibited mitosis, yet had little effect on cytokinin content, whereas chloramphenicol and cycloheximide affected growth of the roots and their cytokinin content.

B. Root Temperature

Elevated root temperatures over a three-day period increased the synthesis of protein nitrogen in leaves of squash without affecting root growth (Vinokur, 1963). Shoot growth and fruit set of sultana vines were greater at root temperatures of 30°C than 20°C (Woodham and Alexander, 1966). Skene and Kerridge (1967) found qualitative differences in the cytokinins of xylem sap from sultana vines grown at root temperatures of 30°C and 20°C, a cytokinin nucleotide being absent from the 30°C samples. Although these workers doubted whether temperature had affected cytokinin production per unit weight of roots, the results of one of their experiments shows an increased export of cytokinins from the roots of plants grown at the higher temperature. This is associated with a larger root system and a higher bleeding rate.

C. Growth Retardant

The growth retardant CCC [(2-chloroethyl)trimethylammonium chloride] is regarded as an inhibitor of gibberellin biosynthesis in many plants (see Lang, 1970). When applied to the roots of grape vines, the cytokinin content of the sap was increased (Skene, 1968, 1970). Skene (1970) suggested that CCC directly affected cytokinin synthesis at the root tip, and Woolley and Wareing (1972c) proposed that the results may provide an example of a feedback regulatory mechanism between cytokinins and gibberellins. It is interesting to speculate whether the elevated cytokinin levels have any bearing on the effects of CCC on fruit set in grapes (Coombe, 1965, 1967); synthetic cytokinins also affect set in grapes (Weaver *et al.*, 1965). However, it seems more likely that CCC exerts its effects on set by reducing shoot growth and thereby reducing competition between developing berries and young shoots for available nutrients (Skene, 1969; Coombe, 1970).

D. Senescence

Senescence of the whole plant as an orderly developmental programme, and the concept of a signal for senescence have been discussed by Carr and Pate (1967). How this signal might operate in annual species is illustrated by the following examples.

Sitton and co-workers (1967) found that the cytokinin content of xylem sap of sunflowers increased during the exponential growth phase, and decreased dramatically when growth ceased and flowering had commenced. The results suggest that a reduction in the supply of cytokinins from roots to leaves is one of the factors leading to senescence. These workers are also of the opinion that the growing fruits take over as centres of cytokinin synthesis at the time the supply from the roots declines. By virtue of cytokinin's ability to attract metabolites (Mothes *et al.*, 1959), the flow of assimilates is directed from leaves to fruits, leading to further ageing of the shoot. It is well known that flower and fruit removal delays the ageing process (Molisch, 1938; Leopold *et al.*, 1959). However, other results allow slightly different interpretations, one point of contention being the origin of cytokinins in the developing fruits.

Beever and Woolhouse (1973) examined the effect of floral induction on cytokinin activity of the xylem sap of the short day plant *Perilla frutescens*. The concentration and amount of cytokinin in the sap increased rapidly after exposure to short days, the highest levels being present during the period of flower and fruit formation. Activity declined to a low level in sap from plants with mature fruit. Whilst senescence of the plant could be associated with the eventual decline in cytokinin transport from root to shoot, the results also implicate the root as a potential source of cytokinins for developing fruits. Furthermore, the observations of Wareing and Seth (1967) suggest that cytokinins may be diverted to developing fruits at the expense of leaves. These workers assessed the effects of developing fruits on levels of senescence-delaying factors in extracts of roots and leaves of French beans. No differences were detected in the roots, but leaves contained more activity in the absence of fruits than in their presence. Indirect evidence for low levels of endogenous cytokinins in leaves during fruiting was provided by Fletcher (1969), who observed BAP to be effective in delaying senescence of primary leaves on intact plants of French bean only after the onset of fruiting.

Irrespective of whether there is a sequential production of cytokinins by roots and fruits, or whether root-produced cytokinins are diverted to developing fruits, the net effect would be a lowered availability of cytokinins for the leaves, leading to their senescence. The onset of senescence is seen to be a complex interaction between the organs of the plant, a part of this interaction being mediated by cytokinins of root origin.

E. Stress

Various stresses applied to the root system also precipitate physiological changes reminiscent of the ageing process, as well as affecting the quantities of cytokinins moving from root to shoot. Temporary water deficiency induced by drought (Itai and Vaadia, 1965; Vaadia and Itai, 1969), and saline and osmotic stresses (Vaadia and Itai, 1969) reduce the levels of cytokinins in xylem sap in sunflowers, and reduce the capacity of tobacco leaf discs to incorporate (^{14}C)L-leucine into protein (Ben-Zioni *et al.*, 1967).

These adverse effects of stress are partially counteracted by cytokinins. Pretreatment of water-stressed sugar beets with BAP prevented the declines in RNA and protein levels that otherwise were associated with the stress (Shah and Loomis, 1965). Similarly, pretreatment of stressed tobacco leaf discs with kinetin prior to incubation with labelled leucine partially restored incorporation (Ben-Zioni *et al.*, 1967; Vaadia and Itai, 1969). The results suggest a lower endogenous level of cytokinins in stressed discs, although Mizrahi and co-workers (1971) subsequently were unable to demonstrate an effect of saline stress on levels of extractable cytokinins in tobacco leaves. The stress may not have been as severe. Alternatively, Mizrahi considers that activity could have been released from biologically-inactive bound forms in the stressed discs during extraction. This possibility requires investigation, as it is not unlikely that bound forms of cytokinins exist in leaves. Itai and Vaadia (1971) detected a reduction in the cytokinin content of xylem sap from tobacco leaves when the shoots were exposed to enhanced evaporative demands, and found evidence for inactivation of cytokinins in the leaves. The rapid increase in cytokinin content of excised mature poplar leaves on exposure to red light is suggestive of release from a bound or inactive form (Hewett and Wareing, 1973c). Stress also induces other modifications to the hormone balance of the plant. Discussion of the important role that abscisic acid plays in mediating the plant's response to stress is dealt with in a recent review by Livne and Vaadia (1972).

F. Waterlogging

The shoots of plants subjected to prolonged flooding of the root system exhibit characteristic symptoms indicative of a drastic imbalance of endogenous growth substances (see Phillips, 1964a). Symptoms such as epinasty of leaves, and formation of adventitious roots near the base of the stem appear to be caused by elevated levels of auxin in the shoot (Phillips, 1964b). Phillips (1964a) proposed that the root system served as a centre for the oxidative inactivation of excess shoot-synthesized auxin.

Waterlogging also reduces growth of the shoots and induces chlorosis of the lower leaves, symptoms which in part may be related to reduced quantities of gibberellin (Reid *et al.*, 1969) and cytokinin (Burrows and Carr, 1969) moving from root to shoot. The decline in cytokinin content of sunflower sap paralleled a decline in metabolic activity of the root apices up to a flooding time of 72 hours (Burrows and Carr, 1969). Thereafter, a drastic drop in the cytokinin content of the sap coincided with a large increase in the number of blackened root apices unable to reduce tetrazolium, which were probably dead.

Exogenous gibberellin only partially overcame the reduced shoot growth of flooded tomato plants (Reid and Crozier, 1971); treatment with BAP relieved many of the symptoms of flooding injury (Railton and Reid, 1973). Leaf chlorophyll levels were maintained, epinastic curvatures of petioles were reduced, no adventitious roots formed, and stem growth was greater than that of water-logged control plants. Railton and Reid consider that cytokinin may have a more important role in shoot growth than is commonly appreciated.

In addition, the relief by BAP of symptoms usually associated with excess auxin or ethylene levels (see references in Railton and Reid, 1973) suggests to these workers that cytokinins transported to the shoot could also be involved in maintaining normal growth through an effect on other hormones in the shoot.

G. Photoperiod; Light

Reports on the cytokinin content of xylem sap in relation to photoperiod are too few in number to draw any general conclusions on causal relationships between root cytokinins and morphogenetic changes induced by daylength. In the short day plant *Xanthium strumarium*, cytokinins in sap of plants grown in long days appear to predominate in the nucleotide form (Van Staden and Wareing, 1972). This fraction declined after exposure to short days; the free base/riboside fraction remained virtually unchanged. On the other hand, the cytokinin content of xylem sap from *Perilla frutescens*, another short day plant, increased under inductive conditions (Beever and Woolhouse, 1973). It is not clear whether differences in the cytokinin content of mature leaves of *X. strumarium* under long days and short days (Van Staden and Wareing, 1972) are due to an effect of daylength on cytokinin synthesis within the leaf, or are a reflection of inter-conversions within the leaf of cytokinins imported from the root system. The rapid effects of red light on changes in the cytokinin content of excised poplar leaves (Hewett and Wareing, 1973c) indicate that interconversions may occur within leaves.

Cytokinin levels in extracts of the roots of *Solanum andigena* are higher under long days than short days (Woolley and Wareing, 1972c). Long days also increase the tendency for lateral buds to develop as orthotropic leafy shoots rather than as diageotropic stolons when plants are decapitated and given the appropriate hormone treatments (see Booth, 1963; Woolley and Wareing, 1972a, c). As synthetic cytokinins promote the formation of leafy shoots on rootless cuttings of *S. andigena* (Woolley and Wareing, 1972b), the higher levels of cytokinins in roots of plants grown in long days are in keeping with the supposition that cytokinins synthesized by the root system determine the form of lateral shoots in this species.

Light quality also appears to influence cytokinin production by roots. Poplar plants exposed to fluorescent lighting for three weeks contained less cytokinins in their xylem sap than plants illuminated by fluorescent plus incandescent lights (Hewett and Wareing, 1973c).

H. Season

Cytokinins in the rising spring sap of woody species (Loeffler and van Overbeek, 1964; Nitsch and Nitsch, 1965a; Reid and Burrows, 1968; Skene and Antcliff, 1972; Jones, 1973) may fill a particular role in initiating processes associated with emergence from winter rest, as well as participating in events leading to normal growth and development of the emerging shoots. Processes stimulated by synthetic cytokinins in woody species include: bursting of dormant buds (Chvojka *et al.*, 1962; Weaver, 1963; Pieniazek, 1964), stem elongation and leaf production in isolated apple shoots (Jones, 1967), retention of inflorescences on grape vine cuttings in the absence of roots (Mullins, 1967, 1968), fruit set in grape vines (Weaver *et al.*, 1965).

Studies on xylem sap extracted by suction enable observations on cytokinin levels to be carried out on field material throughout the whole year. Luckwill and Whyte (1968) reported that the cytokinin content of xylem sap extracted by suction from apple stems remained low during winter, increased in early spring, reached a maximum level about the time of full bloom, and thereafter decreased (Fig. 6). As the disappearance of activity coincided with the cessation of extension growth, Luckwill and White suggested that shoot extension may be associated with cytokinins in the sap. Increases in activity at the time of budburst are also suggestive of a causal relationship between cytokinins and budburst, but as Luckwill and Whyte pointed out, such a conclusion is not supported by pomological evidence. It is the character of the scion variety, not of the rootstock that determines time of budburst in apples. Clearly, changes in the balance of all growth regulators in the bud must be considered.

Furthermore, evidence has already been given (Section III) indicating that cytokinins accumulate in xylem sap of isolated stems of grape vine (Skene, 1972b) and poplar (Hewett and Wareing, 1973a) independently of roots during

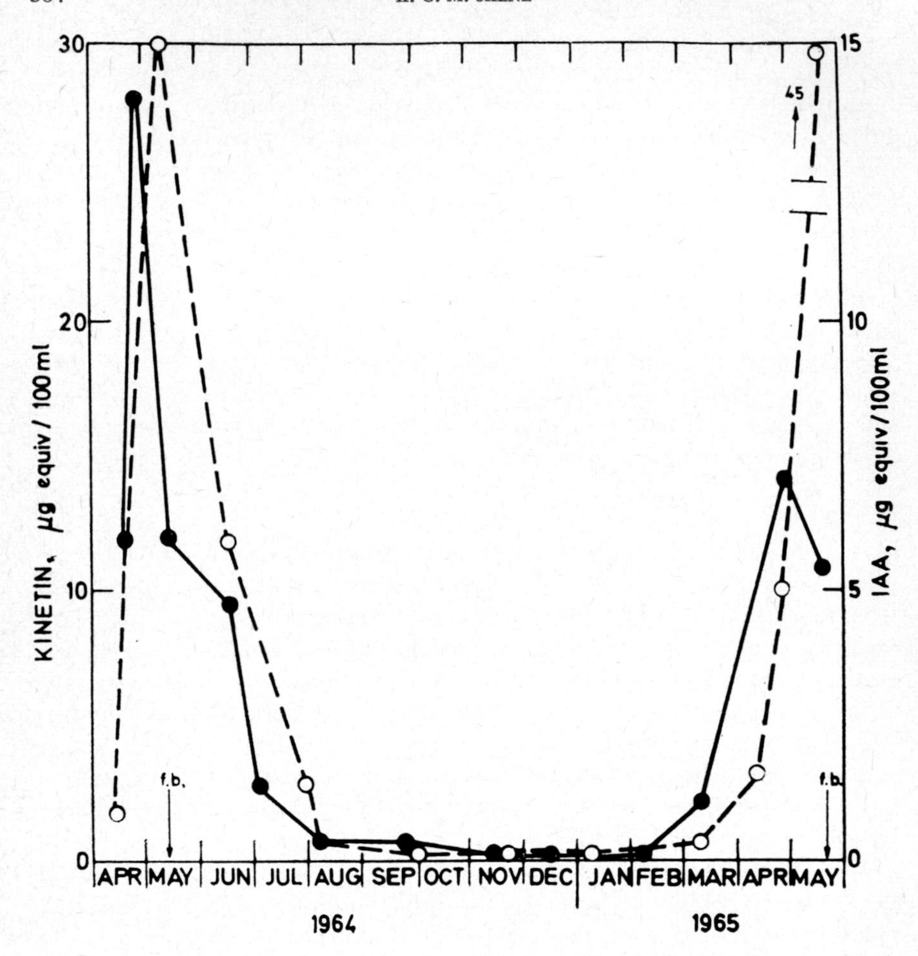

FIG. 6. Seasonal variation in concentration of hormones K and A in the xylem sap of seedling apple trees. f.b. = full bloom. ●—● K as μg equivalents kinetin ○—○ A as μg equivalents IAA. (From Luckwill and Whyte, 1968; by permission of the authors.)

cold treatment. In fact, the cytokinin content of poplar buds, as well as activity in the xylem sap increases during both natural and artificial chilling (Hewett and Wareing, 1973a). As buds on excised stems, when forced to burst at 20°C in late winter showed similar rises in their cytokinin content prior to budburst, it is clear that cytokinins can also increase in buds independently of a root system. It is not known whether this cytokinin is produced by the bud or accumulated from the stem. The fact that in the field, maximum cytokinin content of xylem sap is attained three weeks before maximum levels are reached in the buds suggests the latter alternative.

Thus it is proposed that in the intact woody perennial, high levels of cytokinins are already present in the above-ground parts in late winter–early spring before sap flow reaches a maximum. However, some of this accumulated cytokinin in stems and buds in the field may originate in the roots. Roots of many deciduous species exhibit some growth in winter (see references in Richardson, 1958), and Hewett and Wareing (1973a) observed that xylem sap and buds on field material achieved higher levels of cytokinin activity than did sap and buds from artificially chilled excised stems.

As growth of the emerging shoot continues, with an attendant depletion of accumulated cytokinin from the stem (Skene, 1972b), it is envisaged that cytokinins of root origin assume increasing importance.

Elevated levels of cytokinins in xylem sap also appear to be implicated in the release of dormancy in flower buds of *Coffea arabica* (Browning, 1973). Dormancy release in this species is usually associated with the relief of a temporary water stress. The pattern of changes in cytokinin content of buds and xylem sap following irrigation or rainfall suggests uptake into buds from the xylem, although buds may have produced some of the cytokinin. Browning is of the opinion that the relief of water stress causes cytokinins to be released directly from the wood into the xylem stream. Transport of cytokinins from roots to flower buds may be important later.

I. Rootstocks

Observations on the cytokinin production of grape vine rootstocks (Skene and Antcliff, 1972) indicate a possible role of root cytokinins in fruit set. The results are also consistent with the hypothesis that rootstock–scion relationships are in part mediated by hormones produced by roots. Skene and Antcliff (1972) investigated the cytokinin content of bleeding sap from sultana vines growing on their own roots or on the nematode-resistant rootstocks Salt Creek or 1613. They compared levels of activity with yields of fruit obtained from the same groups of vines. Vines were on soils free from plant parasitic nematodes.

Sultana vines on their own roots and on the Salt Creek rootstock yielded equally (Table II) and there were no differences in the cytokinin concentrations of their sap (Table III). However, for the season considered, yield was depressed by the 1613 rootstock, which also contained less cytokinins in its sap, both on a concentration basis and in terms of total activity passing to the shoot each day. The lower yield of sultana vines on 1613 roots was primarily due to fewer berries both on the vine and in each bunch (Table II). It was not possible to decide whether the observed differences in berry number were a consequence of fewer flowers being initiated, or because a lower proportion of flowers in each bunch developed into berries. However, synthetic cytokinins can affect both processes in grape vines (M. G. Mullins, personal communication; Weaver

TABLE II. Effect of rootstock on yield of sultana scions. (After Skene and Antcliff, 1972)

	Sultana	Salt Creek	1613	LSD (P = 0·05)
Fruit Weight (kg/vine)	45·6	45·9	30·9	11·9
Berry Weight (g/berry)	1·71	1·96	1·98	0·12
Berry Number/vine	26,700	23,660	15,820	7,000
Berry Number/bunch	327	281	252	50

TABLE III. Cytokinin content of bleeding sap from sultana grape vines, and from the rootstocks Salt Creek and 1613, when bearing sultana scions (After Skene and Antcliff, 1972)

Dilution assayed (ml sap/flask)	Cytokinin concentration (μg/l, kinetin equivalents)			
	Sultana	Salt Creek	1613	Mean
10	23	29	5	19
50	21	22	4	16
Mean	22	26	5	

et al., 1965), suggesting that the lower yield of the sultana scion on 1613 roots was causally related to the reduced levels of cytokinins passing from these roots.

V. Concluding Remarks and Some Unresolved Problems

Conclusions on the physiological significance of cytokinins in xylem sap are supported by the summation of evidence from several avenues of research. One group of results implicates the root tip as a site of cytokinin synthesis. Other work has shown that at least some of the inhibitory effects of root removal on shoot metabolism or development are specifically alleviated by exogenous applications of cytokinins. In other cases, the effects of unfavourable root environments on shoot performance are counteracted by cytokinin applications. Finally, there is a limited number of examples of correlations between the cytokinin content of xylem sap and development of the above-ground parts.

Evidence from all the above sources provides a strong case for the contention that root cytokinins regulate the protein metabolism of leaves. Other evidence, although not unequivocal, suggests that root cytokinins are involved in many aspects of shoot development, including the regulation of the activities of photosynthetic enzymes, the growth of lateral buds, the initiation of floral primordia and their continued development. Cytokinins of root origin may regulate cell division and growth of the stem in part through an effect on the activities of other hormones (e.g. Railton and Reid, 1973). The contribution of root cytokinins to the hormone level of developing fruits remains an important and elusive question.

Despite these advances in our knowledge of root cytokinins, many questions remain unanswered. The indisputable fact is that xylem sap contains high levels of cytokinins. It is still not known whether they originate exclusively from root tips, and we are ignorant of the actual amounts moving in the transpiration stream. Extrapolation from a decapitated bleeding root system to the intact plant is not without its dangers.

A further question is the degree to which the shoot controls the production and export of cytokinins from roots. The development of roots and shoots is mutually interdependent, so it is perhaps not surprising that treatments perceived by the shoot, such as photoperiod (Beever and Woolhouse, 1973) or light quality (Hewett and Wareing, 1973c) eventually affect the cytokinin content of xylem sap. However, the apparently rapid effects on cytokinin production by roots of stresses applied directly to the shoot (Itai and Vaadia, 1971) may be harder to explain.

Whilst unfavourable root environments depress the amounts of cytokinin detected in xylem sap, an effect on cytokinin content of leaves under the same circumstances is yet to be demonstrated. Apparent reductions in *activity*, as indicated by positive responses to exogenous cytokinin (e.g. Ben-Zioni *et al.*, 1967; Tal *et al.*, 1970) may merely reflect a change in the balance of hormones antagonistic to endogenous cytokinin, such as abscisic acid. This word of caution could be applied generally to instances where responses to added cytokinin have been taken as indicative of reduced endogenous levels of the hormone.

Problems of interpretation are partially related to our ignorance of the fate of root cytokinins once they reach the shoot. Their movement in the aerial parts of the plant, as well as the extent to which they are metabolized or converted to storage forms are both relatively unknown quantities.

Whilst it is probable that initially there is a passive movement of cytokinins into leaves with the transpiration stream, there does appear to be competition between the developing organs for available root cytokinins. For instance, leaf senescence can be reversed if tobacco plants are reduced to a single leaf (Mothes and Baudisch, 1958); removal of the apex retards the response to kinetin by the lower leaves of intact *Nicotiana rustica* plants grown under conditions of nutrient deficiency (Kulaeva, 1962). This latter result suggests that the limited supplies of root cytokinins in deficient plants move preferentially to the apex. It also suggests, like the results to be discussed below, that at least some component of cytokinin movement in the shoot is under directional control, and probably ultimately takes place in living tissues.

Acropetal transport of (^{14}C)BAP applied to stems of *Lens culinaris* is suppressed by removal of the apical bud (Pilet, 1968). In fact, radioactivity in decapitated *Cicer arietinum* seedlings moves to the most actively growing buds irrespective of whether they are above or below the point of application of BAP (Guern and Sadorge, 1967). Essentially similar results were obtained by Morris and Winfield (1972). The label from (^{14}C)kinetin, when applied to the roots or upper stem of decapitated pea seedlings accumulated in the axillary buds, whereas pretreatment of the cut stem surface with IAA inhibited transport of label to the buds; instead, activity accumulated in the IAA-treated region of the stem. Application of labelled kinetin to the roots of intact plants resulted in activity accumulating in the apical bud itself. Morris and Winfield suggest that hormone-directed transport of cytokinins is involved in the regulation of lateral bud growth.

Woolley and Wareing (1972a, b) are of the opinion that auxin influences the distribution and metabolism of cytokinins from roots. Auxin applied to the cut upper surface of *Solanum andigena* cuttings prevents the outgrowth of lateral buds and inhibits the accumulation in the lateral buds of label from basally administered (^{14}C)BAP. In the absence of auxin, label (identified as BAP and BAP-riboside) accumulates in the lateral buds prior to their growth as leafy shoots (Woolley and Wareing, 1972b). Related experiments on the conversion of stolons to leafy shoots (Woolley and Wareing, 1972a) also showed a build-up, in the absence of IAA, of activity from labelled BAP in the tip of the induced stolon prior to its conversion to a leafy shoot.

Although Woolley and Wareing were unable to demonstrate that exogenous IAA reduced cytokinin accumulation in lateral buds or stolons by diverting cytokinins to the point of auxin application (c.f. Morris and Winfield, 1972), they found that IAA promoted the formation of an unknown metabolite of BAP, which accumulated in stem tissue. In the absence of IAA, BAP-riboside was the main metabolite formed from BAP. As BAP, its riboside and the unknown metabolite appear to have differing transport characteristics, Woolley and Wareing propose that auxin influences cytokinin distribution by a combination of its effects on cytokinin metabolism and transport.

Whether cytokinins from roots move directly to the apex and other centres of high auxin concentration, or whether they are re-exported from the leaves is not known. They are not necessarily mutually exclusive situations. Bilateral interchange between xylem and phloem of both cytokinins and gibberellins seems to occur very readily (Bowen and Wareing, 1969); similarly, the accumulation of zeatin ribotide in hypocotyls of rootless radish seedlings supplied with labelled zeatin probably resulted from lateral movement from the transpiration stream rather than from basipetal transport from the upper parts of the seedlings (Parker and Letham, 1973). On the other hand, experiments with (^{14}C)BAP indicate that BAP applied to the base of rootless cuttings of *Solanum andigena* rapidly accumulates in the leaf and subsequently is redistributed to other parts of the cutting (Woolley and Wareing, 1972b).

Engelbrecht's studies (1972) on the cytokinin content of rooted leaf cuttings of *Phaseolus vulgaris* indicate that the leaf may have a mechanism for storing root cytokinins in excess of its requirements. She suggests this storage form to be a nucleotide, which in single-leaf cuttings builds up to levels far higher than those in leaves of 15-day-old plants. Other results (see below) indicate that cytokinin glucosides may have similar functions.

Exogenous zeatin is extensively metabolized by plant tissues. Adenine, adenosine, adenosine-5'-monophosphate, zeatin riboside and zeatin riboside-5'-monophosphate were all detected to varying extents in rootless radish seedlings supplied basally with (^{3}H)zeatin (Parker *et al.*, 1972; Parker and Letham, 1973). However, the major metabolite in the cotyledons was identified as a zeatin–glucose complex (raphanatin, 7-glucosylzeatin), and this too is probably a storage form for excess cytokinins. Its distribution is not restricted to the upper parts of the plant. Most of the activity from labelled zeatin applied through the roots to intact radish seedlings was recovered from the roots (Gordon *et al.*, 1974), and here the major metabolite in both roots and cotyledons was also found to be identical with 7-glucosylzeatin.

An endogenous compound with properties resembling a cytokinin glucoside has also been extracted from mature poplar leaves (Hewett and Wareing, 1973b, d). Highest levels were attained during late summer–early autumn, after shoot elongation had ceased, and the levels of other cytokinins in the leaf and the xylem sap had fallen (Hewett and Wareing, 1973d); removal of the apex during midsummer also caused a rise in the level of the glucoside in leaves. Mature poplar leaves contain at least six other cytokinins (Hewett and Wareing, 1973b). Two of these are probably zeatin and its riboside. Another has been identified as 6-(2-hydroxybenzyl)aminopurine riboside (Horgan *et al.*, 1973b) and this is the first isolation of a naturally-occurring cytokinin with an aromatic side chain. It is the main cytokinin present in poplar leaves cultivated in growth chambers and specifically increases after red light treatment (Hewett and Wareing 1973c).

The physiological significance of these diverse forms of cytokinins in leaf tissue is far from established, and we have no direct information on their origin, although the evidence suggests that interconversions between the various cytokinins take place within the leaf itself. Answers to these questions are fundamental to an understanding of the contribution of root cytokinins to the control of plant development.

Acknowledgements

The author wishes to thank Dr. D. S. Letham for providing the results of experiments unpublished at the time this manuscript was prepared. He is also grateful for the helpful criticism of his colleagues.

References

ANDREENKO, S. S., POTAPOV, N. G. and KOSULINA, L. G. (1964). The effect of sap from maize plants grown at various pH levels on growth of carrot callus. *Doklady Akad. Nauk. U.S.S.R.* (Engl. transl.) **155**; 35–37.

BEEVER, J. E. and WOOLHOUSE, H. W. (1973). Increased cytokinin from root system of *Perilla frutescens* and flower and fruit development. *Nature New Biol.* **246**; 31–32.

BEN-ZIONI, A., ITAI, C. and VAADIA, Y. (1967). Water and salt stresses, kinetin and protein synthesis in tobacco leaves. *Plant. Physiol.* **42**; 361–365.

BLUMENFELD, A. and GAZIT, S. (1971). Growth of avocado fruit callus and its relation to exogenous and endogenous cytokinins. *Physiol. Plant.* **25**; 369–371.

BOLLARD, E. G. (1960). Transport in the xylem. *Ann. Rev. Plant. Physiol.* **11**; 141–166.

BOOTH, A. (1963). *In*: "The Growth of the Potato" (J. D. Ivins and F. L. Milthorpe,Eds.); pp. 99–113. Butterworths, London.

BOWEN, M. R. and WAREING, P. F. (1969). The interchange of ^{14}C-kinetin and ^{14}C-gibberellic acid between the bark and xylem of willow. *Planta* **89**; 108–125.

BROWNING, G. (1973). Flower bud dormancy in *Coffea arabica* L. II. *J. hort. Sci.* **48**; 297–310.

BUI-DANG-HA, D. and NITSCH, J. P. (1970). Isolation of zeatin riboside from the chicory root. *Planta* **95**; 119–126.

BURROWS, W. J. and CARR, D. J. (1969). Effects of flooding the root system of sunflower plants on the cytokinin content in the xylem sap. *Physiol. Plant.* **22**; 1105–1112.

BUTTROSE, M. S. and MULLINS, M. G. (1968). Proportional reduction in shoot growth of grapevines with root systems maintained at constant relative volumes by repeated pruning. *Aust. J. biol. Sci.* **21**; 1095–1101.

CARR, D. J. and BURROWS, W. J. (1966). Evidence of the presence in xylem sap of substances with kinetin-like activity. *Life Sci.* **5**; 2061–2077.

CARR, D. J. and PATE, J. S. (1967). Ageing in the whole plant. *Symp. Soc. exp. Biol.* **21**; 559–600.

CARR, D. J. and REID, D. M. (1968). *In*: "Biochemistry and Physiology of Plant Growth Substances" (F. Wightman and G. Setterfield, Eds.) pp. 1169–1185. Runge Press, Ottawa.

CARR, D. J., REID, D. M. and SKENE, K. G. M. (1964). The supply of gibberellins from the root to the shoot. *Planta* **63**; 382–392.

CHIBNALL, A. C. (1939). "Protein Metabolism in the Plant". Yale Univ. Press, New Haven.

CHIBNALL, A.C. (1954). Protein metabolism in rooted runner-bean leaves. *New Phytol.* **53**; 31–37.

CHVOJKA, L., TRAVNICEK, M. and ZAKOURILOVA, M. (1962). The influence of stimulating doses of 6-benzylaminopurine on awakening apple buds and on their consumption of oxygen. *Biol. Plant.*, **4**; 203–206.

COOMBE, B. G. (1965). Increase in fruit set of *Vitis vinifera* by treatment with growth retardants. *Nature, Lond.* **205**; 305–306.

COOMBE, B. G. (1967). Effects of growth retardants on *Vitis vinifera* L. *Vitis* **6**; 278–287.

COOMBE, B. G. (1970). Fruit set in grape vines: the mechanism of the CCC effect. *J. hort. Sci.* **45**; 415–425.

CROZIER, A. and REID, D. M. (1972). *In*: "Plant Growth Substances 1970" (D. J. Carr, Ed.) pp. 414–419. Springer, Berlin.

DELEUZE, G. G., MCCHESNEY, J. D. and FOX, J. E. (1972). Identification of a stable cytokinin metabolite. *Biochem. biophys. Res. Commun.* **48**; 1426–1432.

EINSET, J. W. and SKOOG, F. (1973). Biosynthesis of cytokinins in cytokinin-autotrophic tobacco callus. *Proc. natn. Acad. Sci. U.S.A.* **70**; 658–660.

ENGELBRECHT, L. (1972). Cytokinins in leaf cuttings of *Phaseolus vulgaris* L. during their development. *Biochem. Physiol. Pflanzen*, **163**; 335–343.

FEIERABEND, J. (1969). Der Einfluss von Cytokininen auf die Bildung von Photosynthese-enzymen. *Planta.* **84**; 11–29.

FLETCHER, R. A. (1969). Retardation of leaf senescence by benzyladenine in intact bean plants. *Planta.* **89**; 1–8.

FOX, J. E., CORNETTE, J., DELEUZE, G., DYSON, W., GIERSAK, C., NIU, P., ZAPATA, J. and MCCHESNEY, J. (1973). The formation, isolation and biological activity of a cytokinin 7-glucoside. *Plant. Physiol.* **52**; 627–632.

GOLDACRE, P. L. (1959). Potentiation of lateral root induction by root initials in isolated flax roots. *Aust. J. biol. Sci.* **12**; 388–396.

GORDON, M. E., LETHAM, D. S. and PARKER, C. W. (1974). Regulators of cell division in plant tissues XVII. *Ann. Bot.* **38**; 809-825.

GUERN, J. and SADORGE, P. (1967). Polarité du transport de la 6-benzylaminopurine dans les jeunes plantes étiolées de *Cicer arietinum*. *C. r. hebd. Séanc. Acad. Sci., Paris*, **264**; 2106–2109.

GUERN, J., DOREE, M. and SADORGE, P. (1968). *In*: "Biochemistry and Physiology of Plant Growth Substances" (F. Wightman and G. Setterfield, Eds.) pp. 1155–1167. Runge Press, Ottawa.

GUPTA, G. R. P. and MAHESHWARI, S. C. (1970). Cytokinins in seeds of pumpkin. *Plant Physiol.* **45**; 14–18.

HALL, R. H. (1973). Cytokinins as a probe of developmental processes. *Ann. Rev. Plant Physiol.* **24**; 415–444.

HEWETT, E. W. and WAREING, P. F. (1973a). Cytokinins in *Populus* × *robusta*: changes during chilling and bud burst. *Physiol. Plant.* **28**; 393–399.

HEWETT, E. W. and WAREING, P. F. (1973b). Cytokinins in *Populus* × *robusta* Schneid: a complex in leaves. *Planta* **112**; 225–233.

HEWETT, E. W. and WAREING, P. F. (1973c). Cytokinins in *Populus* × *robusta* (Schneid): light effects on endogenous levels. *Planta* **114**; 119–129.

HEWETT, E. W. and WAREING, P. F. (1973d). Cytokinins in *Populus* × *robusta*: qualitative changes during development. *Physiol. Plant.* **29**; 386–389.

HOLM, R. E. and KEY, J. L. (1969). Hormonal regulation of cell elongation in the hypocotyl of rootless soybean: an evaluation of the role of DNA synthesis. *Plant Physiol.* **44**; 1295–1302.

HORGAN, R., HEWETT, E. W., PURSE, J. G., HORGAN, J. M. and WAREING, P. F. (1973a). Identification of a cytokinin in sycamore sap by gas chromatography-mass spectrometry. *Plant Sci. Letters* **1**; 321–324.

HORGAN, R., HEWETT, E. W., PURSE, J. G. and WAREING, P. F. (1973b). A new cytokinin from *Populus robusta*. *Tetrahedron Letters* No. 30; pp. 2827–2828.

ITAI, C. and VAADIA, Y. (1965). Kinetin-like activity in root exudate of water-stressed sunflower plants. *Physiol. Plant.* **18**; 941–944.

ITAI, C. and VAADIA, Y. (1971). Cytokinin activity in water-stressed shoots. *Plant Physiol.* **47**; 87–90.

JONES, O. P. (1967). Effect of benzyl adenine on isolated apple shoots. *Nature, Lond.* 215; 1514–1515.

JONES, O. P. (1973). Effects of cytokinins in xylem sap from apple trees on apple shoot growth. *J. hort. Sci.* **48**; 181–188.

KENDE, H. (1964). Preservation of chlorophyll in leaf sections by substances obtained from root exudate. *Science N.Y.* **145**; 1066–1067.

KENDE, H. (1965). Kinetinlike factors in the root exudate of sunflowers. *Proc. natn. Acad. Sci. U.S.A.* **53**; 1302–1307.

KENDE, H. (1971). The cytokinins. *Int. Rev. Cytol.* **31**; 301–338.

KENDE, H. and SITTON, D. (1967). The physiological significance of kinetin- and gibberellin-like root hormones. *Ann. N.Y. Acad. Sci.* **144**; 235–243.

KLÄMBT, D. (1968). Cytokinine aus *Helianthus annuus*. *Planta*, **82**; 170–178.

KULAEVA, O. N. (1962). The effect of roots on leaf metabolism in relation to the action of kinetin on leaves. *Soviet Plant Physiol.* (Engl. transl.) **9**; 182–189.

KUMAR, D. and WAREING, P. F. (1972). Factors controlling stolon development in the potato plant. *New Phytol.* **71**; 639–648.

LANG, A. (1970). Gibberellins: structure and metabolism. *Ann. Rev. Plant Physiol.* **21**; 537–570.

LENTON, J. R. BOWEN, M. R. and SAUNDERS, P. F. (1968). Detection of abscisic acid in the xylem sap of willow (*Salix viminalis* L.) by gas-liquid chromatography. *Nature, Lond.* **220**; 86–87.

LEOPOLD, A. C., NIEDERGANG–KAMIEN, E. and JANICK, J. (1959). Experimental modification of plant senescence. *Plant Physiol.* **34**; 570–573

LETHAM, D. S. (1966a). Regulators of cell division in plant tissues II. *Phytochem.* **5**; 269–286.

LETHAM, D. S. (1966b). Purification and probable identity of a new cytokinin in sweet corn extracts. *Life Sci.* **5**; 551–554.

LETHAM, D. S. (1966c). Isolation and probable identity of a third cytokinin in sweet corn extracts. *Life Sci.* **5**; 1999–2004.

LETHAM, D. S. (1967). Chemistry and physiology of kinetin-like compounds. *Ann. Rev. Plant Physiol.* **18**; 349–364.

LETHAM, D. S. (1968). *In*: "Biochemistry and Physiology of Plant Growth Substances" (F. Wightman and G. Setterfield, Eds.) pp. 19–31. Runge Press, Ottawa.

LETHAM, D. S. (1973). Cytokinins from *Zea mays*. *Phytochem.* **12**; 2445–2455.

LETHAM, D. S., SHANNON, J. S. and McDONALD, I. R. (1964). The structure of zeatin, a factor inducing cell division. *Proc. Chem. Soc.* pp. 230–231.

LETHAM, D. S. and WILLIAMS, M. W. (1969). Regulators of cell division in plant tissues VIII. *Physiol. Plant.* **22**; 925–936.

LIVNE, A. and VAADIA, Y. (1972). *In*: "Water Deficits and Plant Growth" (T. T. Kozlowski, Ed.) Vol. III; pp. 255–275. Academic Press, New York, London and San Francisco.

LOEFFLER, J. E. and VAN OVERBEEK, J. (1964). *In*: "Régulateurs Naturels de la Croissance Végétale" (J. P. Nitsch, Ed.) pp. 77–82. C.N.R.S., Paris.

LUCKWILL, L. C. and WHYTE, P. (1968). Hormones in the xylem sap of apple trees. *S. C. I. Monograph* No. 31; pp. 87–101.

MACKENZIE, I. A. and STREET, H. E. (1972). The cytokinins of cultured sycamore cells. *New Phytol.* **71**; 621–631.

McDAVID, C. R., SAGAR, G. R. and MARSHALL, C. (1973). The effect of root pruning and 6-benzylaminopurine on the chlorophyll content, $^{14}CO_2$ fixation and the shoot/root ratio in seedlings of *Pisum sativum* L. *New Phytol.* **72**; 465–470.

MIGINIAC, E. (1971). Influence des racines sur le développement végétatif ou floral des bourgeons cotylédonaires chez le *Scrofularia arrguta*: role possible des cytokinines. *Physiol. Plant.* **25**; 234–239.

MILLER, C. O. (1961). A kinetin-like compound in maize. *Proc. natn. Acad. Sci., U.S.A.* **47**; 170–174.

MILLER, C. O. (1965). Evidence for the natural occurrence of zeatin and its derivatives: compounds from maize which promote cell division. *Proc. natn. Acad. Sci., U.S.A.* **54**; 1052–1058.

MILLER, C. O. (1967). Zeatin and zeatin riboside from a mycorrhizal fungus. *Science, N.Y.* **157**; 1055–1057.

MIURA, G. A. and HALL, R. H. (1973). *trans*-Ribosylzeatin. Its biosynthesis in *Zea mays* endosperm and the mycorrhizal fungus, *Rhizopogon roseolus*. *Plant Physiol.* **51**; 563–569.

MIURA, G. A. and MILLER, C. O. (1969). Cytokinins from a varient strain of cultured soybean cells. *Plant Physiol.* **44**; 1035–1039.

MIZRAHI, Y., BLUMENFELD, A., BITTNER, S. and RICHMOND, A. E. (1971). Abscisic acid and cytokinin contents of leaves in relation to salinity and relative humidity. *Plant Physiol.* **48**; 752–755.

MOLISCH, H. (1938). "The Longevity of Plants". Science Press, Lancaster.

MORRIS, D. A. and WINFIELD, P. J. (1972). Kinetin transport to axillary buds of dwarf pea (*Pisum sativum* L.). *J. exp. Bot.* **23**; 346–355.

MOTHES, K. (1960). Über das Altern der Blätter und die Moglichkeit ihrer Wiederverjüngung. Naturwiss enschaften **47**; 337–351.

MOTHES, K. (1973). Some remarks about cytokinins. *Soviet Plant Physiol.* (Engl. transl.), **19**; 863–872.

MOTHES, K. and BAUDISCH, W. (1958). Untersuchungen über die Reversibilität der Ausbleichung grüner Blätter. *Flora* **146**; 521–531.

MOTHES, K. and ENGELBRECHT, L. (1956). Über den Stickstoffumsatz in Blattstecklingen. *Flora* **143**; 428–472.

MOTHES, K. ENGELBRECHT, L. and KULAEVA, O. N. (1959). Über die Wirkung des Kinetins auf Stickstoffverteilung und Eiweisssynthese in isolierten Blättern. *Flora* **147**; 445–465.

MULLINS, M. G. (1967). Morphogenetic effects of roots and of some synthetic cytokinins in *Vitis vinifera* L. *J. exp. Bot.* **18**; 206–214.

MULLINS, M. G. (1968). Regulation of inflorescence growth in cuttings of the grape vine (*Vitis vinifera* L.). *J. exp. Bot.* **19**; 532–543.

NITSCH, J. P. and NITSCH, C. (1965a). Présence d'une phytokinine dans le cambium. *Bull. Soc. bot. France* **112**; 1–10.

NITSCH, J. P. and NITSCH, C. (1965b). Présence de phytokinines et autres substances de croissance dans la sève d'*Acer saccharum* et de *Vitis vinifera*. *Bull. Soc. bot. France* **112**; 11–19.

PARKER, C. W. and LETHAM, D. S. (1973). Regulators of cell division in plant tissues XVI. *Planta* **114**; 199–218.

PARKER, C. W. and LETHAM, D. S. (1974). Regulators of cell division in plant tissues XVIII. *Planta* **115**; 337–344.

PARKER, C. W., LETHAM, D. S., COWLEY, D. E. and MACCLEOD, J. K. (1972). Raphanatin, an unusual purine derivative and a metabolite of zeatin. *Biochem. biophys. Res. Commun.* **49**; 460–466.

PARTHIER, B. (1964). Proteinsynthese in grünen Blättern II. *Flora* **154**; 230–244.

PHILLIPS, I. D. J. (1964a). Root-shoot hormone relations I. *Ann. Bot.* **28**; 17–35.

PHILLIPS, I. D. J. (1964b). Root-shoot hormone relations II. *Ann. Bot.* **28**; 37–45.

PHILLIPS, I. D. J. and JONES, R. L. (1964). Gibberellin-like activity in bleeding sap of root systems of *Helianthus annuus* detected by a new dwarf pea epicotyl assay and other methods. *Planta* **63**; 269–278.

PHILLIPS, D. A. and TORREY, J. G. (1970). Cytokinin production by *Rhizobium japonicum*. *Physiol. Plant.* **23**; 1057–1063.

PIENIAZEK, J. (1964). Kinetin induced breaking of dormancy in 8-month old apple seedlings of Antonovka variety. *Acta Agrobot.* **16**; 157–169.

PILET, P. E. (1968). *In*: "Biochemistry and Physiology of Plant Growth Substances" (F. Wightman and G. Setterfield, Eds.) pp. 993–1004. Runge Press, Ottawa.
PRAKASH, R. and MAHESHWARI, S. C. (1970). Studies on cytokinins in watermelon seeds. *Physiol. Plant.* **23**; 792–799.
RADIN, J. W. and LOOMIS, R. S. (1971). Changes in the cytokinins of radish roots during maturation. *Physiol. Plant.* **25**; 240–244.
RAILTON, I. D. and REID, D. M. (1973). Effects of benzyladenine on the growth of water-logged tomato plants. *Planta* **111**; 261–266.
REID, D. M. and BURROWS, W. J. (1968). Cytokinin and gibberellin-like activity in the spring sap of trees. *Experientia* **24**; 189.
REID, D. M. and CROZIER, A. (1971). Effects of waterlogging on the gibberellin content and growth of tomato plants. *J. exp. Bot.* **22**; 39–48.
REID, D. M., CROZIER, A. and HARVEY, B. M. R. (1969). The effects of flooding on the export of gibberellins from the root to the shoot. *Planta* **89**; 376–379.
RICHARDSON, S. D. (1958). *In*: "The Physiology of Forest Trees" (K. V. Thimann, Ed.) pp. 409–426. Ronald Press, New York.
RICHMOND, A. E. and LANG, A. (1957). Effect of kinetin on protein content and survival of detached *Xanthium* leaves. *Science, N.Y.* **125**; 650–651.
SHAH, C. B. and LOOMIS, R. S. (1965). Ribonucleic acid and protein metabolism in sugar beet during drought. *Physiol. Plant.* **18**; 240–254.
SHELDRAKE, A. R. and NORTHCOTE, D. H. (1968). The production of auxin by tobacco internode tissues. *New Phytol.* **67**; 1–13.
SHIRAI, H., KANATANI, H. and TAGUCHI, S. (1972). 1-Methyladenine biosynthesis in starfish ovary: action of gonad stimulating hormone in methylation. *Science, N.Y.* **175**; 1366–1368.
SHORT, K. C. and TORREY, J. G. (1972a). Cytokinins in seedling roots of pea. *Plant Physiol.* **49**; 155–160.
SHORT, K. C. and TORREY, J. G. (1972b). Cytokinin production in relation to the growth of pea root callus tissue. *J. exp. Bot.* **23**; 1099–1105.
SITTON, D., ITAI, C. and KENDE, H. (1967). Decreased cytokinin production in the roots as a factor in shoot senescence. *Planta* **73**; 296–300.
SKENE, K. G. M. (1967). Gibberellin-like substances in root exudate of *Vitis vinifera*. *Planta* **74**; 250–262.
SKENE, K. G. M. (1968). Increases in the levels of cytokinins in bleeding sap of *Vitis vinifera* L. after CCC treatment. *Science, N.Y.* **159**; 1477–1478.
SKENE, K. G. M. (1969). A comparison of the effects of "Cycocel" and tipping on fruit set in *Vitis vinifera* L. *Aust. J. biol. Sci.* **22**; 1305–1311.
SKENE, K. G. M. (1970). The relationship between the effects of CCC on root growth and cytokinin levels in the bleeding sap of *Vitis vinifera* L. *J. exp. Bot.* **21**; 418–431.
SKENE, K. G. M. (1972a). *In*- "Plant Growth Substances 1970" (D. J. Carr, Ed.) pp. 476–483. Springer, Berlin.
SKENE, K. G. M. (1972b). Cytokinins in the xylem sap of grape vine canes: changes in activity during cold-storage. *Planta* **104**; 89–92.
SKENE, K. G. M. (1972c). The effect of ringing on cytokinin activity in shoots of the grape vine. *J. exp. Bot.* **23**; 768–774.
SKENE, K. G. M. and ANTCLIFF, A. J. (1972). A comparative study of cytokinin levels in bleeding sap of *Vitis vinifera* (L.) and the two grapevine rootstocks, Salt Creek and 1613. *J. exp. Bot.* **23**; 283–293.
SKENE, K. G. M. and KERRIDGE, G. H. (1967). Effect of root temperature on cytokinin activity in root exudate of *Vitis vinifera* L. *Plant Physiol* **42**; 1131–1139.
SKOOG, F. and ARMSTRONG, D. J. (1970). Cytokinins. *Ann. Rev. Plant. Physiol.* **21**; 359–384.
SKOOG, F., STRONG, F. M. and MILLER, C. O. (1965). Cytokinins. *Science, N.Y.* **148**; 532–533.

SMITH, D. L. (1968). The growth of shoot apices and inflorescences of *Carex flacca* Schreb. in aseptic culture. *Ann. Bot.* **32**; 361–370.

SMITH, D. L. (1969). The role of leaves and roots in the control of inflorescence development in *Carex*. *Ann. Bot.* **33**; 505–514.

SMITH, H. and WAREING, P. F. (1964). Gravimorphism in trees III. *Ann. Bot.* **28**; 297–309.

TAL, M., IMBER, D. and ITAI, C. (1970). Abnormal stomatal behaviour and hormonal imbalance in *flacca*, a wilty mutant of tomato I. *Plant Physiol.* **46**; 367–372.

TEGLEY, J. R., WITHAM, F. H. and KRASNUK, M. (1971). Chromatographic analysis of a cytokinin from tissue cultures of crown gall. *Plant Physiol.* **47**; 581–585.

TORREY, J. G. and LOOMIS, R. S. (1967). Auxin-cytokinin control of secondary vascular tissue formation in isolated roots of *Raphanus*. *Am. J. Bot.* **54**; 1098–1106.

VAADIA, Y. and ITAI, C. (1969). *In*: "Root Growth" (W. J. Whittington, Ed.); pp. 65–79. Butterworths, London.

VAN STADEN, J. and WAREING, P. F. (1972). The effect of photoperiod on levels of endogenous cytokinins in *Xanthium strumarium*. *Physiol. Plant.* **27**; 331–337.

VINOKUR, R. L. (1963). Nitrogen metabolism of squash plants in relation to root temperature. *Soviet Plant Physiol.* (Engl. transl.) **10**; 275–278.

WAGNER, H. and MICHAEL, G. (1969). Cytokinin-Bildung in Wurzeln von Sonnenblumen. *Naturwissenschaften* **56**; 379.

WAGNER, H. and MICHAEL, G. (1971). Der Einfluss unterschiedlicher Stickstoffversorgung auf die Cytokininbildung in Wurzeln von Sonnenblumenpflanzen. *Biochem. Physiol. Pflanzen* **162**; 147–158.

WAREING, P. F. and NASR, T. A. A. (1961). Gravimorphism in trees. Effects of gravity on growth and apical dominance in fruit trees. *Ann. Bot.* **25**; 321–340.

WAREING, P. F. and SETH, A. K. (1967). Ageing and senescence in the whole plant. *Symp. Soc. exp. Biol.* **21**; 543–558.

WEAVER, R. J. (1963). Use of kinin in breaking rest in buds of *Vitis vinifera*. *Nature, Lond.* **198**; 207–208.

WEAVER, R. J., VAN OVERBEEK, J. and POOL, R. M. (1965). Induction of fruit set in *Vitis vinifera* L. by a kinin. *Nature, Lond.* **206**; 952–953.

WEISS, C. and VAADIA, Y. (1965). Kinetin-like activity in root apices of sunflower plants. *Life Sci.* **4**; 1323–1326.

WENT, F. W. (1938). Specific factors other than auxin affecting growth and root formation. *Plant Physiol.* **13**; 55–80.

WOODHAM, R. C. and ALEXANDER, D. MCE. (1966). The effect of root temperature on development of small fruiting sultana vines. *Vitis* **5**; 345–350.

WOOLLEY, D. J. and WAREING, P. F. (1972a). The role of roots, cytokinins and apical dominance in the control of lateral shoot form in *Solanum andigena*. *Planta* **105**; 33–42.

WOOLLEY, D. J. and WAREING, P. F. (1972b). The interaction between growth promoters in apical dominance I. *New Phytol.* **71**; 781–793.

WOOLLEY, D. J. and WAREING, P. F. (1972c). The interaction between growth promoters in apical dominance II. *New Phytol.* **71**; 1015–1025.

YOSHIDA, R., ORITANI, T. and NISHI, A. (1971). Kinetin-like factors in the root exudate of rice plants. *Plant and Cell Physiol.* **12**; 89–94.

YOSHIDA, R. and ORITANI, T. (1972). Cytokinin glucoside in roots of the rice plant. *Plant and Cell Physiol.* **13**; 337–343.

ZWAR, J. A., BOTTOMLEY, W. and KEFFORD, N. P. (1963). Kinin activity from plant extracts II. *Aust. J. Biol. Sci.* **16**; 407–415.

Note added in proof

Two recent results indicate that seeds and fruits probably do not have an exclusive dependence on root cytokinins for their development:

1. Fruits and seeds can develop to varying extents on rootless plants (Peterson and Fletcher, 1973). 2. The seeds of pea pods cultured *in vitro* are capable of synthesizing cytokinins independently of a root system (Hahn *et al.*, 1974).

HAHN, H., DE ZACKS, R. and KENDE, H. (1974). Cytokinin formation in pea seeds. *Naturwissenschaften* **61**; 170.

PETERSON, C. A. and FLETCHER, R. A. (1973). Formation of fruits on rootless plants. *Can. J. Bot.* **51**; 1899–1905.

Chapter 18

Water Relations of the Root System

P. E. WEATHERLEY

Botany Department, University of Aberdeen, Aberdeen, Scotland

I. Introduction

In passing from the soil to the transpiring leaves, water must traverse the cells of the root cortex, the endodermis and the stele before entering the tracheae of the xylem. Once in the xylem conduits it is pulled up to the veins of the leaf by relatively simple mechanical forces. The passage of water across the root cells is however far from simple and many aspects of it are still not well understood. Likewise the passage of water from the soil into the cells of the root surface has often been regarded as a simple process—the films of water round the soil particles being continuous with those on the root surface. However there is some evidence that the soil: root interface has properties peculiar to itself which may have implications for the water relations of the whole plant.

II. Movement from Epidermis to Xylem

A. The Root as an Osmometer

It has long been known that the root in some respects acts like an osmometer. The tissues are differentially permeable and it was shown half a century ago (Sabinin 1925) that exudation from the stump of a decapitated root system is defined by the equation:

$$J_v = L(\pi_x - \pi_e) \tag{1}$$

where J_v is the rate of flow through the system (rate of exudation), π_x and π_e are the osmotic pressures of the xylem sap and the external medium respectively. L is a constant, the hydraulic conductivity coefficient. This has been confirmed by the elegant experiments of Arisz *et al.*, (1951) and many others since. Of course as water moves into the xylem, the sap would become diluted and π_x and the rate of exudation would fall were it not for a continual transfer of solutes (mainly ions) into the tracheae of the xylem from surrounding cells, these ions having their ultimate origin in the external medium. More recently, the possible permeability of the root cells to the solutes in the medium and sap has been taken into account by inserting σ, the reflexion coefficient, into the equation. Thus

$$J_v = L\sigma(\pi_x - \pi_e) \tag{2}$$

If the barrier between medium and sap is perfectly semipermeable, $\sigma = 1$. Values of σ less than unity indicate leakage of the osmotic solutes through the barrier. In fact for root systems σ is often close to unity (Ginsburg and Ginzburg, 1970).

In the transpiring plant, water is drawn up the xylem tracheae and because of the resistance to flow across the root cortex and stele a hydrostatic tension arises in the tracheae of the root. It is in response to this tension (T) that water moves from the external medium to the stele. Thus

$$J_v = LT \tag{3}$$

and if the pathway were perfectly semipermeable and the osmotic and hydrostatic forces acted along the same pathway, equations (2) and (3) can be summed:

$$J_v = L(\pi_x - \pi_e + T) \tag{4}$$

The flow in response to a difference in osmotic pressure as represented by equation (2) has in the past been referred to as "active" absorption, presumably because it depends on a continuous secretion of solutes into the xylem, whilst the tension-induced flow (equation 3) has been called "passive" absorption. This distinction is not particularly useful since the nature of the movement is similar whether induced by hydrostatic tension or solutes. The motive force is the difference of water potential between the root surface and the xylem tracheae, and if the root is behaving like a perfect osmometer it makes no difference whether the water potential in the xylem is lowered by the presence of solutes or a hydrostatic tension. With the exuding plant or the intact plant transpiring very slowly, T is small and π_x large. With rapid transpiration T is large and the ingress of water so great that the solution in the xylem tracheae becomes very dilute indeed π_x can be lower than π_e and the osmotic and tension components act in "opposite" directions.

If in addition to the osmotic pathway across the root tissues, there was a free flow pathway with no osmotic barrier in it, then a tension in the xylem would lead to an additional mass flow of solution from medium to tracheae. In this case the total flow becomes:

$$J_v = L(\pi_x - \pi_e + T) + L^1 T \tag{5}$$

where L^1 is the hydraulic conductivity coefficient of the mass flow pathway. The existence of $L^1 T$ is impossible to demonstrate in the intact plant since the measurement of π_x requires decapitation and this eliminates T. The artificial application of a pressure gradient across a decapitated root system is open to the criticism that a suction applied to the stump is applied to the whole cross section of stem and not to the xylem tracheae alone and in any case the drop of pressure must necessarily be limited to less than one bar. Mees and Weatherley (1957a, b) attempted to compare osmotic and hydrostatic pressure differences by applying positive pressures to the medium surrounding the decapitated root system. Their results indicate that there was an "all-through" mass flow pathway, L^1 being about $\frac{1}{3}$ L. These experiments have been criticized (Briggs, 1967) but whilst a further examination of possible errors would be worth while, a different experimental approach is difficult to devise.

B. Active Transport

So far the radial flux of water across the root tissues has been regarded as passive, i.e. moving down a gradient of water potential, the water potential in the xylem always being lower than that in the medium. An "active" flux implies a movement against the operational gradient of water potential. This has been investigated by a number of workers using detopped plants. The method is to alter the difference of osmotic pressure between medium and xylem by altering the osmotic pressure of the medium and measuring the corresponding change of rate of exudation from the stump of the stem. If the data fit equation (2) the flux is passive; J_v is plotted against $(\pi_x - \pi_e)$ when a straight line should be obtained passing through the origin, i.e. no flux when there is no gradient across the tissues. The classic work in this field is that of Arisz *et al.*, (1951). Using tomato plants they found that exudation fitted equation (2) but there was a small fraction which appeared to be non-osmotic. A similar result was obtained by House and Findlay (1966) for maize (see Fig. 1). However this does not necessarily imply an active component. As the authors themselves point out, the assumption is made that π_x in equation (2), i.e. the osmotic pressure in the absorbing region of the root, is identical with the osmotic pressure of the sap emerging from the cut end of the stem or root. It is, however, possible that solutes are abstracted from the xylem sap as it moves upwards and so the emergent sap is more dilute than that in the operative region. Such a dilution has been demonstrated in barley plants by Oertli (1966). It would therefore be unwarranted to conclude from the rather small intercepts such as featured in Fig. 1 that there is any active water transfer.

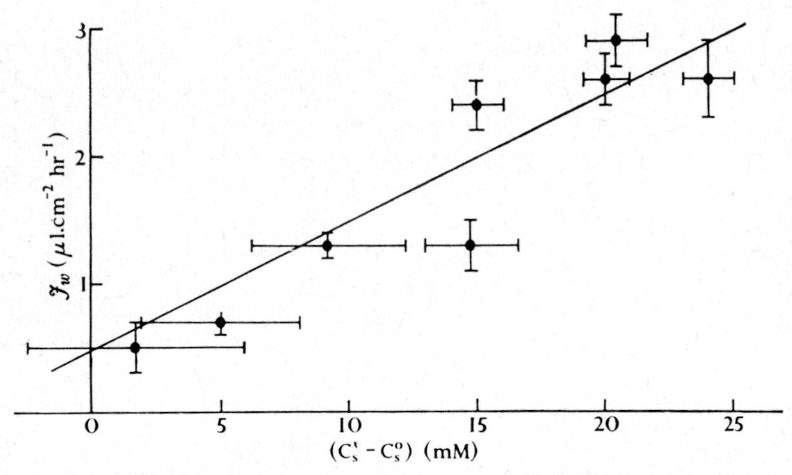

FIG. 1. The relation between rate of exudation, J_w, and the difference in osmolarity between exudate and external medium ($C_s^x - C_s^o$). Vertical and horizontal bars indicate $\pm$ S.E. (from House and Findlay, 1966: by permission of the Clarendon Press, Oxford).

In this connexion the work of Ginsburg and Ginzburg (1970) is particularly interesting since they used root "sleeves" i.e. lengths of root from which the steles had been pulled out, the rupture being at the endodermis and a cylinder of cortex remaining. The sleeves were irrigated internally and externally with solutions of known osmotic pressure. With this arrangement no "abstraction error" could occur. Even so when J_v was plotted against $\Delta\pi$ straight lines were obtained which, when extrapolated, gave a significant J_v with $\Delta\pi = 0$. However this was rejected by the authors as incontrovertible evidence for an active water transport and a model system has been put forward (Ginsburg, 1971) in which water is envisaged as moving in the symplasm and the iso-osmotic flow ($\Delta\pi = 0$) results from a difference in values for σ at the outer and inner surfaces of the sleeves. To sum up the present position: iso-osmotic flow can occur across the root cortex and does not necessarily depend on an intact endodermis. It is small as a fraction of the total flux and always seems to be associated with a flux of solutes of one sort or another (Anderson *et al.*, 1970; Tyree, 1970). It does not appear to be a manifestation of an active water flux, the existence of which has not yet been unequivocally demonstrated, and the importance of which in the normal life of the plant is doubtful.

C. The Pathway of Water Movement

Three possibilities present themselves and are illustrated in Fig. 2 (a, b and c): (1) Water might move from vacuole to vacuole down a stepwise gradient of

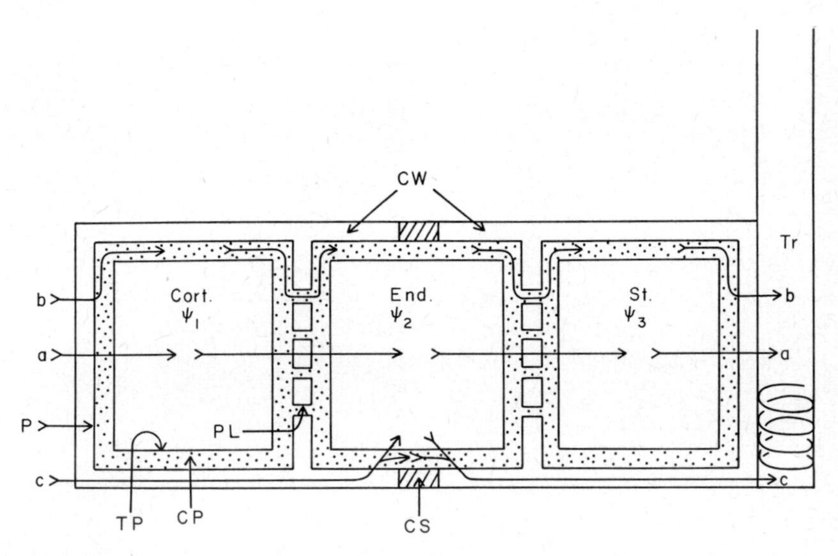

FIG. 2. Diagram showing three pathways for water movement across the root cortex, (a) vacuole to vacuole, (b) movement in the symplast, (c) movement in the cell walls. Cort. = Cortical cell; End. = Endodermal cell; St. = Stelar cell; Tr. = Trachea; CP = Cytoplasm; CS = Casparian strip; CW = Cell Wall; P = Plasmalemma; PL = Plasmodesma; TP = Tonoplast.

vacuolar water potential ($\Psi_1 - \Psi_3$). At each cell the water must cross the tangential cell wall, plasmalemma, cytoplasm and tonoplast on entry and exit. (2) A second possible pathway is the symplast. Here water would enter through the plasmalemma of the epidermal cells and thence traverse the whole cortex and stele within the symplast, moving from cell to cell through the plasmodesmata and leaving via the plasmalemma of the parenchyma cells adjacent to the tracheae. This pathway would thus entail passage across only two cell membranes. (3) Water might move in the cell walls in response to a gradient of hydrostatic pressure. This pathway is interrupted at the endodermal cylinder by the Casparian strips. At this point water must enter the endodermal cells and move through the cytoplasm and perhaps the vacuole and emerge on the inside of the endodermis through the plasmalemma into the cell wall system of the stele (see also p. 415).

With this pathway then, the only membranes through which water need move in its radial passage to the tracheae are those of the endodermis.

Thus we are presented with three pathways in parallel. Water demonstrably moves through the cell walls (De Lavison, 1910; Strugger, 1938–9; Tanton and Crowdy, 1972) and since the symplast and membranes are to some degree permeable to water, movement must also occur through them too (see also p. 429). But what we want to know is the relative flow in each, in other words, what are the relative resistances of the three channels? There is strong indirect evidence that flow from vacuole to vacuole can only constitute a minor part of the total. Following the argument of Briggs (1967) if J_v is the rate of swelling or contraction of the cortical cells when immersed in osmotica and the difference in water potential between the cell wall and vacuole of each cell is $\Delta\Psi$ bars then:

$$J_v = P\Delta\Psi \text{ or } \Delta\Psi = \frac{J_v}{P}$$

where $J_v = \text{cm sec}^{-1}$ and P is the permeability coefficient (cm sec^{-1} bar^{-1}).

Now when water is moving *through* a tissue it will encounter two resistances at each cell, one on entering and the other on leaving, so that if there are n cells across the cortex, the total drop in water potential will be

$$\Delta\Psi = \frac{J_v 2n}{P} \tag{6}$$

A value for J_v of transpiring plants can be obtained from Brouwer (1953) who measured the rate of uptake by a single root of a transpiring bean plant and assuming a root diameter of 1 mm, a value of $J_v = 10^{-5}$ cm sec^{-1} is obtained. Again the values of P for individual plant cells vary widely. Values as high as 10^{-5} cm sec^{-1} bar^{-1} (Dainty and Hope, 1959) and as low as 10^{-6} cm sec^{-1} bar^{-1} (Stadelmann, 1963) have been obtained. Taking the highest value for P, i.e. 10^{-5} cm sec^{-1} bar^{-1}, Brouwer's rate of uptake would require a value of $\Delta\Psi$ of 2 bars/cell. If there were as few as five cells along a radius through cortex and stele, $\Delta\Psi$ would be 10 bars between medium and trachea. This drop of potential seems rather unlikely and if Stadelmann's value for P is used, say 10^{-6} cm sec^{-1} bar^{-1}, $\Delta\Psi$ would be ten times greater—clearly an impossible figure. Looked at in another way, the resistance to flow of the whole cortex is no greater than that of a single layer of cells as measured by osmotic influx or efflux with single cells. Thus it may be concluded that either the membranes of root cells are exceptionally permeable, or that water travels across the cortex without crossing the membranes, e.g. through the cell walls or symplast, crossing only two membranes, those of the endodermis in the former case, or on entry to and exit from the symplast in the latter.

Briggs (1967) lends support to the cell wall as a perfectly feasible pathway by calculating the difference of pressure between medium and tracheae that would be needed to drive water through the radial walls of the cortex. He obtains a figure for the distance apart of the microfibrils of cellulose and applies Poiseuille's equation. He concludes that the pressure drop across a radial chain of five cortical cells would be only about 1·4 bars.

As pointed out above, if the resistance of the cell walls were small, the site of the main resistance might reside in the endodermis. The work of Ginsburg and Ginzburg (1970) is interesting in this connexion. They used cortical sleeves in which the endodermis was ruptured and could not therefore play any part in the movement of water. They measured the flux of water across the sleeves in response to differences of osmotic pressure and obtained values of Lp of approximately 10^{-6} cm sec^{-1} bar^{-1}; in other words similar to a root system with an intact endodermis. This suggests that the endodermis is not the site of the main resistance. Furthermore unlike the intact root the cell wall system could, in my view, have played little part as a pathway since in their system there was no difference of hydrostatic pressures across the cortex and hence no mass flow in the walls. (NB: in a root with an intact endodermis there will be a gradient of hydrostatic tension in the cell walls of the cortex irrespective of whether the water potential in the stele is lowered by the presence of solutes or hydrostatic pressure; the outer surface of the endodermis could not discriminate between the two). Ginsburg and Ginzburg concluded that the movement of water was in the symplast and considered that the flow through the cell wall was inconsistent with their finding of a reflexion coefficient of unity and a marked inhibition of the flux by DNP, KCN and CCCP. This inhibition they attribute to an effect on movement within the symplast since movement of water across cell membranes is rather insensitive to inhibitors. A measurement of the water flux, across the sleeves produced by a difference of hydrostatic pressure, would have been interesting since this would induce a flow through the cell walls. Such an experiment is hardly practicable though, since the sleeves are too delicate to withstand the necessary pressure difference.

The position with regard to the radial pathway for water seems to be that the fraction moving from vacuole to vacuole is small because the resistance of the plasma membrane is too great judged from permeability studies on other cells. It may be that the cells of the root cortex are exceptionally permeable to water but until this is measured with parallel measurements of Lp we cannot be certain. Meantime it would appear more likely that most of the flow is in the cell walls or symplast or both (Tyree, 1970).

It is not easy to decide unequivocally between them. With the root sleeves of Ginsburg and Ginzburg, flow through the cell walls will be small, but the high value of Lp eliminates the likelihood of vacuole to vacuole movement and so we are left with the symplast as the pathway. With intact roots on the other hand there could be a gradient of hydrostatic tension in the cell wall system leading to a flow of water from the epidermis to the endodermis. Passage through the endodermis would mean crossing the plasma membrane at both entry and exit and it is here that one would expect the main resistance to reside and this too would be the inhibitor-sensitive step (though the possibility that the resistance of the cell-wall system could be inhibitor-sensitive cannot be entirely ruled out). Perhaps the similarity of the values of Lp for intact segments and sleeves points to the symplast as the major pathway in both, but with the more complex intact system, cell wall movement cannot be excluded.

The mechanism of the movement of water, i.e. diffusion, laminar flow, the role of protoplasmic streaming, etc. has not been included in the scope of the present contribution, but reference on this subject may be made to Tyree (1970), Clarkson *et al.*, (1971) and Tanton and Crowdy (1972).

III. Root Resistance of the Transpiring Plant

The investigations described so far concern the exuding root system, i.e. when the flux of water is in response to a difference of osmotic pressure and is necessarily rather slow. Under these conditions the relationship between flux and $\Delta\pi$ is demonstrably linear. However, with transpiring plants the resistance of the root system seems to be a function of the flux itself. The water potential in the xylem of the root of an intact plant cannot be measured directly, but if the root offers the major resistance to water flow within the plant (Tinklin and Weatherley 1966), the water potential of the transpiring leaf may be taken as approximating to that of the xylem in the root. It has been found (Macklon and Weatherley, 1965; Tinklin and Weatherley, 1966; Stoker and Weatherley 1971) that when the transpiration is varied by changing the humidity of the airstream in a wind-tunnel, the relationship between rate of transpiration and leaf water potential far from being linear, has three phases (Fig. 3). At very low rates of transpiration there is a steep rise in depression of leaf water potential ($\Delta\Psi_l$) with increasing rate of transpiration (exudation would presumably fall in this phase), but above a certain rate of transpiration a plateau value of $\Delta\Psi_l$ is reached (at between 6–9 atm for cotton and sunflower) beyond which there is little further rise in $\Delta\Psi_l$ over a wide range of transpiration rate. Lastly, at very high rates of transpiration there is a linear rise in $\Delta\Psi_l$ with increasing transpiration. It has long been known (Kramer, 1933) that killing a root system decreases its resistance and, as shown in Fig. 3, a killed root system has not only less resistance but gives a linear plot between $\Delta\Psi_l$ and rate of transpiration, indicating that the hydraulic resistance coefficient is a constant. In other words it behaves like a straightforward physical resistance. Judging from Fig. 3, it is tempting to think that the third phase of the living root system coincides with an extrapolation of the line obtained with the dead root system. However, the significance of such a situation is at present not understood.

There is also no clear explanation of the plateau phase. A fall in resistance with increasing rate of transpiration has long been established (Jost, 1916; Köhnlein, 1930) whilst Brewig (1936) and Brouwer (1953, 1954) have shown that at low rates of flow most of the water is taken up by the tip region of the root whereas at high rates it is the basal region which is the main absorbing zone. This locates the phenomenon but provides no explanation. Brouwer suggested that an increase in flow would be accompanied by a lower turgidity in the root cells and this might increase their hydraulic conductivity. However

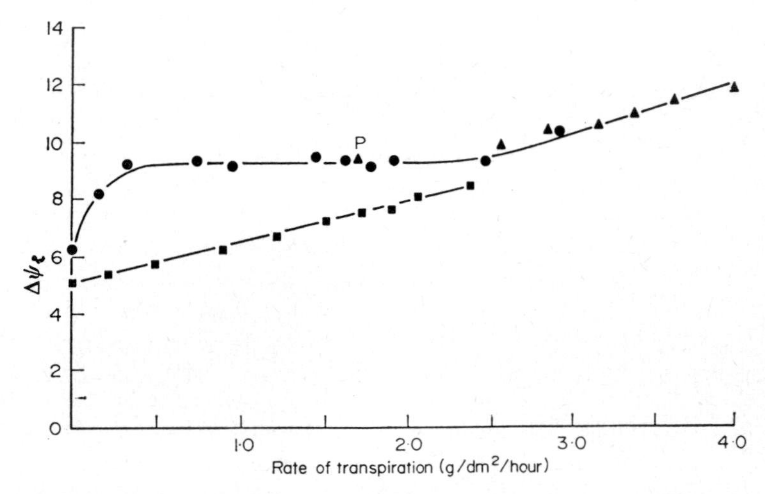

FIG. 3. Relationship between depression of water potential in the leaves ($\Delta\Psi_l$) and the rate of transpiration. Plants with living root systems (● and ▲) and with heat killed root systems (■). (After Stoker and Weatherley, 1971: by permission of Blackwell Sci. Pub.).

this could not offer an explanation for our plateau phase since the $\Delta\Psi$ in the xylem, and hence the turgidity of the cortical cells, remains constant throughout. Thought of in terms of a variable resistance it means that as the transpirational flow increases, there is a proportionate fall in resistance. This fall seems to be instantaneous for when the water content of one of the leaves was continuously monitored using a β gauge, no change was recorded when the plant was suddenly subjected to a threefold increase in rate of transpiration (Stoker, 1968; Weatherley, 1974). Also these authors found that at temperatures of 10°C and 0°C the adjustment of resistance with increasing flow was lessened but still remained large. Perhaps these facts can be interpreted in terms of the structure of the pathway, the maintenance of which needs metabolic energy and which becomes more conductive as the flow of water within it increases. This latter property is also manifest by a simple physical system, the rotameter flowmeter, the resistance of which remains constant as the rate of flow through it increases (Tinklin and Weatherley, 1966). However, the equivalence to the rotameter in physiological and cytological terms is difficult to see and no other testable hypothesis has yet been put forward.

IV. The Resistance of the Soil

So far the above experiments with transpiring plants were carried out with the plants in water culture, i.e. with the roots in a medium at virtually zero water potential. However, as shown by Macklon and Weatherley (1965) and Tinklin and Weatherley (1968) values of $\Delta\Psi_l$ much higher than the plateau values (6–9 bars) were obtained when the plants were rooted in sand or soil. Tinklin and Weatherley grew *Ricinus communis* plants in boxes of sand or soil in which a water table was maintained at various levels below the root system. The boxes were placed in a windtunnel and the rate of transpiration varied by altering the humidity of the air stream. It was found that if the water table was high (say 15 cm below the roots) $\Delta\Psi_l$ remained at a plateau value of 6 bars irrespective of the rate of transpiration, thus the plant behaved as if the roots were surrounded by free water. However, when the water table was 25 cm below the roots, $\Delta\Psi_l$ remained at a value of 6 atm provided the transpiration rate was low (0·4 g dm^{-2} hr^{-1}), but when the rate of transpiration was raised threefold, $\Delta\Psi_l$ rose steeply to a value of 12 bars, a steady value being attained in about 8 h (see Fig. 4). On reducing the rate of transpiration $\Delta\Psi_l$ returned to 6 bars.

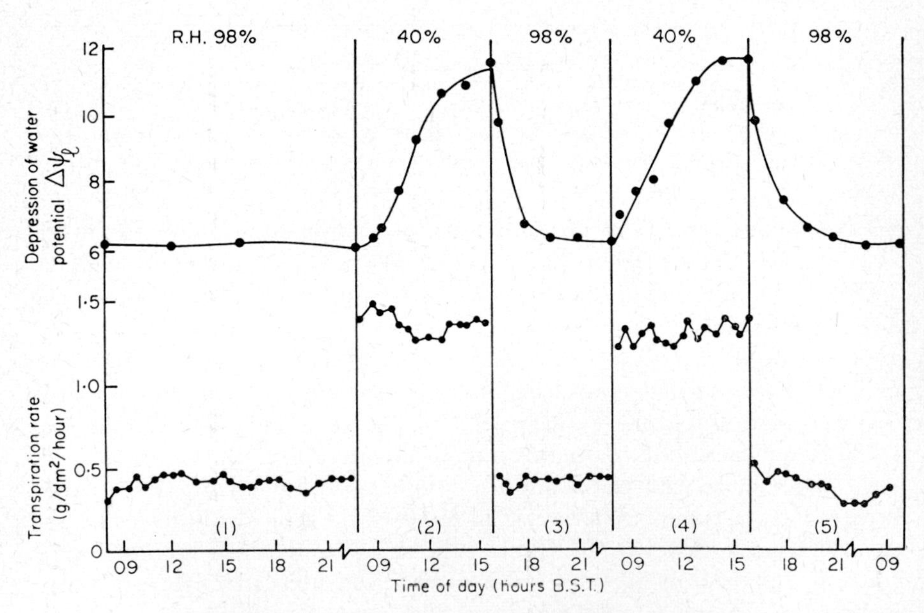

FIG. 4. Effects of repeated sudden changes of relative humidity on the rate of transpiration and leaf water potential of sand rooted plants. The water table was 25 cm below the rooting zone. (From Tinklin and Weatherley, 1968: by permission of Blackwell Sci. Pub.).

This rise in $\Delta\Psi_l$ is interpreted as reflecting a rise in the $\Delta\Psi$ of the water in the sand at the root surface and this in turn was thought to arise because of the low water conductivity of the sand at this height above the water table. Thus the value of $\Delta\Psi_l$ of 12 bars represented a soil moisture tension of 6 bars in the perirhizal zones with an additional drop of 6 bars from root surface to leaf (= plateau value). When the rate of transpiration was reduced, $\Delta\Psi_l$ returned to a value of 6 bars, this recovery representing the rewetting of the perirhizal zones. Similar patterns of results were obtained with soil of various textures.

The interpretation of these results has been criticized by Newman (1969) who argues that, on theoretical grounds, the hydraulic resistance of the soil round the roots would not be great enough to cause any appreciable drop of water potential in the rhizosphere (= perirhizal zones) unless the water potential of the soil approached the wilting point; in these experiments it certainly did not. Newman's explanation of Tinklin and Weatherley's results is that, at the higher rates of transpiration, the soil in the rooting zone as a whole becomes depleted of water owing to the resistance to flow in the soil between the rooting zone and the water table or the adjacent masses of moist soil. This so called pararhizal resistance can be considerable because of the greater distances through which the water has to travel. On this interpretation the rise in $\Delta\Psi_l$ with increased transpiration (Fig. 4) represents the rise in $\Delta\Psi_s$ (depression of water potential of the soil) of the whole mass of the soil in the rooting zone and the fall in $\Delta\Psi_l$ on reducing the transpiration represents the rewetting of this soil by movement from the adjacent soil masses. This is entirely different from the hypothesis of Tinklin and Weatherley where the changes in $\Delta\Psi_l$ reflect redistribution of water *within* the rooting zone.

Some recent work by Faiz (1973) suggests that Newman's explanation is not correct. Sunflower plants were grown in sand or soil above a fixed water table as in the experiments of Tinklin and Weatherley. However, in these experiments the container was divided into upper and lower sections separated by a diaphragm of nylon gauze attached to the bottom of the upper section. Both containers were filled with sand or soil and pressed together so that there was capillary continuity between the sand in the two containers. Psychrometers were fixed in the paper section to monitor $\Delta\Psi_s$ within the rooting zone. A seedling was planted in the upper section four or five days prior to the experiment. In this time the roots ramified densely throughout the upper section, but could not penetrate the gauze into the lower section. The lower section was stood in a pan of water and a constant water table maintained.

When the water table was maintained at a little over 15 cm below the gauze $\Delta\Psi_l$ remained at about 7 bars even at a high rate of transpiration. This indicates that upward movement of water to the rooting zone was not hindered by the gauze. The results obtained when the water table was 25 cm below the gauze are shown in Fig. 5. At a low rate of transpiration $\Delta\Psi_l$ remained at 8 bars—a plateau value indicating the absence of any significant hydraulic resistance in the sand. When the rate of transpiration was increased $\Delta\Psi_l$ rose steadily over a period of 48 h to a value of 15 bars. At this point the lower section was removed and the upper section was closed in a polythene bag and kept in darkness at a high humidity to reduce the rate of transpiration to a minimum. A sharp fall in $\Delta\Psi_l$ followed and after about 12 h an equilibrium value of approximately 7 atm was

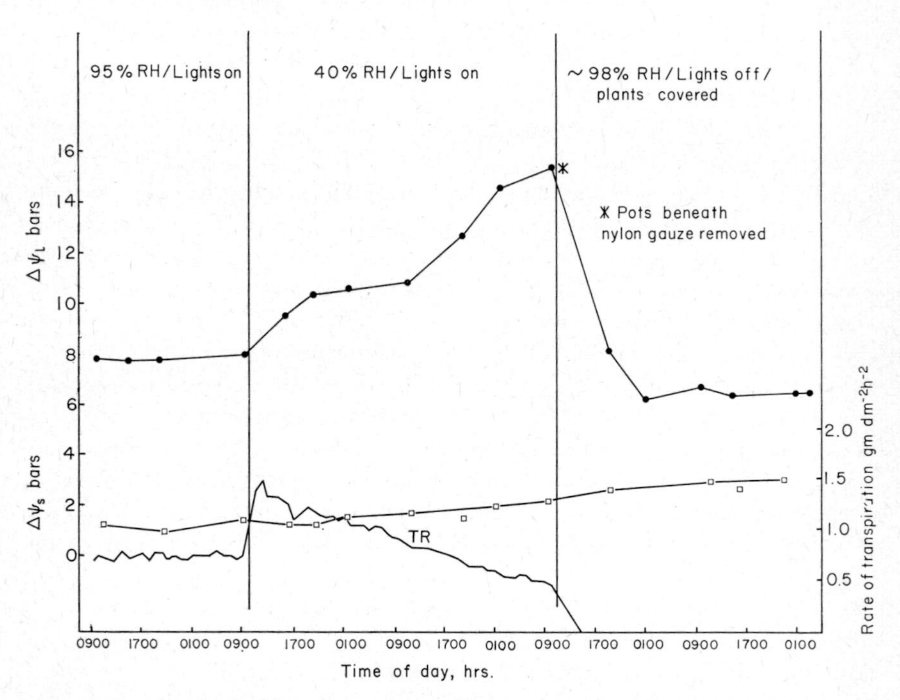

FIG. 5. Results of a double pot experiment demonstrating that after the development of a water stress ($\Delta\Psi_l$) resulting from a period of rapid transpiration, recovery can take place when continued flow into the rooting zone is prevented. For further explanation see text. (After Faiz, 1973).

attained. Clearly this recovery was to do with events in the upper section and not due to a continued supply of water from the lower section. Further it will be seen that the $\Delta\Psi$ of the sand in the rooting zone rose steadily throughout the experiment from an initial value of 1 bar to a final one of 3 bars. During the final recovery period, it will be noted that the fall in $\Delta\Psi_l$ was accompanied by a slight *rise* in $\Delta\Psi_s$ consistent with an internal adjustment within the rooting zone. Similar results were obtained with garden soil.

Subsequent examination of the root system in these experiments revealed a dense growth of roots evenly distributed throughout the container so that any perirhizal gradients in water potential must be very steep. However as Newman pointed out, on theoretical grounds such steep gradients are not to be expected, and further experiments of the above pattern (Faiz, 1973) argue against their existence. In these experiments two more parameters were measured. First, $\Delta\Psi_s$ was measured at different depths both above and below the gauze and secondly, an estimate of the total surface area of the root system was made from measurements of length and diameter of representative samples. When the water table was 23 cm below the gauze and a high rate of transpiration maintained, $\Delta\Psi_l$ reached a value of 13 bars in 12 h and thereafter remained steady for the following 22 h. During this period $\Delta\Psi_s$ of the sand in the rooting zone was uniformly 2·5 bars down to a level 3·5 cm above the gauze, whilst a value of 1 bar was recorded immediately above the gauze and zero immediately below the gauze. Thus the pararhizal gradient including the gauze was 2·5 bars in 3·5 cm = 0·7 bar cm^{-1}. Now, since these figures refer to a steady state, the volume flow upwards from the lower container into the rooting zone equals the flow into the roots, and the relative velocities of pararhizal flow and perirhizal flow will be inversely proportional to the surface areas through which flow is occurring. In this experiment the surface area of the gauze was 177 cm^2 and the estimated surface area of the roots was 16023 cm^2. Thus relative velocity is 16023 ÷ 177 = 91 and assuming that the whole surface area of the roots was absorbing, it can be concluded that the velocity of flow into the roots was only a nintieth that of the parathizal flow with the gradient of potential in the perirhizal zones correspondingly smaller. If the pararhizal gradient was, as shown above 0·7 bar cm^{-1}, the perirhizal gradient will be of the order of 10^{-2} bar cm^{-1} and since the roots are less than 1 cm apart, the drop of water potential across the hypothetical perirhizal zones would probably be less than 10^{-2} bar, an insignificant value relative to the demonstrated difference in Ψ between the bulk sand in the rooting zone and the root surface. In this experiment this difference amounts to 4·5 bars, arrived at in the following way. The total difference in water potential between the soil in the rooting zone and the leaf is 10·5 bars (13 — 2·5) .The drop of potential within the plant is 6 bars (plateau value). Thus the drop in water potential between the bulk soil and the root surface is 4·5 bars (10·5 — 6). In an experiment of the same kind using garden soil the following steady state values were obtained under conditions of rapid transpiration. $\Delta\Psi_l$ = 15 bars: soil in rooting zone = 2 atm, thus total drop within the system = 13 bars (15 — 2). Drop in potential within the plant = 6 bars (plateau value) ∴ apparent drop in potential within the soil = 7 bars (13 — 6).

We are thus faced with a large difference in water potential between the surface of the root and the bulk soil in the rooting zone (a distance of only a few millimetres) which is difficult to explain in terms of perirhizal gradients. It is therefore suggested (Faiz, 1973) that the site of the large drop in potential outside the root is the soil: root interface. It seems quite possible that the interface between a particulate medium, such as the soil, and elongated cylindrical objects, such as root hairs or the root surface, might have special properties; for example with an unsaturated soil the water absorbing area of the root may be restricted to wedges of water around points of contact between root and soil particles. It can easily be imagined that the lower the water content of the soil and the more rapid the rate of absorption by the root surface, the more restricted may these areas of contact become.

Another factor which would undoubtedly increase the resistance of the soil: root interface is the contraction of the roots with increasing hydrostatic tension in the plant thus possibly leaving a gap between soil and root surface. The occurrence of such a gap has been envisaged by Phillip (1957), Bonner (1959) and De Roo (1969), whilst shrinkage of roots with increasing water stress was observed long ago by MacDougal (1936) and more recently by Huck *et al* (1970) and Faiz (1973). Also Cruiziat (1972) found that for any degree of water stress in the plant the relative water content of the roots was lower than that of the shoots, whilst Faiz (1973) has found that with sand rooted sunflower plants the relative water content of the roots fell from 94% to 70% during a period of rapid transpiration and this was accompanied by a reduction in root diameter of 19%. Evidence for the development of a gap between soil and root surface has also been provided by Faiz (1973). He grew sunflower plants in sand contained in polythene bags and the water stress in the plant was monitored by a β gauge fitted to one of the leaves. During a period of rapid transpiration, the water stress in the leaves steadily increased. When the bag was gently squeezed the β gauge recorded an immediate fall in stress, presumably due to the elimination of the gaps by disturbing the sand around the roots. A similar result was obtained when potted plants were briefly shaken on a mechanical shaker. The formation of such perirhizal gaps would presumably not initiate a rise in $\Delta\Psi_l$ above the plateau value. A rise in $\Delta\Psi_l$ would first occur because of the resistance of the soil: root interface and this would lead to a contraction of the roots and the formation of gaps which would lead to a secondary increase in resistance.

V. Summary and Conclusions

Movement of water through the root is in response to a difference in water potential between the soil and the tracheae resulting from a lowering of water potential in the latter. This lowering of water potential arises either from a secretion of ions into the xylem or by a lowering of pressure due to the transpirational pull. These represent the old so-called "active" and "passive" absorptions. In fact, they are probably largely passive (down the potential gradient) in both cases. There is some evidence which at first sight points to an active fraction, i.e. a continued flow through the root when there is no difference in water potential between medium and tracheae. However, closer examination leads to the conclusion that this flow could result from a solute movement of one sort or another. In any case it is a small fraction of the total flow and is a negligible factor in the normal functioning of the root.

With regard to the pathway followed by the water there are three possibilities. It could move either from vacuole to vacuole along a gradient of water potential, in the cell wall system in response to a gradient of hydrostatic pressure or in the symplast. No doubt it moves along all three, but is one predominant? There is strong evidence that little flow occurs from vacuole to vacuole, for the resistance of the whole cortex and stele is little more than that expected of a single layer of cells judging from cell membrane permeability measurements. It is less easy to discriminate between the apoplast and symplast, and conceivably both play a part.

Whichever pathway the water follows, the cells impose a resistance to flow. This resistance is very sensitive to environmental factors such as aeration and temperature and, rather curiously, to the flux of water itself. The faster water moves through the root the lower is the root resistance and in some cases the fall in resistance is proportional to the increase in flux so that the difference in water potential across the root remains constant in the face of wide changes in the rate of transpiration. This adjustment does not seem to be "metabolic" in that it is not inhibited by low temperature and is immediate. No satisfactory mechanism for it has as yet been proposed.

For many plants the resistance of the root system is the major one within the plant and is therefore the main factor controlling the water stress developing in the plant for a given rate of transpiration. This is only true however if the soil is more or less saturated with water. Quite modest soil moisture tensions can give rise to much larger tensions within the plant and this has been attributed to a high hydraulic resistance developing in the zones of soil immediately surrounding the roots. However, such high resistances are not expected on theoretical grounds and this is supported by measurements of gradients of water potential in the soil at various distances from the roots. But the fact remains that there is a considerable drop in water potential adjacent to the root surface and it is proposed that soil : root interface can present a high resistance barrier to flow of water into the root. The tensions in the plant which arise from this high resistance can cause contraction of the roots, leading to a gap forming between the soil and the root surface and thus a secondary increase in resistance may arise.

References

Anderson, W. P., Aikman, D. P. and Meiri, A. (1970). Excised root exudation: a standing gradient osmotic flow. *Proc. Roy. Soc.* (*London*) *Ser. B.*, **174**; 445–458.

Arisz, W. H., Helder, R. J., and Van Nie, R. (1951). Analysis of the exudation process in tomato plants. *J. exp. Bot.* **2**; 257–297.

Bonner, J. (1959). Water transport. *Science, N.Y.*, **129**; 447–450.

Briggs, G. E. (1967). "Movement of Water in Plants." Botanical Monographs, Blackwell Scientific publications, Oxford and Edinburgh.

Brewig, A. (1936). Die Regulationserscheinungen bei der Wasseraufnahme und die Wasserleitgeschwindigkeit in *Vicia faba* Wurzeln. *Jb. wiss. Bot.*, **82**; 803–828.

Brouwer, R. (1953). Water absorption by the roots of *Vicia faba* at various transpiration strengths. II Causal relation between suction tension, resistance and uptake. *Proc. K. Ned. Akad. Wet.* **C56**; 129–136.

Brouwer, R. (1954). The regulating influence of transpiration and suction tension on the water and salt uptake by the roots of intact *Vicia faba* plants. *Acta Bot. Neerl.*, **3**; 264–312.

Clarkson, D. T., Robards, A. W. and Sanderson, J. (1971). The tertiary endodermis in barley roots: fine structure in relation to radial transport of ions and water. *Planta* (*Berl.*), **96**; 292–305.

Cruiziat, P. (1972) Contribution à l'étude des réserves en eau de la plante. Thése, Fac. Sci., Paris.

Dainty, J. and Hope, A. B. (1959) The water permeability of cells of *Chara Australis* R. Br. *Aust. J. biol. Sci.* **12,** 136–345.

De Lavison, J. de Rufz (1910) Du mode de pénétration de quelques sels dans la plante vivante. *Refue gén. Bot.* **22**; 225.

De Roo, H. C. (1969). Water stress gradients in plants and soil-root systems. *Agron. J.* **61,** 511–515.

Faiz, S. M. A. (1973) Soil-root water relations. Ph.D. thesis, University of Aberdeen.

Ginsburg, H. (1971). Model for iso-osmotic flow in plant roots. *J. theor. Biol.*, **32**; 147–158.

Ginsburg, H. and Ginzburg, B. Z. (1970). Radial water and solute flows in roots of *Zea mays*. I. Water flow. *J. exp. Bot.*, **21**; 580–592.

House, C. R. and Findlay, N. (1966). Water transport in isolated maize roots. *J. exp. Bot.*, **17**; 344–354.

Huck, M. G., Klepper, B. and Taylor, H. M. (1970). Diurnal variations in root diameter. *Plant Physiol.*, **45**; 529–530.

Jost, L. (1916). Versuche über die Wasserleitung in der Pflanzen. *Z. Bot.*, **8**; 1–55.

Köhnlein, E. (1930). Untersuchungen über de Hohe des Wurzelwiderstandes und die Bedeutung aktiver Wurzeltätigkeit fur die Wasserversorgung der Pflanzen. *Planta* (*Berl.*), **10**; 381–423.

Kramer, P. J. (1933). The intake of water through dead root systems and its relation to the problem of absorption by transpiring plants. *Am. J. Bot.* **20**; 481–492.

Macdougal, D. T. (1936). Studies in tree growth by the dendrographic method. *Carnegie Inst. Wash. Publ.* 642.

Macklon, A. E. S. and Weatherley, P. E. (1965). Controlled environment studies of the nature and origins of water deficits in plants. *New Phyt.*, **64**; 414–427.

Mees, G. C. and Weatherley, P. E. (1957a). The mechanism of water absorption by roots. I. Preliminary studies on the effects of hydrostatic pressure gradients. *Proc. Roy. Soc.* (*London*) **B147**; 367–380.

Mees, G. C. and Weatherley, P. E. (1957b). Idem. II. The role of hydrostatic pressure gradients across the cortex. *Proc. Roy. Soc.* (*London*), **B147**; 381–391.

Newman, E. I. (1969). Resistance to water flow in soil and plant. II. A review of experimental evidence on the rhizosphere resistance. *J. Appl. Ecol.*, **6**; 261–272.

Oertli, J. J. (1966). Active water transport in plants. *Physiol. Plant.* **19**; 809–817.

Phillip, J. R. (1957). The physical principles of soil water movement during the irrigation cycle. *Proc. Int. Congr. Irrig. Drain*, **8**; 125–154.

Sabinin, D. A. (1925). On the root systems as an osmotic apparatus. *Bull. Inst. Recherche biol. Univ. Perm.* (*Molotov*) **4**; Suppl. 2; 129–136.

STADELMANN, E. (1963). Vergleich und Umrechnung von Permeabilitäts—konstanten fur Wasser. *Protoplasma*, **57**; 660–678.

STOKER, R. (1968). "Studies in the Water Relations of the Transpiring Plant with Emphasis on the Role of the Roots." Ph.D. thesis, University of Aberdeen.

STOKER, R. and WEATHERLEY, P. E. (1971). The influence of the root system on the relationship between the rate of transpiration and depression of leaf water potential. *New Phytol.*, **70**; 547–554.

STRUGGER, S. (1938–9). Die lumineszenzmikroscopische Analyse des Transpirationsstromes in Parenchymen. *Flora*, (Jena), **133**; 56.

TANTON, T. W. and CROWDY, S. H. (1972). Water pathway in higher plants. II. Water pathways in roots. *J. exp. Bot.* **23**; 600–618.

TINKLIN, R. and WEATHERLEY, P. E. (1966). On the relationship between transpiration rate and leaf water potential. *New Phytol.* **65**; 509–517.

TINKLIN, R. and WEATHERLEY, P. E. (1968). The effect of transpiration rate on the leaf water potential of sand and soil rooted plants. *New Phytol.* **67**; 605–615.

TYREE, M. T. (1970). The symplast concept. A general theory of symplast transport according to the thermodynamics of irreversible processes. *J. theor. Biol.* **26**; 181–214.

WEATHERLEY, P. E. (1974). The hydraulic resistance of the root system. *In*: "Structure and Function of Primary Root Tissue." (Ed. J. Kolek). pp. 297–308. Symposium Proceedings, Czechoslovakia 1971. Slovak Acad. Sci., Bratislava.

Chapter 19

The Endodermis, Its Structural Development and Physiological Role

D. T. CLARKSON AND A. W. ROBARDS

Agricultural Research Council, Letcombe Laboratory and Department of Biology, University of York.

I. Introduction

For many purposes a root axis can be envisaged as a central core of vascular tissue separated from an outer collar of collecting cells (the cortex) by the endodermis (Van Tieghem and Douliot, 1886). By virtue of both its structure and position the endodermis is well fitted to influence the movement of substances between the vascular stele and cortex. Substances entering or leaving the stele must enter endodermal protoplasts, either by crossing their plasma membranes or by numerous plasmodesmata linking the endodermis with neighbouring cells in the cortex and pericycle. These constraints are determined, in the State I endodermis, by the deposition of hydrophobic bands, in the radial walls of endodermal cells, known as Casparian bands (or strips) (Kroemer, 1903; De Lavison, 1910; Priestley and North, 1922).

Although greatly varying in appearance the same basic features of the endodermis occur in all roots of vascular plants except the Lycopodiaceae (Mager, 1907), most stems and some leaves. The evolution of its structure occurred early in the development of vascular plants; the endodermis of fossil horsetails of carboniferous times (see Walton, 1953) appears much like its modern angiosperm counterpart. When roots and stems undergo secondary thickening resulting from cambial activity, the endodermis may first stretch (Harrison-Murray and Clarkson, 1973) and then undergo radial division (Van Wisselingh, 1926; Scott, 1928; Bond, 1930) to accommodate the enlarging stele. These observations suggest that there may be some general necessity for a barrier between the vascular elements and the outside environment.

One might imagine that electron microscopy would yield a great deal of new information about the structure of the endodermis. This is not entirely the case. The endodermis was a favourite subject for study by anatomists at the end of the last century and for the first few decades of the present one (e.g. Kroemer, 1903; Mylius, 1913), and modern microscopy has not added greatly to these early and impressive observations (see Priestley and North, 1922 and Van Fleet, 1961 for general reviews), except to show that the endodermis is a highly perforated structure, and that, in spite of its massive appearance at later stages of its development, its cellulosic walls and suberin lamellae are crossed by large numbers of plasmodesmata (Clarkson *et al.*, 1971). Failure to appreciate this fact may have been the cause of the widespread misunderstanding about the significance of suberization and the later stages of endodermal differentiation (e.g. Priestley and North, 1922). Briefly, it has been held that the formation of suberin lamellae effectively seals off access to the stele from the cortex, drastically reducing the transport of material between them; the direct consequence of this view is that roots are said to have an absorbing zone, confined to the apical parts of root axes and branches, in which most of the water and nutrients used by the plant are taken up. We shall show that, in fact, this is not the case in two root systems of contrasting type.

Experiments and observations of several kinds, mainly in barley but also in some other species, will be used to illustrate the role of the endodermis. Since the technical details of these experiments have been described elsewhere only brief notes on the methods and conditions used are given as footnotes to tables and figures.

II. General Features of the Endodermis in Roots

The endodermis is a layer of cells, forming a continuous sheath around the vascular stele, in which four developmental stages can be usually recognized, viz. Pro-endodermis, States I, II, and III. (These last three stages are also referred to as primary, secondary and tertiary. This terminology will be avoided here because it is not synonymous with the corresponding stages of plant cell wall development, and can, therefore, be misleading).

A. Pro-endodermis

Van Fleet (1961) has shown that those cells recently initiated by the apical meristem, which are destined to become the endodermis, can be distinguished histochemically at a very early stage by their high phenol content. The ability to synthesize and elaborate phenolic compounds is characteristic of endodermal cells and it is interesting to see that this feature is established towards the end of cell division prior to elongation and differentiation. At somewhat greater distances from the meristem (usually within 200 μm) the pro-endodermis can be recognized distinctly as the innermost rank of cells having tangential axes longer than radial axes—in marked contrast to the underlying pericycle where the reverse is generally true. Although recognizable at this stage, the radial walls of the cells lack Casparian bands.

In seminal root axes of barley at a distance of 4–5 mm from the tip, the cytoplasm adjacent to the radial walls of pro-endodermal cells becomes dense with membraneous material and ribosomes; microtubules are commonly seen arranged in the transverse plane and parallel to the radial wall (Robards *et al.*, 1973). Shortly afterwards (5–7 mm proximal), the inner part of the radial wall takes on the characteristic appearance of a Casparian band.

B. State I

The formation of a State I endodermal cell is marked by the appearance of Casparian bands in the transverse and radial (anticlinal) longitudinal walls. These bands result from the deposition of lipidic and polyphenolic substances within the frame-work of the existing primary wall (Priestley and North, 1922; Van Fleet, 1961). They are, therefore, part of the wall and not, as implied by some authors (Fahn, 1967), deposited on the wall surface. Although the nature of the substances in the Casparian bands remains to be determined precisely, cytochemical methods suggest that suberin and lignin are the major components. Attempts to plasmolyse State I endodermal cells in hypertonic solutions (Fig. 1) show that there is a very tight binding between the protoplast and the cell wall in the region of the Casparian band (Bryant, 1934; Bonnett, 1968). Early reports that this binding was produced by the presence of abundant plasmodesmata (Scott, 1963) through the Casparian bands appear to have been ill-founded,

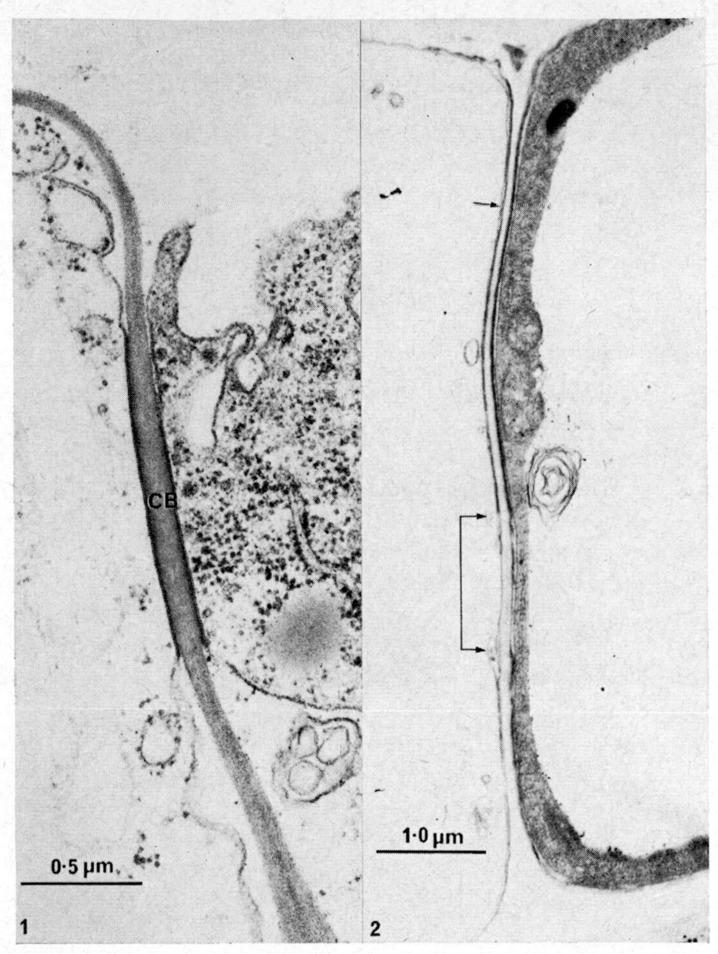

FIG. 1. Radial walls of State I endodermal cells of barley from a transverse section taken 4 cm from the root tip. The cells have been plasmolysed, demonstrating the strong attachment of the plasmalemma to the Casparian band (CB). The roots shown in this, and subsequent, micrographs were fixed in glutaraldehyde followed by osmium tetroxide; dehydration was through an acetone series, or using the diffusion dehydration method; embedding was in epoxy resin. Full details of processing schedules are given in Robards *et al.*, (1973). The Casparian band is revealed as a slightly thicker, slightly more electron-opaque, region of the apposed primary walls.

FIG. 2. Radial walls of State II endodermal cells of barley from a section taken 8 cm from the root tip. The thin, electron-opaque, suberin lamella lines the cell wall of both cells (single arrow). The suberin lamella is much thinner, or missing, along the surface of the Casparian band (indicated by bracketing arrows). As soon as endodermal cells develop suberin lamellae, they become much more difficult to fix and embed satisfactorily—doubtless owing to the presumptive purpose of the suberin lamella in impeding the flow of liquids from the apoplast into the symplast. A more highly magnified view of the suberin lamella, and its stratification, is shown in Fig. 4a.

and recent electron microscopical evidence has shown a very close relationship between the relatively homogenous structure of the Casparian band and the adjacent plasmalemma (Bonnett, 1968; Robards *et al.*, 1973). Indeed, Bonnett has reported that, if the protoplast is plasmolysed so severely that it does separate from the Casparian band, then the membrane cleaves through its mid-line (the presumptive hydrophobic zone). The bonding between wall and membrane is, therefore, an extremely tight and unusual one, possibly involving some association between proteins in the wall and in the outer layer of the plasmalemma.

The appearance of the Casparian bands in the anticlinal walls of different species varies considerably: they may be considerably less than 1·0 μm in the radial direction (e.g. barley), or may occupy virtually the whole of the radial wall > 10·0 μm (e.g. marrow): they do not always extend across the whole width of the wall, although the wall is usually thicker in the region of the band than elsewhere. In longitudinal view the band is often seen to be slightly convoluted; this is probably an artefact arising from differential expansion and contraction (Priestley and North, 1922). Lipid or lignin stains and reagents have been used to demonstrate the band at the optical microscope level; in the electron microscope, normal processing and staining methods reveal it as having a more homogenous structure and slightly more opaque appearance than the rest of the wall. Some early workers (e.g. Priestley and North, 1922) considered that the band was produced by extra-cellular events (e.g. at the point of interaction of inward-diffusing air, and outward diffusing polyphenols), but it seems clear from our own observations that it is, in fact, formed by processes involving cytoplasmic synthesis and secretion.

The formation of State I endodermal cells tends to be synchronized, usually taking place within a few millimetres of the root tip. In some roots no further endodermal wall development takes place, and the endodermis may eventually become lost or occluded when periderm formation occurs. However, in many plants further development to States II and III may take place and, if it does, it tends to be asynchronous.

C. State II

The State II endodermal cell is typified by the presence of a continuous layer of lipidic material—again generally considered to be suberin—between the plasmalemma and the cell wall (Fig. 2). Sometimes this layer is absent or particularly thin over the Casparian bands (Mylius, 1913; Bond, 1930). The suberin is laid down as lamellae in most instances although histochemical tests for suberin have sometimes shown positive results for endodermal cells in which suberin lamellae could not be seen in the electron microscope (e.g. marrow; Harrison-Murray and Clarkson, 1973).

D. State III

State III cells have a relatively thick cellulosic wall deposited over the suberin lamella (Fig. 3). The transition from State II to State III is not always sharply marked: in some cases cellulosic and suberin lamellae may be deposited alternately, while in others the cellulose wall is clearly defined from the outer suberin

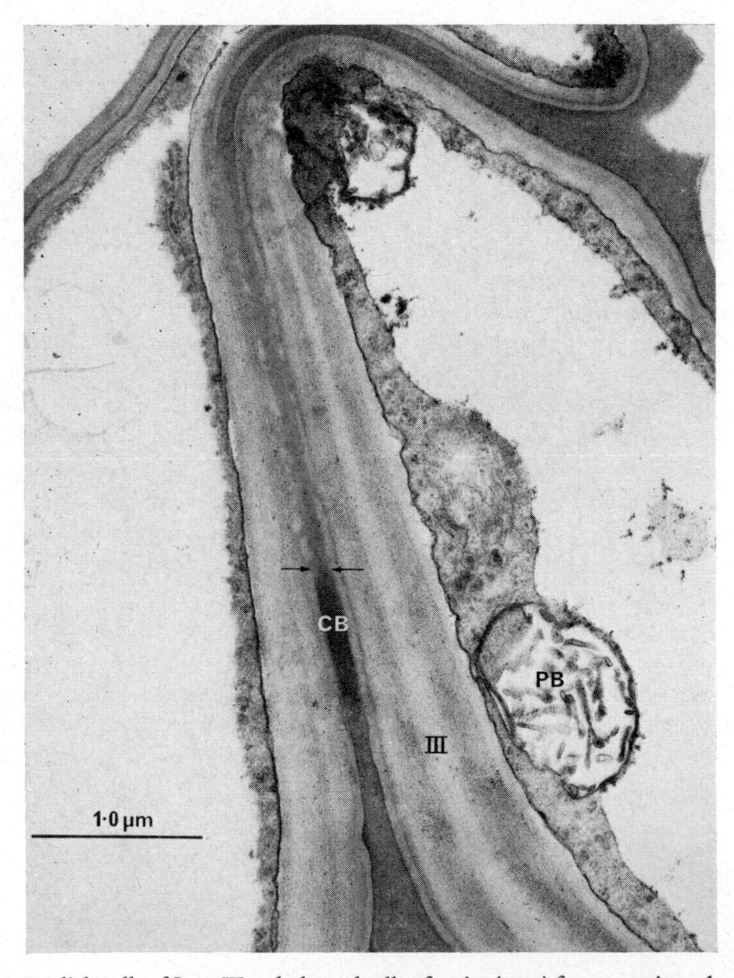

FIG. 3. Radial walls of State III endodermal cells of maize (corn) from a section taken 22 cm from the root tip. The Casparian band of the State I cells is still clearly visible (CB) as are the suberin lamellae (arrowed). The massive cellulosic wall (III) characteristic of the State III cell shows the typically graminaceous situation of having far greater thickness on the inner tangential walls than on the outer tangential walls. This State III cell wall is probably the equivalent of a normal secondary wall. Paramural bodies (PB) are sometimes seen in these cells: their possible function is unknown.

lamella. The cellulosic thickening of the State III cell normally appears in one of two forms: it may be evenly deposited around the cell ('O' type); or, as particularly evident in many graminaceous roots, it may be far thinner on the outer tangential wall than on all other walls ('C' type).

E. Passage Cells

The literature commonly refers to endodermal "passage cells"; these are cells which remain in the State I condition while those around them proceed towards State III. It is usually the case that the later stages of endodermal development in roots are more rapid in the cells external to the phloem than in those over the xylem poles. (A possible explanation lies in the transport of suberin and lignin precursors, *via* the phloem, outwards, with consequent opportunity for earlier differentiation in the adjacent cells.)

The endodermal cells next to the xylem poles are the last to develop State II thickening and, in some cases, they may never acquire it at all, e.g. in roots of members of the Iridaceae. They remain, therefore, as permanent passage cells; in other species, while the development to State II may be delayed in these cells, it does occur eventually. In the latter case, which is by far the most common, passage cells are only temporary features of the endodermis; their possible physiological significance is considered later.

F. Plasmodesmata

At all stages of development the walls of endodermal cells are traversed by numerous plasmodesmata grouped together in pit fields. In barley, fully expanded cells have approximately six hundred thousand plasmodesmata per square millimeter of inner tangential wall and about half this frequency on the outer tangential and radial walls. Tanton and Crowdy (1972) arrived at similar frequencies for the outer tangential wall in their study of wheat roots. It is of great importance to realize that these plasmodesmata are continuous through the suberin lamella (Fig. 4) and tertiary cellulosic wall of the endodermis in roots (Robards *et al.*, 1973) and in the endodermis-like mestome, or bundle sheath, in graminaceous leaves (O'Brien and Carr, 1970). This is not the appropriate place to discuss plasmodesmatal ultrastructure in detail, but the subject is clearly of importance in relation to ion transport. As will emerge later in our discussion, the structure (desmotubule) which traverses the wall cavity between adjacent cells is of crucial importance in the functioning of plasmodesmata.

III. Role of the Endodermis at Different Stages of its Development

A. Pro-endodermis

Between the time at which the pro-endodermis is produced from the meristem and the time that the Casparian band is deposited, there is a period during which

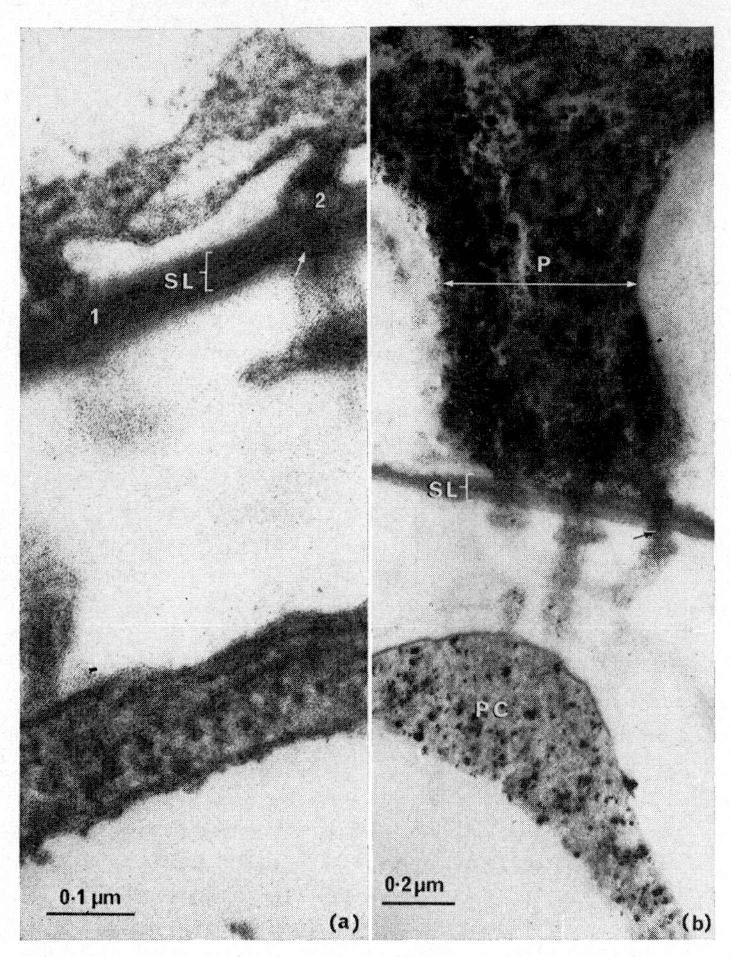

FIG. 4. Plasmodesmata through the inner tangential wall of endodermal cells of barley. Ordinary thin sections commonly give the impression that the suberin lamellae are continuous across the plasmodesmata (e.g. plasmodesma "1" in 4a). This is not so, as borne out by the rather thinner sections illustrated here. The plasmodesmata narrow through the suberin lamellae ("2" in 4a; arrowed plasmodesma in 4b), but remain continuous. The thinner sections, however, reduce the probability that the whole length of a single plasmodesma will be contained within the thickness of a section. Therefore, only short lengths of oblique plasmodesmata are seen. Figure 4a is from a section taken 16 cm from the root tip; Figure 4b is also from 16 cm, but the roots had been growing in a solution containing lanthanum hydroxide in colloidal form. The wide pit (P) through the thick cellulosic State III cell wall leads to the plasmodesmata which traverse the suberin lamellae and the wall of the pericycle cell (PC). (SL—Suberin lamellae).

the young cells give no clear indication of any special deposition or structure associated with the wall. In barley, the Casparian bands are seen from approximately 5 mm behind the tip. There is, therefore, a short zone in which the typical structural modifications associated with a normal endodermal cell are missing. In this zone the contents of protoxylem pole cells in the pericycle may begin to degenerate. At this stage, however, the metaxylem elements are thin-walled and retain their cell contents intact. The question arises, therefore, whether the pro-endodermis controls the movement of materials into the vascular tissue to any significant extent.

It has been shown that uranyl ions (Wheeler and Hanchey, 1971; Robards and Robb, 1972) and lanthanum ions (Nagahashi *et al.*, 1974) taken up by young root tissues are largely associated with cell walls and that the extent of their radial penetration indicates the apoplast volume which is directly accessible from the external solution; particles of lanthanum hydroxide in colloidal solution can also be used for this purpose. The use of this latter compound is preferable because it eliminates the undoubtedly toxic effects of the polyvalent cations (e.g. Levan, 1945 and Clarkson, 1965). Figure 5 shows that particles of lanthanum hydroxide are present on both sides of the endodermis in equal abundance at a distance of 3 mm from the root tip in barley. The result strongly suggests that, at the point where some of the xylem elements have already started to lose their contents and are becoming conductive, material can reach the vascular elements by mass flow in the apoplast.

In this zone, any control over entry into the living xylem must reside in the plasmalemmata of the xylem elements themselves (cf. Anderson and House, 1967).

On the other hand, there would seem to be no way of preventing the direct movement of substances from the free space into the protoxylem pole cells once their cell contents have been lost. Clearly, this is a matter for further investigation if, as seems likely, these small xylem elements are conductive.

The strong attachment of the plasmalemma to the radial walls in plasmolysis experiments, which is such a prominent feature of the State I endodermal cells, was not found to occur in advance of visible evidence of the deposition of the Casparian band (Robards *et al.*, 1973).

B. State I

The formation of the Casparian band and its firm attachment to the plasmalemma prevents direct access to the stele through the apoplast. This fact has been recognized for many years. As early as 1910 De Lavison described the endodermis as "une membrane vivante" around the central cylinder and noted that coloured salts which did not penetrate cells to any great extent were not able to move beyond this point and enter the stele. Three kinds of more recent evidence support this view of the endodermis as the inner limit of the free space.

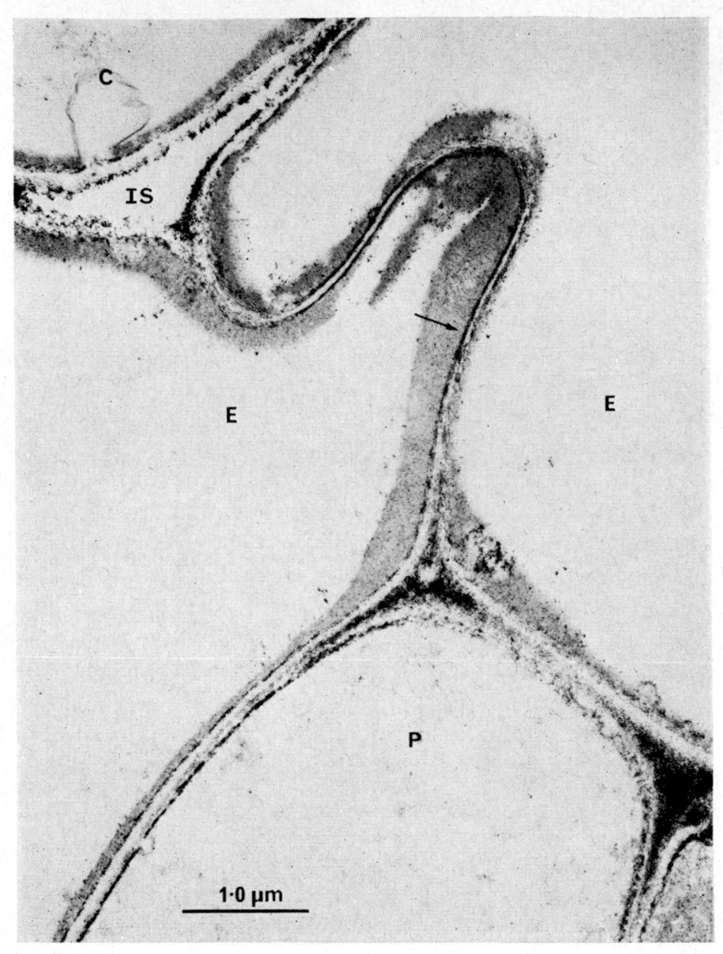

FIG. 5. Micrograph from a section taken 3 mm behind the root apex of a barley plant which had been grown in a colloidal solution of lanthanum hydroxide. The Casparian band has not yet developed in the endodermal cell radial walls (arrowed), and the lanthanum tracer has penetrated from the cell walls of the cortical cells (C), through those of the endodermis (E), and into those of the pericycle (P) without restriction. (IS—intercellular space). This section was not stained, and therefore electron-opacity is confined to the tracer.

Figure 6 shows that the movement of lanthanum hydroxide is brought to an abrupt halt midway down the radial walls of the endodermis. The blockage corresponds with the position of the Casparian band.

The second line of evidence comes from the distribution of ^{45}Ca in barley roots visualized by microautoradiography before and after exchange for 10 minutes in unlabelled calcium chloride solution. This treatment removes between 50–60% of the calcium from whole roots (e.g., Clarkson and Sanderson,

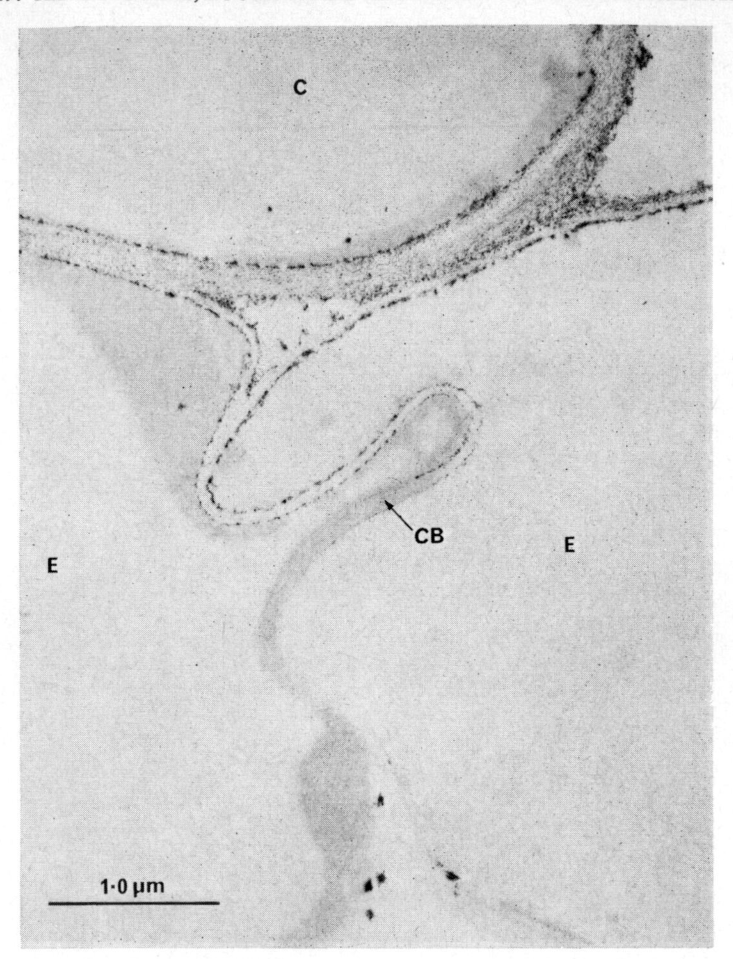

FIG. 6. Micrograph of a section taken 50 mm behind the root apex of a barley plant which had been grown in a colloidal solution of lanthanum hydroxide (exactly as Fig. 5). The Casparian band has now developed and, while the cortical cell walls and the outer part of the endodermal radial wall, have become labelled, the Casparian band has effectively precluded further centripetal movement. (C—cortical cell; E—endodermal cell; CB—position of Casparian band).

1971a). Quantitative measurements of autoradiographs of frozen sections show that exchange removes ^{45}Ca principally from the cortical and epidermal cells (Table I). The decline in the ^{45}Ca in the stele is more likely to have been due to movement into the transpiration stream during the 10 minute period, than to exchange.

The role of the State I endodermis can be seen clearly if the uptake of water and nutrients is observed through portions of seminal axes from which the cortex

TABLE I. The effect of short-term ion exchange on the distribution of ^{45}Ca in tissues of barley seminal root axis, 1 cm from the root tip.

Tissue	Tissue concentration of labelled calcium		Reduction during exchange (%)
	Before exchange	After 10 min exchange	
	(mM) $\pm$ S.E.		
Epidermis	1·66 $\pm$ 0·14	0·62 $\pm$ 0·17	63
Mid-cortex	1·52 $\pm$ 0·16	0·56 $\pm$ 0·14	63
Endodermis	3·02 $\pm$ 0·17	1·49 $\pm$ 0·15	51
Pericycle	3·15 $\pm$ 0·16	2·56 $\pm$ 0·26	19
Xylem parenchyma	3·41 $\pm$ 0·22	2·50 $\pm$ 0·16	27
Central xylem	1·70 $\pm$ 0·16	1·00 $\pm$ 0·10	41

Short lengths of intact root of Barley cv. Midas, 16 days-old, were treated for 2 h at 20°C with a nutrient solution containing 0·1 mM $CaCl_2$ labelled with 2μCi ml^{-1} ^{45}Ca. The apparatus used for this purpose is described in Clarkson and Sanderson (1971b). Segments were washed in de-ionized water and half of them treated with unlabelled 0·1 mM $CaCl_2$ for 10 mins. Segments then were cut from the plant and frozen in liquid nitrogen. Frozen section (10 μm) and autoradiographs were prepared as described in Sanderson (1972). Quantitative estimates of calcium concentration were made by reference to autoradiographs of disks, containing known volumes of the uptake medium, as described in Sanderson (1972).

has been stripped from the stele of an intact root (Fig. 7). In these experiments, stripping caused fractures in the radial walls of the endodermis in the position of the Casparian band (Clarkson and Sanderson, 1974b) and the endodermis was destroyed. In this condition, the stele admitted water at a greatly increased rate for 4–6 h (Fig. 7) until repair mechanisms began to seal it off once more. It was also found that the specific activities of ^{32}P and ^{85}Sr were the same in the transpiration stream from the injured segment as in the external solution, suggesting that the capability for ion concentration and discrimination was lost when the endodermis was damaged.

The foregoing evidence suggests that the plasmalemma of the endodermal cell constitutes a major resistance to water and solute flow into the stele and also that it can exclude colloidal substances, acting rather like an ultrafilter. Materials may also enter and leave endodermal cells through plasmodesmata. For substances which can move freely in the symplast, the principal permeability barrier in their journey towards the stele must be the plasmalemma of the cortical or epidermal cell which they cross first. It is generally agreed that the total amount of water or solute delivered to the stele may be the sum of two components, one of which moves through the symplast and the other moves through the cell walls, by-passing the cortex, and crossing the plasmalemma of the endodermal cells. The relative amounts of a given substance carried in these two pathways will determine the extent to which its movement into the stele will be affected by the next stage of endodermal development.

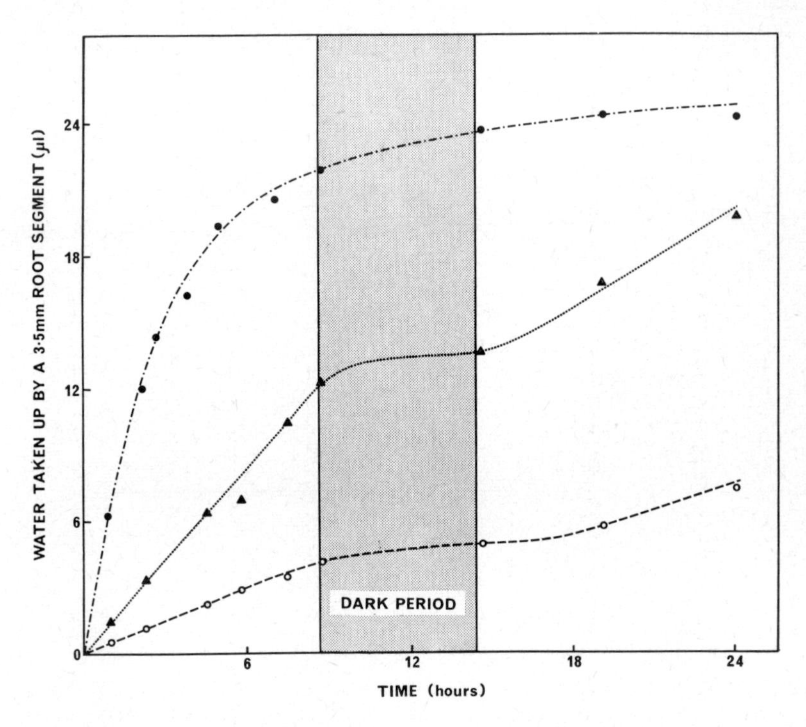

FIG. 7. Uptake of water by segments of barley seminal root axis attached to an intact plant. The water uptake at 20°C was measured using a micropotometer system as described in Clarkson and Sanderson (1971b). Symbols: closed circles—stripped stele approx 1 cm from the root tip; triangles—undamaged root 5 cm from the root tip; open circles—undamaged root > 40 cm from the root tip. (Taken from Clarkson and Sanderson, 1974b).

C. State II

As soon as suberin lamellae begin to make their appearance in the endodermis of barley and marrow, the movement of water (Fig. 8) and calcium into the stele begins to decline. Figure 9 shows that progress towards complete suberization occurs over different distances from the tip in the three types of members in the barley root system. These differences correspond closely to the declining ability of short segments of intact root to translocate calcium to the shoot (Robards *et al.*, 1973). In nodal axes, where suberization is long delayed, a much greater length of root can translocate calcium.

Once direct access to the plasmalemma of the endodermal cell has been sealed off by the deposition of suberin lamellae in the walls, direct absorption from the free space is greatly diminished. These results suggest that much of the water and nearly all of the calcium entering the vascular tissue of a barley root are absorbed directly from the free space by the endodermis.

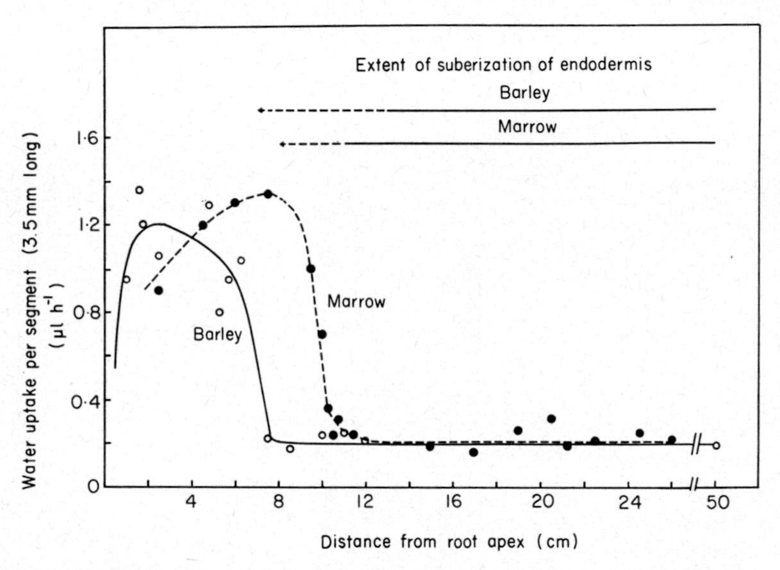

FIG. 8. A comparison of water uptake by 3·5 mm segments of intact root and suberization of the endodermis in barley (*Hordeum vulgare*, c.v. Midas) and marrow (*Cucurbita pepo*, c.v. Greenbush). Water uptake measured using micropotometer system (Clarkson and Sanderson, 1971b) at 20°C. Plants received illumination at 20×10^3 lux with a relative humidity of 70%. Length of axis where the endodermis is only partly suberized is shown as broken lines. (From Graham, *et al.*, 1974).

The alternative pathway into the endodermis *via* the plasmodesmata is not affected by the suberin lamella. Plasmodesmatal pores are continuous through the lamella and the frequency of their occurrence is not reduced (Robards *et al.*, 1973). Tables II and III show that translocation (of phosphate by segments of intact barley root and of potassium by segments of marrow root) is not markedly altered over the zone where water uptake (Fig. 8) and calcium translocation (Fig. 9) are so greatly reduced. This evidence suggests that phosphate and potassium gain access to the vascular tissue from the symplast which is not disturbed by the State II endodermis.

Suberization of the endodermis effectively differentiates, therefore, between those ions reaching the stele mainly from the apoplast, e.g. calcium, and those which enter it *via* the symplast, e.g. phosphate, potassium.

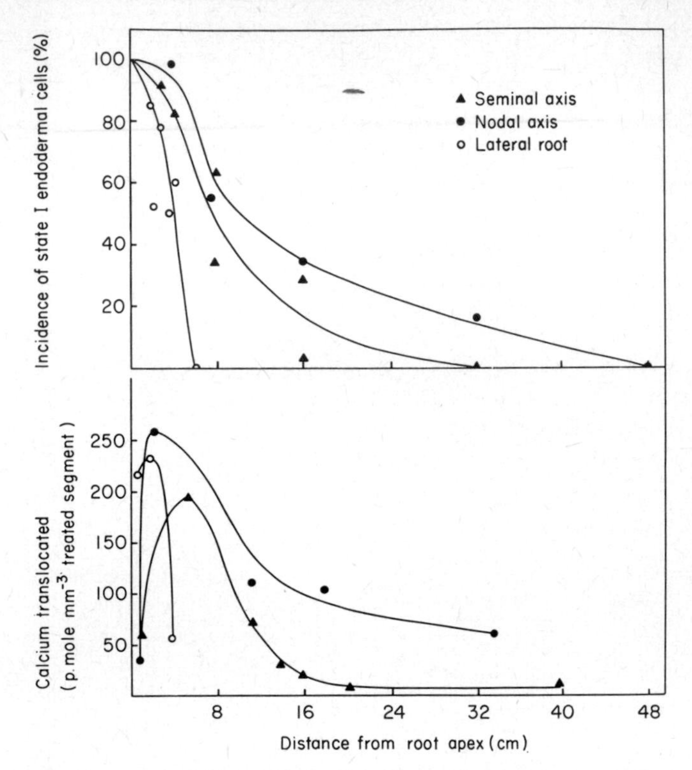

FIG. 9. A comparison of the distribution of unsuberized endodermal cells (State I) and the ability for calcium translocation in 3 types of root member in barley c.v. Midas. Estimates of State I cells made from thin sections examined by electron microscopy. Calcium translocation from 3·5 mm segments of intact root was measured at 20°C using the technique of Clarkson and Sanderson (1971b).

D. State III

Further thickening of the cellulosic walls of the endodermis does little to alter the situation created by the formation of suberin lamellae. Water uptake in barley and marrow continues at a slow rate (Fig. 8) and the rate of phosphate entry into the stele of barley is virtually unchanged (Sanderson and Clarkson, 1973).

IV. Water and Solute Movement through Plasmodesmata in the Endodermis

In barley seminal root axes it was found that the frequency of plasmodesmata in the inner tangential wall of the endodermis did not change once cells had

TABLE II. Uptake and translocation of labelled phosphate by segments of intact seminal root axis from 3-week-old barley plants c.v. Maris Badger.

Distance from root tip	Condition of endodermis	P translocated from segment	Total P uptake by segment
(cm)		(p moles seg^{-1} day^{-1})	
1	State I 100%	53	448
12	State I 30%	35	189
	States II & III 70%		
19	State I $< 5\%$	61	230
	States II & III $> 95\%$		
> 40	State III 100%	241	576

Short segments of intact seminal root axis were treated with a nutrient solution containing 3 μm KH_2PO_4 + 0·25 μCi ml^{-1} ^{32}P for 24 h at 20°C as described in Russell and Sanderson (1967). Estimated condition of the endodermis at the zones stated from data in Robards *et al.* (1973).

TABLE III. Uptake and translocation of labelled potassium by segments of intact seminal root axis from 10-day-old marrow plants, c.v. Greenbush.

Distance from root tip	Condition of endodermis	K translocated from segment	Total K uptake by segment
(cm)		(n moles mm^{-3} h^{-1})	
1	State I	6·72	11·83
15	State I/II	8·00	14·20
50	State III	5·17	10·73

Short segments of intact seminal root axis were treated for 6 h at 20°C with a nutrient solution containing 0·5 mM KCl plus 5 μCi ml^{-1} ^{42}K. Data and methods described fully in Harrison-Murray and Clarkson (1973).

completed elongation (Robards *et al.*, 1973). The constancy of both plasmodesmatal numbers and phosphate entry into the stele supports the suggestion that the former provide the pathway for the latter. If it is assumed that the observed uptake of water from older parts of the root axis also crosses the endodermal walls through plasmodesmata, it is possible to calculate both the rate of flow through them and the pressure difference across their ends necessary to support this flow (Clarkson *et al.*, 1971). It was found that the pressure difference required was not improbably high, being of the order of 0·09 to 2·4 bar, but certain rather surprising conclusions are forced on us if we calculate the speed at which water must be moving through the plasmodesmatal channels. If, as calculated earlier (Clarkson *et al.*, 1971), $3{\cdot}78 \times 10^{-11}$ mm^3 of water pass through a plasmodesma of $3 \cdot 98 \times 10^{-14}$ mm^3 volume in each second, we see that there must be nearly a thousand volume changes per plasmodesmatal channel per second, or a rate of movement of 500 μm per second down a channel of only 10 nm diameter. These flow rates are scarcely credible, and yet the plasmodesmata appear, in electron micrographs, to be the only likely channel for water movement in these old parts of the root axis.

The estimates of water uptake and plasmodesmatal frequency are accurate to within 20% so that major weaknesses in observations must reside in the effective dimensions of the plasmodesmatal pore itself. The crucial dimension in the above calculations was the radius of the desmotubule which runs through the centre of the pore; it is possible that shrinkage of this structure, in common with many others, may occur regularly during processing for electron microscopy. If its radius was actually 50% greater than we observe it to be, then the flow rate through a plasmodesma would be reduced to 250 volume changes per second.

There are two other ways around this difficulty; the first is that the suberized walls have a small, but finite, permeability to water which accounts for much of the flow. The second is that there is an additional channel through the plasmodesmatal cavity. In electron micrographs there appears to be a tight junction between the end of the desmotubule and the plasmalemma lining the pore (Fig. 10). If, *in vivo*, this were not a tight junction, a second pathway might be opened up. With the desmotubule intimately associated with the endoplasmic reticulum (Burgess 1971; Robards, 1971; Withers and Cocking 1972) and the general cytoplasm connected through the outer part of the pore, a system capable of supporting bidirectional movement can be envisaged. It must be admitted, however, that evidence on electrical coupling between cells in leaves and roots (Spanswick 1972) suggests that the resistance of plasmodesmatal junctions between cells is too high for there to be cores of cytoplasm, the diameter of the plasmodesmatal cavity, running from cell to cell. If it is accepted, however, that much of the cross section of a plasmodesma is occupied by the desmotubule and blocked off from the cytoplasm at large by the association with the endoplasmic reticulum, a small ring of cytoplasm may remain around the desmotubule through which water and solutes could move.

The foregoing speculation is prompted by the fact that cells in the cortex separated from the stele by a State III endodermis continue to live and respire at a steady rate. We must assume therefore that they receive carbohydrate from the phloem and that this carbohydrate reaches them by diffusion. If all of the plasmodesmata in the endodermis have water flowing through them at the rate calculated above, it is hard to see how carbohydrate can cross the endodermis in the opposite direction.

Much the same situation must exist, but in reverse, in the leaves of grasses where the vascular bundles are surrounded by a suberized sheath of cells known as the mestome (O'Brien and Carr, 1970). If water leaving the vein is constrained to pass through plasmodesmata in the mestome it is far from clear how sucrose can pass from the mesophyll to the phloem in the symplast; the suberin lamellae in the walls would appear to preclude its diffusion across the mestome in the apoplast.

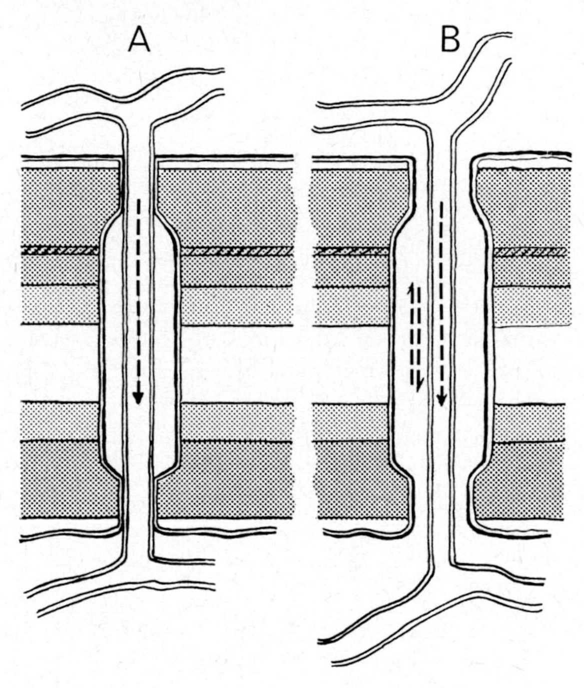

Fig. 10. Possible pathways through the wall cavity created by a plasmodesma. A. Situation envisaged by Robards (1971) and seen frequently in electron micrographs of fixed tissue. B. Hypothetical situation *in vivo* discussed in the text.

V. Other Functions of the Endodermis

There are, in addition to the functions described above, several other ways in which the endodermis acts as a defence for the vascular system. Firstly, in roots it can act as a protection against desiccation of phloem. The upper layers of the soil are frequently subjected to drying and high temperatures which may cause the death of the cortex at the base of the root. Using a technique for labelling segments of intact roots, Clarkson *et al.*, (1968) showed that, when the cortex was destroyed in this way, the entry of ^{32}P into the stele ceased in the dried zone. However, younger portions of the root below the desiccated zone grew normally, indicating that no disturbance had occurred in downward movement of carbohydrate in the phloem. In these conditions it is likely that the remnants of the plasmodesmatal channels became blocked with callose or some other structural carbohydrate material. The process also largely prevents the loss of water from the root to the dry soil at the base of the plant.

There is also evidence that the endodermis can protect the vascular tissue from invasion by some parasites and pathogens. The fungus causing the "take all" disease in wheat, *Gaeumannomyces graminis*, encounters some resistance from the endodermis in its invasion of the vascular tissue (Fellows, 1928; Clarkson *et al.*, 1975). In this instance, as in many others where damage is sustained by the root, there is considerable oxidation of the polyphenolic compounds in the endodermis which leads to dark brown discolouration. There has been a suggestion that these substances, e.g. scopoletin, may possess antibiotic properties (Goodwin and Pollock, 1954). It is also thought that they render the protoplasts unpalatable to browsing nematodes, many species of which never enter the vascular tissue and clearly avoid feeding in the endodermis even where it is in a State I condition, e.g. *Pratylenchus fallax* on wheat and barley roots (Corbett, 1972).

Roots growing in certain soils are frequently bathed by solutions containing particles in colloidal solution. We have seen earlier that, because of the interruption of the apoplast by the Casparian band, such substances cannot enter the xylem. If they were to do so there is a genuine possibility that the vessels and leaf veins could become "silted" up by the continued arrival and deposition of particles.

The functions of the endodermis-like structures in leaves (Schwendener, 1890) and stems must be similar to those in roots. O'Brien and Carr (1970) point out that the bundle sheath in maize, oat and wheat may impose a restriction on the movement of water out of the xylem into the free space of the mesophyll, and, by doing so, ensure a uniform distribution of water over the entire surface of the leaf.

VI. "Absorbing Zones" of Roots: A Re-assessment

The anatomical observations of Priestley and North (1922) and Scott (1928) provided the basis for a firm belief that, once the endodermis became completely suberized, access to the stele from the external environment was eliminated; this view is still held by some authors, e.g. Bar-Yosef (1970).

Thus it was thought the effective uptake of mineral nutrition of the plant was restricted to an apical unsuberized absorbing zone. So complete was the faith in this, that Priestley and North (1922) questioned Mylius's (1913) observation that protoplasts from the suberized endodermis could be plasmolysed by hypertonic solutions albeit rather slowly, and considered his conclusion that the cells remained in some way permeable to water was mistaken. The evidence on phosphate and potassium translocation reported in this paper, and the results of other workers (e.g. Bowen, 1969, 1970; Burley *et al.*, 1970), suggest that he was not. For those substances which normally enter the endodermal cells by way of the symplast, suberization seems to be of little consequence, and because of this all parts of the root system of a cereal plant can make a significant contribution of these elements to the shoot. Although a slow rate of water absorption is characteristic of older portions of roots, such suberized zones may represent a large fraction of the root volume; thus they may contribute as much water to the shoot as the more rapidly absorbing younger zones (Graham *et al.*, 1974). It is not strictly accurate, therefore, to describe the rapidly absorbing part of the root as the "absorbing zone" for water.

The pattern of calcium translocation in barley and marrow (Harrison-Murray and Clarkson, 1973) does, however, correspond broadly with the notion of the "absorbing zone" as envisaged by earlier workers. It has also been shown in barley that translocation of iron to shoots is associated only with the apical zones of axes and lateral branches (Clarkson and Sanderson, 1974a).

Recent observations by Ferguson (1974) suggest that the suberization of the hypodermis in *Zea mays* may be of greater significance in preventing the translocation of phosphate along the root axis than the suberization of the endodermis. Suberization at the root periphery probably places a restriction on the entry of materials into the symplast and should be an important factor in considerations of structure and function in roots.

Acknowledgements

We would like to thank John Sanderson and Margaret Jackson for their excellent collaboration in the experiments and observations which we have reviewed in this article, Dr. R. Scott Russell for his encouragement of this work and the Agricultural Research Council for financial assistance to AWR.

References

ANDERSON, W. P. and HOUSE, C. R. (1967). A correlation between structure and function in the root of *Zea Mays*. *J. exp. Bot.* **18**; 544–555.

BAR-YOSEF, B. (1970). Fluxes of P and Ca into intact Corn Roots and their Dependence on Solution Concentration and Root Age. *Plant and Soil.* **35**; 589–600.

BOND, G. (1930). The occurrence of cell division in the endodermis. *Proc. Roy. Soc. Edin.*, **50**; 38–50.

BONNETT, JR., H. T. (1968). The root endodermis: fine structure and function. *J. Cell Biol.*, **37**; 109–205.

BOWEN, G. D. (1969). The uptake of orthophosphate and its incorporation into organic phosphates along roots of *Pinus radiata*. *Aust. J. Biol. Sci.* **22**; 1125–1135.

BOWEN, G. D. (1970). Effects of soil temperature on root growth and on phosphate uptake along roots of *Pinus radiata*. *Aust. J. Soil Res.*, **8**; 31–42.

BRYANT, A. E. (1934). A demonstration of the connection of the protoplasts of the endodermal cells with the Casparian strips in the roots of barley. *New Phytol.*, **33**; 231.

BURGESS, J. (1971). Observations on Structure and Differentiation in Plasmodesmata. *Protoplasma*, **73**; 83–95.

BURLEY, W. J., NWOKE, F. I. O., LEISTER, G. L. and POPHAM, R. A. (1970). The relationship xylem maturation to the absorption and translocation of ^{32}P. *Am. J. Bot.* **57**; 504–511.

CLARKSON, D. T. (1965). The effect of aluminium and some other trivalent metal cations on cell division in the root apices of *Allium cepa*. *Ann. Bot.*, **29**; 309–315.

CLARKSON, D. T. and SANDERSON, J. (1971a). Inhibition of the uptake and long-distance transport of calcium by aluminium and other polyvalent cations. *J. exp. Bot.*, **23**; 837–851.

CLARKSON, D. T. and SANDERSON, J. (1971b). Relationship between the anatomy of cereal roots and the absorption of nutrients and water. *Agricultural Research Council Letcombe Laboratory, Report* 1970; 16–25.

CLARKSON, D. T. and SANDERSON, J. S. (1974a). The uptake of iron and its distribution in the root tissue of barley. *Agricultural Research Council Letcombe Laboratory Report* 1973; 15–19.

CLARKSON, D. T. and SANDERSON, J. S. (1974b). The endodermis and its development in barley roots as related to radial migration of ions and water. pp. 87–100. *In*: "Structure and Function of Primary Root Tissues". (J. Kolek, Ed.). Slovak Academy of Sciences, Bratislava.

CLARKSON, D. T., SANDERSON, J. and RUSSELL, R. S. (1968). Ion uptake and root age. *Nature, Lond.* **220**; 805–806.

CLARKSON, D. T., ROBARDS, A. W. and SANDERSON, J. (1971). The tertiary endodermis in barley roots: fine structure in relation to radial transport of ions and water. *Planta (Berl.)*, **96**; 292–305.

CLARKSON, D. T., DREW, M. C., FERGUSON, I. B., and SANDERSON, J. (1975). The effect of the "take-all" fungus, *Gaeumannomyces graminis*, on the transport of ions by wheat plants. *Physiol. Pl. Pathol.* **5**; (In press).

CORBETT, D. C. M. (1972). The effect of *Pratylenchus Fallax* on Wheat, Barley and Sugar Beet Roots. *Nematologica*, **18**; 303–308.

DE LAVISON, J. DE R. (1910). Du mode de penetration de quelques sels dans la plante vivante. Rôle de l'endoderme. *Rev. Gen. de Bot.* **22**; 225–241.

FAHN, A. (1967) "Plant Anatomy". Pergamon Press (Oxford).

FELLOWS, H. (1928). Some chemical and morphological phenomena attending infection of the wheat plant by *Ophiobolus graminis*. *J. Agric. Sci.*, **37**; 674–61.

FERGUSON, I. B. (1974). Ion uptake and translocation by the root system of maize. *Agricultural Research Council Letcombe Laboratory Report* 1973; 13–15.

GOODWIN, R. H. and POLLOCK, B. M. (1954). Studies on Roots. I. Properties and Distribution of fluorescent constituents of *Avena* roots. *Am. J. Bot.*, **41**; 516–520.

GRAHAM, J., CLARKSON, D. T. and SANDERSON, J. (1974). Water uptake by the roots of marrow and barley plants. *Agricultural Research Council Letcombe Laboratory Report* 1937; 9–12.

HARRISON–MURRAY, R. S. and CLARKSON, D. T. (1973). Relationships between structural developement and the absorption of ions by the root system of *Cucurbita Pepo*. *Planta* (Berl.) **114**; 1–16.

KROEMER, K. (1903). Wurzelhaut, Hypodermis und Endodermis der Angiospermenwurzel. *Bibl. Bot.*; **59** 1–159.

LEVAN, A. (1945). Cytological reactions induced by inorganic salt solutions. *Nature, Lond.*, **156**; 751–2.

MAGER, H. (1907). Beitrage zur Anatomie der physiologiochen Scheiden der Pteridophyten. *Bibl. Bot.* **14**; (66).

MYLIUS, G. (1913). Das Polyderm. *Bibl. Bot.*, **18**; (79) 1–119.

NAGAHASHI, G., THOMSON, W. W. and LEONARD, R. T. (1974). The Casparian strip as a barrier to the movement of lanthanum in corn roots. *Science, N.Y.*, **183**; 670–671.

O'BRIEN, T. P. and CARR, D. J. (1970). A suberized layer in the cell walls of the bundle sheath of grasses. *Aust. J. Biol. Sci.*, **23**; 275–287.

PRIESTLEY, J. H. and NORTH, E. E. (1922). Physiological Studies in Plant Anatomy. III. The structure of the endodermis in relation to its function. *New Phytol.* **21**; 111–139.

ROBARDS, A. W. (1971). The Ultrastructure of Plasmodesmata. *Protoplasma*, **72**; 315–323.

ROBARDS, A. W. and ROBB, M. E. (1972). Uptake and binding of uranyl ions by barley roots. *Science, N.Y.*, **178**; 980–982.

ROBARDS, A. W., JACKSON, S. M., CLARKSON, D. T. and SANDERSON, J. (1973). The structure of barley roots in relation to the transport of ions into the stele. *Protoplasma*, **77**; 291–312.

RUSSELL, R. S. and SANDERSON, J. (1967). Nutrient uptake by different parts of the intact roots of plants. *J. exp. Bot.*, **18**; 491–508.

SANDERSON, J. (1972). Micro-autoradiography of diffusible ions in plant tissues—problems and methods. *J. Micros.*, **96**; 245–254.

SANDERSON, J. and CLARKSON, D. T. (1973). Quantitative microautoradiography of accumulation of phosphorus-32 in tissues from young and mature zones of the barley root. *Agricultural Research Council Letcombe Laboratory Report* 1972; 7–10.

SCHWENDENER, S. (1890). Die Mestomscheiden der Gramineenblatter. *Sitzungsber. Konig. Preuss, Akad. Wiss.* 405–426.

SCOTT, F. M. (1963). Root hair zone of soil-grown roots. *Nature Lond.* **199**; 1009–1010.

SCOTT, L. I. (1928). The root as an absorbing organ. II. Delimitation of the absorbing zone. *New Phytol.* **27**; 141–174.

SPANSWICK, R. M. (1972). Electrical coupling between cells of higher plants: A Direct Demonstration of intercellular communication. *Planta* (*Berl.*), **102**; 215–227.

TANTON, T. W. and CROWDY, S. H. (1972). Water pathways in higher plants. II. Water pathways in roots. *J. exp. Bot.* **23**; 600–18.

VAN FLEET, D. S. (1961). Histochemistry and function of the endodermis. *Bot. Rev.*, **27**; 165–221.

VAN TIEGEM, P. and DOULIOT, H. (1886). Sur la polystelie. *Ann. Sci. Nat. Bot.* **7**; Ser 3; 275–322.

VAN WISSELINGH, C. (1926). Beitrag zur Kenntnis der inneren Endodermis. *Planta* (*Berl.*), **2**; 27–43.

WALTON, J. (1953). "An Introduction to the Study of Fossil Plants" 2nd edition. Adam and Charles Black, London.

WHEELER, H. and HANCHEY, P. (1971). Pinocytosis and membrane dilation in uranyl-treated plant roots. *Science, N.Y.*, **171**; 68–71.

WITHERS, L. A. and COCKING, E. C. (1972). Fine structural studies on spontaneous and induced fusion of higher plant protoplasts. *J. Cell Sci.* **11**; 59–75.

Chapter 20

Ion Transport through Roots

W. P. ANDERSON*

Department of Botany, University of Liverpool, Liverpool, England.

I. Introduction

It has long been recognized that the root system of higher plants is the chief organ for absorption of nutrient salts and water in the ordinary terrestrial species. There is also a complex internal circulation of ions within a plant. Figure 1 shows how a radioactive ion, initially supplied to half the root system, will circulate around the plant, presumably moving upward to the leaves in the

* Present Address: Director's Unit, Research School of Biological Sciences, Australian National University, Canberra A.C.T. Australia.

xylem translocation stream and downward again from the leaves in the phloem. Growth substances, particularly cytokinins and gibberellins, are also translocated upward to the shoot in the xylem from the chief synthesis sites in the root meristems. Apart from these hormones the solutes in xylem sap are mainly inorganic, except in the legumes where significant amounts of nitrogen-rich amino acids are translocated in the xylem from the nitrogen-fixing nodules on the root.

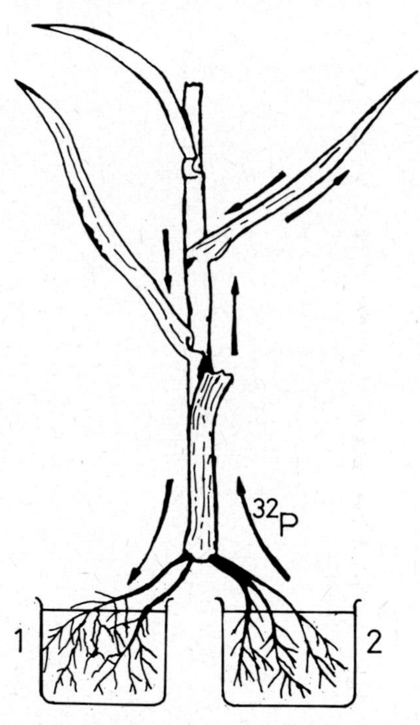

FIG. 1. Diagram showing the distribution of ^{32}P within a plant. The isotope was originally supplied to the nutrient solution in container 2, and is then found at a later time in container 1. (after Biddulph and Biddulph; 1959, Sci. Am. **200**, 44–49).

This article will briefly review what is known about ion uptake and transport by higher plant roots, with particular emphasis on the results obtained from excised root exudation studies, the preparation in which the best definitive work can be done. This has been a favourite experimental svstem since the earliest days (e.g. Priestley, 1920) and the basic mechanism of xylem exudation from excised roots has long been recognized (Priestley, 1920; Arisz *et al.*, 1951). More recently, root pressure exudation has been characterized in a formalism based upon irreversible thermodynamics (House and Findlay, 1966), and the possibilities of two osmotic barriers (Ginsburg, 1971) and of longitudinal variation in the transport parameters (Anderson *et al.*, 1970) have been considered.

II. Root Anatomy

Any exposition of ion transport through the root requires a basic knowledge of root anatomy. The most fundamental point about ion transport through the root to be appreciated is that secretion or translocation from the root to the shoot of either ions or water depends upon the integrated functioning of all the cells in the root.

Figure 2 shows the cross-section of a typical root in a zone where ion absorption occurs. The outermost, epidermal layer of cells and the cortical cells are separated from the inner region of the stele which contains the vascular elements of

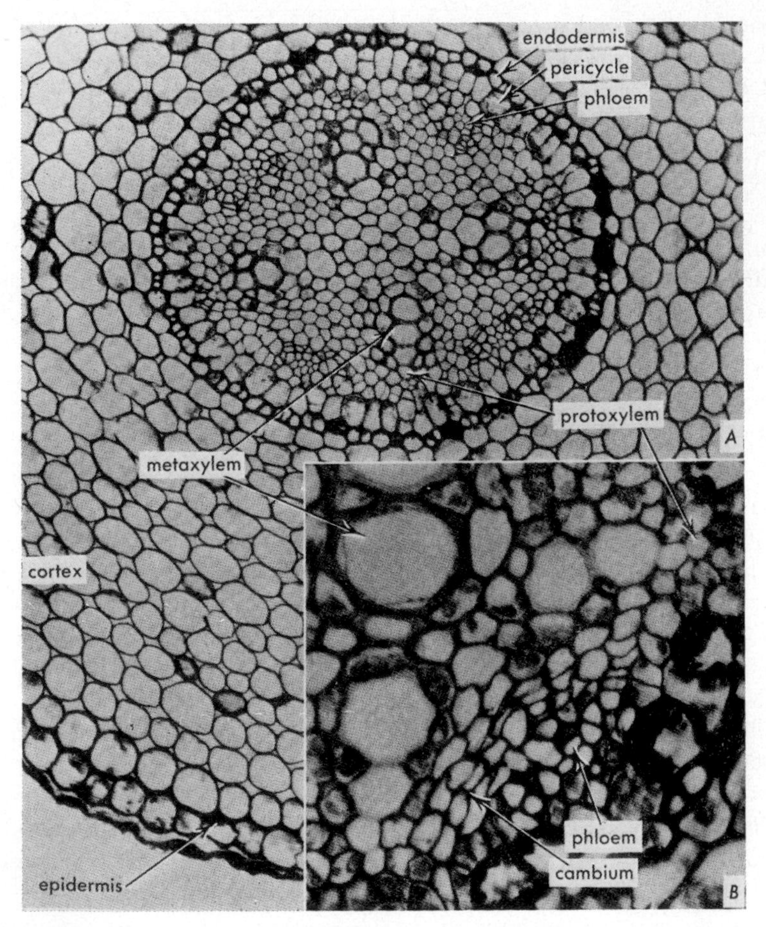

FIG. 2. Cross sections of the root of strawberry (*Fragaria*) in the primary state of growth. Reproduced from Esau, K. "Anatomy of Seed Plants." Wiley and Sons Inc. New York-London, 1961.

xylem and phloem, by the endodermis. Endodermal cells develop characteristic bands on the radial walls as a result of suberin deposition at a very early stage in their ontogeny, often within 1–3 mm from the root tip meristem. This so-called Casparian band effectively isolates the apoplasm of the cortex from that of the stele and thus any transport from the external medium to the xylem of the root is forced to cross cell membranes at some point (see also p. 416). This is in accord with the classical enunciation of endodermal function as set out by Crafts and Broyer (1938) and Van Fleet (1961). Further, this necessity for water ultrafiltration through cell membranes may help to explain the high cohesive strength of the water in the xylem columns, so often observed in field trials on xerophytic species where xylem water tensions of the order of − 100 atm have been reported.

The epidermal and cortical cells of the typical root have the normal complement of organelles, but are highly vacuolate; in *Zea mays* primary roots for example, some 80% of the cortical cell volume is vacuole. In the average 25 μm diameter cortical cell the peripheral cytoplasm layer is only 1 μm thick. The cytoplasm of the epidermal and cortical cells is interconnected through the plasmodesmata into a symplasm (Arisz, 1956). Figure 3 shows a diagram of plasmodesmatal structure, taken from Robards (1971) who has produced perhaps the most complete survey of the ultrastructure. There is general agreement among anatomists that the plasmalemma is continuous from one cell to the next, as shown in this diagram, although there is still some controversy as to whether the desmotubule is or is not a strand of endoplasmic reticulum. A discussion of ion transport through the symplasm will be given in a later section of this article.

It is clear from the work of Helder and Boerma (1969) and Clarkson *et al.* (1971) that the cortical symplasm extends through the endodermis to the stele. The parenchyma cells of the stele are also connected by plasmodesmata in a symplasm. This may be an important consideration in those upper regions of the root where tertiary thickening and lignification have occurred; cell to cell communication through the cell walls, highly impermeable following lignification, will be much restricted and the symplasm pathway seems likely to be the only viable one under these conditions.

In most species the root xylem parenchyma cells are apparently quite ordinary, although Anderson and House (1967) did observe in *Zea* that they seem to be less vacuolate and more densely cytoplasmic than the average. Läuchli (to be published) notes that in soybean roots the parenchyma cells adjacent to the xylem look like the transfer cells discussed in a recent review by Pate and Gunning (1972); convoluted walls and extensive plasmalemma surface area in these cells lead one to suppose that they may be specialized in some transport function, perhaps in secreting ions to the xylem vessels. Läuchli and Epstein (1971), Läuchli *et al.* (1971) and Pitman (1972) have given evidence to suggest that xylem ion transport in the roots of *Zea* and *Hordeum vulgare* may indeed involve active secretion of ions by the xylem parenchyma into the vessels.

The general consensus is that the early metaxylem, through which translocation of salts and water from the young absorbing zones of the root can be shown to

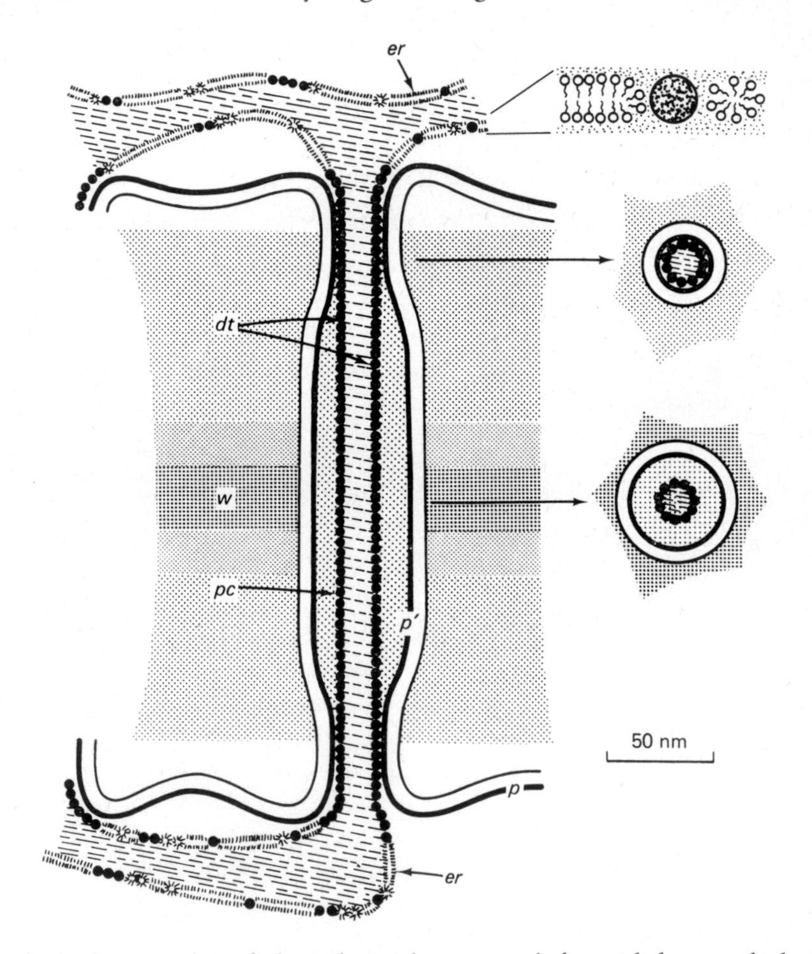

FIG. 3. An interpretation of plasmodesmatal structure. *dt* desmotubule; *er* endoplasmic reticulum; *p* plasmalemma; *p'* plasmalemma through plasmodesmatal canal; *pc* plasmodesmatal canal; various layers of the cell wall, *w*, are represented by stippling. Reproduced from Robards (1971).

occur (in *Zea* for example it is easily demonstrated by simply observing the freshly cut end of a young root under a dissecting microscope), is fully mature within a few millimetres of the root tip (see also pp. 82). Full maturity in the metaxylem implies that the vessels elements have completed secondary wall deposition with fully developed pits, and have completely perforated end walls. The mature metaxylem vessel is therefore a dead, inert and open channel through which the translocation stream flows, impelled from below by root pressure (see later) and from above by transpirational tension. However there is a considerable body of opinion which holds that in some species at least, the metaxylem still contains viable cytoplasm in the absorbing zones of the root. Scott (1949; 1965) has evidence that the metaxylem in *Ricinus communis* may contain cytoplasm up to 20 cm from the root tip. In *Zea mays* primary roots, both Anderson and House (1967) and Higinbotham *et al.* (1973) have electron micrographs to demonstrate that some of the outer ring of early metaxylem elements (through which the xylem exudate flows) contain cytoplasm and have imperfectly perforate end walls at 10–15 cm from the root tip. The difficulty is that it is almost impossible to establish whether these immature vessels exude; if they do, we must surely be forced to accept as a mechanism something like the test-tube hypothesis proposed by Hylmö (1953). In contrast, Läuchli (to be published) has made a very careful examination of young barley roots and can find no cytoplasm-containing xylem beyond 3 mm from the root tip.

III. Ion Transport Through Excised Roots

A. Exudate Collection Experiments

The most common preparation for this type of experiment is shown in Fig. 4 and full details can be found in many papers (e.g. House and Findlay, 1966). The rate of production of xylem exudate fluid can be continuously monitored by measuring the column height in the collecting capillary, and this rate is usually normalized over unit area of root surface, or over unit fresh weight of root system, to give a volume flux ($cm^3\ cm^{-2}s^{-1}$) or ($cm^3\ gFW^{-1}s^{-1}$). The collected exudate fluid can be assayed chemically and the concentration of any ion, j, can be determined. Let us call it C_j (moles 1^{-1}). Then the *flux* of that ion, j, transported into the xylem and exuded from the cut end of the excised root, J_j, (moles $cm^{-2}s^{-1}$) or (moles $gFW^{-1}s^{-1}$) will be given by equation (1).

$$J_j = J_v \cdot C_j \tag{1}$$

Therefore by this simple technique, the rate of transport of any ion to the xylem stream can be easily and unequivocally determined, but it should be noted that there is no implication that the ion has been necessarily transported all the way from the external solution to the xylem. There is evidence (e.g. Hodges and Vaadia, 1964b; Jarvis and House, 1970; Anderson *et al.*, 1971) to suggest that ion supply to the xylem stream may be either wholly or partly derived from endogenous reservoirs within the root, presumably from the cortical cell vacuoles which constitute the largest single volume fraction of the root.

However, if seedlings are grown in a culture medium and the excised roots from them are then allowed to exude while bathed in an identical medium, it seems evident that the root tissue will be everywhere in flux equilibrium and in this case the ion fluxes exuded from the cut end of the root will be a good measure of the net ion transport from the external medium. The data in Table I have been

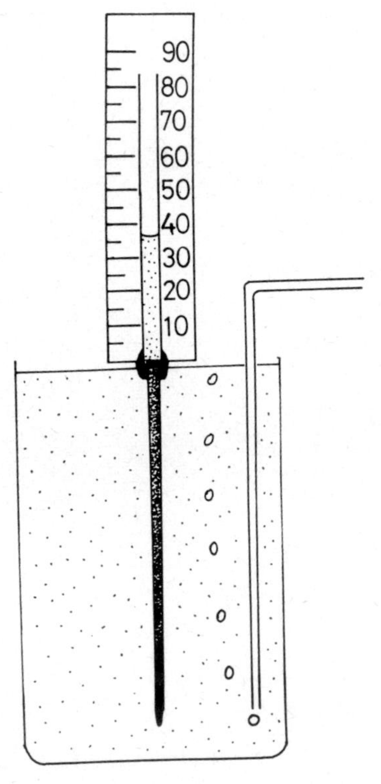

FIG. 4. Diagram of the normal method of observing excised root exudation. The scale to measure column height is accurately aligned with the bottom of the water column at the start of the experiment.

collected from root preparations where these conditions have been met. Note that there is a very definite accumulation of ions into the xylem stream in all the species investigated.

Xylem exudation from these roots is of course chiefly an exudation of water, and the generally held opinion is that this water is driven through the root from the external medium to the xylem stream by *root pressure*, i.e. by the osmotic pressure difference between the external solution and the xylem fluid. Root pressure exudation from excised root systems, or its equivalent in the intact, non-transpiring, plant (guttation from the leaves), has been known for many years,

TABLE I. Exudation data from excised root preparations

Material/Preparation	Water Flux	Exudate Concn.	Exudate Ion Flux
Zea mays			
bathed in 1 mM KCl*(1)	7·22[a]	20·4[b], 25·1[c]	147·3[d], 181·2[e]
10 mM KCl*(1)	6·92[a]	28·4[b], 30·7[c]	196·5[d], 212·4[e]
50 mM KCl*(1)	3·60[a]	54·3[b], 57·6[c]	195·5[d], 207·4[e]
0·1 mM K_2SO_4**(2)	1·94[a]	13·8[b], 2·9[f]	26·7[d], 5·6[g]
10 mM K_2SO_4**(2)	1·66[a]	25·2[b], 6·2[f]	41·8[d], 10·3[g]
bathed in 0·1X solution (3)	—	14·7[b], 0·8[h] 1·4[j] 5·6[k]	
bathed in 1X solution (3)	—	19·4[b], 1·2[h] 3·3[j] 1·4[m] 9·7[k], 3·1[c]	
Ricinus communis			
in mM KNO_3*(4)	6·76[n]	12·2[b]	82·4[p]
1mM KNO_3***(4)	4·72[n]	7·5[b]	35·4[p]
5 mM KNO_3*(4)	4·33[n]	15·6[b]	67·5[p]
5 mM KNO_3***(4)	3·39[n]	16·4[b]	55·6[p]
Sinapis alba			
in 1 mM KCL*(5)	0·2[q]	4·4[b], 5·1[c] 0·6[j]	
in 10 mM KCl*(5)	0·16[q]	12·4[b], 13·2[c] 1·3[j]	
Triticum aestivum			
in distilled water (6)	0·56[q]	22·1[k]	
Avena fatua			
in 1mM KCl*(7)	1·66[a]	5·9[b]	9·8[d]
10 mM KCl*(7)	1·11[a]	10·4[b]	11·5[d]
50mM KCl*(7)	0·55[a]	63·7[b]	35·3[d]
Avena sativa			
in 1mM KCl*(7)	3·32[a]	5·0[b]	16·6[d]
10 mM KCl*(7)	1·94[a]	8·4[b]	16·3[d]
50 mM KCl*(7)	1·11[a]	45·3[b]	50·2[d]
Helianthus annuus			
in complete nutrient solution (8)	33·3[q]	0·5[f]	
Allium cepa			
in mM KCl*(9)	3·6[a]	12·5[b], 5·6[c]	45·0[d] 20·2[e]
10 mM KCl*(9)	3·3[a]	12·3[b], 9·1[c]	40·6[d], 30·0[e]

The superscripts are as follows: [a] $\times 10^{-4}$ μ_l $cm^{-2}s^{-1}$; [b]mM K^+; [c]mM Cl^-; [d] $\times 10^{-1}$ pmoles K^+ $cm^{-2}s^{-1}$; [e] $\times 10^{-1}$ pmoles Cl^- $cm^{-2}s^{-1}$; [f]mM SO_4^{--}; [g] $\times 10^{-1}$ pmoles SO_4^{--} $cm^{-2}s^{-1}$; [h]mM Na^+; [j]mM Ca^{2+}; [k]mM NO_3^-; [m]mM Mg^{2+}; [n]μl $gFW^{-1}s^{-1}$; [p]nmoles K^+ $gFW^{-1}s^{-1}$; [q]$\times 10^{-3}$ μ_l $plant^{-1}s^{-1}$; *0·1 mM $CaCl_2$ added; **0·1 mM $CaSO_4$ added; ***2·5 mM $CaCl_2$ added.

The source references are: (1) House and Findlay (1966); (2) Anderson and Collins, (1969); (3) Davis and Higinbotham (1969); (4) Minchin and Baker (1973); (5) Collins and Kerrigan (1974); (6) Lundegardh, H. (1950); (7) Collins, J. C. (personal communication); (8) Pettersson, S. (1966); (9) Hay and Anderson (1972).

but the best modern formalism of the process, based upon the the terminology of irreversible thermodynamics, is due to House and Findlay (1966). These authors showed that the volume flow of exudate, or more strictly the water flow — an academic distinction rather than a practical one —, may be described by equation (2)

$$J_v = L_p\sigma\,(C^x - C^e) + \phi_0 \qquad (2)$$

where L_p is the hydraulic conductivity (or water permeability) of the root from the external solution to the xylem stream, σ is the reflexion coefficient (a measure of how osmotically good is the barrier across which the osmotic pressure is developed) C^x and C^e are respectively the total osmolarities of xylem fluid and external solution, and ϕ_0 is a small component of water transport which flows in the absence of any apparent osmotic gradient.

This non-osmotic component of root pressure exudation, ϕ_0, has variously been interpreted as an active water flux (most recently by Ginsburg and Ginzburg, 1971) or as an electro-osmotic water flux (Tyree, 1973) but it seems most likely that it is chiefly the consequence of applying a model [resulting in equation (2)], which is too simple. The over-simplification is two fold; firstly, it is certain that the water permeability of the root, L_p, varies as one progresses from the apex to the more mature regions, and secondly it is by no means certain that there is only one osmotic barrier between the external solution and the xylem stream. The first of these modifications leads to the possibility of there being a standing gradient osmotic flow along the root and has been dealt with in some detail by Anderson *et al.* (1970). The second yields a possible two osmotic barriers in series problem (Curran and McIntosh, 1962) and has been discussed for the root, by Ginsburg (1971); see Weatherley (this volume p. 397).

B. Radio-active Tracer Experiments

Only those tracer experiments which have direct information about ion transport into the root xylem and thence to the shoot will be discussed here; there is simply not sufficient space to go into the wealth of information available on the accumulation of ions into the root cells *per se*. There are many recent reviews which deal with this aspect of ion transport (e.g. Anderson, 1972; Epstein, 1972; Higinbotham, 1973; Anderson, 1973; Clarkson, 1974).

Perhaps the best type of experiment to lead on from studies of root pressure exudation, is the sort performed by Hodges and Vaadia (1964a, b). Briefly, these workers added radiotracer ^{36}Cl to the bathing medium of onion roots of different salt status and then assayed the collected exudate for ^{36}Cl activity; the root tissue was also assayed for ^{36}Cl at the end of the exudation time. Figure 5 is a reproduction of a graph taken from their work; note that in all cases there is an approximately linear rise in exudate activity, although the rate of increase

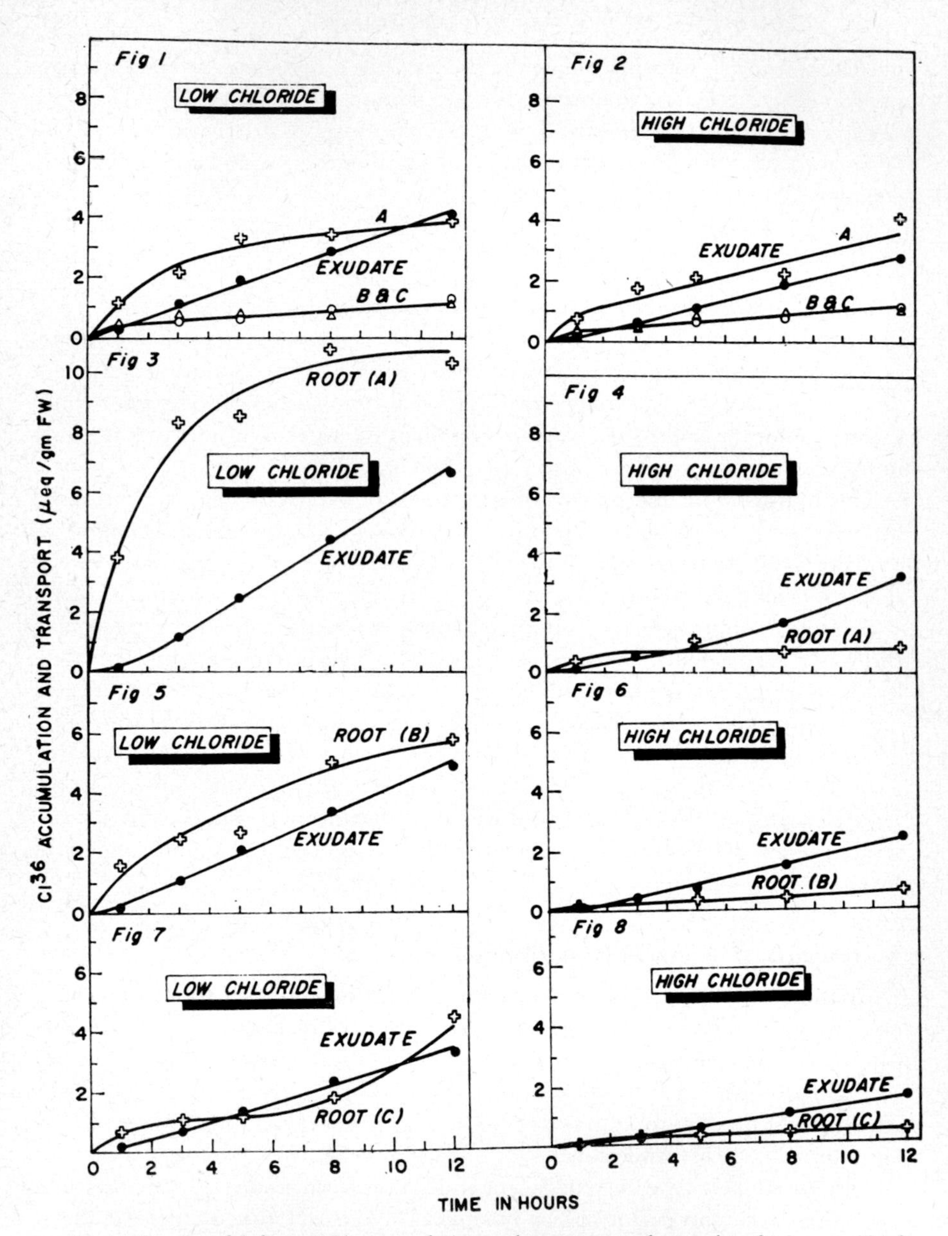

FIG. 5. Figs. 1 and 2 show ^{36}Cl accumulation and transport to the exudate for low and high chloride roots, with all root zones exposed to 2 mM KCl plus 0·3 mM $CaSO_4$. Figs. 3 and 4 give accumulation and transport to exudate with only zone A exposed to ^{36}Cl: other zones in $CaSO_4$. Figs. 5 and 6 are as for 3 and 4 with root zone B exposed to isotope. Figs. 7 and 8 are as for 3 and 4 with root zone C exposed to isotope. Zone A is the apical, zone B the middle and zone C the basal zone. (From Hodges and Vaadia, 1964a).

is much higher in low salt roots than in high salt roots. The chief implication to be drawn from these data, as far as this article is concerned, is that the Cl^- supply to the xylem stream is not a straightforward, direct passage from external solution to the xylem stream, but that there is significant exchange of Cl between the symplasmic through-put to the xylem and the vacuoles of the cortical cells. In the high salt roots there is a larger vacuolar pool of unlabelled Cl^-, and therefore there is a correspondingly greater isotopic dilution of the ^{36}Cl symplasmic through-put, resulting in a lower rate of appearance of radioactivity in the exudate.

There is other evidence, again in onion but also in maize (Anderson *et al.*, 1971), that a large proportion of the ion supply to the xylem stream may be derived from the cortical cell vacuoles. In this work, onion roots and maize seedlings were separated into two batches and were grown in either a nitrate medium or in a chloride medium. After excision, the roots were then allowed to exude from the alternative medium to the one in which they were grown. In both treatments and for both species, it was found that the exudate, collected for up to 24 h after excision and solution-switch, contained, as the predominant anion, the anion supplied during growth, which of course was not present in the external solution during exudation. Again the implication is that the cortical cell vacuoles are providing ions to the xylem stream. Unpublished data from this laboratory confirm essentially the same situation for K^+; the rate of appearance of ^{42}K in the exudate of excised maize roots is slow and can most readily be interpreted in terms of isotopic dilution of the symplasmic through-put by exchange with K^+ already present in the cortical cell vacuoles.

The overall flux balance has been formally given by Pitman (1971) and the basic picture can be represented as in Fig. 6 which is reproduced from his work. In understanding this figure it should be realized that here the whole cortical symplasm is taken as a perfect cytoplasmic continuity; the case for this assumption will be dealt with in a later section of this article. Note however that the data described in the preceding paragraphs can be readily understood on the basis of Fig. 6; briefly, the fluxes ϕ_{cv} and ϕ_{vc} are of sufficient magnitude to much affect the isotopic or indeed ionic composition of the throughput to the xylem, ϕ_x.

Recently, Läuchli *et al.* (1973) have produced some interesting data on the effect of the protein synthesis inhibitor, cycloheximide, on the transport of K^+ in excised barley roots. They find that respiration is not affected, that K^+ uptake into the cortical cell vacuoles is not affected, but that K^+ transport through the symplasm into the xylem is reduced to zero. The explanation offered by the authors is that continuing protein synthesis is needed to maintain the plasmodesmata in their functional state between one cortical cell and the next. Cycloheximide effectively causes the plasmodesmata to close and symplasmic transport is therefore reduced to a negligible rate.

C. The Common Radial Transport Model

Radial transport of ions across a root from the external solution to the xylem stream can be perhaps most easily discussed by considering separately the respective roles of epidermis, cortex, endodermis and stele. It may be helpful to refer to Fig. 7 which gives a diagrammatic representation of the topics brought out below.

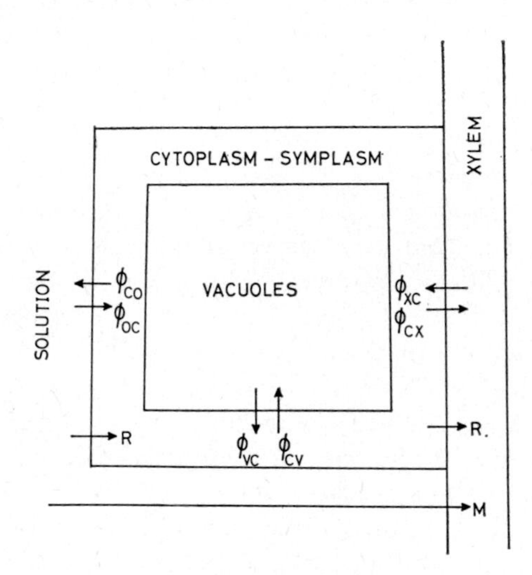

FIG. 6. A schematic representation of the flux situation in the root (after Pitman, 1971).

1. *The Epidermis and Cortex*

In young regions of a root, particularly in hydroponically cultured seedlings, the epidermis has essentially no cuticle and the cell wall offers almost unrestricted passage to ion diffusion into the free space of the root cortex. Ion uptake also occurs at the plasmalemma of the epidermal cells, and the uptake may be magnified by the increased membrane area brought about by the formation of root hairs. Ions accumulated into the epidermal cytoplasm will then migrate, probably by simple diffusion (see later) through the plasmodesmatal connections between the epidermis and the first layer of cortical cells, and so be delivered into the cortical symplasm proper. Ions which diffuse through the walls into the free space of the cortex will be transported across the cortical cell plasmalemma and will therefore also enter the cortical symplasm.

The original discussion of ion transport through the symplasm was given by Arisz (1956) who stressed that the important feature of the transport is that cell to cell passage of ions takes place, almost without restriction, through the plasmodesmata. Circulation of the ions around the peripheral cytoplasm layer of any one cell is aided by cyclosis of the cytoplasm. More recently, ion transport through the symplasm has been discussed on a quantitative basis by Tyree

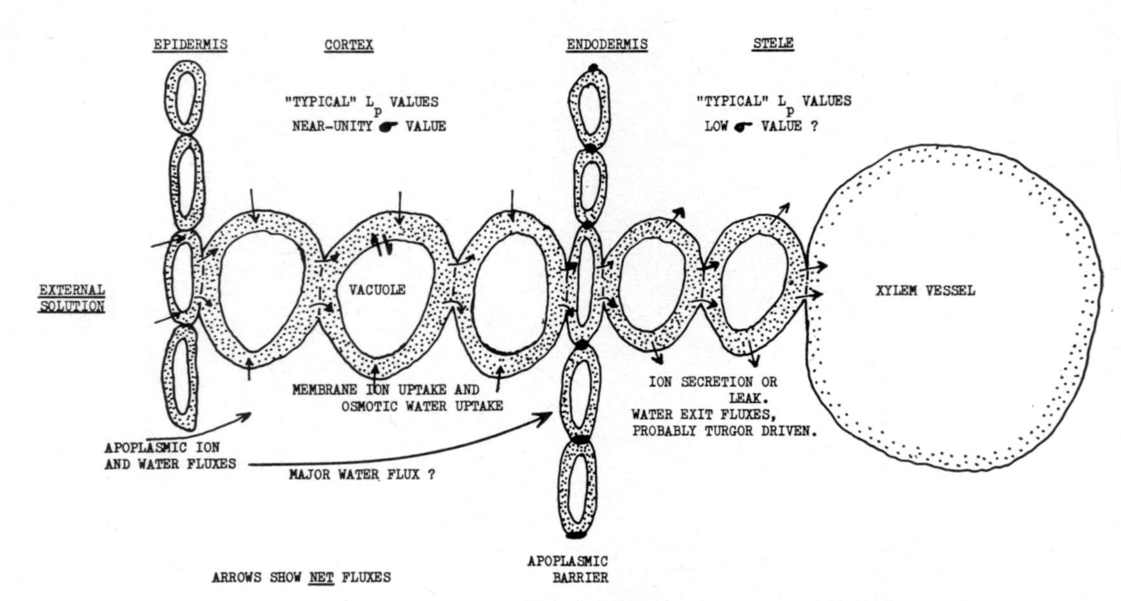

FIG. 7. A schematic representation of the chief pathways for ion and water movement across the root. This diagram embodies the most commonly held view of radial transport; the endodermis acts as an apoplasmic barrier between the cortical free space and the apoplasm of the stele. The parenchyma of the stele may or may not actively secrete ions to the xylem.

(1970). He has demonstrated, on what appear to be sound grounds, that: (i) cell-to-cell movement is chiefly through the plasmodesmata and only a negligible flux of all molecular species (with the possible exception of those which rapidly penetrate membranes, water being perhaps one such) cross the membrane bounded regions of the cell–cell junction; (ii) diffusion is the chief mechanism of transit across the plasmodesmata; (iii) diffusion, and perhaps cyclosis, in the peripheral cytoplasm layer of any single cell is sufficiently rapid to ensure perfect mixing, so that plasmodesmatal transit is the rate-controlling event for symplasmic transport once the ions are in the symplasm. Note that this last statement refers only to transport within the symplasm; overall transport from the external solution to the root xylem stream may be rate controlled by membrane transport rates during loading or unloading of the symplasm.

Finally it should be remembered from the evidence given in Section B that under certain circumstances the symplasm may be chiefly loaded from the cortical cell vacuoles; the generalized flux picture is as given in Fig. 6 and as discussed by Pitman (1971). Since it is clear that the cortical cell vacuolar contents will depend on the past nutritional history of the root, care must be taken to account for this effect when the growth medium and the experimental medium are different.

2. *The Endodermis*

Not much space will be given to this topic because it is very extensively reviewed elsewhere in this volume (Clarkson and Robards, p. 415). To understand the role of the endodermis in the transport function of the root, it is only necessary to appreciate that the Casparian bands which form a continuous, suberized layer surrounding the endodermal cells effectively restrict all apoplasmic movement from the free space of the cortex to the free space of the stele. The Casparian band forms at a very early stage in root ontogeny, at about the same time as the differentiation of most of the vascular elements in the stele (see also p. 417). It is noteworthy that similar structures effectively isolate the free space of other secretory tissues from the ground tissue, e.g. in the salt glands (Ziegler and Lüttge, 1966) and in the nitrogen fixing root nodules of the legumes (Gunning *et al.*, 1974). Thus, although a purist may argue that there is no really definitive evidence that the Casparian band in the young endodermis is an effective apoplasmic barrier, there is in fact much circumstantial, *prima facie*, evidence to suggest that the apoplasm of secretory tissue is isolated from the apoplasm of the ground tissue.

Passage of ions from the cortex to the stele must therefore occur through the endodermis. This is, of course, the traditional role ascribed to this layer of cells by Crafts and Broyer (1938), and later supported by Van Fleet (1961). There is good evidence that the cortical symplasm is indeed continuous through plasmodesmata in both the outer and inner tangential walls of the endodermis (see

Helder and Boerma, 1969; Clarkson *et al.*, 1971), and the assumption is that radial movement of ions in the cortical symplasm, following uptake at the plasmalemma of the epidermal and outer cortical cells, continues in the symplasm through the endodermis and into the stele.

3. *The Stele*

There is a certain amount of on-going controversy about the exact function of the stele with respect to ion-loading of the xylem stream. There are two main topics which are in dispute; first, the functioning of the parenchyma cells of the stele, and in particular of the xylem parenchyma, and second the function and structure of the xylem vessels in the young absorbing regions of the root.

The classic exposition of the role of the stele is due to Crafts and Broyer (1938) who assumed that the cells were "leaky" because low oxygen tension within the stele so reduced cellular respiration rates that membrane integrity and regulatory ion carrier systems could not be maintained. Whether or not the parenchyma cells of the stele are "leaky", it now seems clear that it is not the result of anaerobiosis. Hall *et al.* (1971) measured the respiration rate in freshly isolated maize root steles and found the rate to be 192 $\mu l\,O_2\,g^{-1}\,h^{-1}$, a value which seems quite sufficient to allow the cells to maintain themselves and conduct net transport of ions, although it is true that *in situ* respiration rates are not available. The respiration rate of isolated cortex of maize roots, measured in the same study, is reported as higher, but if it is remembered that a gram fresh weight of stele contains proportionately far more inert material (thickened walls and vessels) than does a gram of cortical tissue, it is possible that the parenchyma respire just as rapidly as does the cortex. Finally, there is the report by Bowling (1973) of his attempt to measure the gradient of O_2 partial pressure across the sunflower root. He found the range to be 151 mm Hg O_2 at the epidermis to 130 mm Hg at the pericycle and 127 mm Hg at the protoxylem, not a very large gradient of partial pressure. Fiscus and Kramer (1970) have other evidence that oxygen tension within the maize root stele *in situ* is sub-optimal, but suggest that the cells are adapted to it.

There is little doubt that the root stele is not under great anaerobic stress, but are the cells in fact "leaky" to ions? Certainly ions leave the parenchyma of the stele and enter the xylem stream, but the question is whether they leak passively from the parenchyma or are secreted actively into the xylem. The first alternative has received recent support from Baker (1973a, b) who has demonstrated that ^{86}Rb efflux from isolated steles is enhanced by low temperature and by the respiratory uncoupler carbonyl cyanide *m*-chlorophenylhydrazone (CCCP). Such effects are not consistent with active secretion of K^+ from the parenchyma of the stele.

In contrast, other recent work has been interpreted as evidence in favour of parenchymal secretion of ions to the xylem. Läuchli and Epstein (1971) and

Läuchli *et al.*, (1971) have reached this conclusion on the basis of the results from several experimental techniques, radioactive pulse labelling, electron probe analysis of ion concentration profiles across the root and inhibitor studies. They conclude that ion transport to the xylem requires two carrier-mediated membrane events—uptake at the plasmalemma of the outer cortical cells and secretion at the plasmalemma of the xylem parenchyma.

A similar conclusion has been reached independently by Pitman (1972) where the evidence is a comprehensive flux analysis for ^{36}Cl in excised barley roots after a variety of pre-treatments and with or without the addition of CCCP. Xylem exudation is inhibited by CCCP but the plasmalemma and tonoplast effluxes in the cortex are not. Since the cortical vacuoles could therefore supply the symplasm in pre-loaded tissue even after the application of CCCP, Pitman (1972) argues that the observed cessation of xylem exudation must be due to CCCP inhibition of active ion secretion by the xylem parenchyma.

In considering differential inhibitor effects on uptake and exudation it must always be borne in mind that the inhibitor may be exercising its effect through acting on the symplasmic transport. Although it is likely that diffusion is the chief mechanism of symplasmic ion transport, it is still reasonable to suppose that the continued functioning of the plasmodesmata as effective transit channels may require continuing energy supply. Indeed there is evidence that certain plant hormones (e.g. cytokinin and abscisic acid) may act on the plasmodesmata in barley roots (Cram and Pitman, 1974). Furthermore, the protein synthesis inhibitor, cycloheximide, is thought to affect xylem ion exudation by acting at the plasmodesmata; cycloheximide has no affect on K^+ uptake (into the cortical cells) nor on respiration, but K^+ transport to the xylem in barley roots is inhibited (Läuchli *et al.*, 1973). The explanacion offered is that continuing protein synthesis is needed for symplasmic ion transport, possibly to maintain the plasmodesmatal pores in their natural state.

D. The Common Longitudinal Model

It has been known for many years that different regions of a plant root absorb and translocate ions at different rates. In general the older regions are less permeable and the first, most obvious, correlation is with rather gross anatomical alterations, tertiary thickening of the endodermis and stele, and cuticulization of the epidermis leading in many cases to the formation of an identifiable exodermis. Concomitant decreases in water permeability are also observed (see also p. 397).

Although these structural changes are of much importance, there are other, more subtle factors involved. For example, in many cereals little Ca reaches the conducting tissues of the older parts of the seminal root axes in which the tertiary endodermis is well developed (Clarkson *et al.*, 1968), while K is still transported in the same regions. The reason for this difference is not perfectly clear, but may be partly explained by the cell wall pathway being the chief route for Ca movement, with the K moving mainly in the symplasm. In younger parts of the root, the behaviour of the two ions differs less. Phosphate, like potassium, can apparently traverse the older regions of the root, but the pattern of uptake again differs from that of K. Sodium uptake proceeds quite rapidly into the xylem of the young regions of *Zea mays* primary roots, but with increasing distance from the root tip, discrimination against sodium absorption relative to potassium increases markedly (Shone *et al.*, 1969).

These various alterations in the pattern of ion uptake as one proceeds along the root from the tip give rise, in a steady state situation, to a standing gradient osmotic flow along the xylem vessels (see Anderson *et al.*, 1970). There will in general be a maintained gradient of salt concentration along the vessel and the superposition of the standing osmotic gradient model and the radial model, discussed earlier, seems to provide a satisfactory explanation of what is known about ion transport along the xylem in excised roots.

E. Hormone Effects on Ion Transport

Recently there has been a great deal of interest in the effects of various plant hormones and growth regulators on ion uptake into roots and ion transport from the root to the shoot along the xylem. Table II contains a summary of work published to date with appropriate literature citations where those interested may find further details.

It may be worth noting that in all such work, the hormone levels applied externally to the tissues are many orders of magnitude greater in concentration than the natural, endogenous, levels. To what extent this may distort the hormone effects is a matter which requires further examination. However, from Table II it can be seen that there are in general two effects, firstly on the rate of xylem ion transport from the root and secondly on the concentration of ions accumulated in the root tissue, predominantly in the cortical cell vacuoles.

Few have attempted to interpret these observations on any hard mechanistic model, but one factor is sure: many of the hormones, in particular abscisic acid (ABA) and indole acetic acid (IAA), are known to have dramatic effects on the water permeability of cells, which accounts in part for the observed increases in fluid exudation rates. However it does seem that ion transport is also directly affected by the hormones, but as yet no proposals for the action have been made.

IV. Electrophysiology of Roots

A. Introductory Theory

This section will be very brief; for a more complete exposition of the relevant theory written for botanists, see Dainty (1962) or Higinbotham (1973). The most fundamental point to grasp is that an ion is subjected to *two* physical driving

TABLE II. Hormone effects on ion and water transport through excised roots.

Material/Treatment	Water Flux	Exudate Concn.	Tissue Concn.
Zea mays			
Control (1)	4·64[a]	22·1[b], 1·8[c] 23·1[d]	69·4[b], 2·9[c] 41·5[d]
Kinetin (1)	1·02[a]	16·1[b], 2·9[c] 15·4[d]	94·2[b], 2·5[c] 69·3[d]
ABA (1)	7·34[a]	20·5[b], 1·1[c] 17·2[d]	91·5[b], 3·7[c] 84·0[d]
Zea mays			
Control (2)	100%		100%[e]
ABA (2)	30%		130%[e]
Hordeum vulgare			
Control (2)		0·61[f], 0·69[g]	
ABA (2)		0·1[f], 0·1[g]	
Lycopersicon esculentum			
Control (3)	0·58[h]	0·73[j]	
Kinetin (3)	0·21[h]	1·14[j]	
ABA (3)	1·18[h]	0·74[j]	
IAA (3)	0·76[h]	0·62[j]	
Pisum sativum			
Control after 1 h (4)		1·61[f]	
Gibberellin after 1 h		1·54[f]	
Control after 4 h		1·65[f]	
Gibberellin after 4 h		1·59[f]	
Control after 5·5 h		1·71[f]	
Gibberellin after 5·5 h		2·38[f]	

The superscripts are: [a] $\times 10^{-4}$ μl $cm^{-2}s^{-1}$; [b]mM K^+; [c]mM Ca^{2+}; [d]mM Cl^-; [e]accumulation into tissue; [f]nmoles K^+ $gFW^{-1}s^{-1}$; [g]nmoles Cl^- $gFW^{-1}s^{-1}$; [h]μl $gDW^{-1}s^{-1}$: [j]atm.

The source references are: (1) Collins and Kerrigan (1974); (2) Cram and Pitman (1972); (3) Tal and Imber (1971); (4) Lüttge *et al.* (1968).

forces to make it move from one side of a membrane, artificial or biological, to the other. Firstly it is driven by the concentration gradient, as is a non-electrolyte, and secondly, and equally important, it is driven by any electric field there may be across the membrane. The electric field is usually defined as the electric potential difference across the membrane (the *membrane potential*, ΔE) divided by the membrane thickness. This definition is the so-called constant field assumption which is commonly used as an acceptable simplification in membrane ion transport work.

It follows, since an ion is subjected to what amounts to two driving forces, that the equilibrium condition for an ion across a membrane is not necessarily that the concentration of that ion is equal on either side of the membrane. Concentration equality is a special equilibrium condition which only pertains if the membrane is short circuited so that there is zero membrane potential. Otherwise, the equilibrium condition is contained in the so-called *Nernst equation*, which is, for univalent ions,

$$\Delta E = 58 \log_{10} \frac{C_1}{C_2} \text{ (mV)}$$

where C_1 and C_2 are the ion concentrations on either side of the membrane. Thus if K^+ say, is 10 mM on side 1 and 1 mM on side 2 the K^+ may yet be in equilibrium across the membrane if the membrane potential is 58 mV with side 1 biased negative with respect to side 2.

There is commonly an accumulation of ions in the interior of plant cells, the internal concentrations of many cations and anions being in the range 1–100 times those of the external solution. However, as has just been explained, the concentration difference, by itself, is no indication that an ion is out of equilibrium across the membrane. In order to reach that decision, for a single membrane and assuming that all ions move independently, one must insert the appropriate values in the Nernst equation and check the equality. Since the membrane potentials of plant cells are almost invariably biased with the interior of the cell negative, it is therefore the case that all accumulated anions are far from equilibrium and most be transported actively inward by the expenditure of metabolic energy; of the cations, K^+ is often close to equilibrium and Na^+ is often far from equilibrium in the opposite sense to the anions and must therefore be transported actively outward.

Where a cell membrane is permeable to several ion species, the membrane potential can be predicted by the use of the so-called Goldman equation, providing that there is no other potential generating event taking place in the membrane. Such a Goldman potential is sometimes referred to as a diffusion potential, because it arises solely from the tendency of the ions to diffuse through the membrane, down their individual concentration gradients. However, the membrane is more permeable to some ion species than to others, and these tend to diffuse more rapidly, leading to a slight charge separation which results in a membrane potential to retard the fast ions and accelerate the slow ones. This diffusion potential or Goldman potential, can be derived for univalent salts and has the following form

$$\Delta E_m = E_G = 58 \log_{10} \left\{ \frac{P_k C_1{}^k + P_{Na} C_1{}^{Na} + P_{Cl} C_2{}^{Cl}}{P_k C_2{}^k + P_{Na} C_2{}^{Na} + P_{Cl} C_1{}^{Cl}} \right\}$$

where the P's are permeability coefficients for the individual ion species across the membrane. For a more complete derivation, see Dainty (1966).

For a number of years it was thought that the membrane potential measured in a plant cell arose solely from the diffusion effect outlined above and could be described by a Goldman equation fit. Indeed in animal cells in many of the best studied examples the membrane potential is still described by the Goldman potential. However, there has recently been increasingly good evidence in a number of plant cells that the Goldman potential is only one component of the membrane potential, and that to it must be added a second component which arises as a direct result of ion transport systems at work in the membrane. Such transport systems are usually known as electrogenic ion pumps, electrogenic because they contribute directly by their action to the observed membrane potential. Thus for many (perhaps all) plant cell membrane potentials, the complete membrane potential will be given by

$$\Delta E_m = E_G + E_p$$

where E_p is the electrogenic pump potential.

B. The Electrogenic Pump

An ion pump is generally considered to be a membrane-located protein complex, which on the fluid mosaic membrane model occupies a space in the predominantly lipid matrix like an iceberg in the sea. The complex runs all the way through the membrane and is in contact with the aqueous solution on either side. Ions are attached to the complex at some binding site on one side and are released at the other, possibly by some change in the quaternary structure of the protein. In addition, many of these ion pumps have the capability of energy transduction, usually by hydrolysing ATP or possibly by utilizing the reducing power provided by NADH or some other reduced intermediate. These energy inputs will be necessary to cause the quaternary structural changes in those cases where ions are being transported against their electrochemical gradients, i.e. when there is active transport.

For our purposes, we may make three broad classifications of ion pumps: (i) neutral salt pumps; (ii) unidirectional pumps; (iii) exchange pumps, which may or may not have 1:1 stoichiometry. A neutral salt pump, e.g. a pump which transports KCl as such across the plant cell membrane, is perhaps the least likely from the chemist's viewpoint, although evidence has been given to suggest it may exist in red beet storage tissue (Poole, 1966). It is obvious that such a pump will not be electrogenic, because it transports what is essentially a neutral, non-electrolyte, KCl.

A unidirectional pump, e.g. a pump which transports only Cl^- say, will be ideally electrogenic. On every cycle of the pump net electric current will be transferred, because there will be direct coupling of net movement of electric charge, carried on the Cl^- ions, to the expenditure of metabolic energy. Thus the action of such a pump will contribute directly to the membrane potential. Net electric current will be transferred, and, as this current must be balanced across the passive resistance of the membrane by an equal but opposite current, it follows from Ohm's law that the membrane will be polarized by the ohmic potential. Since the observed membrane potential in a higher plant cell is hyper-polarized above the Goldman potential, it follows that the electrogenic pump must extrude cations or accumulate anions.

Finally, there is perhaps the most likely situation, an exchange pump but not necessarily with 1:1 stoichiometry. It is obvious that such a pump is somewhere between cases (i) and (ii). If the pump were 1:1 coupled then it is neutral; for every cation transported inward, a cation is transported outward to give zero net charge transfer, which is the equivalent of a neutral salt pump. However, if the pump is 3:2 coupled say, rather than 1:1, then the pump will transfer net charge during one time-average cycle. In this latter situation the pump will be electrogenic and make a direct contribution to the observed membrane potential.

The whole topic of the electrogenic pump can be thought through a little more rigorously than is indicated here, and a more complete account is to be found in Higinbotham and Anderson (1974).

C. Excised Root Exudate Potentials

An electropotential will be recorded between two electrodes, one placed in the xylem exudate fluid of an excised root and the other in the solution bathing the root. Values of several tens of millivolts are observed, the exudate being negative. It is generally thought that this potential arises at the membrane(s) which every ion must cross to reach the xylem. Extensive studies of xylem exudate potentials have been made (Table III). It was on this preparation of excised root exudate potential, rather than on a single cell preparation, that the first evidence for an electrogenic pump in plants was obtained. Higinbotham *et al.* (1970) have argued, from data on the xylem exudate potential of *Zea mays* roots published by Davis and Higinbotham (1969), that a component of the potential is due to an electrogenic pump. They found a fairly rapid depolarization of the exudate potential after poisoning the tissue with CN^-, with subsequent repolarization to something close to the original value after the CN^- had been washed out. This effect is too rapid to be due to alterations in the diffusion potential, but is expected of an electrogenic pump component of potential which will become zero as soon as the pump ceases its activity upon inhibition of respiration. A similar effect was observed independently by Shone (1969) with DNP inhibition again in *Zea mays*, and a similar conclusion was reached.

D. Root Cell Electropotentials

By using a finely-drawn microelectrode (2-4 μm tip diameter) mounted on a micro-manipulator, it is possible to impale a single cell so that the membrane

TABLE III. Xylem exudate electropotentials of roots

Material and Conditions	Exudate potential (mV) (Exudate negative)
Zea mays	
bathed in 0·1 mM KCl (1)	50
0·3 mM KCl (1)	48
1·0 mM KCl (1)	30
3·0 mM KCl (1)	25
10·0 mM KCl (1)	18
Zea mays	
bathed in 0·2 mM KCl (2)	58
0·2 mM KCl, with 10^{-5} DNP added (2)	32
Zea mays	
bathed in 1X solution (3)	49
1X solution with 1 mM CN added (3)	32
Ricinus communis	
bathed in 1/10 Stout and Arnon's solution (4)	58
Helianthus annuus	
bathed in 1/10 culture solution (5)	48
bathed in full strength culture solution	34

The source references are: (1) Dunlop and Bowling (1971); (2) Shone (1969); (3) Davis and Higinbotham (1969); (4) Bowling and Spanswick (1964); (5) Bowling, (1966). The solution described as 1X has the following composition: 1 mM KCl; 1 mM $Ca(NO_3)_2$; 0·25 mM $MgSO_4$; NaH_2PO_4 and Na_2HPO_4 to give 1 mM Na^+ and a pH of 5·5 to 5·7.

seals around the electrode tip and the cell apparently suffers little damage. Then the electropotential between this intracellular electrode and the bathing solution can be measured with suitable circuitry. Figure 8 shows the profile of potential across a root cross section obtained by this technique by Bowling (1973) on sunflower. Although the values he obtains seem low in the light of other measurements, nevertheless this work highlights the fact that the cortical symplasm forms an equipotential network across the cortex and indeed into the endodermis, as one would expect if indeed the plasmodesmata are a low resistance pathway for ion movement from cell to cell.

There is a direct measurement of the plasmodesmatal resistance available to us in Spanswick (1972). He inserted a microelectrode in a cortical cell of *Zea* root and a second electrode in an adjacent cell and passed current to obtain a direct resistance measurement of the coupling between the two cells. The low value he found confirmed that the plasmodesmata are an effective channel for ion movement from cell to cell.

More recently, Anderson *et al.* (to be published) have striking evidence that the electropotential of *Zea* root and *Pisum sativum* root cells contain a large component due to an electrogenic pump. Respiration was inhibited by light–dark

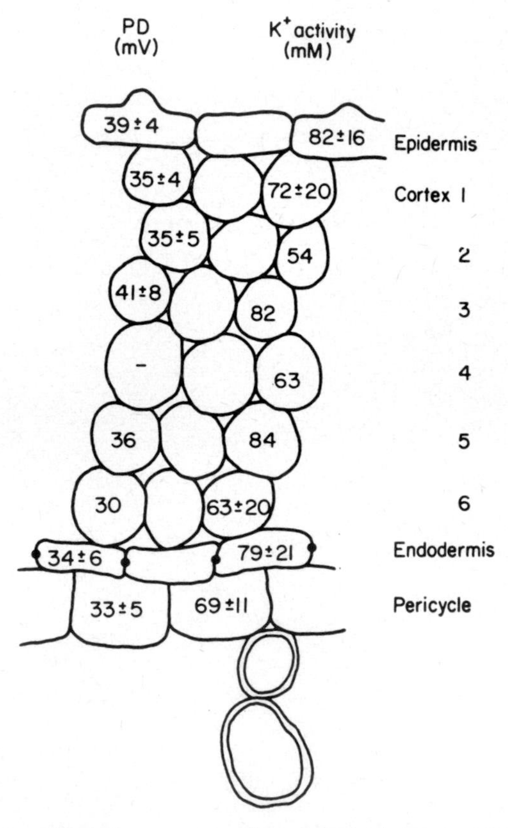

Fig. 8. Electropotential profile across a sunflower root. The K^+ activities also shown were obtained with a K^+-sensitive microelectrode. (Reproduced from Bowling, 1973).

switching of tissue bathed in solution saturated with a 19:1 v:v mixture of CO and O_2 and very rapid depolarization–repolarization of the potential was observed. The tissue could be taken through many cycles, and a fair correlation was established between the time for depolarization and the time for ATP turnover in the cell, the most obvious implication being that the electrogenic pump in this tissue is ATP powered, but other possibilities (such as reducing power in NADH) cannot be excluded. Slayman *et al.* (1973) have recent excellent evidence that the electrogenic pump in *Neurospora crassa* is ATP driven, and in this case believe that H^+ extrusion from the fungal cell is the electrogenic event. In higher plant roots there is at present no definitive evidence of the ion involved in electrogenesis, but if it is anion uptake then the carrier must be very non-specific; changing the external anion composition seems to have little effect on the cell membrane potential. Perhaps the most likely candidate in the higher plant root is again H^+ extrusion, perhaps on a pump exchanging for K^+ but not 1:1 coupled. Variable stoichiometry on such a pump, in response to external conditions, and indeed to internal conditions such as organic acid concentration, might make an interesting feedback mechanism for regulation of ion uptake by the root.

Acknowledgement

I should like to pay tribute to Professor N. Higinbotham for the many insights into ion transport given to me in our conversations at Washington State University, Pullman, Washington.

References

Anderson, W. P. (1972). Ion transport in the cells of higher plant tissues. *A. Rev. Pl. Physiol.*, **23**; 51–72.

Anderson, W. P. (Ed.) (1973). "Ion Transport in Plants." Academic Press, London, New York and San Francisco.

Anderson, W. P., Aikman, D. P. and Meiri, A. (1970). Excised root exudation—a standing gradient osmotic flow. *Proc. Roy. Soc. (London)*. **B174**; 445–458.

Anderson, W. P. and Collins, J. C. (1969). The exudation from excised maize roots bathed in sulphate media. *J. exp. Bot.*, **20**; 72–80.

Anderson, W. P., Goodwin, L. and Hay, R. K. M. (1974). Evidence for vacuole involvement in xylem ion supply in the excised primary roots of two species, *Zea mays* and *Allium cepa*. *In*: "Structure and Function of Primary Root Tissues." (Ed. J. Kolek). pp. 379–388 Slovak Academy of Sciences, Bratislava.

Anderson, W. P., Hendrix, D. L. and Higinbotham, N. The effect of CN and CO on the electrical potential and resistance of cell membranes. (Submitted to Plant Physiol.).

Anderson, W. P. and House, C. R. (1967). A correlation between structure and function in the root of *Zea mays*. *J. exp. Bot.*, **18**; 544–555.

Arisz, W. H. (1956). Significance of the symplast theory for transport across the root. *Protoplasma.*, **46**; 5–62.

Arisz, W. H., Helder, R. J. and Van Nie, R. (1951). Analysis of the exudation process in tomato plants. *J. exp. Bot.*, **2**; 257–297.

Baker, D. A. (1973a). The radial transport of ions in maize roots. *In*: "Ion Transport in Plants." (Ed. W. P. Anderson). pp. 511–517. Academic Press, London, New York and San Francisco.

Baker, D. A. (1973b). The effect of CCCP on ion fluxes in the stele and cortex of maize roots. *Planta (Berl.)*. **112**; 293–299.

BOWLING, D. J. F. (1966). Active transport of ions across sunflower roots. *Planta (Berl.).* **69**; 377–382.

BOWLING, D. J. F. (1973a). Measurement of a gradient of oxygen partial pressure across the intact root. *Planta (Berl.).* **111**; 323–328.

BOWLING, D. J. P. (1973b). The origins of the trans-root potential and the transfer of ions to the xylem of sunflower roots. *In*: "Ion Transport in Plants." (Ed. W. P. ANDERSON). pp. 483–491. Academic Press. London, New York and San Francisco.

BOWLING, D. J. F. and SPANSWICK, R. M. (1964). Active transport of ions across the root of *Ricinus communis. J. exp. Bot.*, **15**; 422–427.

CLARKSON, D. T. (1974). "Ion Transport and Cell Structure in Plants." McGraw Hill, G.B.

CLARKSON, D. T., ROBARDS, A. W. and SANDERSON, J. (1971). The tertiary endodermis in barley roots: fine structure in relation to radial transport of ions and water. *Planta (Berl.).* **96**; 292–305.

CLARKSON, D. T., SANDERSON, J. and RUSSELL, R. S. (1968). Ion uptake and root age. *Nature, Lond.* **220**; 805–806.

COLLINS, J. C. and KERRIGAN, A. P. (1974). The effects of kinetin and abscisic acid on water and ion transport in isolated maize roots. *New Phytol.* **73**; 309–314.

CRAFTS, A. S. and BROYER, T. C. (1938). Migration of the salts and water into xylem of the roots of higher plants. *Am. J. Bot.*, **24**; 415–431.

CRAM, W. J. and PITMAN, M. G. (1972). The action of abscisic acid on ion uptake and water flow in plant roots. *Aust. J. Biol. Sci.*, **25**; 1125–1132.

CURRAN, P. G. and MACINTOSH, J. R. (1962). A model for biological water transport. *Nature, Lond.*, **193**; 347–348.

DAINTY, J. (1962). Ion transport and electrical potentials in plant cells. *A. Rev. Pl. Physiol.*, **13**; 379–402.

DAVIS, R. F. and HIGINBOTHAM, N. (1969). Effects of external cations and respiratory inhibitors on electrical potential of the xylem exudate of excised corn roots. *Plant Physiol.*, **44**; 1383–1392.

DUNLOP, J. and BOWLING, D. J. F. (1971). The movement of ions into the xylem exudate of maize roots. II. A comparison of the electrical potential and electro-chemical potentials of ions in the exudate and in the root cells. *J. exp. Bot.*, **22**; 445–452.

EPSTEIN, E. (1972). "Mineral Nutrition of Plants." Wiley. New York.

FISCUS, E. L. and KRAMER, P. J. (1970). Radial movement of oxygen in plant roots. *Plant Physiol.*, **45**; 667–669.

GINSBURG, H. (1971), Model for iso-osmotic water flow in plant roots. *J. theor. Biol.*, **32**; 147–158.

GINSBURG, H. and GINZBURG, B. Z. (1971). Evidence for active water transport in a corn root preparation. *J. Mem. Biol.*, **4**; 29–41.

GUNNING, B. E. S., PATE, J. S., MINCHIN, F. R. and MARKS, I. (1974). Quantitative aspects of transfer cell structure and function. *Soc. Exp. Biol. Symp.*, **28**; (In press).

HALL, J. L., SEXTON, R. and BAKER, D. A. (1971). Metabolic changes in washed, isolated steles. *Planta (Berl.).*, **96**; 54–61.

HAY, R. K. M. and ANDERSON, W. P. (1972). Characterization of exudation from excised roots of onion, *Allium cepa*. II. Chemical composition of exudate. *J. exp. Bot.*, **23**; 585–590.

HELDER, R. J. and BOERMA, J. (1969). An electron-microscopical study of the plasmodesmata in the roots of young Barley seedlings. *Acta Bot. Neerl.*, **18**; 99–107.

HIGINBOTHAM, N. (1973). Electropotentials of plant cells. *A. Rev. Pl. Physiol.*, **24**; 25–46.

HIGINBOTHAM, N. and ANDERSON, W. P. (1974). Electrogenesis in higher plant cells. *Can. J. Bot.* (In press).

HIGINBOTHAM, N., DAVIS, R. F., MERTZ, S. M. and SHUMWAY, L. K. (1973). Some evidence that radial transport in corn roots is into living vessels. *In*: "Ion Transport in Plants." (Ed. W. P. ANDERSON). pp. 493–506. Academic Press. London, New York and San Francisco.

HIGINBOTHAM, N., GRAVES, J. S. and DAVIS, R. F. (1970). Evidence for an electrogenic ion transport pump in cells of higher plants. *J. Mem. Biol.*, **3**; 210–222.

HODGES, T. K. and VAADIA, Y. (1964a). Uptake and transport of radiochloride and tritiated water by various zones of onion roots of different chloride status. *Plant Physiol.*, **39**; 104–108.

HODGES, T. K. and VAADIA, Y. (1964b). Chloride uptake and transport in roots of different salt status. *Plant Physiol.*, **39**; 109–114.

HOUSE, C. R. and FINDLAY, N. (1966). Water transport in isolated maize roots. *J. exp. Bot.*, **17**; 344–354.

HYLMO, B. (1953). Transpiration and ion absorption. *Physiol. Plant.* **6**; 333–405.

JARVIS, P. and HOUSE, C. R. (1970). Evidence for symplastic ion transport in maize roots. *J. exp. Bot.*, **21**; 83–90.

LÄUCHLI, A. and EPSTEIN, E. (1971). Lateral transport of ions into xylem of corn roots. 1. Kinetics and energetics. *Plant Physiol.*, **48**; 111–117.

LÄUCHLI, A., KRAMER, D., PITMAN, M. G. and LÜTTGE, U. (1974). Ultrastructure of xylem parenchyma cells in barley roots in relation to ion transport to the xylem. *Planta* (*Berl.*) **119**; 85-99.

LÄUCHLI, A., LÜTTGE, U. and PITMAN, M. G. (1973). Ion uptake and transport through barley seedlings: differential effects of cycloheximide. *Z. Naturforsch.* **28c**; 431–434.

LÄUCHLI, A., SPURR, A. R. and EPSTEIN, E. (1971). Lateral transport of ions into the xylem of corn roots. II. Evaluation of a stelar pump. *Plant Physiol.*, **48**; 118–124.

LUNDEGARDH, H. (1950). Translocation of salts and water through wheat roots. *Physiol. Plant.* **3**; 103–151.

LÜTTGE, U., BAUER, K. and KÖHLER, D. (1968). Frühwirkungen von Gibberellinsäure auf Membran-transporte in jungen Erbsenplanzen. *Biochim. Biophys. Acta.* **150**; 452–459.

MINCHIN, F. R. and BAKER, D. A. (1973). The influence of calcium on potassium fluxes across the root of *Ricinus communis*. *Planta* (*Berl.*). **113**; 97–104.

PATE, J. S. and GUNNING, B. E. S. (1972). Transfer cells. *A. Rev. Pl. Physiol.*, **23**; 173–196.

PETTERSSON, S. (1966). Artifically induced water and sulfate transport through sunflower roots. *Physiol. Plant.* **19**; 581–601.

PITMAN, M. G. (1971). Uptake and transport of ions in barley seedlings. 1. Estimation of chloride fluxes in cells of excised roots. *Aust. J. Biol. Sci.*, **24**; 407–421.

PITMAN, M. G. (1972). Uptake and transport of ions in barley seedlings. 2. Evidence for two active stages in transport to the shoot. *Aust. J. Biol. Sci.*, **25**; 243–257.

POOLE, R. J. (1966). *J. gen. Physiol.* **49,** 551–563.

PRIESTLEY, J. H. (1920). The mechanism of root pressure. *New Phytol.*, **19**; 189–200.

ROBARDS, A. W. (1971). The ultrastructure of plasmodesmata. *Protoplasma.*, **72**; 315–323.

SCOTT, F. M. (1949). Plasmodesmata in xylem vessels. *Bot. Gaz.*, **110**; 492–495.

SCOTT, F. M. (1965). The anatomy of plant roots. *In*: "Ecology of Soil-borne Pathogens." (Eds. K. F. Baker and W. C. Syndcr). pp. 145–153. Univ. Calif. Press. Berkeley.

SHONE, M. G. T. (1969). Origins of the electrical potential difference between the xylem sap of maize roots and the external solution. *J. exp. Bot.*, **20**; 698–716.

SHONE, M. G. T., CLARKSON, D. T. and SANDERSON, J. (1969). The absorption and translocation of sodium by maize seedlings. *Planta* (*Berl.*). **86**; 301–314.

SLAYMAN, C. L., LONG, W. S. and LU, C. Y.-H. (1973). The relationship between ATP and an electrogenic pump in the plasma membrane of *Neurospora crassa*. *J. Mem. Biol.*, **14**; 305–338.

SPANSWICK, R. M. (1972). Electrical coupling between cells of higher plants: a direct demonstration of intercellular communication. *Planta* (*Berl*). **102**; 215–227.

TAL, M. and IMBER, D. (1971). Abnormal stomatal behaviour and hormonal imbalance in Flacca, a wilty mutant tomato. III. Hormonal effects on water status in the plant. *Plant Physiol.*, **47**; 849–850.

TYREE, M. T. (1970). The symplast concept. A general theory of symplastic transport according to the thermodynamics of irreversible processes. *J. Theoret. Biol.*, **26**; 181–214.

TYREE, M. T. (1973). An alternative explanation for the apparently active water exudation in excised roots. *J. exp. Bot.*, **24**; 33–37.

VAN FLEET, D. S. (1961). Histochemistry and function of the endodermis. *Bot. Rev.*, **27**; 165–220.

ZIEGLER, H. and LÜTTGE, U. (1966). Die Salzdrusen von *Limonium vulgare*. I. Die Feinstruktur. *Planta* (*Berl.*)., **70**; 193–206.

Part III

ROOTS IN RELATION TO THE SOIL MICROFLORA

Chapter 21

Legume Root Nodule Initiation and Development

P. J. DART

Soil Microbiology Department, Rothamsted Experimental Station, Harpenden, Herts., England.

I. Host–Rhizobium Specificity

Root nodules differ in many ways from other plant structures in their mode of development and this prompted early speculations that they were insect or fungal induced galls, modified roots, or even storage buds. The bacteria-like bodies that Woronin (1867) found in lupin nodules were isolated and grown by Beijerinck (1888) and shown to be the causative organism. At the same time others in Europe and North America showed that these organisms were widespread in soil and that different plants may require different strains of bacteria to nodulate them. This culminated in the cross-inoculation concept, that certain groups of plants tend to be reciprocally nodulated by rhizobia isolated from their nodules, and that rhizobia nodulating one group would not readily nodulate others. Eight major groups were recognized by 1932 (Fred *et al.*, 1932)—alfalfa—*Rhizobium meliloti*; clover—*R. trifolii*; pea—*R. leguminosarum*; bean—*R. phaseoli*; lupin—*R. lupini*; soybean—*R. japonicum*; cowpea—*Rhizobium* sp.; lotus—*Rhizobium* sp.—with several minor groups. Nodulation across these groups occurs occasionally with *R. japonicum*, *R. lupini*, *R. phaseoli* and cowpea rhizobia (e.g. see Vincent, 1974), and much wider promiscuity may result when heterologous combinations are left together for longish periods (e.g. Wilson, 1944). The cowpea group is the largest and here reciprocal relationships between plants and rhizobia are less clearly defined. Generally infection across cross-inoculation groups results in ineffective nodule development. A *Rhizobium* strain capable of effectively nodulating cowpeas also effectively nodulates the non-legume *Trema cannabina* (family *Ulmaceae*). Leghaemoglobin is not formed and the nodule structure differs from that of legumes in that the nodule has a single central vascular trace surrounded by invaded cells. In some of these, bacteroids are enclosed by membrane envelopes, but otherwise the bacteria remain enclosed in infection thread structures which fill the whole host cell (Trinick, 1973 and personal communication).

Detailed serological analysis of the different *Rhizobium* spp has shown cross-reactions among slow growing strains of rhizobia, and reactions among the fast growing rhizobia which mirror their relatedness on physiological characteristics, and DNA composition (Vincent, 1974). There are relationships between the DNA of *R. japonicum* and *R. lupinii*, and *R. trifolii* with *R. leguminosarum* (Gibbins and Gregory, 1972), and affinities between the exopolysaccharides of *R. leguminosarum*, *R. trifolii* and *R. phaseoli* which differ considerably from that of *R. meliloti* (Zevenhuizen, 1973; Vincent, 1974). However there are differences in composition of the exopolysaccharides between strains within species, e.g. *R. trifolii* (Hepper, 1972), which have no obvious relationship to their nodulating abilities. The red pigmented organism which nodulates only *Lotononis* spp however has distinctly different antigenic affinities (Vincent and Humphreys, personal communication) and DNA composition (Godfrey, 1972) from the other groups of organisms. However there seem to be no obvious common, cross-reacting antigens between the host plant and *Rhizobium* which mirror their nodulation affinities (Charudattan and Hubbell, 1973). Thus no differences have been demonstrated between strains which readily define their differences in nodulating ability. Perhaps more detailed analyses of the structure of the polysaccharides will reveal where the specificities reside.

Plant glycoproteins seem to be involved in moderating the production and activity of polygalacturonases secreted by various fungi with a specificity that correlates well with resistance or susceptibility of the host to the fungus (e.g. Fisher *et al.*, 1973). Similar proteins are widespread in legumes (Toms, 1971), and phytohaemagglutinins from *Phaseolus vulgaris* bind to *R. phaseoli* and are released from the root at discrete sites (Hamblin and Kent, 1973). Rhizobia readily grow in the rhizospheres of plants which they do not nodulate, but such proteins may be involved in the "recognition process" between plant and bacterium, i.e. a specific type of binding of *Rhizobium* to the root may be necessary to trigger later events in the infection process.

Large variations in nodulating ability exist between different lines and species of various legumes. Both the host and *Rhizobium* genomes affect the time taken to form nodules and their number and effectiveness in fixing nitrogen. Plant breeding can influence the ability of a plant to nodulate, and lines have been released which are difficult to nodulate. Most commercial cultivars of *Phaseolus vulgaris* do not nodulate on the primary root, but the "primitive", Mexican, cultivar 120 does. Grafting generally does not affect the nodulation characteristics of a root. Some single gene host mutants are non-nodulating with one or more *Rhizobium* strains; others are difficult to nodulate, or the nodules formed are ineffective (see Nutman, 1969). Red clover seedlings occasionally nodulate with *R. leguminosarum* (e.g. Kleczkowska *et al.*, 1944). Crossing and breeding can produce lines in which all plants still nodulate effectively with *R. trifolii*, and up to 90% of them with *R. leguminosarum* but none with *R. meliloti* or *R. japonicum* (Hepper, 1973a).

Most of the 12,000 or so species of legumes have still not been examined for nodulation (see Allen and Allen, 1961); most careful study has been made of potentially agronomically useful herbs. Nodules have been found on about 90% of plants examined in the sub-families Mimosoideae and Papilionatae, but only on about 23% of the Caesalpinoideae examined. Non-nodulating roots are often brown, perhaps because of tannins or polyphenols. Many tropical forests are dominated by legume trees, but little is known of their nodulation status or whether nodules confer an advantage (e.g. Döbereiner and Campelo, 1974).

II. Nodule Form

Morphologically, nodules can be divided rather arbitrarily into (a) elongate with continuing apical meristematic activity and a basically cylindrical structure,

(b) spherical nodules which usually have several discrete meristematic foci, and (c) collar nodules as in lupins where the meristematic activity becomes displaced to the lateral peripheries of the nodule so that the nodule grows around the root. Nodules with apical meristems may later develop a branched or even coralloid structure through dichotomy of the meristem. In spherical nodules such as soybean, meristematic activity declines after the initial 14 days or so of nodule development, with later growth largely from expansion of bacteroid filled cells. However, activity restarts when the demand for fixed nitrogen increases during pod formation and fill (Day, 1972). Nodule shape is plant determined and is usually consistent within genera, although lupins may form collar nodules on older, thicker roots, and elongate ones on finer, lateral roots.

Nodules vary in the firmness of their attachment to the root with no obvious relationship with nodule form; for elongate nodules it is substantial in clovers and fragile in *Cajanus cajan* (pigeon pea), and for spherical nodules it is substantial in *Arachis hypogaea* (peanut) and fragile in *Centrosema pubescens.*

III. Development of Infection Threads

The best known mechanism for entry of *Rhizobium* into roots is via root hairs through formation of an infection thread. Most observations have been made on small seeded *Trifolium* plants (with 16 species studied) and reliable observations have been recorded for only 34 species in other genera. In a further 41 species, infection threads have been found in sections of nodules; perhaps these infection threads also originated in root hairs. For 29 species, of which the best known examples are lupins and peanut, no infection thread structures have been found in nodules or root hairs (see Dart, 1974).

A. Rhizosphere Events

The first interaction between *Rhizobium* and host occurs in the rhizosphere with *Rhizobium* growth stimulated over that in soil, and over that of other micro-organisms. There seems to be some selection for the strain which will nodulate the legume, particularly in agar culture (e.g. Purchase and Nutman, 1957; Nutman and Ross, 1970). The mechanism for this is unknown, but could be related to the stimulation of *Rhizobium* growth afforded by substances such as biotin and thiamin exuded from the root (e.g. Graham, 1963). For *Pisum sativum* some selection results from the preponderance of homoserine in the ninhydrin positive fraction of the seedling root exudate. *Rhizobium leguminosarum* grows well using homoserine as a sole N and C source but *R. phaseoli* and *R. trifolii* do not (van Egeraat, 1972).

However the growth patterns of *Rhizobium* strains in the rhizosphere seem unable to account for either cross-inoculation specificity or the competition between strains able to nodulate a given legume. In the latter case one strain may preponderate in the rhizosphere by a ratio $10^4:1$ but only form 40% of the nodules (Robinson, 1969). For soybeans the competitive ability of some strains changes with the soil temperature (Weber and Miller, 1972). This host selection can sometimes operate to advantage in the field, with effective strains forming most of the nodules; this is very much dependent on the site and strains involved (e.g. Sherwood and Masterson, 1974). While the overall numbers of different strains of rhizobia on a root can be counted we know nothing of their localized distribution in relation to the curled root hairs which are the sites of infection. Some legume roots have a prominent "mucigel" layer which may enclose the root hairs and which is produced partly by the root cap (Dart and Mercer, 1964a) and perhaps partly by older epidermal cells and root hairs (see Head, 1964; Bell and McCully, 1970). Rhizobia are often concentrated within this layer, but can also be excluded by it from the root (Figs 1 and 2).

Serological analysis (of fewer than 10 species!) suggested that a nodule, particularly from plants grown in soil, usually contains only one *Rhizobium* strain although adjacent nodules may be formed by another strain. However for *Trifolium subterraneum* grown in tube culture and inoculated with two strains, some 3% of the nodules contained two strains (Vincent, 1954); for soybean two strains may be found in as many as 10% of the nodules (Skrdleta, 1970). The mechanism for excluding other virulent strains is obscure. Since only bacteria near the infection thread tip in clovers appear to multiply, this could "dilute out" any bacteria other than those dividing. Possibly only a small, single strain colony develops at the site of infection. When large numbers of rhizobia of both strains are well distributed over the root, as in agar tube culture, then the likelihood of more than one strain in a nodule apparently increases. Perhaps separate infections are involved in the nodules containing two strains.

Some nodules harbour other organisms as well as *Rhizobium*; *Klebsiella pneumoniae* has been isolated from surface sterilized soybean and clover nodules (Evans *et al.*, 1972). A fast growing aerobe belonging to the Rhizobiaceae was also isolated from soybean nodules along with the slow growing *R. japonicum*. It could also be readily recovered in large numbers from *Vigna unguiculata* and *Arachis hypogaea* nodules and could even enter *V. unguiculata* nodules three weeks after they had formed (Jansen van Rensburg and Strijdom, 1972a, b). An organism identified as *Protaminobacter ruber*, red coloured on yeast mannitol agar, was also isolated from *Lotononis bainesii* nodules, along with the normal red coloured *Rhizobium* nodulating this host (Diatloff, personal communication).

The first host response to *Rhizobium* is a marked curling and branching of root hairs. Very young root hairs are not effected; uninoculated root hairs and inoculated non-legume root hairs grow straight (e.g. Ward, 1887; Fahraeus, 1957; Nutman, 1959; Haack, 1964; Dart, 1971). Already formed hairs may become distorted up until about 10 days after their initiation (Nutman, personal communication). If inoculation is delayed, older zones of the root become progressively unavailable for nodule formation (Nutman, 1949; Dart and Pate, 1959; Skrdleta, 1970). The degree of distortion and number of hairs involved depends on the legume and *Rhizobium* strain (Nutman, 1959) with up to 60% of hairs

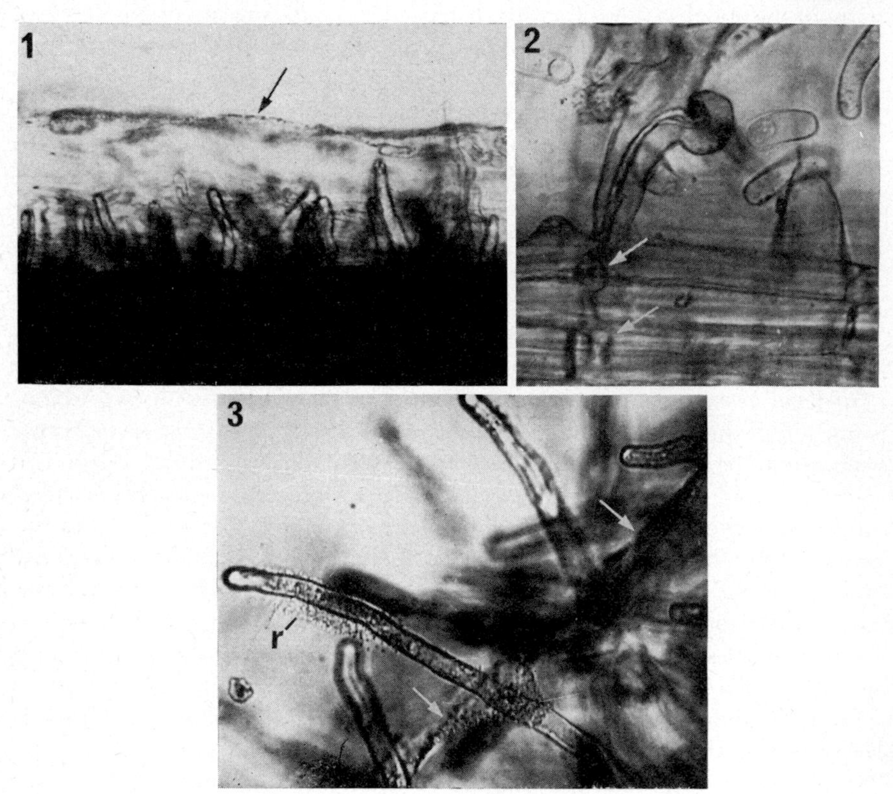

FIGS. 1–3. FIGS. 1 and 3. Mucigel layer (arrows) enclosing root hairs on a *Trifolium subterraneum* seedling. In Fig. 3 colonies of rhizobia (r) are present on a root hair and on the surface of the mucigel. Fig. 1. ×150, Fig. 3. ×380. FIG. 2. Double infection thread arising in a lateral branch of a *T. subterraneum* root hair. The two threads join and then expand characteristically as they cross the epidermal and adjacent cortical cell walls (arrows). ×280.

distorted on peas (Haack, 1964). Non-nodulating combinations generally produce less marked reactions (e.g. Yao and Vincent, 1969), although red clover root hairs are as readily curled by *R. leguminosarum* as by *R. trifolii* (Hepper, personal communication). Yao and Vincent (personal communication) found that branching can be produced without marked curling, but curling is closely linked with infection. However infections can occur in either lateral branches (Fig. 3) or at the original tip of the root hair (Nutman, 1959).

The factor inducing curling is released by *Rhizobium* in pure culture into the medium, is dialysable and heat labile, but when added to clover roots becomes more heat stable, suggesting a linking of a *Rhizobium* and root produced factor to become the active component. Clover root hairs are deformed by sterile culture filtrates of lucerne nodule bacteria and vice versa but not as markedly as with the homologous combination (McCoy, 1932; Yao and Vincent, 1969; Solheim and Raa, 1973). The amount of root hair curling does not seem to be a limiting factor in infection—Polymixin β increased curling of clover roots with no effect on infection (Darbyshire, 1964) and pre-treatment of *Medicago sativa* roots with nitrate before inoculation reduced curling markedly with much less effect on infection (Munns, 1968c).

All stages of infection of *M. sativa*, including root hair curling and thread growth, seem sensitive to nitrate (0·02–0·5 mM); fewer threads form and more abort. Nitrate, and particularly nitrite, also inhibit root hair initiation (Darbyshire, 1966). Although the rapid initial phase of nodulation is inhibited by nitrate, the subsequent nodulation pattern in the presence of nitrate is similar for plants without (Munns, 1968b, c). For cowpeas, ammonium nitrate additions at sowing inhibit nodulation of the primary root but not when given 4 days after sowing, although nodules are not visible until 7 days (Pate and Dart, 1961). Only nitrate in direct contact with the root surface has a major effect on nodule formation. Adding IAA can alleviate some of the effects of nitrate on curling and nodule formation (Valera and Alexander, 1965a; Munns, 1968d), although IAA added alone did not stimulate curling or infection (Sahlman and Fahraeus, 1962; Darbyshire, 1964). IAA is produced by *Rhizobium* in the legume rhizosphere (e.g. Kefford *et al.*, 1960). Very small amounts of combined nitrogen (10–20 μg N/plant) may even stimulate initial infections of clover seedlings, but 40 μg and above tend to inhibit (Darbyshire, 1966).

The pH and calcium level in the soil or root medium also interact to influence infection. Increasing acidity inhibits nodulation, with the response dependent on the species, variety, *Rhizobium* strain and Ca supply (e.g. Andrew and Norris, 1961; Robson and Loneragan, 1970). Generally legumes will grow with a combined nitrogen source at pH's which inhibit nodulation.

Some species nodulate at pH 4–4·5, e.g. Kudzu-*Pueraria phaseoloides* (Norris, 1965), *Trifolium africanum* (Small, 1968), but others such as *Medicago* spp nodulate little below pH 5·5. The initiation of root hair curling of *M. sativa* is restricted at pH 4·5 at all Ca levels and limited by low levels of Ca at pH 5·2. Curling, once started, is not stopped by lowering the pH, but nodule initiation remains sensitive after curling is completed and requires a further 12 h or so at an equitable pH, i.e. a stage subsequent to curling is also pH sensitive (Munns, 1968a, 1970). Infection in peas is sensitive to acidity at 2–4 days after inoculation (Lie, 1969). Although acidity reduces *Rhizobium* numbers in the rhizosphere, a heavy inoculum did not overcome the sensitivity of infection to low pH in the rhizosphere. Calcium is required at the site of infection during the infection thread initiation for nodules to form (Munns, 1970) independent of any effect on root growth and root hair production. Magnesium does not substitute for Ca (Lowther and Loneragan, 1968; Munns, 1970).

B. Thread Initiation

Rhizobium rods often attach "end on" to root hairs and other parts of the root epidermis of legumes (Dart, 1971; Nutman *et al.*, 1973) and non-legumes (Menzel *et al.*, 1972), but the significance of this is not known. *Rhizobium* apparently produces cellulose fibrils in pure culture; possibly such fibrils are involved in establishing the required affinity between *Rhizobium* and the host wall (Deinema and Zevenhuizen, 1971). As Ward (1887) so beautifully illustrated, infection usually occurs at a kink or distorted part of the hair associated with the curling. Very occasionally apparently "straight" root hairs are infected (Fahraeus, 1957; Nutman, 1959). Perhaps the physical constraint of the kink in the hair maintains a high concentration of the hormones produced by the entrapped rhizobia. Auxin and the root hair curling factor would both be presumably required—auxin to enable stretching of the existing wall, and allow new growth which is characteristically distorted by the curling factor. Nodules form on some roots apparently lacking root hairs e.g. the adventitious roots of the water plant *Neptunia oleracea* (Schaede, 1940) and it is presumed that the thread starts in an epidermal cell. This sometimes also occurs in clovers (Nutman, 1959).

Rhizobia can sometimes be seen swimming freely inside the root hair with or without infection thread formation (see Dart, 1974). This suggests that the hair wall is "fragile" at the point of initiation of the infection thread. Perhaps a localized wall softening allows entry of *Rhizobium* into the host cytoplasm and this initiates cell wall deposition to repel the invader; if *Rhizobium* growth is too rapid to be completely contained, the growing infection thread results. One mechanism for host–*Rhizobium* specificity could operate at this stage—lysis of incompatable rhizobia or the rhizobia could trigger collapse of the root hair cytoplasm. It has also been postulated that *Rhizobium* might specifically induce host root, pectolytic enzyme activity which would condition the root hair wall for infection (e.g. Fahraeus and Ljunggren, 1959; Ljunggren, 1969); this is not well supported by experiment (e.g. Bonish, 1973; and see Dart, 1974 for discussion). *Rhizobium* in culture does not produce cellulase or pectolytic enzymes (e.g. Ljunggren, 1969).

The first sign of infection is a hyaline spot or swelling in the root hair wall, accompanied by a greater "opacity" of the associated cytoplasm and increased cytoplasmic streaming. After about 3 h the refractile sheath of the infection thread becomes visible and this then grows at about 7 μm/h for *Trifolium* spp., similar to the previous rate of growth of the root hair. The nucleus is double the size of the uninfected root hair's nucleus and has a large nucleolus. Unless the nucleus remains close to the thread tip at all stages of development growth stops, although it may sometimes renew if the nucleus returns. If the thread begins in a lateral branch it may sometimes follow the nucleus towards the hair tip away from the root and then usually abort. Within the clover infection thread the rhizobia are usually aligned in a single row longitudinally. The growth is often irregular at the point of entry, sometimes branching and rejoining, before becoming a straighter and smoother tube once it has passed beyond the distorted part of the hair. Sometimes further irregular growth occurs, particularly on infection threads that abort, as though wall material is being deposited to contain escaping rhizobia. The thread is quite rigid, remaining unchanged when the root hair wall is accidentally broken. Two infection threads may start in a hair (Fig. 3) and very rarely three (Nutman, 1959, 1962; Haack, 1964). The thread wall contains the same histochemical components as the root hair tip (McCoy, 1932). A 15 minute film of the infection process in clovers is available (Nutman *et al.*, 1973).

C. Numbers of Infection Threads

Only a small proportion of the root hairs are infected, e.g. an estimated 2·8% for peas (Haack, 1964). Not all infections give rise to nodules (Ward, 1887), the proportion depending on the *Rhizobium* strain and plant—1·4–32% of infections on *Trifolium* spp. were successful (Nutman, 1959). Whether the proportion changes with plant age is unknown, as infections are more difficult to find in older, thicker roots. Many infections abort (cease to grow) before reaching the base of the root hair, others abort at the base of the hair and in the root cortex. The nucleus associated with aborted threads often rounds up and becomes smaller (Fahraeus, 1957).

The number of infections is not obviously related to the number of nodules a clover cultivar forms (Roughley *et al.*, 1970b) or the effectiveness of the association in nitrogen fixation, but is affected by the invasiveness of the strain (Nutman, 1962). Infections begin 3–20 days after germination depending on the species. The numbers increase almost exponentially until the first nodule forms when the rate of infections slows markedly. Formation of a lateral root can also slow the rate of infection (Nutman, 1962; Lim, 1963; Roughley *et al.*, 1970a, b).

Infections do not form at random, but are restricted initially to broad zones down the root, within which smaller groupings of infections occur with several within a diameter of 250 μm. This is not related to any obvious cytological feature. Later infections reduce the demarcation between the original zones. Nodules usually develop in acropetal succession, their location unrelated to the initial site of infection. On clovers fewer infections seem to occur in soil than in agar or water culture, although their distribution and the effect of the first nodule is similar (Lim, 1961).

For the few plants studied, the optimum temperature for nodule formation is about 20°C for temperate species and 27–30°C for tropical plants, but the response away from the optimum temperature varies greatly with the host (see Gibson, 1971). Few or no nodules form at root temperatures below 7°C or above 36°C. At 6–7°C and at 36°C the start of infection of *Trifolium* species is much delayed, and then continues at a much slower rate, even though root hairs and rhizobia are plentiful. At 7°C the formation of the first nodule had little effect on the infection rate. The *T. subterraneum* cultivar Cranmore was not infected at all by 40 days, and in the root zone where nodules usually formed, lateral roots were produced instead (Roughley and Dart, 1970). The rate of infection begins at similar times and continues at similar rates between 18 and 30°C (Roughley *et al.*, 1970a, b; Kumarasinghe, personal communication).

IV. Nodule Number

After the initial phase of nodulation, on many plants subsequent nodule formation is slower. For *Medicago sativa* (Munns, 1968b) and *Trifolium subterraneum* (Day, 1972) new nodules form in groups on the newer roots. For plants such as soybean where root growth is not continuous, nodules form on or close to the primary root at the "crown" of the root in good conditions for infection. Nodulation on peripheral, secondary, roots usually begins during early pod fill, but it is not known whether this is dependent on new root and root hair growth. Do some infections for later nodulations remain latent, i.e. infection occurs but does not proceed immediately to nodule initiation? Support for this comes from the appearance on overwintered white clover plants of new nodules on old, brown-coloured roots before any noticeable new season, white-coloured root growth (Masterson and Murphy, 1975). Delayed inoculation experiments indicate that old root hairs do not become infected. A few new root hairs sometimes develop among old root hairs (Nutman, personal communication).

Nodule formation often restricts lateral and secondary root production in the root zone bearing nodules, with "normal" root development below this zone. This suggests that the same hormones—produced in the nodules or induced by *Rhizobium* in the roots—are involved in lateral root and nodule initiation. Nodule formation restricts overall root growth as well (e.g. Dart and Pate, 1959).

Excising effective *Trifolium pratense* nodules from the root stimulated further nodule production (Nutman, 1952). Such experiments are difficult to interpret because of wound hormone production. Existing nodules exert some control over further nodulation because the number of nodules formed on some legumes is relatively small. There is also an inverse correlation between nodule number and the volume of nodule tissue on a clover root under some growth conditions (Nutman, 1967). However for other legumes such as *Vigna radiata* or *V. mungo* temperature has an overriding influence on nodule number. At 33°C nodules continue to form over the root system, as do tertiary roots, resulting in hundreds of nodules per plant. At 27°C and 21°C progressively fewer, larger nodules form. Temperature has only a small effect on root weight (Islam and Dart, unpublished observations).

Any control by existing nodules of further nodulation could be mediated by hormone production in the nodule, perhaps by the rhizobia (e.g. gibberellic acid —Radley, 1961; Katznelson and Cole, 1965; Dullaart and Duba, 1970, or ethylene—Day, 1972) or by the activity of nodules as powerful sinks. Thus nodules would sequester either carbohydrate (Small and Leonard, 1969; Minchin and Pate, 1973; Gunning *et al.*, 1974) and/or hormones (abscisic acid or cytokinin? —Hocking *et al.*, 1972) leaving too low a concentration of these substances for nodule initiation in the root proximal to the already formed nodules. Increasing the CO_2 concentration around the leaves (Wilson, 1933) or the light intensity (Day, 1972) stimulates nodulation of clover roots. The reason why ineffective nodules are often more numerous on roots may be that they have a much lower utilization of photosynthate since they are not fixing nitrogen. In the effective pea nodule, 47% of the carbon input returned to the shoot in the form of amino acids produced by N_2 fixation (Minchin and Pate, 1973).

V. Other Modes of Infection

Infection threads have not yet been found in *Arachis hypogaea* root hairs or nodules (Allen and Allen, 1940; Chandler and Dart, 1973). *Arachis hypogaea* nodules arise (perhaps exclusively) at root junctions where a tuft of hair-like structures with thick walls occurs. This results in the nodules forming in a very regular pattern along the root in four well defined lines opposite the xylem points on the primary root. Sections of young nodule primordia show meristematic activity and intercellular zooglaeal strands containing rhizobia, but no infection threads or obvious intracellular rhizobia. Rhizobia are distributed in the young nodule by the host cell division (Fig. 4) and this results in virtually all the central tissue containing rhizobia. A few files of large duct cells, continuous with the uninvaded cortex, run through the bacteroid zone (Fig. 5 and Chandler and Dart, 1973). *Aeschynomene indica* nodules are similar, forming on both stem and root near emerging lateral roots, where root hairs are absent (Arora, 1954).

This pattern of intercellular infection, and dissemination by cell division probably exists for other nodules which apparently do not contain infection threads, and where uninvaded cells are absent or uncommon in the bacteroid tissue, e.g. *Lupinus* spp. (Haack, 1961; and see Dart, 1974). A consequence of this mode of infection may be an ability to form nodules with a wider variety of strains. *Arachis hypogaea* is quite promiscuous, nodulating readily with strains of the cowpea miscellany, *R. japonicum*, *R. phaseoli*, *R. lupini* and occasionally with *R. trifolii* and *R. meliloti* (Allen and Allen, 1940; Saric, 1963). Some rhizobia can thus form nodules which have infection threads (e.g. *Ornithopus sativus* and *Vigna* spp.) and others without (*Lupinus* spp. and *Arachis hypogaea*).

The *Arachis* mode of infection has similarities with crown gall development where *Agrobacterium* remains intercellular.

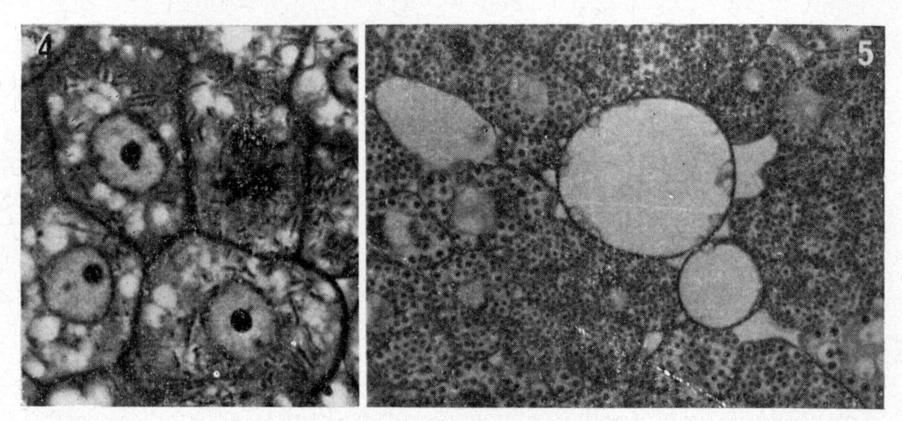

FIGS. 4 and 5. Effective *Arachis hypogaea* nodules. FIG. 4. Young nodule in which all the central cells are invaded. The rod to club shaped rhizobia are spread by host cell division. ×1090. FIG. 5. Older nodule with spherical bacteroids and large, uninvaded duct cells. ×510.

VI. Nodule Development

A. Initiation

It takes about 24 h for an infection thread to grow from the root hair into the cortex of clover roots. The infection thread passes close to the cortical cell nuclei which enlarge, presumably with associated DNA synthesis. Often the cells traversed also enlarge and adjacent cells may divide (e.g. Libbenga and Harkes, 1973). Although threads may divide and occasionally join, the overall direction of their growth is towards the stele (Fig. 9). Meristematic activity occurs in advance of the thread in *Ornithopus sativum* (Haack, 1961), and in *Pisum sativum* (e.g. Libbenga and Harkes, 1973) where activity is localized in the inner cortex (Figs 6–11). On reaching this zone, the thread divides and ramifies through the cells releasing rhizobia. Nodules usually have a cortical origin with pericyclic involvement in production of vascular tissue, although it is claimed that *Arachis hypogaea* and *Aeschymonene indica* nodules arise in the pericycle (Allen and Allen, 1940; Arora, 1954). Some nodules arise in the outer cortex (e.g. *Phaseolus vulgaris*, *Cyamopsis tetragonolobus*, *Lupinus albus*).

Infection threads may penetrate the cortex without stimulating nodule formation, suggesting that foci (cells with the right hormonal balance?) exist in the root for nodule initiation, and that these are not necessarily connected to the

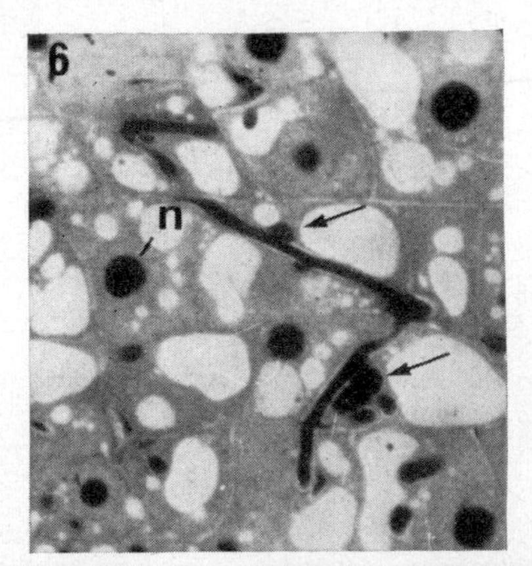

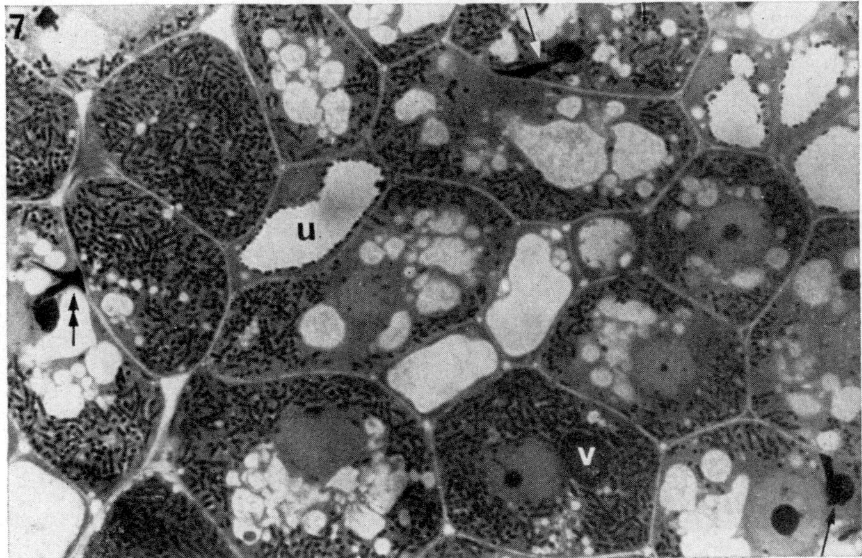

FIGS. 6 and 7. FIG. 6. Differentiating zone in a young *T. subterraneum* nodule showing infection threads crossing cells and releasing vesicles containing rhizobia into the cells (arrows). n—nucleus with prominent dense staining nucleolus. ×720. FIG. 7. *T. subterraneum* nodule showing multiplication of rhizobia after their release. The infection threads (arrows) have terminal vesicles of zooglaea still attached; one thread has divided after entering the cell (double arrow). A vesicle of zooglaea (z) remains in another cell packed with rhizobia. The vacuoles of the uninvaded cells (u) are lined with a dense staining product, possibly tannin. ×850.

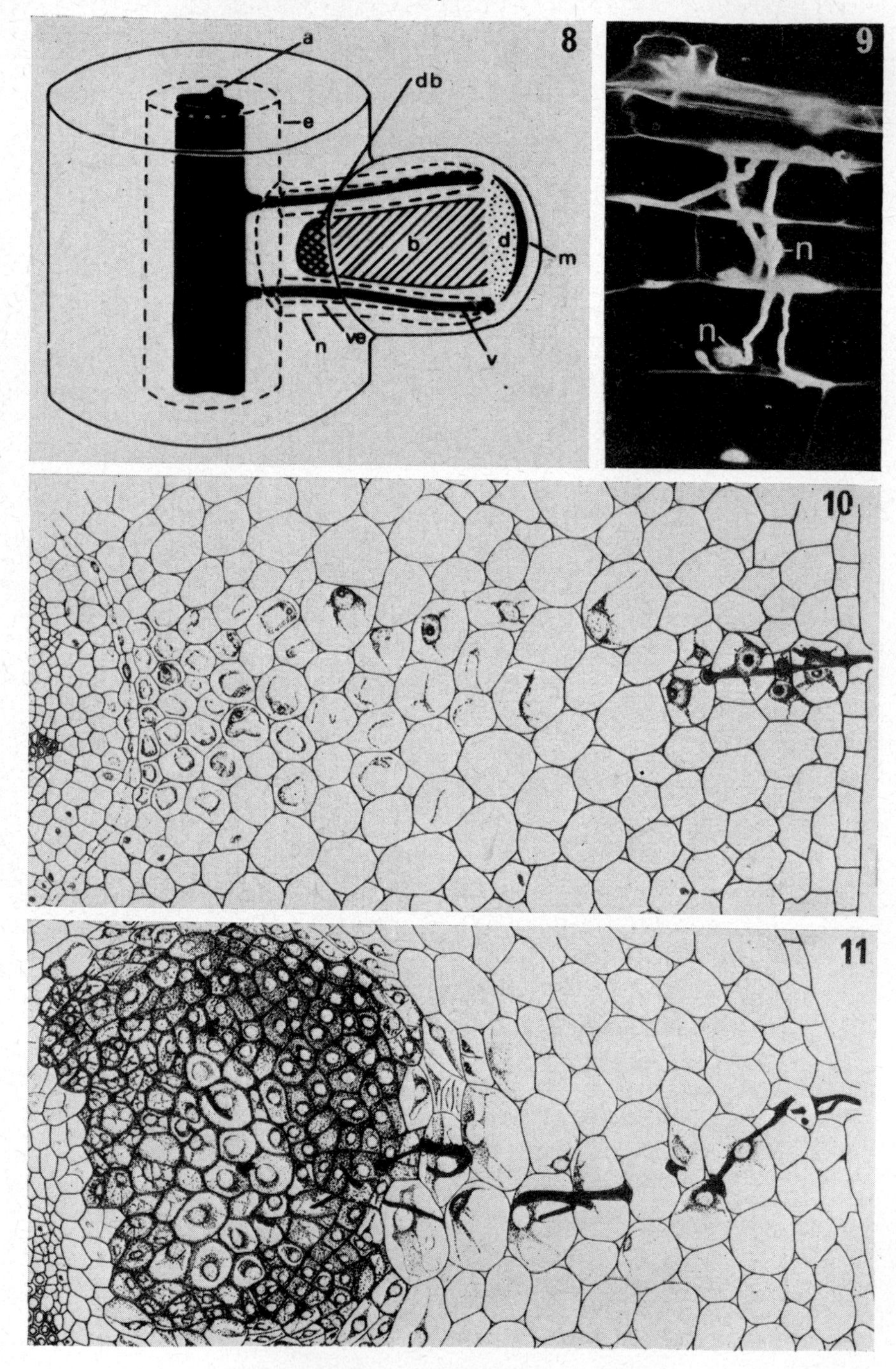
8
a
db
e
b
d
m
n
ve
v
9
n
n
10
11

sites of the infection. On many legumes, young nodules and bacteroid containing cells have nuclei that are disomatic or polyploid or have 4–16 c levels of DNA (see Libbenga and Torrey, 1973, for discussion and details). They suggested that cytokinin (see Phillips and Torrey, 1972) and perhaps also auxin production by *Rhizobium* in the infection thread induces endoreduplication of chromosomes in some cortical cells, which may be uninvaded and in advance of the thread, and these then go into polyploid divisions under the same stimulus.

Nodules of *Pisum sativum*, *P. arvense*, *Ulex europaeus*, *Phaseolus vulgaris*, *Glycine max*, *Lupinus lueteus* contain higher levels of auxin on a fresh weight basis that the rest of the root (see Dullaart, 1970, for references). The high auxin level may result from a slower rate of breakdown in the host cells of the nodule rather than to production by the rhizobia—IAA levels are higher than in the root but indole carboxylic acid levels may be lower (Dullaart, 1967).

The response to hormones of cortical tissue explants of pea roots suggests an analogous sequence may occur in nodule initiation. Adding auxin to the explant medium induces an increase in cell size but no DNA synthesis or cell division. If 0·01 ppm kinetin is also present, two rounds of DNA synthesis occur, starting some 24 h after excision; the first is an endoreduplication or doubling of the chromosome complement, and the second the normal process of cell division preceding mitosis (Libbenga and Torrey, 1973). Cell division begins 48 h after excision. Explants from within 2 mm of the root tip have 75% diploid mitoses, but the proportion of polyploid divisions increases as explants are taken further from the tip, being 85% polyploid for explants taken 5–6 mm from the root tip. Nodules are not initiated close to the root tip.

Thus in nodule initiation the infection thread is suggested to perform a similar function to the exogenous cytokinin and auxin for the explant. The thread-produced hormones induce the polyploidy and cell division, and it is in these cells that the released rhizobia survive to develop into bacteroids. Nodule initiation may also perhaps involve cortical cells that were already polyploid before infection. Some uninvaded nuclei of the nodule cortex may be polyploid but most appear to be diploid. *Rhizobium* produced auxin and perhaps enhanced

---

FIGS. 8–11. FIG. 8. Diagram of elongate nodule with an apical meristem (m), differentiating zone (d) where cells are invaded by infection threads, bacteroid zone (b), and degenerating zone (db). The nodule cortex contains the nodule endodermis (n) arising from the root endodermis (e) and merging with the meristem, the nodule vascular traces (v) joined to the central root stele (a). Each trace is enclosed by a vascular bundle endodermis (ve) also joined to the root endodermis. After Bond (1948). FIG. 9. Section of a pea root showing infection threads crossing the cortex, branching and passing close to the nuclei (n). FIGS. 10 and 11. Drawings of sections of pea roots showing nodule initiation. Figure 9 shows the infection thread initiating cytoplasmic activity in the inner root cortex in advance of the thread. In Fig. 10 the nodule initial has started to differentiate an apical meristematic region. Figures 9–11 by courtesy of Libbenga and Harkes (1973).

endogenous levels, would seem to be involved in the cell hypertrophy that accompanies nodule development. The requirement for adequate, but not inhibitory, auxin levels is indicated by the nodulation pattern of *Trifolium subterraneum* var. Woogenellup. Roots of this variety have a higher tryptophan content than others, and the ability of various *Rhizobium* strains to nodulate Woogenellup seems inversely related to their ability to produce auxin in culture from tryptophan. Exposing the roots to light also increases nodulation, perhaps by increasing IAA oxidation (Gibson, 1968, 1969).

The role of cotyledons and the plant tops in nodulation seems to vary with the species, perhaps reflecting different levels of hormones, and different mobilization and transport of materials to the root. Sections of *Pisum sativum* roots showed that over 83% of the 400 nodules and all lateral roots examined, originated opposite protoxylem points (Phillips, 1971a); where nodules appeared to arise opposite the phloem, the initial cell divisions apparently occurred opposite the protoxylem (Libbenga and Harkes, 1973). Removing a cotyledon before germination resulted in the formation of primary root nodules closer to the cotyledons, and doubled the proportion of nodules formed midway between xylem and phloem. There was a significant increase in the percentage of lateral roots bearing nodules for those lateral roots with most vascular connection to the excised cotyledon position, over lateral roots mainly connected to the remaining cotyledon. Phillips suggested that an inhibitor was translocated in the phloem from the cotyledons to the roots. This is possibly abscisic acid which can inhibit nodule initiation subsequent to the infection process and also inhibit the polyploid divisions in pea root segments normally stimulated by cytokinin addition (Phillips, 1971b).

The results are difficult to reconcile with those of Libbenga *et al.* (1973) who extracted a factor from the stele of *P. sativum* which promoted cell division of root cortical explants. Possibly this fraction contains cytokinin produced in the root tip and carried to the root cortex via the xylem. Hence the higher endogenous level of cytokinin opposite the protoxylem points would promote cell division there, with the phloem factor restricting cell division in its vicinity.

Some factors necessary for infection and nodulation of *Trifolium* spp. seedlings are supplied by the cotyledons since their removal up to 8 days after germination reduces infection and prevents nodulation; these infections tend to form in a zone closer to the root tip (Nutman, personal communication). Similarly for *Glycine max*, excision of one or both cotyledons up to 7 days after germination markedly reduced nodulation. Nodulation was stimulated in plants with or without cotyledons by an extract of leaves or cotyledons (Raggio and Raggio, 1956; Yatazawa and Yoshida, 1965). *Phaseolus vulgaris* seedlings however nodulated even when cotyledons were removed before germination (Valera and Alexander, 1965b).

B. Meristematic Activity

The pattern of meristematic activity in young nodules soon differentiates it from that in roots. In the elongate nodules meristematic activity becomes localized in the initial away from the root stele, before the nodule emerges from the root, forming a hemispherical cap of uninvaded cells, with large nuclei and nucleoli, which are presumably polyploid. As the nodule grows through the root cortex the relatively small meristematic cells become elongate transverse to the nodule long axis, producing towards the root a zone of enlarging, mostly uninvaded, vacuolating cells. On the sides of the young nodule, transverse divisions among the uninvaded cortical cells differentiate a procambial strand which at this stage may be unconnected to the nodule meristem or root stele. The strand grows towards the meristem and root, there linking with cells produced by pericyclic division, thus connecting the incipient vascular trace to the stele (Bond, 1948).

Although the young nodule initial may contain dividing cells which are infected, this is rarer in older, elongate nodules. However in the nodules such as *Phaseolus vulgaris* and soybean, meristematic activity with divisions in all planes is spread through the nodule initial, with dissemination of rhizobia through infection threads and host cell division. After about 14 days, cell divisions become localized in a cambium-like layer, 2–3 cells wide, or in small localized groups, at the edge of the nodule bacteroid zone. Other meristematic regions in the cortex give rise to vascular traces (McCoy, 1929; Bergersen and Briggs, 1958).

In *Arachis hypogaea* and *Lupinus* spp. where rhizobia seem to be entirely spread by cell division, meristematic activity initially spreads throughout the nodule with all the central cells invaded. Later, meristematic activity becomes localized at the edge of the bacteroid zone.

C. Cortex and Vascular Tissue

The mature nodule has an uninvaded cortex of very large, vacuolate, thick-walled cells within which lie a nodule endodermis connected to the root endodermis, and internal to this sometimes a layer of scleroids, as in soybean, and several vascular traces (Figs 8 and 12). Outside the endodermis may be a periderm layer, with the endodermis perhaps acting as a phellogen. The nodule endodermis envelops the nodule merging with the meristem at the apex of cylindrical nodules, and usually has cells with thickened suberized walls. In spherical nodules the endodermis is believed to surround the whole nodule (e.g. Fraser, 1942).

Nodules may have one to five vascular connections with the root, depending on the species, and seem mostly to be connected to a single protoxylem point in the root. The vascular bundles branch dichotomously at the base of the nodule, with further branching in the nodule cortex keeping pace with nodule growth, reaching as many as 126 traces for a year-old *Sesbania grandiflora* nodule (Harris *et al.*, 1949). Each bundle has its own endodermis, with Casparian thickening, which is joined to the root endodermis (Fig. 8). The arrangement of cell types

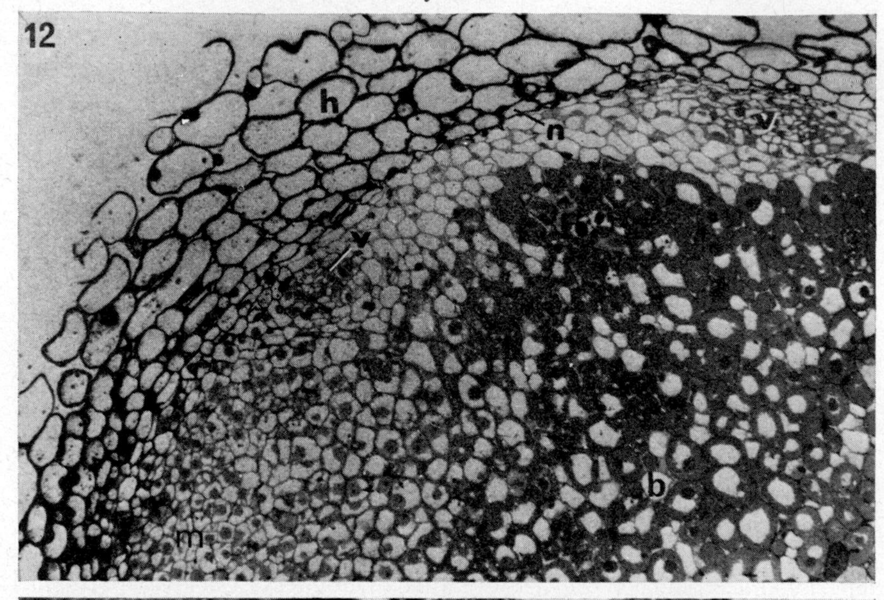
12
h
n
v
v
b
m

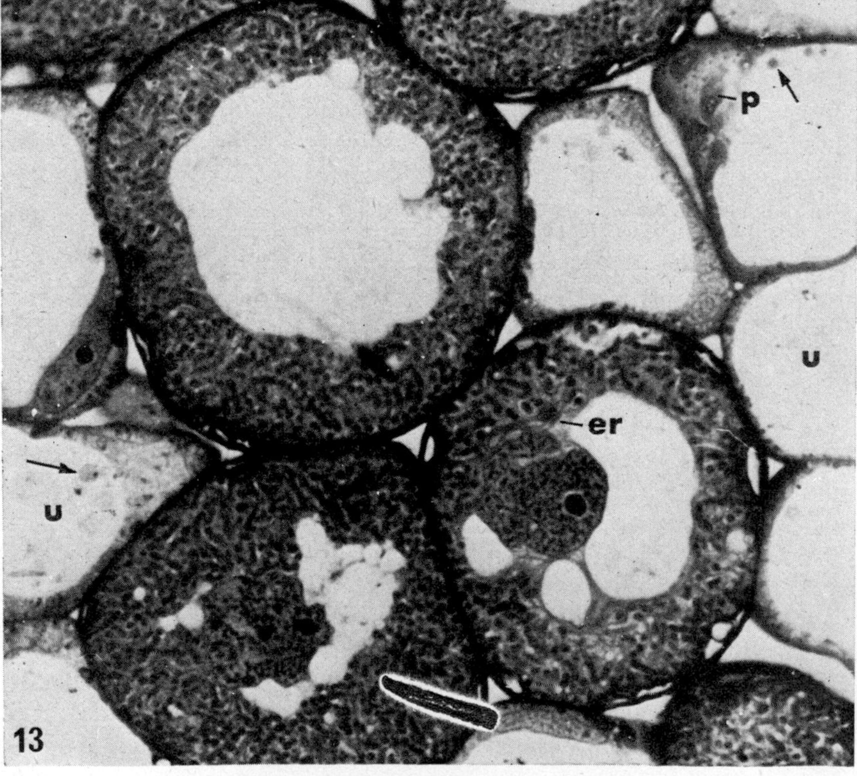
p
u
er
u
13

within the bundles seems to vary with the position in the nodule. For peas, xylem lies closest to the bacteroid zone with phloem outside it. The xylem has spiral and occasionally scalariform thickening.

Surrounding the phloem and xylem are pericycle cells with dense cytoplasmic contents. Out of 71 genera examined, 8 in the Papillionatae had pericyclic cells developed into transfer cells. These were all elongate or collar nodules with continuing apical or lateral meristematic activity. Spherical nodules such as soybean lacked the transfer cells, as did nodules developed in waterlogged conditions which fixed nitrogen poorly. The development of the wall ingrowth varied with the species; for *Trifolium repens* transfer cells developed when nitrogen fixation started, after xylem and phloem had differentiated, occurring in the vascular tissue opposite the bacteroid zone. Transfer cells were prominent at the junction of the root stele with the nodule vascular traces. In peas the wall in-growths increased the transfer cell surface by 2–3 times (Gunning *et al.*, 1974).

Transfer cells are depicted as mediating the symplastic flow of sugars from the phloem to the bacteroid tissues; for peas this is estimated as 10·3 mg sugar for every milligram of fixed nitrogen exported. Secondly, transfer cells are believed to secrete amino acids coming to them from the bacteroid zone into the xylem, thus lowering the water potential of the bundle apoplast, and inducing an influx of water, probably through the "open" procambial end of the bundle near the nodule meristem. This results in the solutes being moved out of the nodule in the xylem. Xylem sap from the pea and *Vicia faba* nodules contains glutamine, asparagine and aspartic acid at much higher concentrations than in the nodule (Pate *et al.*, 1969; Gunning *et al.*, 1974). Other nodules such as soybean, without transfer cells, also concentrate the products of nitrogen fixation in their xylem sap for export out of the nodule.

Several undifferentiated, uninvaded cortical cells with little cytoplasm lie between the vascular bundles and the bacteroid zone and are continuous with channels of similar cells running through the bacteroid filled cells (Figs 12, 13). There are many plasmodesmata between these uninvaded cells and the bacteroid filled cells (Figs 14, 15). Intercellular air-filled spaces also penetrate from the edge of the nodule to its interior (e.g. Bergersen and Goodchild, 1973a).

FIGS. 12 and 13. FIG. 12. Young effective *Vicia faba* nodule showing large thick walled outer cortex or husk cells (h), nodule endodermis (n) and inner cortex with vascular traces in cross section (v). There is a rapid increase in size from cells of the meristem (m) to bacteroid filled cells (b). × 110. FIG. 13. Young bacteroids in host cells of a *Vicia faba* nodule. Sheets of endoplasmic reticulum (er) are prominent in one cell. The invaded host cell nuclei have become amoeboid, but still retain a prominent nucleolus. Small platelets of starch lie opposite the large, empty, intercellular spaces. Two files of uninvaded cells (u) lie adjacent to the invaded cells. They have large vacuoles with inclusions (arrows), and large plastids (p), some with small starch grains. × 1110.

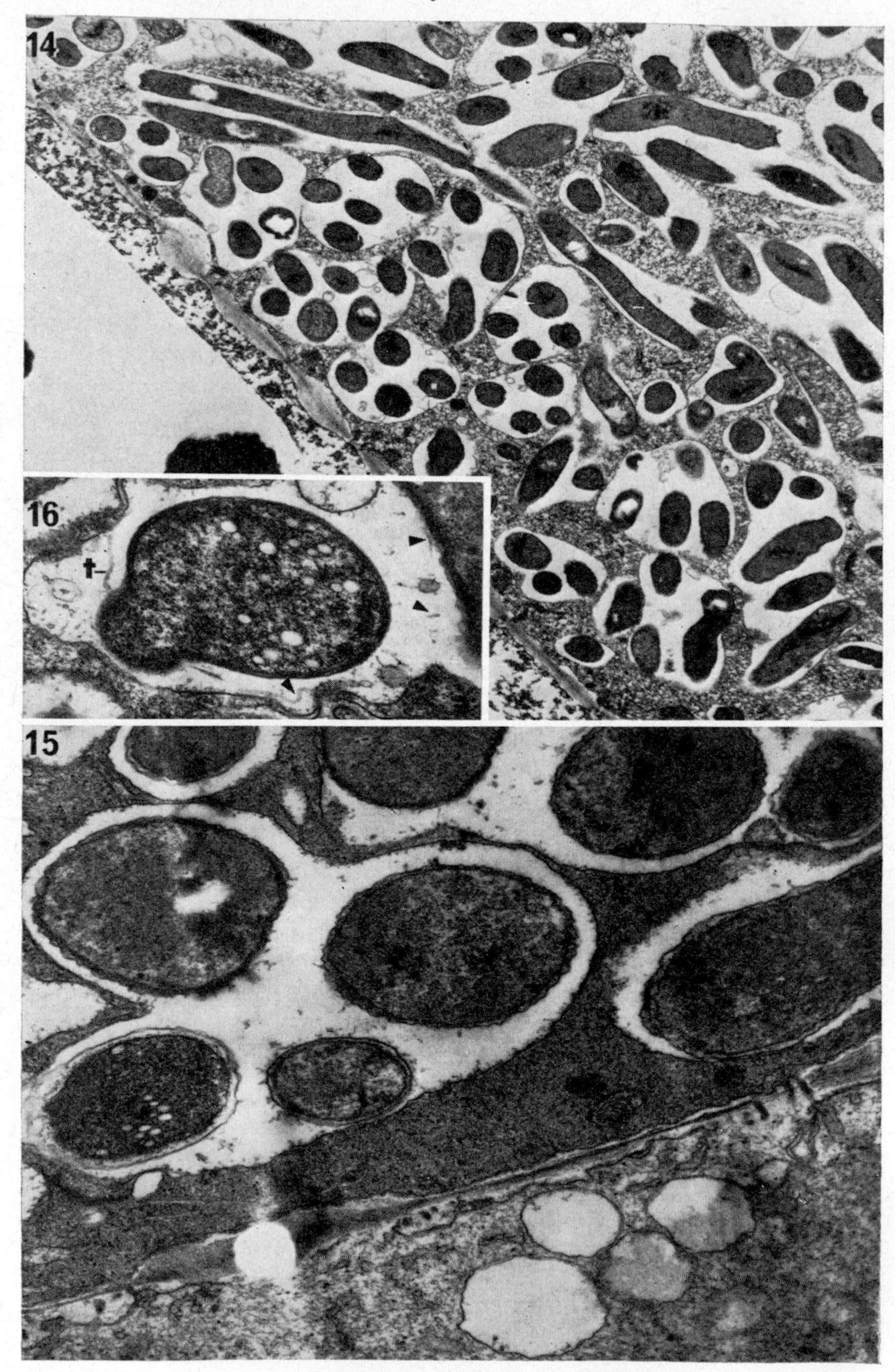
14
16
15

Cortical cells usually contain starch—the amount varying greatly between species and depending on the physiological state of the nodule—and sometimes large crystals, presumably of calcium oxalate. *Arachis hypogaea* nodules may also contain prominent protein bodies in the inner cortex.

D. Cell Invasion and Rhizobium Release

Bacteria are released from infection threads, usually terminally, as vesicles of zooglaea enclosing *Rhizobium* rods which often contain glycogen and poly-β-hydroxybutyrate inclusions. The bacteria escape from the zooglaea singly or in small groups to lie free in the cytoplasm (Fig. 7), and most images show them enclosed by an electron transparent space bordered by a membrane envelope. Much host cell synthetic activity accompanies the infection. There are marked increases in the number of dictyosomes and rough endoplasmic reticulum—often organized in parallel arrays (Fig. 13) or whorls—and free polyribosomes. Many large, empty vesicles, similar in size to those enclosing the rhizobia, abound. It is as though the plant's response to the "invader" is to wall it off in the infection thread, and when this fails, to enclose it in a membrane which is perhaps a lytic vacuole (Truchet and Coulomb, 1973). The entry of *Rhizobium* can be viewed as an endocytotic process, but the continued multiplication of the bacteria in a compatible host, necessitates a continuing reaction by the host cell. New material is incorporated into the membrane envelopes by accretion of small vesicles and also apparently by coalescence of the envelope with larger, empty uninvaded vacuoles. The small vesicles are possibly derived from endoplasmic reticulum, and directed to the envelope by the microtubules occasionally found near the groups of rhizobia just released from infection threads. *Rhizobium* multiplies within the membrane envelopes, and these also divide so that the host cell comes to be filled with bacteria predominantly enclosed singly in envelopes.

Small tubules have been observed in the space between envelope and bacteroid during this phase of *Rhizobium* multiplication in *Lotus*, *Astragalus*, soybean and *Phaseolus vulgaris* nodules (e.g. Fig. 16). The tubules are occasionally seen to connect the envelope membrane with the bacteroid, and may mediate transfer of material between host and bacteroid.

FIGS. 14–16. FIG. 14. Effective *Vigna radiata* nodule formed by the same strain as in the *A. hypogaea* nodules in Figs. 13–15. Several rhizobia, mostly long and rod shaped, are enclosed in each membrane envelope. Several pits with plasmodesmata are present between the invaded cell with its dense cytoplasm and the adjacent uninvaded cell with much less dense cytoplasm. Note the electron dense deposits often found in the vacuoles. ×8,300. FIG. 15. *Lotus corniculatus* nodule showing several bacteroids enclosed in each membrane envelope. The cytoplasm of the invaded cell is much more electron dense than the adjacent uninvaded cell. Pit areas with several plasmodesmata link the cells. ×25,000. FIG. 16. *Lotus corniculatus* nodule, showing a tubular connection between bacteroid and membrane envelope (t). Several other tubules are present in the envelope space e.g. arrows. ×28,300.

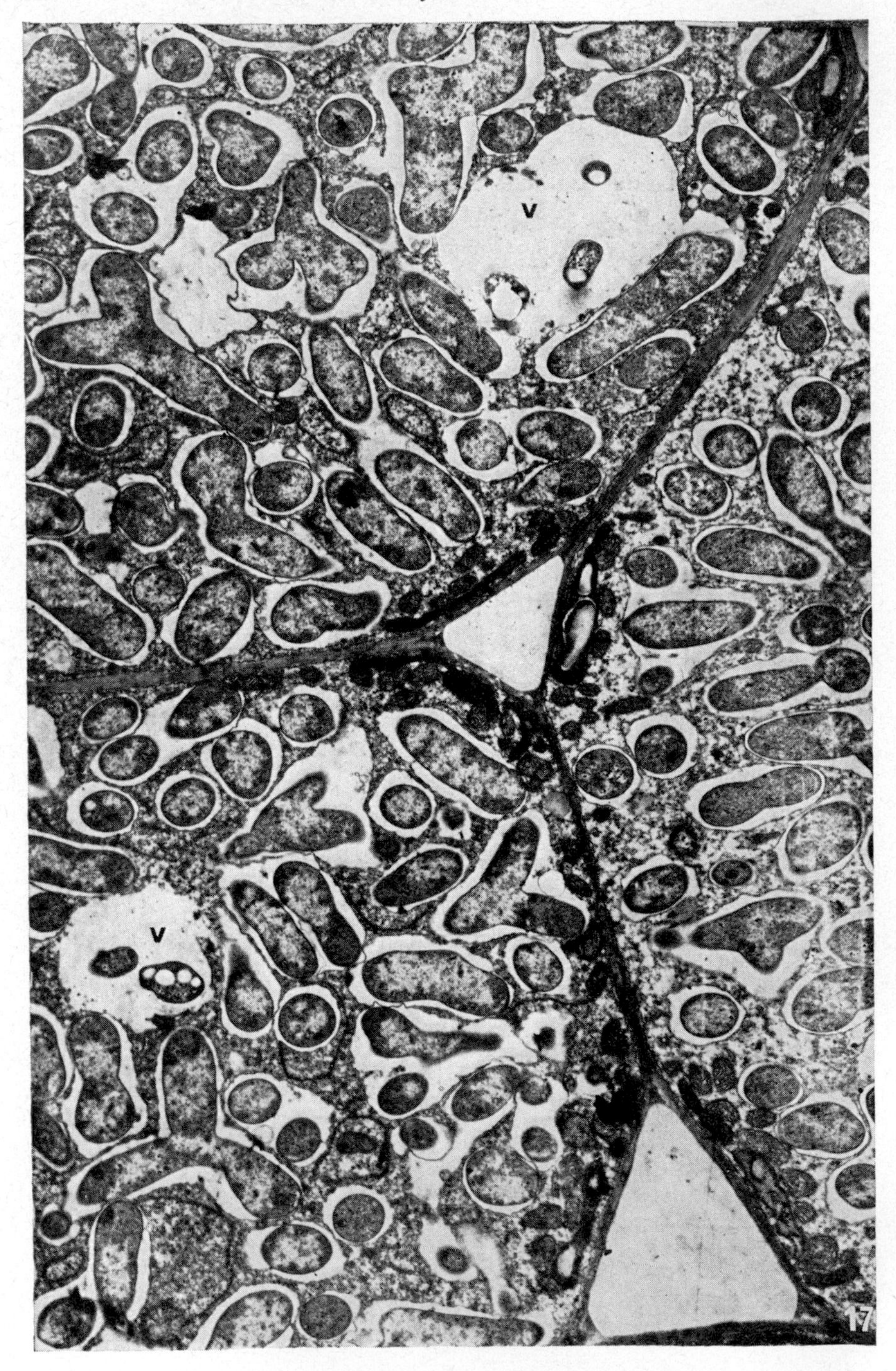

V
V
17

At the same time, plastids, usually containing starch and much phytoferritin, and most of the mitochondria, become lined up at the cell periphery adjacent to the intercellular spaces, perhaps because there the pO_2 is favourable for their activity (Figs 13, 17, 18). A decrease in the amount of phytoferritin accompanies the synthesis of nitrogenase and leghaemoglobin, both iron-containing proteins. The nucleus with its large nucleolus remains central in the cell, often adjacent to a single, large vacuole. The amount of endoplasmic reticulum and ribosomes per unit volume of cell decreases, but since the cell is enlarging at the same time, it is not immediately obvious whether there is an absolute decrease in their amount per cell.

Certainly *Rhizobium* dissemination through the cell is accompanied by an increase in the density of the host cytoplasmic matrix. This does not occur in uninvaded cells, and is perhaps related to the leghaemoglobin (Lb) synthesis occurring at this time. Assays for peroxidase activity at both light and electron microscope level; examination of unstained, thin sections with filtered light of wavelength 400–420 nm (which covers the Soret absorption band of aldehyde fixed Lb); and electron microscope microanalysis, all support the proposal that Lb is located in the cytoplasm of invaded cells outside the membrane envelopes (Dart, 1968, 1969; Dart and Chandler, 1972). It has been suggested from autoradiography of ^{59}Fe in *Ornithopus sativus* nodules (Dilworth and Kidby, 1968) and from peroxidase activity in soybean nodules (Bergersen and Goodchild, 1973b) that Lb is located in the space between bacteroid and membrane envelope.

The spreading of rhizobia through the host cell is accompanied by a large increase in their size and loss of inclusion granules. For plants in the clover, medic, and pea cross inoculation group the rhizobia remain predominantly singly enclosed in their envelopes, and may develop considerable pleomorphy—the well known club, x and y shapes (Figs 13, 17). These bacteroids have a dispersed nucleoid, and few ribosomes, and sometimes intracytoplasmic membranes (e.g. Dart and Mercer, 1963, 1964b; Jordan *et al.*, 1963; Grilli, 1963; Dixon, 1964; Mosse, 1964). The fine structural image is consistent with a decline in the amount of DNA and RNA per bacteroid—as found in *Lupinus* nodules (Dilworth and Williams, 1967) and their loss of ability to divide perhaps results. Inclusion granules are rare in these mature, nitrogen fixing bacteroids. Later in some plants such as peas, the older bacteroids accumulate PHBA granules. *Trifolium subterraneum* bacteroids may have prominent crystalline inclusions (Chandler and Dart,

---

FIG. 17. Electron micrograph of an effective pea nodule showing pleomorphic bacteroids with large dispersed nucleoids, enclosed mainly singly in membrane envelopes. Envelopes apparently containing two rhizobia are thought to represent cross sections through a single Y-shaped bacteroid. Rhizobia in two vesicles (v) have not expanded and retain prominent poly-β-hydroxybutyrate granules absent from the bacteroids. Plastids and mitochondria are mostly confined to areas adjacent to intercellular spaces. $\times$7,900.

unpublished). In *Lupinus* spp., bacteroids are smaller, and rod shaped but enclosed singly in envelopes (e.g. Jordan and Grinyer, 1965; Dart and Mercer, 1966; Kidby and Goodchild, 1966). In *Lotus* spp. bacteroids are often enclosed several per envelope and there is a slight increase in overall size, although they remain rod shaped (Fig. 15, Dart, 1969). Lotus bacteroids often contain inclusions, including polyphosphate and poly-β-hydroxybutyrate (Figs 15 and 16; Craig and Williamson, 1972; Craig *et al.*, 1972). For plants in the soybean, *Phaseolus* and many in the cowpea cross inoculation groups, there is less pleomorphy and the bacteroid size increases are usually in length. Continued *Rhizobium* division results in several per membrane envelope packet (Fig. 14). Such bacteroids as they age accumulate a great deal of PHBA and other inclusions, possibly glycogen (e.g. Goodchild and Bergersen, 1966; Dart and Mercer, 1966). Leghaemoglobin synthesis in soybean nodules starts two days before nitrogenase activity can be detected (four days after nodules first appear), while the bacteroids are still predominantly enclosed singly in their membrane envelopes (Bergensen and Goodchild, 1973b).

E. Starch in Nodules

Starch is often abundant in the nodule cortex, but within the bacteroid zone of effective clover, medic and pea nodules it is largely confined to a band, 2–3 cells wide, which is filled with rhizobia not yet in their mature bacteroid form. Some starch may reappear in the senescent zone at the base of the nodule. In *Vicia faba* nodules starch has a more widespread distribution. Uninvaded cells in the bacteroid zone contain occasional starch grains (Fig. 13). Starch may be prominent in the young nodule, but is absent from cells with mature bacteroids. It is not known whether starch levels have a diurnal fluctuation, or whether the starch can provide an energy source for nitrogenase activity so that the latter has little diurnal periodicity. In soybean, nitrogenase activity responds to changes in light intensity (Bergersen, 1970). In nodules developing at low root temperatures, or in ineffective nodules, starch accumulates in all tissue (e.g. Dart and Mercer, 1965a; Roughley *et al.*, 1974) as though the nodule remained a strong sink, but had insufficient nitrogenase activity to utilize the energy source. As the nitrogenase enzyme itself functions at these temperatures a block in enzyme synthesis may be involved as few bacteroids develop. Alternatively translocation of fixed nitrogen out of the nodule may be inhibited and some end product repression may limit nitrogenase levels.

F. Host Modification of Bacteroid Development

The pattern of bacteroid development by a given *Rhizobium* strain can be very much influenced by the host plant. In *Lupinus* spp. the bacteroids contain few inclusion granules, and are enclosed singly in membrane envelopes, whereas in *Ornithopus sativus* nodules formed by the same *Rhizobium* strain, bacteroids often contain PHBA and are enclosed several per envelope (Kidby and Goodchild,

1966). Other *Rhizobium* strains form bacteroids enclosed several per envelope in *Lotus* nodules, but in *Astragalus glycyphyllos* the same strains form greatly enlarged and pleomorphic bacteroids which mainly occur singly in envelopes (Dart, 1969).

Bacteroids of peanut nodules are strikingly different from those of *Vigna* spp. formed by the same strain. In peanuts the rhizobia are initially rod shaped and spread through the host cell enclosed singly in membrane envelopes. The rhizobia then become enlarged and pleomorphic and finally enlarge still further into spherical bacteroids 3–5 μm in diameter. The nucleoid is condensed at the centre, and the plasma membrane in older bacteroids develops many intracytoplasmic membranes (Fig. 18). By contrast nodules on *Vigna* spp. have prominent infection threads, and the bacteroids, although more elongate than rhizobia in infection threads, remain rod shaped and several are enclosed in each membrane envelope (Fig. 14). The bacteroids also become packed with PHBA and have no obvious intracytoplasmic membranes.

Cold and hot temperatures also modify the pattern of bacteroid development in *Trifolium subterraneum* nodules, with more bacteria failing to develop into the bacteroid form, and more envelope packets containing more than one *Rhizobium* (e.g. Roughley, 1970; Pankhurst and Gibson, 1973).

G. Nodule Longevity

The degenerate zone of a nodule usually has a green or brown colour associated with the breakdown of leghaemoglobin. Elongate and spherical nodules have different patterns of senescence. In the elongate nodule with continuing, apical, meristematic activity, senescence begins at the base of the nodule after a period dependent on both the growth conditions and the species, and runs concurrently with new tissue development. The bacteroids round up, clump and lyse along with the breakdown of the host organelles, and a loss of host cell wall rigidity with consequent folding and collapse. Rod shaped, untransformed rhizobia from infection threads, intercellular spaces and the occasional groups left untransformed amongst the bacteroids, multiply during this process, but then are mainly lysed themselves. Later in the life of the nodule the meristematic activity ceases and the whole nodule degenerates.

Occasionally some nodules appear to escape from a control which governs growth of their neighbours, resulting in rapid growth, with much dichotomous branching of the meristem to produce a large coralloid structure with many active lobes, and degenerate bases. Such nodule masses in *Cicer arientinum* may be more than 2 cms in diameter; a "normal" nodule adjacent being less than 5 mm long (Fig. 19).

Degeneration is influenced by the environmental conditions and is promoted by interference with the carbohydrate supply, desiccation of the root system, addition of combined nitrogen and high temperature treatment. Should the adverse condition be relieved, meristematic activity may be restored and bacteroid

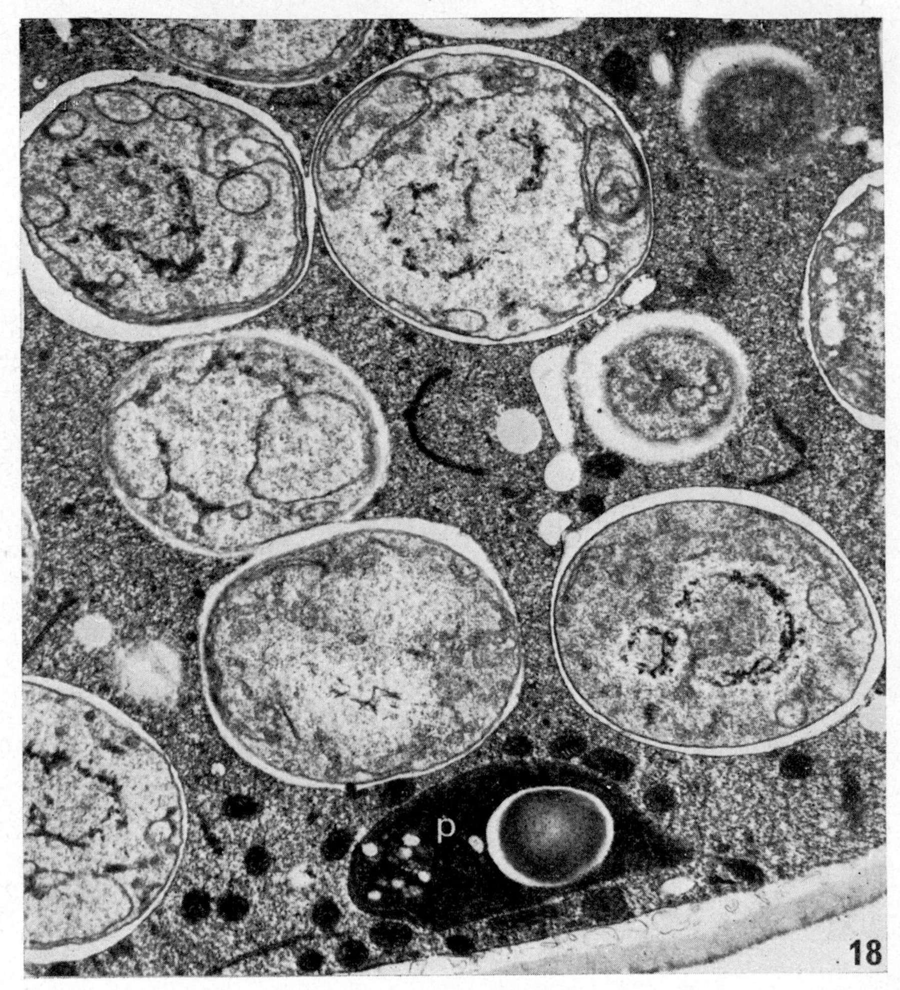

FIG. 18. Effective *Arachis hypogaea* nodule. Electron micrograph showing the large spherical bacteroids enclosed singly in membrane envelopes. The bacteroids contain intracytoplasmic membranes, and a dense staining, condensed nucleoid. Note the electron density of the host cytoplasm. P—plastid with starch grain surrounded by many mitochondria. ×14,200.

development and nitrogen fixation restart (Islam and Dart, unpublished observations). A similar process obtains in overwintering nodules in many plants. The nodule ceases activity and the bacterial tissue degenerates until the spring or wet season when meristematic activity restarts (Pate, 1958; Bergersen *et al*., 1963; Masterson and Murphy, 1975). Some nodules persist in this fashion for up to 6 years. Such nodules have a well-developed periderm, and may be up to 4 cm long and weigh over 1 gm fresh weight (e.g. Jimbo, 1927; Pate, 1961).

Plant-produced ethylene can also hasten senescence, particularly for nodules on clover plants grown inside cotton wool stoppered tubes. These plants grow poorly compared to plants grown in open sand culture because nodule senescence begins very early in the life cycle—as soon as 10–14 days after formation—and proceeds rapidly. Sand grown nodules have little senescence by 35 days. Removing the plant-produced ethylene by adsorption onto activated charcoal, or by recycling the enclosed atmosphere through mercuric perchlorate, retards senescence and increases nitrogenase activity and plant growth almost to the level of the sand grown plants, where presumably the C_2H_4 can readily escape from the tissues. In nodules where C_2H_4 is removed or can escape, most bacteroids remain as an enlarged rod form in contrast with the untreated tube grown nodule where bacteroids rapidly become oval (Day, 1972).

In round nodules degeneration, once started, usually proceeds rapidly, but occasionally the centre of the nodule may degenerate while the periphery remains red and active. Nodules formed by some *Rhizobium phaseoli* strains on *Phaseolus vulgaris* senesce at flowering, with new nodules forming during pod fill (Day and Dart, unpublished observations). On *Pisum sativum*, nodules also degenerate rapidly after flowering. Nitrogenase activity of soybean nodules decreases at flowering, but then increases during pod and seed development (Day, 1972). Hormones are implicated in this change of nodule activity associated with flowering because competition with the developing fruits for photosynthate has not yet begun.

VII. Ineffective Nodules

The failure to establish a symbiosis can be attributed to the *Rhizobium* strain, to the host genotype, and more often to the combination of these, and to adverse environmental conditions. A few *Rhizobium* strains seem to be genuinely ineffective on any host, others nodulate one host ineffectively and another effectively (e.g. Helz *et al.*, 1927; Vincent, 1954). Some hosts have a genotype which blocks effective nodulation with a particular strain (see Nutman, 1969), and others seem to be unable to produce any effective nodules (e.g. Holl, 1973). Although ineffective nodules vary in structure, each association has a consistent pattern of development. Nodule number and size varies greatly with the association—sometimes fewer are formed than with an effective strain (see Nutman, 1969).

In the first ineffective association described, the large pea nodules had cells filled with rod shaped bacteria with no bacteroids (Nobbe and Hiltner, 1893). A common type of ineffectiveness in many plants occurs when bacteroids develop in an apparently normal sequence but which break down rapidly (e.g. Chen and Thornton, 1940; Bergersen, 1955). In one pea variety leghaemoglobin may be formed, along with stable bacteroids, but these apparently do not synthesize nitrogenase (Holl, 1973). A range of ineffective structures, some determined by single, recessive, genes (Nutman, 1969), occur on clovers. In the "simplest", a nodule shaped growth is induced consisting of small uninvaded parenchyma-like cells, with no vascular tissue and no obvious sign of *Rhizobium* in any sections (Chandler *et al.*, 1974; Pankhurst, 1974). In another, a few cells are invaded and then fill with zooglaea; the nodule then grows for a limited extent without any further invasion. In another host mutant the bacteria may be released and some spread through the cell, but these then invoke a breakdown of host cytoplasmic organization before bacteroid development can occur. In another type bacteria remain trapped in large infection thread vesicles surrounded by a massive cell wall; some bacteria escape and may form normal bacteroids, but most remain in groups which become enclosed by a loose polysaccharide matrix, and later by what appears to be massive deposition of fibrous, host wall material which stains as for lignin; the host cytoplasm then becomes completely disorganized (Chandler *et al.*, 1974).

A riboflavin-requiring, auxotrophic mutant of *R. trifolii* formed ineffective nodules on red clover in which rhizobia were spread through the cells as rods but bacteroid development was limited. Adding riboflavin to the plant growth medium induced normal nodule development (Pankhurst *et al.*, 1972).

Many antibiotic and antimetabolite resistant strains form ineffective nodules. Five patterns of nodule development were found for 15 ineffective mutants of *R. trifolii*, but apart from some strains releasing internal antigens more readily, the mutants had similar antigenic characteristics to the parent strains. Some mutants differed quantitatively in the proteins of the cell envelopes. There was no obvious relationship between patterns of resistance and nodule development on *Trifolium pratense*. Seven mutants formed tumour-like growths with no rhizobia or infection threads visible; in two strains, release from the infection threads was blocked; for three other mutants the released rhizobia did not develop into bacteroids, and for one strain abnormally large, very pleomorphic bacteroids were formed (Pankhurst, 1974).

Molybdenum deficiency also produces ineffective nodules, in which bacteroid degeneration is rapid (e.g. Anderson and Thomas, 1946; Mullens, 1965). Cowpeas grown in low light intensity, or short days, or cold temperatures may also form ineffective nodules with an otherwise effective strain (Dart and Mercer, 1965a; Doku, 1970; Day, 1972). *Trifolium subterraneum* nodules formed at 7°C or 30°C by strains effective at 19°C may also be ineffective. The response to temperature depends on the host line; one line formed effective nodules with one strain at 19°C but those formed at 15°C were much less effective (Roughley, 1970; Roughley and Dart, 1970b; Pankhurst and Gibson, 1973; Roughley *et al.*, 1975). Leghaemoglobin formation and bacteroid development in such nodules is very restricted; much starch accumulates in the nodules.

VIII. Nodule Roots

Some nodules produce roots under "natural" conditions. In *Sesbania grandiflora* up to 18 roots may emerge from a large multilobed nodule, developing from within the endodermis around the nodule vascular bundles, emerging at right angles to them (Harris *et al.*, 1949). In *Caragana arborescens* the roots usually develop from the apical ends of the nodule vascular traces (Allen *et al.*, 1955). *Trifolium pratense* nodules very occasionally produce a small root at the end of the nodule (Nutman, 1956). In several clover species, particularly *Trifolium subterraneum* var. Tallarook, and *Medicago sativa*, nodule roots can be induced by transferring nodulated plants to a temperature of 35°C for a week or more (Day and Dart, 1971b) (Figs 20–23). The bacteroid zone tissue degenerates and normal meristematic activity is replaced by a disorganized callus type of growth from which emerge the roots, from the end of the vascular traces, some 7–10 days later. The roots have a central vascular trace and root hairs which are curled by *Rhizobium*, but their length, as for other nodule roots, is limited to about 2 cms. Bacteroids do not redevelop. If young *T. subterraneum* nodules are placed at the high temperature they occasionally grow rapidly without producing roots or fixing nitrogen (Fig. 24). Others develop a rapidly dividing meristem producing nodules with up to 34 lobes.

IX. Excised Roots and Tissue Culture Associations

Excised roots of *Phaseolus vulgaris*, soybean, *Trifolium pratense*, *T. repens*, *Melilotus alba*, *Medicago sativa*, and *Lotus corniculatus*, but not *Pisum sativum*, nodulate when fed on an organic medium containing sucrose, glycine, thiamine, pyridoxine, nicotinic acid and myoinositol. Not all roots nodulate. The nodules are red, and fix nitrogen (e.g. Raggio *et al.*, 1957; Valera and Alexander, 1965b; Molina and Alexander, 1967; Higashi *et al.*, 1971; Hepper, 1973b). Such systems have shown that combined nitrogen fed through the base of the root has little effect on nodulation, contrary to the drastic inhibition that occurs when it is in contact with the root surface (Raggio *et al.*, 1965; Cartwright, 1967a). Nodulation of *P. vulgaris* is improved by adding a piece of hypocotyl (e.g. Bunting and Horrocks, 1964; Hough *et al.*, 1966). Ethylene produced by the tissue inhibits nodulation if not allowed to diffuse away. Three percent CO_2 also decreases nodulation suggesting that a small amount of ethylene may be needed for nodule formation (Grobbelaar *at al.*, 1971). Adding auxins to the root medium inhibited nodulation of *P. vulgaris* but antiauxins at low concentration were stimulating. Gibberellic acid at 1 ppm caused a 6-fold reduction in nodulation with no reduction in root growth (Cartwright, 1967b).

Associations have been established between *Rhizobium japonicum* and soybean root callus tissue with nitrogenase activity. Initially, experiments involved cells

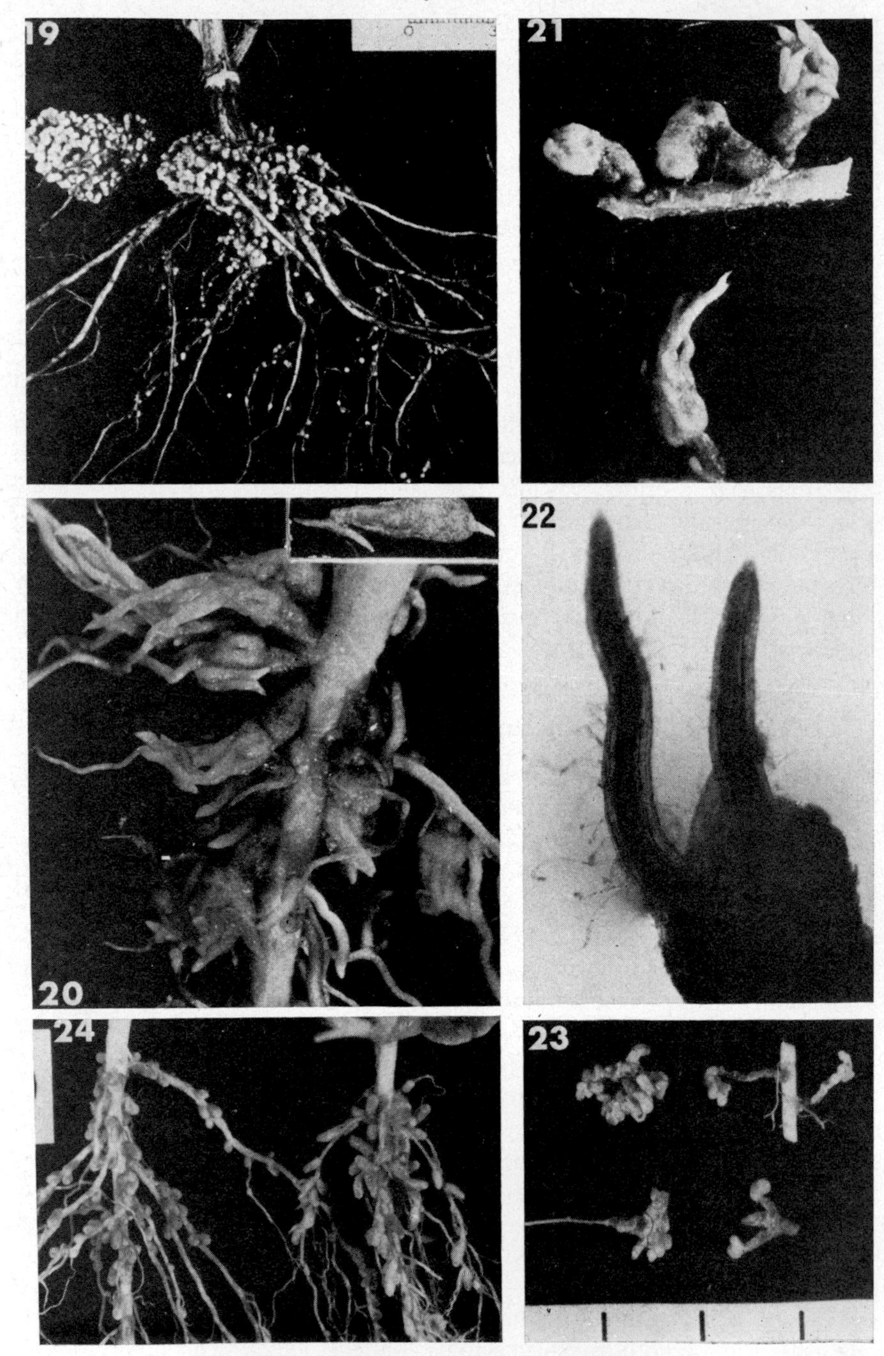
19
21
20
22
24
23

grown in liquid culture in a klinostat (Holsten *et al.*, 1971), but latterly cells growing in clumps in agar have given more reproducible results (Child and La Rue, 1974; Phillips, 1974). Infection threads were not found although intercellular zooglaea containing rhizobia were present. A group of cells in the tissue clump apparently become filled with *Rhizobium*. Infection began with a characteristic darkening of the callus due to formation of lignin-like material (Velicky and Le Rue, 1967). Nitrogenase activity was detectable 5–7 days after inoculation and was dependent on the *R. japonicum* strain; an active association was also formed with cowpea *Rhizobium*. Activity was not increased by lowering the pO_2 (Phillips, 1974).

X. Environment and Effects on Nodule Formation

Extreme root temperatures inhibit nodule growth and their efficiency in fixing nitrogen. Nodules formed by some *Rhizobium* strains are much more tolerant than others of such temperatures. One strain, effective at 21 and 27°C with soybean was quite ineffective at 33°C although the nodules grew and contained much leghaemoglobin. Not all red nodules are actively fixing nitrogen (Dart *et al.*, 1974).

Inorganic combined nitrogen in the root medium usually inhibits nodule formation and reduces the amount of nitrogen fixed by the plant. The extent of this inhibition varies with the host and the *Rhizobium* strain, being absent at some levels of combined nitrogen for some associations (e.g. Pate and Dart, 1961; Dart *et al.*, 1974; Summerfield, Huxley, Eaglesham, Day and Dart, unpublished observations). This shows the importance of choosing the legume–*Rhizobium* association carefully; some will be more capable of building up soil nitrogen to higher levels than others, and will fix nitrogen when soil levels are high, e.g. when used in a rotation following a cereal heavily N-fertilized.

Effectively nodulated cowpeas grown in good environmental conditions yield significantly more grain than plants liberally and continuously supplied with inorganic combined nitrogen at up to 240 ppm N as $NH_4^+ + NO_3^-$

FIGS. 19–24. FIG. 19. Effectively nodulated *Cicer arietinum* root showing two large clusters of nodule tissue apparently developed from single nodules, along side smaller "normal" sized nodules. The scale is 3 cm. FIGS. 20–22. *Trifolium subterraneum* var Tallarook root, with rooted nodules induced by transferring nodulated plants to a root temperature of 35°C. Inset shows *T. pratense* nodule with a single root similarly induced. FIG. 21. Shows the apical, callous-like growth which develops before the roots form. FIG. 22. Nodule roots showing the small root cap, central stele and root hairs. FIG. 23. *Medicago sativa* nodules induced by high temperature treatment to form terminal callus-like growths and a root. Distance between scale marks is 1 cm. FIG. 24. Five weeks old *T. subterraneum* plants nodulated by an effective *R. trifolii* strain. The root on the left was kept at a constant temperature of 20°C, and for on the other the much larger, ineffective nodules were formed at 20°C and transferred at 18 days to 35°C.

(Summerfield and others, unpublished observations). This suggests that nodules are capable of supplying all the nitrogen plants need for high yields. For soybean, nitrogen fertilizing in the field generally has little or no effect on yields of nodulated plants (see Ham *et al.*, 1975). Addition of combined nitrogen to nitrogen fixing nodules of *Trifolium subterraneum*, *Medicago truncatula* (Dart and Mercer, 1965b), cowpea (Day and Dart, 1971a), soybean and peanut (Hardy *et al.*, 1973) induces rapid nodule senescence and restricts their nitrogenase activity.

Vigna radiata and *V. mungo* grow poorly with nutrients supplied in mineral form only. Addition of a 10% loam soil greatly increased growth and nodulation. Growth and subsequent nodulation in soils varied with the amount and type of organic matter present. Cowpea (*V. unguiculata*), *Phaseolus vulgaris* and pigeon pea (*Cajanus cajan*) growth and nodulation were also greatly stimulated by the presence of organic matter. The cause of the stimulation is obscure, but perhaps the organic matter has hormonal activity, or else adsorbs some plant product which can inhibit growth. Nutritional effects do not seem to be involved (Dart *et al.*, 1975).

XI. In Conclusion

Nodules are complex structures with a precisely controlled development mediated by the interaction of two different genomes. Some single gene changes can modify this. As neither partner can readily produce the *Rhizobium* located nitrogenase or the leghaemoglobin of the host cytoplasm on its own, some genetic exchange or unmasking of the other partner's genome apparently takes place. The plant may provide growth factors essential for the production of nitrogenase by *Rhizobium*, as the enzyme can be induced in pure culture (Dilworth, Scowcroft and Gibson, personal communication). The interaction starts early in the association as host nuclear changes are apparent from the initial infection stage. *Rhizobium* morphology only changes significantly when released into the host cytoplasm. Small tubular connections are found between the bacteria and host and perhaps these mediate the transfer of metabolites (Dart, 1969).

Nodule development also seems to involve endogenous plant hormones—some such as abscisic acid, perhaps directed to the nodule by its strong sink effect (e.g. Small and Leonard, 1969; Hocking *et al.*, 1972), and some such as auxin, where plant tissue levels are increased by the presence of *Rhizobium*, interacting with *Rhizobium* produced cytokinins, auxins, gibberellins and possibly ethylene. Perhaps one can usefully alter this balance to produce more nodules or place nodules on the root so that they are less affected by factors such as soil temperature. Although plants can compensate to some extent for low nodule number by increase in nodule size (e.g. Nutman, 1967), under adverse environmental conditions, plants with most nodules fare best.

One is also looking for nodule associations which can both stave off senescence and compete with fruits or vegetation as sinks for carbohydrate, thus ensuring an adequate nitrogen supply. Nodules usually reduce overall root growth, perhaps because they are strong sinks, but perhaps because of their hormone metabolism. A small root system is a disadvantage when soil moisture and nutrients are limiting, so that breeding plants which nodulate well and retain an adequate root system should be our objective.

References

ALLEN, E. K. and ALLEN, O. N. (1961). The scope of nodulation in the Leguminosae. *In*: "Recent Advances in Botany," pp. 585–588. University of Toronto Press.

ALLEN, E. K., GREGORY, K. F. and ALLEN, O. N. (1955). Morphological development of nodules on *Caragana arborescens* LAM. *Can. J. Botany* **33**; 139–148.

ALLEN, O. N. and ALLEN, E. K. (1940). Response of the peanut plant to inoculation with rhizobia, with special reference to morphological development of the nodules. *Bot. Gaz.* **102**; 121–142.

ANDERSON, A. J. and THOMAS, M. P. (1946). Plant responses to molybdenum as a fertilizer. I. Molybdenum and symbiotic nitrogen fixation. *Council Sci. Ind. Res. Australia. Bull.* **198**; pp. 44.

ANDREW, C. S. and NORRIS, D. O. (1961). Comparative responses to calcium of five tropical and four temperate pasture legume species. *Aust. J. Agric. Res.* **12**; 40–55.

ARORA, N. (1954). Morphological development of the root and stem nodules of *Aeschynomene indica* L. *Phytomorphology* **4**; 211–216.

BEIJERINCK, M. W. (1888) Die bacterien der Papilionaceenknollchen. *Bor. Ztg.* **46**; 726–735, 741–750, 757–771, 781–790, 797–804.

BELL, J. K. and MCCULLY, M. E. (1970). A histological study of lateral root initiation and development in *Zea mays. Protoplasma* **70**; 179–205.

BERGERSEN, F. J. (1955). The cytology of bacteroids from root nodules of subterranean clover (*Trifolium subterraneum* L.). *J. Gen. Microbiol.* **13**; 411–419.

BERGERSEN, F. J. (1970). The quantitative relationship between nitrogen fixation and the acetylene reduction assay. *Aust. J. Biol. Sci.* **23**; 1015–1025.

BERGERSEN, F. J. and BRIGGS, M. J. (1958). Studies on the bacterial component of soybean root nodules: cytology and organisation in the host tissue. *J. Gen. Microbiol.* **19**; 482–490.

BERGERSEN, F. J. and GOODCHILD, D. J. (1973a). Aeration pathways in soybean root nodules. *Aust. J. Biol. Sci.* **26**; 729–740.

BERGERSEN, F. J. and GOODCHILD, D. J. (1973b) Cellular location and concentration of leghaemoglobin in soybean root nodules *Aust. J. Biol. Sci.* **26**; 741–756.

BERGENSEN, F. J., HELY, F W. and COSTIN, A. B. (1963) Overwintering of clover nodules in alpine conditions. *Aust. J. Biol. Sci.* **16**; 920–921.

BOND, L. (1948). Origin and developmental morphology of root nodules of *Pisum sativum*. *Bot Gaz.* **109**; 411–433.

BONISH, P. M. (1973). Pectolytic enzymes in inoculated and uninoculated red clover seedlings. *Pl. Soil* **39**; 319–328.

BUNTING, A. H. and HORROCKS, J. (1964). An improvement in the Raggio technique for obtaining nodules on excised roots of *Phaseolus vulgaris* L. in culture. *Ann. Bot.* **28**; 229–237.

CARTWRIGHT, P. M. (1967a) .The effect of combined nitrogen on the growth and nodulation of excised roots of *Phaseolus vulgaris. Ann. Bot.* **31**; 309–321.

CARTWRIGHT, P. M. (1967b). The effect of growth regulators on the growth and nodulation of excised roots of *Phaseolus vulgaris. Wiss. Zeit. Univ. Rostock.* **4/5**; 537–538.

CHANDLER, M. R. and DART, P. J. (1973). Structure of nodules of peanut and *Vigna* spp. *Rothamsted Exp. Sta. Rept for* 1972, *Pt.* 1, 85.

CHANDLER, M. R., DART, P. J. and NUTMAN, P. S. (1974). The fine structure of hereditarily ineffective red clover nodules. *Rothamsted Exp. Sta. Rept for* 1973, *Pt.* 1, 83–84.

CHARUDATTAN, R. and HUBBELL, D. H. (1973). The presence and possible significance of cross-reactive antigens in *Rhizobium*-legume associations. *Antonie van Leeuwenhoek. J. Microbiol. Serol.* **39**; 619–627.

CHEN, H. K. and THORNTON, H. G. (1940). The structure of "ineffective" nodules and its influence on nitrogen fixation. *Proc. Roy. Soc. Ser. B.* **129**; 208–229.

CHILD, J. J. and LA RUE, T. A. (1974). A simple technique for the establishment of nitrogenase in soybean callus culture. *Plant Physiol.* **53**; 88–90.

CRAIG, A. S. and WILLIAMSON, K. I. (1972). Three inclusions of rhizobial bacteroids and their cytochemical character. *Arch. Mikrobiol.* **87**; 165–171.

CRAIG, A. S., GREENWOOD, R. M. and WILLIAMSON, K. I. (1972). Ultrastructural inclusions of rhizobial bacteroids of *Lotus* nodules and their taxonomic significance. *Ardh. Mikrobiol.* **89**; 23–32.

DARBYSHIRE, J. F. (1964). A Study of the Initial Stages of infection of Clovers by Nodule Bacteria. Ph.D. thesis, University of London, pp. 149.

DARBYSHIRE, J. F. (1966). Studies on the physiology of nodule formation. IX. The influence of combined nitrogen, glucose, light intensity and day length on root hair infection in clover. *Ann. Bot.* **30**; 623–638.

DART, P. J. (1968). Localization of peroxidase activity in legume root nodules. "4th European Regional Conf. Electron Microscopy, Rome," 69–70.

DART, P. J. (1969). Nodule and *Rhizobium* fine structure. *Rothamsted Exp. Sta. Rept. for* 1968, Pt. 1; 89–90.

DART, P. J. (1971). Scanning electron microscopy of plant roots. *J. exp. Bot.* **22**; 163–168.

DART, P. J. (1974). The infection process *In*: "Biological Nitrogen Fixation" (A. Quispel, Ed.); 381–429. North Holland Publishing Co., Amsterdam.

DART, P. J. and CHANDLER, M. R. (1972). The site of iron in nodules. *Rothamsted Exp. Sta. Rept. for* 1971, *Pt.* 1 : 99.

DART, P. J., DAY, J. M., ISLAM, R. and DÖBEREINER, J. (1975). Symbiosis in tropical grain legumes—some effects of temperature and the composition of the rooting medium. *In*: "Symbiotic Nitrogen Fixation in Plants" (Ed. P. S. Nutman); 361–384. Cambridge University Press.

DART, P. J. and MERCER, F. V. (1963). Development of the bacteroid in the root nodule of barrel medic (*Medicago tribuloides* Desr.) and subterranean clover (*Trifolium subterraneum* L.) *Arch. Mikrobiol.* **46**; 382–401.

DART, P. J. and MERCER, F. V. (1964a). The legume rhizosphere. *Arch. Mikrobiol.* **47**; 344–378.

DART, P. J. and MERCER, F. V. (1964b). Fine structure changes in the development of the nodules of *Trifolium subterraneum* L. and *Medicago tribuloides* Desr. *Arch. Mikrobiol.* **49**; 209–235.

DART, P. J. and MERCER, F. V. (1965a). The effect of growth temperature, level of ammonium nitrate, and light intensity on the growth and nodulation of cowpea (*Vigna sinensis* Endl. ex Hassk). *Aust. J. Agric. Res.*, **16**; 321–345.

DART, P. J. and MERCER, F. V. (1965b). The influence of ammonium nitrate on the fine structure of nodules of *Medicago tribuloides* Desr. and *Trifolium subterraneum* L. *Arch. Mikrobiol.* **51**; 233–257.

DART, P. J. and MERCER, F. V. (1966). Fine structure of bacteroids in root nodules of *Vigna sinensis, Acacia longifolia, Viminaria juncea* and *Lupinus augustifolius*. *J. Bact.* **91**; 1314–1319.

DART, P. J. and PATE, J. S. (1959). Nodulation studies in legumes. III. The effects of delaying the inoculation on the seedling symbiosis of barrel medic (*Medicago tribuloides* Desr.). *Aust. J. Biol. Sci.* **12**; 427–444.

DAY, J. M. (1972) "Studies on the Role of Light in Legume Symbioses." Ph.D. thesis, University of London, p. 264.

DAY, J. M. and DART. P. J. (1971a). Effect of temperature and combined nitrogen on nodule structure. *Rothamsted Exp. Sta. Rept. for* 1970, *Pt.* 1, pp. 84–85.

DAY, J. M. and DART, P. J. (1971b). Root formation by legume nodules. *Rothamsted Exp. Sta. Rept. for* 1970, *Pt.* 1; 85.

DEINEMA, M. H. and ZEVENHUIZEN, L. P. T. M. (1971) Formation of cellulose fibrils by gram-negative bacteria and their role in bacterial flocculation. *Arch. Mikrobiol.* **78**; 42–57.

DILWORTH, M. J. and KIDBY, D. K. (1968). Localization of iron and leghaemoglobin in the legume root nodule by electron microscope autoradiography. *Exp. Cell. Res.* **49**; 148–159.

DILWORTH, M. J. and WILLIAMS, D. C. (1967). Nucleic acid changes in bacteroids of *Rhizobium lupini* during nodule development. *J. Gen. Microbiol.* **48**; 31–36.

DIXON, R. O. D. (1964). The structure of infection threads, bacteria and bacteroids in pea and clover root nodules. *Arch. Mikrobiol.* **48**; 166–178.

DÖBEREINER, J. and CAMPELO, A. B. (1975). Importance of legumes and their contribution to tropical agriculture *In*: "Treatise on Dinitrogen (N_2) Fixation" (Ed. R. W. F. Hardy) Vol. III. In press. John Wiley and Sons Inc., New York.

DOKU, E. V. (1970). Effect of day length and water on the nodulation of cowpea (*Vigna unguiculata* L. Walp) in Ghana. *Exp. Agric.* **6**; 13–18.

DULLAART, J. (1967). Quantitative estimation of indoleacetic acid and indolecarboxylic acid in the root nodules and roots of *Lupinus luteus* L. *Acta. Bot. Neerl.* **16**; 222–230.

DULLAART, J. (1970). The bioproduction of indole-3-acetic acid and related compounds in root nodules and roots of *Lupinus luteus* L. and by its rhizobial symbiont. *Acta. Bot. Neerl.* **19**; 573–618.

DULLAART, J. and DUBA, L. I. (1970). Presence of gibberellin-like substances and their possible role in auxin bioproduction in root nodules and roots of *Lupinus luteus* L. *Acta. Bot. Neerl.* **19**; 877–883.

VAN EGERAAT, A. W. S. M. (1972). Pea-root exudates and their effect upon root nodule bacteria. *Meded. Landbouwhogeschool Wageningen*, 72–27.

EVANS, M. J., CAMPBELL, N. E. R. and HILL, S. (1972). Asymbiotic nitrogen-fixing bacteria from the surface of nodules and roots of legumes. *Can. J. Microbiol.* **18**; 13–21.

FAHRAEUS, G. (1957). The infection of clover root hairs by nodule bacteria studied by a simple glass slide technique. *J. Gen. Microbiol.* **16**; 374–381.

FAHRAEUS, G. and LJUNGGREN, H. (1959). The possible significance of pectic enzymes in root hair infection by nodule bacteria. *Physiol. Plant.* **12**; 145–154.

FISHER, M. L., ANDERSON, A. J. and ALBERSHEIM, P. (1973). Host-pathogen interactions. VI. A single plant protein efficiently inhibits endopolygalacturonases secreted by *Colleototrichum lindemuthianum* and *Aspergillus niger*. *Plant Physiol.* **51**; 489–491.

FRASER, H. L. (1942). The occurrence of endodermis in leguminous root nodules and its effect upon nodule function. *Proc. Roy. Soc. Edinburgh, Ser. B.* **61**; 328–343.

FRED, E. B., BALDWIN, I. L. and MCCOY, E. (1932). "Root Nodule Bacteria and Leguminous Plants." University Wisconsin Press, Madison.

GIBBINS, A. M. and GREGORY, K. F. (1972). Relatedness among *Rhizobium* and *Agrobacterium* species determined by three methods of nucleic acid hybridization. *J. Bact.* **111**; 129–141.

GIBSON, A. H. (1968). Nodulation failure in *Trifolium subterraneum* L. c.v. Woogenellup (syn. Marrar). *Aust. J. Agric. Res.* **19**; 907–918.

GIBSON, A. H. (1969). Indole acetic acid as a factor in legume nodulation. *Proc.* 11*th Int. Botan. Congr., Seattle*, p. 70.

GIBSON, A. H. (1971). Factors in the physical and biological environment affecting nodulation and nitrogen fixation by legumes. *Pl. Soil*, special volume; 139–152.

GODFREY, C. A. (1972). The carotenoid pigment and desoxyribonucleic acid base ratio of a *Rhizobium* which nodulates *Lotononis bainesii* Baker. *J. Gen. Microbiol.* **72**; 399–402.

GOODCHILD, D. J. and BERGERSEN, F. J. (1966). Electron microscopy of the infection and subsequent development of soybean nodule cells. *J. Bact.* **92**; 204–213.

GRAHAM, P. H. (1963). Vitamin requirements of root nodule bacteria. *J. Gen. Microbiol.* **30**; 245–248.

GRILLI, M. (1963). Osservazioni sui rapporti tra cellule ospiti e rizobia nei tubercoli radicali di Pisello (*Pisum sativum*). *Caryologia* **16**; 561–594.

GROBBELAAR, N., CLARKE, B. and HOUGH, M. C. (1971). The nodulation and nitrogen fixation of isolated roots of *Phaseolus vulgaris* L. III. The effect of carbon dioxide and ethylene. *Pl. Soil*, special volume; 215–224.,

GUNNING, B. E. S., PATE, J. S., MINCHIN, F. R. and MARKS, I. (1974). Quantitative aspects of transfer cell structure and function. *Symp. Soc. Exp. Biol.* **24**; 87–126.

HAACK, A. (1961). Über den urspring der Wurzelknollchen von *Ornithopus sativus* L. und *Lupinus albus* L. *Z. Bakteriol. Parisitenk. Abt. II*, **114**; 577–589.

HAACK, A. (1964). Über den Einfluss du Knollchenbakterien auf die Wurzelhaare von Leguminosen und Nichtleguminosen. *Z. Bakteriol. Parisitenk. Abt. II*, **117**; 343–366.

HAM, G. E., LAWN, R. J. and BRUN, W. A. (1975) Influence of inoculation, nitrogen fertilizers, and photosynthetic source-sink manipulations on field grown soybeans. *In*: "Symbiotic Nitrogen Fixation in Plants." (Ed. P. S. Nutman); 239–254. Cambridge University Press.

HAMBLIN, J. and KENT, S. P. (1973). A possible role of phytohaemagglutinin in *Phaseolus vulgaris* L. *Nature, Lond.* **245**; 28–30.

HARDY, R. W. F., BURNS, R. C. and HOLSTEN, R. D. (1973). Applications of the acetylene-ethylene assay for measurement of nitrogen fixation. *Soil Biol. Biochem.* **5**; 47–81.

HARRIS, J. O., ALLEN, E. K. and ALLEN. O. N. (1949). Morphological development of nodules of *Sesbania grandiflora* poir. with reference to the origin of nodule rootlets. *Am. J. Botany*, **36**; 651–661.

HEAD, G. C. (1964). A study of exudation from the root hairs of apple roots by time-lapse cine-photomicrography. *Ann. Bot.* **28**; 495–498.

HELZ, G. E., BALDWIN, H. and FRED, E. B. (1927). Strain variations and host specificity of the root-nodule bacteria of the pea group. *J. Agr. Res.* **35**; 1039–1055.

HEPPER, C. M. (1972). Composition of extra-cellular polysaccharides of *Rhizobium trifolii*. *Antonie van Leeuwenhoek. J. Microbiol. Serol.* **38**; 437–445.

HEPPER, C. M. (1973a). Genetics of red clover nodulation. *Rothamsted Exp. Sta. Rept. for* 1972, *Pt.* 1; 82.

HEPPER, C. M. (1973b). Nodulation of excised roots of red clover. *Rothamsted Exp. Sta. Rept. for* 1972, *Pt.* 1, p. 84.

HIGASHI, S., ABE, M. and YAMANE, G. (1971). The influence of plant hormones on the formation of infection thread into root hair of *Trifolium repens* L. by *Rhizobium trifolii* K.102 *Rep. Fac. Sci. Kagoshima Univ.* (*Earth Sci. Biol.*) no. 4.

HOCKING, T. J., HILLMAN, J. R. and WILKINS, M. B. (1972). Movement of abscisic acid in *Phaseolus vulgaris* plants. *Nature New Biology*, **235**; 124–125.

HOLL. F. B. (1973). A nodulating strain of *Pisum* unable to fix nitrogen. *Plant Physiol.* **51**; Suppl.; 35.

HOLSTEN, R. D., BURNS, R. C., HARDY, R. W. F. and HEBERT, R. R. (1971). Establishment of symbiosis between *Rhizobium* and plant cells in vitro. *Nature, Lond.* **232**; 173–175.

HOUGH, M. C., CLARKE, B. and GROBBELAAR, N. (1966). The influence of the hypocotyl and vitamin B-12 on the nitrogen fixation of isolated bean roots. *Phyton* (Buenos Aires), **23**; 15–19.

JANSEN VAN RENSBURG, H. and STRIJDOM, B. W. (1972a). A bacterial contaminant in nodules of leguminous plants. *Phytophylactica* **4**; 1–8.

JANSEN VAN RENSBURG, H. and STRIJDOM, B. W. (1972b). Information on the mode of entry of a bacterial contaminant into nodules of some leguminous plants. *Phytophylactica* **4**; 73–78.

JIMBO, T. (1927). Physiological anatomy of the root-nodule of *Wistaria sinensis*. *Proc. Imp. Acad. Japan*, **3**; 164–166.

JORDAN, D. C., GRINYER, I. and COULTER, W. W. (1963). Electron microscopy of infection threads and bacteria in young root nodules of *Medicago sativa*. *J. Bact.* **86**; 125–137.

JORDAN, D. C. and GRINYER, I. (1965). Electron microscopy of the bacteroids and root nodules of *Lupinus luteus*. *Can. J. Microbiol.* **11**; 721–725.

KATZNELSON, H. and COLE, S. E. (1965). Production of gibberellin-like substances by bacteria and actinomycetes. *Can. J. Microbiol.* **11**; 733–741.

KEFFORD, N. P., BROCKWELL, J. and ZWAR, J. A. (1960). The symbiotic synthesis of auxin by legumes and nodule bacteria and its role in nodule development. *Aust. J. Biol. Sci.* **13**; 456–467.

KIDBY, D. K. and GOODCHILD, D. J. (1966). Host influence on the ultrastructure of root nodules of *Lupinus luteus* and *Ornithopus sativus*. *J. Gen. Microbiol.* **45**; 147–152.

KLECZKOWSKA, J., NUTMAN, P. S. and BOND, G. (1944). Note on the ability of certain strains of rhizobia from peas and clover to infect each others host plants. *J. Bact.* **48**; 673–675.

LIBBENGA, K. R. and HARKES, P. A. A. (1973). Initial proliferation of cortical cells in the formation of root nodules in *Pisum sativum* L. *Planta (Berl.)* **114**; 17–28.

LIBBENGA, K. R., VAN IREN, F., BOGERS, R. J., and SCHRAAG-LAMERS, M. F. (1973). The role of hormones and gradients in the initiation of cortex proliferation and nodule formation in *Pisum sativum* L. *Planta (Berl.)* **114**; 29–39.

LIBBENGA, K. R. and TORREY, J. G. (1973). Hormone-induced endoreduplication prior to mitosis in cultured pea root cortex cells. *Am. J. Bot.* **60**; 293–299.

LIE, T. A. (1969). The effect of low pH on different phases of nodule formation in pea plants. *Pl. Soil* **31**; 391–406.

LIM, G. (1961). "Microbiological Factors Influencing Infection of Clover by Nodule Bacteria." Ph.D. thesis, London University.

LIM, G. (1963). Studies on the physiology of nodule formation. VIII. The influence of the size of the rhizosphere population of nodule bacteria on root hair infection in clover. *Ann. Bot.* **27**; 55–67.

LJUNGGREN, H. (1969). Mechanism and pattern of *Rhizobium* invasion into leguminous root hairs. *Physiol. Plant.* Supplement 5.

LOWTHER, W. L. and LONERAGAN, J. F. (1968). Calcium and nodulation in subterranean clover (*Trifolium subterraneum* L.). *Plant Physiol.* **43**; 1362–1366.

McCOY, E. F. (1929). A cytological and histological study of the root nodules of the bean, *Phaseolus vulgaris* L. *Z. Bakteriol. Parisitenk., Abt. II*, **79**; 394–412.

McCOY, E. F. (1932). Infection by *Bact. radicicola* in relation to the microchemistry of the host cell walls. *Proc. Roy. Soc. (London), Ser. B.* **110**; 514–533.

MASTERSON, C. L. and MURPHY, P. M. (1975). Application of the acetylene reduction technique to the study of nitrogen fixation by white clover in the field. *In*: "Symbiotic Nitrogen Fixation in Plants." (Ed. P. S. Nutman); 299–316. Cambridge University Press.

MENZEL, G., UHLIG, H. and WEICHSEL, G. (1972). Über die besiedlung der wurzeln einiger leguminosen und Nichtleguminosen mit Rhizobien und auderen boden bakterien. *Z. Bakteriol. Parisitenk. Abt. II*, **127**; 348–358.

MINCHIN, F. R. and PATE, J. S. (1973). The carbon balance of a legume and the functional economy of its root nodules. *J. exp. Bot.* **24**; 259–271.

MOLINA, J. A. E. and ALEXANDER, M. (1967). The effect of antimetabolites on nodulation and growth of leguminous plants. *Can. J. Microbiol.* **13**; 819–827.

MOSSE, B. M. (1964). Electron microscope studies of nodule development in some clover species. *J. Gen. Microbiol.* **36**; 49–66.

MULLENS, R. (1965). "Effect of Molybdenum Deficiency on Ultrastructure of Root Nodules. M.Sc. thesis, University of Sydney.

MUNNS, D. N. (1968a). Nodulation of *Medicago sativa* in solution culture. I. Acid-sensitive steps. *Pl. Soil* **28**; 129–146.

MUNNS, D. N. (1968b). Nodulation of *Medicago sativa* in solution culture. II. Compensating effects of nitrate and of prior nodulation. *Pl. Soil* **28**; 246–257.

MUNNS, D. N. (1968c) Nodulation of *Medicago sativa* in solution culture. III. Effects of nitrate on root hairs and infection. *Pl. Soil.* **29**; 33–47.

MUNNS, D. N. (1968d). Nodulation of *Medicago sativa* in solution culture. IV. Effects of indole-3-acetate in relation to acidity and nitrate. *Pl. Soil.* **29**; 257–262.

MUNNS, D. N. (1970). Nodulation of *Medicago sativa* in solution culture. V. Calcium and pH requirements during infection. *Pl. Soil* **32**; 90–102.

NOBBE, F. and HILTNER, L. (1893). Werden die knöllchenbesitzenden Leguminosen befähigt, den freien atmosphärischen Stickstoff für sich zu verwerten? *Landw. Vers. Sta.* **42**; 459–478.

NORRIS, D. O. (1965). Acid production by *Rhizobium*. A unifying concept. *Pl. Soil* **22**; 143–166.

NUTMAN, P. S. (1949). Physiological studies on nodule formation. II. The influence of delayed inoculation on the rate of nodule formation in red clover. *Ann. Bot.* **13**; 261–283.

NUTMAN, P. S. (1952). Studies on the physiology of nodule formation. III. Experiments on the excision of root tips and nodules. *Ann. Bot.* **14**; 79–101.

NUTMAN, P. S. (1956). The influence of the legume in root-nodule symbiosis. A comparative study of host determinants and functions. *Biol. Rev.* **31**; 109–151.

NUTMAN, P. S. (1959a). Some observations on root-hair infection by nodule bacteria. *J. exp. Bot.* **10**; 250–263.

NUTMAN, P. S. (1962). The relation between root-hair infection by *Rhizobium* and nodulation in *Trifolium* and *Vicia*. *Proc. Roy. Soc.* (*London*) *Ser. B.* **156**; 122–137.

NUTMAN, P. S. (1967). Varietal differences in the nodulation of subterranean clover. *Aust. J. Agric. Res.* **18**; 381–425.

NUTMAN, P. S. (1969). Genetics of symbiosis and nitrogen fixation in legumes. *Proc. Roy. Soc.* (*London*) *Ser. B.* **172**; 417–437.

NUTMAN, P. S., DONCASTER, C. C. and DART. P. J. (1973). "Infection of clover by root-nodule bacteria". Black and white, 16 mm, optical sound track film available from The British Film Institute, 81, Dean Street, London, W1V 6AA.

NUTMAN, P. S. and ROSS, G. J. S. (1970). *Rhizobium* in the soils of the Rothamsted and Woburn Farms. *Rothamsted Exp. Sta. Rept. for* 1969, *Pt.* 2, 148–167.

PANKHURST, C. E. (1974). Ineffective *Rhizobium trifolii* mutants examined by immune-diffusion, gel-electrophoresis and electron microscopy. *J. Gen. Microbiol.* **82**; 405–413.

PANKHURST, C. E. and GIBSON, A. H. (1973). *Rhizobium* strain influence on disruption of clover nodule development at high root temperature. *J. Gen. Microbiol.* **74**; 111–222.

PANKHURST, C. E., SCHWINGHAMER, E. A. and BERGERSEN, F. J. (1972). The structure and acetylene-reducing activity of root nodules formed by a riboflavin-requiring mutant of *Rhizobium trifolii*. *J. Gen. Microbiol.* **70**; 161–177.

PATE, J. S. (1958). Nodulation studies in legumes. II. The influence of various environmental factors on symbiotic expression in the vetch (*Vicia sativa* L.) and other legumes. *Aust. J. Biol. Sci.* **11**; 496–515.

PATE, J. S. (1961). Perennial nodules on native legumes in the British Isles. *Nature, Lond* **192**; 376–377.

PATE, J. S. and DART, P. J. (1961) Nodulation studies in legumes. IV. The influence of inoculum strain and time of application of ammonium nitrate on symbiotic response. *Pl. Soil* **15**; 329–346.

PATE, J. S., GUNNING, B. E. S. and BRIARTY, L. G. (1969). Ultrastructure and functioning of the transport system of the leguminous root nodule. *Planta* (*Berl.*) **85**; 11–34.

PHILLIPS, D. A. (1971a). Abscisic acid inhibition of root nodule initiation in *Pisum sativum*. *Planta* (*Berl.*) **100**; 181–190.

PHILLIPS, D. A. (1971b). A cotyledonary inhibitor of root nodulation in *Pisum sativum*. *Physiol. Plant.* **25**; 482–487.

PHILLIPS, D. A. (1974). Factors affecting the reduction of acetylene by *Rhizobium*–soybean cell associations in vitro. *Plant Physiol.* **53**; 67–72.

PHILLIPS, D. A. and TORREY, J. G. (1972). Studies on cytokinin production by *Rhizobium*. *Plant Physiol.* **49**; 11–15.

PURCHASE, H. F. and NUTMAN, P. S. (1957). Studies on the physiology of nodule formation. VI. The influence of bacterial numbers in the rhizosphere on nodule initiation. *Ann. Bot.* **21**; 439–454.

RADLEY, M. (1961). Gibberellin-like substances in plants. *Nature, Lond* **191**; 684–685.

RAGGIO, M. and RAGGIO, N. (1956). Relacion entre cotiledones y nodulacion y factores que la afectan. *Phyton* (Buenos Aires) **7**; 103–119.

RAGGIO, M., RAGGIO, N. and TORREY, J. G. (1957). The nodulation of isolated leguminous roots. *Am. J. Bot.* **44**; 325–334.

RAGGIO, M., RAGGIO, N. and TORREY, J. G. (1965). The interaction of nitrate and carbohydrates in rhizobial root nodule formation. *Plant Physiol.* **40**; 601–606.

ROBINSON, A. C. (1969). Competition between effective and ineffective strains of *Rhizobium trifolii* in the nodulation of *Trifolium subterraneum*. *Aust. J. Agric. Res.* **20**; 827–841.

ROBSON, A. D. and LONERAGAN, J. F. (1970). Nodulation and growth of *Medicago truncatula* on acid soils. I. Effect of calcium carbonate and inoculation level on the nodulation of *Medicago truncatula* on a moderately acid soil. *Aust. J. Agric. Res.* **21**; 427–434.

ROUGHLEY, R. J. (1970). The influence of root temperature, *Rhizobium* strain and host selection on the structure and nitrogen-fixing efficiency of the root nodules of *Trifolium subterraneum*. *Ann. Bot.* **34**; 631–646.

ROUGHLEY, R. J. and DART, P. J. (1970a). Root temperature and root-hair infection of *Trifolium subterraneum* L. c.v. Cranmore. *Pl. Soil* **32**; 518–520.

ROUGHLEY, R. J. and DART, P. J. (1970b). Growth of *Trifolium subterraneum* L. selected for sparse and abundant nodulation as affected by root temperature and *Rhizobium* strain. *J. exp. Bot.* **21**; 776–786.

ROUGHLEY, R. J., DART, P. J., NUTMAN, P. S. and RODRIGUEZ-BARRUECO, C.(1970a). The influence of root temperature on root-hair infection of *Trifolium subterraneum* L. by *Rhizobium trifolii* Dang. *Proc. 9th Int. Grassland Cong.* 451–455.

ROUGHLEY, R. J., DART, P. L., NUTMAN, P. S. and CLARKE, P. A. (1970b). The infection o *Trifolium subterraneum* root hairs by *Rhizobium trifolii*. *J. exp. Bot.* **21**; 186–194.

ROUGHLEY, R. J., DART, P. J. and DAY, J. M. (1975). The structure of *Trifolium subterraneum* L. root nodules. II. The influence of sub optimal temperatures on nodule development. MS submitted to *J. exp. Bot.*

SAHLMAN, K. and FAHRAEUS, G. (1962). Microscopic observations on the effect of indole-3-acetic acid upon root hairs of *Trifolium repens*. *K. LantbrHogsk. Annlr.* **28**; 261–268.

SARIC, Z. (1963a). Specificity of strains of the nodule bacteria *Rhizobium* sp. isolated from peanuts in some parts of Yugoslavia. *J. Sci. Agric. Res.* (Beograd) **16**; 60–67.

SCHAEDE, R. (1940). Die Knöllchen der adventiven Wasserwurzeln von *Neptunia oleracea* und ihre Bakteriensymbiose. *Planta* (*Berl.*) **31**; 1–21.

SHERWOOD, M. and MASTERSON, C. L. (1974). Importance of using the correct test host in assessing the effectiveness of indigenous populations of *Rhizobium trifolii*. *Ir. J. Agric. Res.* **13**; 101–108

SKRDLETA, V. (1970). Competition for nodule sites between two inoculum strains of *Rhizobium japonicum* as affected by delayed inoculation. *Soil Biol. Biochem.* **2**; 167–171.

SMALL, J. G. C. (1968). Physiological studies on the genus *Trifolium* with special reference to the South African species. IV. Effect of calcium and pH on growth and nodulation. *S. African J. Agric. Sci.* **11**; 441–458.

SMALL, J. G. C. and LEONARD, D. A. (1969). Translocation of C^{14}-labelled photosynthate in nodulated legumes as influenced by nitrate nitrogen. *Am. J. Bot.* **56**; 187–194.

SOLHEIM, B. and RAA, J. (1973). Characterisation of the substances causing deformation of root hairs of *Trifolium repens* when inoculated with *Rhizobium trifolii*. *J. Gen. Microbiol.* **77**; 241–247.

TOMS, G. C. (1971). Phytohaemagglutinins. *In*: "Chemotaxonomy of the Leguminosae" (Eds. J. B. Harborne, D. Boulter and B. L. Turner), pp. 376–462. Academic Press, London and New York.

TRINICK, M. J. (1973), Symbiosis between *Rhizobium* and the non-legume, *Trema aspera*. *Nature, Lond.* **244**; 459–460.

TRUCHET, G. and COULOMB, Ph. (1973). Mise en évidence et évolution dy système phytolysosomal dans les cellules des différentes zones de nodules radiculaires de Pois (*Pisum sativum* L.). Notion d'hétérophagie. *J. Ultrastruct. Res.* **43**; 36–57.

VALERA, C. L. and ALEXANDER, M. (1965a). Reversal of nitrate inhibition of nodulation by indolyl-3-acetic acid. *Nature, Lond.* **206**; 326.

VALERA, C. L. and ALEXANDER, M. (1965b). Nodulation factor for *Rhizobium*-legume symbiosis. *J. Bacteriol.* **89**; 1134–1139.

VELICKY, I. and LA RUE, T. A. (1967). Changes in soybean root culture induced by *Rhizobium japonicum*. *Naturwissenschaften* **54**; 96–97.

VINCENT, J. M. (1954). The root-nodule bacteria of pasture legumes. *Proc. Linnean Soc. N.S. Wales* **79**; IV-XXXII.

VINCENT, J. M. (1974). Root-nodule symbioses with *Rhizobium In*: "Biological Nitrogen Fixation" (Ed. A. Quispel) 265–341. North Holland Publishing Co., Amsterdam.

WARD, H. M. (1887). On the tubercular swellings on the roots of *Vicia faba*. *Phil. Trans. Roy. Soc. (London) Ser. B.* **178**; 539–562.

WEBER, D. F. and MILLER, V. L. (1972). Effect of soil temperature on *Rhizobium japonicum* serogroup distribution in soybean nodules. *Agron. J.* **64**; 796–798.

WILSON, J. K. (1944). Over five hundred reasons for abandoning the cross inoculation groups of the legumes. *Soil Sci.* **58**; 61–69.

WILSON. P. W. (1933). Relation between carbon dioxide and elemental nitrogen assimilation in leguminous plants. *Soil Sci.* **35**; 145–165.

WORONIN, M. (1867). Observations sur certaines excroissances que presentent les racines de l'aune et du lupin des jardins. *Ann. Sci. Nat., Bot., Ser.* 5, **7**; 73–86.

YAO, P. Y. and VINCENT, J. M. (1969). Host specificity in the root hair "curling factor" of *Rhizobium* spp. *Aust. J. Biol. Sci.* **22**; 413–423.

YATAZAWA, M. and YOSHIDA, S. (1965). Role of the cotyledon of soybean seedlings on root nodule formation. *Nippon Dojo-Hirygaku Zasshi* **36**; 263–267.

ZEVENHUIZEN, L. P. T. M. (1973). Methylation analysis of acidic exopolysaccharides of *Rhizobium* and *Agrobacterium*. *Carbohyd. Res.* **26**; 409–419.

Chapter 22

Root Nodules in Non-legumes

J. H. BECKING

Institute for Atomic Sciences in Agriculture, Wageningen, The Netherlands

I. Introduction

A number of dicotyledonous non-leguminous plants of phylogenetically unrelated families and genera (Hutchinson, 1973) possess root nodules with the capacity to fix molecular nitrogen. The fixation of nitrogen is due to the presence of special micro-organisms within the root nodules. In most cases, the endophytes of these non-leguminous plants belong to the Actinomycetales, but in some non-leguminous symbioses described recently the endophyte appears to be a *Rhizobium* species. In the latter type, cross-inoculation of the *Rhizobium* endophyte of the non-legume with some legumes proved to be possible.

The present report surveys the occurrence of nodulated non-leguminous plants within the angiosperms, their geographic distribution, the types of nodulation and symbioses, and the amount of nitrogen fixed by the root nodules. Some biochemical properties related to the nitrogen fixation process, the initiation and development of the root nodules and the growth requirements of the endophytes will be discussed also. Special attention will be paid to the structure and fine structure of the actinomycete endophyte in relation to the origin of nodules and the taxonomy of the organism.

As will be shown in the following paragraphs the nitrogen-fixing capacity of these non-legumes is of considerable economic significance. These non-leguminous plants have a world-wide distribution and occur in all kinds of climates and ecological sites. Some of these plants are woody, but others are shrubs or even herbs. Nearly all of them are active colonizers of bare soil and therefore may contribute considerably to the cover and nitrogen status of these exposed sites.

II. Taxonomic Aspects of Nodule-bearing Non-legumes

Nodule-bearing non-leguminous dicotyledons can be separated into two main groups: those having an actinomycete symbiosis and those with a *Rhizobium* symbiosis. Moreover, there is a miscellaneous group probably related to the second group, which comprises a number of plants unrelated phylogenetically. Root nodulation and nitrogen fixation of the latter group have been claimed only recently by one research group (Farnsworth and Clawson, 1972; Clawson *et al.*, 1972) and the root nodules are described to be inhabited by "*Rhizobium*-like" organisms.

A. Actinomycetes Symbioses

The actinomycete symbioses, undoubtedly the largest and the best-known group comprises 6 plant orders, 7 families and about 14 genera, the latter depending on the genus delimination of this group. The plant taxa of these symbioses are presented in Table I, in which the phylogenetic sequence of orders and families

TABLE I. Non-leguminous nodule-bearing dicotyledons with actinomycete symbioses

Genus	Family	Order
Coriaria	Coriariaceae (23)[a]	Coriariales (5)[a]
Dryas, *Purshia*, *Cercocarpus*	Rosaceae (24)	Rosales (6)
Casuarina	Casuarinaceae (67)	Casuarinales (18)
Myrica, *Comptonia*	Myricaceae (59)	Myricales (14)
Alnus	Betulaceae (61)	Fagales (16)
Elaeagnus, *Hippophaë*, *Shepherdia*	Elaeagnaceae (200)	Rhamnales (44)
Ceanothus, *Discaria*, *Colletia*	Rhamnaceae (201)	Rhamnales (44)

[a] Serial numbers according to Hutchinson's (1973) phylogenetic system of orders and families.

according to Hutchinson (1973) is given. There is, however, a departure from Hutchinson's system in the table since the Casuarinales (18) are placed in front of the Myricales (14) and Fagales (16). This was done because the Casuarinales and Myricales have representatives with the same type of nodulation, i.e. in both types the apex of each nodule lobe gives rise to a normal but negative geotropic root (see p. 536). In addition, in Engler's (1964) system of phylogenetic relationships the Casuarinales are regarded as a primitive order with connections to the Gymnospermae and for this reason they were placed at the beginning of their system.

The genus *Alnus* of the Fagales (16) seems to be a rather isolated genus, although its relation to the Myricales (14) is quite evident according to Hutchinson's (1973) system, the six other genera of the Betulaceae, including *Betula*, *Corylus*, *Carpinus*, etc., are non-nodulated. The relationship between the Elaeagnaceae (200) and

Table II. Genera, well-known species and present distribution of non-leguminous nodule-bearing dicotyledons with actinomycete symbioses

Genus	Number of nodulated species reported	Total number of species[a]	Distribution
Coriaria *C. myrtifolia* *C. japonica* *C. arborea*	12	15	Warm temperate and tropical regions of Northern and Southern hemisphere. Mediterranean region (Europe, Africa), Asia, Japan, New Zealand, Central and South America (Chile).
Dryas *D. drummondii*	3	4	Northern temperate zone to Arctic zone and mountains: Europe, Asia and North America.
Purshia *P. tridentata* *P. glandulosa*	2	2	West coast of North America.
Cercocarpus *C. betuloides* *C. montanus* *C. paucidentatus*	3	20	Western North America, down to Mexico.
Casuarina *C. equisetifolia* (littoral) *C. cunninghamiana* *C. montana* (montane) *C. junghuhniana*	18	45	Tropical and subtropical. Australia, New Caledonia, Indo-Malayan Archipelago, Pacific Isl., India, East coast of Africa, Madagascar. Widely introduced elsewhere.
Myrica *M. gale* *M. faya*	20	35	Temperate, subarctic, subtropical, tropical (montane) Western and Northern Europe.

TABLE II (*continued*)

M. cerifera *M. pensylvanica* *M. javanica* (montane) *M. pilulifera*			Siberia, Asia, South-East Asia, North America, South America (Venezuela), Africa (Cape Province).
Comptonia *C. peregrina* (syn. *M. asplenifolia*)	1	1	East coast of North America
Alnus *A. glutinosa* *A. incana* *A. crispa* *A. japonica* *A. jorullensis*	33	35	Northern temperate zone: Europe, Siberia, North America, nearly to the Arctic Circle, Near East and Central Asia, Japan. Temperate South America: Andes, Argentina, Chile.
Elaeagnus *E. macrophylla* *E. angustifolia*	14	45	Temperate. Northern hemisphere: Asia, Europe, North America. South-East Asia (montane).
Hippophaë *H. rhamnoides* *H. salicifolia*	1	3	Temperate: Europe, Asia, from Himalayas (*H. salicifolia*) to Arctic Circle. Common on coastal sand dunes and on river gravels in the Alps.
Shepherdia *S. canadensis* *S. argentea*	2	3	Temperate to Subarctic. North America.
Ceanothus *C. velutinus* *C. prostratus*	31	55	Temperate: North America. Mainly Pacific coast region to South Mexico.
Discaria *D. toumatou*	1	10	Extra-tropical South America: South Brazil, Uruguay, New Patagonia. Australia (1 sp. *D. pubescens*) and New Zealand (1 sp. *D. toumatou*).
Colletia species unknown[b]	2	17	Temperate and subtropical South America.

[a]mainly according to Willis (1973); [b]Bond (1975).

Rhamnaceae (201), both belonging to the Rhamnales (44) is also quite evident; both families contain 5 genera and have a large number of nodulated species. The relationship in Hutchinson's (1973) system between the Coriariales (5) and Rosales (6) is also evident, because of their subsequent sequence: both families, the Coriariaceae (23) (a single family in the order) and the Rosaceae (24) (one of the 3 families in the order) have a number of nodulated genera and species. In addition, in Hutchinson's system the Rosaceae are placed in front of the other families of the Rosales, indicating its closer relationship to the Coriariales.

Of the 14 non-leguminous plant genera with actinomycete symbioses some well-known species are listed in Table II and a rough sketch of their distribution is given. This table gives also information of the number of species within each genus (according to the author's best estimate) and the number of species so far reported to have root nodulation. It is clear from this table that the largest number of root-nodulated species are *Alnus* (33), *Ceanothus* (31), *Myrica* (20), *Casuarina* (18) *Coriaria* (12), whereas the largest number of species within the genus are in the sequence *Ceanothus* (55), *Casuarina* (45), *Elaeagnus* (45), *Alnus* (35), *Myrica* (35), *Cercocarpus* (20) and *Coriaria* (15).

An extensive list covering all the non-leguminous plant species with actinomycete symbioses so far reported has been presented by Rodriguez-Barrueco (1969), while Uemura (1971) gives a local list of such non-legumes in Japan.

B. Rhizobium and Related Symbioses

Non-leguminous nodule-bearing dicotyledons with *Rhizobium* symbioses or supposed *Rhizobium* symbioses are listed in Table III. The first convincing evidence of a *Rhizobium* symbioses in a non-legume is the recent discovery of Trinick (1973) who found root nodulation in *Trema cannabina* var. *scabra* (not *T. aspera*—personal communication M. J. Trinick) of the Ulmaceae (68) belonging to the Urticales (19). These root nodules were large and pink in colour and showed active nitrogen fixation when assayed with the acetylene reduction test. From the root nodules a "cowpea-type" of *Rhizobium* could be isolated, which after inoculation produced root nodules in *Trema* plants growing in green-house conditions, but also was effective in producing root nodulation in three species of legumes, i.e. *Vigna sinensis*, *Macroptilium* (syn. *Phaseolus*) *atropurpureum* and *M. lathyroides*. It is curious that, in Hutchinson's (1973) scheme of phylogenetic relationships, the Urticales (19) follow the Casuarinales (18) and the latter order is also known to contain many nodulated species. However, it must be stressed that, apart from the fact that root nodules in the Casuarinales are initiated by actinomycete species, they show a completely different nodulation type (see p. 536). The close proximity of both orders may, therefore, be accidental.

Trinick (1973) mentioned that the *Trema* endophyte is the first *Rhizobium* known to effectively nodulate a non-legume. This statement is not true, however, since similar experiments were reported by Sabet (1946) and Mostafa and Mahmoud (1951). The latter authors observed root nodulation in Zygophyllaceae (132) of the Malpighiales (34) in the genera *Zygophyllum*, *Fagonia* and *Tribulus* growing in poor sandy soils of the Egyptian deserts. Isolates from these root nodules were *Rhizobia*-like and could produce root nodulation in some legumes such as *Trifolium alexandrinum* and *Arachis hypogaea*. But this work was criticized by Allen and Allen (1950), who studied the anatomy of the root nodules of *Tribulus cistoides* in a green-house plant raised from imported seed. The latter authors found no evidence of an endophyte within these root nodules and they considered that the nodular structures were starch-storage organs. In view of Trinick's work, it is highly desirable that root nodulation of Zygophyllaceae should be re-investigated and preferably using plants grown in soil from its natural habitats.

Recently, Farnsworth and Clawson (1972) and Clawson *et al.* (1972) enumerated a number of plant species growing on rangeland and forest soils of Northern Utah, U.S.A., whose roots were nodulated and which were found to possess nitrogen-fixing capacity as shown by the acetylene reduction assay. The main species was *Artemisia ludoviciana* of the Asteraceae (Compositae), whose root nodules showed an ethylene production surpassing that of some legumes growing in the same soil. These authors found such type of root nodulation in five unrelated orders of dicotyledonous plants. According to Hutchinson's serial numbers of phylogenetic orders which indicate mutual relationships, these orders are: Violales (26), Cactales (31), Umbellales (72), Asterales (76) and Boraginales (81). Nothing definite is known about the nature of the endophyte inhabiting these root nodules, except that it is stated that a "*Rhizobium*-like" organism could be isolated from the root nodules and brought in pure culture. The isolate, however, failed to nodulate some of these plants grown under laboratory conditions. Since all evidence of root nodulation and nitrogen fixation of these plants is from one research group, it is desirable that these claims should be confirmed by other workers in this field.

III. Evidence of Nitrogen Fixation

Nitrogen fixation by the root-nodule system is usually assessed in three different ways, i.e. in growth experiments, in $^{15}N_2$ fixation tests, or with the acetylene reduction test.

A. Growth Experiments

Nodulated non-leguminous plants are grown from their seedling stage in nitrogen-poor soil or in water or sand cultures lacking a nitrogen source. Appropriate blanks of non-nodulated (i.e. non-inoculated) plants are included in the

TABLE III. Non-leguminous nodule-bearing dicotyledons with *Rhizobium* or supposed *Rhizobium* symbioses

Genus and Species	Family	Order	Legume cross-inoculation	Reference
Trema *T. cannabina* var. *scabra*	Ulmaceae (68)[a]	Urticales (19)[a]	*Vigna sinensis* *Macroptilium (Phaseolus) atropurpureum* *M. (Ph.) lathyroides* *Vigna marina*	Trinick (1973) (Original classification as *Trema aspera* was wrong, *T. aspera* is non-nodulated.) Trinick (1975)
Zygophyllum *Z. album* *Z. coccineum* *Z. decumbens* *Z. simplex*	Zygophyllaceae (132)	Malpighiales (34)	*Trifolium alexandrinum* *Arachis hypogaea*	Sabet (1946), Mostafa and Mahmoud (1951)
Fagonia *F. arabica*				
Tribulus *T. alatus*				

(continued)

Artemisia *A. ludoviciana* *A. michauxiana*	Asteraceae (320) (Compositae)	Asterales (76)	Unknown, only stated '*Rhizobium*-like' organism.	Farnsworth and Clawson (1972) Clawson *et al.* (1972)
Chrysothamnus *Ch. viscidiflorus*				
Lomatuim *L. triternatum*	Apiaceae (311) (Umbelliferae)	Umbellales (72)		
Mertensia *M. brevistyla*	Boraginaceae (338)	Boraginales (81)		
Viola *V. praemorsa*	Violaceae (99)	Violales (26)	—	
Opuntia *O. fragilis*	Cactaceae (113)	Cactales (31)		

[a] Serial numbers according to Hutchinson's (1973) phylogenetic system of orders and families.

series and their N-content is subtracted from that found in the nodulated plants. It is usually advisable to have a series of non-nodulated plants supplied with a combined nitrogen source for comparison of the yield with the nodulated plants growing at the expense of molecular nitrogen (i.e. fixed N). At the end of the experiment, or at certain regular intervals during its course, the nodulated plants are harvested and their dry weight and N-content determined. As an example of such an approach the dry matter production and total nitrogen increase of *Alnus glutinosa* plants grown in nitrogen-free nutrient solution is given in Fig. 1. In 48 weeks (one growing season), the alder plants accumulated

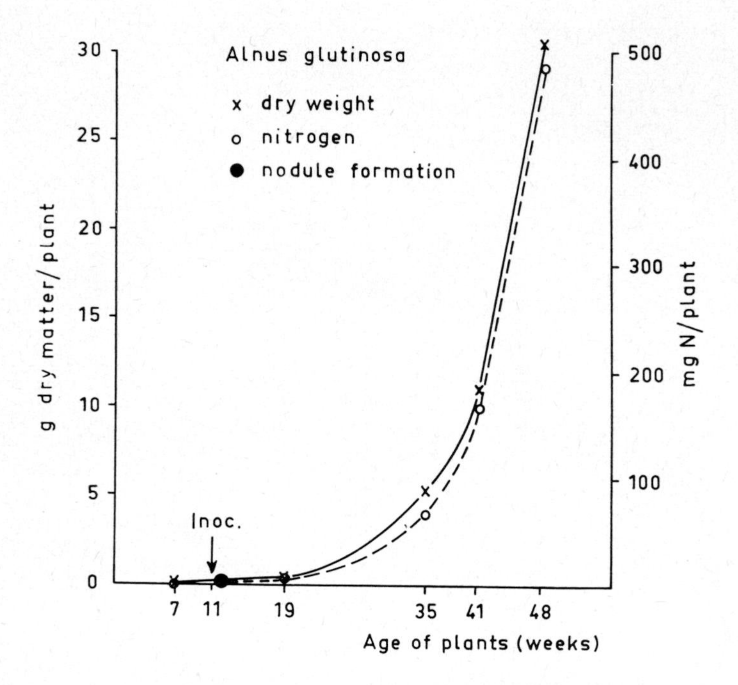

FIG. 1. Dry matter and nitrogen increase of *Alnus glutinosa* plants grown in nitrogen-free nutrient solution. At an age of 11 weeks the plants were inoculated and at point (●) a root nodule appeared.

30 g dry matter representing 500 mg of N per plant. As shown in Fig. 2, half of the nitrogen was present in the leaves indicating that leaf fall in alder in the winter season can contribute considerably to the N-status and the cycling of nitrogen in soil. We have measured similar amounts of nitrogen fixation in other non-leguminous plants (*Hippophaë rhamnoides*, *Myrica gale* and *Casuarina* species). Russell and Evans (1966) found that *Ceanothus velutinus* growing in a nitrogen-free medium fixed 286 mg N per plant in 18 weeks; by extrapolation this would give

about 760 mg N fixed per plant in 48 weeks. This value is close to the value obtained by us for alder.

One of the advantages of growth experiments is that they measure overall effects over rather long growing periods and thus eliminate the variability in

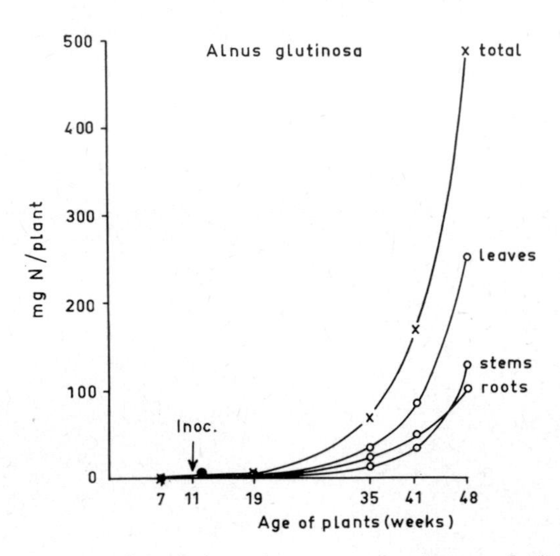

FIG. 2. Nitrogen contents of total plants, leaves, stems and roots of *Alnus glutinosa* plants grown in nitrogen-free nutrient solution. At an age of 11 weeks the plants were inoculated and at point (●) a root nodule appeared.

activity of tissues likely to occur in short-termed experiments. Moreover, they require rather simple chemical analyses and no expensive and/or complicated equipment.

B. $^{15}N_2$-fixation Tests

Excised or intact root nodules or parts of nodules containing the endophyte are exposed to an atmosphere containing ^{15}N labelled N_2. Analyses are made of the ^{15}N enrichment of the total nitrogen in the nodular tissue or in other parts of the plant. At low enrichment values an increased sensitivity of the determination is obtained by extracting the tissue with acid, e.g. 3 N HCl, which yields only the acid soluble N-compounds.

The ^{15}N enrichment can be measured by mass spectrometry requiring an expensive apparatus or by the recently developed, less expensive, emission spectrometry for ^{15}N (Faust, 1960, 1965, 1967; Cook *et al.*, 1967; Leicknam *et al.*, 1968; Goleb and Middelboe, 1968; Perschke and Proksch, 1971). In non-legumes this method has been used mainly to demonstrate qualitatively the presence or absence of nitrogen fixation, but this method can also be used for quantitative estimation of the amount of nitrogen fixed by the system.

Figure 3 shows the increase of ^{15}N-content in excised root nodules of alder (*Alnus glutinosa*) and in excised nodulated roots of bean (*Phaseolus vulgaris*) exposed to an atmosphere containing labelled $^{15}N_2$. It is evident that a very young root nodule of *Alnus* does not fix nitrogen, but that a mature root nodule

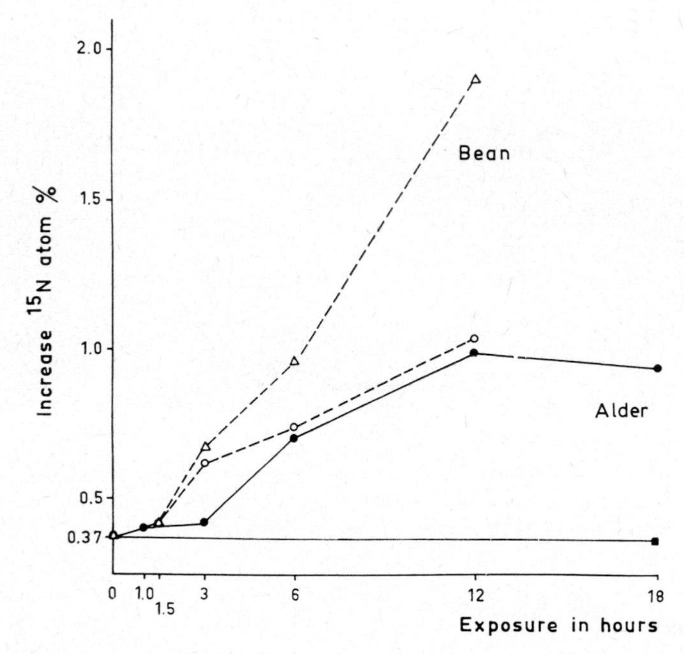

FIG. 3. Increase in amount of ^{15}N found in excised root nodules of alder (*Alnus glutinosa*) and excised nodule-bearing roots of bean (*Phaseolus vulgaris*) exposed to an atmosphere containing 19 % vol. N_2 (labelled with 35·0 atom % ^{15}N), 21% vol. O_2, 3% vol. CO_2, and 57% vol. Ar. Bean: root △ ; root nodules ○; Alder: mature root nodules ●; very young root nodule ■.

fixed nitrogen for 12 hours after excision. Such prolonged nitrogenase activity after nodule excision is unknown for leguminous root nodules and is probably due to the woody structure and the presence of reserve substances in the non-leguminous root nodule.

C. Acetylene Reduction Tests

Nitrogenase activity of the tissue can also be measured with the acetylene reduction assay. This method has been independently developed by Dilworth (1966) and by Schöllhorn and Burris (1966, 1967). All nitrogen-fixing systems so far examined reduce acetylene (C_2H_2) to ethylene (C_2H_4). The nitrogenase system apparently functions in a similar fashion for each substrate reduced. The assay is based on the similarity in the rate of electron activation and transfer to either N_2 or C_2H_2 as substrate assuming that:

$$N_2 + 6\,H^+ \xrightarrow{6e} 2\,NH_3 \qquad \text{and}$$

$$C_2H_2 + 2\,H^+ \xrightarrow{2e} C_2H_4 \qquad \text{then}$$

theoretically a ratio of 1 N_2:3 C_2H_2 or 1 N:1·5 C_2H_2 can be expected. Indeed, in experiments with blue-green algae the theoretical value has been closely approached (Stewart *et al.*, 1968). For non-legumes Hardy *et al.* (1973) gave as average conversion factor C_2H_2/N_2 a value of 2·4.

Acetylene and ethylene can be measured readily by gas chromatography, with an increase in sensitivity over ^{15}N methods of some 10^3–10^4-fold. Compared to a mass spectrometer and the recently developed emission spectrometrical method for ^{15}N determination, a gas chromatograph is a rather inexpensive instrument and easy to handle. Moreover, the acetylene gas is inexpensive and the acetylene reduction test can be easily applied under field conditions. For the latter reason it is obviously the superior method for field work.

Table IV shows results from some field experiments with excised root nodules of some non-leguminous nitrogen-fixing plants. The nitrogenase activity of the *Casuarina rumphiana* root nodules was measured in the field in the Botanical Gardens of Bogor, Java, Indonesia. The activity found (0·8–0·9 nmoles C_2H_4 per min per mg dry weight of nodule) is one of the highest ever observed in a non-legume. Usually, the activity of the root nodules of non-leguminous plants is a factor about 10 times lower. In acetylene reduction tests care should be taken that the acetylene concentration (in air) applied to the tissue produces maximal nitrogenase activity. Experiments with root nodules of *Alnus glutinosa* of the same age growing in nitrogen-free nutrient solution indicated that nitrogenase activity with 20% vol. acetylene was higher than with 10 or 30% vol. acetylene (Fig. 4). Obviously with 10% vol. acetylene in the atmosphere no maximal acetylene reduction was obtained and 30% vol. acetylene was clearly inhibitory.

IV. Nitrogen Increments in the Field

Nitrogen-fixation figures, in particular those obtained by the acetylene reduction test, have been used to estimate the nitrogen increments in the field per square unit of area and in unit of time. For such a quantitative evaluation, it is necessary to know in addition to the nitrogen-fixing capacity of the root nodules, their number and the nodule mass, i.e. the quantity of nodule material per plant, the number of plants (trees) per square unit of area and the diurnal fluctuations in

TABLE IV. Acetylene reduction tests with some non-leguminous plants.

Plant species	nmoles C_2H_4/ min/mg N nodules	nmoles C_2H_4/ min/mg fresh wt nodules	nmoles C_2H_4/ min/mg dry wt nodules	nmoles N min/mg N nodules[a]
Alnus glutinosa	12·3–20·7	0·074–0·124	0·27–0·45	8·2–13·8
Casuarina rumphiana	58·1–67·3	0·22–0·25	0·79–0·91	38·8–44·8
Myrica cerifera	1. 18·0–30·3	1. 0·079–0·13[b]	1. 0·29–0·48[b]	1. 12·0–20·2
(Silver and Mague, 1970)	2. 4·1–11·7			2. 2·8–7·8
Myrica javanica	10·1–10·5	0·041–0·043	0·150–0·156	6·8–7·0

[a] Assuming 0·3 mole N fixed per mole of C_2H_4 formed. [b] Converted by assuming: N content of fresh wt = 0·44% and fresh wt/dry wt = 3·6.

nitrogen-fixing capacity of the root nodules. When these data are available, it is possible to calculate the nitrogen fixation in kg N per ha per year. Some values obtained in this way with non-leguminous plants possessing an actinomycete symbiosis are presented in Table V.

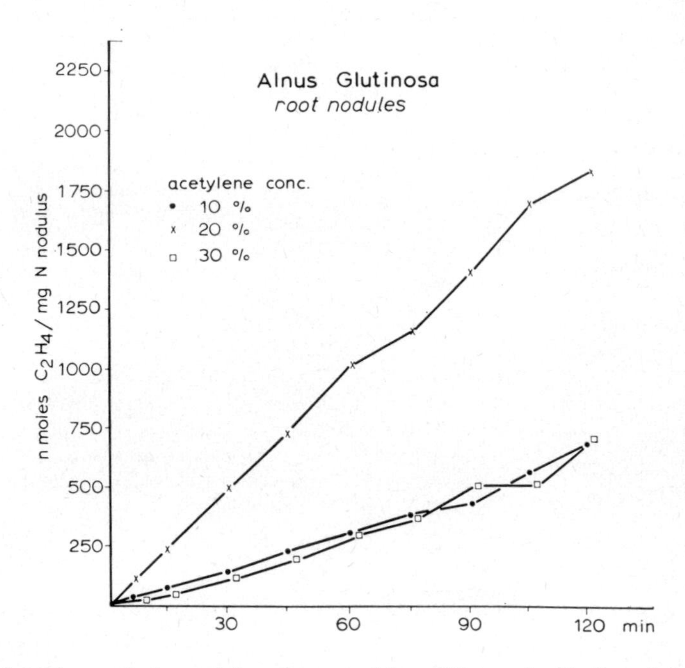

FIG. 4. Acetylene-reducing activity of root nodules of 17-month-old *Alnus glutinosa* plants grown on nitrogen-free nutrient solution. The excised root nodules were incubated at 20°C in air with 10, 20, or 30% vol. acetylene.

V. Physiological Aspects of Nitrogen Fixation

A. Mineral Requirements

The mineral nutrients for non-leguminous plants are broadly the same as for other plants. Only those minerals whose function is related to the nitrogen fixation process will be discussed here.

1. *Molybdenum*

All nitrogen-fixing systems, including the symbiotic ones, have a requirement for molybdenum. Although Mo is essential also for some other biochemical transformations, e.g. nitrate reduction, higher amounts of molybdenum are required in nitrogen fixation than for other processes (Becking, 1962).

TABLE V. Field measurements of N_2 fixation by some non-legumes

Non-legume	N_2 fixation kg N/ha/year	Reference
Casuarina equisetifolia	229	Dommergues, 1963
Myrica cerifera	3·4	Silver and Mague, 1970
M. rubra (3–15 years old)	15–25[a]	Uemura, 1971
Alnus crispa (50 years old)	61·5	Crocker and Major, 1955
A. crispa	40·0	Crocker and Dickson, 1957
A. crispa (5 years old)	157	Lawrence, 1953, 1958
A. glutinosa (0–7 years old) (1 plant/m²)	28	Virtanen, 1957
(5 plants/m²)	100	Virtanen, 1962
A. glutinosa (12 years old)	26–48	Delver and Post, 1968
A. glutinosa	56	Akkermans, 1971
A. incana	40	Ovington, 1956
A. rubra	139	Tarrant *et al.*, 1969; Tarrant and Trappe 1971
A. rubra	up to 300	Zavitkowski and Newton, 1968; 1971
A. rugosa	193	Daly, 1966
A. rugosa (natural stand)	85	Voigt and Steucek, 1969
Hippophaë rhamnoides (0–3 years old)	27	Stewart and Pearson, 1967
(13–16 years old)	179	
H. rhamnoides (1–2 years old)	2	Akkermans, 1971
(10–15 years old)	15	
Ceanothus spp. (natural shrub community)	60	Delwiche *et al.*, 1965
Dryas drummondii	18–36	Lawrence *et al.*, 1967
D. drummondii and some *Shepherdia canadensis*	61·5	Lawrence and Hulbert, 1950

[a] Data from comparison of *Myrica*-pine stands and pure pine stands; N_2 fixation determined by substracting total amount N pine stand from total amount N *Myrica*-pine stand.

A molybdenum requirement for the growth of *Alnus glutinosa* plants growing on molecular nitrogen was found by Becking (1961a, b). Nodulated alder plants growing on nitrogen-poor, molybdenum-deficient soil showed considerably less growth than plants growing in the same soil supplied with molybdenum (Fig. 5). As is obvious from the dry weight yields and the nitrogen content, the plant supplied with Mo showed an increase in dry weight of 3·2 times and an

FIG. 5. Six-month-old *Alnus glutinosa* plants growing in pots with molybdenum-deficient soil. All plants were inoculated and nodulated. *Left*, two plants without added molybdenum; *right*, two plants with 150 μg Mo as $Na_2MoO_4 \cdot 2H_2O$ per pot. Note the pale colour of the leaves of the molybdenum-deficient plants. $\times 0{\cdot}14$.

TABLE VI. Effect of molybdenum on yield and nitrogen fixation by alder plants grown in molybdenum-deficient soil.

Molybdenum[a] addition (μg per pot)	Mean shoot length (cm)	Fresh weight of shoots (g per pot)	Dry matter of shoots (g per pot)	Total nitrogen of shoots (mg per pot)	Per cent nitrogen increase
0	18·2	11·2 ± 2·0	2·38 ± 0·64	46·4 ± 8·8	—
150	33·7	34·4 ± 0·1	7·58 ± 0·25	218·3 ± 8·6	370

[a] Applied as $Na_2MoO_4 \cdot 2H_2O$.

increase in nitrogen content of 370% (Table VI). Molybdenum analyses of the tissue parts showed that most of the molybdenum present was in the root nodules (Table VII). Also alder plants growing in the field showed about 6 times more Mo in root nodules than in roots bearing them. It was remarkable that the seeds of

TABLE VII. Molybdenum contents of leaves, stems, roots and nodules of *Alnus glutinosa* plants grown in Mo-deficient soil in the absence or presence of added molybdenum.

Plant tissue	Molybdenum content in ppm of dry matter	
	I[a]	II[b]
Leaves	0·01	0·27
Stems	0·14	1·89
Roots	0·24	2·62
Nodules	2·00[c]	17·30

The molybdenum-deficient soil (pH = 5·2) from Groningen contained approx. 0·006 ppm available Mo (*Aspergillus niger* method, Hewitt and Hallas, 1951). [a]I.: no added Mo. [b]II.: 150 μg Mo (as $Na_2MoO_4 \cdot 2H_2O$) supplied per pot containing 500 g of soil. [c]Approx. value since the sample was very small.

Alnus from field trees contained about 10 times more Mo than leaves, stems and roots (Table VIII).

A molybdenum requirement for growth of *Alnus* on atmospheric nitrogen was also shown by Hewitt and Bond (1961) in water culture experiments. In addition, using the same method a Mo requirement for the growth of *Casuarina cunninghamiana* and *Myrica gale* was established (Hewitt and Bond, 1961; Bond and Hewitt, 1961). In the water culture experiments, the deficiency symptoms

TABLE VIII. Nitrogen and molybdenum contents of different parts of a 12 metre high *Alnus glutinosa* tree growing at natural site[a] in an Alder grove at Wageningen, The Netherlands.

Part of plant	Nitrogen % of dry matter	Mo ppm of dry matter[b]
Leaves (young)	3·84	0·25
Stems and branches	0·92 (0·73–1·11)	0·18
Roots	1·76	0·28
Nodules	2·37	1·71
Seeds	3·69	2·10

[a] The soil was a sandy loam (pH = 5·3) containing approx. 0·2 ppm available Mo (*Aspergillus niger* method, Hewitt and Hallas, 1951). [b] Determined chemically with potassium rhodanide, after reduction of any hexavalent molybdenum by stannous chloride.

were again reduced growth and N-content, chlorosis and leaf scorching of the older leaves. In water culture there was neither an increased number nor a reduction in size of the root nodules. By contrast, in the molybdenum-deficient plants grown in soil there were very large numbers of small root nodules distributed over the whole root system (Fig. 6). Similar symptoms are known for legumes infected with an ineffective *Rhizobium* strain or for effective *Rhizobium* strains on molybdenum-deficient legumes.

2. *Cobalt*

The importance of cobalt for the growth of nodulated legumes is well known (Ahmed and Evans, 1960, 1961; Delwiche *et al.*, 1961). A similar function might be anticipated in non-legumes and, indeed, a cobalt requirement for nitrogen fixation was demonstrated in *Alnus glutinosa*, *Casuarina cunninghamiana* and *Myrica gale* by Bond and Hewitt (1962) and Hewitt and Bond (1966). Deficiency in Co resulted in symptoms of severe nitrogen deficiency. No cobalt requirement was observed in non-nodulated plants supplied with combined nitrogen.

Root nodules from plants supplied with cobalt were relatively rich in vitamin B_{12} analogues. Bond *et al.* (1965) found, from the *Euglena gracilis* assay, that the vitamin B_{12} content of various non-legume root nodules from plants supplied with cobalt was of the same order as that of legume nodules. Root nodules of *Alnus glutinosa*, *Casuarina cunninghamiana*, *Myrica gale*, and *Hippophaë rhamnoides* contained about 130–300 ng vitamin B_{12} per g of fresh tissue, although some values were lower, i.e. 55–63 ng vitamin B_{12} per g of fresh tissue. The results showed considerable variation with the extraction procedure followed. The vitamin B_{12} content of *Coriaria myrtifolia* root nodules was much lower, i.e.

28–53 ng per g of fresh tissue. Roots of these non-leguminous plants, tested simultaneously, contained negligible amounts of vitamin B_{12}.

Earlier, Kliewer and Evans (1962) showed the presence of vitamin B_{12} coenzyme in the root nodules of legumes and of *Alnus rubra* (=*Alnus oregona*).

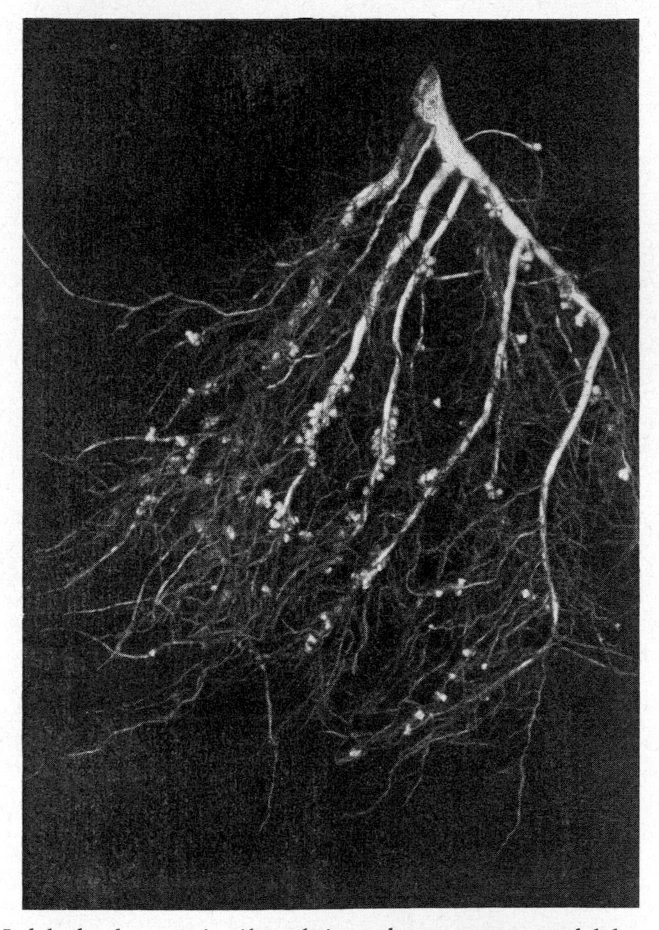

FIG. 6. Nodule development in *Alnus glutinosa* plants grown on molybdenum-deficient soil. A large number of small nodules were formed, which were uniformly distributed over the entire root system. ×0·9.

However, the published B_{12} coenzyme values were incorrect; unfortunately, a dilution factor was omitted (pers. comm. Kliewer and Evans, 1963). The vitamin B_{12} coenzyme content of the root nodules of a two-year-old *Alnus rubra* plant should be 62 nmoles, not 174 nmoles, per g of fresh tissue.

In more recent experiments Russell *et al.* (1968) observed that cobalt addition to the nutrient solution increased the dry matter production and N-content of *Alnus rubra* plants. Also the vitamin B_{12} content of the root nodules was significantly increased by a comparatively small cobalt (0·05 ppm) addition. Increasing amounts of Ni, Fe, and Mn in the nutrient solution interfered, however, with the incorporation of Co into vitamin B_{12} compounds.

3. *Copper*

Copper was also observed to stimulate the growth of *Alnus glutinosa* plants on a nitrogen-free mineral medium (Bond and Hewitt, 1967), but the differences between plus and minus copper plants failed to be statistically significant for *Casuarina cunninghamiana* growing under the same experimental conditions. The copper level, which stimulated nitrogen fixation in alder was 0·02 ppm of copper as $Cu\ (SO_4)_2$.

B. Soil Factors

Many non-leguminous nodulated plants are found in very poor soils. In general, they play an active role in colonizing of virgin soils developed after rock weathering or soil exposed in recently deglaciated areas (e.g. *Alnus crispa* and *Dryas drummondii*) or they occupy a prominent role in covering bare soil after land slides, soil erosion or volcanic activity (*Comptonia peregrina, Myrica javanica*). Plant succession in recently deglaciated areas at Glacier Bay in Alaska showed the succession of *Dryas drummondii*, *Shepherdia canadensis* and *Alnus crispa*; after a certain N-status of the soil had been built up, *Populus trichocarpa* appeared as has been described by Crocker and Major (1955) and Lawrence (1953, 1958). The same situation holds for the tropics, where *Myrica javanica* colonizes soil of recently erupted lava streams or forest destroyed by volcanic activity (Van der Pijl, 1938).

Soil pH is also not limiting for the occurrence of nodulated non-leguminous plants. In the temperate region *Myrica gale* and *Alnus glutinosa* occur in peat bogs often under appreciable acidic soil conditions, whereas *Hippophaë rhamnoides* inhabiting the sand dunes near the coast can grow under considerable alkaline conditions. In the tropics *Casuarina equisetifolia* grows preferentially on coastal sands temporarily flooded by the sea. Experiments with nutrient solutions have also confirmed the wide ecological pH range of nodulated non-legumes. Bond *et al.* (1954) noted that about half of the inoculated plants of *Myrica gale* and *Alnus glutinosa* nodulated at pH = 4·2, whereas *Hippophaë rhamnoides* and red clover did not. However, growth at this pH of non-nodulated plants supplied with combined nitrogen was satisfactory for all the species and it was adequate for *Myrica* and *Alnus* even at pH = 3·3. The effect of pH is thus primarily an effect on nodule initiation in *Hippophaë*, a situation not unlike that found in certain legumes. Once nodulation had occurred, the effects of pH on subsequent plant growth showed the following optima during the first season's growth in N-free nutrient medium: *Alnus glutinosa*, pH 4·2–5·4 (Bond *et al.*, 1954); *Myrica gale*, pH 5·4 (Bond, 1951); *Hippophaë rhamnoides* pH 6·3 (Bond *et al.*, 1954). These pH optima correlate fairly well with the acidity of the typical habitats of these species.

C. Effect of Combined Nitrogen

1. *Ammonium-N Sources*

The presence of combined nitrogen increased the weight of the root nodules per plant of *Alnus glutinosa* and *Myrica gale* at the lower concentrations of nitrogen (10 mg NH_4–N per litre). Nodule development was continually depressed as the levels of combined nitrogen increased to 50 and 100 mg NH_4–N per litre (MacConnell and Bond, 1957).

Nodulated plants of *Alnus glutinosa* and *Myrica gale* were cultivated by Stewart and Bond (1961) in the presence of ^{15}N-labelled ammonium ion (10, 50 and 100 mg NH_4–N per litre) in water culture. The root nodules fixed atmospheric nitrogen despite the presence of ammonium in the root medium, though fixation per unit weight of nodule tissue formed was slightly less than in nitrogen-free nutrient solution. In *Alnus glutinosa*, but not in *Myrica gale*, fixation per plant was enhanced in the presence of a low ammonium level of 10 mg NH_4–N per litre because of the greater nodule development. At the highest ammonium level of 100 mg NH_4–N per litre, which exceeded the plant's requirements, fixation per plant was similar to that in the nitrogen-free solution, but represented only 24 to 45% of the total nitrogen accumulation by the plants. It was concluded that, under field conditions, some fixation of atmospheric nitrogen will always occur when nodules are present.

Growth of *Myrica gale* and *Casuarina cunninghamiana* in the presence of combined nitrogen at rates of 10, 50, 100, 150 mg NH_4–N per litre was significantly better than that in nitrogen-free medium. High levels of combined nitrogen caused a depression in the number of nodules formed and in nodule dry weight as a percentage of total plant weight. Nodule dry weight was appreciable at the highest nitrogen levels of 100 and 150 mg NH_4–N per litre, and an increase in dry weight per nodule occurred in the presence of combined nitrogen (Stewart, 1963).

2. *Nitrate-N Sources*

Pizelle (1965) studied the effect of NO_3–N on the nodulation and nitrogen fixation of young *Alnus glutinosa* plants, whereby the root system of each plant was divided over two compartments with and without combined nitrogen. The results confirmed earlier experiments of MacConnell and Bond (1957) and Stewart and Bond (1961) that combined nitrogen depressed nodule formation. However, this was not a local effect, since the non-nodulated part of the root system supplied with combined nitrogen suppressed nodule formation in the other compartment without combined nitrogen. The plants receiving combined nitrogen showed better growth than those growing on nitrogen-free nutrient solution. Later, in a more detailed study Pizelle (1966) found virtually the same results. He attributed the outcome to two opposite reactions: first, a stimulative effect of the total nitrogen status of the plant on nodule development; and secondly, a local depressive effect of combined nitrogen on nodule formation in roots in contact with a nutrient solution containing combined nitrogen.

D. Nitrogen Compounds and Transfer of Sugar and Nitrogen Nodule-plant

Leaf *et al.* (1958) determined the free amino acids and related soluble N-compounds in *Alnus glutinosa* root nodules with chromatographic methods. Citrulline was the predominant amino acid present accompanied by smaller amounts of aspartic, glutamic and γ-amino-butyric acids, arginine and other constituents. When molecular nitrogen labelled with ^{15}N was applied, the highest atom per cent of ^{15}N was found in glutamic acid, followed by citrulline or aspartic acid. Ammonia contained a smaller portion of ^{15}N than these compounds and arginine showed only a very small enrichment. When citrulline was degraded to ammonia and ornithine, the ammonia liberated was richer in ^{15}N than even glutamic acid.

In *Myrica gale* these authors (Leaf *et al.*, 1959) found asparagine the predominant amino acid. Glutamine and various other amino acids were present in smaller amounts. The highest enrichment with ^{15}N was in the amide-N of glutamine and the next highest was in the amide-N of asparagine and amino-N of glutamic acid. These data support the view that the fixed nitrogen passes through the form of ammonia before entry into organic combination.

Wheeler and Bond (1970) studied the free amino acids in the root nodules of nine species of non-leguminous plants belonging to seven genera. In species of *Myrica* (*M. gale*, *M. cerifera*, *M. pilulifera*, *M. cordifolia*) and also in *Hippophaë rhamnoides*, *Elaeagnus angustifolia*, *Ceanothus velutinus* var. *laevigatus* and *Casuarina cunninghamiana*, the pattern of amino acids resembled that previously reported in the nodules of *Myrica gale* (Leaf *et al.*, 1959). Asparagine was the predominant amino acid in terms of nitrogen content but substantial amounts of glutamine were often present. In *Alnus inokumai* citrulline was prominent, as previously shown for the nodules of *Alnus glutinosa* (Leaf *et al.*, 1958), though now amides were present in substantial quantities as well. In nodules of *Coriaria myrtifolia* glutamic acid, glutamine and arginine were the chief amino acids.

Bond (1956) showed that in *Alnus glutinosa* plants a substantial enrichment in ^{15}N is detectable in the shoot within 6 hours from the beginning of the exposure of the root nodules to ^{15}N-labelled N_2 gas. By ringing the nodulated plants over a short zone at shoot base, thereby removing the tissues external to the xylem, the upward movement of the fixed nitrogen was not interfered with. He concluded that the fixed nitrogen, which is probably in organic form, can be translocated in the xylem by the transpiration stream, and that this is most likely its normal route.

Stewart (1962) made a quantitative study of the fixation of nitrogen and its transfer from the nodule to the remainder of the plant in *Alnus glutinosa* plants during their first season. Fixation per plant reached a maximum in late August (30·5 mg N per plant in 12 days), but fell rapidly with the onset of autumn. Fixation per unit dry weight of nodule tissue was greatest in young nodules and of the same order as that of nodulated legumes. Throughout the growth season there was a steady transfer from the nodules of some 90% of the nitrogen fixed.

Wheeler (1969) found that maximal rates of nitrogen fixation occurred about mid-day in the root nodules of first-year plants of *Alnus glutinosa* and *Myrica gale* growing under natural illumination, but at constant temperature. The nitrogen content of the sap exuding from the stumps of decapitated plants and the level of soluble nitrogen in the nodules were also highest around mid-day, the same being true of the rate of respiration in detached nodules. The hypothesis that the mid-day period must have the maximum ingress of photosynthates into the nodules proved not to be valid, since the analyses of "soluble" and "reserve" carbohydrate in the nodules showed no accumulation during the mid-day period.

In later experiments Wheeler (1971) followed the translocation of ^{14}C-labelled photosynthates to the nodules of first-year alder plants growing under the same conditions as mentioned previously. It was shown that there is a maximum influx of new photosynthates at the time of the mid-day peak in fixation. Analysis of fluctuations in the levels of the main sugars present in the nodules at different times of day, and a study of the effects of interrupting supplies of photosynthates to the root system by stem ringing, suggested that a substantial part of the nodule carbohydrate is unavailable to support N_2-fixation and that maximal rates of fixation can be attained only when new photosynthate enters the nodules in quantity. No evidence was obtained for an autonomous element in the mid-day maximum in fixation when plants were kept in continuous darkness; a fall in acetylene reduction to an insignificant level 24 hours after darkening was accompanied by a fall in nodule sugars, notably in the amount of sucrose present. The rate of acetylene reduction by the root nodules of *Alnus glutinosa* reached a maximum at about 25°C, and at this temperature, was some six times faster than at 15°C.

E. Effect of Gases on Nitrogen Fixation

Various gases are known to influence, or inhibit, nitrogen fixation in legumes and free-living nitrogen-fixing organisms (Wilson, 1958; Bergersen *et al.*, 1961; Mortenson, 1962). Some of these gases have been tested on the symbiotic system of non-leguminous nitrogen fixation.

1. *Oxygen*

MacConnell (1959) applied different levels of oxygen after inoculation of *Alnus glutinosa* plants in water culture and observed that the number of nodules was progressively reduced as the oxygen tension was lowered from the normal

21% vol. Nodulated alder plants were considerably more sensitive to oxygen level than non-nodulated plants. Therefore, oxygen supply seems to be of special importance in the development and function of alder root nodules. Later, Bond (1961) showed with the ^{15}N-method that nitrogen fixation was small at low levels of oxygen, but rose with increasing oxygen supply and attained a maximum at 12% vol. O_2 in *Hippophaë rhamnoides* and 20 to 25% vol. O_2 in *Casuarina cunninghamiana* and *Myrica gale*. Results with *Alnus glutinosa* closely resembled those with *Hippophaë rhamnoides*; the optimum fixation occurred at 12% vol. O_2. Application of 40% vol. O_2 virtually eliminated nitrogen fixation in the root nodules of *Alnus glutinosa* but not in *Casuarina cunninghamiana, Myrica gale* and *Hippophaë rhamnoides* (Bond, 1959; 1961).

2. *Hydrogen*

Bond (1960) found an inhibition of nitrogen fixation by hydrogen in *Alnus glutinosa, Casuarina cunninghamiana*, and *Myrica gale* which was similar to that in peas and soybeans. The application of 60% vol. H_2 to the gas mixture resulted in an inhibition of fixation of 77 to 86% in the legumes and 84 to 88% in the non-legumes. The exposure of *Alnus glutinosa* nodules to 20 and 60% vol. H_2 showed an inhibition of 52 and 80% in the nitrogen fixation, respectively.

3. *Carbon Monoxide*

The effect of carbon monoxide on nitrogen fixation by non-legumes was studied by Bond (1960) with the ^{15}N-method in detached root nodules of *Alnus glutinosa* and *Myrica gale*; parallel tests with peas were included. The inhibition of nitrogen fixation on application of 0·1 and 1·0% vol. CO in the gas mixture was 31 to 66% and 98% respectively. Carbon monoxide also strongly inhibited N_2-fixation in peas. The inhibition of nitrogen fixation by carbon monoxide was about the same in non-legumes as in legumes. As will be pointed out later, most workers could not detect haemoglobin-like compounds in non-leguminous root nodules. If this is true, the mechanism of action of carbon monoxide in root nodules of non-legumes must be different from that in legumes.

VI. Biochemical Properties of Root Nodules

A. Presence of Haemoglobin

It is well known that haemoglobin plays a prominent role in legumes in the symbiotic process of nitrogen fixation (Bergersen, 1960). Smith (1949) could not demonstrate haemoglobin in *Alnus glutinosa* root nodules. Bond reported that haemoglobin was not detected spectroscopically by Smith (*cf* Bond, 1951, *loc. cit.*, p. 457, footnote) in root nodules of *Myrica gale* and that the intensive red colour of the bog-myrtle root nodule is due to anthocyanin pigments as also observed in *Alnus* and *Casuarina* (Bond, 1958).

Egle and Munding (1951) studied haemin compounds in root nodules of *Alnus glutinosa*, *Hippophaë rhamnoides*, *Myrica gale*, *Encephalartos altensteinii*, *Podocarpus neriifolius*, and *Lupinus angustifolius*. Using a spectroscope method, absorption bands with maxima at 557·2 and 526·2 nm could be observed in root-nodule extracts treated with sodium dithionite containing traces of pyridine. These absorption bands were attributed to the presence of haemoglobin, observed as "pyridine haemochromogen". The spectroscopic evidence was positive for the root nodules of *Lupinus*, *Alnus*, *Hippophaë* and *Myrica*, but no absorption could be observed in the root-nodule extracts of *Encephalartos* and *Podocarpus*, nor in the extracts of the root-cortex tissue of all the species. It was concluded that *Alnus*, *Hippophaë* and *Myrica* root nodules contained haemoglobin. Estimates indicated, however, that its quantity on fresh weight basis was only about half of that found in the *Lupinus* root nodules tested simultaneously.

In a search for haemoglobin in nodular tissue from *Alnus glutinosa*, *Myrica gale*, *Hippophaë rhamnoides* and *Casuarina cunninghamiana*, Davenport (1960) observed by spectroscopic examination, especially in *Casuarina*, an absorption band at 562 nm due to deoxygenated haemoglobin. This haemoglobin was not extractable by aqueous solvents in contrast to haemoglobin from leguminous nodules simultaneously tested. Apparently it was firmly bound to cell debris in non-legumes and could not be removed from this. *Alnus* and *Myrica* root nodules also contained a bound form of haemoglobin, similar to that in *Casuarina*.

Haemoglobin was not detected by us in root nodules of several *Alnus* species (Becking, unpublished observations). The red colour of these nodules and root tips, especially in nitrogen-deficient plants, was caused by pigments of the anthocyanin type as already observed by Bond (1958). Also Moore (1964) did not detect haemoglobin with a microspectroscope in nodule slices of *Alnus rugosa*, *Elaeagnus commutata*, *Shepherdia canadensis*, and *Hippophaë rhamnoides*.

B. Presence of Poly-β-hydroxybutyrate

Free-living, nitrogen-fixing bacteria such as *Azotobacter* spp. and *Beijerinckia* spp., as well as many other bacteria, are rich in the reserve substance poly-β-hydroxybutyrate. The presence of this substance has been demonstrated in the root nodules of leguminous plants, but it was absent from the root nodules of non-leguminous plants (*Alnus glutinosa* and *A. incana*) so far tested (Schlegel, 1962).

Transmission electron microscopy of *Alnus glutinosa* root nodules indicated the presence of lipid-like reserve substances in the so-called bacteroids or bacteria-like cells of the *Alnus* endophyte (Becking *et al.*, 1954). These lipids are most probably triglycerids of branched fatty acids, principally C_{15} and C_{16}, as has been found in the sclerotal cells of related actinomycete species (Lechevalier *et al.*, 1973) and in *Streptomyces* species (Asselineau, 1966).

C. Growth Substances

1. *Auxin-like Substances*

As will be mentioned in Section VII. B, root-nodule initiation in non-legumes is associated with a pronounced curling of root hairs and deformation of their tips prior to the infection process. In parallel with observations made in legumes, it is likely that an auxin-like substance is responsible for such an effect. Thimann (1936, 1939) and Chen (1938) were the first to connect these effects with the production of indole-3-acetic acid from tryptophan by the bacteria in the legume symbiosis. The presence of growth substances was confirmed by many other workers Thornton, 1946; Nutman, 1959, 1958; Kefford *et al.*, 1960; Saheman and Fahraeus, 1962). Other workers attribute, however, the deforming principle of root hairs to an extracellular polysaccharide obtainable from preparations of *Rhizobia* (Hubbell, 1970), or a substance separable into two fractions, one with nucleic acid properties and one which is either protein or polysaccharide (Solheim and Raa, 1973). According to the latter authors the deforming principle was heat-labile in filtrates of bacterial cultures, but heat-stable in the inoculated root medium.

Nothing is known about the substance producing curling of the root hairs of non-leguminous roots, except that our own preliminary tests showed that the substance is diffusible and gives a positive response in the *Avena*-coleoptile bioassay. This would indicate that indoleacetic acid (IAA) is involved. In any case IAA was shown to be present in appreciable quantities in *Alnus glutinosa* root nodules. Dullaart (1970) estimated spectrofluorometrically, in the acid ether-soluble fraction of methanol extracts of *Alnus glutinosa* root nodules, indol-3yl-acetic acid (IAA) in a concentration of 0·6–1·0 mg per kg fresh weight of nodules. Using the same method, alder roots had an IAA concentration 5–10 times lower than the root nodule. The related substance indol-3yl-carboxylic acid (ICA) was only found in the low concentration of 0·05–0·15 mg per kg fresh tissue and exclusively in the roots, not in the nodule. This situation was similar to that observed in *Lupinus luteus* (Dullaart, 1967). A significant observation relating IAA and root nodulation was made by Clarke *et al.* (1959) who found that ether extracts of tomato tumor tissue contained both IAA and ICA, but normal tissue only ICA.

Silver *et al.* (1966) measured by a bioassay method the IAA concentration in root nodules of *Myrica cerifera, Casuarina cunninghamiana* and *Alnus serulata.* They found no detectable IAA in the so-called nodule roots with negative geotropic behaviour in *Myrica* and *Casuarina*, but an appreciable amount of IAA in the normal nodulated root of both species and in *Alnus* root nodules. They estimated that the latter species contained 20 mg IAA per kg fresh weight in its root nodules; this figure is about 20–30 times higher than that given by Dullaart (1970). According to Silver *et al.* (1966) the low IAA concentration in the root nodules of both *Myrica* and *Casuarina* is due to an active IAA oxidase system in the root nodules. It was suggested that the negative geotropic response of the roots associated with the nodules of *Myrica* and *Casuarina* can be explained by this activity producing a suboptimal IAA concentration.

In legume root nodules IAA was able to counteract to some extent the inhibitory effect of nitrate on root nodulation (Tanner and Anderson, 1963, 1964; Valera and Alexander, 1965) and a similar effect was observed in *Alnus glutinosa* (Pizelle, 1970).

2. *Cytokinins*

Cytokinins undoubtedly play a role in the nodulation process in legumes and non-legumes. Arora *et al.* (1959) found that kinetin was able to "pseudo-nodulate" tobacco roots in excised root culture. This observation was preceded by an earlier observation by Allen *et al.* (1953), who found pseudo-nodulation in legumes by the application of 2-bromo-3,5-dichlorobenzoic acid. Phillips and Torrey (1972) observed that cytokinin was released by two *Rhizobium* species grown in a culture medium and that the amount of cytokinin released during the logarithmic phase of growth of these strains would be sufficient to initiate the cortical cell division to form a root nodule in the pea plant. The cytokinin substance released was identified as a zeatin-like compound. In later experiments (Short and Torrey, 1972) it was shown that the zeatin-like compound produced polyploid divisions and the differentiation of tracheary element in pea root segments. Most of the cytokinin was found in the terminal 0–1 mm root tip, which contained 44 times more cytokinin on fresh weight basis than the next 1–5 mm root segment. Apart from free cytokinin, cytokinin bound to transfer RNA was also found to be present. There was, however, in the root apices 27 times more free cytokinin present than bound cytokinin.

Cytokinin was also observed to be produced by *Corynebacterium fascians* giving the fasciation disease in pea tissue (Thimann and Sachs, 1966). Moreover, Becking (1971) found that this substance was involved in the leaf symbiosis of *Psychotria*. In the latter plant species in which the endophyte is probably a *Chromobacterium* species, the cytokinin produced gives a pronounced chlorophyll retention around the leaf nodules in the senescent leaf-nodulated leaves. The substance was diffusible producing the same effect in detached etiolated oat leaf blades. Although there is no direct proof that cytokinin plays a role in the nodulation process of non-leguminous plants it is likely that it does so. This is based on the visual observation that alder tissue may have an abundance of cytokinin; by contrast with many non-legumes without root nodulation, alder trees (but also *Myrica* species) do not show yellowing of leaves when they become sensescent in autumn. Alder leaves drop when they are green and become black when lying on the forest floor. The same holds for *Myrica* leaves growing in *Myrica* bogs or at other sites. The absence of yellowing in the senescence pattern is likely to be due to the effect of cytokinin, since it is the only known compound exhibiting a chlorophyll-retention effect. In addition, in experiments in which we tried to grow the endophyte within root-nodule callus tissue or excised root nodules, the inoculation of sterile alder plants in test tubes with macerated root-nodule tissue sometimes resulted in abundant development of pseudo-nodules on the alder roots. These root nodules were mostly free of endophyte. This phenomenon can be explained by the action of cytokinin which is known to produce pseudo-nodulation. Indeed, Rodriguez-Barrueco and Bermudez de Castro (1973) found that the application of kinetin or 2-iso-pentenyladenine (2IP) to the rooting medium of sterile non-inoculated alder plants growing in test tubes, caused the profuse formation of pseudonodules. Histologically these root nodules were composed of unorganized tissue originating from root cortex. The nodules also contained groups of tracheids giving the impression of unorganized vascular bundles.

D. Other Substances

Seidel (1972) reported that root nodules of *Alnus glutinosa* produced a red coloured pigment, which, when excreted into the environment, showed bactericide action against a number of common pathogenic bacteria. The substance was not characterized; it was suggested to be a "streptomycin-like" substance probably because the alder endophyte is actinomycetal. The substance was also able to lyse bacterial colonies on membrane filters used for the purification of sewage water.

E. Nodule Breis and Cell-free Extracts

The destructive effect of nodule fragmentation on nitrogen fixation has been reported for legumes (Aprison *et al.*, 1954), but later tissue breis of leguminous root nodules has been found to continue nitrogen fixation for some time and the nitrogenase of leguminous root nodules has been isolated in a relatively pure form (Koch *et al.*, 1967; Klucas *et al.*, 1968). The nitrogenase activity in cell-free extracts was inhibited by phenolic substances, but these could be removed by treatment of the ascorbate-buffered homogenates with solid polyvinyl-pyrrolidone (PVP). Bond (1964), Sloger and Silver (1965, 1966) and Sloger (1968) did some experiments with fragmented root nodules or nodule breis of non-legumes. Generally, nitrogen fixation by the breis was much lower than that by intact nodules. The presence of a reducing agent (sodium dithionite) and the absence of O_2 was essential during homogenization, but O_2 was required, presumably for the production of ATP, during exposure of the homogenate to $^{15}N_2$ or acetylene. This phenomenon was also noted by Bergensen (1966a, b) for soybean nodule breis. Although it could be expected that the use of PVP to sequester phenolic compounds would increase the activity of the homogenized root nodules of non-legumes and should lead to the preparation of an active non-leguminous subcellular nitrogenase system, no reports of such systems have been published in literature.

VII. Root Nodules

A. Morphology and Structure

In non-legumes two types of root nodules can be distinguished, i.e. an *Alnus* and a *Myrica/Casuarina* root nodule type. The *Alnus* type of root nodule is found in representatives of the Betulaceae (*Alnus* species), Elaeagnaceae, Rhamnaceae,

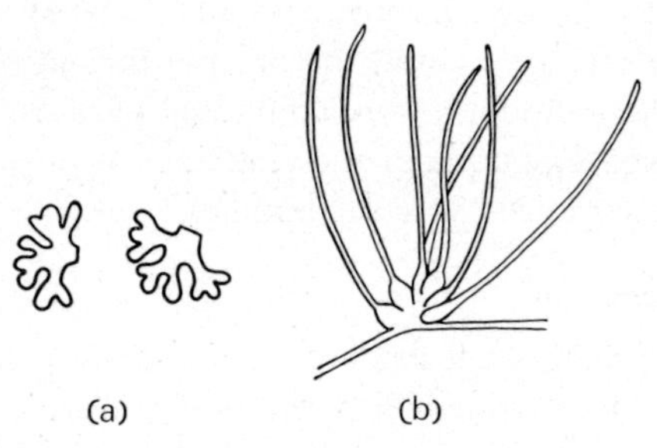

FIG. 7. Root-nodule types of non-leguminous plants. (a) *Alnus*-type; (b) *Casuarina/Myrica*-type. $\times 0{\cdot}7$.

Coriariaceae, and Rosaceae. The root nodules are composed of modified, often dichotomously branched roots of arrested growth (Fig. 7a) which usually have a coralloid appearance (Fig. 8b, c). By contrast the apex of each root-nodule lobe of the *Myrica/Casuarina* type gives rise to a normal, but negatively geotropic root (Fig. 7b). Because of the negative geotropic growth of the apical meristem, these root nodules become clothed with upward growing rootlets (Fig. 8a, 7b). As mentioned earlier (p. 534) Silver *et al.* (1966) attributed the negative geotropic curvature of nodular rootlets in *Myrica* and *Casuarina* to a low auxin concentration resulting from a very active indoleacetic acid oxidase system. The activity of this enzyme was far less in non-nodulated root tissue, which had an auxin content within the customary range of 10 mg IAA per kg fresh weight of tissue.

Intermediate types of nodule occur occasionally in *Alnus rubra* plants inoculated with the *Alnus glutinosa* endophyte; here root nodules are produced in which some nodular tips give rise to negative geotropic growth (Becking, 1966, 1968, 1970a). The same has been observed in *Alnus jorullensis* plants inoculated with the *Alnus glutinosa* endophyte (Rodriguez-Barrueco, 1966). It is likely, from pictures published by Ziegler (1962, Tafel X, Abb. 2a, b), that *Comptonia peregrina* (syn. *Myrica asplenifolia*) of the Myricaceae has root nodules resembling those of the *Alnus type*.

The root nodules of non-legumes are perennial and for this reason increase in size every season. Under field conditions *Alnus glutinosa* root nodules may attain

FIG. 8. Root-nodule types of non-leguminous plants. (a) Root nodule of *Casuarina equisetifolia* showing the feature that the apex of each nodule lobe gives rise to a negative geotropic root. ×0·8. (b) Coralloid root nodules of *Alnus glutinosa*. Detached and divided root nodules showing the dichotomous branching of the nodule lobes. ×1·0. (c) Coralloid root nodules of *Alnus glutinosa in situ* on the root. ×0·7. (d) Root nodules of *Alnus rubra* produced by an *Alnus glutinosa* inoculum show nodular lobes with negatively geotropic rootlets like root nodules of the *Casuarina/Myrica* type. ×1·0.

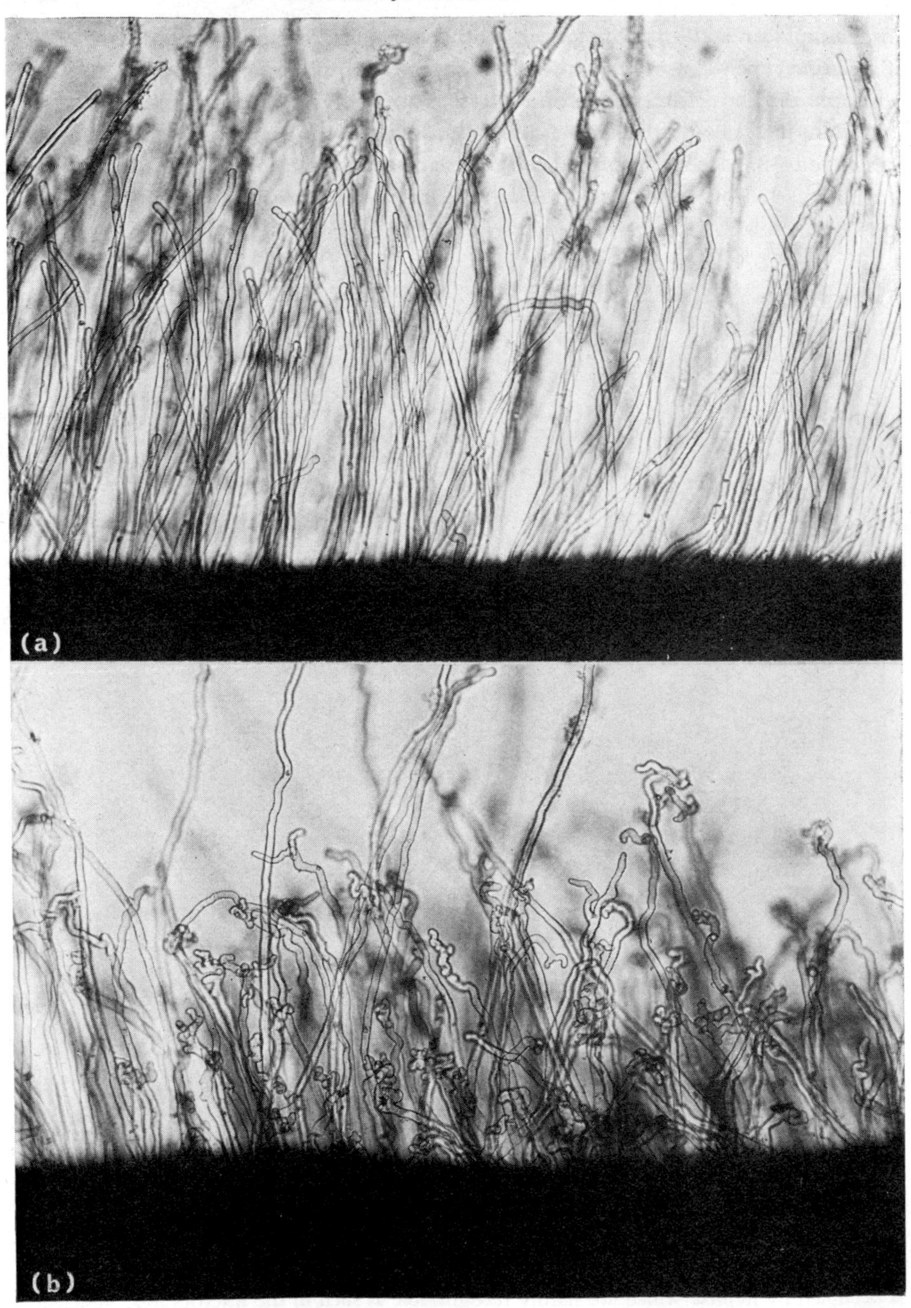

FIG. 11. Root-hair deformation of *Alnus glutinosa* prior to the infection process. (a) Alder root with straight unicellular and bicellular root hairs. (b) Alder root with root hairs showing

Longitudinal sections of root nodules indicate that the endophyte is present in the cortical cells immediately behind the apical meristem; from this site newly formed cells produced by the meristem are invaded. A gradient is thus established in which the infected host cells and the endophyte are youngest near the apex and oldest near the base of the root nodules. There are morphological changes in both the host cell and the endophyte which accompany ageing. The infected cells near the meristem are rather small in size and contain only the hyphal form of the endophyte. In a zone below this tissue the host cells become enlarged and they contain, in addition to the centrally located hyphal form of the endophyte, "so-called" vesicular structures at the tips of the hyphae close to the host-cell wall (Fig. 10). This zone, with enlarged host cells containing the endophyte predominantly in the vesicular form is probably responsible for nitrogen fixation (see p. 548). Below this zone there is another zone where numerous host cells are found with "bacteria-like" endophytic structures (see p. 547) and where vesicles are disintegrated by the host. There is no sharp boundary between these tissue zones but the frequency of the bacteria-like endophyte structures is far greater towards the base of the nodule. Finally, a tissue zone near the base of the nodule can be distinguished, where the host cells themselves begin to disintegrate and for this reason the root nodules are usually narrower near the base than at the apex (Fig. 7a).

B. Infection Process and Root-nodule Initiation

So as to study endophyte penetration into *Alnus glutinosa* we used a technique similar to that of Fahraeus (1957) working with legumes, i.e. observing root-hair growth and endophyte penetration under the microscope at low magnification (Becking, 1966, 1968, 1970a). The experiments showed that root hairs which were initially straight became deformed and curved by the action of the endophyte (Fig. 11a, b). The curling of the root hairs was probably due to the excretion of growth substances or other products by the endophyte as discussed earlier (p. 533). Root-hair curling is a prerequisite for root-nodule formation as only at these sites on the root surface is a nodule formed (Fig. 12). On the other hand, root hair curling does not always give rise to a root nodule because the hair may abort or a lateral root primordium may not be present (see p. 543). The first event during infection is when the endophyte invades the root hair at the tip; this frequently swells and may become deformed or branched (Fig. 13). At the site of endophyte penetration a pocket is formed and an infection thread proceeds into the host cell by invagination of the cell wall and host cytoplasmic membranes in much the same way as observed in legumes by Mosse (1964) (see also p. 467).

---

curling due to the production of growth substances by the endophyte in the rhizosphere. Only at places where root hairs are curved will a root nodule form. Micrograph of living, unstained preparation. ×100.

Moreover, as in legumes, the growth of infection thread is associated with the host-cell nucleus (see N, Fig. 13) which keeps in close proximity to the infection thread tip as it extends.

At a very early stage of development, the site of root-nodule formation is apparent as a slight thickening of the main root (Fig. 12). In longitudinal section,

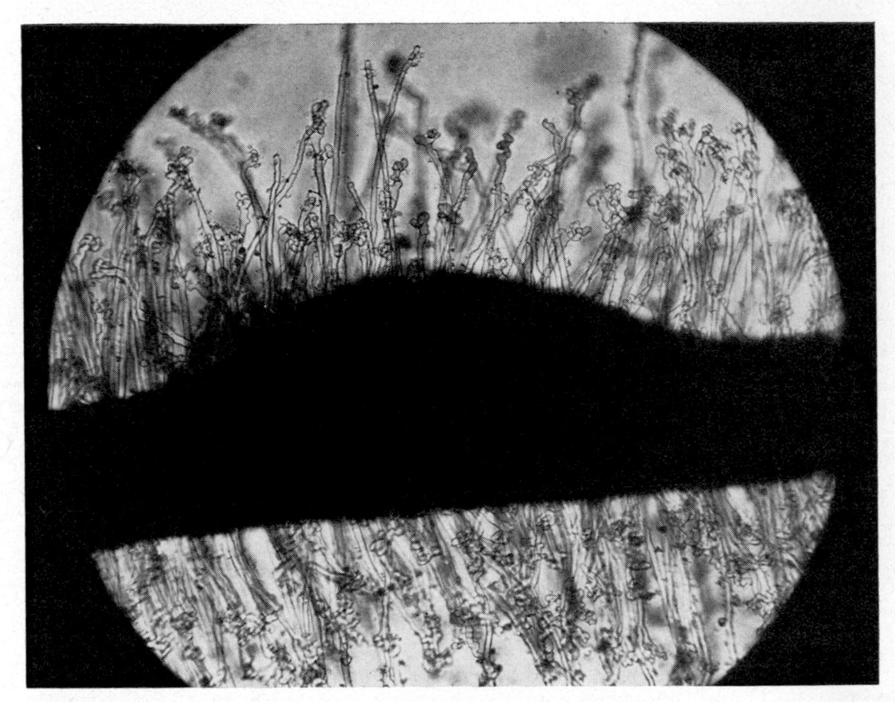

FIG. 12. Root-nodule development in *Alnus glutinosa* in its initial stage. Note the curling of the root hairs due to excretion of growth substances by the endophyte. The thickening of the root indicates the formation of a so-called "pre-nodule" on the main root. Micrograph of living, unstained preparation. ×90.

nodules like this can be seen to be only a few cell layers thick (Fig. 14). This stage of development is more properly described as a "pre-nodule", since the true root nodule is produced eventually by a lateral root primordium and not by the main root. In the "pre-nodule", the endophyte is already present in its vesicular form in a cortical parenchyma cell-layer of only 3–4 cells thick (Fig. 14). Host cells containing bacteria-like endophytic structures were not observed in the main-root cortex. In a later stage of root-nodule development, the infection progresses in a lateral root emerging at the infection site (Fig. 14) and not in the main root. Erythrosin and methylene green stained paraffin thin-sections have indicated that the developing primordium is actually a lateral root meristem originating from protoxylem and protophloem strands within the pericycle of the parent root. The thin sections showed clearly that two endodermal layers staining green with the methylene-green dye were present, one of the main root and the other of the lateral root (see Becking 1968, colour plate 7). Thus, by contrast with

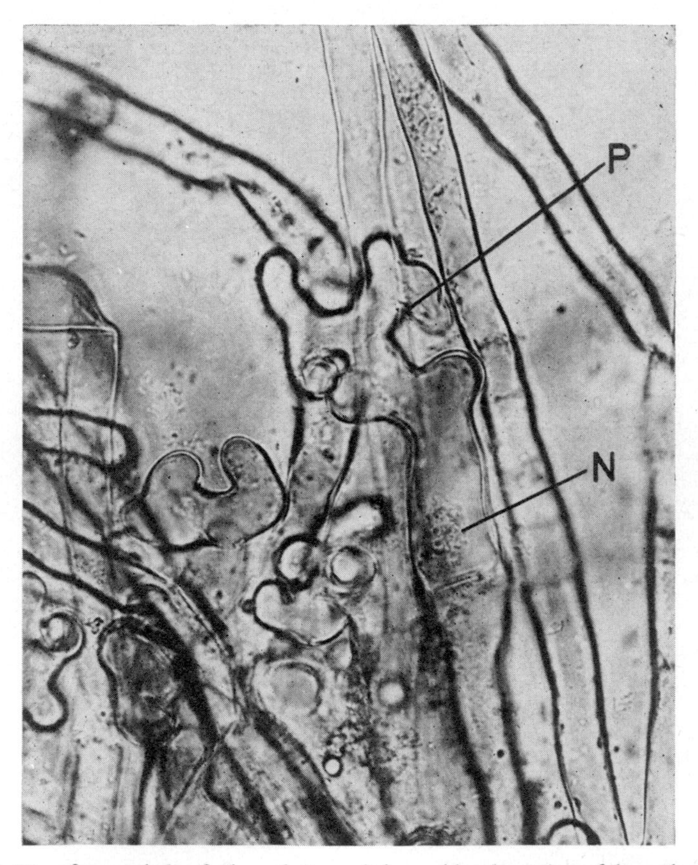

FIG. 13. Tip of a root hair of *Alnus glutinosa* deformed by the action of the endophyte. At the tip of the root hair a pocket (P) is formed by invagination of the host-cell wall and from here an infection thread develops inside the host cell which is accompanied during growth by the host-cell nucleus (N). Micrograph of living, unstained preparation; phase contrast ×780.

legumes, root nodules of non-legumes with actinomycete symbiosis are modified lateral roots. A further contrast is seen in the "pre-nodule" stage. It is likely root nodules can only occur at places where there are dormant lateral root primordia. So far, there is no indication that these lateral root primordia are induced *de novo* by the endophyte present in the "pre-nodule". Lateral root primordia are spaced

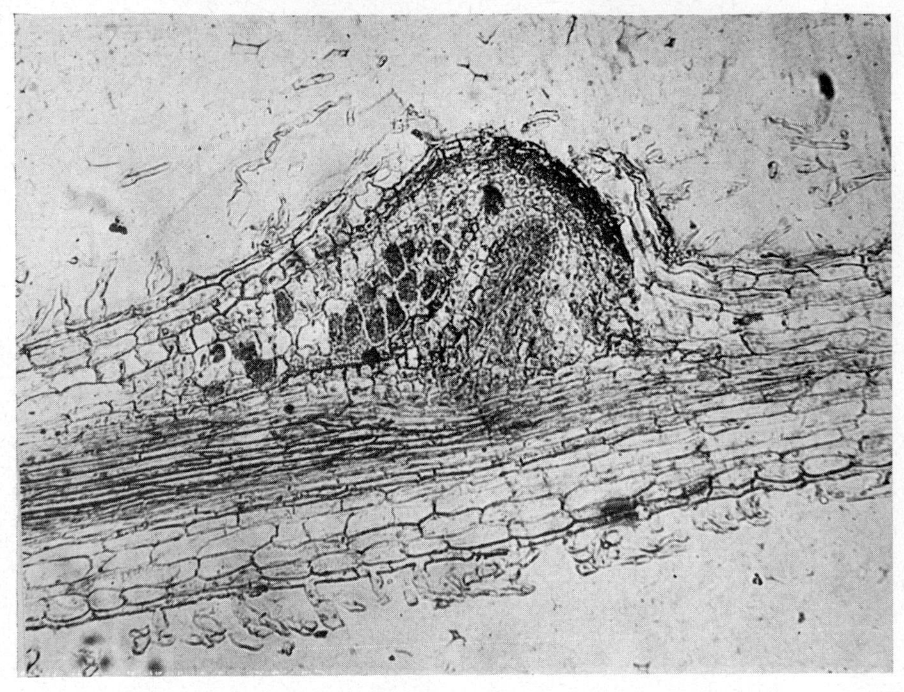

FIG. 14. Longitudinal section through a so-called 'pre-nodule' of *Alnus glutinosa*. In this very early stage there is cell division in the cortex of the main root and the endophyte is already present in its vesicular form. In a later stage of root nodule development the infection is taken over by the side-root primordium (visible in the micrograph), which produces the actual root nodule. Micrograph of fixed preparation, stained with erythrosin and methylene green. ×130.

regularly along the root and the presence of the endophyte may serve to stimulate their further development into a root nodule.

Cortical parenchyma cells containing the endophyte have bigger nuclei and are larger than cells of non-infected tissue (Fig. 10). This suggests that the infected tissue is polyploid as it is in some legumes (Wipf, 1939; Wipf and Cooper, 1938, 1940; Trolldenier, 1959; and Kodama, 1967a, 1968). In nodule tissue from *Alnus glutinosa*, thin section revealed mitotic divisions accumulated in the presence of colchicine (0·5%) and separate chromosomes were visible (see Fig. 9). *Alnus glutinosa* has a somatic chromosome number of 28 or 56 chromosomes (Wetzel, 1929; Woodworth, 1931); other *Alnus* species have 28, 42 or 56 chromosomes (Darlington and Wylie, 1961). Recently, Kodama (1967b, 1970) investigating Japanese alder species found for *Alnus pendula* 2n = 28, for *A. hirsuta* 2n = 56, and for *A. sieboldiana* (syn. *A. firma*) 2n = 112. In our experiments it was rather difficult to ascertain whether polyploidy had occurred in the infected root-nodule tissue because the chromosome number in *Alnus glutinosa* is variable due to natural polyploidy and so far we have not examined root tips and nodular tissue of the same plant. Furthermore, treatment of nodules with colchicine seldom gives rise to mitotic activity in the nucleus of host cells containing the endophyte but these nuclei are often masked by the endophyte. Some preliminary micro-densitometer measurements on thin sections of Feulgen-stained *Alnus glutinosa* nuclei suggest that the infected tissue is indeed polyploid since the nuclei contained more DNA than those of adjacent non-infected tissue. On the other hand, Kodama (1967b, 1970) counting chromosome numbers in three species of Japanese alder found no difference in chromosome number in root nodule cells and root tip cells, but from his studies it is evident that he examined only the apical meristem of the root nodules and not the infected cells themselves.

VII. Endophytic Organisms

A. Cross-inoculation Groups

In non-legumes two main types of endophytic organisms can be distinguished, i.e. actinomycete species and a *Rhizobium* species.

1. *Actinomycete Endophyte*

On the basis of morphology and cultural characteristics the actinomycete endophytes of non-legumes can be placed in one family, Frankiaceae, with a single genus, *Frankia* (Becking, 1970b; see also Bergey's Manual of Determinative Bacteriology, 1974). A further classification in species was made according to cross-inoculation incompatibilities in much the same way as cross-inoculation groups are used to distinguish *Rhizobium* spp. Using this criterion, ten *Frankia* species could be recognized, i.e. *Frankia alni*, *F. elaeagni*, *F. ceanothli*, *F. discariae*, *F. brunchorstii*, *F. casuarinae*, *F. coriariae*, *F. dryadis*, *F. purshiae*, and *F. cercocarpi*. These *Frankia* species differed to some extent also in their morphological characteristics (see Becking, 1970b, Table 1; Becking, 1974).

We have refrained from splitting the classification of the genus *Frankia* into more taxa, although there are some arguments for doing so. For instance, Bond (1962) observed that the endophyte of *Coriaria myrtifolia* did not nodulate in *Coriaria japonica* plants, and that there were also specific endophyte adaptations within the genus *Myrica*. Experiments with nodular material from *Myrica gale* tested on *M. cerifera*, *M. californica*, *M. cordifolia*, *M. pilulifera*, and *M. rubra* showed that these associations, and also some reciprocal crosses, gave less nodule formation or less effective nitrogen fixation (Gardner and Bond, 1966; Mackintosh and Bond, 1970). Similarly, Becking (1966, 1968) found reduced compatibility between the endophyte of *Alnus glutinosa* and *A. rubra*, an observation confirmed by Rodriguez-Barrueco and Bond (1968). The same non-adaptive relation was observed by Rodriguez-Barrueco (1966) between the endophyte of *A. glutinosa* and *A. jorullensis*. Also Mackintosh and Bond (1970) observed reduced symbiotic performance or effectiveness of inocula of one *Alnus* species when tested on some other *Alnus* species. Especially the combinations *A. jorullensis* inoculum on *A. incana*, and *A. jorullensis* on *A. sieboldiana* produced poor results, pointing again to a specific adaptation of the endophyte in these *Alnus* species. Although less pronounced, other cross-inoculation combinations between alder species gave less satisfactory results.

On the other hand, as in the *Rhizobium* symbiosis of non-legumes, the barriers between cross-inoculation groups may be overcome in some cases. Claims have been made that *Alnus glutinosa* inoculum can produce nodules in *Myrica gale*, but that the reciprocal combination apparently does not give root nodulation (C. Rodriguez-Barrueco, personal communication). These cross-inoculations between species of a different genus were initiated by Rodriguez-Barrueco, because he observed that most soils containing the alder endophyte could also produce root nodulation in *Myrica gale* plants (Rodriguez-Barrueco, 1968).

In spite of this obvious exception, it can be stated that the barriers between the cross-inoculation groups are maintained in most instances. Arguments could be put forward to subdivide the *Frankia* species within one plant genus, because of specific adaptations. The non-identity of the endophyte in different non-legumes is also evident in the case of sympatric occurrence, where one species is nodulated and another not. For instance, in Eurasia all *Alnus* species are nodulated, but not the sympatric species of *Dryas* which often occurs in the same habitats, while both *Dryas* and *Alnus* species are nodulated in Northern North-America. Moreover, the marked difference in acid tolerance of the endophytes of *Alnus glutinosa*, *Myrica gale* and *Hippophaë rhamnoides* for nodule initiation also indicates that the endophytes of these species are not identical (Bond *et al.*, 1954).

2. *Rhizobium Endophyte*

The symbiosis with *Rhizobium* has so far been found in only one non-leguminous species, *Trema cannabina* var. *scabra*, of the Ulmaceae (Trinick, 1973)*. The *Trema* endophyte could be isolated from nodular tissue of the host plant by the normal procedures for isolating *Rhizobium* from legume root nodules. The strain isolated (NGR 231) proved to be a slow-growing *Rhizobium* species, which could also produce effective nodulation on three legume species, i.e. in *Vigna sinensis*, *Macroptilium atropurpureum* (formerly *Phaseolus atropurpureus*), and *M. lathyroides* (formerly *P. lathyroides*) (Table III). The nodules produced on the latter plants were pink, large and typical of nodules produced by effective "cowpea-type" rhizobia. Fifteen other species of legumes tested with the *Trema* endophyte, e.g. *Lablab purpureum* (formerly *Dolichos lablab*), *Crotalaria anagyroides*, *Glycine max*, *Acacia farnesiana*, *Phaseolus vulgaris*, *Trifolium pratense*,

* Trinick (personal communication) informed me that the original classification as *T. aspera* was wrong and that other *Trema* species growing at the same locality (New Guinea), including *T. aspera* were non-nodulated.

and *Medicago sativa*, failed to produce root nodulation. Nodules produced on *Trema* in response to inoculation with strain NGR 231 were in cross section also pink or reddish coloured, but they lack haemoglobin (C. A. Appleby, personal communication). Hence, it is evident that the endophyte of *Trema cannabina* var. *scabra* is a real *Rhizobium* species, cross-inoculable with some legumes and showing a close affinity to *Rhizobium* of the "cowpea-group".

B. Cytology and Classification

1. *Actinomycete Endophyte*

In the older literature various claims have been made concerning the identity and nature of the endophyte of non-legumes such as *Alnus*, *Myrica*, *Elaeagnus*, *Casuarina* or *Ceanothus*; it has been described as a fungus (Woronin, 1866; Brunchorst, 1886; Frank, 1887, 1891; Arzberger, 1910), a bacterium (Hiltner, 1896; Spratt, 1912; McLuckie, 1923; Koslova *et al.*, 1966) or a species of *Plasmodiophora* (Möller, 1885; Schröter, 1889; Yendo and Takase, 1932; Hawker and Fraymouth, 1951). It is now well established that the endophyte is an actinomycete. Although filamentous, the organism is definitely prokaryotic since it lacks a nuclear membrane (Becking *et al.*, 1954). Besides this, a relation to fungi is excluded because of the narrow hyphae which are characteristic of actinomycetes being 0·2–0·3 μm in diameter. In addition, the hyphae contain plasmalemmosomes associated with cross-wall formation; these are found commonly in Streptomycetaceae and Actinomycetaceae (Glauert and Hopwood, 1960; Chen, 1962, 1964), and in the related Mycobacteria (Imaeda and Ogura, 1963), but also in some gram-positive bacteria (Fitz-James, 1960; Giesbrecht, 1960).

The morphology of the actinomycetous structures within the host cells has been extensively studied by transmission electron microscopy in the *Alnus glutinosa* endophyte (Becking *et al.*, 1964; Gardner, 1965) and in the *Hippophaë rhamnoides* endophyte (Gatner and Gardner, 1970). The endophyte hyphae are branched and septate, but branching is not necessarily correlated with septum formation. Both light and electron microscopy (including scanning electron microscopy) have indicated that the hyphae bear terminal swellings which are called vesicles. These vesicular structures are spherical, measuring 3·0–5·0 μm in diameter, in the *Alnus* and the *Elaeagnus* endophyte, but are more club-shaped in the endophyte of *Casuarina*, *Myrica* and *Dryas*.

In addition to vesicular structures, polyhedral-shaped, bacteria-like cells are found in most endophyte species; the latter are called by some authors "bacteroids" or "granulae". Our observations (Becking *et al.*, 1964) have shown that the vesicular structures are always found in living host cells, whereas the bacteria-like endophytic cells occur solely in dead host cells. It is likely that host cells containing bacteria-like structures were alive when first infected but died during the infection process. The vesicular structures are always situated at the tips of the hyphae and within the host cell near the periphery of the hyphal mass close to the host-cell wall, as is obvious from light micrographs (see Fig. 10) and transmission electron micrographs (Fig. 15). The vesicles have a very elaborate internal struc-

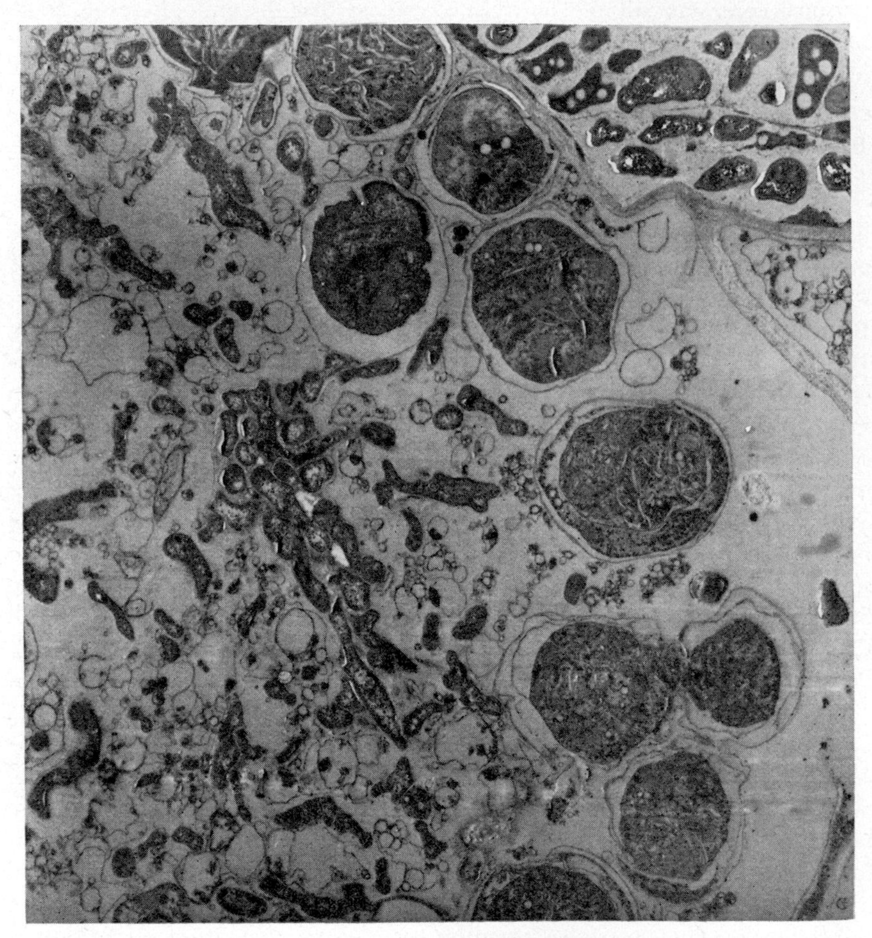

FIG. 15. *Alnus glutinosa* endophyte in host cells showing vesicular structures near the periphery of the hyphal mass close to the host-cell wall. The host cell is still alive as evident from the cytoplasmic membranes of host origin enclosing each vesicle and the endoplasmic reticulum in between the hyphal mass. At the top right corner bacteria-like endophyte cells are visible in a host cell lacking host-cell organelles. Therefore this host cell is dead. Transmission electron micrograph. × 8,300.

ture with many membranes and incomplete cell walls (Fig. 16). The vesicle probably serves as a site of nitrogen fixation as it possesses a strong tetrazolium-reducing capacity (Akkermans, 1971). Moreover, we obtained evidence that parts of *Alnus* nodules containing many vesicles had a higher capacity for acetylene reduction per unit tissue volume or host cell number than tissue parts containing less of these structures. Vesicles are often surrounded by host-cell

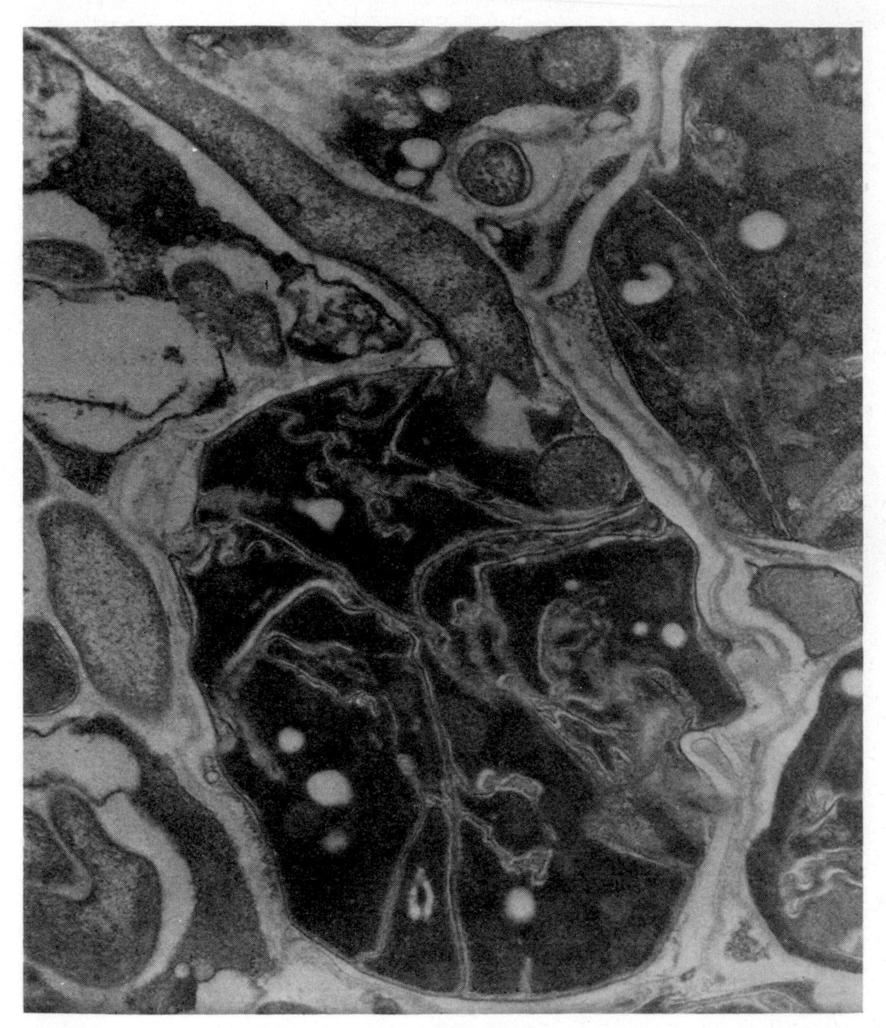

FIG. 16. Section through a vesicle of the *Alnus glutinosa* endophyte showing the point of attachment of the hypha with the vesicle and the very complicated internal structure of the vesicle with many internal and partly incomplete cell walls and their associated membranes. Transmission electron micrograph. × 33,500.

mitochondria (Fig. 17) suggesting an active exchange or cross-feeding of substances between vesicle and host cytoplasm. This exchange is probably nitrogen from vesicle to host cytoplasm and carbohydrate in the reverse direc-

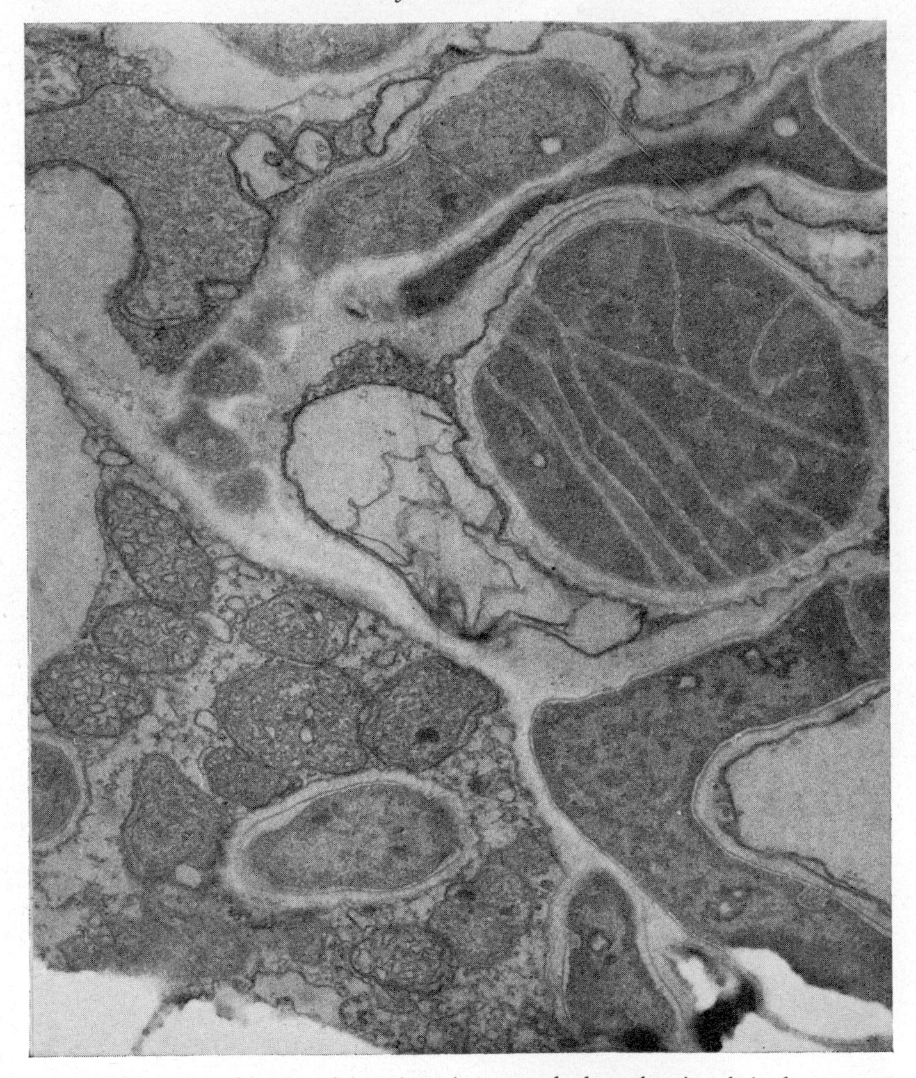

FIG. 17. Vesicle and hypha of the *Alnus glutinosa* endophyte showing their close contact with host-cell mitochondria indicating metabolic interaction or cross-feeding between host and endophyte. Transmission electron micrograph. $\times$24,000.

tion. Older vesicles, or vesicles in older tissue parts, are resorbed, disintegrate and loose their contents by the action of the host cytoplasm as shown in transmission electron micrographs (Fig. 18).

Vesicles may be regarded as a kind of sporangia as there is occasionally formation of a complete cell-wall within the vesicle. The vesicle then tends to fall apart

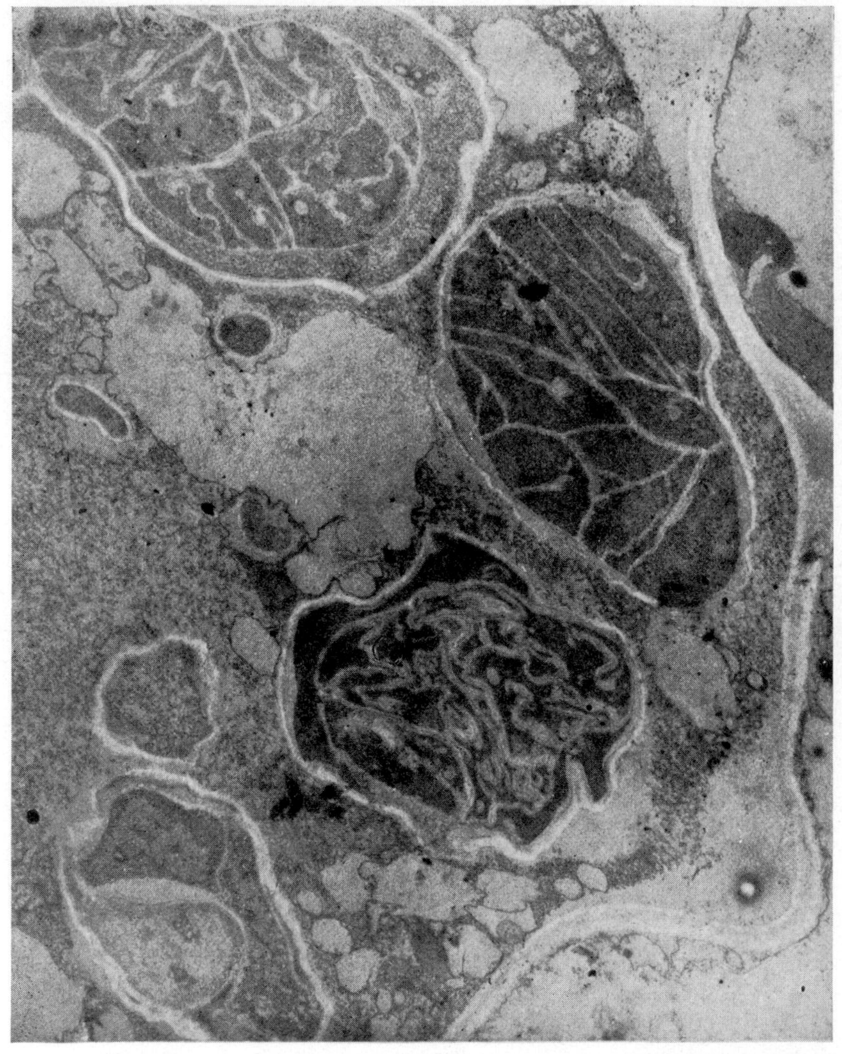

FIG. 18. Of the three vesicles visible in the micrograph within the living host cell, the lower one is in the process of becoming disintegrated by the active host cytoplasm. Transmission electron micrograph. ×14,500.

in particles resembling the "bacteria-like" cells. Such a situation has been demonstrated in some scanning (Fig. 19) and transmission electron micrographs (Becking *et al.*, 1964, Fig. 29). Bacteria-like cells usually develop from hyphae which become broader and vacuolated and finally differentiate in small thick-walled and rigid polyhedral structures. The formation of such enlarged hyphae

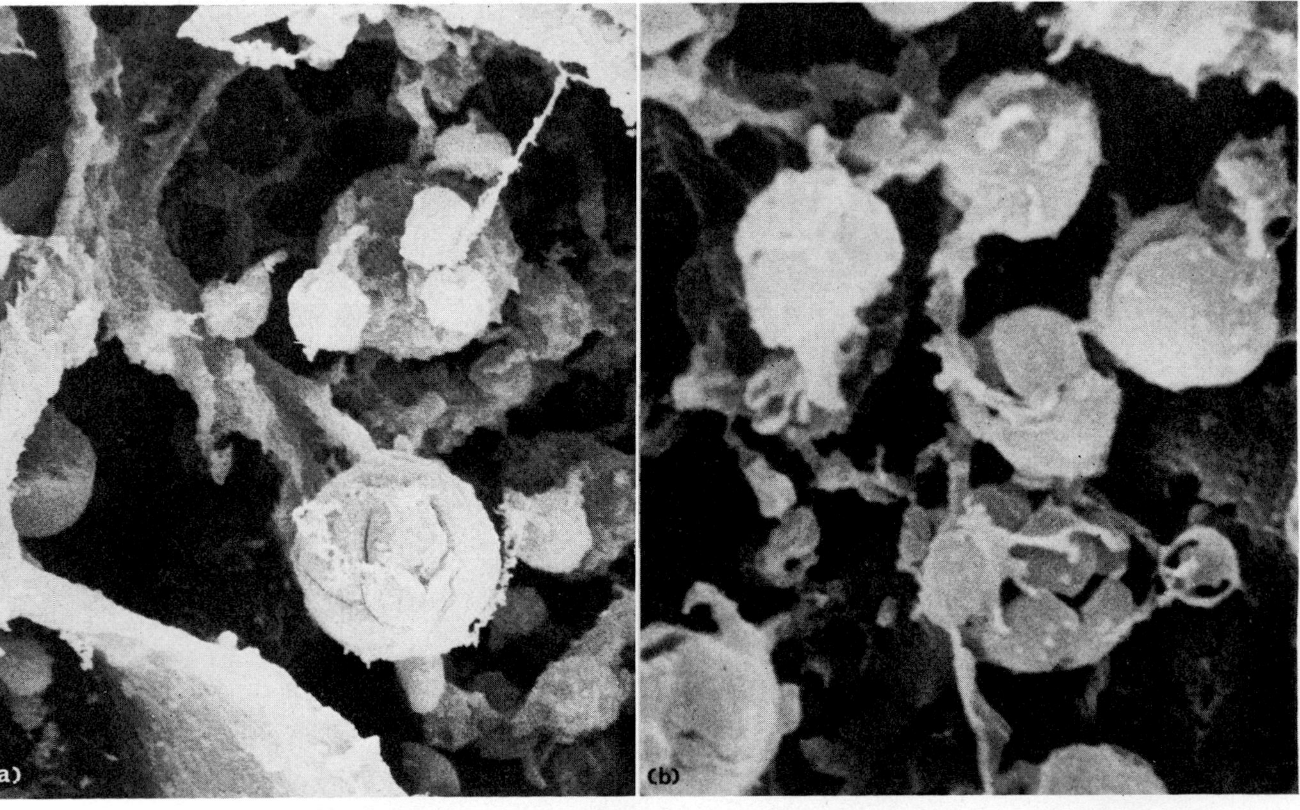

FIG. 19. Vesicles of the *Elaeagnus umbellata* endophyte showing stages of segregation of the vesicle into small endophytic particles reminiscent of bacteria-like endophyte cells. Such a process seldom occurs, because vesicles have mostly incomplete cell walls. (a) Small cracks can be seen at the outer surface of the vesicle, (b) A vesicle has broken up into particles. Scanning electron micrograph. ×3,500.

has already been observed by Käppel and Wartenberg (1958), who called these structures "Spindeln". These broad vacuolated hyphae and the formation of endophytic vegetative cells within them were once thought to be unique features of *Frankia* species. However, recently Lechevalier *et al.* (1973) have described similar structures, together with the formation of vegetative cells within these structures in another actinomycete, *Chainia olivacea.* Lechevalier *et al.* called these structures sclerotia and found experimentally that the sclerotia are more heat resistant and therefore help the organism to survive under adverse conditions. Similarly, we think that the bacteria-like cells of *Frankia* (see Fig. 15), which resemble the *Chainia* sclerotal cells in having a rigid structure with a thick cell wall containing substantial amounts of lipids, function in the survival and dispersal of the endophyte in soil.

We have analysed the cell wall of the endophyte of *Alnus glutinosa* root nodules. The endophyte was obtained from the host tissue by a filtration technique of tissue homogenates using nylon filter cloth of different mesh. The tissue fraction which sedimented on cloth of 10 μm mesh consisted mainly of vesicular clusters of the endophyte. The endophyte clusters were broken by ultra-sonic treatment to subcellular fragments. The cell-walls obtained were thoroughly washed with phosphate buffer prior to hydrolysis and cell-wall analyses. The hydrolysates were analysed by paper chromatography following the chemical methods given by Lechevalier and Fekete (1971) for the separation and classification of actinomycetes into genera. The hydrolysates indicated that meso-DAP (DAP = 2,6-diaminopimelic acid), glycine, arabinose, and galactose were cell-wall constituents as well as some lysine and aspartic acid. Reference strains of actinomycetes containing other forms of DAP were simultaneously tested for cell-wall constituents in order to confirm the technique and recovery. Although other forms of DAP (L-, 3-OH DAP) were found in the reference strains, only meso-DAP was present in the alder organism. According to the classification of actinomycetes based on major cell-wall constituents (Lechevalier *et al.* 1966; Lechevalier and Fekete, 1971) the alder endophyte showed affinities to the cell-wall types II (*Micromonospora*, *Actinoplanes*) and IV (*Nocardia*). The presence of lysine and aspartic acid also suggests a relationship with the anaerobic actinomycetes of the genus *Actinomyces*. The vesicle formation in the *Alnus* organism is reminiscent of sporangia formation in representatives of the actinomycetes of cell-wall type II, e.g. Actinoplanaceae, and it is likely that the endophyte within the root nodule is micro-aerophilic. The differences in cell-wall constituents with other actinomycetes support the separation of these endophytic organisms in a separate family Frankiaceae.

2. *Rhizobium Endophyte*

The cytology and ultra-structure of the *Trema cannabina* var. *scabra* endophyte has not yet been studied. There is no reason to anticipate that it will be different from other rhizobia, especially since the close relationship with the "cowpea-type" *Rhizobium* has been established and cross inoculation ability with legumes has been observed. So far as the picture of the internal structure of the *Trema* root nodule published by Trinick (1973, Fig. 1b) indicates, the internal organization of this nodule is different from that of a legume and conforms more with that of a non-legume. For this reason a more thorough anatomical study of the *Trema cannabina* var. *scabra* nodule and its comparison with a nodule produced by the same organism in a legume, would be very worthwhile.

C. Growth Requirements and Isolation of the Endophyte

1. Actinomycete Endophyte

So far all attempts to isolate the symbiotic actinomycetes of non-legumes have failed. Claims that the endophyte had been isolated by normal plating procedures (Von Plotho, 1941; Uemura, 1952a, b, 1961; Fiuczek, 1959; Niewiarowska, 1961; Danilewicz, 1965; Allen *et al.*, 1966; Wollum *et al.*, 1963) are subject to criticisms, since in most of these studies alternative possibilities of infection were not considered, or the re-infection tests were made under non-sterile conditions, or Koch's postulate was not satsified. We tested the endophyte from *Alnus* on a wide spectrum of media under both aerobic and anaerobic conditions, but failed to obtain growth. However, by other methods we could obtain the endophyte of alder in apparently pure culture. First, by cultivating *in vitro* alder root nodule tissue on a nutrient agar medium formulated for plant-tissue cultures (Becking, 1965). Surface sterilized nodule slices were tested for sterility on the usual media for bacteria growth (Plate Count Agar, Oxoid, etc.) and the contaminant-free tissues plated on plant-tissue medium in tubes. These tissue fragments grew until quite large amounts of root-nodule callus tissue had formed (Fig. 20). The callus tissue still contained the endophyte as evidenced by effective root-nodule formation on sterile alder seedlings growing in test tubes. However, in further explants of the callus tissue the endophyte was often lost, because of insufficient transmission in the newly developed callus cells (Fig. 21). Moreover, if nodules on sterile alder seedlings were produced by crushed tissue of the explants, the nodules were either ineffective or did not contain the endophyte (Becking, 1965). Another method was more profitable in purifying and maintaining the endophyte. Suspensions of crushed surface-sterilized root nodules were plated, for verification of freedom from contaminants, on media for bacterial growth. Suspensions which did not produce any growth were diluted in serial dilutions and inoculated on sterile alder seedlings growing in test tubes. The highest dilution producing root nodulation was expected to contain only one or a very few cells of the endophyte. This is in agreement with the normal procedure of serial dilutions for obtaining pure cultures of bacteria, where the highest dilution producing growth is assumed to contain only one or a very few cells of the dominant organism. The only difference with the normal

FIG. 20. Root-nodule fragments of *Alnus glutinosa* containing the endophyte placed on a nutrient agar medium for the growth of plant callus tissue *in vitro*. As evident, small tissue fragment of 1–2 mm diameter can produce root-nodule callus tissue up to 4–5 cm in diameter. $\times 1{\cdot}0$.

dilution procedure for acquiring pure cultures of bacteria was that the host plant, and not an artificial medium, was used as substrate. Using the dilution method, we obtained "isolates" of the alder organism, which according to our view represented a pure culture of the organism and which could fulfil the postulate of Koch because root nodulation was obtained by inoculation. After passage through the plant the isolates were tested again for contamination; in none of these tests did we observe any growth of the endophyte on artificial media.

The absence of growth of the actinomycete endophyte on artificial media does not mean that the endophyte is an obligate symbiont. Transmission electron micrographs of the rhizosphere of the sterile alder plants infected in the way described above show that, without doubt, the endophyte can grow in the

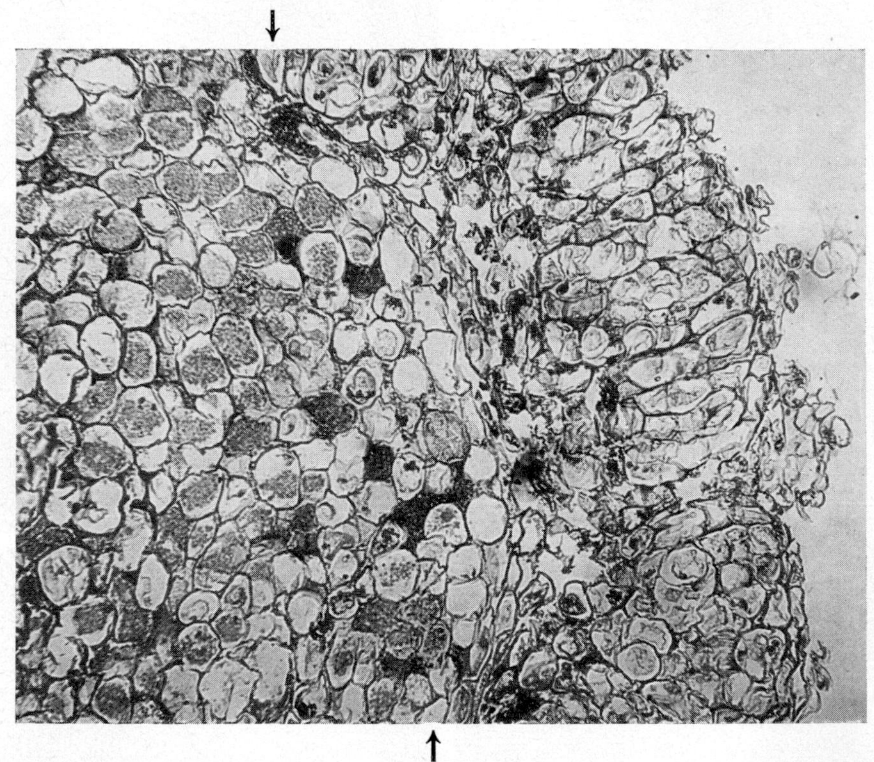

FIG. 21. Section through a callus-producing root nodule fragment of *Alnus glutinosa* showing that the transmission of the endophyte to the newly developed tissue is poor. The arrows indicate the transitional zone between original tissue and the newly developed callus tissue. Micrograph of fixed, unstained preparation; phase contrast. $\times 140$.

mucilaginous layer covering the root and root hairs prior to infection (Fig. 22). Thus, it must be possible to grow the endophyte outside the plant, if its environmental and nutritional requirements are sufficiently known.

2. Rhizobium Endophyte

The growth requirements of the *Rhizobium* endophyte of *Trema cannabina* var. *scabra* are the same as that known for *Rhizobium* of the "cowpea-group".

IX. Concluding Remarks

In the present treatise it has been shown that many non-leguminous plants with root nodulation are of economic importance by increasing the nitrogen status of

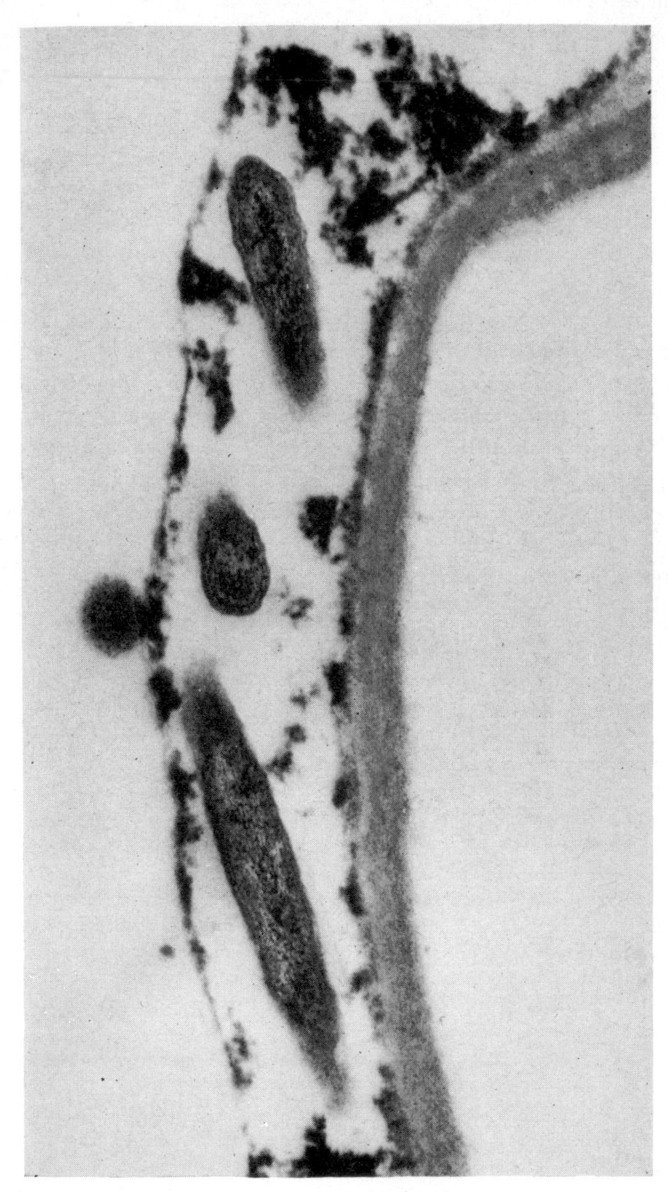

FIG. 22. *Alnus glutinosa* endophyte growing in the mucilaginous layer covering the root and root hair surface. Transmission electron micrograph. $\times$ 35,500.

soil. Therefore, like legumes they can be used for improving soil fertility. As legumes are exploited in agriculture, these non-legumes, being perennial, may be used beneficially in forestry. By their independence of combined nitrogen, many of these nodulated non-legumes play an active role as pioneer vegetation in soil colonization after erosion, land slides or retreat of glaciers. In some of these non-leguminous plants (*Alnus crispa*, *A. rubra*) nitrogen increments of 150–300 kg N per ha per year have been measured, values certainly comparable with figures published for legumes.

Two main topics in non-legumes will require more profound attention in the near future. First, as discussed in the previous section, the actinomycete endophyte of non-legumes has not yet been grown in pure culture outside the host. Because the endophyte is obviously not an obligate symbiont, research on the nutritional requirements of the endophyte should be intensified. Secondly, in spite of recent progress obtained in the purification and characterization of the nitrogenase of legumes, comparatively little work has been done on the nitrogenase in sub-cellular fractions of non-leguminous root nodules. The few experiments performed with homogenates of non-leguminous root nodules can only be considered to represent some crude initial steps and a wide field is still open for further investigations.

References

AHMED, S. and EVANS, H. J. (1960). Cobalt: A micronutrient element for growth of soybean plants under symbiotic conditions. *Soil Sci.* **90**; 205–210.

AHMED, S. and EVANS, H. J. (1961). The essentiality of cobalt for soybean plants grown under symbiotic conditions. *Proc. Nat. Acad. Sci. U.S.A.* **47**; 24–36.

AKKERMANS, A. D. L. (1971). "Nitrogen fixation and nodulation of *Alnus* and *Hippophaë* under natural conditions." Ph.D. Thesis, Univ. Leiden, The Netherlands, 84 pp.

ALLEN, E. K. and ALLEN, O. N. (1950). The anatomy of nodular growths on the roots of *Tribulus cistoides* L., *Proc. Soil Sci. Soc. Amer.* **14**; 179–183.

ALLEN, E. K., ALLEN, O. N. and NEWMAN, S. A. (1953). Pseudonodulation of leguminous plants induced by 2-bromo-3,5 dichlorobenzoic acid. *Am. J. Bot.* **40**; 429–435.

ALLEN, J. D., SILVESTER, W. B. and KALIN, M. (1966). *Streptomyces* associated with root nodules of *Coriaria* in New Zealand. *N.Z. J. Bot.* **4**; 57–65.

APRISON, M. H., MAGEE, W. E. and BURRIS, R. H. (1954). Nitrogen fixation by excised soybean root nodules. *J. biol. Chem.* **208**; 29–39.

ARORA, N., SKOOG, F. and ALLEN, O. N. (1959). Kinetin-induced pseudonodules on tobacco roots. *Proc. Soil. Sci. Soc. Am.* **46**; 610–613.

ARZBERGER, E. G. (1910). The fungous root-tubercles of *Ceanothus americanus*, *Elaeagnus argentea*, and *Myrica cerifera*. Missouri Bot. Gardens, 21st Ann. Rep. 60–102.

ASSELINEAU, J. (1966). "The Bacterial Lipids." Hermann, Paris.

BECKING, J. H. (1961a). Molybdenum and symbiotic nitrogen fixation by alder (*Alnus glutinosa* Gaertn.). *Nature, Lond.* **192**; 1204–1205.

BECKING, J. H. (1961b). A requirement of molybdenum for the symbiotic nitrogen fixation in alder (*Alnus glutinosa* Gaertn.). *Pl. Soil* **12**; 217–228.

BECKING, J. H. (1962). Species differences in molybdenum and vanadium requirements and combined nitrogen utilization by *Azotobacteriaceae*. *Pl. Soil* **16**; 171–201.

BECKING, J. H. (1965). *In vitro* cultivation of alder root-nodule tissue containing the endophyte. *Nature, Lond.* **207**; 885–887.

BECKING, J. H. (1966). Interactions nutritionelles plantes-actinomycetes. Rapport général. *Ann. Inst. Pasteur (Paris)* **Suppl. 111**; 295–302.

BECKING, J. H. (1968). Nitrogen fixation by non-leguminous plants. Symposium "Nitrogen in Soil", Groningen, May 17–19, 1967. *In*: "Stikstof", Dutch Nitrogen Fertilizer Review, **12**; 47–74.

BECKING, J. H. (1970a). Plant-endophyte symbiosis in non-leguminous plants. 2nd Conference on Global Impacts of Applied Microbiology, Addis Ababa, Ethiopia, November 5–12, 1967. *Pl. Soil* **32**; 611–654.

BECKING, J. H. (1970b). Frankiaceae fam. nov. (Actinomycetales) with one new combination and six new species of the genus *Frankia* Brunchorst 1886, 174. *Int. J. syst. Bacteriol.* **20**; 201–220.

BECKING, J. H. (1971). The physiological significance of the leaf nodules of *Psychotria*. Proc. Technical Meetings on Biological Nitrogen Fixation of the Int. Biol. Programme (Section PP–N), Prague and Wageningen, 1970, *Pl. Soil*, Special vol.; 361–374.

BECKING, J. H. (1974). Family Frankiaceae. *In*: "Bergey's Manual of Determinative Bacteriology". 8th ed.; xxvi + 1246 pp. Williams and Wilkins Co., Baltimore, U.S.A.

BECKING, J. H., DE BOER, W. E. and HOUWINK, A. L. (1964). Electron microscopy of the endophyte of *Alnus glutinosa*. *Antonie van Leeuwenhoek*, **30**; 343–376.

BERGERSEN, F. J. (1960). Biochemical pathways in legume root nodule nitrogen fixation. *Bacteriol. Rev.* **24**; 246–250.

BERGERSEN, F. J. (1966a). Nitrogen fixation in breis of soybean root nodules. *Biochim. Biophys. Acta.* **115**; 247–249.

BERGERSEN, F. J. (1966b). Some properties of nitrogen-fixing breis from soybean root nodules. *Biochim. Biophys. Acta* **130**; 304–312.

BERGERSEN, F. J., BURRIS, R. H. and WILSON, P. W. (1961). Biochemical studies on soybean nodules. *In*: "Recent Advances in Botany", (Ed. D. L. Bailey), pp. 589–593. Univ. Toronto Press, Canada.

BOND, G. (1951). The fixation of nitrogen associated with the root nodules of *Myrica gale* L., with special reference to its pH relation and ecological significance. *Ann. Bot.* **15**; 447–459.

BOND, G. (1956). Some aspects of translocation in root nodule plants. *J. exp. Bot.* **7**; 387–394.

BOND, G. (1958). Symbiotic nitrogen fixation by non-legumes. *In*: "Nutrition of the Legumes", (Ed. E. G. Hallsworth), pp. 216–231. Butterworths Sci. Publ., London.

BOND, G. (1959). Fixation of nitrogen in non-legume root-nodule plants. *In*: "Utilization of of Nitrogen and its Compounds by Plants", (Ed. H. K. Porter), pp. 59–72. Cambridge Univ. Press, London.

BOND, G. (1960). Inhibition of nitrogen fixation in non-legume root nodules by hydrogen and carbon monoxide. *J. exp. Bot.* **11**; 91–97.

BOND, G. (1961). The oxygen relation of nitrogen fixation in root nodules. *Z. allg. Mikrobiol.* **1**; 93–99.

BOND, G. (1962). Fixation of nitrogen in *Coriaria myrtifolia*. *Nature, Lond.* **193**; 1103–1104.

BOND, G. (1964). Isotopic investigations of nitrogen fixation in non-legume root nodules. *Nature, Lond.* **204**; 600–601.

BOND, G. (1975). The results of the I.B.P. survey of root-nodule formation in non-leguminous angiosperms. *In* "Nitrogen Fixation in the Biosphere", (Ed. P. S. Nutman), Intern. Biol. Programme, PP-N Section, Intern. Synthesis Meeting, Edinburgh, September 1973.

BOND, G. and HEWITT, E. J. (1961). Molybdenum and the fixation of nitrogen in *Myrica* root nodules. *Nature, Lond.* **190**; 1033–1034.

BOND, G. and HEWITT, E. J. (1962). Cobalt and the fixation of nitrogen by root nodules of *Alnus* and *Casuarina*. *Nature, Lond.* **195**; 94–95.

BOND, G. and HEWITT, E. J. (1967). The significance of copper for nitrogen fixation in nodulated *Alnus* and *Casuarina* plants. *Pl. Soil* **27**; 447–449.

BOND, G., FLETCHER, W. W. and FERGUSON, T. P. (1954). The development and function of the root nodules of *Alnus, Myrica* and *Hippophaë*. *Pl. Soil* **5**; 309–323.

BOND, G., ADAMS, J. F. and KENNEDY, E. H. (1965). Vitamin Bizanalogues in non-legume root nodules. *Nature, Lond.* **207,** 319–320.

BRUNCHORST, J. (1886). Über einige Wurzelanschwellungen, besonders diejenigen von *Alnus* und den Elaeagnaceen. *Unters. bot. Inst. Tübingen* **2**; 151–117.

CHEN, H. K. (1938). Production of growth substance by clover nodule bacteria. *Nature, Lond.* **142**; 753–754.

CHEN, P. L. (1962). The fine structure of *Streptomyces cinnamonensis*. *In*: "Electron Microscopy", (Ed. S. S. Brees Jr.), pp. UU-5. Academic Press, New York, London and San Francisco.

CHEN, P. L. (1964). The membrane system of *Streptomyces cinnamonensis*. *Am. J. Bot.* **51**; 125–132.

CLARKE, G., DYE, M. H. and WAIN, R. L. (1959). Occurrence of 3-indolylacetic and 3-indolecarboxylic acids in tomato crown gall tissue extracts. *Nature, Lond.* **184**; 825–826.

CLAWSON, M. A., FARNSWORTH, R. B. and HAMMOND, M. (1972). New species of nodulated non-legumes on range and forest soils. *Agron. Abstr.;* 138.

COOK, C. B., GOLEB, J. A. and MIDDELBOE, V. (1967). Optical nitrogen-15 analysis of small nitrogen samples using a noble gas to sustain the discharge. *Nature, Lond.* **216**; 475–476.

CROCKER, R. L. and DICKSON, B. A. (1957). Soil development on the recessional moraines of the Herbert and Mendenhall glaciers, South-eastern Alaksa. *J. Ecol.* **45**; 169–185.

CROCKER, R. L. and MAJOR, J. (1955). Soil development in relation to vegetation and surface age at Glacier Bay, Alaska. *J. Ecol.* **43**; 427–448.

DALY, G. T. (1966). Nitrogen fixation by nodulated *Alnus rugosa*. *Can. J. Bot.* **44**; 1607–1621.

DANILEWICZ, K. (1965). Symbiosis in *Alnus glutinosa* (L.) Gaertn. *Acta Microbiol. Polon.* **14**; 321–326.

DARLINGTON, C. D. and WYLIE, A. P. (1961) "Chromosome Atlas of Flowering Plants," George Allen and Unwin Ltd. London, 2nd ed., 519 pp.

DAVENPORT, H. E. (1960). Haemoglobin in the root nodules of *Casuarina cunninghamiana*. *Nature, Lond.* **186**; 653–654.

DELVER, P. and POST, A. (1968). Influence of alder hedges on the nitrogen nutrition of apple trees. *Pl. Soil* **28**; 325–336.

DELWICHE, C. C., JOHNSON, C. M. and REISENAUER, H. M. (1961). Influence of cobalt on nitrogen fixation by *Medicago*. *Plant Physiol.* **36**; 73–78.

DELWICHE, C. C., ZINKE, P. J. and JOHNSON, C. M. (1965). Nitrogen fixation by *Ceanothus*. *Plant Physiol.* **40**; 1045–1047.

DILWORTH, M. J. (1966). Acetylene reduction by nitrogen-fixing preparations from *Clostridium pasteurianum*. *Biochim. Biophys. Acta* **127**; 285–294.

DOMMERGUES, Y. (1963). Evaluation du taux de fixation de l'azote dans un sol dunair reboisé en filao (*Casuarina equisetifolia*). *Agrochimica* **7**; 335–340.

DULLAART, J. (1967). Quantitative estimation of indoleacetic acid and indolecarboxylic acid in root nodules and roots of *Lupinus luteus* L., *Acta Bot. Neer.* **16**; 222–230.

DULLAART, J. (1970). The auxin content of root nodules and roots of *Alnus glutinosa*, *J. exp. Bot.* **21**; 975–984.

EGLE, K. and MUNDING, H. (1951). Über den Gehalt an Häminkörpern in den Wurzelknöllchen von Nicht-Leguminosen. *Naturwissenschaften* **38**; 548–549.

ENGLER, A. (1964), "Syllabus der Pflanzenfamilien," Vol. **2**, Angiospermen, 12. Aufl., Bearbeitet von H. Melchior, 666 pp. Borntraeger, Berlin.

FAHRAEUS, G. (1957). The infection of clover root hairs by nodule bacteria studied by a simple glass slide technique. *J. gen. Microbiol.* **16**; 374–381.

FARNSWORTH, R. B. and CLAWSON, M. A. (1972). Nitrogen-fixation by *Artemisia ludoviciana* determined by acetylene-ethylene gas assay. *Agron. Abstr.;* 96.

FAUST, H. (1960). Zur spektroskopischen ^{15}N-Bestimmung unter Verwendung eines Stufenfilters. *Z. Anal. Chem.* **175**; 9–18.

FAUST, H. (1965). Zur Probenchemie von Stickstoffverbindungen für die emissionsspektrometrische Isotopenanalyse des Stickstoffs. *Isotopenpraxis* **1**; 62–65.

FAUST, H. (1967). Probenchemie ^{15}N-markierter Stickstoffverbindungen im Mikro- bis Nanomolbereich für die emissionsspektrometrische Isotopenanalyse. *Isotopenpraxis* **3**; 100–103.

FITZ-JAMES, P. C. (1960). Participation of the cytoplasmic membrane in the growth and spore formation of Bacilli. *J. Biophys. biochem. Cytol.* **8**; 507–528.

FIUCZEK, M. (1959). Fixation of atmospheric nitrogen in pure cultures of *Streptomyces alni* (In Polish). *Acta Microbiol. Polon.* **8**; 283–287.

FRANK, B. (1887). Sind die Wurzelanschwellungen der Erle und Elaeagnaceen Pilzgallen? *Ber. Deut. bot. Ges.* **5**; 50–58.

FRANK, B. (1891). Über die auf Verdauung von Pilzen abzielende Symbiose der mit endotrophen Mykorhizen begabten Pflanzen, sowie der Leguminosen und Erlen. *Ber. Deut. bot. Ges.* **9**; 244–253.

GARDNER, I. C. (1965). Observations on the fine structure of the endophyte of the root nodules of *Alnus glutinosa* (L.) Gaertn. *Arch. Mikrobiol.* **51**; 365–383.

GARDNER, I. C. and BOND, G. (1966). Host plant-endophyte adaptation in *Myrica*. *Naturwissenschaften* **53**; 161.

GATNER, E. M. S. and GARDNER, I. C. (1970). Observations on the fine structure of the root nodule endophyte of *Hippophaë rhamnoïdes* L. *Arch. Mikrobiol.* **70**; 183–196.

GIESBRECHT, P. (1960). Über "organisierte" Mitochondrien und andere Feinstrukturen von *Bacillus megaterium*. *Z.bl Bakteriol. Parasitenk. I. Abt. Orig.* **179**; 538–581.

GLAUERT, A. M. and HOPWOOD, D. A. (1960). The fine structure of *Streptomyces coelicolor*. I. The cytoplasmic membrane system. *J. biophys. biochem. Cytol.* **7**; 479–488.

GOLEB, J. A. and MIDDELBOE, V. (1968). Optical nitrogen-15 analysis of small nitrogen samples with a mixture of helium and xenon to sustain the discharge in an electrodeless tube. *Anal. Chim. Acta* **43**; 229–234.

HARDY, R. W. F., BURNS, R. C. and HOLSTEN, R. D. (1973). Applications of the acetylene-ethylene assay for measurement of nitrogen fixation. *Soil Biol. Biochem.* **5**; 47–81.

HAWKER, L. E. and FRAYMOUTH, J. (1951). A re-investigation of the root-nodules of species of *Elaeagnus*, *Hippophaë*, *Alnus* and *Myrica*, with special reference to the morphology and life histories of the causative organisms. *J. gen. Microbiol.* **5**; 369–386.

HEWITT, E. J. and BOND, G. (1961). Molybdenum and the fixation of nitrogen in *Casuarina* and *Alnus* root nodules. *Pl. Soil* **14**; 159–175.

HEWITT, E. J. and BOND, G. (1966). The cobalt requirement of non-legume root nodule plants. *J. exp. Bot.* **17**; 480–491.

HEWITT, E. J. and HALLAS, D. G. (1951). The use of *Aspergillus niger* (Van Tiegh.) M strain as a test organism in the study of molybdenum as plant nutrient. *Pl. Soil.* **3**; 366–408.

HILTNER, L. (1896). Über die Bedeutung der Wurzelknöllchen von *Alnus*. *Landw. Vers. Sta.* **46;** 153–161.

HUBBELL, D. H. (1970). Studies on the root hair 'curling factor' of *Rhizobium*. *Bot. Gaz.* **131**; 337–342.

HUTCHINSON, J. (1973). "The Families of Flowering Plants (arranged according to a new system based on their probable phylogeny)," Clarendon Press, Oxford, 3rd ed., 968 pp.

IMAEDA, T. and OGURA, M. (1963). Formation of intracytoplasmic membrane system of Mycobacteria related to cell division. *J. Bacteriol.* **85**; 150–163.

KÄPPEL, M. and WARTENBERG, H. (1958). Der Formenwechsel des *Actinomyces alni* Peklo in den Wurzeln von *Alnus glutinosa* Gaertner. *Arch. Mikrobiol.* **30**; 46–63.

KEFFORD, N. P., BROCKWELL, J. and ZWAR, J. A. (1960). The symbiotic synthesis of auxin by legumes and nodule bacteria and its role in nodule development. *Aust. J. Biol. Sci.* **13**; 456–467.

KLIEWER, M. and EVANS, H. J. (1962). B_{12} coenzyme content of the nodules from legumes, alder and of *Rhizobium meliloti*. *Nature, Lond.* **194**; 108–109.

KLUCAS, R. V., KOCH, B., RUSSELL, S. A. and EVANS, H. J. (1968). Purification and some properties of the nitrogenase from soybean (*Glycine max* Merr.) nodules. *Plant Physiol.* **43**; 1906–1912.

KOCH, B., EVANS, H. J. and RUSSELL, S. (1967). Reduction of acetylene and nitrogen gas by breis and cell-free extracts of soybean root nodules. *Plant Physiol.* **42**; 466–468.

KODAMA, A. (1967a). Cytological studies on root nodules of some species in leguminosae II. *Bot. Mag.* (*Tokyo*) **80**; 92–99.

KODAMA, A. (1967b). Karyological studies on root nodules of three species of *Alnus*. *Bot. Mag.* (*Tokyo*) **80**; 230–232.

KODAMA, A. (1968). Cytological studies on root nodules of some species in leguminosae III. Origin of tetraploid root nodules in *Astragalus sinicus*. *Bot. Mag.* (*Tokyo*) **81**; 459–463.

KODAMA, A. (1970). Cytological and morphological studies on the plant tumors. II. Root nodules of three species of *Alnus*. *J. Sci. Hiroshima Univ. Ser. B.* **13**; 261–264.

KOSLOVA, E. I., BADUMYAN, L. S. and VENDILO, M. V. (1966). Properties of bacteria from the nodules of sea buckthorn. *Mikrobiologiya* **35**; 699–706 (in Russian, seen in English translation *Microbiology* **35**; 591–597.

LAWRENCE, D. B. (1953). Development of vegetation and soil in southeastern Alaska with special reference to the accumulation of nitrogen. *Final Rep., Off. Naval Res. Project Nr.* **160–183**; Washington; 39 pp.

LAWRENCE, D. B. (1958). Glaciers and vegetation in southeastern Alaska. *Am. Sci.* **46**; 89–122.

LAWRENCE, D. B. and HULBERT, L. (1950). Growth stimulation to adjacent plants by lupine and alder on recent glacier deposits in southeastern Alaska. *Bull. Ecol. Soc. Am.* **31**; 58.

LAWRENCE, D. B., SCHOENIKE, R. A., QUISPEL, A. and BOND, G. (1967). The role of *Dryas drummondii* in vegetation development following ice recession at Glacier Bay, Alaska, with special reference to its nitrogen fixation by root nodules. *J. Ecol.* **55**; 793–813.

LEAF, G., GARDNER, I. C. and BOND, G. (1958). Observations of the composition and metabolism of the nitrogen-fixing root nodules of *Alnus*. *J. exp. Bot.* **9**; 320–331.

LEAF, G., GARDNER, I. C. and BOND, G. (1959). Observations on the composition and metabolism of the nitrogen-fixing root nodules of *Myrica*. *Biochem. J.* **72**; 662–667.

LECHEVALIER, H., LECHEVALIER, M. P. and BECKER, B. (1966). Comparison of the chemical composition of cell-walls of Nocardiae with that of other aerobic actinomycetes. *Int. J. syst. Bacteriol.* **16**; 151–160.

LECHEVALIER, M. P. and FEKETE, E. (1971). Chemical methods as criteria for separation of Actinomycetes into genera. Workshop sponsored by Subcommittee on Actinomycetes of the Amer. Soc. for Microbiol., 23 pp.

LECHEVALIER, M. P., LECHEVALIER, H. A. and HEINTZ, C. E. (1973). Morphological and chemical nature of the sclerotia of *Chainia olivacea* Thirumalachar and Sukapure of the order *Actinomycetales*. *Int. J. syst. Bacteriol.* **23**; 157–170.

LEICKNAM, J. P., MIDDELBOE, V. and PROKSCH, G. (1968). Analyse isotopique de l'azote par spectrométrie optique pour de faibles teneurs en ^{15}N. *Anal. Chim. Acta.* **40**; 487–502.

MACCONNELL, J. T. (1959). The oxygen factor in the development and function of the root nodules of alder. *Ann. Bot.* **23**; 261–268.

MACCONNELL, J. T. and BOND, G. (1957). A comparison of the effect of combined nitrogen on nodulation in non-legumes and legumes. *Pl. Soil* **8**; 378–388.

MACKINTOSH, A. H. and BOND, G. (1970). Diversity in the nodular endophytes of *Alnus* and *Myrica. Phyton* **27**; 79–90.

MCLUCKIE, J. (1923). Studies in symbiosis. IV. The root-nodules of *Casuarina cunninghamiana* and their physiological significance. *Proc. Linn. Soc. N.S.W.* **48**; 194–204.

MÖLLER, H. (1885). *Plasmodiophora alni. Ber. Deut. bot. Ges.* **3**; 102–105.

MOORE, A. W. (1964). Note on non-leguminous nitrogen-fixing plants in Alberta. *Can. J. Bot.* **42**; 952–955.

MORTENSON, L. E. (1962). Inorganic nitrogen assimilation and ammonia incorporation. *In*: "Bacteria" vol. **3**; (Eds. I. C. Gunsales and R. Y. Stanier), pp. 119–166. Academic Press, New York, London and San Francisco.

MOSSE, B. (1964). Electron-microscope studies of nodule development in some clover species. *J. gen. Microbiol.* **36**; 49–66.

MOSTAFA, M. A. and MAHMOUD, M. Z. (1951). Bacterial isolates from root nodules of Zygophyllaceae. *Nature, Lond.* **167**; 446–447.

MOWRY, H. (1933). Symbiotic nitrogen fixation in the genus *Casuarina. Soil Sci.* **36**; 409–425.

NIEWIAROWSKA, J. (1961). Morphologie et physiologie des Actinomycetes symbiotiques des *Hippophaë. Acta Microbiol. Polon.* **10**; 271–286.

NUTMAN, P. S. (1958). The physiology of nodule formation. *In*: "Nutrition of the Legumes", (Ed. E. G. Hallsworth), pp. 87–107. Butterworths Sci. Publ., London.

NUTMAN, P. S. (1959). Some observations on root-hair infection by nodule bacteria. *J. exp. Bot.* **10**; 250–263.

OVINGTON, J. D. (1956). Studies on the development of woodland conditions under different trees. IV. The ignition loss, water, carbon and nitrogen content of the mineral soil. *J. Ecol.* **44**; 171–179.

PERSCHKE, H. and PROKSCH, G. (1971). Analysis of ^{15}N abundance in biological samples by means of emission spectrometry. *In*: "Nitrogen-15 in Soil-Plant Studies," pp. 223–225. F.A.O./I.A.E.A., Vienna.

PHILLIPS, D. A. and TORREY, J. G. (1970). Cytokinin production by *Rhizobium japonicum. Physiol. Plant.* **23**; 1057–1063.

PHILLIPS, D. A. and TORREY, J. G. (1972). Studies on cytokinin production by *Rhizobium. Plant Physiol.* **49**; 11–15.

PIZELLE, G. (1965). L'azote minéral et la nodulation de l'aune glutineux (*Alnus glutinosa*): Observations sur des plantes cultivées avec systèmes racinaires compartimentés. *Bull. Ecole Nat. Sup. Agron. Nancy* **7**; 55–63.

PIZELLE, G. (1966). L'azote minéral et la nodulation de l'aune glutineux (*Albus glutinosa*). II. Observations sur l'action inhibitrice de l'azote minéral à l'égard de la nodulation. *Ann. Inst. Pasteur (Paris)* **Suppl. 3**; 259–264.

PIZELLE, G. (1970). Effects comparés de l'acide β-indolyl-acétique et de l'azote nitrique sur la nodulation de l'aune glutineux (*Alnus glutinosa*). *Bull. Acad. Soc. Lorraines Sci.* **9**; 174–178.

RODRIGUEZ-BARRUECO, C. (1966). Fixation of nitrogen in root nodules of *Alnus jorullensis* H.B. and K. *Phyton.* **23**; 103–110.

RODRIGUEZ-BARRUECO, C. (1968). The occurrence of the root-nodule endophytes of *Alnus glutinosa* and *Myrica gale* in soils. *J. gen. Microbiol.* **52**; 189–194.

RODRIGUEZ-BARRUECO, C. (1969). The occurrence of nitrogen-fixing root nodules on non-leguminous plants. *Bot. J. Linn. Soc.* **62**; 77–84.

Rodriguez-Barrueco, C. and Bermudez de Castro, F. (1973). Cytokinin-induced pseudonodules on *Alnus glutinosa. Physiol. Plant.* **29**; 277–280.

Rodriguez-Barrueco, C. and Bond, G. (1968). Nodule endophytes in the genus *Alnus. In*: "Biology of Alder", (Eds. J. M. Trappe, J. F. Franklin, R. F. Tarrant and G.M. Hansen), pp. 185–192. Pacific Northwest Forest and Range Exp. Sta., Portland, Oregon, U.S.A.

Russell, S. A. and Evans, H. J. (1966). The nitrogen-fixing capacity of *Ceanothus velutinus. For. Sci.* **12**; 164–169.

Russell, S. A., Evans, H. J. and Mayeux, P. (1968). The effect of cobalt and certain other trace elements on the growth and vitamin B_{12} content of *Alnus rubra. In*: "Biology of Alder," (Eds. J. M. Trappe, J. F. Franklin, R. F. Tarrant and G. M. Hansen) pp. 259–271. Pacific Northwest Forest and Range Exp. Sta., Portland, Oregon, U.S.A.

Sabet, Y. S. (1946) Bacterial root nodules in Zygophyllaceae. *Nature, Lond.* **157**; 656–657.

Sahlman, K. and Fåhraeus, G. (1962). Microscopic observations on the effect of indole-3-acetic acid upon root hairs of *Trifolium repens. Kgl. Lantbrukshögskol. Ann.* **28**; 261–268.

Schlegel, H. G. (1962). Die Isolierung von Poly-β-hydroxybuttersäure aus Wurzelknöllchen von Leguminosen. *Flora* **152**; 236–240.

Schöllhorn, R. and Burris, R. H. (1966). Study of intermediates in nitrogen fixation. *Fed. Proc.* **25**; 710.

Schöllhorn, R. and Burris, R. H. (1967). Acetylene as a competitive inhibitor of N_2 fixation. *Proc. Nat. Acad. Sci. U.S.A.* **58**; 213–216.

Schröter, J. (1889). *In*: "Cohn's Krytogamen-Flora von Schlesien," vol. **3**; Erste Hälfte; p. 134. J. U. Kern's Verlag, Breslau, Germany.

Seidel, K. (1972). Exsudat-Effekt der Rhizodamnien von *Alnus glutinosa* Gaertner. *Naturwissenschaften* **59**; 366–367.

Short, K. C. and Torrey, J. G. (1972). Cytokinins in seedling roots of pea. *Plant. Physiol.* **49**; 155–160.

Silver, W. S., Bendana, F. E. and Powell, R. D. (1966). Root nodule symbiosis II. The relation of auxin to root geotropism in roots and root nodules of non-legumes. *Physiol. Plant.* **19**; 207–218.

Silver, W. S. and Mague, T. (1970). Assessment of nitrogen fixation in terrestrial environments in field conditions. *Nature, Lond.* **227**; 378–379.

Sloger, C. (1968), "Nitrogen fixation by tissues of leguminous and non-leguminous plants." Ph.D. thesis, Univ. Florida, U.S.A. 96 pp.

Sloger, C. and Silver, W. S. (1965). Note on nitrogen fixation by excised root nodules and nodular homogenates of *Myrica cerifera* L. *In*: "Non-heme Iron Proteins: Role in Energy Conversion". (Ed. A. San Pietro) pp. 299–302. Antioch Press, Yellow Springs, Ohio, U.S.A.

Sloger, C. and Silver, W. S. (1966). Nitrogen fixation by excised root nodules and nodular homogenates of *Myrica cerifera* L. *Abstracts 9th Int. Congr. Microbiol.*, Moscow, 285.

Smith, J. D. (1949). The concentration and distribution of haemoglobin in the root nodules of leguminous plants. *Biochem. J.* **44**; 585–591.

Solheim, B. and Raa, J. (1973). Characterization of the substances causing deformation of root hairs of *Trifolium repens* when inoculated with *Rhizobium trifolii. J. gen. Microbiol.* **77**; 241–247.

Spratt, E. R. (1912). The morphology of the root tubercles of *Alnus* and *Elaeagnus* and the polymorphism of the organism causing their formation. *Ann. Bot.* **26**; 119–128.

Stewart, W. D. P. (1962). A quantitative study of fixation and transfer of nitrogen in *Alnus. J. exp. Bot.* **13**; 250–256.

Stewart, W. D. P. (1963). The effect of combined nitrogen on growth and nodule development of *Myrica* and *Casuarina. Z. allg. Mikrobiol.* **3**; 152–156.

STEWART, W. D. P. and BOND, G. (1961). The effect of ammonium nitrogen in *Alnus* and *Myrica*. *Pl. Soil* **14**; 347–359.

STEWART, W. D. P., FITZGERALD, G. P. and BURRIS, R. H. (1968). Acetylene reduction by nitrogen-fixing blue-green algae. *Arch. Mikrobiol.* **62**; 336–348.

STEWART, W. D. P. and PEARSON, M. (1967). Nodulation and nitrogen-fixation by *Hippophaë rhamnoides* L. in the field. *Pl. Soil* **26**; 348–360.

TANNER, J. W. and ANDERSON, I. C. (1963). An external effect of inorganic nitrogen in root nodulation. *Nature, Lond.* **198**; 303–304.

TANNER, J. W. and ANDERSON, I. C. (1964). External effect of combined nitrogen on nodulation. *Plant Physiol.* **19**; 1039–1043.

TARRANT, R. F., LU, K. C., BOLLEN, W. B. and FRANKLIN, J. F. (1969). Nitrogen enrichment of two forest ecosystems by red alder (*Alnus rubra*). *U.S.D.A. Forest Serv. Res. Paper PNW-76. Pacific Northwest Forest and Range Exp. Sta., Portland, Oregon, U.S.A.*, 8 pp.

TARRANT, R. F. and TRAPPE, J. M. (1971). The role of *Alnus* in improving the forest environment. *Pl. Soil* **Spec. vol.**; 335–348.

THIMANN, K. V. (1936). On the physiology of the formation of nodules on legume roots. *Proc. Nat. Acad. Sci. U.S.*A. **22**; 511–514.

THIMANN, K. V. (1939). The physiology of nodule formation. *Trans. 3rd Comm. Int. Soc. Soil Sci.* **A**; 24–281

THIMANN, K. V. and SACHS, T. (1966). The role of cytokinins in the "fasciation" disease caused by *Corynebacterium fascians*. *Am. J. Bot.* **53**; 731–739.

THORNTON, H. G. (1946). The nodule bacteria of leguminous plants. *Rept. Rothamsted Exp. Sta. 1939–1945*; 74–82.

TRINICK, M. J. (1973). Symbiosis between *Rhizobium* and the non-legume *Trema aspera*. *Nature, Lond.* **244**; 459–460.

TRINICK, M. J. (1975). Rhizobium symbiosis with a non-legume. *In* "International Symposium on Nitrogen Fixation: Interdisciplinary Discussions", (Eds. W. E. Newton and C. J. Nyman), Washington State University, Pullman, Washington, June 3–7, 1974.

TROLLDENIER, G. (1959). Polyploidie und Knöllchenbildung bei Leguminosen. *Arch. Mikrobiol.* **32**; 328–345.

UEMURA, S. (1952a). Studies on the root nodules of alders (*Alnus* spp.). (IV). Experiments on the isolation of actinomycetes from alder root nodules. *Bull. Govt. Forest Exp. Sta. (Japan)* **52**; 1–18.

UEMURA, S. (1952b). Studies on the root nodules of alders (*Alnus* spp.). (V) .Some new isolation methods of *Streptomyces* from alder root nodules. *Bull. Govt. Forest Exp. Sta. (Japan)* **57**; 209–226.

UEMURA, S. (1961). Studies on the *Streptomyces* isolated from alder root nodules. *Sci. Rep. Agr. Forest Fisheries Res. Council (Japan)* **7**; 1–90.

UEMURA, S. (1971). Non-leguminous root nodules in Japan. *Pl. Soil* **Spec. vol.**; 349–360.

VALERA, C. L. and ALEXANDER, M. (1965). Reversal of nitrate inhibition of nodulation by indolyl-3-acetic acid. *Nature, Lond.* **206**; 326.

VAN DER PIJL, L. (1938). The re-establishment of vegetation on Mt. Goentoer (Java). *Ann. Jard. bot. Buitenz.* **48**; 129–152.

VIRTANEN, A. I. (1957). Investigations on nitrogen fixation by the alder. II. Associated culture of spruce and inoculated alder without combined nitrogen. *Physiol. Plant.* **10**; 164–169.

VIRTANEN, A. I. (1962). On the fixation of molecular nitrogen in nature. *Commun. Inst. Forest. Fenniae* **55.22**; 1–11.

VOIGT, G. K. and STEUCEK, G. L. (1969). Nitrogen distribution and accretion in an alder ecosystem. *Soil Sci. Soc. Am. Proc.* **33**; 946–949.

VON PLOTHO, O. (1941). Die Synthese der Knöllchen an den Wurzeln der Erle. *Arch. Mikrobiol.* **12**; 1–18.

WETZEL, G. (1929). Chromosomenstudien bei den Fagales. *Bot. Arch.* **25**; 258–284.

WHEELER, C. T. (1969). The diurnal fluctuation in nitrogen fixation in the nodules of *Alnus glutinosa* and *Myrica gale*. *New Phytol.* **68**; 675–682.

WHEELER, C. T. (1971). The causation of the diurnal changes in nitrogen fixation in the nodules of *Alnus glutinosa*. *New Phytol.* **70**; 487–495.

WHEELER, C. T. and BOND, G. (1970). The amino acids of non-legume root nodules. *Phytochem.* **9**; 705–708.

WILLIS, J. C. (1973). "A Dictionary of the Flowering Plants and Ferns", 8th Ed. (revised by H. K. Airy Shaw), Cambridge Univ. Press, London, 1245 pp + lxvi.

WILSON, E. H. (1958). Asymbiotic nitrogen fixation. *In*: "Handbuch der Pflanzenphysiologie," **vol. 8** (Ed.W. Ruhland) pp. 9–47. Springer Verlag, Germany.

WIPF, L. (1939). Chromosome numbers in root nodules and root tips of certain Leguminosae. *Bot. Gaz.* **101**; 51–67.

WIPF, L. and COOPER, D. C. (1938). Chromosome numbers in nodules and roots of red clover, common vetch and garden pea. *Proc. Nat. Acad. Sci. U.S.A.* **24**; 87–91.

WIPF, L. and COOPER, D. C. (1940). Somatic doubling of chromosomes and nodular infection in certain Leguminosae. *Am. J. Bot.* **27**; 821–824.

WOLLUM II, A. G., YOUNGBERG, C. T. and GILMOUR, C. M. (1966). Characterization of a *Streptomyces* sp. isolated from root nodules of *Ceanothus velutinus* Dougl. *Soil Sci. Soc. Am. Proc.* **30**; 463–467.

WOODWORTH, R. H. (1931). Polyploidy in the Betulaceae. *J. Arnold Arboretum* (Harvard Univ.) **12**; 206–217.

WORONIN, M. (1866). Über die bei der Schwarzerle (*Alnus glutinosa*) und der gewöhnlichen Garten-Lupine (*Lupinus mutabilis*) auftretenden Wurzelanschwellungen. *Mém. Acad. Imp. Sci. St. Pétersbourg, Serie* 7, **10**; No. 6; 1–13.

YENDO, Y. and TAKASE, T. (1932). On the root-nodule of *Elaeagnus*. *Bull. Seric. Silk Industr., Uyeda (Japan)* **4**; 114–134. (Japanese, with English summary).

ZAVITKOVSKI, J. and NEWTON, M. (1968). The effect of combined nitrogen and organic matter on nodulation and nitrogen fixation of red alder. *In*: "Biology of Alder", (Eds. J. M. Trappe, J. F. Franklin, R. F. Tarrant and G. M. Hansen) pp. 209–224. Pacific Northwest Forest and Range Exp. Sta., Portland, Oregon, U.S.A.

ZAVITKOVSKI, J. and NEWTON, M. (1971). Litterfall and litter accumulation in red alder stands in Western Oregon. *Pl. Soil* **35**; 257–268.

ZIEGLER, H. (1962). Die Rhizothamnien bei *Comptonia peregrina* (L.) Coult. *Mitt. Deut. dendrol. Ges.* **61**; Jahrb. 1959/60; 28–31.

Chapter 23

The Genetics of Mycorrhizal Associations between *Amanita muscaria* and *Betula verrucosa*

P. MASON

Institute of Terrestrial Ecology, Unit of Tree Biology, Bush Estate, Penicuik, Midlothian, Scotland.

I. Introduction

Complex populations of microbes living in soil play an indispensable part in the circulation of nutrients, transforming them into forms that are "available" to plants. These changes occur most actively in the rhizosphere and on the rhizoplane where the complexes of microbes contain free-living saprophytes, which utilize decomposing organic matter and parasites. The latter group includes unspecialized microbes which cause damage to their hosts' tissues and others that are highly specialized entering into balanced symbiotic relationships in which the partners usually contribute to the well-being of each other. The microsymbionts include species of the bacterial genus *Rhizobium*, found in legume nodules, actinomycetes in nodules of *Alnus* and *Myrica* and fungi which form mycorrhizas.

The important contributions made by leguminous crops to amounts of soil nitrogen have been sufficient reason for intensive studies of nodule formation and activity. These have demonstrated that (a) strains of *Rhizobium* species have different abilities to form nodules and (b) all nodules are not similarly effective in fixing nitrogen (Nutman, 1969). Further, with the increasing demand for unvarying pure lines of legumes the propensity of different strains of legumes to form nodules has been found to differ appreciably. Evidence has, therefore, accumulated to show that factors in both the host and bacterium control the formation and subsequent activity of legume nodules. To increase plant yield and maximize soil fertility the recognition of these features has led to techniques in which legume seeds are inoculated with highly effective strains of *Rhizobium*.

Although sheathing mycorrhizas (ectomycorrhizas) have been the subject of scientific interest for many decades their formation is still poorly understood. By analogy with the *Rhizobium*–legume complex, factors present within both the host and fungus may control the formation of sheathing mycorrhizas and influence their ability to take up and utilize nutrients. Rosendahl (1942) found that pine seedlings grew better in the presence of *Boletus felleus* than B. *granulatus*. Benecke and Göbl (1974) reported that mycorrhizas formed on seedlings of mountain pine (*Pinus mugo*) by a single symbiont, *Hebeloma mesophaeum*, were more efficient in terms of seedling nutrition than mycorrhizas formed by the natural fungus population from a native site. Lundeberg (1970) found that seedlings of *Pinus sylvestris* formed mycorrhizas with *Boletus bovinus* and *B. granulatus* but not with *B. submentosus* and *Paxillus involutus*. However in tests using more than one isolate of *B. submentosus* he found that mycorrhizas were formed with one but not with the others thus indicating within-species differences. Each fungal isolate can be considered as having a unique character. Using pot culture experiments, Levisohn (1961) suggested that the unequal performance of young trees might be attributable to differing benefits gained from differing strains of the same symbiotic fungal species, a suggestion that may explain the apparent inconsistencies among the performances of some mycorrhizal species, as recorded by different research workers. For example, the drought tolerant fungus, *Cenococcum graniforme* was found to be most beneficial to plant growth by Shemakhanova (1967) but somewhat inhibitory by Lundeberg (1970) and Mejstrik and Krause (1973). In an analogous manner, it has been suggested that genotypic variation within a host species might influence the development of sheathing mycorrhizas (Marx and Bryan, 1971) but as yet there is little substantive supporting evidence. Linnemann (1960) and Wright and Ching (1962) found differences in the frequency of occurrence of mycorrhizas on seedlings of *Pseudotsuga menziesii* derived from different seed sources. Lundeberg (1968) found distinct mycorrhizal differences among different provenances of *Pinus sylvestris*. Approximately 80% of the roots of three progenies became mycorrhizal whereas only 53% of the fourth developed mycorrhizas. Interestingly,

Marx and Bryan (1971) noted that the growth responses of the progenies differed.

Although still scanty, there is evidence suggesting that the formation of sheathing mycorrhizas and the benefit gained by their tree hosts are both controlled genetically. If this is so, the inclusion of mycorrhizal selection into tree breeding programmes might be profitable.

With this in mind a search for a screening test was initiated at the Unit of Tree Biology, the aim being to seek variation within fungal and plant components of the symbiotic complexes. A technique was required to enable large numbers of host seedlings and fungal isolates, from different origins, to be screened singly and in combination, with their interactions being closely monitored; the work was focussed on *Amanita muscaria* and birch (*Betula verrucosa*).

II. Materials and Methods

A. Materials

1. *Plant Material*

Prior to each experiment, after being shaken in 30% hydrogen peroxide for 30 min, seeds of *Betula verrucosa* Ehrh. were washed in sterile distilled water and plated onto water agar. Germination occurred after 7–10 days incubation at room temperature.

2. *The Fungus*

Isolates of *Amanita muscaria* (L. ex Fr.) Pers. ex Hooker were made from sporophore tissue and then maintained on Hagem's agar (Hagem, 1910).

For flask culture experiments, five pieces of agar inoculum (3–5 mm^2) were placed in each 250 ml Erlenmeyer flask containing 80 ml maize seeds and 60 ml Hagem's liquid medium. The flasks were then incubated for 6 weeks at room temperature so as to allow a thick superficial fungal mat to form.

To ensure that cultures of *Amanita muscaria* were growing actively, colonized agar blocks (2–3 mm^2) were transferred to fresh agar 10–12 days before being inoculated to tube culture.

B. Techniques

1. *Flask Culture*

A modified Marx and Zak (1965) technique was used. Wide necked Erlenmeyer flasks (500 ml volume), each containing 250 ml of thoroughly mixed vermiculite/peat (7·8 : 1·2) and 166 ml of modified Melin–Norkrans nutrient solution (Norkrans, 1949), were autoclaved for 30 min. at 121°C. (The modified medium differed from the original by the inclusion of 100 μg/l thiamine hydrochloride and the replacement of sucrose by 10 g/litre glucose).

Flasks were seeded with either three maize grains colonized by *Amanita muscaria* or, in the control series, with three grains autoclaved after colonization by the appropriate isolate. Three days later, following incubation at 23°C during which flasks were monitored for contaminants, two aseptically germinated birch seedlings were planted into each flask. The flasks, arranged in factorial experimental designs, were given continuous light from "Gro-lux" lamps.

After ten weeks incubation, seedlings were harvested when flasks were filled with tap water and agitated so as to loosen the substrate adhering to the roots of the birch seedlings. Subsequently the seedlings were removed and a further attempt was made to release adhering particles. Numerous criteria of growth, including total numbers of roots, numbers of laterals and numbers of mycorrhizal and non-mycorrhizal short roots, were recorded and analysed, care being taken to exclude pseudomycorrhizas. For this, randomly selected short roots were fixed in weak formalin-acetic acid-alcohol (F.A.A.), dehydrated in Johansen's tertiary butyl alcohol series and infiltrated with paraffin wax. Sections cut at 12 μm, after being fixed to glass slides with Haupts adhesive, were stained using Pianeze III B (Simmons and Shoemaker, 1952) and examined.

2. *Tube Culture*

Twelve millilitres of a nutrient agar medium (Pelham and Mason, in *litt*), containing a modification of Ingestad's inorganic salts (Ingestad, 1970; 1971), glucose (10 g/l), and thiamine hydrochloride (50 μg/l) were put in each sterile polystyrene container (Sterilin Ltd., Teddington, Middlesex, U.K.). After gelling as a slope, aseptically germinated birch seedlings were placed on the surface (one plant per tube). Where appropriate, seedlings were inoculated with a piece of agar inoculum measuring 1–2 mm^2, placed in contact with the root system. Subsequently the tubes were arranged in a Trojan square experimental design (Darby and Gilbert, 1958) and incubated at room temperature in continous light.

After 6–8 weeks, the tubes were harvested and growth assessed. Seedlings were carefully removed from the agar medium and adhering fungal mycelium was teased from the roots before randomly selected samples were examined for the presence or absence of sheathing mycorrhizas by the method already described.

III. Results

A. Effect of the Fungus

Table I shows the results obtained from a typical flask experiment in which five isolates of different origin were factorially tested against two seed-lots of birch, one (A) collected from Scotland and the other (B) from Latvia. Each of the twelve factorial combinations plus the uninoculated controls were replicated × 6.

TABLE I. Development of mycorrhizas when the factorial combinations of (i) two seedlots of *Betula verrucosa* and (ii) five isolates of *Amanita muscaria*, were tested (flask culture).

Isolates of *A. muscaria*			Seedlots of *B. verrucosa*	
	Origin		A ex Scotland	B ex Latvia
Accession Number	Country	Associated tree	Percentage of roots with mycorrhiza[a]	
1	U.K.	Birch	54·2 ± 7·9	53·8 ± 10·4
2	U.K.	Birch	51·5 ± 9·2	59·5 ± 11·4
3	U.K.	Birch	15·7 ± 6·5	19·5 ± 7·8
4	U.K.	Pine	43·7 ± 10·6	41·4 ± 6·9
5	U.S.A.	Pine	0·0	0·0

[a] Analysed as angles.

The three birch isolates formed mycorrhizas with seedlings grown from Scottish and Latvian seeds but the percentages of mycorrhizal roots differed significantly. Two isolates transformed 50 to 60% of short roots to mycorrhizas whereas the third stimulated only 15–20%. There was an even more marked contrast between the two pine isolates of *Amanita muscaria*, the one transforming 41–44% short roots whereas the other had no effect.

These effects were confirmed in a tube culture experiment in which two isolates, both from birch, were tested against seed lot A growing on a modified Ingestad's medium containing 6·5 ppm phosphorus. Isolate 6, which spread over entire root systems, induced more than three times as many mycorrhizas as isolate 7 (Table II) which was restricted to the roots near to the bases of stems

TABLE II. Effects of mycorrhiza formation when differing isolates of *Amanita muscaria* were inoculated to seedlings of *Betula verrucosa* grown from one lot of seed (tube culture).

Isolate of *A. muscaria*	Mean number of mycorrhiza per seedling	% mycorrhizas that were branched
6	13·5	18·5
7	4·5	44·4

However, at least 40% of the fewer mycorrhizal short roots formed in association with isolate 7 were branched, some with as many as 10 to 20 branches. In contrast, less than 20% of those formed with isolate 6 were branched, none with more than five branches.

B. Effect of the Host

Although mycorrhizas are formed between *Amanita muscaria* and birch seedlings at both levels of phosphorus (Table III), the results from the data presented show that the number of mycorrhizas were found to depend both on the seed lot characteristics and amounts of phosphorus.

TABLE III. Interacting effects of seed sources and phosphate concentrations on numbers of mycorrhizas produced per seedling of *Betula verrucosa* when inoculated with *Amanita muscaria*, isolate 6 (tube culture)

		Sources of seeds of *B. verrucosa* A ex Scotland	B ex Latvia
Phosphate Concentrations (see text)	Low	13·5	27·5
	High	14·0	13·3

With isolate 6 the response of seed lot A to *Amanita muscaria* was not determined by the amount of available phosphorus there being *c* 13/14 mycorrhizas per seedling at the low and high levels of P respectively. By contrast, seed lot B, although forming the same number of mycorrhizas as seed lot A at the higher phosphorus level, developed more than twice as many with only 6·5 ppm of phosphorus.

The lack of constancy was also found when several assessments of growth, including estimates of root numbers, were made. By combining some of the factional treatments it was found that *Amanita muscaria* stimulated root production more on Latvian than Scottish derived seedlings; increases of 60% in the former and 27% in the latter were recorded (Table IV).

IV. Discussion and Conclusions

Results with A. *muscaria*/*B. verrucosa* confirm that host and fungal genotypes can affect the formation of sheathing mycorrhizas.

TABLE IV. Numbers of roots (mycorrhizal + non-mycorrhizal) produced when *Amanita muscaria* isolate 6, was inoculated to *Betula verrucosa* seedling grown from Latvian and Scottish seed (tube culture).

	Sources of seeds of *B. verrucosa* A ex Scotland	B ex Latvia
Uninoculated controls	30·3	41·7
Inoculated with *A. muscaria*	38·6	65·6

Variation attributable to *Amanita muscaria* was detected using two experimental methods, flask culture and tube culture. In both it was possible to use strictly controlled conditions but the tube culture technique is the more convenient as it is less demanding of space so enabling a large number of treatments to be tested.

In addition to controlling the extent of short root infection, *Amanita muscaria* influenced the pattern of mycorrhizal branching, some isolates stimulating small numbers of repeatedly branched mycorrhizas whereas others transformed greater numbers of sparsely branched ones. Is it possible to obtain crosses between these two types of isolates which would stimulate large numbers of repeatedly branched mycorrhizas?

With the tube culture method there was evidence to suggest that the host responded differently to different isolates of *Amanita muscaria* and that this response was influenced by nutrient status.

So far, the effects recorded in this paper have been obtained only in highly controlled conditions. They need to be checked in the field. Already the evidence strongly suggests that the genetics of sheathing mycorrhiza formation have features in common with the legume/*Rhizobium* complex. For the future there may be reason to consider the development of sheathing mycorrhiza in tree improvement programmes so extending the involvement of fungi beyond the more usual realm of plant pathology.

References

BENECKE, U. and GÖBL, F. (1974). The influence of different mycorrhizae on growth, nutrition and gas-exchange of *Pinus mugo* seedlings. *Pl. Soil* **40**; 21–32.

DARBY, L. A. and GILBERT, N. (1958). The Trojan Square. *Euphytica* **7**; 183–188.

HAGEM, O. (1910). Untersuchungen über norwegische Mucorineen II. *Vidensk. Selsk. Skrift* 1. *Math. naturw.* **Kl. 17.**

INGESTAD, T. (1970). A definition of optimum nutrient requirements in birch seedlings I. *Physiol. Plant.* **23**; 1127–1138.

INGESTAD, T. (1971). A definition of optimum nutrient requirements in birch seedlings II. *Physiol. Plant.* **24**; 118–125.

LEVISOHN, I. (1961). "Researches in Mycorrhiza." Forestry Commission Report on Forest Research March 1960.

LINNEMANN, G. (1960). Rassenunterschiede bei *Pseudotsuga taxifolia* hinsichtlich der Mycorrhiza. *Allg. Forst. Jagdztg.* **131**: 41–47.

LUNDEBERG, G. (1968). The formation of mycorrhizal in different provenances of pine (*Pinus sylvestris* L.). *Svensk. Bot. Tidskr.* **62**; 249–255.

LUNDEBERG, G. (1970). Utilisation of various nitrogen sources, in particular bound soil nitrogen, by mycorrhizal fungi. *Studia Forestalia Suecica* **79**; 1–95.

MARX, D. H. and ZAK, B. (1965). Effect of pH on mycorrhizal formation of slash pine in aseptic culture. *Forest Sci.* **11**; 66–75.

MARX, D. H. and BRYAN, W. C. (1971). Formation of ectomycorrhizae on half-sib progenies of slash pine in aseptic culture. *Forest Sci.* **17**; 488–492.

MEJSTŘIK, V. K. and KRAUSE, H. H. (1973). Uptake of ^{32}p by *Pinus radiata* roots inoculated with *Suillus luteus* and *Cenococcum graniforme* from different sources of available phosphate. *New Phytol.* **72**; 137–140.

NORKRANS, B. (1949). Some mycorrhiza-forming *Tricholoma* species. *Svensk Botan. Tidskr.* **43**; 485–490.

NUTMAN, P. S. (1969). Genetics of symbiosis and nitrogen fixation in legumes. *Proc. Roy. Soc. B.* **172**; 417–437.

ROSENDAHL, R. O. (1942). The effect of mycorrhizal and non-mycorrhizal fungi on the availability of difficulty-soluble potassium and phosphorus. *Soil Sci. Soc. Amer., Proc.* **7**; 477–479.

SHEMAKHANOVA, N. M. (1967). "Mycotrophy of Woody Plants." Isr. Program Sci. Transl., Jerusalem.

SIMMONS, S. A. and SHOEMAKER, R. A. (1952). Differential staining of fungus and host cells using a modification of Pianeze IIIb. *Stain Technology* **27**: 121.

WRIGHT, E. and CHING, K. K. (1962). Effect of seed source on mycorrhizal formation of Douglas fir seedlings. *Northwest Sci.* **36**; 1–6.

Chapter 24

Vesicular-arbuscular Mycorrhizae

J. W. GERDEMANN

Department of Plant Pathology, University of Illinois, Urbana, U.S.A.

I. Introduction

Vesicular-arbuscular (VA) mycorrhizae occur on more plant species and are more widely distributed geographically than any other type of mycorrhiza. They occur on most cultivated crops, native grasses, herbs and shrubs as well as the majority of forest and shade tree species. The fossil record suggests that the VA fungi were present in the subterranean parts of the earliest land plants (Butler, 1939). Despite their near omnipresence the VA mycorrhizae have, until quite recently, received relatively little attention.

The first and some of the best descriptions of VA mycorrhizae were published in the late 1800's and early 1900's. These early works were followed by further descriptions, reports on occurrence, and attempts to culture the endophytes. There was, however, very little experimental research prior to the mid 1950's. At this time new methods were devised for the study of VA mycorrhizae, and the investigations that followed provided firm evidence for their importance in plant nutrition. These developments greatly stimulated interest in the subject. Mosse (1973b) documented the large increase in number of papers on VA mycorrhiza in the previous 5 year period and noted the changes in direction of the research.

A number of general reviews of this subject have recently been published (Boullard, 1968; Gerdemann, 1968, 1970; Harley, 1969; Khan, 1972a; Mosse, 1973b; Nicolson, 1967). The role of VA mycorrhiza in phosphate uptake has also been reviewed (Bieleski, 1973). In this paper I will briefly state some general principles that now seem well established, discuss some of the most recent papers, and indicate some of the implications for both basic and applied research.

II. Occurrence

VA mycorrhizae are found in Bryophytes, Pteridophytes, Gymnosperms and Angiosperms. Meyer (1973) estimated that only about 3% of the phanerogams have ectomycorrhizae. The vast majority of the remaining species have VA mycorrhizae. They are so common that it is far easier to list the plant families in which they have not been found than to compile a list of plant families in which they are known to occur (Gerdemann, 1968). They occur on species in most families with the following exceptions: (1) families that are ectomycorrhizal, primarily Pinaceae, Betulaceae, and Fagaceae; (2) families that are endomycorrhizal with separate endophytes, primarily Orchidaceae, and Ericaceae; (3) certain groups that have been reported to be non-mycorrhizal, primarily families in the order Centrospermae and the families Crucifereae, Fumariaceae, Cyperaceae, Commelinaceae, Urticaceae, and Polygonaceae. However, exceptions have been noted (Gerdemann, 1968) and more recently several members of the Centrospermae in the family Chenopodiaceae (Williams, *et al.*, 1974; Ross and Harper, 1973; Kruckelmann, 1973), several species in the Cyperaceae (Mejstrik, 1972) and Cruciferae (Ross and Harper, 1973; Kruckelmann, 1973) were found to have VA mycorrhizae.

There are also some plant groups in which both ectomycorrhizae and VA mycorrhizae occur; Salicaceae, Juglandaceae, Tiliaceae, Myrtaceae, *Juniperus* and *Chamaecyparis* (Gerdemann, 1968). To this list we must now add Fagaceae (*Quercus*) (Grand, 1969) and Caesalpiniaceae (Redhead, 1968). As more plant species are examined, generalizations about the occurrence of VA mycorrhizae will become more difficult.

Geographically VA mycorrhizae occur from the tropics to the arctic and because of the wide host ranges of the fungi, there are very few natural plant communities that do not contain VA mycorrhizal species. Two possible exceptions are plants growing in aquatic habitats and dense stands of strictly ectomycorrhizal trees that have shaded out all other vegetation. If one considers mycorrhizae of trees, ectomycorrhizae are most common in cool regions. In temperature regions both ectomycorrhizae and VA mycorrhizae are common, and in the tropics trees with VA mycorrhizae predominate (Meyer, 1973).

III. Morphology

VA mycorrhizal infections produce little change in external root morphology. They have an extensive loose hyphal network that extends a considerable distance into the soil; however, unless one takes special precautions in removing plants from soil or utilizes some type of root observation chamber this extensive mycelium will not be seen. VA mycorrhizae can be recognized on relatively thin unsuberized roots by their bright yellow color. This color, however, disappears rapidly when exposed to light.

The internal morphology of VA mycorrhizae is readily observable by clearing and staining techniques (Bevege, 1968; Gerdemann, 1955; Phillips and Hayman, 1970; Trappe *et al.*, 1973). Hyphae produce appressoria on epidermal cells or root hairs in back of the meristematic region. Following infection the fungi colonize epidermal and cortical cells but never invade the endodermis, stele or root meristem. The hyphae may be entirely intracellular or mainly intercellular depending on the host species. For example, *Endogone fasciculata* Thaxter [*Glomus fasciculatus* (Thaxter) Gerd. and Trappe] forms a VA mycorrhiza on tuliptree that is almost entirely intracellular. The same fungus forms a mycorrhiza in maize in which the hyphae are primarily intercellular (Gerdemann, 1965). The hyphae produced by VA fungi are highly variable in size, irregular in shape and non-septate. Septa may form, however, when growing conditions are unfavorable or the fungus is dying. Soon after infection the fungi produce arbuscules within cortical cells. As arbuscules form, the starch within the invaded cell disappears and the nuclei enlarge and at times they also divide. Such changes have also been observed in uninfected cells in close proximity to arbuscules. Arbuscules form by repeated dichotomous branching from a coarse hyphal "trunk" and when mature they resemble complex "little trees". The ultimate branches of the arbuscular hyphae may be less than 1 μm in diameter and the entire structure may nearly fill the cell. Arbuscules are quickly digested starting at the tips and the contents are absorbed by the host (Kaspari, 1973). In cleared and stained mycorrhizae one frequently sees only the coarse hyphal "trunk" of the arbuscule surrounded by stained granular material. After the arbuscules are destroyed the nuclei return to their normal size and starch may reappear.

Vesicles are terminal, ovate to spherical, structures that contain oil droplets. They form either intra- or intercellularly depending on the host species. They may remain thin-walled and function as temporary storage organs for food, or in some species of VA fungi the vesicles become thick-walled and do not differ in any important respect from chlamydospores produced in the soil. Some species of VA fungi form spores within roots in such abundance that the cortex is ruptured. At the other extreme there are some VA fungi (*Gigaspora* spp.) (Gerdemann and Trappe, 1974) which have never been reported to produce vesicles within roots. In this genus, thin-walled distinctive vesicles form in the soil. They are probably not analogous structures to the vesicles that form within roots; however, their function may be similar.

IV. The Fungi

VA mycorrhizae are formed by certain species of Endogonaceae, a family of fungi in the Mucorales. This is a relatively large and diverse family and its taxonomy has recently been revised (Gerdemann and Trappe, 1974). The species that form VA mycorrhizae produce large globose, subglobose, elliptical to ovoid spores that contain globules of oil. In some species the spores are grouped in sporocarps, in others the spores may form either in sporocarps or singly in soil. In other species sporocarps have not been observed and it appears that the spores are always borne singly in the soil or roots. The species that produce large sporocarps are usually found in forests. Such sporocarps are occasioanlly produced on the soil surface but generally they are hypogeous or beneath leaf litter. They can be collected by raking the soil with a truffle fork or by collecting soil samples and soaking them in acid fuchsin (Kessler and Blank, 1972). The staining technique renders the sporocarps bright red and makes them easily detectable. From soil samples collected in a stand of sugar maple in Michigan, Kessler and Blank (1972) calculated there were more than 2,800,000 sporocarps per acre.

Species that produce small sporocarps or single spores are generally separated from soil by wet-sieving and decanting (Gerdemann, 1955; Gerdemann and Nicolson, 1963). Spores have also been collected from soil by a combination of wet-sieving and decanting, and centrifuging in a sucrose density gradient column (Ohms, 1957; Ross and Harper, 1970); and differential sedimentation on gelatine columns (Mosse and Jones, 1968). Sutton and Barron (1972) have described a flotation–adhesion technique for recovery of spores from small soil samples. Soils from colder regions or from forests often contain a high percentage of organic matter which makes it difficult or impossible to use wet-sieving and decanting to collect spores. VA mycorrhizal fungi can be obtained from such soils by collecting mycorrhizae and using them to inoculate a sterilized sandy soil containing little organic matter. A suitable mycorrhizal host is planted and after infection is established, a crop of spores is produced which can easily be wet-sieved from the soil. It is also possible to sample high organic matter soils by transplanting plants from the field into pots of sterilized sandy soil and allowing spores to develop (Gerdemann and Trappe, 1974).

The fungi that form VA mycorrhizae make little independent growth in soil and they probably obtain most organic nutrients from their hosts. Ho and Trappe (1973) found that ^{14}C supplied to mycorrhizal *Festuca* plants was transported to soil-borne spores of the fungus.

It is doubtful if these fungi have been obtained in axenic culture, and they are maintained in pot cultures which are established by carefully collecting spores of a particular species and using them to inoculate a suitable host growing in sterilized soil. Once established such cultures can be maintained indefinitely. Pot cultures contain only one VA fungus; however, many other microorganisms are present. By starting with surface-sterilized spores and growing plant roots under aseptic conditions Ross and Harper (1970) produced pure inoculum for use in their experiments. Pot cultures have proven extremely useful and they are maintained by nearly every investigator who works with VA mycorrhizae. However, progress in research and in its application would be greatly facilitated if we could grow these fungi in axenic culture.

The fungi that form VA mycorrhizae have extremely wide host ranges, and there is no definitive evidence that places any restriction on the host range of any particular species provided the host is capable of forming VA mycorrhiza. This is not to say that there are no adaptations to particular environments. Some species have been found only in forests while others, although they may occur in forests, are much more common in cultivated fields. It is very likely that some VA mycorrhizal fungi will be found to have restricted host ranges or at least species preferences. The extremely wide host ranges of these fungi raises the possibility of interspecific translocation of nutrients through hyphae in the soil between unrelated hosts. Woods and Brock (1964) injected ^{32}P and ^{45}Ca into stumps of red maple and later found the radioisotopes in leaves of 19 other species which were up to 24 feet from the stumps.

V. Effect of VA Mycorrhizae on Nutrition and Plant Growth

It is well established that VA mycorrhizae can increase plant growth, and that the growth improvement is greatest in soils of low fertility (Mosse, 1973b). When growth increases are large, mycorrhizal plants usually contain higher quantities of all essential nutrients. However, this does not necessarily provide evidence that mycorrhizae are more efficient in uptake of a particular nutrient than comparable non-mycorrhizal roots. If increased growth is accompanied by a higher concentration of a nutrient this provides strong evidence that mycorrhizal infection is directly responsible for the increased uptake. In experiments with Keen sour orange and rough lemon, mycorrhizal seedlings grew much larger than non-mycorrhizal seedlings and the above ground parts of mycorrhizal plants contained larger amounts of all minerals that were measured (Kleinschmidt and Gerdemann, 1972); however, only phosphorus and copper were present in higher concentration. All other elements were present in higher concentration in the stunted nonmycorrhizal seedlings. This suggests that the plants' ability to obtain phosphorus and possibly copper were the factors limiting growth, and that mycorrhizae were more efficient in uptake of these nutrients than comparable non-mycorrhizal roots.

A. Phosphorus

Most investigators have found that plants with VA mycorrhiza contain a higher concentration of phosphorus than do comparable non-mycorrhizal plants (Mosse, 1973b). When mycorrhizal onions were exposed to ^{32}P mycorrhiza, segments of the roots contained much higher levels of radioactivity than did comparable non-mycorrhizal segments or root tips (Gray and Gerdemann, 1969). Sanders and Tinker (1971, 1973) concluded that the increased efficiency of mycorrhizae to absorb phosphate could be accounted for by the uptake and transport of phosphate by the hyphae in the soil. Hattingh *et al.* (1973) obtained evidence that the theory is correct. Onion mycorrhizae had high levels of radioactivity when ^{32}P-labeled phosphate was injected into soil 27 mm from the root surface. Non-mycorrhizal roots had little radioactivity. Autoradiography indicated diffusion in the soil of only 7·5 mm or less from the point of application. When the hyphae from mycorrhizal roots were severed, mycorrhizae did not significantly differ in radioactivity from non-mycorrhizal roots. Therefore the external hyphae enable the plant to obtain phosphate from a larger volume of soil extending a greater distance from the root surface.

Schoknecht and Hattingh (in press) using X-ray microanalysis of onion mycorrhizae found that onion cells with arbuscules contained higher levels of phosphorus than adjacent cells without arbuscules. Thus, the external hyphae of VA fungi absorb phosphate, translocate it through the soil and into the root and it is likely that much of it is released in cortical cells containing arbuscules.

Mycorrhizal and non-mycorrhizal plants utilize the same sources of soil phosphate, and there is no evidence that the fungi can dissolve insoluble forms. In soil labeled with ^{32}P the proportion of ^{32}P to total P (specific activity) was nearly identical for mycorrhizal and non-mycorrhizal plants (Sanders and Tinker, 1971; Hayman and Mosse, 1972; Mosse, *et al.*, 1973). Ross and Gilliam (1973) supplied mycorrhizal and non-mycorrhizal soybeans with phosphate sources of varying availability and concluded that the principal source of phosphate utilized by mycorrhizal plants was the one most readily available.

Phosphate is relatively immobile in soil and it is present in the soil solution in very low concentration (Bieleski, 1973). However, if the amount in solution is depleted by roots it is quickly replenished by release from solid-phase forms.

Stout and Overstreet (1950) grew plants in soil containing 1 ppm phosphate in the soil solution. From the phosphorus content of the plants at the end of the experiment they calculated that the phosphate in soil solution was renewed on the average of 10 times a day. The exploitation of a greater volume of soil by mycorrhizae should result in the release of a larger amount of phosphate into the soil solution. This is a likely explanation for the improved growth of mycorrhizal plants given relatively insoluble phosphate (Daft and Nicolson, 1966; Murdoch *et al.*, 1967).

B. Nitrogen

There is no evidence that VA fungi are capable of nitrogen fixation, and I have frequently observed what appeared to be nitrogen deficiency symptoms on mycorrhizal plants when controls grown under the same conditions lacked such symptoms. Baylis (1959, 1967) reported a lower percent nitrogen in mycorrhizal plants of *Griselinia*, *Podocarpus*, *Coprosma*, *Pittosporum* and *Myrsine*. However, Ross and Harper (1970) and Ross (1971) found significantly higher concentrations of nitrogen in mycorrhizal soybeans. The mycorrhizal plants also grew larger and had higher seed yields than the controls. Schenck and Hinson (1973) grew a nodulating soybean variety and a non-nodulating isoline in a soil in which phosphorus was not believed to be limiting. Inoculation with a VA fungus increased yields of the nodulating but not the non-nodulating isoline. The available evidence suggests that mycorrhizal infection may increase the rate of nitrogen fixation by *Rhizobium* in leguminous plants. This is a subject of great importance that deserves intensive study.

C. Other Nutrients

VA mycorrhizal infection can also increase the uptake of other nutrients. Peach seedlings grown in steamed soil and watered with a nutrient solution containing N, P, K, Ca, Mg, B and an iron chelate (Gilmore, 1971) were stunted and had severe zinc deficiency symptoms. Inoculation with an *Endogone* (*Glomus*) species greatly improved growth and eliminated the deficiency symptoms. Three seasons following inoculation, foliage from mycorrhizal plants had 2·7 times more ppm zinc than did foliage from the check plants. VA mycorrhizae have been shown to have increased rates of sulphur uptake (Gray and Gerdemann, 1973). Mycorrhizal and non-mycorrhizal red clover and maize grown in sand and watered with a nutrient solution were given ^{35}S. Mycorrhizae of both species had much greater cpm/mg dry weight than did non-mycorrhizal roots. Jackson *et al.* (1973) working with mycorrhizal and non-mycorrhizal soybeans grown in soil watered every other day with Hoagland's solution found that mycorrhizal plants absorbed significantly more ^{90}Sr than control plants.

Many investigators have analysed mycorrhizal and non-mycorrhizal plants for concentrations of N, K, Ca, Mg, Fe, Cu, Mn, Zn, Na, and B and have obtained inconsistent results (Mosse, 1973b). In some experiments they were present in mycorrhizal plants in greater concentration. In others the concentrations were higher in non-mycorrhizal plants, or the differences were not significant. It is probable that the availability of the element in the soil determines the result. If an element is deficient and limiting plant growth it is likely to occur in higher concentration in mycorrhizal plants than in the controls.

In order to test for increased efficiency of uptake of a particular element by mycorrhizae, all other essential elements should be present at high enough levels so that they are not limiting growth of either mycorrhizal plants or the controls.

D. Water

VA mycorrhizae decrease the resistance to water transport in intact soybean plants (Safir *et al.*, 1971). There was no difference in resistances of stems plus leaves, indicating that the reduction of resistance occurred in the roots (Safir *et al.*, 1972). Addition of Hoagland's nutrient solution essentially eliminated differences in resistance to water transport between mycorrhizal and non-mycorrhizal plants, and it appeared that the decreased resistance to water transport in mycorrhizae was related to enhanced nutrient status of the mycorrhizal plant when grown in low nutrient soil.

Decreased resistance to water transport may be caused by changes in morphology of the host plant. Daft and Okusanya (1973b) found that mycorrhizal infection increased the amount of vascular tissue in tomato, petunia and maize stems. They believed that this was an indirect effect resulting from the greater uptake of phosphorus.

VI. Effect of Soil Fertility on VA Mycorrhizal Infection

VA mycorrhizal infection tends to be most prevalent in soils of moderate or low fertility. Generally, additions of P, N, or complete fertilizers reduce the amount of infection. In pot culture, VA mycorrhizal infection was reduced by high applications of available phosphate (Baylis, 1967; Daft and Nicolson, 1969a, b; Mosse, 1973a).

In field soil, addition of phosphate reduced the degree of mycorrhizal infection in maize (Khan, 1972b). In soils cropped to peanuts there were more spores of *E. gigantea* Nicol. and Gerd. [*Gigaspora gigantea* (Nicol. and Gerd.) Gerd. and Trappe[in soils maintained at low nitrogen levels than in soil maintained at high nitrogen levels (Porter and Beute, 1972). Spore numbers have been shown to be an accurate measure of the amount of VA mycorrhizae (Daft and Nicolson, 1972). In wheatfield soil at Rothamsted, plots that were not fertilized contained the largest number of spores of VA fungi, and the plants had the most abundant mycorrhizal infection (Hayman, 1970). Spore numbers and VA mycorrhiza were reduced in plots receiving N, NP, or NPKNaMg. However, the effect of fertilization can vary depending on the soil type. In a heavy loam soil, fertilization with manure, manure + N, or NPKMgNa reduced the numbers of chlamydospores in the soil (Kruckelmann, 1973). However, in a sandy soil, fertilization with manure compost, manure compost + NPK or NPK all increased the number of spores.

In agar culture, the amount of mycorrhizal infection in *Trifolium parviflorum* Ehrh. increased with the addition of small amounts of phosphate, but large amounts of phosphate decreased infection (Mosse and Philips, 1971). High phosphate concentrations in plants seem to make them resistant to infection (Mosse, 1973a).

In high arsenic soils the rootlets of apple trees were few and stunted, and the intensity of mycorrhizal infection was also reduced. It was suggested that the mycorrhizal deficiency might be partially responsible for the stunting of apple trees growing in high As soil. However, in correlations of this sort, it is difficult to discern cause and effect (Trappe *et al.*, 1973).

The mechanism whereby mycorrhizal infection is reduced in soils containing high levels of phosphate, other essential nutrients, or non-essential ions should be further investigated.

VII. Mycorrhiza Dependency

Mycorrhiza dependency can be defined as the degree to which a plant is dependent on the mycorrhizal condition to produce its maximum growth or yield, at a given level of soil fertility. In order to determine mycorrhiza dependency of a particular species, one would need to compare the growth of mycorrhizal and non-mycorrhizal plants at various fertility levels and determine the level at which mycorrhizal and non-mycorrhizal plants grow equally well. A non-mycorrhizal species, if such exists, would have no mycorrhiza dependency. At the other extreme there are some citrus varieties that are highly mycorrhiza dependent (Kleinschmidt and Gerdemann, 1972). These varieties, when non-mycorrhizal, are severely stunted in highly fertile nursery soils, and it is doubtful that fertility levels can be achieved at which the non-mycorrhizal and mycorrhizal plants grow equally well.

Most plant species probably fall between these extremes. At low nutrient levels one can assume that all plants that normally produce VA mycorrhiza are benefited by mycorrhizal infection. However, as nutrient levels, particularly phosphate levels, are increased a point is reached at which the mycorrhizal and non-mycorrhizal plants grow equally well. If this point could be clearly determined for each plant species, this information could be of considerable practical importance. Soil fertility is relative, and it can only be defined in relation to a particular plant species, or variety. In addition one needs to take into considera-

beneficial to their hosts than others (Gilmore, 1971; Mosse, 1972). It is also probable that spore populations in some soils are low. When virgin deserts containing little vegetation are first cultivated and planted with citrus seed, the plants are stunted and chlorotic and resemble citrus seedlings grown in methyl-bromide fumigated soil (D. A. Newcomb, personal communication). However, scattered groups of seedlings are healthy, suggesting that they may be growing in soil that was previously occupied by a mycorrhizal shrub.

Increases in growth have also been obtained by inoculating plants prior to transplanting in non-sterile soil. Mosse *et al.* (1969) inoculated onions and *Liquid-ambar styraciflua* L. by growing them temporarily in a soil containing a high spore population of a VA mycorrhizal fungus. Ten and 9 weeks, respectively, after transplanting into non-sterilized soil the inoculated plants of both species were larger than the controls. Inoculated onion plants also had more mycorrhizae than the controls.

Many horticultural plants are now grown in containers in sterilized soil, bark, quartz sand, or various types of synthetic media. Artificial inoculation of such plants could be of considerable practical importance. F. F. Hendrix (personal communication) inoculated woody ornamentals by placing small amounts of inoculum directly beneath the cuttings in the rooting beds. Much better rooting was obtained on the inoculated cuttings. According to Hendrix it is necessary to fertilize certain container-grown nursery stock with up to 90% of the toxicity level in order to obtain satisfactory growth. He believes that if the cuttings are mycorrhizal at the time they are transplanted to containers that fertilizer rates could be lowered to more reasonable levels.

Flowering annual plants also are commonly started in containers of sterilized soil. Daft and Okusanya (1973b) showed that flower production started earlier and was more profuse on mycorrhizal petunias. Comparisons should be made of mycorrhizal and non-mycorrhizal annuals after they are transplanted to home owners' gardens. Would the mycorrhizal plants become established more rapidly and bloom earlier and more profusely? If so, could one achieve the same effect with non-mycorrhizal plants grown under high phosphate levels?

The degree to which plants became mycorrhizal in the field can be altered by chemical treatment or cultural practices. Fumigation with the nematocide 1,2-dibromo-3-chloropropane (DBCP) or 1,3-dichloropropane resulted in increased mycorrhizal infection of cotton roots (Bird *et al.*, 1974), whereas application of the insecticide aldrin reduced mycorrhizal infection and chlamy-dospore numbers (Kruckelmann, 1973). Fields planted to *Beta vulgaris* L. or *Solanum tuberosum* L. continuously for 16 years contained few spores compared to those planted to *Zea mays* L., *Avena sativa* L., *Triticum aestivum* L. or *Secale cereale* L. (Kruckelmann, 1973). The roots of *Beta vulgaris* were almost free of mycorrhizal infection which undoubtedly accounts for the low spore number. However, the roots of *Solanum tuberosum* had about 25% infection. The volume of soil occupied by the root systems of various crops probably also influences the number of spores produced. It is conceivable that planting fields for a period of years to crops that produce few spores might adversely affect the growth of the crop that follows them. Conversely, planting to a crop that produces an abundance of spores might result in increased growth of the crop to follow, particularly if that crop has a high mycorrhiza dependency.

Soils are often said to be "toxic" to plants following heat-treatment or fumigation. Toxins can be produced by heat-sterilization (Rovira and Bowen, 1966) and such toxins can be leached out and their effects demonstrated in sterile sand. The toxins were believed to be organic compounds and they were neutralized by certain bacteria and by a number of common soil-borne fungi. Many of these organisms are airborne and would colonize sterilized soil exposed to air.

Soil fumigation has become a common practice in nurseries, and problems with poor plant growth are sometimes associated with it. If the problem persists for a period of months and if it can be overcome or partially corrected by fertilizer applications, then it is likely that the stunting is caused primarily by the killing of mycorrhizal fungi rather than soil toxins. The subject of "soil-toxicity" following sterilization should be re-evaluated in the light of present knowledge of VA mycorrhizal fungi.

Citrus seedlings grown in fumigated nursery soils or in heat-treated or fumigated soil in the greenhouse are commonly stunted and chlorotic. For many years this problem was attributed to soil toxicity. The major cause of the problem has now been shown to be inadequate nutrition brought about by the killing of VA mycorrhizal fungi (Kleinschmidt and Gerdemann, 1972). In order to control pathogenic nematodes and fungi, citrus nursery soils are routinely fumigated and stunting has become a major problem. Heavy fertilization partially overcomes stunting. However, citrus varieties vary in their mycorrhiza dependency. At high fertility levels some varieties grow moderately well in the non-mycorrhizal condition, while the more dependent varieties remain stunted and some barely survive. Inoculation of such soils would solve the problem of stunting, and also fertilization rates could be greatly reduced. Inoculum can be produced in quantity on the roots of living plants and it could be used for large scale inoculation. It should be produced, however, under very carefully controlled conditions in order to avoid the possibility of contamination with pathogens. Such inoculum could be placed in the row at the time of planting. However, seed inoculation offers many advantages. Smaller amounts of inoculum would be needed, and no alteration in present planting practices would be required. Sievings containing chlamydospores of *Glomus fasciculatus*, root debris, and soil organic matter were attached to Brazilian sour orange seed with methyl-cellulose. This method of inoculation proved successful in greenhouse and field experiments (Hattingh and Gerdemann, in press).

If the VA fungi could be grown in axenic culture both basic and applied research would be greatly facilitated. Here is a problem for someone with great patience. Now that other "obligate" parasitic fungi have been obtained in pure culture the chances of success may be greater.

References

BALTRUSCHAT, H. and SCHÖNBECK, F. (1972). Untersuchungen über den Einfluss der endotrophen Mycorrhiza auf die Chlamydosporenbildung von *Thielaviopsis basicola* in Tabakwurzeln. *Phytopathol. Z.* **74**; 358–361.

BALTRUSCHAT, H., SIKORA, R. A. and SCHÖNBECK, F. (1973). Effect of VA-mycorrhiza (*Endogone mosseae*) on the establishment of *Thielaviopsis basicola* and *Meloidogyne incognita* in tobacco. Abst. 2nd Int. Congr. Plant Pathol. Abst. 0661.

BAYLIS, G. T. S. (1959). The effect of vesicular-arbuscular mycorrhizas on growth of *Griselinia littoralis* (Cornaceae). *New Phytol.* **58**; 274–280.

BAYLIS, G. T. S. (1967). Experiments on the ecological significance of phycomycetous mycorrhizas. *New Phytol.* **66**; 231–243.

BAYLIS, G. T. S. (1970). Root hairs and phycomycetous mycorrhizas in phosphate-deficient soil. *Plant. Soil.* **33**; 713–716.

BAYLIS, G. T. S. (1972). Minimum levels of available phosphorus for non-mycorrhizal plants. *Plant Soil.* **36**; 233–234.

BEVEGE, D. I. (1968). A rapid technique for clearing tannins and staining intact roots for detection of mycorrhizas caused by *Endogone* spp., and some records of infection in Australian plants. *Trans. Brit. Mycol. Soc.* **51**; 808–810.

BIELESKI, R. L. (1973). Phosphate pools, phosphate transport and phosphate availability. *A. Rev. Plant Physiol.* **24**; 225–252.

BIRD, G. W., RICH, J. R. and GLOVER, S. U. (1974). Increased endomycorrhizae of cotton roots in soil treated with nematocides. *Phytopathology* **64**; 48–51.

BOULLARD, B. (1968). "Les Mycorrhizes." Masson et Cie, Paris, 135 pp.

BUTLER, E. J. (1939). The occurrences and systematic position of the vesicular-arbuscular type of mycorrhizal fungi. *Trans. Brit. Mycol. Soc.* **22**; 274–301.

DAFT, M. J. and NICOLSON, T. H. (1966). The effect of *Endogone* mycorrhiza on plant growth. *New Phytol.* **65**; 343–350.

DAFT, M. J. and NICOLSON, T. H. (1969a). The effect of *Endogone* mycorrhiza on plant growth. II. Influence of soluble phosphate on endophyte and host in maize. *New Phytol.* **68**; 945–952.

DAFT, M. J. and NICOLSON, T. H. (1969b). Effect of *Endogone* mycorrhiza on plant growth. III. Influence of inoculum concentration on growth and infection in tomato. *New Phytol.* **68**; 953–963.

DAFT, M. J. and NICOLSON, T. H. (1972). Effect of *Endogone* mycorrhiza on plant growth. IV. Quantitative relationships between the growth of the host and the development of the endophyte in tomato and maize. *New Phytol.* **71**; 287–295.

DAFT, M. J. and OKUSANYA, B. O. (1973a). Effect of *Endogone* mycorrhiza on plant growth. V. Influence of infection on the multiplication of viruses in tomato, petunia and strawberry. *New Phytol.* **72**; 975–983.

DAFT, M. J. and OKUSANYA, B. O. (1973b). The effect of *Endogone* mycorrhiza on plant growth. VI. Influence of infection on the anatomy and reproductive development in four hosts. *New Phytol.* **72**; 1333–1339.

FOX, J. A. and SPASOFF, L. (1972). Interaction of *Heterodera solanacearum* and *Endogone gigantea* on tobacco. Abstr. *J. Nematology* **4**; 224–225.

GERDEMANN, J. W. (1955). Relation of a large soil-borne spore to phycomycetous mycorrhizal infections. *Mycologia* **47**; 619–632.

GERDEMANN, J. W. (1964). The effect of mycorrhiza on the growth of maize. *Mycologia* **56**; 342–349.

GERDEMANN, J. W. (1965). Vesicular-arbuscular mycorrhizae formed on maize and tuliptree by *Endogone fasciculata*. *Mycologia* **57**; 562–575.

GERDEMANN, J. W. (1968). Vesicular-arbuscular mycorrhiza and plant growth. *A. Rev. Phytopathol.* **6**; 397–418.

GERDEMANN, J. W. (1970). The significance of vesicular-arbuscular mycorrhizae in plant nutrition *In*: "Root Diseases and Soil-Borne Pathogens" (Eds. T. A. Toussoun, R. V. Bega and P. E. Nelson) pp. 125–129. University of California Press, Berkeley, Los Angeles, London.

GERDEMANN, J. W. and NICOLSON, T. H. (1963). Spores of mycorrhizal *Endogone* species extracted from soil by wet sieving and decanting. *Trans. Brit. Mycol. Soc.* **46**; 235–244.

GERDEMANN, J. W. and TRAPPE, J. M. (1974). The Endogonaceae in the Pacific Northwest. Mycologia Memoir No. 5. 76 pp.

GILMORE, A. E. (1971). The influence of endotrophic mycorrhizae on the growth of peach seedlings. *J. Am. Soc. Hort. Sci.* **96**; 35–38.

GRAND, L. F. (1969). A beaded endotrophic mycorrhiza of northern and southern red oak. *Mycologia* **61**; 408–409.

GRAY, L. E. and GERDEMANN, J. W. (1969). Uptake of phosphorus-32 by vesicular-arbuscular mycorrhizae. *Plant Soil* **30**; 415–422.

GRAY, L. E. and GERDEMANN, J. W. (1973). Uptake of sulphur-35 by vesicular-arbuscular mycorrhizae. *Plant Soil.* **39**; 687–689.

HARLEY, J. L. (1969). "The Biology of Mycorrhiza" 334 pp. Leonard Hill, London.

HATTINGH, M. J., and GERDEMANN, J. W. Inoculation of Brazilian sour orange seed with an endomycorrhizal fungus. *Phytopathology* (In press).

HATTINGH, M. J., GRAY, L. E. and GERDEMANN, J. W. (1973). Uptake and translocation of ^{32}P-labeled phosphate to onion roots by endomycorrhizal fungi. *Soil Sci.* **116**; 383–387.

HAYMAN, D. S. (1970). *Endogone* spore numbers in soil and vescular-arbuscular mycorrhiza in wheat as influenced by season and soil treatment. *Trans. Brit. Mycol. Soc.* **54**; 53–63.

HAYMAN, D. S. and MOSSE, B. (1972). Plant growth responses to vesicular-arbuscular mycorrhiza. III. Increase uptake of labile P from soil. *New Phytol.* **71**; 41–47.

HO, I. and TRAPPE, J. M. (1973). Translocation of ^{14}C from *Festuca* plants to their endomycorrhizal fungi. *Nature New Biol.* **244**; 30–31.

JACKSON, N. E., MILLER, R. H. and FRANKLIN, R. E. (1973). The influence of vesicular-arbuscular mycorrhizae on uptake of ^{90}Sr. from soil by soybeans. *Soil Biol. Biochem.* **5**; 205–212.

KASPARI, H. (1973). Elektronenmikroskopische Untersuchung zur Feinstruktur der endotrophen Tabakmykorrhiza. *Arch. Mikrobiol.* **92**; 201–207.

KESSLER, K. L., JR. and BLANK, R. W. (1972). *Endogone* sporocarps associated with sugar maple. *Mycologia* **64**; 634–638.

KHAN, A. G. (1972a). Mycorrhizae and their significance in plant nutrition. *Biologia.* Special supplement. April, 1972. pp. 42–78.

KHAN, A. G. (1972b). The effect of vesicular-arbuscular mycorrhizal associations on growth of cereals. I. Effects on maize growth. *New Phytol.* **71**; 613–619.

KLEINSCHMIDT, G. D. and GERDEMANN, J. W. (1972). Stunting of citrus seedlings in fumigated nursery soils related to the absence of endomycorrhizae. *Phytopathology* **62**; 1447–1453.

KRUCKELMANN, H. W. (1973). "Die vesikulär-arbuskuläre mykorrhiza und ihre beeinflussung in landwirtschaftlichen Kulturen." Diss. Naturwiss. Fakultät Tech. Universität. Carolo-Wilhelmina, Braunschweig. 1–56.

MARX, D. H. (1972). Ectomycorrhizae as biologic deterrents to pathogenic root infections. *A. Rev. Phytopathol.* **10**; 429–454.

MARX, D. H. (1973). Mycorrhizae and feeder root diseases. *In*: "Ectomycorrhizae, Their Ecology and Physiology," (Eds. G. C. Marx and T. T. Kozlowski) pp. 351–382. Academic Press, New York, London and San Francisco.

MEJSTRIK, V. K. (1972). Vesicular-arbuscular mycorrhizas of the species of a *Molinietum coeruleae* L. I. Association: The ecology. *New Phytol.* **71**; 883–890.

MEYER, F. H. (1973). Distribution of ectomycorrhizae in native and man-made forests. *In*: "Ectomycorrhizae Their Ecology and Physiology," (Eds. G. C. Marks and T. T. Kozlowski) pp. 79–105. Academic Press, New York, London and San Francisco.

MOSSE, B. (1972). Effects of different *Endogone* strains on the growth of *Paspalum notatum*. *Nature, Lond.* **239**; 221–223.

MOSSE, B. (1973a). Plant growth responses to vesicular-arbuscular mycorrhiza. IV. In soil given additional phosphate. *New Phytol.* **72**; 127–136.

MOSSE, B. (1973b). Advances in the study of vesicular-arbuscular mycorrhiza. *A. Rev. Phytopathol.* **11**; 171–196.

MOSSE, B. and HAYMAN, D. S. (1971). Plant growth responses to vesicular-arbuscular mycorrhiza. II. In unsterilized field soils. *New Phytol.* **70**; 29–34.

MOSSE, B., HAYMAN, D. S. and ARNOLD, D. J. (1973). Plant growth responses to vesicular-arbuscular mycorrhiza. V. Phosphate uptake by three plant species from P-deficient soils labelled with ^{32}P. *New Phytol.* **72**; 809–815.

MOSSE, B., HAYMAN, D. S. and IDE, G. J. (1969). Growth responses of plants in unsterilized soil to inoculation with vesicular-arbuscular mycorrhiza. *Nature, Lond.* **224**; 1031–1032.

MOSSE, B. and JONES, G. W. (1968). Separation of *Endogone* spores from organic soil debris by differential sedimentation on gelatin columns. *Trans. Brit. Mycol. Soc.* **51**; 604–608.

MOSSE, B. and PHILLIPS, J. M. (1971). The influence of phosphate and other nutrients on the development of vesicular-arbuscular mycorrhiza in culture. *J. Gen. Microbiol.* **69**; 157–166.

MURDOCH, C. L., JACKOBS, J. A. and GERDEMANN, J. W. (1967). Utilization of phosphorus sources of different availability by mycorrhizal and non-mycorrhizal maize. *Plant Soil* **27**; 329–334.

NICOLSON, T. H. (1967). Vesicular-arbuscular mycorrhiza—a universal plant symbiosis. *Sci. Progr., Oxford* **55**; 561–581.

OHMS, R. E. (1957). A flotation method for collecting spores of a phycomycetous mycorrhizal parasite from soil. *Phytopathology* **47**; 751–752.

PHILLIPS, J. M. and HAYMAN, D. S. (1970). Improved procedures for clearing roots and staining parasitic and vesicular-arbuscular mycorrhizal fungi for rapid assessment of infection. *Trans. Brit. Mycol. Soc.* **55**; 158–161.

PORTER, D. M. and BEUTE, M. K. (1972). *Endogone* species in roots of Virginia type peanuts. Abst. *Phytopathology* **62**; 783.

REDHEAD, J. F. (1968). Mycorrhizal associations in some Nigerian forest trees. *Trans. Brit. Mycol. Soc.* **51**; 377–387.

ROSS, J. P. (1971). Effect of phosphate fertilization on yield of mycorrhizal and non-mycorrhizal soybeans. *Phytopathology* **61**; 1400–1403.

ROSS, J. P. (1972). Influence of *Endogone* mycorrhiza on Phytophthora rot of soybean. *Phytopathology* **62;** 896–897.

ROSS, J. P. and GILLIAM, J. W. (1973). Effect of *Endogone* mycorrhiza on phosphorus uptake by soybeans from inorganic sources. *Soil Sci. Soc. Amer. Proc.* **37**; 237–239.

ROSS, J. P. and HARPER, J. A. (1970). Effect of *Endogone* mycorrhiza on soybean yields. *Phytopathology* **60**; 1552–1556.

ROSS, J. P. and HARPER, J. A. (1973). Hosts of a vesicular-arbuscular *Endogone* species. *J. Elisha Mitchell Sci. Soc.* **89**; 1–3.

ROVIRA, A. D. and BOWEN, G. D. (1966). The effects of microorganisms upon plant growth. II. Detoxication of heat-sterilized soils by fungi and bacteria. *Plant Soil* **25**; 129–142.

RUEHLE, J. L. (1973). Nematodes and forest trees—types of damage to tree roots. *A. Rev. Phytopathol.* **11**; 99–118.

SAFIR, G. R., BOYER, J. S. and GERDEMANN, J. W. (1971). Mycorrhizal enhancement of water transport in soybean. *Science, N.Y.* **172**; 581–583.

SAFIR, G. R., BOYER, J. S. and GERDEMANN, J. W. (1972). Nutrient status and mycorrhizal enhancement of water transport on soybean. *Plant Physiol.* **49**; 700–703.

SANDERS, F. E. and TINKER, P. B. (1971). Mechanism of absorption of phosphate from soil by *Endogone* mycorrhizas. *Nature, Lond.* **233**; 278–279.

SANDERS, F. E. and TINKER, P. B. (1973). Phosphate flow into mycorrhizal roots. *Pestic. Sci.* **4**; 385–395.

SCHENCK, N. C. and HINSON, K. (1973). Response of nodulating and non-nodulating soybeans to a species of *Endogone* mycorrhiza. *Agron. J.* **65**; 849–850.

SCHÖNBECK, F. and SCHINZER, U. (1972). Untersuchungen über den Einfluss der endotrophen Mycorrhiza auf die TMV-Läsionenbildung in *Nicotiana tabacum* L. var. Xanthi-nc. *Phytopathol. Z.* **73**; 78–80.

STOUT, P. R. and OVERSTREET, R. (1950). Soil chemistry in relation to inorganic nutrition of plants. *A. Rev. Plant. Physiol.* **1**; 305–342.

SUTTON, J. C. and BARRON, G. L. (1972). Population dynamics of *Endogone* spores in soil. *Can. J. Bot.* **50**; 1909–1914.

TRAPPE, J. M., STAHLY, E. A., BENSON, N. R. and DUFF, D. M. (1973). Mycorrhizal deficiency of apple trees in high arsenic soils. *Hort. Science* **8**; 52–53.

WILLIAMS, S. E., WOLLUM, A. G. II, and ALDON, E. F. (1974). Growth of *Atriplex cahescens* (Pursh) Nutt. Improved by formation of vesicular-arbuscular mycorrhizae. *Soil Sci. Soc. Amer. Proc.* 38; 962–965.

WOODS, F. W. and BROCK, K. (1964). Interspecific transfer of Ca-45 and P-32 by root systems. *Ecology* **45**; 886–889.

YOUNG, J. L., HO, I., and TRAPPE, J. M. (1972). Endomycorrhizal invasion and effect on free amino acids content of corn roots. *Agron. Abstr.* p. 102.

Author Index

Numbers in italic indicate the page on which references are given in full.

D

L

S

Subject Index

E

F

G

H

U

V

W

X

Z